Modeling of Chemical Kinetics and Reactor Design

A. Kayode Coker, Ph.D.

Lecturerer and Consultant, AKC Technology

Gulf Professional Publishing
an imprint of Butterworth-Heinemann

Boston Oxford Johannesburg Melbourne New Delhi Singapore

Dedication

To my wife Victoria and the boys Akin and Ebun,
love and thanks.

Gulf Professional Publishing is an imprint of Butterworth–Heinemann.

Copyright © 2001 by Butterworth–Heinemann

 A member of the Reed Elsevier group

Recognizing the importance of preserving what has been written, Butterworth–Heinemann prints its books on acid-free paper whenever possible.

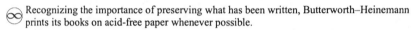

GLOBAL RELEAF 2000 Butterworth–Heinemann supports the efforts of American Forests and the Global ReLeaf program in its campaign for the betterment of trees, forests, and our environment.

Library of Congress Cataloging-in-Publication Data

Coker, A. Kayode
    Modeling of chemical kinetics and reactor design / A. Kayode Coker. -- 2nd ed.
      p. cm.
    Includes bibliographical references and index.
    ISBN 0-88415-481-5
    1. Chemical processes--Mathematical models. 2. Chemical reactors--Mathematical models. I. Title
models. I. Title
    TP155.7 .C655 2001
    660'.281--dc21   2001023630

British Library Cataloguing-in-Publication Data
A catalogue record for this book is available from the British Library.

The publisher offers special discounts on bulk orders of this book.
For information, please contact:

Manager of Special Sales
Butterworth–Heinemann
225 Wildwood Avenue
Woburn, MA 01801-2041
Tel: 781-904-2500
Fax: 781-904-2620

Library
University of Texas
at San Antonio

For information on all Gulf Professional publications available, contact our World Wide Web home page at: http://www.gulfpp.com

10 9 8 7 6 5 4 3 2 1

Printed in the United States of America

Acknowledgments

I wish to express my gratitude to the following for giving their time in proofreading various sections of the text: Drs. A. A. Adesina, L. M. Rose, C. J. Mumford, and J. D. Jenkins. I am indebted to Emeritus Professor Octave Levenspiel for his encouragement and advice in some chapters of the text, and to Drs. Waldram and Singh for their comments, suggestions on safety in reaction engineering, and the inclusion of HEL safety photographs in the text. I would also like to thank Mr. Ed Steve for his comments and suggestions on scale-up of reactors, and Mr. Joseph Rivera for some of the figures in the text. I wish to express my gratitude to Drs. A. Bakker, J. B. Fasano, and V. V. Ranade for contributing to Chapter 10 (Computational Fluid Dynamics and Computational Fluid Mixing). I would like to acknowl-edge the following companies for the use of their materials: Arthur D. Little, HEL, M. W. Kellogg Ltd., Stone & Webster, Fauske & Associates, Inc., Simulation Sciences Inc., Chemineer, PROCEDE, and Absoft Corporation.

I would like to express my gratitude to the following institutions for permission to reproduce their materials: Institution of Chemical Engineers (U.K.), the American Institute of Chemical Engineers and Chemical Engineering—a publication of Chemical Week Associates. I am also indebted to those whose work was drawn.

I thank Dr. E. L. Smith for his comments and suggestions during the preparation of some of the chapters in the text and wish him a happy retirement. It has been a pleasure to have learned so much from him during his tenure at Aston University.

Sincere gratitude to Tim Calk of Gulf Publishing Company for his direction and editing of the book, to Danette DeCristofaro and Jerry Hayes of ExecuStaff for their excellent production of the book, to Mr. Phil Carmical and Ms. Jennifer Plumley of Butterworth-Heinemann for the production of the CD-ROM.

My thanks to Ahmed Mutawa of Saudi Aramco Shell Refinery (SASREF), an excellent student in the short course program for developing conversion table software for the book.

In gratitude to the Almighty father, the Omnipotence, Omniscience, and Omnipresence.

A. Kayode Coker, Ph.D.

Nippon Petroleum FCC Unit—Japan. *(Courtesy M. W. Kellogg Ltd.)*

Contents

Preface

SCOPE

This valuable reference volume conveys a basic understanding of chemical reactor design methodologies that incorporate both scale-up and hazard analysis. It shows readers how to select the best reactor for any particular chemical reaction, how to estimate its size, and how to obtain the best operating conditions.

An understanding of chemical reaction kinetics and the design of chemical reactors is very important to the chemist and the chemical engineer. Engineers share interests in fluid mechanics and transport phenomena, while the chemist deals with the kinetics and mechanisms of reactions. The chemical engineer combines the knowledge of these subjects for the better understanding, design, and control of the reactor. The recent accidents that have occurred in the chemical process industries with inherent fatalities and environmental pollution have imposed greater demands on chemical engineers. Consequently, chemical reactor design methodologies must incorporate both control and hazard analysis. However, the design of chemical reactors is still essential for its proper sizing, and is included in various types of process simulators. In an industrial problem, it is essential to select the best type of reactor for any particular chemical reaction. Additionally, it is necessary to estimate its size and determine the best operating conditions. The chemical engineer confronted with the design of various reactor types often depends on the scale of operation and the kinetics.

Many excellent texts have appeared over the years on chemical reactor design. However, these texts often lack sections on scale-up, biochemical reactor design, hazard analysis, and safety in reactor design methodology. The purpose of this book is to provide the basic theory and design and, sometimes, computer programs (Microsoft Excel spreadsheet and software) for solving tedious problems. This speeds up the work of both chemists and engineers in readily arriving at a solution. The following highlights some of the subjects that are covered in this text.

MIXING

An important unit operation in chemical reaction engineering, mixing, finds application in petrochemicals, food processing, and biotechnology. There are various types of fluid mixing such as liquid with liquid, gas with liquid, or solids with liquid. The text covers micromixing and macromixing, tracer response and residence time distribution (RTD), heat transfer, mixing fundamentals, criteria for mixing, scale of segregation, intensity of segregation, types of impellers, dimensional analysis for liquid agitation systems, design and scale-up of mixing pilot plants, the use of computational fluid dynamics (CFD) in mixing, and heat transfer in agitated vessels.

BIOCHEMICAL REACTION

This is an essential topic for biochemists and biochemical engineers. Biochemical reactions involve both cellular and enzymatic processes, and the principal differences between biochemical and chemical reactions lie in the nature of the living systems. Biochemists and biochemical engineers can stabilize most organic substances in processes involving microorganisms.

This chapter discusses the kinetics, modeling and simulation of biochemical reactions, types and scale-up of bioreactors. The chapter provides definitions and summary of biological characteristics.

CHEMICAL REACTOR MODELING

This involves knowledge of chemistry, by the factors distinguishing the micro-kinetics of chemical reactions and macro-kinetics used to describe the physical transport phenomena. The complexity of the chemical system and insufficient knowledge of the details requires that reactions are lumped, and kinetics expressed with the aid of empirical rate constants. Physical effects in chemical reactors are difficult to eliminate from the chemical rate processes. Non-uniformities in the velocity, and temperature profiles, with interphase, intraparticle heat, and mass transfer tend to distort the kinetic data. These make the analyses and scale-up of a reactor more difficult. Reaction rate data obtained from laboratory studies without a proper account of the physical effects can produce erroneous rate expressions. Here, chemical reactor flow models using mathematical expressions show how physical processes interact with chemical processes. The proposed model must represent the flow behavior of an actual reactor, which is realistic enough to give useful information for its design and analysis. The text reviews different reactor flow models.

SAFETY IN CHEMICAL REACTION

Equipment failures or operator errors often cause increases in process pressures beyond safe levels. A high increase in pressure may exceed the set pressure in pipelines and process vessels, resulting in equipment rupture and causing major releases of toxic or flammable chemicals. A proper control system or installation of relief systems can prevent excessive pressures from developing. The relief system consists of the relief device and the associated downstream process equipment (e.g., knock-out drum, scrubber, absorbers, and flares) that handles the discharged fluids. Many chemical reactions (e.g., polymerization, sulphonation, nitration) in the chemical process industry result in runaway reactions or two-phase flow. This occurs when an exothermic reaction occurs within a reactor. If cooling no longer exists due to a loss of cooling water supply or failure of a control system (e.g., a valve), then the reactor temperature will rise. As the temperature rises, the reaction rate increases, leading to an increase in heat generation. This mechanism results in a runaway reaction. The pressure within the reactor increases due to increased vapor pressure of the liquid components and gaseous decomposition products as a result of the high temperature. Runaway reactions can occur within minutes for large commercial reactors and have resulted in severe damage to a complete plant and loss of lives. This text examines runaway reactions and two-phase flow relief.

SCALE-UP

The chemical engineer is concerned with the industrial application of processes. This involves the chemical and microbiological conversion of material with the transport of mass, heat and momentum. These processes are scale-dependent (i.e., they may behave differently in small and large-scale systems) and include heterogeneous chemical reactions and most unit operations. The heterogeneous chemical reactions (liquid-liquid, liquid-gas, liquid-solid, gas-solid, solid-solid) generate or consume a considerable amount of heat. However, the course of such chemical reactions can be similar on both small and large scales. This happens if the mass and heat transfer processes are identical and the chemistry is the same. Emphasis in this text is on dimensional analysis with respect to the following:

- Continuous chemical reaction processes in a tubular reactor.
- Influence of back mixing (macromixing) on the degree of conversion and in continuous chemical reaction operation.
- Influence of micro mixing on selectivity in a continuous chemical reaction process.
- Scale-up of a batch reactor

AN INTEGRATING CASE STUDY—AMMONIA SYNTHESIS

This book briefly reviews ammonia synthesis, its importance in the chemical process industry, and safety precautions. This case study is integrated into several chapters in the text. See the Introduction for further details.

Additionally, solutions to problems are presented in the text and the accompanying CD contains computer programs (Microsoft Excel spreadsheet and software) for solving modeling problems using numerical methods. The CD also contains colored snapshots on computational fluid mixing in a reactor. Additionally, the CD contains the appendices and conversion table software.

A. Kayode Coker, Ph.D.

Introduction

Increased collaboration between chemical reaction engineering and chemistry disciplines in recent years has produced significant advances in kinetic and thermodynamic modeling of processes. Additionally, improvement in analytical chemistry techniques, the formulation of mathematical models, and the development of computational tools have led to a deeper understanding of complex chemical reaction kinetics, particularly in mixtures with large numbers of compounds. Activities in both academic and industrial research organizations have enabled these groups to review the state of the art and cooperate with the overall objectives of improving the safety, yields, and quality of the products. Also, the final commitment to the production of any chemical product often depends on its profitability and other economic factors.

Chemical kinetics mainly relies on the rates of chemical reactions and how these depend on factors such as concentration and temperature. An understanding of chemical kinetics is important in providing essential evidence as to the mechanisms of chemical processes. Although important evidence about mechanisms can be obtained by non-kinetic investigations, such as the detection of reaction intermediates, knowledge of a mechanism can be confirmed only after a detailed kinetic investigation has been performed. A kinetic investigation can also disprove a mechanism, but cannot ascertain a mechanism.

Kinetic investigations cover a wide range from various viewpoints. Chemical reactions occur in various phases such as the gas phase, in solution using various solvents, at gas-solid, and other interfaces in the liquid and solid states. Many techniques have been employed for studying the rates of these reaction types, and even for following fast reactions. Generally, chemical kinetics relates to the studies of the rates at which chemical processes occur, the factors on which these rates depend, and the molecular acts involved in reaction mechanisms. Table 1 shows the wide scope of chemical kinetics, and its relevance to many branches of sciences.

Table 1
Some branches of science to which kinetics is relevant [1]

Branch	Applications of kinetics
Biology	Physiological processes (e.g., digestion and metabolism), bacterial growth
Chemical engineering	Reactor design
Electrochemistry	Electrode processes
Geology	Flow processes
Inorganic chemistry	Reaction mechanisms
Mechanical engineering	Physical metallurgy, crystal dislocation mobility
Organic chemistry	Reaction mechanisms
Pharmacology	Drug action
Physics	Viscosity, diffusion, nuclear processes
Psychology	Subjective time, memory

CASE STUDY

As an introduction to the modeling of chemical kinetics and reactor design, consider the manufacture of ammonia. The synthesis of ammonia is performed on a large scale with over 100 million tons produced each year. Computer simulation of the plant is increasingly employed as the first stage in identifying which parameters control the conversion rate, the product purity, the energy expended, and the production rate. The economic considerations that affect the reduction of costs with increased efficiency and profitability are high. The principal licensors of ammonia synthesis are ICI, Braun, and M.W. Kellogg. Figure 1 shows a typical ammonia plant.

Ammonia is one of the largest volume inorganic chemicals in the chemical process industries. Its major applications are in the production of fertilizers, nitrates, sulfates, phosphates, explosives, plastics, resins, amines, amides, and textiles. The fertilizer industry is the largest user of ammonia, and large quantities must be stored to meet the demand and maintain constant production levels. Ammonia may be stored in very large insulated tanks at pressure near ambient; in large spheres at a moderate pressure, refrigerated to reduce the pressure; and at ambient temperature but higher pressure, corresponding

Figure 1. Tugas ammonia plant—Gemlik, Turkey. *(Courtesy M. W. Kellogg Ltd.)*

to the vapor pressure at ambient temperature. Table 2 shows the raw materials requirements, yield, and properties of ammonia.

Transportation is by railroad tank vehicle, by tank truck, or by pipeline. In this case, transportation at ambient temperature is the best choice. The choice of storing ammonia at an ambient temperature liquid or partially refrigerated liquid or an ambient pressure liquid depends mostly on economic factors. One of the factors that determines the storage method is the quantity of ammonia to be stored.

Ammonia is toxic and flammable, although the lower flammable limit is quite high and fires in ammonia facilities are rare. However, spillage from storage vessel or transfer piping must be considered and adequate precautions taken to minimize its effect. Storage tanks must have adequate vents so the pressure cannot rise above safe levels, and are diked to prevent the spread of liquid in case of a spill. For ambient pressure storage, the vents must be large in area and operate at pressures only slightly above ambient pressure. Alternatively, for ambient temperature, the vents are smaller, but operate at much higher pressure. At ambient temperature, ammonia in common with other liquified gases must have sufficient ullage space in the tank to allow for expansion when the temperature rises. Otherwise, liquid loaded into

Table 2
Raw materials requirements and yield

Raw materials required per ton of ammonia:

Natural gas: 810 m^3 Yield: 85%

Properties: Colorless liquid or gas with a very pungent odor. Soluble in water, ethyl alcohol, and ether.

Molecular weight: 17.03   Vapor density (air = 1): 0.597

Density at 20°C (gas): 0.771

Melting point: –77.7°C   Boiling point: –33.4°C

Autoignition temperature: 651°C

Critical temperature, T_C: 132.4°C

Critical pressure, P_C: 111.3°C

Critical volume, V_C: 72.4 cm^3/gmol

Critical compressibility factor Z: 0.243

the tank from refrigerated storage will expand to cause the tank to be full of liquid as it heats. The tank vent would then open and subsequently leak liquid ammonia to the atmosphere. Training personnel for the hazardous nature of anhydrous ammonia and how to handle emergency conditions in the plant is essential.

The American Conference of Government Industrial Hygienists (ACGIH) has established threshold doses called the threshold limit values (TLVs) for many chemical agents. The TLV is the airborne concentration that corresponds to conditions where no adverse effects are normally expected during a worker's lifetime. The exposure is assumed to occur only during normal working hours, eight hours per day and five days per week.

TLVs are calculated using ppm (parts per million by volume), mg/m^3 (mg of vapor per cubic meter of air). For vapor, mg/m^3 is converted to ppm by the equation:

$$C_{ppm} = \text{Conc. in ppm} = \frac{22.4}{M_W}\left(\frac{T}{273}\right)\left(\frac{1}{P}\right), \frac{mg}{m^3}$$

$$= 0.08205\left(\frac{T}{P \bullet M_W}\right), \frac{mg}{m^3}$$

where T = temperature in K
       P = absolute pressure in atm
       M_W = molecular weight in gm/gm – mole

Table 3 gives the threshold limit value (TLV) and permissible exposure level (PEL) of ammonia.

Table 3

Substance	Threshold Limit Value (TLV), time weighted ppm	Value average, mg/m^3 at 25°C	U.S. Safety and Administration Permissible Exposure Level (PEL), ppm	OSHA Exposure, mg/m^3 at 25°C
Ammonia	25	18	25	18

The raw material source for ammonia plants is natural gas for producing hydrogen by steam reforming. An alternative raw material is naphtha, which also requires partial oxidation. Hydrogen streams from catalytic reformers are another source of hydrogen. However, the volumes available are negligible to meet the requirements of an average size ammonia plant. Nitrogen is obtained by the liquefaction of air or from producer gas mixed with hydrogen in the mole ratio of 3:1.

The synthesis of ammonia is divided into four stages. In stage 1, the natural gas undergoes catalytic reforming to produce hydrogen from methane and steam. The nitrogen required for ammonia production is introduced at this stage. Stage 2 involves the "synthesis gas" (syngas) that is purified by removing both carbon monoxide and carbon dioxide. Stage 3 is the compression of the syngas to the required pressure. Stage 4 is the ammonia loop. A typical feed stock for ammonia synthesis is 0.17 million standard cubic meter per day (6 Mscfd) of natural gas at a temperature of 16°C and a pressure of 23.4 barg. Table 4 shows its composition.

Natural gas is desulfurized because sulfur has an adverse effect on the catalysts used in the reforming and synthesis reactions. After desulfurization and scrubbing, the natural gas is mixed with super-heated steam at 23 barg and 510°C. Nitrogen is supplied from the air, which is fed to the secondary reformer at 20 barg and 166°C. Table 5 shows the composition of air.

Table 4
Natural gas feed

Component	Mole %
Carbon dioxide (CO_2)	2.95
Nitrogen (N_2)	3.05
Methane (CH_4)	80.75
Ethane (C_2H_6)	7.45
Propane (C_3H_8)	3.25
Butane (C_4H_{10})	2.31
Pentane (C_5H_{12})	0.24

Used with permission from Simulation Sciences Inc.

Table 5
Air feed

Component	Mole %
Oxygen (O_2)	21.0
Nitrogen (N_2)	78.05
Argon (Ar)	0.95

STAGE 1: CATALYTIC REFORMING

After the removal of sulfur, the primary steam reformer converts about 70% of the hydrocarbon feed into synthesis gas. Methane is mixed with steam and passed over a nickel catalyst. The main reforming reactions are:

$$CH_4 + H_2O \rightleftharpoons CO + 3H_2$$

$$CO + H_2O \rightleftharpoons CO_2 + H_2$$

The catalytic steam hydrocarbon reforming process produces raw synthesis gas by steam reforming under pressure. The reactions are endothermic, thus the supply of heat to the reformer is required to maintain the desired reaction temperature. The gases leaving the reformer are CH_4, 6 mol/%; CO, 8%; CO_2, 6%; H_2, 50%; and H_2O, 30%. The operating pressure is between 20–35 bar, and the gases leaving the reformer contain about 6% CH_4. This represents approximately 30% of the original natural gas input. Figure 2 shows the process flowsheet of catalytic reforming.

In the secondary reformer, air is introduced to supply the nitrogen required for the 3:1 hydrogen H_2 and nitrogen N_2 synthesis gas. The heat of combustion of the partially reformed gas supplies the energy to reform the remaining hydrocarbon feed. The reformed product steam is employed to generate steam and to preheat the natural gas feed.

STAGE 2: SHIFT AND METHANATION CONVERSION

The shift conversion involves two stages. The first stage employs a high-temperature catalyst, and the second uses a low-temperature catalyst. The shift converters remove the carbon monoxide produced in the reforming stage by converting it to carbon dioxide by the reaction

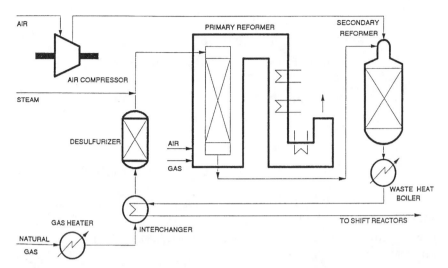

Figure 2. Catalytic reforming flowsheet. *(Used with permission of Simulation Sciences Inc.)*

$$CO + H_2O \rightleftharpoons CO_2 + H_2$$

The reaction produces additional hydrogen for ammonia synthesis. The shift reactor effluent is cooled and the condensed water is separated. The gas is purified by removing carbon dioxide from the synthesis gas by absorption with hot carbonate, Selexol, or methyl ethyl amine (MEA). After purification, the remaining traces of carbon monoxide and carbon dioxide are removed in the methanation reactions.

$$CO + 3H_2 \rightleftharpoons CH_4 + H_2O$$

$$CO_2 + 4H_2 \rightleftharpoons CH_4 + 2H_2O$$

Figure 3 illustrates the shift and methanation conversion. The resulting methane is inert and the water is condensed. Thus purified, the hydrogen-nitrogen mixture with the ratio of $3H_2 : 1N_2$ is compressed to the pressure selected for ammonia synthesis.

STAGE 3: COMPRESSION PROCESS

The purified synthesis gas is cooled and the condensed water is removed. The syngas is then compressed in a series of centrifugal

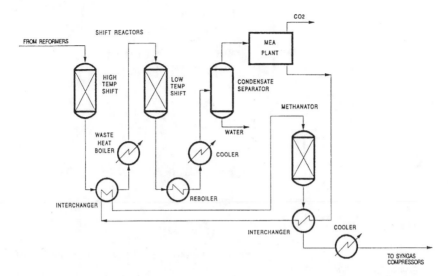

Figure 3. Shift and methanation process flow diagram. *(Used with permission of Simulation Sciences Inc.)*

compressors with interstage cooling to a pressure of 150 bar. The centrifugal compressors are driven by steam turbines using steam generated in the plant itself. This reduces the overall power consumption. Coker [2] illustrates the design of a centrifugal compressor. Figure 4 shows the compressor with interstage cooling.

STAGE 4: CONVERSION UNIT

The compressed synthesis gas is dried, mixed with a recycle stream, and introduced into the synthesis reactor after the recycle compressor. The gas mixture is chilled and liquid ammonia is removed from the secondary separator. The vapor is heated and passed into the ammonia converter. The feed is preheated inside the converter prior to entering the catalyst bed. The reaction occurs at 450–600°C over an iron oxide catalyst. The ammonia synthesis reaction between nitrogen, N_2, and hydrogen, H_2, is

$$N_2 + 3H_2 \rightleftharpoons 2NH_3$$

The reaction is an equilibrium reaction that is exothermic. Lower temperatures favor the production of ammonia. High pressures in

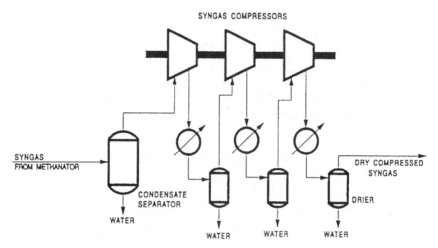

Figure 4. Compressors with interstage cooling. *(Used with permission of Simulation Sciences Inc.)*

excess of 21 bar are required to achieve sufficient conversion. Conversions of 20%–25% ammonia per pass are achieved. However, the conversion of hydrogen per pass is still less than 30%, therefore, the process requires a large recycle of unreacted gases. The converted vapor product is cooled by ammonia refrigeration in the primary separator to condense the ammonia product. A purge is removed from the remaining gases to prevent the buildup up of inerts (in particular, CH_4 and Ar) in the synthesis reactor. Figure 5 shows the process flow diagram of the conversion and Figure 6 illustrates the complete ammonia plant process flow diagram.

Recently, the price of ammonia has nearly doubled as global supplies have been tightened and are now in line with the demands. Ammonia process licensors are employing new technologies that can be retrofitted to existing plants to increase the capacity by 20%–40% [3]. A wide range of newer more reactive catalysts are now replacing the iron-based catalysts. These catalysts are found to be advantageous in operating at lower synthesis pressures. Iron-titanium metals, ruthenium-alkali metals, or ruthenium promoted by potassium and barium on activated carbon have exhibited high efficiency. The raw material hydrogen must be free from the oxides of carbon, which degrade the catalyst activity. Additionally, phosphorus, sulfur, and arsenic compounds tend to poison the catalyst in the subsequent reaction.

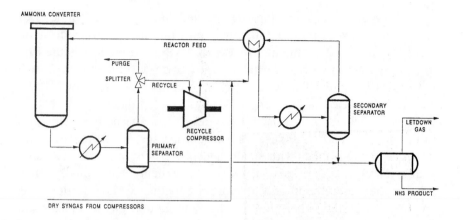

Figure 5. Process flow diagram of the conversion unit. *(Used with permission of Simulation Sciences Inc.)*

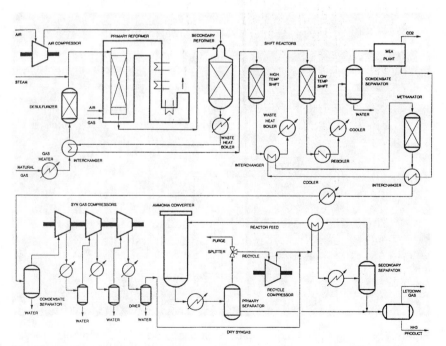

Figure 6. A complete ammonia plant process flow diagram. *(Used with permission of Simulation Sciences Inc.)*

M. W. Kellogg has developed a new technology in the synthesis of ammonia. They employ a ruthenium on graphite as the catalyst on Kellogg Advanced Ammonia Process (KAAP). The process is the first to employ a non-iron based catalyst and was co-developed with British Petroleum Ventures. The KAAP has been commercialized since 1994, and has been used in an increasing number of projects.

Process technology licensors have developed alternative techniques to the primary and secondary reformer processes. These technologies integrate process units with steam and power systems, thereby using heat exchange networks to capture waste heat. Additionally, they provide the energy required for reforming methane. M.W. Kellogg has employed a system where the desulfurized natural gas and steam are first divided into two streams and heated. The mixed feed is then fed to a tubular reforming exchanger and an autothermal reformer. Enriched air at 600°C is then passed to the autothermal reformer and the effluent at 1,000°C flows to the shell side of the reforming heat exchanger. In the autothermal reformer, which contains conventional secondary

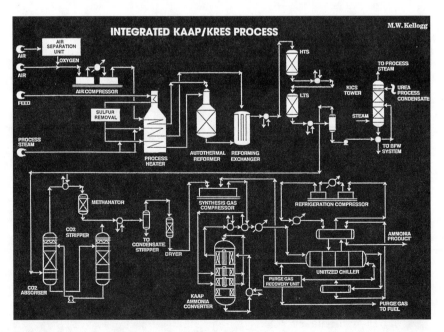

Figure 7. Kellogg's new ruthenium-catalyst based advanced ammonia process combined with the reforming exchange system. *(Used with permission of Chemical Engineering.)*

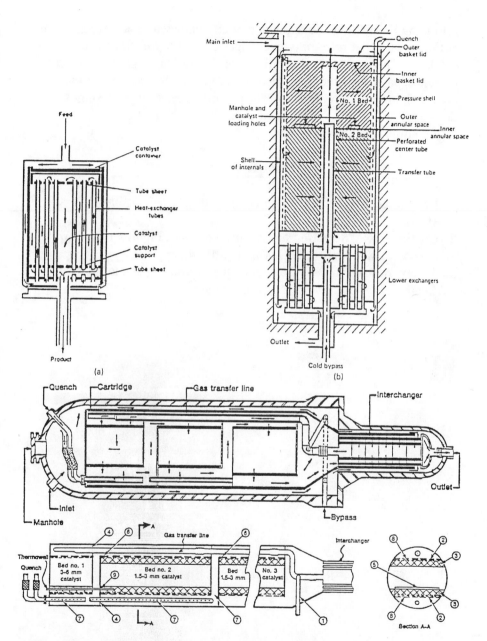

Figure 8. Designs of ammonia synthesis converters (a) Principle of the autothermal ammonia synthesis reactor; (b) Radial flow converter with capacities of 1,800 tpd; (c) Horizontal three-bed converter and detail of the catalyst cartridge. (*Source: Walas, M. S.,* Chemical Process Equipment, Selection and Design, *Butterworth Series in Chemical Engineering, 1988.*)

reforming catalyst, the feed gas is partially oxidized. The mixed stream is then sent to the reforming exchanger consisting of tubes filled with catalysts. This is designed to minimize the buildup of pressure and to expand separately without any constraint. Finally, the heat for reforming comes from an autothermal reformer effluent. Figure 7 shows the designs features in an integrated system. Figure 8 shows a selection of ammonia reactors.

REFERENCES

1. Laidler, K. J. *Chemical Kinetics,* 3rd ed., Harper Collins Publisher, 1987.
2. Coker, A. K., "Selecting and Sizing Process Compressors," *Hydrocarbon Processing,* pp. 39–47, July 1994.
3. Shaley, A. and Ondrey, G., "Ammonia's on the Upswing," *Chemical Engineering,* pp. 30–35, November 1996.

CHAPTER ONE

Reaction Mechanisms and Rate Expressions

INTRODUCTION

The field of chemical kinetics and reaction engineering has grown over the years. New experimental techniques have been developed to follow the progress of chemical reactions and these have aided study of the fundamentals and mechanisms of chemical reactions. The availability of personal computers has enhanced the simulation of complex chemical reactions and reactor stability analysis. These activities have resulted in improved designs of industrial reactors. An increased number of industrial patents now relate to new catalysts and catalytic processes, synthetic polymers, and novel reactor designs. Lin [1] has given a comprehensive review of chemical reactions involving kinetics and mechanisms.

Conventional stoichiometric equations show the reactants that take part and the products formed in a chemical reaction. However, there is no indication about what takes place during this change. A detailed description of a chemical reaction outlining each separate stage is referred to as the mechanism. Mechanisms of reactions are based on experimental data, which are seldom complete, concerning transition states and unstable intermediates. Therefore, they must to be continually audited and modified as more information is obtained.

Reaction steps are sometimes too complex to follow in detail. In such cases, studying the energy of a particular reaction may elucidate whether or not any intermediate reaction is produced. Energy in the form of heat must be provided to the reactants to enable the necessary bonds to be broken. The reactant molecules become activated because of their greater energy content. This change can be referred to as the activated complex or transition state, and can be represented by the curve of Figure 1-1. The complex is the least stable state through

1

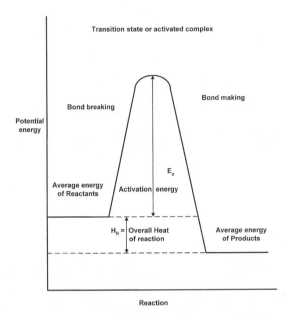

Figure 1-1. Potential energy curve for an exothermic reaction.

which the reactants pass before forming the products. The amount of energy required to raise the reactant molecules to this state is known as the activation energy, E_a. This energy is important in determining the rate at which a reaction proceeds.

The use of a catalyst affects the rate of reaction by enabling the products to form by an alternative route. Each stage has lower activation energy than the uncatalyzed reaction.

Once the reactants have absorbed sufficient energy to cross over this peak, energy is then released as the new bonds are made in yielding the stable products. For reactions at constant pressure, the difference between the amount of energy provided to break the bonds of the reactants and that evolved during the formation of new molecules is termed the enthalpy of reaction, ΔH_R. When more energy is evolved than absorbed, the reaction is exothermic, that is, ΔH_R is negative as shown in Figure 1-1. Alternatively, when less energy is evolved than absorbed, the reaction is endothermic, that is, ΔH_R is positive as indicated by Figure 1-2.

Products formed through an exothermic reaction have a lower energy content than the reactants from which they are formed. Alternatively, products formed via an exothermic process have a higher

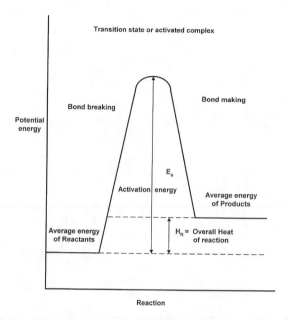

Figure 1-2. Potential energy curve for an endothermic reaction.

energy content than their reactants. In general, exothermic compounds are more stable than endothermic compounds.

There are cases where the activated complex exists as an unstable intermediate. This is observed in reaction profile as a trough in the activated peak of the curve. This produces a double hump and as the minimum in the trough is more marked, that is, as the intermediate becomes more stable, it becomes more difficult to separate the intermediate from the reaction mixture during the course of the reaction. Figure 1-3 shows the curve of an unstable intermediate.

Generally, all practical reactions occur by a sequence of elementary steps that collectively constitute the mechanism. The rate equation for the overall reaction is developed from the mechanism and is then used in reactor design. Although there are cases where experimental data provide no information about intermediate chemical species, experimental data have provided researchers with useful guidelines in postulating reaction mechanisms. Information about intermediate species is essential in identifying the correct mechanism of reaction. Where many steps are used, different mechanisms can produce similar forms of overall rate expression. The overall rate equation is the result

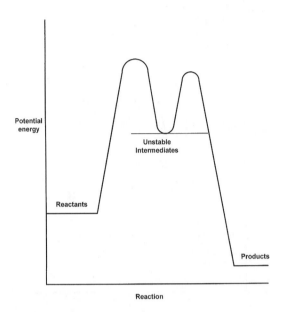

Figure 1-3. Potential energy curve for reactions showing the unstable intermediate state.

of the correct mechanism and is developed in terms of concentrations of the reactants and products. In the case of complex chemical reactions, the overall rate equation may be erroneous for reactor design. Therefore, assumptions are employed to make a satisfactory kinetic representation resulting in the design of a reliable reactor.

Chemists and engineers interpret the mechanisms in different ways. The chemist defines the reaction mechanism as how the electron densities around the molecule move in order to provide charged areas, allowing the second reactant to attach because of induced opposite charge. The activated complex has a modified electron structure that results in part of the complex being weakly attached, thereby making detachment possible. The overall rearrangement of the charges around the molecules gives the product of the reaction. The chemical engineer, on the other hand, often views the mechanism in terms of its reaction steps, where each step changes from one distinct chemical species to another. This reduces the reaction mechanism so that it can be treated quantitatively.

The following describes many types of reaction mechanisms with a view toward developing their overall rate expressions.

TYPICAL REACTION MECHANISMS

CRACKING OF ALKANES (Paraffins C_nH_{2n+2})

Pyrolysis of alkanes is referred to as cracking. Alkanes from the paraffins (kerosene) fraction in the vapor state are passed through a metal chamber heated to 400–700°C. Metallic oxides are used as a catalyst. The starting alkanes are broken down into a mixture of smaller alkanes, alkenes, and some hydrogen.

$$\text{Alkanes} \xrightarrow{400-700^\circ C} \text{Smaller alkanes} + \text{alkenes} + H_2$$

For example:

$$2CH_3CH_2CH_3 \xrightarrow{400-700^\circ C} CH_4 + CH_3CH = CH_2 + CH_2 = CH_2 + H_2 \quad (1\text{-}1)$$
$$\phantom{2CH_3CH_2CH_3 \xrightarrow{xx}CH_4 + }\text{Propene}\text{Ethene}$$

Mechanism

The reaction proceeds via a free radical mechanism with homolytic fission occuring between the carbon-carbon atoms. The mechanism of reactions for the cracking of propane is:

Chain initiation:

$$1. \quad CH_3CH_2 - CH_3 \rightarrow CH_3CH_2^* + CH_3^* \qquad (1\text{-}2)$$

Chain propagation:

$$2. \quad CH_3CH_2^* \rightarrow CH_2 = CH_2 + H^* \qquad (1\text{-}3)$$
$$3. \quad CH_3CH_2CH_3 + H^* \rightarrow H_2 + CH_3\dot{C}HCH_3 \qquad (1\text{-}4)$$
$$4. \quad CH_3CH_2CH_3 + CH_3^* \rightarrow CH_4 + CH_3\dot{C}HCH_3 \qquad (1\text{-}5)$$

Chain termination:

$$5. \quad 2CH_3CH_2^* \rightarrow CH_2 = CH_2 + CH_3CH_3 \qquad (1\text{-}6)$$
$$6. \quad 2CH_3\dot{C}HCH_3 \rightarrow CH_3CH = CH_2 + CH_3CH_2CH_3 \qquad (1\text{-}7)$$

SULFUR DIOXIDE OXIDATION

The overall stoichiometric equation is:

$$SO_2 + \frac{1}{2}O_2 \rightleftharpoons SO_3 \qquad (1\text{-}8)$$

Mechanism

Vanadium pentoxide, V_2O_5, is used as a catalyst in the oxidation of sulfur dioxide. The mechanism involves oxidation-reduction of V_2O_5 that exists on the support at operating conditions in the molten state. The mechanism of reaction is:

1. $SO_2 + 2V^{5+} + O^{2-} \rightleftharpoons SO_3 + 2V^{4+}$ $\qquad (1\text{-}9)$

2. $\frac{1}{2}\left[O_2 + V^{4+} \rightleftharpoons V^{5+} + O_2^-\right]$ $\qquad (1\text{-}10)$

3. $\frac{1}{2}\left[O_2^- + V^{4+} \rightleftharpoons V^{5+} + 2O^-\right]$ $\qquad (1\text{-}11)$

4. $O^- + V^{4+} \rightleftharpoons V^{5+} + O^{2-}$ $\qquad (1\text{-}12)$

AMMONIA SYNTHESIS

The overall stoichiometric equation is:

$$\frac{1}{2}N_2 + \frac{3}{2}H_2 \rightarrow NH_3 \qquad (1\text{-}13)$$

Studies of ammonia synthesis on iron catalyst suggest that the reaction occurs through surface imine radicals.

Mechanism

The following elementary steps are:

1. $N_{2(g)} \rightarrow 2N_{(ads)}$ $\qquad (1\text{-}14)$

2. $H_{2(g)} \rightarrow 2H_{(ads)}$ $\qquad (1\text{-}15)$

3. $N_{(ads)} + H_{(ads)} \rightarrow NH_{(ads)}$ $\qquad (1\text{-}16)$

4. $NH_{(ads)} + H_{(ads)} \rightarrow NH_{2(ads)}$ (1-17)

5. $NH_{2(ads)} + H_{(ads)} \rightarrow NH_{3(ads)}$ (1-18)

6. $NH_{3(ads)} \rightarrow NH_{3(g)}$ (1-19)

The "ads" denotes the adsorbed species.

AMMONIA OXIDATION

The overall stoichiometric reaction for the oxidation of ammonia to nitric oxide is:

$$4NH_3 + 5O_2 \rightarrow 4NO + 6H_2O \qquad (1-20)$$

This reaction is very rapid and has been difficult to study mechanistically. The direct oxidation of ammonia, NH_3, to nitric oxide, NO, over platinum catalyst is one of the major steps in the manufacture of nitric acid, HNO_3.

Mechanism

1. $O_2 \rightarrow 2O^*$ (1-21)

2. $NH_3 + O^* \rightarrow NH_2OH$ (1-22)

3. $NH_2OH \rightarrow NH^* + H_2O$ (1-23)

4. $NH^* + O_2 \rightarrow HNO_2$ (1-24)

5. $HNO_2 \rightarrow NO + OH^*$ (1-25)

6. $2OH^* \rightarrow H_2O + O^*$ (1-26)

The oxygen is chemisorbed on the catalyst. This then reacts with ammonia to produce a chemisorbed imide radical. The imide reacts with a molecular oxygen to yield nitric oxide.

STEAM REFORMING

Steam reforming is an important process to generate hydrogen for such uses as ammonia synthesis because of the high endothermic heat reaction and its rapidity. High heat fluxes with a direct-fired furnace are required. Although many steps of reactions are possible, the typical reaction steps are as follows:

1. $CH_4 + H_2O \rightarrow 3H_2 + CO$ $\qquad$ (1-27)

2. $CH_4 + 2H_2O \rightarrow 4H_2 + CO_2$ $\qquad$ (1-28)

3. $CO + H_2O \rightarrow CO_2 + H_2$ $\qquad$ (1-29)

4. $CH_4 + CO_2 \rightarrow 2CO + 2H_2$ $\qquad$ (1-30)

5. $CO + H_2 \rightleftharpoons C + H_2O$ $\qquad$ (1-31)

6. $CH_4 \rightleftharpoons C + 2H_2$ $\qquad$ (1-32)

7. $2CO \rightleftharpoons C + CO_2$ $\qquad$ (1-33)

BIOCHEMICAL REACTION: CONVERSION OF GLUCOSE TO GLUCONIC ACID

The fermentation of glucose to gluconic acid involves oxidation of the sugar in the aldehyde group (RCHO), thereby converting it to a carboxyl group (RCOOH). The conversion process is achieved by a micro-organism in the fermentation process. The enzyme glucose oxidase that is present in the micro-organism converts the glucose to gluconolactone. The gluconolactone is further hydrolyzed to form the gluconic acid.

The enzyme glucose oxidase is useful in medicinal applications where glucose or oxygen removal is required. This enzyme prevents the browning reaction, or a Mailland reaction, when dried egg powders are darkened due to a reaction between glucose and protein. Also, the presence of oxygen in the production and storage of orange soft drinks, dried food powders, canned beverages, salad dressing, and cheese slices can be avoided by adding glucose oxidase. Since the activity of the enzyme is maintained for a long time at storage temperature, such enzyme additions increase the shelf-life of food products. This is achieved by the removal of oxygen that diffuses through food packaging [2].

The hydrogen peroxide produced in the glucose oxidase catalyzed reaction has an antibacterial action. The addition of a catalase catalyzes the decomposition of hydrogen peroxide to water and oxygen.

REACTION MECHANISMS

The reaction mechanisms in the fermentation of glucose to gluconic acid are:

1. *Cell growth:*

   Glucose + cells → more cells

   That is:

   $$C_6H_{12}O_6 + \text{cells} \rightarrow \text{more cells} \qquad (1\text{-}34)$$

2. *Glucose oxidation:*

   Glucose + $O_2 \xrightarrow{\text{Glucose oxidase}}$ gluconolactone + H_2O

   That is:

   $$C_6H_{12}O_6 + O_2 \xrightarrow{\text{Glucose oxidase}} C_6H_{10}O_6 + H_2O_2 \qquad (1\text{-}35)$$

3. *Gluconolactone hydrolysis:*

   Gluconolactone + H_2O → gluconic acid

   That is:

   $$C_6H_{10}O_6 + H_2O \longrightarrow C_5H_{11}O_5COOH \qquad (1\text{-}36)$$

4. *Hydrogen peroxide decomposition:*

   $$H_2O_2 \xrightarrow{\text{Catalase}} H_2O + \frac{1}{2}O_2 \qquad (1\text{-}37)$$

Analysis of the rate equation and kinetic model of the conversion of glucose to gluconic acid is discussed in Chapter 11.

ELEMENTARY AND NON-ELEMENTARY REACTIONS

Consider the reaction between hydrogen and bromine in the gas phase:

$$H_2 + Br_2 \rightarrow 2HBr \qquad (1\text{-}38)$$

Bodenstein and Lind [3] first studied the thermal reaction over the temperature range of 500–600 K. The relative reaction rates of hydrogen and bromine and the formation of hydrogen bromide are:

$$\left(-r_{H_2}\right) = \left(-r_{Br_2}\right) = \frac{1}{2}\left(+r_{HBr}\right) \qquad (1\text{-}39)$$

and, if the reaction took place in a single step, then the rate expression may be represented as:

$$\left(+r_{HBr}\right) = k\,C_{H_2}C_{Br_2} \qquad (1\text{-}40)$$

The slope of a concentration-time curve to define the rate expression can be determined. However, experimental studies have shown the reaction cannot be described by simple kinetics, but by the relationship:

$$\left(+r_{HBr}\right) = \frac{k_1 C_{H_2} C_{Br_2}^{0.5}}{k_2 + \dfrac{C_{HBr}}{C_{Br_2}}} \qquad (1\text{-}41)$$

where k_1 and k_2 are the rate constants. The reaction between hydrogen and bromine is an example of a non-elementary reaction. The following steps account for the rate expression:

$$Br_2 \rightarrow 2Br^* \qquad (1\text{-}42)$$

$$Br^* + H_2 \rightarrow HBr + H^* \qquad (1\text{-}43)$$

$$H^* + HBr \rightarrow Br^* + H_2 \qquad (1\text{-}44)$$

Br^* and H^* being highly reactive intermediates.

TYPES OF INTERMEDIATE

STABLE INTERMEDIATES

Stable intermediates are those where concentration and lifespan are comparable to those of stable reactants and products. An example is the reaction between methane and oxygen in the gas phase at 700 K and 1 atmosphere. The overall reaction is:

$$CH_4 + 2O_2 \rightarrow CO_2 + 2H_2O \qquad (1\text{-}45)$$

Successive single reactions are:

$$CH_4 + O_2 \rightarrow CH_2O + H_2O \qquad (1\text{-}46)$$
$$(Formaldehyde)$$

$$CH_2O + O_2 \rightarrow CO + H_2O_2 \qquad (1\text{-}47)$$

$$CO + \frac{1}{2}O_2 \rightarrow CO_2 \qquad (1\text{-}48)$$

$$H_2O_2 \rightarrow H_2O + \frac{1}{2}O_2 \qquad (1\text{-}49)$$

In these reactions, the stable intermediates are CH_2O, CO, and H_2O_2.

ACTIVE CENTERS

Reactions that are catalyzed by solids occur on the surfaces of the solids at points of high chemical activity. Therefore, the activity of a catalytic surface is proportional to the number of active centers per unit area. In many cases, the concentration of active centers is relatively low. This is evident by the small quantities of poisons present (material that retards the rate of a catalytic reaction) that are sufficient to destroy the activity of a catalyst. Active centers depend on the interatomic spacing of the solid structure, chemical constitution, and lattice structure.

Generally, active centers are highly reactive intermediates present in very small concentrations with short lifespans. For example, in the case of

$$H_2 + Br_2 \rightarrow 2HBr \qquad (1\text{-}50)$$

$$H^* \text{ and } Br^*$$

and in the case of

$$CH_3CHO \rightarrow CH_4 + CO \qquad (1\text{-}51)$$

$$CH_3^*, CHO^*, CH_3CO^*$$

TRANSITION STATE INTERMEDIATES

Each elementary step proceeds from reactants to products through the formation of an intermediate called the transition state. Such intermediates cannot be isolated, as they are species in transit. The act of reaction will involve the breaking or making of a chemical bond, whereby transition state intermediates are formed (see also "Transition State Theory").

THE ARRHENIUS EQUATION AND THE COLLISION THEORY

THE ARRHENIUS EQUATION

Generally, the rate of reaction depends on three principal functions: temperature, pressure, and composition. However, as a result of phase rule and thermodynamics, there is a relationship between temperature, pressure, and composition. This relationship can be expressed as:

$$r_i = f(\text{temperature, composition}) \tag{1-52}$$

Consider the reaction:

$$A + B \rightarrow C + D \tag{1-53}$$

Here, a molecule of C is formed only when a collision between molecules of A and B occurs. The rate of reaction r_C (that is, rate of appearance of species C) depends on this collision frequency. Using the kinetic theory of gases, the reaction rate is proportional to the product of the concentration of the reactants and to the square root of the absolute temperature:

$$r_C \propto C_A C_B T^{0.5} \tag{1-54}$$

The number of molecules reacting per unit time is smaller than the number of binary collisions between A and B. Also, temperature is known to have a much greater effect on the reaction rate than one would expect from $T^{0.5}$. For binary collisions between A and B to result in a reaction, the collision must involve energies of translation and vibration that are in excess of energy E, known as the activation energy. The fraction of collisions having energies in excess of E is represented by $e^{-E/RT}$, which can now be substituted in Equation 1-54 to give:

$$r_C \propto e^{-E/RT} C_A C_B T^{0.5} \tag{1-55}$$

The effect of temperature in $T^{0.5}$ is small compared with its effect in $e^{-E/RT}$; therefore, $T^{0.5}$ can be combined with the proportionality constant resulting in:

$$r_C = k_0 e^{-E/RT} C_A C_B \tag{1-56}$$

Generally, $r_i = f_1$ (temperature) f_2 (composition) and at a given temperature:

$$r_i = k f_2 \text{(composition)} \tag{1-57}$$

where

$$\boxed{k = k_0 e^{-E/RT}} \tag{1-58}$$

where k = reaction rate constant or velocity constant
k_0 = frequency factor or preexponential factor
E = activation energy, J/mol or cal/mol
R = gas constant = 8.314 J/mol•K = 1.987 cal/mol•K
T = absolute temperature, K

Equation 1-58 is referred to as the Arrhenius equation.

Effect of Temperature on Reaction Rates

We can evaluate the effect of temperature on the reaction rate from the Arrhenius equation, $k = k_0 e^{-E/RT}$, as:

$$\ln k = \ln k_0 - \frac{E}{RT} \tag{1-59}$$

When plotting experimentally determined reaction rate constants as a function of temperature (i.e., ln k against 1/T), a straight line is obtained with –E/R equal to the slope and the intercept as ln k_0. Figure 1-4 shows the linear relationship between the reaction rate constant and the temperature.

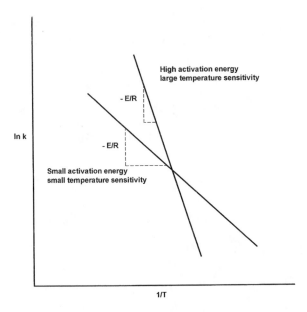

Figure 1-4. Reaction rate temperature dependence.

THE COLLISION THEORY

The collision theory for bimolecular reactions assumes that a chemical reaction occurs when two molecules collide with enough energy to penetrate the molecular van der Waals repulsive forces, thus combining together. For the bimolecular collisions of unlike molecules A, the collision number is:

$$Z_{AA} = \sigma_A^2 n_A^2 \left(\frac{4\pi kT}{M_A} \right)^{0.5} = \sigma_A \frac{N^2}{10^6} \left(\frac{4\pi kT}{M_A} \right)^{0.5} C_A^2$$

$$= \frac{\text{number of collisions of A with A}}{\text{sec} \cdot \text{cm}^3} \tag{1-60}$$

For bimolecular collisions of unlike molecules in a mixture of A and B,

$$A + B \rightarrow \text{Products}$$

$$Z_{AB} = \left(\frac{\sigma_A + \sigma_B}{2}\right)^2 n_A \cdot n_B \left(8\pi kT \cdot \frac{M_A + M_B}{M_A \cdot M_B}\right)^{0.5}$$

$$= \left(\frac{\sigma_A + \sigma_B}{2}\right)^2 \frac{N^2}{10^6} \left(8\pi kT \cdot \frac{M_A + M_B}{M_A . M_B}\right)^{0.5} C_A C_B \qquad (1\text{-}61)$$

where σ = diameter of a molecule, cm

$\quad\quad$ M = (molecular weight)/N, mass of a molecule, gm

$\quad\quad$ N = 6.023×10^{23} molecules/mol, Avogadro's number

$\quad\quad$ C_A = concentration of A, mol/l

$\quad\quad$ C_B = concentration of B, mol/l

$\quad\quad$ n_A = $NC_A/10^3$, number of molecules of A/cm^3

$\quad\quad$ n_B = $NC_B/10^3$, number of molecules of B/cm^3

$\quad\quad$ k = R/N = 1.30×10^{-16} erg/K, Boltzmann constant

The rate equation is given by

$$\left(-r_A\right) = -\frac{1}{V}\frac{dN_A}{dt} = kC_A C_B$$

$$= \left(\begin{array}{c}\text{collision rate} \\ \text{mol}/l\cdot\text{sec}\end{array}\right)\left(\begin{array}{c}\text{fraction of collision} \\ \text{involving energies} \\ \text{in excess of E}\end{array}\right) \qquad (1\text{-}62)$$

where E is the minimum energy.

$$= Z_{AB} \cdot \frac{10^3}{N} e^{-E/RT}$$

$$= \left(\frac{\sigma_A + \sigma_B}{2}\right)^2 \frac{N}{10^3} \left(8\pi kT \cdot \frac{M_A + M_B}{M_A M_B}\right)^{0.5} C_A C_B e^{-E/RT} \qquad (1\text{-}63)$$

where $e^{-E/RT}$ represents the fraction of collisions involving molecules with the necessary activation energy E.

TRANSITION STATE THEORY

The transition state theory describes reactants combining to form unstable intermediates called activated complexes, which rapidly decompose into products. This is represented by the following reaction steps:

$$A + B \rightarrow P \qquad\qquad (1\text{-}64)$$

$$A + B \rightleftharpoons AB^* \rightarrow P \qquad\qquad (1\text{-}65)$$

      Transition state intermediate

The overall rate of reaction depends on the rate of decomposition of AB* to product P.

The collision theory considers the rate to be governed by the number of energetic collisions between the reactants. The transition state theory considers the reaction rate to be governed by the rate of the decomposition of intermediate. The formation rate of the intermediate is assumed to be rapid because it is present in equilibrium concentrations.

CHAIN REACTIONS

Atoms and free radicals are highly reactive intermediates in the reaction mechanism and therefore play active roles. They are highly reactive because of their incomplete electron shells and are often able to react with stable molecules at ordinary temperatures. They produce new atoms and radicals that result in other reactions. As a consequence of their high reactivity, atoms and free radicals are present in reaction systems only at very low concentrations. They are often involved in reactions known as chain reactions. The reaction mechanisms involving the conversion of reactants to products can be a sequence of elementary steps. The intermediate steps disappear and only stable product molecules remain once these sequences are completed. These types of reactions are referred to as open sequence reactions because an active center is not reproduced in any other step of the sequence. There are no closed reaction cycles where a product of one elementary reaction is fed back to react with another species. Reversible reactions of the type $A + B \rightleftharpoons C + D$ are known as open sequence mechanisms. The chain reactions are classified as a closed sequence in which an active center is reproduced so that a cyclic reaction pattern is set up. In chain reaction mechanisms, one of the reaction intermediates is regenerated during one step of the reaction. This is then fed back to an earlier stage to react with other species so that a closed loop or cycle is formed. The intermediate species are intermittently formed by the reaction, and the final products are the result of a cyclic repetition of the intervening processes. Chain reactions play active roles in industrial processes.

Bodenstein [4] first suggested the idea of a chain reaction by postulating that ions such as chlorine, Cl_2^+, are chain carriers. Later, Nernst [5] proposed his mechanism for the hydrogen in chlorine reaction that gave rise to the idea that organic free radicals are important in reaction mechanisms. Taylor [6] investigated the reactions of hydrogen atoms with various substances. He proposed a reaction between a hydrogen atom and ethane. The resulting ethyl radical plays an important role in hydrocarbon reactions. Rice and Herzfeld [7] postulated reaction schemes involving the participation of free radicals in the pyrolysis of organic compounds. Techniques such as spectroscopy, electron-spin resonance spectroscopy, and mass spectrometry have confirmed the validity of these types of reaction mechanism. They confirmed that free radicals are important in many reactions, and thus act as chain carriers. Chain reactions are involved in such processes as combustion, polymerization, and photochemical processes.

A chain reaction consists of three main steps:

1. Initiation (or activation)
2. Propagation (closed sequence steps)
3. Termination

The reaction between hydrogen and bromine is the first step in the initiation reaction in which the chain carriers are formed. The thermal hydrogen bromine reaction begins with the initiation reaction:

$$Br_2 \rightarrow Br^* + Br^* \qquad \text{(1-66)}$$

This is followed by chain propagation reactions:

$$Br^* + H_2 \rightarrow HBr + H^* \qquad \text{(1-67)}$$

$$H^* + Br_2 \rightarrow HBr + Br^* \qquad \text{(1-68)}$$

$$H^* + HBr \rightarrow H_2 + Br^* \qquad \text{(1-69)}$$

The termination reaction is the removal of the carrier from the system:

$$Br^* + Br^* \rightarrow Br_2 \tag{1-70}$$

The Rice-Herzfeld mechanisms of reaction are:

1. *Initiation*
   - Free radicals are formed by scission of the weakest bond in the molecule.
2. *Propagation*
   - One or both of the radicals formed in the initiation step abstracts a hydrogen atom from the parent compound to form a small saturated molecule and a new free radical.
   - The new free radical stabilizes itself by splitting out a simple molecule such as olefin or CO:

$$RCH_2 - CH_2^* \rightarrow R^* + CH_2 = CH_2$$

3. *Termination*
   - The chain is broken by a combination or disproportionation reaction between the two radicals.

Employing mechanistic equations based on the Rice-Herzfeld postulation yields:

1. *Initiation*

$$M \xrightarrow{\ k_1\ } R^* + R_1^* \tag{1-71}$$

2. *Propagation*

$$R^* + M \xrightarrow{\ k_2\ } R_2 + R^*H \tag{1-72}$$

$$R_2 \xrightarrow{\ k_3\ } R^* + P_1 \tag{1-73}$$

3. *Termination*

$$R^* + R^* \xrightarrow{\ k_4\ } P_2 \tag{1-74}$$

$$R^* + R_2 \xrightarrow{\ k_5\ } P_3 \tag{1-75}$$

$$R_2 + R_2 \xrightarrow{\ k_6\ } P_4 \tag{1-76}$$

The following mechanism has been postulated for the gas phase decomposition of acetaldehyde:

$$CH_3CHO \xrightarrow{\text{Thermal decomposition}} CH_4 + CO$$

Acetaldehyde   Methane Carbon monoxide

Initiation

$$CH_3CHO \xrightarrow{k_1} CH_3^* + CHO^* \qquad (1\text{-}77)$$

This thermal initiation generates two free radicals by breaking a covalent bond. The aldehyde radical is long-lived and does not markedly influence the subsequent mechanism. The methane radical is highly reactive and generates most reactions.

Propagation

$$CH_3^* + CH_3CHO \xrightarrow{k_2} CH_4 + CH_3CO^* \qquad (1\text{-}78)$$

$$CH_3CO^* \xrightarrow{k_3} CO + CH_3^* \qquad (1\text{-}79)$$

The propagation reactions use a methyl radical and generate another. There is no net consumption, and a single initiation reaction can result in an indefinite number of propagation reactions.

Termination

$$CH_3^* + CH_3^* \xrightarrow{k_4} C_2H_6 \qquad (1\text{-}80)$$

The following mechanism has been postulated for the decomposition of ethane into ethylene and hydrogen. The overall rate expression is first order in ethane.

Initiation

$$C_2H_6 \xrightarrow{k_1} CH_3^* + CH_3^* \qquad (1\text{-}81)$$

Chain transfer

$$CH_3^* + C_2H_6 \xrightarrow{k_2} CH_4 + C_2H_5^* \tag{1-82}$$

Propagation

$$C_2H_5^* \xrightarrow{k_3} C_2H_4 + H^* \tag{1-83}$$

$$H^* + C_2H_6 \xrightarrow{k_4} H_2 + C_2H_5^* \tag{1-84}$$

Termination

$$H^* + C_2H_5^* \xrightarrow{k_5} C_2H_6 \tag{1-85}$$

High temperatures, electromagnetic radiation, ionizing radiation, or highly reactive chemical initiators often generate active centers.

POLYMERIZATION KINETICS

Many polymerization reactions proceed by free radical mechanisms. The following is a sequence of elementary steps for the polymerization reaction:

1. Dissociation of initiator (I) into two radical fragments

$$I \xrightarrow{k_D} R^* + R^* \tag{1-86}$$

2. Addition of initiator monomer. M

$$R^* + M \xrightarrow{k_i} RM^* \tag{1-87}$$

3. Chain propagation

$$RM^* + M \xrightarrow{k_P} RMM^* \tag{1-88}$$

$$RMM^* + M \xrightarrow{k_P} RMMM^* \tag{1-89}$$

$$RMMM^* + M \xrightarrow{k_P} RMMMM^* \tag{1-90}$$

4. Chain termination
  (a) Radical combination

$$R(M)_m^* + R(M)_n \xrightarrow{k_{tc}} R(M)_m (M)_n R \tag{1-91}$$

  (b) Radical disproportionation

$$R(M)_m^* + R(M)_n \xrightarrow{k_{tD}} \text{saturated polymer} \atop + \text{ unsaturated polymer} \tag{1-92}$$

CATALYTIC REACTIONS

ENZYME CATALYZED REACTIONS

Enzymes are proteins of high molecular weight and possess exceptionally high catalytic properties. These are important to plant and animal life processes. An enzyme, E, is a protein or protein-like substance with catalytic properties. A substrate, S, is the substance that is chemically transformed at an accelerated rate because of the action of the enzyme on it. Most enzymes are normally named in terms of the reactions they catalyze. In practice, a suffice -*ase* is added to the substrate on which the enzyme acts. For example, the enzyme that catalyzes the decomposition of urea is *urease,* the enzyme that acts on uric acid is *uricase,* and the enzyme present in the micro-organism that converts glucose to gluconolactone is *glucose oxidase.* The three major types of enzyme reaction are:

Soluble enzyme–insoluble substrate
Insoluble enzyme–soluble substrate
Soluble enzyme–soluble substrate

The study of enzymes is important because every synthetic and degradation reaction in all living cells is controlled and catalyzed by specific enzymes. Many of these reactions are the soluble enzyme–soluble substrate type and are homogeneous in the liquid phase.

The simplest type of enzymatic reaction involves only a single reactant or substrate. The substrate forms an unstable complex with the enzyme that decomposes to give the product species or, alternatively, to generate the substrate.

Using the Bodenstein steady state approximation for the intermediate enzyme substrate complexes derives reaction rate expressions for enzymatic reactions. A possible mechanism of a closed sequence reaction is:

$$\underset{k_2}{\overset{k_1}{\rightleftharpoons}} \quad E + S \overset{k_1}{\underset{k_2}{\rightleftharpoons}} ES^* \tag{1-93}$$

Enzyme Substrate enzyme-substrate complex

$$ES^* \xrightarrow{k_3} E + P \tag{1-94}$$

where E = enzyme
$\quad\quad\;\;$ S = substrate
$\quad\;$ ES* = enzyme-substrate complex
$\quad\quad\;\;$ P = product of the reaction

The stoichiometry of the reaction may be represented as:

$$S \rightarrow P \tag{1-95}$$

The net rate of an enzymatic reaction is usually referred to as its velocity, V, represented by:

$$V = \frac{dC_P}{dt} = k_3 C_{ES^*} \tag{1-96}$$

The concentration of the complex can be obtained from the net rate of disappearance:

$$\left(-r_{SE^*}\right)_{net} = -\frac{dC_{SE^*}}{dt} = k_2 C_{SE^*} + k_3 C_{SE^*} - k_1 C_S C_E \tag{1-97}$$

Using the steady state approximation,

$$\frac{dC_{SE^*}}{dt} = k_1 C_S C_E - k_2 C_{SE^*} - k_3 C_{SE^*} \cong 0 \tag{1-98}$$

or

$$C_{SE^*} = \frac{k_1 C_S C_E}{k_2 + k_3}$$

(1-99)

Substituting Equation 1-99 into Equation 1-96 gives:

$$V = \frac{k_1 \cdot k_3 \cdot C_E \cdot C_S}{k_2 + k_3}$$

(1-100)

From the material balance, the total concentration of the enzyme in the system, C_{ET}, is constant and equal to the sum of the concentrations of the free or unbounded enzyme, C_E, and the enzyme-substrate complex, C_{SE^*}, that is:

$$C_{ET} = C_E + C_{SE^*}$$

(1-101)

For the substrate C_S, the total concentration of the substrate in the system, C_{ST}, is equal to the sum of the concentration of the substrate and the enzyme substrate complex C_{SE^*}

$$C_{ST} = C_S + C_{SE^*}$$

(1-102)

In laboratory conditions, $C_{ST} \gg C_{ET}$, since C_{SE^*} cannot exceed C_{ET}. This shows that $C_{ST} \approx C_S$.

Rearranging Equation 1-101 gives:

$$C_E = C_{ET} - C_{SE^*}$$

(1-103)

Using $C_{ST} \approx C_S$ in Equation 1-102 and substituting Equation 1-103 into Equation 1-98 gives:

$$\frac{dC_{SE^*}}{dt} = k_1 C_{ST}\left(C_{ET} - C_{SE^*}\right) - k_2 C_{SE^*} - k_3 C_{SE^*} \cong 0$$

$$k_1 C_{ST} C_{ET} - k_1 C_{ST} C_{SE^*} - k_2 C_{SE^*} - k_3 C_{SE^*} = 0$$

(1-104)

or

$$C_{SE^*} = \frac{k_1 C_{ST} C_{ET}}{k_1 C_{ST} + k_2 + k_3}$$

(1-105)

Substituting Equation 1-105 into Equation 1-96 gives

$$V_O = \frac{dC_P}{dt} = \frac{k_1 k_3 C_{ST} C_{ET}}{k_1 C_{ST} + k_2 + k_3} \tag{1-106}$$

Equation 1-106 predicts that the initial rate will be proportional to the initial enzyme concentration, if the initial substrate concentration is held constant. If the initial enzyme concentration is held constant, then the initial rate will be proportional to the substrate concentration at low substrate concentrations and independent of the substrate concentration at high substrate levels. The maximum reaction rate for a given total enzyme concentration is

$$V_{max} = k_3 C_{ET} \tag{1-107}$$

Equation 1-106 can be rearranged to

$$\boxed{V_o = \frac{V_{max} C_{ST}}{C_{ST} + \left(\dfrac{k_2 + k_3}{k_1} \right)}} \tag{1-108}$$

Equation 1-108 can be considered as the Michaelis-Menten equation, where K_m is the Michaelis constant and represented as

$$K_m = \frac{k_2 + k_3}{k_1} \tag{1-109}$$

Equation 1-108 then becomes

$$V_o = \frac{V_{max} C_{ST}}{C_{ST} + K_m} \tag{1-110}$$

Rearranging Equation 1-110 gives

$$\frac{1}{V_o} = \frac{1}{V_{max}} + \frac{K_m}{V_{max} C_{ST}} \tag{1-111}$$

Equation 1-111 is known as the Lineweaver-Burk or reciprocal plot. If the data fit this model, a plot of $1/V_o$ versus $1/C_{ST}$ will be linear with a slope K_m/V_{max} and the intercept $1/V_{max}$.

At low substrate concentration, Equation 1-110 becomes

$$V_o \approx \frac{V_{max}C_{ST}}{K_m} \qquad\qquad (1\text{-}112)$$

At high substrate concentration,

$$C_{ST} \gg K_m$$

and Equation 1-110 becomes

$$V_O \approx V_{max}$$

Figure 1-5 shows a plot of $1/V_o = 1/V_{max} + K_m/V_{max}C_{ST}$.

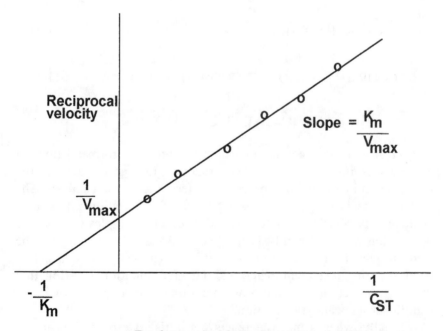

Reciprocal substrate concentration

Figure 1-5. Lineweaver-Burk plot.

ACID-BASE CATALYSIS: HOMOGENEOUS LIQUID PHASE

A catalyst is defined as a substance that influences the rate or the direction of a chemical reaction without being consumed. Homogeneous catalytic processes are where the catalyst is dissolved in a liquid reaction medium. The varieties of chemical species that may act as homogeneous catalysts include anions, cations, neutral species, enzymes, and association complexes. In acid-base catalysis, one step in the reaction mechanism consists of a proton transfer between the catalyst and the substrate. The protonated reactant species or intermediate further reacts with either another species in the solution or by a decomposition process. Table 1-1 shows typical reactions of an acid-base catalysis. An example of an acid-base catalysis in solution is hydrolysis of esters by acids.

$$R_1COOR_2 + H_2O \overset{Acid}{\rightleftharpoons} R_1COOH + R_2OH \qquad (1\text{-}113)$$

Mechanism

$$R_1COOR_2 + H^+ \rightleftharpoons \left(R_1COOR_2H^+\right)^* \qquad (1\text{-}114)$$

$$\left(R_1COOR_2H^+\right)^* + H_2O \rightleftharpoons R_1COOH + R_2OH + H^+ \qquad (1\text{-}115)$$

AUTOCATALYTIC REACTIONS

There are many reactions in which the products formed often act as catalysts for the reaction. The reaction rate accelerates as the reaction continues, and this process is referred to as autocatalysis. The reaction rate is proportional to a product concentration raised to a positive exponent for an autocatalytic reaction. Examples of this type of reaction are the hydrolysis of several esters. This is because the acids formed by the reaction give rise to hydrogen ions that act as catalysts for subsequent reactions. The fermentation reaction that involves the action of a micro-organism on an organic feedstock is a significant autocatalytic reaction.

Normally, when a material reacts, its initial rate of disappearance is high and the rate decreases continuously as the reactant is consumed. However, in autocatalytic reaction, the initial rate is relatively slow

Table 1-1

Examples of acid-base catalysis in aqueous solution

Types of catalysis	Brief title of rearction	Equation of reaction
Specific acid	Inversion of cane sugar	$C_{12}H_{22}O_{11} + H_2O = C_6H_{12}O_6 + C_6H_{12}O_6$
	Hydrolysis of acetals	$R_1CH(OR_2)_2 + H_2O = R_1 \cdot CHO + 2R_2OH$
	Hydration of unsaturated aldehydes	$CH_2{:}CH \cdot CHO + H_2O = CH_2OH \cdot CH_2 \cdot CHO$
Specific base	Cleavage diacetone-alcohol	$CH_3CO \cdot CH_2 \cdot C(OH)(CH_3)_2 = 2(CH_3)_2 \cdot CO$
Specific acid and base	Hydrolysis of esters	$R_1 \cdot COOR_2 + H_2O = R_1 \cdot COOH + R_2OH$
General acid	Decomposition of acetaldehyde hydrate	$CH_3 \cdot CH(OH)_2 = CH_3 \cdot CHO + H_2O$
General base	Decomposition of nitramide	$NH_2NO_2 = N_2O + H_2O$
General acid and base	Halogenation exchange racemization of ketones	$R \cdot CO \cdot CH_3 + X_2 = R \cdot CO \cdot CH_2X + XH$

since little or no product is formed. The rate then increases to a maximum as the products are formed and then decreases to a low value as the reactants are consumed. Consider the following mechanism for an autocatalytic reaction:

$$2A \xrightarrow{\;k_1\;} B + C \tag{1-116}$$

$$A + B \xrightarrow{\;k_2\;} AB \tag{1-117}$$

$$2AB \xrightarrow{\;k_3\;} 3B + C \tag{1-118}$$

where AB is an intermediate complex.

The observed rate of reaction is the rate of formation of species:

$$\left(+r_C\right) = k_1 C_A^2 + k_3 C_{AB}^2 \tag{1-119}$$

The concentration of the reaction intermediate AB may be determined by using the steady state approximation for intermediates,

$$\left(+r_{AB}\right) = k_2 C_A C_B - 2k_3 C_{AB}^2 \approx 0 \tag{1-120}$$

Therefore,

$$C_{AB} = \left(\frac{k_2}{2k_3} \cdot C_A C_B \right)^{0.5} \tag{1-121}$$

Substituting Equation 1-121 into Equation 1-119 gives

$$\left(+r_C\right) = k_1 C_A^2 + \frac{k_2}{2} C_A C_B \tag{1-122}$$

Therefore, the mechanism results in a rate expression in which species B is responsible for the autocatalytic behavior.

GAS-SOLID CATALYTIC REACTIONS

Consider a gaseous reactant flowing through a bed of solid catalyst pellets. The physical steps involved are the transfer of the component gases up to the catalyst surface, diffusion of reactants into the interior of the pellet, diffusion of the products back to the exterior surface, and finally the transfer of the products from the exterior surface to the main stream. Interpreting the experimental results requires minimizing the resistance offered by each of these physical processes and focusing on the chemical aspects of the reaction. The chemical procedures involve activated adsorption of reactants with or without dissociation, surface reactions on active sites, and activated desorption of the products. The uncatalyzed reaction also takes place in the main gas stream simultaneously with the surface reaction.

In industrial applications of kinetics, an understanding of the mechanisms of chemical reactions is essential. This is helpful in establishing the optimum operating conditions in relation to parameters such as temperature, pressure, feed composition, space velocity, and the extent of recycling and conversion. Yang and Hougen [7] have established procedures in planning and correlating experimental data for gaseous reactions catalyzed by solids. They provided methods for eliminating, minimizing, or evaluating the temperatures and concentration gradients in gas films and catalyst pellets. Hougen and Watson [8] have developed rate equations for various mechanisms that may occur in gaseous reactions when catalyzed by solid surfaces. The following illustrates a gas-solid catalytic reaction:

$$A \underset{x}{\rightleftharpoons} R$$

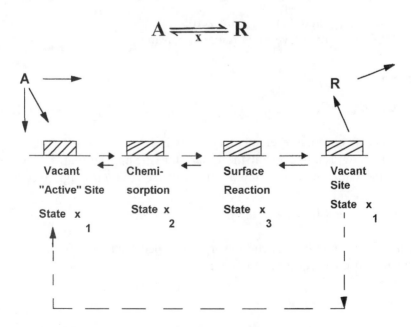

Adsorption Step: $A + x_1 \overset{k_1}{\underset{k_{-1}}{\rightleftharpoons}} x_2$ (1-123)

Surface Reaction: $x_2 \overset{k_2}{\underset{k_{-2}}{\rightleftharpoons}} x_3$ (1-124)

Desorption Step: $x_3 \overset{k_3}{\underset{k_{-3}}{\rightleftharpoons}} R + x_1$ (1-125)

The rate equations are:

A: $\left(-r_A\right)_{net} = k_1 C_A C_{x_1} - k_{-1} C_{x_2}$ (1-126)

R: $\left(+r_R\right)_{net} = k_3 C_{x_3} - k_{-3} C_R C_{x_1}$ (1-127)

x_2: $\left(+r_{x_2}\right)_{net} = k_1 C_A C_{x_1} + k_{-2} C_{x_3} - \left(k_{-1} + k_2\right) C_{x_2}$ (1-128)

x_3: $\left(+r_{x_3}\right)_{net} = k_2 C_{x_2} + k_{-3} C_{x_1} C_R - \left(k_{-2} + k_3\right) C_{x_3}$ (1-129)

Assume:

1. $(+r_{x_2})_{net}$ and $(+r_{x_3})_{net}$ are zero
2. Number of active sites remains constant, C_T.

Thus,

$C_T = C_{x_2} + C_{x_3} + C_{x_1}$ (1-130)

Usually, one of the elementary steps is rate controlling (that is, it is very slow relative to all the other steps). Suppose that $A + x_1 \rightarrow x_2$ is the rate-controlling step and the reverse reaction is ignored, then

$\left(-r_A\right)_{net} \cong k_1 C_A C_{x_1}$ (1-131)

Under these conditions, the desorption process and surface reactions are in a pseudo-equilibrium state:

$k_2 C_{x_2} \cong k_{-2} C_{x_3}$ (1-132)

$$k_3 C_{x_3} \cong k_{-3} C_{x_1} C_R \tag{1-133}$$

and

$$C_T = C_{x_1} + C_{x_2} + C_{x_3} \tag{1-134}$$

Rearranging Equations 1-132 and 1-133 in terms of C_{x_2} and C_{x_3} and substituting them in Equation 1-134 gives

$$C_T = C_{x_1} + \frac{k_{-2}}{k_2} \cdot \frac{k_{-3}}{k_3} C_{x_1} C_R + \frac{k_{-3}}{k_3} C_{x_1} C_R$$

Therefore,

$$C_{x_1} = \frac{C_T}{1 + K_T C_R} \tag{1-135}$$

where

$$K_T = \left(\frac{k_{-2}}{k_2} \cdot \frac{k_{-3}}{k_3} + \frac{k_{-3}}{k_3} \right) \tag{1-136}$$

Substituting Equation 1-136 into Equation 1-131 to eliminate C_{x_1}, yields

$$(-r_A)_{net} = \frac{(k_1 C_T) C_A}{1 + K_T C_R} \tag{1-137}$$

or

$$(-r_A)_{net} = \frac{K_4 C_A}{1 + K_T C_R} \tag{1-138}$$

If the reverse step in adsorption is not ignored, then

$$(-r_A)_{net} = \frac{K_4 C_A - K_5 C_R}{1 + K_T C_R} \tag{1-139}$$

GUIDELINES TO FORMULATING
REACTION MECHANISM

Guidelines have been proposed within which the kineticist works in developing a reaction mechanism. Edward et. al [9] have collated the most extensive collection of guidelines (see Chapter 3), which enable a kineticist to judge the feasibility of a proposed mechanism. They also allow the elimination of a proposed sequence of reactions as being unreasonable from the experimental data. The following are the guidelines to formulating reaction mechanism.

1. The most fundamental basis for mechanistic speculation is a complete analysis of the reaction products. It is essential to obtain a complete quantitative and qualitative analysis for all products of the reaction. Such analyses can provide essential clues as to the identity of the reaction intermediates.

2. The atomic and electronic structure of the reactants and products may point to the nature of possible intermediate species. The structural arrangement of atoms in the molecules that react must correspond at the instant of reaction to interatomic distances appropriate for the formation of new species.

3. All of the elementary reactions involved in a mechanistic sequence must be feasible with respect to bond energies. Bond energy considerations also may conclude that highly endothermic elementary reactions will be slow processes because of the large activation energies normally associated with these reactions.

4. The number of elementary reactions employed must be sufficient to provide a complete path for the formation of all observed products.

5. All of the intermediates produced by the elementary reactions must be consumed by other elementary reactions so that there will be no net production of intermediate species.

6. The majority of known elementary steps are bimolecular, the remainder being unimolecular or termolecular. The ammonia synthesis reaction is known to occur by a number of steps rather than as $N_2 + 3H_2 \rightarrow 2NH_3$.

7. A postulated mechanism for a reaction in the forward direction must also hold true for the reverse reaction. Three corollaries of this guideline should be borne in mind when postulating a reaction mechanism. First, the rate-limiting step for the reverse

reaction must be the same as that for the forward reaction. Second, the reverse reaction cannot have a molecularity greater than three, just as the forward reaction is so limited. As a consequence, the ammonia decomposition reaction $2NH_3 \rightarrow N_2 + 3H_2$ cannot occur as a simple bimolecular process. Third, if the reaction rate expression for the forward reaction consists of two or more independent terms corresponding to parallel reaction paths, there will be the same number of independent terms in the rate expression for the reverse reaction. At equilibrium, both the total rate of the forward reaction is equal to the total rate of the reverse reaction, and the forward rate of each path is equal to the reverse rate of that particular path.

8. Transitory intermediates involving highly reactive species do not react preferentially with one another to the exclusion of their reaction with stable species.

9. When the overall order of a reaction is greater than three, the mechanism probably has one or more equilibria and intermediates prior to the rate-determining step.

10. Inverse orders arise from rapid equilibria prior to the rate-determining step.

11. Whenever a rate law contains non-integers orders, there are intermediates present in the reaction sequence. When a fractional order is observed in an empirical rate expression for a homogeneous reaction, it is often an indication that an important part of the mechanism is the splitting of a molecule into free radicals or ions.

12. If the magnitude of the stoichiometric coefficient of a reactant exceeds the order of the reaction with respect to that species, there are one or more intermediates and reactions after the rate-determining step. Before applying this rule, the stoichiometric equation must be formulated for the reaction such that all coefficients are integers.

13. If the order of a reaction with respect to one or more species increases as the concentration of those species increases, it is an indication that the reaction may be proceeding by two or more parallel paths.

14. If there is a decrease in the order of a reaction with respect to a particular substance as the concentration of that species increases, the dominant form of that species in solution may be undergoing a change caused by the change in concentration.

A decrease in reaction order with respect to hydrogen ion concentration with increasing acidity has frequently been observed for reactions involving weak acids.

TESTING KINETIC MODELS

Testing kinetic models involves the following procedures:

1. Propose possible mechanisms involving elementary steps.
2. Assume that the rate equations for each elementary step can be written by inspection of the stoichiometric equation:

$$A + B^* \rightarrow AB^* \tag{1-140}$$

or

$$(-r_A) = (-r_B)^* = kC_A C_B^* \tag{1-141}$$

3. Assume that after a short initial period that the rates of concentration change of all active centers are zero. For example:

$$A + B^* \xrightarrow{\ k_1\ } AB^* \tag{1-142}$$

$$AB^* + A \xrightarrow{\ k_2\ } A_2 + B^* \tag{1-143}$$

$$(-r_{B^*}) = k_1 C_A C_{B^*} \tag{1-144}$$

$$(+r_{B^*}) = k_2 C_A C_{B^*} \tag{1-145}$$

The net rate of disappearance is

$$(-r_{B^*})_{net} = -\frac{dC_B}{dt} = k_1 C_A C_B - k_2 C_A C_{AB^*} \tag{1-146}$$

From the above assumption, $(-r_{B^*})_{net} = 0$, steady state approximation.

This assumption derives rate equations from which terms involving the concentrations of active centers can be eliminated.

4. If component i takes place in more than one elementary step, then

$$\left(\pm r_i\right) = \sum_{\text{All elementary steps}} \left(\pm r_i\right) \tag{1-147}$$

For example, at 480°C the gaseous decomposition of acetaldehyde has an order with respect to concentration of 3/2. The main reaction is $CH_3CHO = CH_4 + CO$

A possible reaction mechanism is:

Initiation

1. $CH_3CHO \xrightarrow{k_1} CH_3^* + CHO^*$

Propagation

2. $CH_3^* + CH_3CHO \xrightarrow{k_2} CH_4 + CH_3CO^*$

3. $CH_3CO^* \xrightarrow{k_3} CH_3^* + CO$

Termination

4. $CH_3^* + CH_3^* \xrightarrow{k_4} C_2H_6$

It is assumed that for each initiation there are many propagation cycles before termination. The main reaction is therefore given by the addition of the propagation steps alone, which gives the correct stoichiometric equation. A small amount of ethane, C_2H_6, is expected due to the termination reaction.

The rate expressions are:

$$\left(+r_{CH_3CO^*}\right)_{net} = \frac{dC_{CH_3CO^*}}{dt} = k_2 C_{CH_3^*} C_{CH_3CHO} - k_3 C_{CH_3CO^*} \tag{1-148}$$

$$\left(+r_{CH_3^*}\right)_{net} = \frac{dC_{CH_3^*}}{dt} = k_1 C_{CH_3CHO} + k_3 C_{CH_3CO^*}$$
$$- k_2 C_{CH_3^*} C_{CH_3CHO} - 2 k_4 C_{CH_3^*}^2 \tag{1-149}$$

Using the steady state approximation,

$$\left(+r_{CH_3CO^*}\right)_{net} = 0 \text{ and } \left(+r_{CH_3^*}\right)_{net} = 0$$

The overall rate equation is:

$$\left(-r_{CH_3CHO}\right)_{net} = -\frac{dC_{CH_3CHO}}{dt}$$

$$= k_1 C_{CH_3CHO} + k_2 C_{CH_3^*} C_{CH_3CHO} \tag{1-150}$$

From Equations 1-148 and 1-149, $2k_4 C_{CH_3^*}^2 = k_1 C_{CH_3CHO}$ and

$$C_{CH_3^*} = \left(\frac{k_1}{2k_4} C_{CH_3CHO}\right)^{0.5} \tag{1-151}$$

Substituting Equation 1-151 into Equation 1-150 gives

$$\left(-r_{CH_3CHO}\right) = -\frac{dC_{CH_3CHO}}{dt}$$

$$= k_1 C_{CH_3CHO} + k_2 \left(\frac{k_1}{2k_4} \bullet C_{CH_3CHO}\right)^{0.5} C_{CH_3CHO} \tag{1-152}$$

If $k_2 \left(\dfrac{k_1}{2k_4} C_{CH_3CHO}\right)^{0.5} \gg k_1 C_{CH_3CHO}$, then Equation 1-152 becomes

$$\left(-r_{CH_3CHO}\right) = -\frac{dC_{CH_3CHO}}{dt}$$

and

$$-\frac{dC_{CH_3CHO}}{dt} = \frac{dC_{CH_4}}{dt} = \left(\frac{k_2^2 k_1}{2k_4}\right)^{0.5} C_{CH_3CHO}^{3/2} \tag{1-153}$$

This agrees with experimental results.

The activation energy that is expected according to a relevant chain reaction mechanism can be determined if each elementary rate constant is expressed with a temperature dependence according to the Arrhenius equation,

$$K_i = A_i \exp\left(\frac{-E_i}{RT}\right) \tag{1-154}$$

Substituting Equation 1-154 into the predicted rate law allows the temperature dependence of the overall reaction to be predicted. In the decomposition of acetaldehyde, it follows that

$$
\begin{aligned}
K_{obs} &= A_{obs} \exp\left(\frac{-E_a}{RT}\right) = k_2\left(\frac{k_1}{2k_4}\right)^{0.5} \\
&= A_2 \exp\left(\frac{-E_2}{RT}\right)\left(\frac{A_1 \exp(-E_1/RT)}{2A_4 \exp(-E_4/RT)}\right)^{0.5}
\end{aligned} \tag{1-155}
$$

Separating the variables gives

$$E_a = E_2 + \frac{1}{2}(E_1 - E_4) \quad \text{and} \quad A_{obs} = A_2\left(\frac{A_1}{2A_4}\right)^{0.5} \tag{1-156}$$

The activation energy for an overall chain reaction can be smaller than E_1, the activation energy for the initiation step.

CHAIN LENGTH

The efficiency of a chain reaction depends on the number of recycles of the active center for each initiation. The chain length can be defined as:

$$\text{Chain length} = \frac{\text{the rate of the overall chain reaction}}{\text{the rate of the initiation reaction}}$$

For example, for the decomposition of acetaldehyde:

$$\text{Chain length} = \frac{k_2 \left(\dfrac{k_1}{2k_4}\right)^{0.5} C_{CH_3CHO}^{3/2}}{k_1 C_{CH_3CHO}}$$

$$= \left(\frac{k_2^2}{2k_1k_4}\right)^{0.5} C_{CH_3CHO}^{0.5}$$

(1-157)

In this case, the chain length increases with the reactant concentration.

Example 1-1

The Hydrogenation of ethylene can be written as $C_2H_4 + H_2 = C_2H_6$. It is suggested that the reaction sequence is:

Initiation

$$C_2H_4 + H_2 \xrightarrow{\ k_1\ } C_2H_5^* + H^*$$

(1-158)

Closed sequence

$$H^* + C_2H_4 \xrightarrow{\ k_2\ } C_2H_5^*$$

(1-159)

$$C_2H_5^* + H_2 \xrightarrow{\ k_3\ } C_2H_6 + H^*$$

(1-160)

Termination

$$C_2H_5^* + H^* \xrightarrow{\ k_4\ } C_2H_6$$

(1-161)

Derive a rate expression for the overall reaction listing any assumptions made.

Solution

The overall rate of reaction of $C_2H_5^*$ is

$$\left(+r_{C_2H_5^*}\right)_{net} = k_1 C_{C_2H_4} C_{H_2} + k_2 C_{C_2H_4} C_{H^*} - k_3 C_{C_2H_5^*} C_{H_2}$$
$$- k_4 C_{C_2H_5^*} C_{H^*} \tag{1-162}$$

The overall rate of reaction of C_{H^*} is

$$\left(+r_{H^*}\right)_{net} = k_1 C_{C_2H_4} C_{H_2} + k_3 C_{C_2H_5^*} C_{H_2} - k_2 C_{C_2H_4} C_{H^*}$$
$$- k_4 C_{C_2H_5^*} C_{H^*} \tag{1-163}$$

The overall rate of reaction of C_{H_2} is

$$\left(-r_{H_2}\right)_{net} = k_1 C_{C_2H_4} C_{H_2} + k_3 C_{C_2H_5^*} C_{H_2} \tag{1-164}$$

Assuming that $(+r_{C_2H_5^*})_{net}$ and $(+r_{H^*})_{net}$ are zero, equating Equations 1-162 and 1-163 gives:

$$k_1 C_{C_2H_4} C_{H_2} + k_2 C_{C_2H_4} C_{H^*} - k_3 C_{C_2H_5^*} C_{H_2} - k_4 C_{C_2H_5^*} C_{H^*}$$
$$= k_1 C_{C_2H_4} C_{H_2} + k_3 C_{C_2H_5^*} C_{H_2} - k_2 C_{C_2H_4} C_{H^*} - k_4 C_{C_2H_5^*} C_{H^*} \tag{1-165}$$

Therefore,

$$2 k_2 C_{C_2H_4} C_{H^*} = 2 k_3 C_{C_2H_5^*} C_{H_2} \tag{1-166}$$

$$\frac{k_2 C_{C_2H_4}}{k_3 C_{H_2}} = \frac{C_{C_2H_5^*}}{C_{H^*}} \tag{1-167}$$

From Equation 1-164

$$\left(-r_{H_2}\right)_{net} = k_1 C_{C_2H_4} C_{H_2} + k_3 C_{C_2H_5^*} C_{H_2}$$

Substituting Equation 1-167 into Equation 1-164 gives

$$\left(-r_{H_2}\right)_{net} = k_1 C_{C_2H_4} C_{H_2} + \frac{k_2}{k_3} \cdot \frac{C_{C_2H_4}}{C_{H_2}} \cdot C_{H^*} k_3 C_{H_2}$$

$$= k_1 C_{C_2H_4} C_{H_2} + k_2 C_{C_2H_4} C_{H^*}$$

$$(1\text{-}168)$$

The overall rate of reaction of $C_{C_2H_4}$ is

$$\left(-r_{C_2H_4}\right)_{net} = k_1 C_{C_2H_4} C_{H_2} + k_2 C_{C_2H_4} C_{H^*} \qquad (1\text{-}169)$$

$$C_{H^*} = \frac{\left(-r_{C_2H_4}\right)_{net} - k_1 C_{C_2H_4} C_{H_2}}{k_2 C_{C_2H_4}} \qquad (1\text{-}170)$$

Substituting Equation 1-170 into Equation 1-168 yields

$$\left(-r_{H_2}\right)_{net} = k_1 C_{C_2H_4} C_{H_2} + k_2 C_{C_2H_4} \left\{ \frac{\left(-r_{C_2H_4}\right)_{net} - k_1 C_{C_2H_4} C_{H_2}}{k_2 C_{C_2H_4}} \right\}$$

$$= k_1 C_{C_2H_4} C_{H_2} + \left(-r_{C_2H_4}\right)_{net} - k_1 C_{C_2H_4} C_{H_2}$$

$$\left(-r_{H_2}\right)_{net} = \left(-r_{C_2H_4}\right)_{net} \qquad (1\text{-}171)$$

$$\left(-r_{H_2}\right)_{net} = \left(-r_{C_2H_4}\right)_{net} = k_1 C_{C_2H_4} C_{H_2} + k_2 C_{C_2H_4} C_{H^*} \qquad (1\text{-}172)$$

but

$$C_{H^*} = \frac{k_3}{k_2} \cdot \frac{C_{C_2H_5^*} C_{H_2}}{C_{C_2H_4}}$$

Since $(+r_{C_2H_5^*})_{net}$ and $(+r_{H^*})_{net}$ are zero

$$k_1 C_{C_2H_4} C_{H_2} + k_2 C_{C_2H_4} C_{H^*} - k_3 C_{C_2H_5^*} C_{H_2} - k_4 C_{C_2H_5^*} C_{H^*} = 0 \quad (1\text{-}173)$$

$$k_1 C_{C_2H_4} C_{H_2} - k_2 C_{C_2H_4} C_{H^*} + k_3 C_{C_2H_5^*} C_{H_2} - k_4 C_{C_2H_5^*} C_{H^*} = 0 \quad (1\text{-}174)$$

Adding Equations 1-173 and 1-174 gives

$$2k_1 C_{C_2H_4} C_{H_2} - 2k_4 C_{C_2H_5^*} C_{H^*} = 0 \tag{1-175}$$

Subtracting Equations 1-173 from Equation 1-174 yields

$$2k_2 C_{C_2H_4} C_{H^*} - 2k_3 C_{C_2H_5^*} C_{H_2} = 0 \tag{1-176}$$

$$k_1 C_{C_2H_4} C_{H_2} = k_4 C_{C_2H_5^*} C_{H^*} \tag{1-177}$$

and

$$k_2 C_{C_2H_4} C_{H^*} = k_3 C_{C_2H_5^*} C_{H_2} \tag{1-178}$$

From Equation 1-177

$$C_{C_2H_5^*} = \frac{k_1 C_{C_2H_4} C_{H_2}}{k_4 C_{H^*}} \tag{1-179}$$

Substituting Equation 1-179 into Equation 1-178 gives

$$k_2 C_{C_2H_4} C_{H^*} = k_3 \cdot C_{H_2} \cdot \frac{k_1 C_{C_2H_4} C_{H_2}}{k_4 C_{H^*}}$$

Therefore,

$$C_{H^*}^2 = \frac{k_3}{k_2} \cdot \frac{k_1}{k_4} C_{H_2}^2$$

where

$$K = \frac{k_3}{k_2} \cdot \frac{k_1}{k_4}$$

Hence,

$$C_{H^*} = K^{0.5}C_{H_2} \tag{1-180}$$

Substituting Equation 1-160 into Equation 1-168 gives

$$\left(-r_{H_2}\right)_{net} = k_1 C_{C_2H_4} C_{H_2} + k_2 C_{C_2H_4} K^{0.5}C_{H_2}$$
$$= C_{C_2H_4} C_{H_2}\left\{k_1 + k_2 K^{0.5}\right\} \tag{1-181}$$

Example 1-2

The overall reaction $N_2 + O_2 \rightarrow 2NO$ involves the elementary steps:

$$O^* + N_2 \xrightarrow{k_1} NO + N^*$$

$$N^* + O_2 \xrightarrow{k_2} NO + O^*$$

Write down equations for $(-r_{O^*})_{net}$ and $(-r_{N^*})_{net}$.
Assuming these rates are equal to zero and $C_{O^*} + C_{N^*} = C_T$, a constant, derive an expression for $(+r_{NO})_{net}$ in terms of C_T, C_{N_2}, and C_{O_2}.

Solution

The net rates of disappearance of the free radicals $(-r_{O^*})_{net}$ and $(-r_{N^*})_{net}$ are:

$$\left(-r_{O^*}\right)_{net} = k_1 C_{O^*} C_{N_2} - k_2 C_{N^*} C_{O_2} \tag{1-182}$$

and

$$\left(-r_{N^*}\right)_{net} = k_2 C_{N^*} C_{O_2} - k_1 C_{O^*} C_{N_2} \tag{1-183}$$

Assuming that $(-r_{O^*})_{net}$ and $(-r_{N^*})_{net}$ are zero, then

$$k_1 C_{O^*} C_{N_2} = k_2 C_{N^*} C_{O_2} \tag{1-184}$$

and

$$C_{O^*} + C_{N^*} = C_T \tag{1-185}$$

Therefore,

$$C_{O^*} = C_T - C_{N^*} \tag{1-186}$$

Substituting Equation 1-186 into Equation 1-184 gives

$$k_1 C_{N_2}\left(C_T - C_{N^*}\right) = k_2 C_{N^*} C_{O_2}$$

$$k_1 C_{N_2} C_T - k_1 C_{N_2} C_{N^*} = k_2 C_{N^*} C_{O_2}$$

Therefore,

$$C_{N^*} = \frac{k_1 C_{N_2} C_T}{k_1 C_{N_2} + k_2 C_{O_2}} \tag{1-187}$$

The net rate of formation of nitric oxide is:

$$\left(+r_{NO}\right)_{net} = k_1 C_{O^*} C_{N_2} + k_2 C_{N^*} C_{O_2} \tag{1-188}$$

$$= k_1 C_{N_2}\left(C_T - C_{N^*}\right) + k_2 C_{N^*} C_{O_2}$$

$$= k_1 C_{N_2}\left\{C_T - \frac{k_1 C_{N_2} C_T}{k_1 C_{N_2} + k_2 C_{O_2}}\right\} + k_2 C_{O_2}\left\{\frac{k_1 C_{N_2} C_T}{k_1 C_{N_2} + k_2 C_{O_2}}\right\}$$

$$= k_1 C_T C_{N_2} + \left(k_2 C_{O_2} - k_1 C_{N_2}\right)\left(\frac{k_1 C_{N_2} C_T}{k_1 C_{N_2} + k_2 C_{O_2}}\right)$$

$$= \frac{2 k_1 k_2 C_T C_{O_2} C_{N_2}}{k_1 C_{N_2} + k_2 C_{O_2}} \tag{1-189}$$

Example 1-3

The Thermal decomposition of ethane to ethylene, methane, butane, and hydrogen can be expressed by the following mechanism.

$$C_2H_6 \xrightarrow{\ k_1\ } 2CH_3^* \tag{1-190}$$

$$CH_3^* + C_2H_6 \xrightarrow{\ k_2\ } CH_4 + C_2H_5^* \tag{1-191}$$

$$C_2H_5^* \xrightarrow{\ k_3\ } C_2H_4 + H^* \tag{1-192}$$

$$H^* + C_2H_6 \xrightarrow{\ k_4\ } H_2 + C_2H_5^* \tag{1-193}$$

$$2C_2H_5^* \xrightarrow{\ k_5\ } C_4H_{10} \tag{1-194}$$

Derive a rate law for the formation of ethylene C_2H_4 assuming $k_3 \ll k_5$. The free radicals are CH_3^*, and $C_2H_5^*$.

Solution

Applying the steady state equations for the free radicals H^*, CH_3^*, and $C_2H_5^*$, the rate of formation of ethylene C_2H_4 for a constant volume batch reactor is:

$$\left(+r_{C_2H_4}\right) = \frac{dC_{C_2H_4}}{dt} = k_3 C_{C_2H_5^*} \tag{1-195}$$

The rate of formation of the free radicals are:

$$\left(+r_{CH_3^*}\right) = \frac{dC_{CH_3^*}}{dt} = k_1 C_{C_2H_6} - k_2 C_{CH_3^*} C_{C_2H_6} = 0 \tag{1-196}$$

$$\left(+r_{C_2H_5^*}\right) = \frac{dC_{C_2H_5^*}}{dt} \tag{1-197}$$

$$= k_2 C_{CH_3^*} C_{C_2H_6} + k_4 C_{H^*} C_{C_2H_6} - k_5 C_{C_2H_5^*}^2 = 0$$

$$\left(+r_{H^*}\right) = \frac{dC_{H^*}}{dt} = k_3 C_{C_2H_5^*} - k_4 C_{H^*} C_{C_2H_6} = 0 \tag{1-198}$$

From Equation 1-198 $k_3 C_{C_2H_5^*} - k_4 C_{H^*} C_{C_2H_6} = 0$. Therefore,

$$C_{H^*} = \frac{k_3}{k_4} \cdot \frac{C_{C_2H_5^*}}{C_{C_2H_6}} \tag{1-199}$$

From Equation 1-196, $k_1 C_{C_2H_6} - k_2 C_{CH_3^*} C_{C_2H_6} = 0$. Therefore,

$$C_{CH_3^*} = \frac{k_1}{k_2} \tag{1-200}$$

From Equation 1-197, $k_2 C_{CH_3^*} C_{C_2H_6} + k_4 C_{H^*} C_{C_2H_6} - k_5 C_{C_2H_5^*}^2 = 0$. Therefore,

$$k_5 C_{C_2H_5^*}^2 = k_2 C_{CH_3^*} C_{C_2H_6} + k_4 C_{H^*} C_{C_2H_6} \tag{1-201}$$

Substituting Equations 1-199 and 1-200 into Equation 1-201 gives

$$k_5 C_{C_2H_5^*}^2 = k_2 \cdot C_{C_2H_6} \cdot \frac{k_1}{k_2} + k_4 C_{C_2H_6} \cdot \frac{k_3}{k_4} \cdot \frac{C_{C_2H_5^*}}{C_{C_2H_6}} \tag{1-202}$$

$$k_5 C_{C_2H_5^*}^2 = k_1 C_{C_2H_6} + k_3 C_{C_2H_5^*}$$

Assuming $k_3 \ll k_5$, therefore,

$$C_{C_2H_5^*} = \left(\frac{k_1}{k_5} \cdot C_{C_2H_6}\right)^{0.5} \tag{1-203}$$

Substituting Equation 1-203 into Equation 1-195 gives

$$\left(+r_{C_2H_4}\right) = k_3 \left(\frac{k_1}{k_5} C_{C_2H_6}\right)^{0.5} \tag{1-204}$$

Example 1-4

Houser and Lee* studied the pyrolysis of ethly nitrate using a stirred flow reactor. They proposed the following mechanism for the reaction:

Initiation

$$C_2H_5ONO_2 \xrightarrow{k_1} C_2H_5O^* + NO_2 \tag{1-205}$$

Propagation

$$C_2H_5O^* \xrightarrow{k_2} CH_3^* + CH_2O \tag{1-206}$$

$$CH_3^* + C_2H_5ONO_2 \xrightarrow{k_3} CH_3NO_2 + C_2H_5O^* \tag{1-207}$$

Termination

$$2C_2H_5O^* \xrightarrow{k_4} CH_3CHO + C_2H_5OH \tag{1-208}$$

The rate of reaction, mols/(ksec)(m^3), was measured in a stirred tank reactor in terms of concentration mols/m^3. The results are shown in Table 1-2.

Find a rate equation consistent with the proposed mechanism and verify it against the data. Assume that the initiation and termination steps are relatively slow.

Table 1-2

Ethyl nitrate concentration $\dfrac{mols}{m^3}$	Rate $\dfrac{mols}{(ksec)m^3}$
0.0975	13.4
0.0759	12.2
0.0713	12.1
0.2714	23.0
0.2346	20.9

J. Phys. Chem., 71, 3422, 1967.

Solution

The free radicals are $C_2H_5O^*$ and CH_3^*. The net rate of disappearance of $C_2H_5ONO_2$ is:

$$\left(-r_{C_2H_5ONO_2}\right)_{net} = k_1 C_{C_2H_5ONO_2} + k_3 C_{CH_3^*} C_{C_2H_5ONO_2} \tag{1-209}$$

The net rates of disappearance of the free radicals, $C_2H_5O^*$ and CH_3^* are:

$$\left(-r_{C_2H_5O^*}\right)_{net} = k_2 C_{C_2H_5O^*} + k_4 C^2_{C_2H_5O^*}$$
$$- k_1 C_{C_2H_5ONO_2} - k_3 C_{CH_3^*} C_{C_2H_5ONO_2} = 0 \tag{1-210}$$

$$\left(-r_{CH_3^*}\right)_{net} = k_3 C_{CH_3^*} C_{C_2H_5ONO_2} - k_2 C_{C_2H_5O^*} = 0 \tag{1-211}$$

From Equation 1-210

$$k_2 C_{C_2H_5O^*} + k_4 C^2_{C_2H_5O^*} - k_1 C_{C_2H_5ONO_2} - k_3 C_{CH_3^*} C_{C_2H_5ONO_2} = 0 \tag{1-212}$$

From Equation 1-211

$$k_3 C_{CH_3^*} C_{C_2H_5ONO_2} - k_2 C_{C_2H_5O^*} = 0 \tag{1-213}$$

$$k_3 C_{CH_3^*} C_{C_2H_5ONO_2} = k_2 C_{C_2H_5O^*} \tag{1-214}$$

Substituting equation 1-214 into Equation 1-212 gives

$$k_2 C_{C_2H_5O^*} + k_4 C^2_{C_2H_5O^*} - k_1 C_{C_2H_5ONO_2} - k_2 C_{C_2H_5O^*} = 0$$

and

$$C_{C_2H_5O^*} = \left(\frac{k_1}{k_4} \cdot C_{C_2H_5ONO_2}\right)^{0.5} \tag{1-215}$$

From Equation 1-214

$$\begin{aligned}C_{CH_3^*} &= \frac{k_2}{k_3} \cdot \frac{C_{C_2H_5O^*}}{C_{C_2H_5ONO_2}}\\[2mm]&= \frac{k_2}{k_3} \cdot \frac{1}{C_{C_2H_5ONO_2}}\left(\frac{k_1}{k_4} \cdot C_{C_2H_5ONO_2}\right)^{0.5}\\[2mm]&= \frac{k_2}{k_3}\left(\frac{k_1}{k_4} \cdot \frac{1}{C_{C_2H_5ONO_2}}\right)^{0.5} \end{aligned} \tag{1-216}$$

For ethyl nitrate, Equation 1-209 becomes

$$\begin{aligned}\left(-r_{C_2H_5ONO_2}\right)_{net} &= \left\{k_1 + k_3 \cdot \frac{k_2}{k_3}\left(\frac{k_1}{k_4} \cdot \frac{1}{C_{C_2H_5ONO_2}}\right)^{0.5}\right\}C_{C_2H_5ONO_2}\\[2mm]&= \left\{k_1 + k_2\left(\frac{k_1}{k_4}\right)^{0.5} \cdot \frac{1}{\left(C_{C_2H_5ONO_2}\right)^{0.5}}\right\}C_{C_2H_5ONO_2}\end{aligned} \tag{1-217}$$

where k_1 and k_4 are small in comparison with k_2, therefore,

$$\left(-r_{C_2H_5ONO_2}\right)_{net} = kC^{0.5}_{C_2H_5ONO_2} \tag{1-218}$$

Equation 1-218 can be expressed in the form $Y = AX^B$.
Using natural logarithm, Equation $Y = AX^B$ gives

$$\ln(Y) = \ln A + B \bullet \ln(X) \tag{1-219}$$

If we perform a regression analysis on Equation 1-219, the slope gives the order of the reaction and the rate constant is determined by the intercept. Appendix A illustrates a developed computer program that performs the regression analysis of equations for any given set of data. The results of the linearized regression analysis from the computer program give slope B as 0.486, and intercept A as 42.7. Table 1-3 shows the results of the estimated reaction rate of ethyl nitrate with a correlation coefficient = 0.9979.

Figure 1-6 shows plots of the regression model and the experimental results. Equation 1-218 can now be expressed as:

$$\left(-r_{C_2H_5ONO_2}\right) = 42.7 \, C^{0.486}_{C_2H_5ONO_2}$$

This confirms the half order of the rate expression.

Table 1-3

Ethyl nitrate concentration $\dfrac{mols}{m^3}$	Reaction rate $\dfrac{mols}{(ksec)m^3}$	
	Actual	Estimated
0.0975	13.4	13.8
0.0759	12.2	12.2
0.0713	12.1	11.8
0.2714	23.0	22.7
0.2346	20.9	21.1

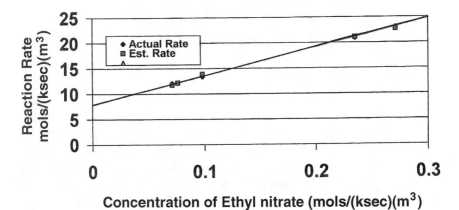

Plots of Ethyl nitrate concentration vs. Reaction Rate mols/(ksec)(m³)

Figure 1-6. Reaction rate of ethyl nitrate as a function of concentration.

Example 1-5

Determine the Michaelis-Menten Parameters V_{max} and K_m for the reaction:

$$\text{urea} + \text{urease} \underset{k_2}{\overset{k_1}{\rightleftharpoons}} [\text{urea} \bullet \text{urease}]^* + H_2O$$

$$[\text{urea} \bullet \text{urease}]^* + H_2O \xrightarrow{k_3} 2NH_3 + CO_2 + \text{urease}$$

The rate of reaction is given as a function of urea concentration in Table 1-4.

Solution

Using Equation 1-220

$$V_O = \frac{V_{max} \bullet C_{urea}}{C_{urea} + K_m} \tag{1-220}$$

and rearranging gives

Table 1-4

$C_{urea}, \left(\dfrac{kmol}{m^3}\right)$	$V_{urea}, \left(\dfrac{kmol}{m^3 \bullet s}\right)$
0.6	1.80
0.4	1.45
0.2	1.07
0.02	0.54
0.01	0.36
0.005	0.19
0.002	0.085
0.001	0.06

$$\frac{1}{V_O} = \frac{1}{V_{max}} + \frac{K_m}{V_{max}} \bullet \frac{1}{C_{urea}} \qquad (1\text{-}221)$$

A plot of the reciprocal reaction rate versus the reciprocal urea concentration should give a straight line with an intercept $1/V_{max}$ and the slope K_m/V_{max}. Figure 1-7 gives a plot of $1/V_O$ versus $1/C_{urea}$ from the developed linear regression method. The results of the computer program give the intercept $1/V_{max} = 1.1741$ and, therefore, the maximum rate is $V_{max} = 0.852 (kmol/m^3.s)$. The slope $K_m/V_{max} = 0.0167$ and $K_m = 0.0142$ ($kmol/m^3$). Substituting K_m and V_{max} in Equation 1-220 gives

$$V_O = \frac{0.852\,C_{urea}}{0.0142 + C_{urea}}, \quad \frac{kmol}{m^3 \bullet s}$$

Example 1-6
Pyrolysis of Ethane C_2H_6 (See Example 1-3)

The decomposition of ethane has been extensively studied and several mechanisms have been postulated. The main products of the reaction are ethylene, C_2H_4, and hydrogen, H_2. However, there are small amounts of methane, CH_4, butane, C_4H_{10}, and other products.

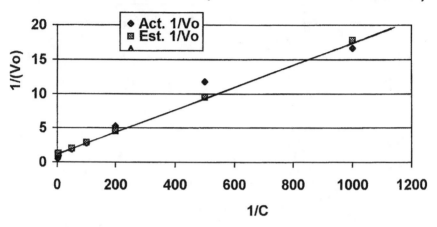

Figure 1-7. Lineweaver-Burk plot of $(1/C_{urea})$ versus $(1/V_o)$.

The overall reaction is:

$$C_2H_6 = C_2H_4 + H_2$$

The reaction is first order in the temperature range of 550–640°C and at pressure in excess of 50 mm Hg. The rate law is given by

$$\left(-r_{C_2H_6}\right) = -\frac{dC_{C_2H_6}}{dt} = kC_{C_2H_6}$$

(1) Develop a mechanism to account for the first-order rate law for the case where only methane is formed as a byproduct. Assume that the chain length is large.

Mechanism

Initiation

1. $C_2H_6 \xrightarrow{\ k_1\ } CH_3^* + CH_3^*$ (1-222)

2. $CH_3^* + C_2H_6 \xrightarrow{\ k_2\ } CH_4 + C_2H_5^*$ (1-223)

Propagation

3. $C_2H_5^* \xrightarrow{k_3} C_2H_4 + H^*$ (1-224)

4. $H^* + C_2H_6 \xrightarrow{k_4} C_2H_5^* + H_2$ (1-225)

Termination

5. $H^* + C_2H_5^* \xrightarrow{k_5} C_2H_6$ (1-226)

Applying the rate expressions to Equations 1-222, 1-223, 1-224, 1-225 and 1-226, and using the steady state approximation for CH_3^*, $C_2H_5^*$, and H^*, for a constant volume batch reactor yields:

$$\left(-r_{C_2H_6}\right)_{net} = -\frac{dC_{C_2H_6}}{dt} = k_1 C_{C_2H_6} + k_2 C_{CH_3^*} C_{C_2H_6}$$
$$+ k_4 C_{H^*} C_{C_2H_6} - k_5 C_{H^*} C_{C_2H_5^*} \tag{1-227}$$

$$\left(-r_{CH_3^*}\right)_{net} = -\frac{dC_{CH_3^*}}{dt} = k_2 C_{CH_3^*} C_{C_2H_6} - 2k_1 C_{C_2H_6} = 0 \tag{1-228}$$

$$\left(-r_{C_2H_5^*}\right)_{net} = -\frac{dC_{C_2H_5^*}}{dt} = k_3 C_{C_2H_5^*} + k_5 C_{C_2H_5^*} C_{H^*}$$
$$- k_2 C_{CH_3^*} C_{C_2H_6} - k_4 C_{H^*} C_{C_2H_6} = 0 \tag{1-229}$$

$$\left(-r_{H^*}\right)_{net} = -\frac{dC_{H^*}}{dt} \tag{1-230}$$
$$= k_4 C_{H^*} C_{C_2H_6} + k_5 C_{H^*} C_{C_2H_5^*} - k_3 C_{C_2H_5^*} = 0$$

Rearranging Equations 1-228, 1-229, and 1-230 gives

$$\frac{dC_{CH_3^*}}{dt} = 2k_1 C_{C_2H_6} - k_2 C_{CH_3^*} C_{C_2H_6} = 0 \tag{1-231}$$

$$\frac{dC_{C_2H_5^*}}{dt} = k_2 C_{CH_3^*} C_{C_2H_6} + k_4 C_{H^*} C_{C_2H_6}$$

$$-k_3 C_{C_2H_5^*} - k_5 C_{C_2H_5^*} C_{H^*} = 0 \tag{1-232}$$

$$\frac{dC_{H^*}}{dt} = k_3 C_{C_2H_5^*} - k_4 C_{H^*} C_{C_2H_6} - k_5 C_{H^*} C_{C_2H_5^*} = 0 \tag{1-233}$$

Adding Equations 1-231, 1-232, and 1-233 gives

$$k_1 C_{C_2H_6} = k_5 C_{H^*} C_{C_2H_5^*} \tag{1-234}$$

This shows that the rate of initiation equals the rate of termination. Thus,

$$C_{H^*} = \frac{k_1 C_{C_2H_6}}{k_5 C_{C_2H_5^*}} \tag{1-235}$$

Substituting Equation 1-235 into Equation 1-233 gives

$$k_3 C_{C_2H_5^*} - k_4 C_{C_2H_6} \cdot \frac{k_1 C_{C_2H_6}}{k_5 C_{C_2H_5^*}} - k_5 C_{C_2H_5^*} \cdot \frac{k_1 C_{C_2H_6}}{k_5 C_{C_2H_5^*}} = 0$$

$$k_3 k_5 C_{C_2H_5^*}^2 - k_1 k_4 C_{C_2H_6}^2 - k_1 C_{C_2H_6} = 0 \tag{1-236}$$

The rate of production of ethylene C_2H_4 is:

$$\left(+r_{C_2H_4}\right) = \frac{dC_{C_2H_4}}{dt} = k_3 C_{C_2H_5^*} \tag{1-237}$$

Introducing the chain length,

$$\text{Chain length} = \frac{k_3 C_{C_2H_5^*}}{k_1 C_{C_2H_6}} \tag{1-238}$$

If the chain length is large, it follows that

$$k_3 C_{C_2H_5^*} \gg k_1 C_{C_2H_6} \tag{1-239}$$

and Equation 1-236 becomes

$$k_3 k_5 C_{C_2H_5^*}^2 - k_1 k_4 C_{C_2H_6}^2 = 0 \tag{1-240}$$

Thus,

$$C_{C_2H_5^*} = \left(\frac{k_1 k_4}{k_3 k_5} \right)^{0.5} C_{C_2H_6} \tag{1-241}$$

Substituting Equation 1-241 into Equation 1-237 gives

$$\left(+r_{C_2H_4} \right) = \frac{dC_{C_2H_4}}{dt} = k_3 \left(\frac{k_1 k_4}{k_3 k_5} \right)^{0.5} C_{C_2H_6}$$

$$= \left(\frac{k_1 k_3 k_4}{k_5} \right)^{0.5} C_{C_2H_6} \tag{1-242}$$

The rate of decrease in ethane C_2H_6 is:

$$\left(-r_{C_2H_6} \right) = -\frac{dC_{C_2H_6}}{dt}$$

$$-\frac{dC_{C_2H_6}}{dt} = k_1 C_{C_2H_6} + k_2 C_{CH_3^*} C_{C_2H_6} + k_4 C_{H^*} C_{C_2H_6} - k_5 C_{H^*} C_{C_2H_5^*}$$

which together with Equation 1-232 yields

$$\left(-r_{C_2H_6} \right) = -\frac{dC_{C_2H_6}}{dt} = k_1 C_{C_2H_6} + k_3 C_{C_2H_5^*} \tag{1-243}$$

From Equations 1-237, 1-239, and 1-242, it follows that

$$\left(-r_{C_2H_6}\right) = \left(+r_{C_2H_4}\right)$$

$$-\frac{dC_{C_2H_6}}{dt} = \frac{dC_{C_2H_4}}{dt} = k_3 C_{C_2H_5^*} = \left(\frac{k_1 k_3 k_4}{k_5}\right)^{0.5} C_{C_2H_6}$$

$$= K^{0.5} C_{C_2H_6} \tag{1-244}$$

This mechanism explains the first order dependence of the reaction.

(2) Develop now another mechanism accounting for the formation of both methane and butane.

If other possible termination steps are considered, then it can be inferred that recombination or disproportionation of ethyl radicals, $C_{C_2H_5^*}$ is more likely than recombination of C_{H^*} and $C_{C_2H_5^*}$. Step 5 should then be replaced by

5a. $C_2H_5^* + C_2H_5^* \rightarrow C_4H_{10}$

and

5b. $C_2H_5^* + C_2H_5^* \rightarrow C_2H_4 + C_2H_6$

This also accounts for the production of the small amount of butane. If the reaction mechanism were steps 1, 2, 3, 4, 5a, and 5b, then applying the steady state approximations would give the overall order of reaction as 1/2.

Assuming that the initiation reaction is second order, then step 1 becomes

1a. $C_2H_6 + C_2H_6 \rightarrow 2CH_3^* + C_2H_6$

The reaction mechanism now becomes

Initiation

1a. $C_2H_6 + C_2H_6 \xrightarrow{\ k_{1a}\ } 2CH_3^* + C_2H_6$ $\hspace{2cm}$ (1-245)

2. $CH_3^* + C_2H_6 \xrightarrow{\ k_2\ } CH_4 + C_2H_5^*$ $\hspace{2cm}$ (1-246)

Propagation

3. $C_2H_5^* \xrightarrow{k_3} C_2H_4 + H^*$ (1-247)

4. $H^* + C_2H_6^* \xrightarrow{k_4} C_2H_5^* + H_2$ (1-248)

Termination

5a. $C_2H_5^* + C_2H_5^* \xrightarrow{k_{5a}} C_4H_{10}$ (1-249)

5b. $C_2H_5^* + C_2H_5^* \xrightarrow{k_{5b}} C_2H_4 + C_2H_6$ (1-250)

Applying the steady state approximation gives

$$\left(+r_{CH_3^*}\right) = \frac{dC_{CH_3^*}}{dt}$$

$$\frac{dC_{CH_3^*}}{dt} = 2k_{1a}C_{C_2H_6}^2 - k_2C_{CH_3^*}C_{C_2H_6} = 0$$ (1-251)

$$\left(+r_{C_2H_5^*}\right) = \frac{dC_{C_2H_5^*}}{dt}$$

$$\frac{dC_{C_2H_5^*}}{dt} = k_2C_{CH_3^*}C_{C_2H_6} - k_3C_{C_2H_5^*} + k_4C_{H^*}C_{C_2H_6}$$

$$-2\left(k_{5a} + k_{5b}\right)C_{C_2H_5^*}^2 = 0$$ (1-252)

and

$$\left(+r_{H^*}\right) = \frac{dC_{H^*}}{dt}$$

$$\frac{dC_{H^*}}{dt} = k_3C_{C_2H_5^*} - k_4C_{H^*}C_{C_2H_6} = 0$$ (1-253)

Adding Equations 1-251, 1-252, and 1-253 gives

$$C_{C_2H_5^*} = \left(\frac{k_{1a}}{k_{5a}+k_{5b}}\right)^{0.5} C_{C_2H_6} \qquad (1\text{-}254)$$

Therefore,

$$-\frac{dC_{C_2H_6}}{dt} = \left(+r_{C_2H_4}\right) = \frac{dC_{C_2H_4}}{dt}$$

$$\frac{dC_{C_2H_4}}{dt} = k_3 C_{C_2H_5^*} = k_3 \left(\frac{k_{1a}}{k_{5a}+k_{5b}}\right)^{0.5} C_{C_2H_6} \qquad (1\text{-}255)$$

This mechanism once again predicts the first order dependence of the reaction.

REFERENCES

1. Bailey, J. E. and Ollis, D. F. *Biochemical Engineering Fundamentals,* 2nd ed., McGraw-Hill Chemical Engineering Series, 1988.
2. Lin, K. H. "Part 1, Applied Kinetics," *Industrial and Engineering Chemistry,* Vol. 60, No. 5, pp. 61–82, May 1968.
3. Bodenstein, M. and Lind, S. C. *Physik. Chem,* 57, 168, 1907.
4. Bodenstein, M. Z. *Elektrochem,* 85, 329, 1913.
5. Nernst, W. Z. *Elektrochem,* 24, 335, 1918.
6. Taylor, H. S. *Trans. Faraday Soc.,* 21, 5601, 1926.
7. Rice, F. O. and Herzfeld, K. F. *J. Am. Chem. Soc.,* 56, 284, 1934.
8. Yang, K. H. and Hougen, O. A. "Determination of Mechanism of Catalyzed Gaseous Reactions," *Chemical Engineering Process,* p. 146, March 1950.
9. Hougen, O. A. and Watson, K. M. *Ind. Eng. Chem.,* 35, 529, 1943.

CHAPTER TWO

Thermodynamics of Chemical Reations

INTRODUCTION

The two main principles involved in establishing conditions for performing a reaction are chemical kinetics and thermodynamics. Chemical kinetics is the study of rate and mechanism by which one chemical species is converted to another. The rate is the mass in moles of a product produced or reactant consumed per unit time. The mechanism is the sequence of individual chemical reaction whose overall result yields the observed reaction. Thermodynamics is a fundamental of engineering having many applications to chemical reactor design.

Some chemical reactions are reversible and, no matter how fast a reaction takes place, it cannot proceed beyond the point of chemical equilibrium in the reaction mixture at the specified temperature and pressure. Thus, for any given conditions, the principle of chemical equilibrium expressed as the equilibrium constant, K, determines how far the reaction can proceed if adequate time is allowed for equilibrium to be attained. Alternatively, the principle of chemical kinetics determines at what rate the reaction will proceed towards attaining the maximum. If the equilibrium constant K is very large, for all practical purposes the reaction is irreversible. In the case where a reaction is irreversible, it is unnecessary to calculate the equilibrium constant and check the position of equilibrium when high conversions are needed.

Both the principles of chemical reaction kinetics and thermodynamic equilibrium are considered in choosing process conditions. Any complete rate equation for a reversible reaction involves the equilibrium constant, but quite often, complete rate equations are not readily available to the engineer. Thus, the engineer first must determine the temperature range in which the chemical reaction will proceed at a reasonable rate (possibly in the presence of any catalyst that may have been developed for the reactions). Next, the values of the equilibrium constant, K, in this temperature range must be computed using the principles of thermodynamics. The equilibrium constant of the reaction depends only on the temperature and is used to determine the limit to which the reaction can proceed under the conditions of temperatures, pressure, and reactant compositions that appear most suitable.

CHEMICAL EQUILIBRIUM

Consider a single homogeneous phase of one component of unchanging composition. If it undergoes an isothermal reversible change and does work, then from the first law of thermodynamics:

$$q = dU + pdV \text{ or } dU = q - pdV \tag{2-1}$$

From the second law of thermodynamics:

$$dS = \frac{q}{T}, \text{ reversible change}$$

$$q = TdS \tag{2-2}$$

Combining the first and second laws (Equations 2-1 and 2-2) gives:

$$dU = TdS - pdV \tag{2-3}$$

Now consider a homogeneous phase containing different substances or components. Its phase contains:

- n_1 moles of component 1
- n_i moles of component i
- n_k moles of component k

For a constant composition, it is known that $dU = TdS - pdV$. For a variable composition,

$$U = U(S, V, n_1, \ldots n_i \ldots n_k) \tag{2-4}$$

Performing a partial differentiation on Equation 2-4 yields

$$dU = \left(\frac{\partial U}{\partial S}\right)_{V,n_i} dS + \left(\frac{\partial U}{\partial V}\right)_{S,n_i} dV + \sum_{i=1}^{i=k}\left(\frac{\partial U}{\partial n_i}\right)_{S,V,n_j,\,i\neq j} dn_i \qquad (2\text{-}5)$$

If the number of moles are held constant, then from Equation 2-3

$$\left(\frac{\partial U}{\partial S}\right)_{V,n_i} = T \quad \text{and} \quad \left(\frac{\partial U}{\partial V}\right)_{S,n_i} = -p \qquad (2\text{-}6)$$

The chemical potential μ_i is defined by

$$\mu_i \equiv \left(\frac{\partial U}{\partial n_i}\right)_{S,V,n_j} \qquad (2\text{-}7)$$

Equation 2-5 thus becomes

$$dU = TdS - pdV + \sum_{i=1} \mu_i dn_i \qquad (2\text{-}8)$$

The Gibbs function is expressed as:

$$G = U + pV - TS \qquad (2\text{-}9)$$

Differentiating Equation 2-9 gives

$$dG = dU + pdV + Vdp - TdS - SdT \qquad (2\text{-}10)$$

Substituting Equation 2-3 into Equation 2-10 yields

$$dG = TdS - pdV + pdV + Vdp - TdS - SdT \qquad (2\text{-}11)$$

$$dG = Vdp - SdT \qquad (2\text{-}12)$$

Since Equation 2-12 holds for a system of constant composition, it is expressed as

$$G = G(T, P, n_1, n_i \ldots n_{k)} \qquad (2\text{-}13)$$

Partial differentiation of Equation 2-13 gives

$$dG = \left(\frac{\partial G}{\partial T}\right)_{P,n_i} dT + \left(\frac{\partial G}{\partial P}\right)_{T,n_i} dP + \sum_{i=1}^{k}\left(\frac{\partial G}{\partial n_i}\right)_{T,P,n_j} dn_i \qquad (2\text{-}14)$$

If the compositions are held constant, then

$$\left(\frac{\partial G}{\partial T}\right)_{P,n_i} = -S, \quad \left(\frac{\partial G}{\partial P}\right)_{T,n_i} = V \qquad (2\text{-}15)$$

And again, μ_i is defined as

$$\mu_i \equiv \left(\frac{\partial G}{\partial n_i}\right)_{T,P,n_j} \qquad (2\text{-}16)$$

Substituting Equations 2-15 and 2-16 into Equation 2-14 gives

$$dG = -S\,dT + V\,dP + \sum_{i=1}^{i=k}\mu_i\,dn_i \qquad (2\text{-}17)$$

Other state functions are the enthalpy function:

$$H = U + pV \qquad (2\text{-}18)$$

and the Helmholtz function:

$$A = U - TS \qquad (2\text{-}19)$$

The chemical potential μ can also be expressed in terms of H and A. The complete set of equations are:

$$dU = T\,dS - p\,dV + \sum \mu_i\,dn_i \qquad (2\text{-}20)$$

$$dG = V\,dp - S\,dT + \sum \mu_i\,dn_i \qquad (2\text{-}21)$$

$$dH = T\,dS + V\,dp + \sum \mu_i\,dn_i \qquad (2\text{-}22)$$

$$dA = -S\,dT - p\,dV + \sum \mu_i\,dn_i \qquad (2\text{-}23)$$

where

$$\mu_i = \left(\frac{\partial U}{\partial n_i}\right)_{S,V,n_j} = \left(\frac{\partial U}{\partial n_i}\right)_{T,P,n_j} = \left(\frac{\partial H}{\partial n_i}\right)_{S,P,n_j} = \left(\frac{\partial A}{\partial n_i}\right)_{T,V,n_j} \qquad (2\text{-}24)$$

CRITERIA FOR EQUILIBRIUM

For a reversible change, $dS = q/T$. If we consider a closed system undergoing a reversible change at $dT = 0$, $dP = 0$, then the only form of work is pdV. From the first law:

$$dU = q - pdV \qquad (2\text{-}1)$$

From the second law of thermodynamics:

$$dS = \frac{q}{T}, \text{ is a reversible change}$$

$$q = TdS \qquad (2\text{-}2)$$

Combining the first and second laws gives:

$$dU = TdS - pdV \qquad (2\text{-}3)$$

That is

$$dU - TdS + pdV = 0 \qquad (2\text{-}25)$$

Defining

$$G = U + PV - TS \qquad (2\text{-}9)$$

Differentiating gives

$$dG = du + pdV + Vdp - TdS - SdT \qquad (2\text{-}11)$$

Restricting this to the case of isothermal and isobaric, $dP = 0$, $dT = 0$, then

$$dG = dU + pdV - TdS \qquad (2\text{-}26)$$

Thus, for any small reversible displacement at equilibrium, dG = 0, if dP = 0, dT = 0. Similarly:

If dT = 0, dV = 0 then dA = 0
If dP = 0, dS = 0 then dH = 0
If dV = 0, dS = 0 then dU = 0

REACTION EQUILIBRIUM

Considering any generalized reversible chemical reaction, such that at dT = 0 and dP = 0:

$$aA + bB \rightleftharpoons cC + dD \tag{2-27}$$

If the reaction mixture is large enough that the mole numbers corresponding to the stoichiometric numbers react, then the compositions remain unchanged. If these mole numbers react at equilibrium, then the overall change in Gibbs function is

$$dG = c\mu_C + d\mu_D - a\mu_A - b\mu_B = 0 \quad \text{at equilibrium} \tag{2-28}$$

Since

$$\mu_A = \frac{\partial G_A}{\partial n_A} \tag{2-29}$$

The chemical potential as a function of composition can be expressed as

$$\mu = \mu_i^o + RT \ln a_i \tag{2-30}$$

where a_i is the activity of i. Introducing Equation 2-30 into the change in the Gibbs statement, and separating the standard state terms on the right side, gives

$$RT \ln \frac{a_C^c \cdot a_D^d}{a_A^a \cdot a_B^b} = -\left\{ c\mu_C^o + d\mu_D^o - a\mu_A^o - b\mu_B^o \right\} \tag{2-31}$$

$$\frac{a_C^c \cdot a_D^d}{a_A^a \cdot a_B^b} = \exp \left\{ \frac{-\left(c\mu_C^o + d\mu_D^o - a\mu_A^o - b\mu_B^o \right)}{RT} \right\} \tag{2-32}$$

$$= \exp\left\{\frac{-\sum \upsilon_i \mu_i}{RT}\right\} = K \tag{2-33}$$

The right side is the equilibrium constant, K. The left side contains the fugacities of the reactants in the mixture at the equilibrium composition. For gases, the right side is independent of pressure and composition and is equivalent to

$$ag_A^o + bg_B^o - cg_C^o - dg_D^o \tag{2-34}$$

where g_A^o is the specific Gibbs function of pure A at the standard state conditions, that is

$$g_A^o = h_A^o - Ts_A^o \tag{2-35}$$

IDEAL GAS MIXTURES

For an ideal gas mixture, $a_i = p_i$, which is expressed as

$$\frac{p_C^c \cdot p_D^d}{p_A^a \cdot p_B^b} = K_p \tag{2-36}$$

where p is the partial pressure. This is related to the total pressure as the total pressure, P_T, multiplied by the mole fraction of the component in the mixture. That is:

$$p_i = y_i P_T \tag{2-37}$$

Substituting Equation 2-37 into Equation 2-36 yields

$$\frac{y_C^c \cdot y_D^d}{y_A^a \cdot y_B^b} \, P_T^{(c+d-a-b)} = K_p \tag{2-38}$$

REAL GASES—IDEAL GASEOUS SOLUTION

In many cases, the assumption of ideal gases is not justified and it will be essential to determine fugacities. An example of such reactions is ammonia synthesis where the operating pressure may be as high as 1,500 atm.

The real gas can be expressed as

$$\frac{f_C^c \cdot f_D^d}{f_A^a \cdot f_B^b} = K_f \tag{2-39}$$

The fugacity in Equation 2-39 is that of the component in the equilibrium mixture. However, fugacity of only the pure component is usually known. It is also necessary to know something about how the fugacity depends on the composition in order to relate the two, therefore, assumptions about the behavior of the reaction mixture must be made. The most common assumption is that the mixture behaves as an ideal solution. In this case, it is possible to relate the fugacity, f, at equilibrium to the fugacity of the pure component, f', at the same pressure and temperature by

$$f_i = y_i f' \tag{2-40}$$

Equation 2-40 is known as the Lewis Randall rule, where f' is the fugacity of pure component i at the same temperature and total pressure as the mixture.

Substituting Equation 2-40 into Equation 2-39 gives

$$\frac{y_C^c \cdot y_D^d}{y_A^a \cdot y_B^b} = K_f \cdot \frac{\left(f_A'\right)^a \cdot \left(f_B'\right)^b}{\left(f_C'\right)^c \cdot \left(f_D'\right)^d} \tag{2-41}$$

where

$$K_y = \frac{y_C^c \cdot y_D^d}{y_A^a \cdot y_B^b} \tag{2-42}$$

The f_A', f_B', f_C', and f_D' are determined for the pure gas at the pressure of the mixture and depend on the pressure and the temperature. In gaseous mixtures, the quantity K_p as defined by Equation 2-38 is used. For an ideal gas reaction mixture, $K_f = K_p$. For a non-ideal system, Equation 2-39 can be used to calculate K_p from the measured equilibrium compositions K_y using Equation 2-42. The composition ratio K_f can then be determined from the equilibrium constant. This step is necessary in determining the equilibrium conversion from free energy data.

The steps in the process are:

1. Calculate the standard change of free energy, ΔG^o.

$$\Delta G^o = -RT \ln K_f \qquad (2\text{-}43)$$

or

$$\Delta G^o = -RT \ln K_p \qquad (2\text{-}44)$$

where ΔG^o is the difference between the free energies of the products and reactants when each is in a given standard state.
2. Determine the equilibrium constant K_f using Equation 2-43.
3. Evaluate K_y from Equation 2-42.
4. Calculate the conversion from K_y.

Steps 1 and 2 require thermodynamic data. Figure 2-1 shows the equilibrium constants of some reactions as a function of temperature. The Appendix at the end of this chapter gives a tabulation of the standard change of free energy ΔG^o at 298 K.

REAL GASES

If ϕ is the fugacity coefficient, and is defined by

$$\phi = \frac{f_i}{p} \qquad (2\text{-}45)$$

Then

$$f_i = \phi p \qquad (2\text{-}46)$$

Substituting Equation 2-45 into Equation 2-39 gives

$$\frac{\phi_C^c \bullet \phi_D^d}{\phi_A^a \bullet \phi_B^b} \bullet \frac{p_C^c \bullet p_D^d}{p_A^a \bullet p_B^b} = K_f \qquad (2\text{-}47)$$

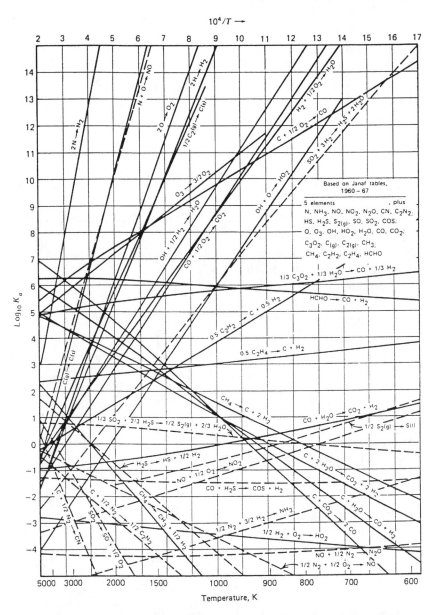

Figure 2-1. Chemical equilibrium constants as a function of temperature. (*Source: M. Modell and R. C. Reid,* Thermodynamics and its Applications, *Prentice-Hall, Inc., Englewood Cliffs, NJ.*)

Thus,

$$\frac{f_C^c \cdot f_D^d}{f_A^a \cdot f_B^b} = K_f = \exp\left(\frac{-\Delta G^o}{RT}\right) \tag{2-48}$$

Replacing f_i by $\phi_i p_i$ gives

$$\frac{\phi_C^c \cdot \phi_D^d}{\phi_A^a \cdot \phi_B^b} \cdot \frac{p_C^c \cdot p_D^d}{p_A^a \cdot p_B^b} = K_f \tag{2-49}$$

Since $p = yP$, Equation 2-49 then becomes

$$\frac{\phi_C^c \cdot \phi_D^d}{\phi_A^a \cdot \phi_B^b} \cdot \frac{y_C^c \cdot y_D^d}{y_A^a \cdot y_B^b} = K_f \, P^{(a+b-c-d)} \tag{2-50}$$

Fugacities are evaluated at the conditions at which the gases exist in the mixture.

LIQUID STATE

Chemical potentials can be expressed as:

$$\mu_i = \mu_i^* + RT \ln \gamma_i x_i \tag{2-51}$$

or

$$\mu_i = \mu_i^* + RT \ln a_i \tag{2-52}$$

where x_i = mole fraction of component i
γ_i = activity coefficient of component i
a_i = activity of component i

Equation 2-52 gives

$$\frac{a_C^c \cdot a_D^d}{a_A^a \cdot a_B^b} = \exp\left(\frac{-\Delta\mu^*}{RT}\right) = K_L \tag{2-53}$$

K_L is not independent of pressure because it is a function of the chemical potentials of pure i at the temperature T and pressure P of the mixture or of the infinite dilution state of i.

DETERMINING THE FUGACITY
AND FUGACITY COEFFICIENT

For any single component fluid at constant temperature and composition,

$$\left(\frac{\partial \mu_i}{\partial p}\right)_{T,n_i,n_j} = \overline{V}_i \qquad (2\text{-}54)$$

or

$$d\mu_i = \overline{V}dp \qquad (2\text{-}55)$$

where μ_i is the chemical potential of the pure gas at temperature T and pressure P. The fugacity f of the gas in which μ^o is a function of the temperature is

$$\mu_i = \mu_i^o + RT \ln f_i \qquad (2\text{-}56)$$

Differentiating Equation 2-56 at constant temperature and composition gives

$$d\mu_i = RTd \ln f_i \qquad (2\text{-}57)$$

Combining Equations 2-55 and 2-57 gives

$$\left[RTd\ln f_i = \overline{V}_i\, dp\right]_T \qquad (2\text{-}58)$$

Subtracting RTd ln p_i from both sides of Equation 2-58 gives

$$RTd\ln\left(\frac{f_i}{p_i}\right) = \overline{V}_i\, dp - RT\, d\ln p_i$$

$$= \overline{V}_i\, dp - RT\, d\ln p_i - RT\, d\ln y_i \qquad (2\text{-}59)$$

Since the composition is constant, $d\ln y_i = 0$ and Equation 2-59 becomes

$$RT\,d\ln\left(\frac{f_i}{p_i}\right) = \left(\overline{V}_i - \frac{RT}{p}\right)dp \tag{2-60}$$

where $\overline{V}_i$ is the partial molar volume. For a single gas, Equation 2-60 reduces to give

$$RT\,d\ln\left(\frac{f}{p}\right) = \left(V - \frac{RT}{p}\right)dp \tag{2-61}$$

where V is the specific volume. If we integrate at constant temperature and constant composition from $p = 0$ to the particular pressure $p = p'$, which is required to calculate the fugacity, the following is obtained:

$$\ln\left(\frac{f_i}{p_i}\right)_{p=p'} - \ln\left(\frac{f_i}{p_i}\right)_{p=0} = \int_0^{p'}\left(\frac{V}{RT} - \frac{1}{p}\right)dp \tag{2-62}$$

$f_i/p_i = 1$ at $p = 0$, which yields

$$\ln\left(\frac{f_i}{p}\right)_{p=p'} = \int_0^{p'}\left(\frac{V}{RT} - \frac{1}{p}\right)dp \tag{2-63}$$

Equation 2-63 gives the fugacity at p' and T in terms of an integral that is evaluated from experimental data.

Equation 2-63 can be expressed in terms of the compressibility factor

$$Z = \frac{pV}{RT} \tag{2-64}$$

and Equation 2-63 now becomes

$$\ln\left(\frac{f}{p'}\right) = \int_0^{p'}\left(\frac{Z-1}{p}\right)dp \tag{2-65}$$

PARTIAL MOLAR QUANTITIES

Considering any extensive property K, the partial molar quantity is defined by

$$\overline{K}_i = \left(\frac{\partial K}{\partial n_i}\right)_{T,P,n_{j,i \ne j}} \tag{2-66}$$

where T, P, n_j are held constant, then

$$K = K(T, P, n_1, n_2, \ldots n_i \ldots n_k) \tag{2-67}$$

Differentiating Equation 2-67 gives

$$dK = \left(\frac{\partial K}{\partial T}\right)_{P,n_i} dT + \left(\frac{\partial K}{\partial P}\right)_{T,n_i} dP + \sum_{i=1}\left(\frac{\partial K}{\partial n_i}\right)_{T,P,n_j} dn_i \tag{2-68}$$

Once again, integrating as in the Gibbs-Duhem equation, yields

$$K = \sum_{i=1} n_i \overline{K}_i \tag{2-69}$$

Similarly,

$$U = \sum_{i=1} n_i \overline{U}_i, \quad H = \sum_{i=1} n_i \overline{H}_i \quad \text{and} \quad G = \sum_{i=1} n_i \overline{G}_i \tag{2-70}$$

For a pure component

$$\overline{U}_i = u_i, \quad \overline{H}_i = h_i \text{ and } \overline{G} = g_i \tag{2-71}$$

The partial molar Gibbs function is given by

$$\left(\frac{\partial G}{\partial n_i}\right)_{T,P,n_j}$$

This is identical with the definition of μ_i in terms of G_i, but this is only true for G because only the pressure P and temperature T hold constant for G. Therefore, in general

$$\mu_i \neq \left(\frac{\partial K}{\partial n_i}\right)_{T,P,n_j} \tag{2-72}$$

This can be expressed as

$$\overline{H}_i = \overline{U}_i + P\overline{V}_i \tag{2-73}$$

and

$$\mu_i = \overline{G}_i \equiv \overline{H}_i - T\overline{S}_i \tag{2-74}$$

From Equation 2-21

$$dG = -SdT + PdV + \sum_{i=1} \mu_i \, dn_i \tag{2-21}$$

Also

$$\left(\frac{\partial G}{\partial p}\right)_{T,n_i,n_j} = V \tag{2-75}$$

Differentiating Equation 2-75 with respect to n gives

$$\frac{\partial^2 G}{\partial n \partial p} = \frac{\partial V}{\partial n} \tag{2-76}$$

But $\dfrac{\partial G}{\partial n_i} = \mu_i$ $\qquad\qquad$ (2-77)

Differentiating Equation 2-77 with respect to P yields

$$\frac{\partial^2 G}{\partial P \partial n_i} = \frac{\partial \mu_i}{\partial P} \tag{2-78}$$

Since the order of differentiating is indifferent, then

$$\left(\frac{\partial \mu_i}{\partial P}\right)_{T,n_i,n_j} = \left(\frac{\partial V}{\partial n_i}\right)_{T,P,n_j} = \overline{V}_i \qquad (2\text{-}79)$$

and similarly

$$\left(\frac{\partial \mu_i}{\partial T}\right)_{P,n_i,n_j} = \left(\frac{-\partial S}{\partial n}\right)_{T,P,n_j} = -\overline{S}_i \qquad (2\text{-}80)$$

Substituting Equation 2-80 into Equation 2-74 gives

$$\mu_i = \overline{H}_i + T\left(\frac{\partial \mu_i}{\partial T}\right)_{P,n_i,n_j} \qquad (2\text{-}81)$$

Rearranging Equation 2-81 yields

$$\left(\frac{\partial\left(\frac{\mu_i}{T}\right)}{\partial T}\right)_{P,n_i,n_j} = \frac{-\overline{H}_i}{T^2} \qquad (2\text{-}82)$$

EFFECT OF TEMPERATURE ON
THE EQUILIBRIUM CONSTANT

Generally,

$$K = \exp\left\{\frac{-\sum v_i \mu^{\circ}}{RT}\right\} \qquad (2\text{-}83)$$

or

$$\ln K = \frac{-1}{R}\sum \frac{v_i \mu^{\circ}}{T} \qquad (2\text{-}84)$$

where υ_i is the stoichiometric number, that is, a, b, c, d. Differentiating Equation 2-84 with respect to T gives

$$\frac{d \ln K}{dT} = \frac{-1}{R} \sum \upsilon_i \frac{d\left(\frac{\mu_i^o}{T}\right)}{dT} \qquad (2\text{-}85)$$

but

$$\frac{d\left(\frac{\mu_i^o}{T}\right)}{dT} = \frac{-h_i^o}{T^2} \qquad (2\text{-}86)$$

Hence,

$$\frac{d \ln K}{dT} = \frac{-1}{RT^2} \sum \upsilon_i h_i^o \qquad (2\text{-}87)$$

where h_i^o is the enthalpy of pure i at temperature T of the mixture for gases at unit fugacity, for liquids pressure P of mixture.

If $\upsilon_i h_i^o$ is replaced by the conventional ΔH^o when the moles represented by stoichiometric numbers react, then

$$\frac{d \ln K}{dT} = \frac{\Delta H^o}{RT^2} \quad \text{van't Hoffs equation} \qquad (2\text{-}88)$$

where ΔH^o is a function of temperature, but can be assumed constant over a small temperature range to enable an equilibrium constant at one T to be deduced from that for a T close to it. Integrating Equation 2-88 gives

$$\boxed{\ln \frac{K_2}{K_1} = \frac{\Delta H^o}{R}\left(\frac{1}{T_1} - \frac{1}{T_2}\right)} \qquad (2\text{-}89)$$

HEATS OF REACTION

If a process involves chemical reactions, heat effects will invariably be present. The amount of heat produced in a chemical reaction depends on the conditions under which the reaction is carried out. The standard heat of reaction is the enthalpy variation when the reaction is carried out under standard conditions using pure components in their most stable state or allotropic form, at standard pressure (1 atm) and temperature (usually, but not necessarily 298 K).

The values for standard heats of reaction may be found in the literature or calculated by thermodynamic methods. The physical state of the reactants and products (e.g. gas, liquid, or solid) must also be specified, if the reaction conditions are such that different states may coexist. For example,

$$H_2(g) + \frac{1}{2}O_2(g) \rightarrow H_2O(g) \quad \Delta H_{298} = -241.6 \text{ kJ} \quad (2\text{-}90)$$

$$H_2(g) + \frac{1}{2}O_2(g) \rightarrow H_2O(l) \quad \Delta H_{298} = -285.6 \text{ kJ} \quad (2\text{-}91)$$

In process design calculations, it is usually more convenient to express the heat of reaction in terms of the enthalpy per mole of product formed or reactant consumed. Since enthalpy is a state function, standard heats of reaction can be used to estimate the ΔH at different temperatures by making a heat balance over a hypothetical process:

$$\Delta H_{rxnT} = \Delta H_{rnxT_o} + \Delta H_{products} - \Delta H_{reactants} \quad (2\text{-}92)$$

where ΔH_{rxnT} = heat of reaction at temperature, T
  ΔH_{rnxT_o} = heat of reaction at a known standard temperature
  $\Delta H_{reactants}$ = enthalpy change to bring reactants from temperature T to standard temperature
  $\Delta H_{products}$ = enthalpy change to bring products from the standard temperature back to reaction temperature T

The specific heats, C_p are usually expressed as a quadratic or a polynomial function of temperature and expressed as:

$$C_{p_i} = A_i + \left(B_i \cdot 10^{-2} \cdot T\right) + \left(C_i \cdot 10^{-5} \cdot T^2\right) \quad (2\text{-}93)$$

or

$$C_{p_i} = A_i + \left(B_i \cdot 10^{-2} \cdot T\right) + \left(C_i \cdot 10^{-5} \cdot T^2\right) + \left(D_i \cdot 10^{-9} \cdot T^3\right) \quad (2\text{-}94)$$

where C_{p_i} is in cal/mol K or J/mol K and T is in degrees Kelvin (1 cal = 4.184 J).

If stoichiometric quantities of the reactants enter and the products leave at the same temperature as represented by the equation:

$$\alpha_1 A + \alpha_2 B \rightarrow \alpha_3 C + \alpha_4 D \tag{2-95}$$

Equation 2-95 represents a reaction at a temperature different from the standard state of 25°C (298 K) in stoichiometric quantities. Figure 2-2 shows the information for calculating the heat of reaction.

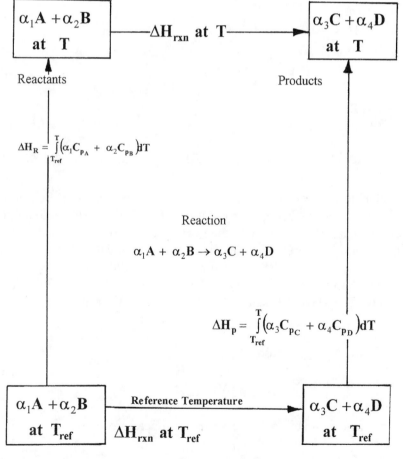

Figure 2-2. Heats of reaction at temperature other than standard condition.

Assuming a steady state process ($\Delta E = 0$), no kinetic (KE) or potential energy (PE) changes and mechanical work done (W) by the system on the surroundings are zero, that is KE, PE, and W = 0, the general energy balance

$$\Delta E = E_2 - E_1 = -\Delta\{(H + KE + PE)_m\} + Q - W \qquad (2\text{-}96)$$

where

$$\Delta E = \left(U + KE + PE\right)_{t2} m_2 - \left(U + KE + PE\right)_{t1} m_1 \qquad (2\text{-}97)$$

reduces to

$$Q = \Delta H = \left\{\Delta H_f^o + \Delta\left(H - H^o\right)\right\}_{products}$$
$$- \left\{\Delta H_f^o + \Delta\left(H - H^o\right)\right\}_{reactants} \qquad (2\text{-}98)$$

or

$$Q = \Delta H^o_{rxnT_{ref}} + \left(\Delta H_{products} - \Delta H_{reactants}\right) \qquad (2\text{-}99)$$

Calculating the heat of reaction is a multi-step process. Beginning with the standard heats of formation at 298 K, first calculate the standard heat of reaction, and then calculate ΔH for the actual system temperature and pressure. The heat of reaction at 298 K, ΔH_{298} is usually referred to as the standard heat of reaction. This can be readily calculated from the standard heats of formation of the reaction components. The standard heat of reaction is expressed as:

$$\Delta H^o_{rxnT_{298}} = \left(\sum \alpha_{products} \cdot \Delta H_{products}\right)$$
$$- \left(\sum \alpha_{reactants} \cdot \Delta H_{reactants}\right) \qquad (2\text{-}100)$$

The heat of reaction at the system temperature is

$$\Delta H_{rxnT} = \Delta H^o_{rxnT_{298}} + \Delta H_{products} - \Delta H_{reactants} \qquad (2\text{-}101)$$

If no phase change occurs, the difference $\Delta H_{products} - \Delta H_{reactants}$ can be written as

$$\Delta H_{products} - \Delta H_{reactants} = \int_{298}^{T} \Delta C_p \, dT \tag{2-102}$$

where C_p is given by

$$\Delta C_p = \Delta a + (\Delta b \cdot T) + (\Delta c \cdot T^2) + (\Delta d \cdot T^3) \tag{2-103}$$

and ΔC_p is in cal/mol K, J/mol K, using T in degrees Kelvin.
The constants for the equation are found by

$$\Delta a = \sum \alpha_{products} \cdot a_{products} - \sum \alpha_{reactants} \cdot a_{reactants} \tag{2-104}$$

$$\Delta b = \sum \alpha_{products} \cdot b_{products} - \sum \alpha_{reactants} \cdot b_{reactants} \tag{2-105}$$

$$\Delta c = \sum \alpha_{products} \cdot c_{products} - \sum \alpha_{reactants} \cdot c_{reactants} \tag{2-106}$$

$$\Delta d = \sum \alpha_{products} \cdot d_{products} - \sum \alpha_{reactants} \cdot d_{reactants} \tag{2-107}$$

So that

$$\int_{298}^{T} \Delta C_p dT = \int_{298}^{T} \left\{ \Delta a + (\Delta b \cdot T) + (\Delta c \cdot T^2) + (\Delta d \cdot T^3) \right\} dT \tag{2-108}$$

Integrating Equation 2-108 yields

$$\int_{298}^{T} \Delta C_p dT = \Delta a (T - 298) + \frac{\Delta b}{2} (T^2 - 298^2)$$

$$+ \frac{\Delta c}{3} (T^3 - 298^3) + \frac{\Delta d}{4} (T^4 - 298^4) \tag{2-109}$$

The heat of reaction at the system temperature is therefore given by

$$\Delta H_{rxnT} = \Delta H_{rxnT_{298}} + \Delta a (T - 298) + \frac{\Delta b}{2} (T^2 - 298^2)$$

$$+ \frac{\Delta c}{3} (T^3 - 298^3) + \frac{\Delta d}{4} (T^4 - 298^4) \qquad (2\text{-}110)$$

ENTHALP, a computer software, was developed to calculate the heat of reaction at the system temperature. To run the program, the user must know the following:

- The standard temperature (298 K) T_o and the system temperature TF.
- The reaction stoichiometric coefficients α_1, α_2, α_3, and α_4.
- Heat capacity constants, and a_1, a_2, a_3, and a_4.
- Standard heats of reaction of each component, dHf_A, dHf_B, dHf_C, and dHf_D.

HEAT CAPACITIES OF GASES

The heat capacity of gases is essential for some process engineering design involving gas-phase chemical reactions. Here, the heat capacities, C_p^o, for gases are required to determine the heat necessary to bring the chemical compound increase to the reaction temperature. The heat capacity of a mixture of gases may be found from the heat capacities of the individual components contained in the mixtures.

The correlation for C_p^o of the ideal gas at low pressure is a third degree polynomial, which is a function of temperature.

$$C_p^o = A + BT + CT^2 + DT^3 \qquad (2\text{-}111)$$

where C_p^o = heat capacity of ideal gas at low pressure, cal/mol K
A, B, C, and D = constants for the chemical compounds
T = temperature, K

The Appendix at the end of this chapter gives values of heat capacity constants.

HEATS OF FORMATION

Heats of formation, ΔH_f^o, for individual chemicals involved in chemical reactions are important in determining the heat of reaction, ΔH_r^o, and associated heating and cooling requirements. If $\Delta H_r^o < 0$, then the chemical reaction is exothermic and will require cooling. If $\Delta H_r^o > 0$, the reaction is endothermic and heating is therefore required.

The correlation of ΔH_f^o for the ideal gas at low temperature is based on a series expansion in temperature and is expressed as:

$$\Delta H_f^o = A + BT + CT^2 \tag{2-112}$$

where ΔH_f^o = heat of formation of ideal gas a low pressure, kcal/gmol
A, B, and C = correlation constants
            T = temperature, K

The Appendix at the end of this chapter shows heats of formation of some compounds.

Example 2-1

Calculate the heat of reaction for the synthesis of ammonia from hydrogen and nitrogen at 1 atm and 155.0°C in $N_2 + 3H_2 \rightarrow 2NH_3$

1. kcal/kmol of N_2 reacted.
2. kJ/mol of H_2 reacted.
3. The true equilibrium constant (K).

The heats of formation, the standard molar entropies at 298 K, are given below.

Component	ΔS_s^o kJ/kmol • K	ΔS_s^o cal/gmol • K	ΔH_f^o kcal/kmol
$N_2(g)$	191.9	45.87	0.0
$H_2(g)$	130.9	31.29	0.0
$NH_3(g)$	192.9	46.10	11,020

The heat capacities, expressed as quadratic function of temperature, are shown below:

Component	C_p, kcal/kmol • K
$N_2(g)$	$6.457 + 1.39 \times 10^{-3}T - 0.069 \times 10^{-6}T^2$
$H_2(g)$	$6.946 - 0.196 \times 10^{-3}T + 0.476 \times 10^{-6}T^2$
$NH_3(g)$	$5.92 - 8.963 \times 10^{-3}T - 1.764 \times 10^{-6}T^2$

Solution

$$N_2 + 3H_2 \rightarrow 2NH_3$$

or

$$\frac{1}{3}N_2 + H_2 \rightarrow \frac{2}{3}NH_3$$

Calculate the heat of reaction at the reference temperature using the heats of formation.

$$\Delta H^o_{rxnT}(T_R) = 2\Delta H^o_{F_{NH_3}} - 3\Delta H^o_{F_{H_2}} - \Delta H^o_{F_{N_2}}$$

The heats of formation of the elements H_2, N_2 are zero at 298 K. Therefore,

$$\Delta H^o_{rxn}(298\,K) = 2(-11,020)\text{ kcal/kmol}$$

$$= -22,040\,\frac{\text{kcal}}{\text{kmol }N_2}\text{ reacted.}$$

The minus sign shows that the heat of reaction is exothermic.

Component	a	b × 10⁻³	c × 10⁻⁶
$N_2(g)$	6.457	1.39	−0.069
$H_2(g)$	6.946	−0.196	0.476
$NH_3(g)$	5.92	8.963	−1.764
Δ	−15.455	17.124	−4.887

Substituting $\Delta H^o_{rxn}(298\,K)$ and the parameters Δa, Δb, and Δc in Equation 2-110 gives

$$\Delta H_{rxnT}(T) = \Delta H^o_{rxnT}(T_R) + \Delta a(T - T_R) + \frac{\Delta b}{2}(T^2 - T_R^2)$$

$$+ \frac{\Delta c}{3}(T^3 - T_R^3)$$

$$\Delta H_{rxn}(428K) = -22,040 - 15.455(428-298)$$

$$+ \frac{17.124 \times 10^{-3}}{2}(428^2 - 298^2)$$

$$- \frac{4.887 \times 10^{-6}}{3}(428^3 - 298^3)$$

$$\Delta H_{rxn}(428K) = -22,040 - 2,009.15 + 808.08 - 84.61$$

$$-23,325.68 \frac{kcal}{kmol\,N_2}$$

1 kcal = 4.184 kJ

$$= -97,594.65 \frac{kJ}{kmol\,N_2}$$

$$\Delta H_{rxn}(428K) = \frac{1\ kg\,mol\,N_2}{3\ kg\,mol\,H_2}\left(-97,594.65 \frac{kJ}{kmol\,N_2}\right)$$

$$= -32,531.55 \frac{kJ}{kg\,mol\,H_2} \quad at\ 428K$$

When the reaction proceeds at a constant temperature T, and the reactants and products remain at the standard state (represented by the superscript o), we can use Equation 2-44 to calculate the true equilibrium constant:

$$\Delta G = -RT \ln K \qquad\qquad (2\text{-}44)$$

$$= \Delta H_r^o - T\Delta S^o$$

where

$$\Delta H_r^o = \Delta H_{rs}^o + \int_{Ts}^{T} \Delta C_p^o\, dT$$

and

$$\Delta S^{o} = \Delta S_{s}^{o} + \int_{Ts}^{T} \frac{\Delta C_{p}}{T} dT$$

Hence,

$$\Delta C_{p} = -15.455 + 17.124 \times 10^{-3} T - 4.887 \times 10^{-6} T^{2} \frac{cal}{mol\ K\ N_{2}}$$

and

$$\Delta S_{s}^{o} = 2(46.10) - 45.87 - 3(31.29)$$

$$= -47.54 \frac{cal}{mol \bullet K}$$

This is the entropy change for a complete conversion to NH_3 where 1 kmol of N_2 is converted. The entropy change at 428 K is:

$$\Delta S^{o} = -47.54 + \int_{298}^{428} \left(\frac{-15.455}{T} + 17.124 \times 10^{-3} - 4.887 \times 10^{-6} T \right) dT$$

$$= -47.54 + \left[-15.455 \ln T + 17.124 \times 10^{-3} T - 4.887 \times 10^{-6} \frac{T^{2}}{2} \right]_{298}^{428}$$

$$= -47.54 + \left[-15.455 \ln \left(\frac{428}{298} \right) + 17.124 \times 10^{-3} (428 - 298) \right.$$

$$\left. - \frac{4.887 \times 10^{-6}}{2} \left(428^{2} - 298^{2} \right) \right]$$

$$= -47.54 - 5.595 + 2.226 - 0.231$$

$$= -51.14 \frac{cal}{mol\ K\ N_{2}}$$

The true equilibrium constant at 428 K is $\Delta G^{\circ} = -RT \ln K = \Delta H^{\circ} - T\Delta S$, that is

$$\ln K = \frac{1}{R}\left(\Delta S^{\circ} - \frac{\Delta H^{\circ}}{T}\right)$$

where

$$\Delta H^{\circ} = -23,326 \frac{cal}{mol\,N_2} \quad \text{and} \quad R = 1.987 \frac{cal}{mol \bullet K}$$

$$\ln K = \frac{1}{1.987}\left[-51.14 - \left(\frac{-23326}{428}\right)\right] = 1.691$$

and

$$K = 5.42$$

The value of K is very sensitive to small errors in ΔH_r° and ΔS° and the equilibrium constant belongs to the reaction equation $N_2 + 3H_2 \rightarrow 2NH_3$. Table 2-1 shows the calculated results of the heats of reaction at standard state (298 K) and at system temperature of 428 K using the software ENTHALP.

Example 2-2

Calculate the heat that must be removed or provided for the gas phase reaction proceeding

$$CO_2(g) + 4H_2(g) \rightarrow 2H_2O(g) + CH_4(g)$$

with 100% conversion and the gas entering at 450°C. The heat of formation at standard conditions and the specific heat capacities are shown below:

Component	Heat of formation ΔH_f(J/gmol)
$CO_2(g)$	−393,513
$H_2(g)$	0.0
$H_2O(g)$	−241,826
$CH_4(g)$	−74,848

Table 2-1
The standard heat of a chemical reaction calculation:
$aA + bB \rightarrow cC + dD$

```
THE STANDARD HEAT OF A CHEMICAL REACTION CALCULATION:aA + bB ---> cC + dD
************************************************************************
INITIAL TEMPERATURE OF THE REACTION, K:                       298.000
FINAL TEMPERATURE OF THE REACTION, K:                         428.000
STOICHIOMETRIC COEFFICIENT OF COMPONENT 1:                        1.0
STOICHIOMETRIC COEFFICIENT OF COMPONENT 2:                        3.0
STOICHIOMETRIC COEFFICIENT OF COMPONENT 3:                        2.0
STOICHIOMETRIC COEFFICIENT OF COMPONENT 4:                         .0
HEAT CAPACITY CONSTANT OF COMP.1, A(1):                    .645700E+01
HEAT CAPACITY CONSTANT OF COMP.1, B(1):                    .138900E+00
HEAT CAPACITY CONSTANT OF COMP.1, C(1):                   -.690000E-02
HEAT CAPACITY CONSTANT OF COMP.2, A(2):                    .694600E+01
HEAT CAPACITY CONSTANT OF COMP.2, B(2):                   -.196000E-01
HEAT CAPACITY CONSTANT OF COMP.2, C(2):                    .476000E-01
HEAT CAPACITY CONSTANT OF COMP.3, A(3):                    .592000E+01
HEAT CAPACITY CONSTANT OF COMP.3, B(3):                    .896300E+00
HEAT CAPACITY CONSTANT OF COMP.3, C(3):                   -.176400E+00
HEAT CAPACITY CONSTANT OF COMP.4, A(4):                    .000000E+00
HEAT CAPACITY CONSTANT OF COMP.4, B(4):                    .000000E+00
HEAT CAPACITY CONSTANT OF COMP.4, C(4):                    .000000E+00
STANDARD HEAT OF FORMATION OF COMP. 1, cal/gmol:          .000000E+00
STANDARD HEAT OF FORMATION OF COMP. 2, cal/gmol:          .000000E+00
STANDARD HEAT OF FORMATION OF COMP. 3, cal/gmol:         -.110200E+05
STANDARD HEAT OF FORMATION OF COMP. 4, cal/gmol:          .000000E+00
THE STANDARD HEAT OF REACTION,cal/g.mol:AT 298.000 K     -.2204000E+05
THE STANDARD HEAT OF REACTION,cal/g.mol:AT 428.000 K     -.2332563E+05
```

ΔC_p : J/gmol • K

Component	Specific heat capacity equation
$CO_2(g)$	$C_p = 26.75 + 4.226 \times 10^{-2}T - 1.425 \times 10^{-5}T^2$
$H_2(g)$	$C_p = 26.88 + 0.435 \times 10^{-2}T - 0.033 \times 10^{-5}T^2$
$H_2O(g)$	$C_p = 29.16 + 1.449 \times 10^{-2}T - 0.202 \times 10^{-5}T^2$
$CH_4(g)$	$C_p = 13.44 + 7.703 \times 10^{-2}T - 1.874 \times 10^{-5}T^2$

Solution

The heat of reaction at the standard state 298 K is:

$$\Delta H_{rxn}(298 \text{ K}) = 2(-241,826) - 74,848 - (-393,513) - 164,987 \text{ J/gmol } CO_2$$

Component	a	b × 10⁻²	c × 10⁻⁵
$CO_2(g)$	26.75	4.226	−1.425
$H_2(g)$	26.88	0.435	−0.033
$H_2O(g)$	29.16	1.449	−0.202
$CH_4(g)$	13.44	7.703	−1.874
Δ	−62.54	4.635	−0.721

$$\Delta C_p = -62.54 + 4.635 \times 10^{-2}\,T - 0.721 \times 10^{-5}\,T^2 \; \frac{J}{mol\,K}$$

$$\Delta H_{rxn}(723K) = \Delta H_{rxn}(298K) + \int_{298}^{723} \Delta C_p dT$$

Substituting $\Delta H^o_{rxn}(298\,K)$ and the parameters Δa, Δb, and Δc in Equation 2-110 gives

$$\Delta H_{rxn}(723K) = \Delta H_{rxn}(298K)$$

$$+ \int_{298}^{723} \left(-62.54 + 4.635 \times 10^{-2}\,T - 0.721 \times 10^{-5}\,T^2\right) dT$$

$$\Delta H_{rxnT}(T) = \Delta H^o_{rxnT}(T_R) + \Delta a(T - T_R) + \frac{\Delta b}{2}\left(T^2 - T_R^2\right)$$

$$+ \frac{\Delta c}{3}\left(T^3 - T_R^3\right)$$

$$\Delta H_{rxn}(723K) = -164,987 + \{-62.54(723 - 298)$$

$$+ \frac{4.635 \times 10^{-2}\left(723^2 - 298^2\right)}{2}$$

$$- \frac{0.721 \times 10^{-5}\left(723^3 - 298^3\right)}{3}\}$$

$$= -164,987 - 26,579.5 + 10,056.21 - 844.70$$

$$= -182,354.99$$

$$= -182,355 \frac{J}{g\,mol\,CO_2}$$

The result shows that $-182,355$ J/gmol CO_2 must be removed. The results of the ENTHALP1 computer program (S.I. units) for the heats of reaction at 298 K and 723 K respectively are shown in Table 2-2.

Example 2-3

Using the standard heats of formation of the reaction components, calculate the standard heat of reaction and the heat of reaction at system temperature of 350°C and 1 atm. The reaction is:

$$\alpha_1 A + \alpha_2 B \rightarrow \alpha_3 C + \alpha_4 D$$

The system temperature $T = 623$ K
Number of components $n = 4$
Reaction coefficients $\alpha_1 = 1, \alpha = 2, \alpha_3 = 1, \alpha_4 = 0$

$$C_p = a + b \times 10^{-2} T + c \times 10^{-5} T^2 + d \times 10^{-9} T^3 \quad \frac{cal}{g \, mol \, K}$$

Table 2-2
The standard heat of a chemical reaction calculation:
aA + bB → cC + dD

THE STANDARD HEAT OF A CHEMICAL REACTION CALCULATION:aA + bB ---> cC + dD	
INITIAL TEMPERATURE OF THE REACTION, K:	298.000
FINAL TEMPERATURE OF THE REACTION, K:	723.000
STOICHIOMETRIC COEFFICIENT OF COMPONENT 1:	1.0
STOICHIOMETRIC COEFFICIENT OF COMPONENT 2:	4.0
STOICHIOMETRIC COEFFICIENT OF COMPONENT 3:	2.0
STOICHIOMETRIC COEFFICIENT OF COMPONENT 4:	1.0
HEAT CAPACITY CONSTANT OF COMP.1, A(1):	.267500E+02
HEAT CAPACITY CONSTANT OF COMP.1, B(1):	.422600E+01
HEAT CAPACITY CONSTANT OF COMP.1, C(1):	-.142500E+01
HEAT CAPACITY CONSTANT OF COMP.2, A(2):	.268800E+02
HEAT CAPACITY CONSTANT OF COMP.2, B(2):	.435000E+00
HEAT CAPACITY CONSTANT OF COMP.2, C(2):	-.330000E-01
HEAT CAPACITY CONSTANT OF COMP.3, A(3):	.291600E+02
HEAT CAPACITY CONSTANT OF COMP.3, B(3):	.144900E+01
HEAT CAPACITY CONSTANT OF COMP.3, C(3):	-.202000E+00
HEAT CAPACITY CONSTANT OF COMP.4, A(4):	.134100E+02
HEAT CAPACITY CONSTANT OF COMP.4, B(4):	.770300E+01
HEAT CAPACITY CONSTANT OF COMP.4, C(4):	-.187400E+01
STANDARD HEAT OF FORMATION OF COMP. 1, J/g mol:	-.393513E+06
STANDARD HEAT OF FORMATION OF COMP. 2, J/g mol:	.000000E+00
STANDARD HEAT OF FORMATION OF COMP. 3, J/g mol:	-.241826E+06
STANDARD HEAT OF FORMATION OF COMP. 4, J/g mol:	-.748480E+05
THE STANDARD HEAT OF REACTION, J/g mol: AT 298.000 K	-.1649870E+06
THE STANDARD HEAT OF REACTION, J/g mol: AT 723.000 K	-.1823550E+06

Component	Heat of formation ΔH_f(cal/gmol)
A	−26,416
B	0.0
C	−48,080
D	0.0

Component	a	b × 10⁻²	c × 10⁻⁵	d × 10⁻⁹
A	6.726	0.04001	0.1283	−0.5307
B	6.952	−0.04576	0.09563	−0.2079
C	4.55	2.186	−0.291	−1.92
D	0	0	0	0
Δ	−16.08	2.23751	−0.61056	−0.9735

Solution

The heat of reaction at the standard state 298 K is:

$$\Delta H_{rxn}(298K) = \alpha_3 \Delta H^o_{FC} + \alpha_4 \Delta H^o_{FD} - \alpha_1 \Delta H^o_{FA} - \alpha_2 \Delta H^o_{FB}$$

$$\Delta H_{rxn}(298K) = 1(-48,080) + 0(0) - 1(-26,416) - 2(0)$$

$$- 21,664 \text{ cal/mol}$$

$$\Delta H_{rxn}(623K) = \Delta H_{rxn}(298K) + \int_{298}^{623} \Delta C_p dT$$

$$= \Delta H^o_{rxn}(298K) + \int_{298}^{623} \left(\Delta a + \Delta b \times 10^{-2} T + \Delta c \times 10^{-5} T^2 + \Delta d \times 10^{-9} T^3 \right) dT$$

Substituting $\Delta H^o_{rxn}(298 \text{ K})$ and the parameters Δa, Δb, and Δc in Equation 2-110 gives

$$\Delta H_{rxnT}(T) = \Delta H^o_{rxnT}(T_R) + \Delta a(T - T_R) + \frac{\Delta b}{2}(T^2 - T_R^2)$$

$$+ \frac{\Delta c}{3}(T^3 - T_R^3) + \frac{\Delta d}{4}(T^4 - T_R^4)$$

$$\Delta H_{rxn}(623) = \Delta H_{rxn}^o(298) + \int_{298}^{623} \left\{ \Delta aT + \frac{\Delta b \times 10^{-2} T^2}{2} + \frac{\Delta c \times 10^{-5} T^3}{3} \right.$$

$$\left. + \frac{\Delta d \times 10^{-9} T^4}{4} \right\}$$

$$= -21,664 + \left\{ -16.08(623 - 298) + \frac{2.23751 \times 10^{-2} \left(623^2 - 298^2\right)}{2} \right.$$

$$\left. - \frac{0.61056 \times 10^{-5} \left(623^3 - 298^3\right)}{3} - \frac{0.9735 \times 10^{-9} \left(623^4 - 298^4\right)}{4} \right\}$$

$$= -21,664 - 5,226 + 3,348.71 - 438.26 - 34.74$$

$$= -24,014.29 \ \frac{cal}{mol \ A}$$

Table 2-3
The standard heat of a chemical reaction calculation:
aA + bB → cC + dD

```
THE STANDARD HEAT OF A CHEMICAL REACTION CALCULATION:aA + bB ---> cC + dD
***********************************************************************************
INITIAL TEMPERATURE OF THE REACTION, K:                          298.000
FINAL TEMPERATURE OF THE REACTION, K:                            623.000
STOICHIOMETRIC COEFFICIENT OF COMPONENT 1:                          1.0
STOICHIOMETRIC COEFFICIENT OF COMPONENT 2:                          2.0
STOICHIOMETRIC COEFFICIENT OF COMPONENT 3:                          1.0
STOICHIOMETRIC COEFFICIENT OF COMPONENT 4:                           .0
HEAT CAPACITY CONSTANT OF COMP. 1, A(1):                        .672600E+01
HEAT CAPACITY CONSTANT OF COMP. 1, B(1):                        .400100E-01
HEAT CAPACITY CONSTANT OF COMP. 1, C(1):                        .128300E+00
HEAT CAPACITY CONSTANT OF COMP. 1, D(1):                       -.530700E+00
HEAT CAPACITY CONSTANT OF COMP. 2, A(2):                        .695200E+01
HEAT CAPACITY CONSTANT OF COMP. 2, B(2):                       -.457600E-01
HEAT CAPACITY CONSTANT OF COMP. 2, C(2):                        .956300E-01
HEAT CAPACITY CONSTANT OF COMP. 2, D(2):                       -.207900E+00
HEAT CAPACITY CONSTANT OF COMP. 3, A(3):                        .455000E+01
HEAT CAPACITY CONSTANT OF COMP. 3, B(3):                        .218600E+01
HEAT CAPACITY CONSTANT OF COMP. 3, C(3):                       -.291000E+00
HEAT CAPACITY CONSTANT OF COMP. 3, D(3):                       -.192000E+01
HEAT CAPACITY CONSTANT OF COMP. 4, A(4):                        .000000E+00
HEAT CAPACITY CONSTANT OF COMP. 4, B(4):                        .000000E+00
HEAT CAPACITY CONSTANT OF COMP. 4, C(4):                        .000000E+00
HEAT CAPACITY CONSTANT OF COMP. 4, D(4):                        .000000E+00
STANDARD HEAT OF FORMATION OF COMP. 1, cal/gmol:              -.264160E+05
STANDARD HEAT OF FORMATION OF COMP. 2, cal/gmol:               .000000E+00
STANDARD HEAT OF FORMATION OF COMP. 3, cal/gmol:              -.480800E+05
STANDARD HEAT OF FORMATION OF COMP. 4, cal/gmol:               .000000E+00
THE STANDARD HEAT OF REACTION, cal/g.molAT 298.000 K          -.2166400E+05
THE STANDARD HEAT OF REACTION, cal/g.molAT 623.000 K          -.2401429E+05
```

Table 2-3 gives the results of the program for the heats of reaction at 298 K and 623 K, respectively

---

Example 2-4

Experimental values calculated for the heat capacity of ammonia from –40 to 1200°C are shown in Table 2-4.
Fit the data by least squares for the following two functions.

$$C_p^o = a + bT + cT^2$$

$$C_p^o = a + bT + cT^2 + dT^3$$

Solution

A computer program, PROG2, was developed to fit the data by least squares of a polynomial regression analysis. The data of temperature (independent variable) versus heat capacity (dependent variable) were inputted in the program for an equation to an nth degree $Y = C_o + C_1 X + C_2 X^2 + C_3 X^3 + \ldots + C_n X^n$. The results of the computer program for the polynomial regressions of: $C_p^o = a + bT + cT^2$ and $C_p^o = a + bT + cT^2 + dT^3$ are:

$$C_p^o = 8.357 + 0.818 \times 10^{-2} T - 0.16807 \times 10^{-5} T^2 \ \frac{cal}{g\,mol\,°C}$$

Table 2-4

T °C	C_p^o cal/gmolT• °C	T °C	C_p^o cal/gmol • °C
–40	8.180	500	12.045
–20	8.268	600	12.700
0	8.371	700	13.310
18	8.472	800	13.876
25	8.514	900	14.397
100	9.035	1,000	14.874
200	9.824	1,100	15.306
300	10.606	1,200	15.694
400	11.347		

Source: Basic Principles and Calculations in Chemical Engineering by David M. Himmelblau, 5th ed., 1989, Prentice Hall International series.

Table 2-5
Polynomial regression analysis for an equation to an nth
degree: $Y = C_0 + C_1 \cdot X + C_2 \cdot X^2 + C_3 \cdot X^3 + \ldots + C_n \cdot X^n$

VALUE OF THE INDEPENDENT AND DEPENDENT VARIABLES	
X-VALUES	Y-VALUES
-40.00	8.18
-20.00	8.27
.00	8.37
18.00	8.47
25.00	8.51
100.00	9.03
200.00	9.82
300.00	10.61
400.00	11.35
500.00	12.05
600.00	12.70
700.00	13.31
800.00	13.88
900.00	14.40
1000.00	14.87
1100.00	15.31
1200.00	15.69

```
POLYNOMIAL OF DEGREE 1
COEFFICIENTS ARE:      .8527405D+01  .6405385D-02

VARIANCE IS:              .6143E-01
ERROR SUM OF SQUARES:     .9215E+00
TOTAL SUM OF SQUARES:     .1215E+03

COEFFICIENT OF DETERMINATION IS:     .9924
CORRELATION COEFFICIENT:        .9962
```

```
POLYNOMIAL OF DEGREE 2
COEFFICIENTS ARE:      .8357130D+01  .8181601D-02 -.1680716D-05

VARIANCE IS:              .5256E-02
ERROR SUM OF SQUARES:     .7358E-01
TOTAL SUM OF SQUARES:     .1215E+03

COEFFICIENT OF DETERMINATION IS:     .9994
CORRELATION COEFFICIENT:        .9997
```

```
POLYNOMIAL OF DEGREE 3
COEFFICIENTS ARE:      .8378698D+01  .7355283D-02  .4074239D-06 -.1235132D-08

VARIANCE IS:              .2325E-02
ERROR SUM OF SQUARES:     .3023E-01
TOTAL SUM OF SQUARES:     .1215E+03

COEFFICIENT OF DETERMINATION IS:     .9998
CORRELATION COEFFICIENT:        .9999
```

```
POLYNOMIAL OF DEGREE 4
COEFFICIENTS ARE:      .8381830D+01  .6463639D-02  .4782197D-05 -.7483072D-08

VARIANCE IS:              .8187E-03
ERROR SUM OF SQUARES:     .9825E-02
TOTAL SUM OF SQUARES:     .1215E+03

COEFFICIENT OF DETERMINATION IS:     .9999
CORRELATION COEFFICIENT:        1.0000
```

and

$$C_p = 8.379 + 0.736 \times 10^{-2} T + 4.074 \times 10^{-5} T^2 - 1.235 \times 10^{-9} \frac{cal}{g \, mol \, °C}$$

Table 2-5 gives the input data and computer results.

REFERENCES

1. Denbigh, K., *The Principles of Chemical Equilibrium,* 3rd ed., Cambridge University Press, 1971.
2. Houghen, O. A. and Watson, K. A., *Chemical Process Principles,* a combined volume consisting of Parts one, two, and three, John Wiley & Sons Inc., 1947.
3. Walas, S. M., *Phase Equilibrium In Chemical Engineering,* Butterworth Publishers, 1985.
4. Himmelblau, D. M., *Basic Principles and Calculations in Chemical Engineering,* 5th ed., Prentice-Hall Int. Series, 1989.
5. Fogler, H. S., *Elements of Chemical Reaction Engineering,* 2nd ed., Prentice-Hall Int. Series, 1986.

Appendix: Heats and Free Energies of Formation

Compound	State†	Heat of formation‡§ ΔH (formation) at 25°C, kcal./mole	Free energy of formation‖¶ ΔF (formation) at 25°C, kcal./mole	Compound	State†	Heat of formation‡§ ΔH (formation) at 25°C, kcal./mole	Free energy of formation‖¶ ΔF (formation) at 25°C, kcal./mole
Aluminum:				**Barium (Cont.):**			
Al	c	0.00	0.00	Ba(NO₃)₂	c	-236.99	-189.94
AlBr₃	c	-123.4			aq, 600	-227.74	
	aq	-209.5	-189.2	BaO	c	-133.0	
Al₄C₃	c	-30.8	-29.0	Ba(OH)₂	c	-225.9	
AlCl₃	c	-163.8			aq, 400	-237.76	-209.02
	aq, 600	-243.9	-209.5	BaO.SiO₂	c	-363	
AlF₃	c	-329		Ba₃(PO₄)₂	c	-992	
	aq	-360.8	-312.6	BaPtCl₆	c	-284.9	
AlI₃	c	-72.8		BaS	c	-111.2	
	aq	-163.4	-152.5	BaSO₃	c	-282.5	
AlN	c	-57.7	-50.4	BaSO₄	c	-340.2	-313.4
Al(NH₄)(SO₄)₂	c	-561.19	-486.17	BaWO₄	c	-402	
Al(NH₄)(SO₄)₂.12H₂O	c	-1419.36	-1179.26	**Beryllium:**			
Al(NO₃)₃.6H₂O	c	-680.89	-526.32	Be	c	0.00	0.00
Al(NO₃)₃.9H₂O	c	-897.59		BeBr₂	c	-79.4	
Al₂O₃	c, corundum	-399.09	-376.87		aq	-142	-127.9
Al(OH)₃	c	-304.8	-272.9	BeCl₂	c	-112.6	
Al₂O₃.SiO₂	c, sillimanite	-648.7			aq	-163.9	-141.4
Al₂O₃.SiO₂	c, disthene	-642.4		BeI₂	c	-39.4	
Al₂O₃.SiO₂	c, andalusite	-642.0			aq	-112	-103.4
3Al₂O₃.2SiO₂	c, mullite	-1874		Be₃N₂	c	-134.5	-122.4
Al₂S₃	c	-121.6		BeO	c	-145.3	-138.3
Al₂(SO₄)₃	c	-820.99	-739.53	Be(OH)₂	c	-215.6	
	aq	-893.9	-759.3	BeS	c	-56.1	
Al₂(SO₄)₃.6H₂O	c	-1268.15	-1103.39	BeSO₄	c	-281	
Al₂(SO₄)₃.18H₂O	c	-2120			aq	....	-254.8
Antimony:				**Bismuth:**			
Sb	c	0.00	0.00	Bi	c	0.00	0.00
SbBr₃	c	-59.9		BiCl₃	c	-90.5	-76.4
SbCl₃	c	-91.3	-77.8		aq	-101.6	
SbCl₅	l	-104.8		BiI₃	c	-24	
SbF₃	c	-216.6			aq	-27	
SbI₃	c	-22.8		BiO	c	-49.5	-43.2
Sb₂O₃	c, I, orthorhombic	-165.4	-146.0	Bi₂O₃	c	-137.1	-117.9
	c, II, octahedral	-166.6		Bi(OH)₃	c	-171.1	
Sb₂O₄	c	-213.0	-186.6	Bi₂S₃	c	-43.9	-39.1
Sb₂O₅	c	-230.0	-196.1	Bi₂(SO₄)₃	c	-607.1	
Sb₂S₃	c, black	-38.2	-36.9	**Boron:**			
Arsenic:				B	c	0.00	0.00
As	c	0.00	0.00	BBr₃	l	-52.7	
AsBr₃	c	-45.9			g	-44.6	-50.9
AsCl₃	l	-80.2	-70.5	BCl₃	g	-94.5	-90.8
AsF₃	l	-223.76	-212.27	BF₃	g	-265.2	-261.0
AsH₃	g	43.6	37.7	B₂H₆	g	7.5	19.9
AsI₃	c	-13.6		BN	c	-32.1	-27.2
As₂O₃	c	-154.1	-134.8	B₂O₃	c	-302.0	-282.9
As₂O₅	c	-217.9	-183.9		gls	-297.6	-280.3
As₂S₃	c	-20	-20	B(OH)₃	c	-260.0	-229.4
	amorphous	-34.76		B₂S₃	c	-56.6	
Barium:				**Bromine:**			
Ba	c	0.00	0.00	Br₂	l	0.00	0.00
BaBr₂	c	-180.38			g	7.47	0.931
	aq, 400	-185.67	-183.0	BrCl	g	3.06	-0.63
BaCl₂	c	-205.25		**Cadmium:**			
	aq, 300	-207.92	-196.5	Cd	c	0.00	0.00
Ba(ClO₃)₂	c	-176.6		CdBr₂	c	-75.8	-70.7
	aq, 1600	-170.0	-134.4		aq, 400	-76.6	-67.6
Ba(ClO₄)₂	c	-210.2		CdCl₂	c	-92.149	-81.889
	aq, 800	....	-155.3		aq, 400	-96.44	-81.2
Ba(CN)₂	c	-48		Cd(CN)₂	c	36.2	
Ba(CNO)₂	c	-212.1		CdCO₃	c	-178.2	-163.2
	aq	....	-180.7	CdI₂	c	-48.40	
BaCN₂	c	-63.6			aq, 400	-47.46	-43.22
BaCO₃	c, witherite	-284.2	-271.4	Cd₃N₂	c	39.8	
BaCrO₄	c	-342.2		Cd(NO₃)₂	aq, 400	-115.67	-71.05
BaF₂	c	-287.9		CdO	c	-62.35	-55.28
	aq, 1600	-284.6	-265.3	Cd(OH)₂	c	-135.0	-113.7
BaH₂	c	-40.8	-31.5	CdS	c	-34.5	-33.6
Ba(HCO₃)₂	aq	-459	-414.4	CdSO₄	c	-222.23	
BaI₂	c	-144.6			aq, 400	-232.635	-194.65
	aq, 400	-155.17	-158.52	**Calcium:**			
Ba(IO₃)₂	c	-264.5		Ca	c	0.00	0.00
	aq	-237.50	-198.35	CaBr₂	c	-162.20	
BaMoO₄	c	-370			aq, 400	-187.19	-181.86
Ba₃N₂	c	-90.7		CaC₂	c	-14.8	-16.0
Ba(NO₂)₂	c	-184.5		CaCl₂	c	-190.6	-179.8
	aq	-179.05	-150.75		aq	-209.15	-195.36

For footnotes see end of table.

1 kal = 4.184 kJ

Source: Adapted from R. H. Perry and C. H. Chilton, Eds., Chemical Engineers' Handbook, © 1973 McGraw-Hill, Inc. Used with permission of McGraw-Hill Book Company

Heats and Free Energies of Formation *(continued)*

Compound	State†	Heat of formation‡§ ΔH (formation) at 25°C., kcal./mole	Free energy of formation‖ ¶ ΔF (formation) at 25°C., kcal./mole
Calcium (*Cont.*):			
$CaCN_2$	c	−85	
$Ca(CN)_2$	c	−43.3	
	aq		−54.0
$CaCO_3$	c, calcite	−289.5	−270.8
	c, aragonite	−289.54	−270.57
$CaCO_3.MgCO_3$	c	−558.8	
CaC_2O_4	c	−332.2	
$Ca(C_2H_3O_2)_2$	c	−356.3	
	aq	−364.1	−311.3
CaF_2	c	−290.2	
	aq	−286.5	−264.1
CaH_2	c	−46	−35.7
CaI_2	c	−128.49	
Ca_3N_2	aq, 400	−156.63	−157.37
$Ca(NO_3)_2$	c	−103.2	−88.2
	c	−224.05	−177.38
	aq, 400	−228.29	
$Ca(NO_3)_2.2H_2O$	c	−367.95	−293.57
$Ca(NO_3)_2.3H_2O$	c	−439.05	−351.58
$Ca(NO_3)_2.4H_2O$	c	−509.43	−409.32
CaO	c	−151.7	−144.3
$Ca(OH)_2$	c	−235.58	−213.9
	aq, 800	−239.2	−207.9
$CaO.SiO_2$	c, wollastonite	−377.9	−357.5
	c, I, pseudowollastonite	−376.6	−356.6
CaS	c	−114.3	−113.1
$CaSO_4$	c, insoluble form	−338.73	−311.9
	c, soluble form α	−336.58	−309.8
	c, soluble form β	−335.52	−308.8
$CaSO_4.\frac{1}{2}H_2O$	c	−376.13	
$CaSO_4.2H_2O$	c	−479.33	−425.47
$CaWO_4$	c	−387	
Carbon:			
C	c, graphite	0.00	0.00
	c, diamond	0.453	0.685
CO	g	−26.416	−32.808
CO_2	g	−94.052	−94.260
CH_4 methane	g	−17.889	−12.140
C_2H_6 ethane	g	−20.236	−7.860
C_3H_8 propane	g	−24.820	−5.614
C_4H_{10} n-butane	g	−29.812	−3.754
C_4H_{10} isobutane	g	−31.452	−4.296
C_5H_{12} n-pentane	g	−35.00	−1.96
	l	−41.36	−2.21
C_5H_{12} 2-methylbutane	g	−36.92	−3.50
	l	−42.85	−3.59
C_5H_{12} 2,2-dimethylpropane	g	−39.67	−3.64
C_6H_{14} n-hexane	g	−39.96	0.05
	l	−47.52	−0.91
C_6H_{14} 2-methylpentane	g	−41.66	−0.96
	l	−48.82	−1.73
C_6H_{14} 3-methylpentane	g	−41.02	−0.29
	l	−48.28	−1.12
C_6H_{14} 2,2-dimethylbutane	g	−44.35	−2.35
	l	−51.00	−2.88
C_6H_{14} 2,3-dimethylbutane	g	−42.49	−0.73
	l	−49.48	−1.44
C_7H_{16} n-heptane	g	−44.89	2.09
	l	−53.63	0.42
C_7H_{16} 2-methylhexane	g	−46.60	0.98
	l	−54.93	−0.47
C_7H_{16} 3-methylhexane	g	−45.96	1.10
	l	−54.35	−0.39
C_7H_{16} 3-ethylpentane	g	−45.34	2.59
	l	−53.77	1.06
C_7H_{16} 2,2-dimethylpentane	g	−49.29	0.09
	l	−57.05	−1.08
C_7H_{16} 2,3-dimethylpentane	g	−47.62	0.16
	l	−55.81	−1.27
C_7H_{16} 2,4-dimethylpentane	g	−48.30	0.72
	l	−56.17	−0.49
C_7H_{16} 3,3-dimethylpentane	g	−48.17	0.63
	l	−56.07	−0.69
C_7H_{16} 2,2,3-trimethylbutane	g	−48.96	0.76
	l	−56.63	−0.43
C_8H_{18} n-octane	g	−49.82	4.14
	l	−59.74	1.77
C_8H_{18} 2-methylheptane	g	−51.50	3.06
	l	−60.98	0.92
C_8H_{18} 3-methylheptane	g	−50.82	3.29
	l	−60.34	1.12
C_8H_{18} 4-methylheptane	g	−50.69	4.00
	l	−60.17	1.86
C_8H_{18} 3-ethylhexane	g	−50.40	3.95
	l	−59.88	1.80

Compound	State†	Heat of formation‡§ ΔH (formation) at 25°C., kcal./mole	Free energy of formation‖ ¶ ΔF (formation) at 25°C., kcal./mole
Carbon (*Cont.*):			
C_8H_{18} 2,2-dimethylhexane	g	−53.71	2.56
	l	−62.63	−0.72
C_8H_{18} 2,3-dimethylhexane	g	−51.13	4.23
	l	−60.40	2.17
C_8H_{18} 2,4-dimethylhexane	g	−52.44	2.80
	l	−61.47	0.89
C_8H_{18} 2,5-dimethylhexane	g	−53.21	2.50
	l	−62.26	0.59
C_8H_{18} 3,3-dimethylhexane	g	−52.61	3.17
	l	−61.58	1.23
C_8H_{18} 3,4-dimethylhexane	g	−50.91	4.97
	l	−60.23	2.86
C_8H_{18} 2-methyl-3-ethylpentane	g	−50.48	5.08
	l	−59.69	3.03
C_8H_{18} 3-methyl-3-ethylpentane	g	−51.38	4.76
	l	−60.46	2.69
C_8H_{18} 2,2,3-trimethylpentane	g	−52.61	4.09
	l	−61.44	2.22
C_8H_{18} 2,2,4-trimethylpentane	g	−53.57	3.13
	l	−61.97	1.51
C_8H_{18} 2,3,3,-trimethylpentane	g	−51.73	4.52
	l	−60.63	2.54
C_8H_{18} 2,3,4-trimethylpentane	g	−51.97	4.32
	l	−60.98	2.34
C_8H_{18} 2,2,3,3,-tetramethylbutane	g	−53.99	4.88
	c	−64.23	2.74
C_2H_4 ethylene	g	12.496	16.282
C_3H_6 propylene	g	4.879	14.964
C_4H_8 1-butene	g	0.280	17.217
C_4H_8 cis-2-butene	g	−1.362	16.007
C_4H_8 trans-2-butene	g	−2.405	15.323
C_4H_8 2-methyl-2-propene	g	−3.343	14.574
C_5H_{10} 1-pentene	g	−5.000	18.787
C_5H_{10} cis-2-pentene	g	−6.710	17.173
C_5H_{10} trans-2-pentene	g	−7.590	16.575
C_5H_{10} 2-methyl-1-butene	g	−8.680	15.509
C_5H_{10} 3-methyl-1-butene	g	−6.920	17.874
C_5H_{10} 2-methyl-2-butene	g	−10.170	14.267
C_2H_2 acetylene	g	54.194	50.000
C_3H_4 methylacetylene	g	44.319	46.313
C_4H_6 1-butyne	g	39.70	48.52
C_4H_6 2-butyne	g	35.374	44.725
C_5H_8 1-pentyne	g	34.50	50.17
C_5H_8 2-pentyne	g	30.80	46.41
C_5H_8 3-methyl-1-butyne	g	32.60	49.12
C_6H_6 benzene	g	19.820	30.989
	l	11.718	29.756
C_7H_8 toluene	g	11.950	29.228
	l	2.867	27.282
C_8H_{10} ethylbenzene	g	7.120	31.208
	l	−2.977	28.614
C_8H_{10} o-xylene	g	4.540	29.177
	l	−5.841	26.370
C_8H_{10} m-xylene	g	4.120	28.405
	l	−6.075	25.730
C_8H_{10} p-xylene	g	4.290	28.952
	l	−5.838	26.310
C_9H_{12} n-propylbenzene	g	1.870	32.810
	l	−9.178	29.600
C_9H_{12} isopropylbenzene	g	0.940	32.738
	l	−9.848	29.708
C_9H_{12} 1-methyl-2-ethylbenzene	g	0.290	31.323
	l	−11.110	27.973
C_9H_{12} 1-methyl-3-ethylbenzene	g	−0.460	30.217
	l	−11.670	26.977
C_9H_{12} 1-methyl-4-ethylbenzene	g	−0.780	30.281
	l	−11.920	27.041
C_9H_{12} 1,2,3-trimethylbenzene	g	−2.290	29.319
	l	−14.013	25.679
C_9H_{12} 1,2,4-trimethylbenzene	g	−3.330	27.912
	l	−14.785	24.462
C_9H_{12} 1,3,5-trimethylbenzene	g	−3.840	28.172
	l	−15.184	24.832
C_5H_{10} cyclopentane	g	−18.46	9.23
	l	−25.31	8.70
C_6H_{12} methylcyclopentane	g	−25.50	8.55
	l	−33.08	7.53
C_7H_{14} ethylcyclopentane	g	−30.38	10.59
	l	−39.09	8.84

Heats and Free Energies of Formation (continued)

Compound	State	Heat of formation ΔH (formation) at 25°C., kcal./mole	Free energy of formation ΔF (formation) at 25°C., kcal./mole
Carbon (Cont.):			
C_6H_{12} cyclohexane	g	−29.43	7.59
	l	−37.34	6.39
C_7H_{14} methylcyclohexane	g	−37.00	6.52
	l	−45.46	4.86
C_8H_{16} ethylcyclohexane	g	−41.06	9.38
	l	−50.73	6.96
CH_4O methanol	g	−48.08	−38.62
	l	−57.04	−39.80
C_2H_6O ethanol	g	−52.23	−40.23
	l	−66.35	−41.76
C_3H_8O n-propanol	g	−61.17	−38.83
	l	−71.87	−39.84
C_3H_8O isopropanol	g	−62.41	−38.20
	l	−74.32	−38.83
$C_4H_{10}O$ n-butanol	g	−67.81	−38.88
	l	−79.61	−40.37
$C_4H_{10}O$ isobutanol	g	−69.05	−38.25
	l	−81.06	−39.36
$C_2H_6O_2$ ethylene glycol	g	−92.53	−71.26
	l	−107.91	−76.44
$C_3H_8O_3$ glycerol	g		
	l	−159.16	−113.65
C_6H_6O phenol	g	−21.71	−6.26
	l	−37.80	−11.02
C_7H_8O cresol		....	−13.17
C_2H_4O ethylene oxide	g	−16.1	−6.94
C_2H_6O dimethyl ether	g	−43.06	−26.06
	l	−51.3	
$C_4H_{10}O$ diethyl ether	l	−65.2	−27.75
CH_2O formaldehyde	g	−28.29	−26.88
C_2H_4O acetaldehyde	g	−39.72	−31.46
C_3H_4O acrolein	g	−20.50	−15.57
	l	−27.97	−16.17
C_3H_6O propionaldehyde	g	−49.15	−33.96
C_4H_8O n-butyraldehyde	g	−52.40	−73.24
C_7H_6O benzaldehyde	g	−9.57	5.85
	l	−21.23	2.24
C_8H_8O p-toluic aldehyde	g	−17.78	4.09
	l	−29.79	0.97
C_2H_2O ketene	g	−14.78	−14.30
	l	−18.78	−13.32
C_3H_6O acetone	g	−51.79	−36.45
	l	−59.32	−37.16
$C_5H_{10}O$ diethylketone	l	−73.8	
CH_2O_2 formic acid	g	−86.67	−80.24
	l	−97.8	−82.7
$\tfrac{1}{2}(CH_2O_2)_2$ bimolecular formic acid	g	−93.85	−81.90
$C_2H_4O_2$ acetic acid	g	−104.72	−91.24
	l	−116.2	−93.56
$C_3H_6O_2$ propionic acid	g	−108.75	−88.27
	l	−121.7	−91.65
$C_2H_4O_3$ hydroxyacetic acid	l	−155.33	−125.57
$C_6H_{10}O_4$ adipic acid	c	−216.19	−163.96
	g	−235.51	−177.17
$C_2H_4O_2$ methyl formate	g	−84.69	−71.37
	l	−95.26	−71.53
$C_4H_6O_2$ methyl acrylate	g	−70.10	−56.78
	l	−82.76	−58.13
$C_4H_8O_2$ ethyl acetate	g	−102.02	−74.93
	l	−110.72	−76.11
$C_5H_{10}O_2$ ethyl propionate	g	−112.36	−77.37
	l	−122.16	−79.16
$C_4H_6O_3$ acetic anhydride	g	−148.82	−119.29
	l	−155.16	−121.75
$C_6H_{10}O_3$ propionic anhydride	g	−147.32	−109.78
	l	−161.53	−113.66
CS_2 carbon disulfide	g	28.11	16.13
COS carbonyl sulfide	g	−33.83	−40.85
C_2N_2 cyanogen	g	73.82	71.02
HCN hydrogen cyanide	g	31.1	27.94
	l	25.2	29.0
	aq, 100	25.2	26.8
C_2H_3N acetonitrile	g	19.81	
CH_5N methylamine	g	−6.7	6.6
C_2H_7N ethylamine	g	−12.24	10.01
C_3H_9N propylamine	g	−16.45	14.38
$C_4H_{11}N$ butylamine	g	−15.60	19.55
$C_6H_{13}N$ hexamethyleneimine	g	−14.37	31.52
	l	−24.90	24.80
CH_2N_2 cyanamide	l	11.18	24.30
	c	9.15	24.18
$C_6H_8N_2$ adiponitrile	c	33.34	61.43
	g	19.19	54.63
$C_6H_{16}N_2$ hexamethylenediamine	g	−30.57	28.91
	l	−27.48	7.34
CH_5N_3 guanidine	c	−30.68	6.33

Compound	State	Heat of formation ΔH (formation) at 25°C., kcal./mole	Free energy of formation ΔF (formation) at 25°C., kcal./mole
Carbon (Cont.):			
$C_3H_6N_6$ melamine	l	−19.33	40.80
CH_3NO formamide	g	−44.64	−36.60
	l	−62.52	27.50
C_2H_7NO ethanolamine	l	−77.55	−46.45
CH_4N_2O urea	c	−79.634	−47.118
Cerium:			
Ce	c	0.00	0.00
CeN	c	−78.2	−70.8
Cesium:			
Cs	c	0.00	0.00
CsBr	c	−97.64	
	aq, 500	−91.39	−94.86
CsCl	c	−106.31	
	aq, 400	−102.01	−101.61
Cs_2CO_3	c	−271.88	
CsF	c	−131.67	
	aq, 400	−140.48	−135.98
CsH	c	−12	−7.30
$CsHCO_3$	c	−230.6	
	aq, 2000	−226.6	−210.56
CsI	c	−83.91	
	aq, 400	−75.74	−82.61
$CsNH_2$	c	−28.2	
$CsNO_3$	c	−121.14	
	aq, 400	−111.54	−96.53
Cs_2O	c	−82.1	
CsOH	c	−100.2	
	aq, 200	−117.0	−107.87
Cs_2S	c	−87	
Cs_2SO_4	c	−344.86	
	aq	−340.12	−316.66
Chlorine:			
Cl_2	g	0.00	0.00
ClF	g	−25.7	
ClO	g	33	
ClO_2	g	24.7	29.5
ClO_3	g	37	
Cl_2O	g	18.20	22.40
Cl_2O_7	g	63	
Chromium:			
Cr	c	0.00	0.00
$CrBr_3$	aq		−122.7
Cr_3C_2	c	−21.008	−21.20
Cr_4C	c	−16.378	−16.74
$CrCl_2$	c	−103.1	−93.8
	aq		−102.1
CrF_2	c	−152	
CrF_3	c	−231	
CrI_2	c	−63.7	
	aq		−64.1
CrO_3	c	−139.3	
Cr_2O_3	c	−268.8	−249.3
$Cr_2(SO_4)_3$	aq		−626.3
Cobalt:			
Co	c	0.00	0.00
$CoBr_2$	c	−55.0	
	aq	−73.61	−61.96
Co_3C	c	9.49	7.08
$CoCl_2$	c	−76.9	−66.6
	aq, 400	−95.58	−75.46
$CoCO_3$	c	−172.39	−155.36
CoF_2	c	−172.98	−144.2
CoI_2	c	−24.2	
	aq	−43.15	−37.4
$Co(NO_3)_2$	c	−102.8	
	aq	−114.9	−65.3
CoO	c	−57.5	
Co_3O_4	c	−196.5	
$Co(OH)_2$	c	−131.5	−108.9
$Co(OH)_3$	c	−177.0	−142.0
CoS	c	−22.3	−19.8
Co_2S_3	c	−40.0	
$CoSO_4$	c	−216.6	
	aq, 400		−188.9
Columbium:			
Cb	c	0.00	0.00
Cb_2O_5	c	−462.96	
Copper:			
Cu	c	0.00	0.00
CuBr	c	−26.7	−23.8
$CuBr_2$	c	−34.0	
	aq	−42.4	−33.25
CuCl	c	−31.4	−24.13
$CuCl_2$	c	−48.83	
	aq, 400	−64.7	
$CuClO_4$	aq	−28.3	1.34
$Cu(ClO_3)_2$	aq, 400		15.4
$Cu(ClO_2)_2$	aq		−5.5

Heats and Free Energies of Formation *(continued))*

Compound	State†	Heat of formation‡§ ΔH (formation) at 25°C., kcal./mole	Free energy of formation‖¶ ΔF (formation) at 25°C., kcal./mole
Copper *(Cont.)*:			
CuI	c	−17.8	−16.66
CuI₂	c	−4.8	
	aq	−11.9	−8.76
Cu₃N	c	17.78	
Cu(NO₃)₂	c	−73.1	
	aq, 200	−83.6	−36.6
CuO	c	−38.5	−31.9
Cu₂O	c	−43.00	−38.13
Cu(OH)₂	c	−108.9	−85.5
CuS	c	−11.6	−11.69
Cu₂S	c	−18.97	−20.56
CuSO₄	c	−184.7	−158.3
	aq, 800	−200.78	−160.19
Cu₂SO₄	c	−179.6	
	aq		−152.0
Erbium:			
Er	c	0.00	0.00
Er(OH)₃	c	−326.8	
Fluorine:			
F₂	g	0.00	0.00
F₂O	g	5.5	9.7
Gallium:			
Ga	c	0.00	0.00
GaBr₃	c	−92.4	
GaCl₃	c	−125.4	
GaN	c	−26.2	
Ga₂O	c	−84.3	
Ga₂O₃	c	−259.9	
Germanium:			
Ge	c	0.00	0.00
Ge₃N₄	c	−15.7	
GeO₂	c	−128.6	
Gold:			
Au	c	0.00	0.00
AuBr	c	−3.4	
AuBr₃	c	−14.5	
	aq	−11.0	24.47
AuCl	c	−8.3	
AuCl₃	c	−28.3	
	aq	−32.96	4.21
AuI	c	0.2	−0.76
Au₂O₃	c	11.0	18.71
Au(OH)₃	c	−100.6	
Hafnium:			
Hf	c	0.00	0.00
HfO₂	c	−271.1	−258.2
Hydrogen:			
H₃AsO₃	aq	−175.6	−153.04
H₃AsO₄	c	−214.9	
	aq	−214.8	−183.93
HBr	g	−8.66	−12.72
	aq, 400	−28.80	−24.58
HBrO	aq	−25.4	−19.90
HBrO₃	aq	−11.51	5.00
HCl	g	−22.063	−22.778
	aq, 400	−39.85	−31.330
HCN	g	31.1	27.94
	aq, 100	24.2	26.55
HClO	aq, 400	−28.18	−19.11
HClO₃	aq	−23.4	−0.25
HClO₄	aq, 660	−31.4	−10.70
HC₂H₃O₂	l	−116.2	−93.56
	aq, 400	−116.74	−96.8
H₂C₂O₄	c	−196.7	
	aq, 300	−194.6	−165.64
HCOOH	l	−97.8	−82.7
	aq, 200	−98.0	−85.1
H₂CO₃	aq	−167.19	−149.0
HF	g	−64.2	−64.7
	aq, 200	−75.75	
HI	g	6.27	0.365
	aq, 400	−13.47	−12.35
HIO	aq	−38	−23.33
HIO₃	c	−56.77	
	aq	−54.8	−32.25
HN₃	c	70.3	78.50
HNO₃	l	−31.99	−17.57
	g	−41.35	−19.05
	aq, 400	−49.210	
HNO₃.H₂O	l	−112.91	−78.36
HNO₃.3H₂O	l	−252.15	−193.70
H₂O	g	−57.7979	−54.6351
	l	−68.3174	−56.6899
H₂O₂	l	−45.16	−28.23
	aq, 200	−45.80	−31.47
H₃PO₂	c	−145.5	
	aq	−145.6	−120.0
H₃PO₃	c	−232.2	
	aq	−232.2	−204.0

Compound	State†	Heat of formation‡§ ΔH (formation) at 25°C., kcal./mole	Free energy of formation‖¶ ΔF (formation) at 25°C., kcal./mole
Hydrogen *(Cont.)*:			
H₃PO₄	c	−306.2	
	aq, 400	−309.32	−270.0
H₂S	g	−4.77	−7.85
	aq, 2000	−9.38	
H₂S₂	l	−3.6	
H₂SO₃	aq, 200	−146.88	−128.54
H₂SO₄	l	−193.69	
	aq, 400	−212.03	
H₂Se	g	20.5	17.0
	aq	18.1	18.4
H₂SeO₃	c	−126.5	
	aq	−122.4	−101.36
H₂SeO₄	c	−130.23	
	aq, 400	−143.4	
H₂SiO₃	c	−267.8	−247.9
H₄SiO₄	c	−340.6	
H₂Te	g	36.9	33.1
H₂TeO₃	c	−145.0	−115.7
	aq	−145.0	
H₂TeO₄	aq	−165.6	
Indium:			
In	c	0.00	0.00
InBr₃	c	−97.2	
	aq	−112.9	−97.2
InCl₃	c	−128.5	
	aq	−145.6	−117.5
InI₃	c	−56.5	
	aq	−67.2	−60.5
InN	c	−4.8	
In₂O₃	c	−222.47	
Iodine:			
I₂	c	0.00	0.00
	g	14.88	4.63
IBr	g	10.05	1.24
ICl	g	4.20	−1.32
ICl₃	c	−21.8	−6.05
I₂O₅	c	−42.5	
Iridium:			
Ir	c	0.00	0.00
IrCl	c	−20.5	−16.9
IrCl₂	c	−40.6	−32.0
IrCl₃	c	−60.5	−46.5
IrF₆	l	−130	
IrO₂	c	−40.14	
Iron:			
Fe	c, α	0.00	0.00
FeBr₂	c	−57.15	
	aq, 540	−78.7	−69.47
FeBr₃	aq	−95.5	−76.26
Fe₃C	c	5.69	4.24
Fe(CO)₅	l	−187.6	
FeCO₃	c, siderite	−172.4	−154.8
FeCl₂	c	−81.9	−72.6
	aq	−100.0	−83.0
FeCl₃	c	−96.4	
	aq, 2000	−128.5	−96.5
FeF₂	aq, 1200	−177.2	−151.7
FeI₂	c	−24.2	
	aq	−47.7	−45
FeI₃	aq	−49.7	−39.5
Fe₄N	c	−2.55	0.862
Fe(NO₃)₂	aq	−118.9	−72.8
Fe(NO₃)₃	aq, 800	−156.5	−81.3
FeO	c	−64.62	−59.38
Fe₂O₃	c	−198.5	−179.1
Fe₃O₄	c	−266.9	−242.3
Fe(OH)₂	c	−135.9	−115.7
Fe(OH)₃	c	−197.3	−166.3
FeO.SiO₂	c	−273.5	
Fe₂P	c	−13	
FeSi	c	−19.0	
FeS	c	−22.64	−23.23
FeS₂	c, pyrites	−38.62	−35.93
	c, marcasite	−33.0	
FeSO₄	c	−221.3	−195.5
	aq, 400	−236.2	−196.4
Fe₂(SO₄)₃	aq, 400	−653.3	−533.4
FeTiO₃	c, ilmenite	−295.51	−277.06
Lanthanum:			
La	c	0.00	0.00
LaCl₃	c	−253.1	
	aq	−284.7	
La₃N₈	c	−160	
LaN	c	−72.0	−64.6
La₂O₃	c	−539	
LaS	c	−148.3	
La₂S₃	c	−351.4	
La₂(SO₄)₃	aq	−972	

Heats and Free Energies of Formation *(continued)*

Compound	State†	Heat of formation‡§ ΔH (formation) at 25°C., kcal./mole	Free energy of formation‖¶ ΔF (formation) at 25°C., kcal./mole
Lead:			
Pb	c	0.00	0.00
PbBr₂	c	−66.24	−62.06
	aq	−56.4	−54.97
PbCO₃	c, cerussite	−167.6	−150.0
Pb(C₂H₃O₂)₂	c	−232.6	
	aq, 400	−234.2	−184.40
PbC₂O₄	c	−205.3	
PbCl₂	c	−85.68	−75.04
	aq	−82.5	−68.47
PbF₂	c	−159.5	−148.1
PbI₂	c	−41.77	−41.47
Pb(NO₃)₂	c	−106.88	
	aq, 400	−99.46	−58.3
PbO	c, red	−51.72	−45.53
	c, yellow	−50.86	−43.88
PbO₂	c	−65.0	−52.0
Pb₃O₄	c	−172.4	−142.2
Pb(OH)₂	c	−123.0	−102.2
PbS	c	−22.38	−21.98
PbSO₄	c	−218.5	−192.9
Lithium:			
Li	c	0.00	0.00
LiBr	c	−83.75	
	aq, 400	−95.40	−95.28
LiBrO₃	aq	−77.9	−65.70
Li₂C₂	c	−13.0	
LiCN	aq	−31.4	−31.35
LiCNO	aq	−101.2	−94.12
LiC₂H₃O₂	aq	−183.9	−160.00
Li₂CO₃	c	−289.7	−269.8
	aq, 1900	−293.1	−267.58
LiCl	c	−97.63	
	aq, 278	−106.45	−102.03
LiClO₃	aq	−87.5	−70.95
LiClO₄	aq	−106.3	−81.4
LiF	c	−145.57	
	aq, 400	−144.85	−136.40
LiH	c	−22.9	
LiHCO₃	aq, 2000	−231.1	−210.98
LiI	c	−65.07	
	aq, 400	−80.09	−83.03
LiIO₃	aq	−121.3	−102.95
Li₃N	c	−47.45	−37.33
LiNO₃	c	−115.350	
	aq, 400	−115.88	−96.95
Li₂O	c	−142.3	
Li₂O₂	c	−151.9	−138.0
	aq	−159	
LiOH	c	−116.58	−106.44
	aq, 400	−121.47	−108.29
LiOH.H₂O	c	−188.92	
Li₂O.SiO₂	gls	−374	
Li₂Se	c	−84.9	
	aq	−95.5	−105.64
Li₂SO₄	c	−340.23	−314.66
	aq, 400	−347.02	
Li₂SO₄.H₂O	c	−411.57	−375.07
Magnesium:			
Mg	c	0.00	0.00
Mg(AsO₄)₂	c	−731.3	
	aq	−749	
MgBr₂	c	−123.9	
	aq, 400	−167.33	−156.94
Mg(CN)₂	aq	−39.7	−29.08
MgCN₂	c	−61	
Mg(C₂H₃O₂)₂	aq	−344.6	−286.38
MgCO₃	c	−261.7	−241.7
MgCl₂	c	−153.220	−143.77
	aq, 400	−189.76	
MgCl₂.H₂O	c	−230.970	−205.93
MgCl₂.2H₂O	c	−305.810	−267.20
MgCl₂.4H₂O	c	−453.820	−387.98
MgCl₂.6H₂O	c	−597.240	−505.45
MgF₂	c	−263.8	
MgI₂	c	−86.8	
	aq, 400	−136.79	−132.45
MgMoO₄	c	−329.9	
Mg₃N₂	c	−115.2	−100.8
Mg(NO₃)₂	c	−188.770	−140.66
	aq, 400	−209.927	−160.28
Mg(NO₃)₂.2H₂O	c	−336.625	
Mg(NO₃)₂.6H₂O	c	−624.48	−496.03
MgO	c	−143.84	−136.17
MgO.SiO₂	c	−347.5	−326.7
Mg(OH)₂	c, ppt.	−221.90	−200.17
	c, brucite	−223.9	−193.3
MgS	c	−84.2	
	aq	−108	

Compound	State†	Heat of formation‡§ ΔH (formation) at 25°C., kcal./mole	Free energy of formation‖¶ ΔF (formation) at 25°C., kcal./mole
Magnesium *(Cont.)*:			
MgSO₄	c	−304.94	−277.7
	aq, 400	−325.4	−283.88
MgTe	c	−25	
MgWO₄	c	−345.2	
Manganese:			
Mn	c, α	0.00	0.00
MnBr₂	c	−91	
	aq	−106	−97.8
Mn₃C	c	1.1	1.26
Mn(C₂H₃O₂)₂	c	−270.3	
	aq	−282.7	−227.2
MnCO₃	c	−211	−192.5
MnC₂O₄	c	−240.9	
MnCl₂	c	−112.0	−102.2
	aq, 400	−128.9	
MnF₂	aq, 1200	−206.1	−180.0
MnI₂	c	−49.8	
	aq	−76.2	−73.3
Mn₅N₂	c	−57.77	−46.49
Mn(NO₃)₂	c	−134.9	
	aq, 400	−148.0	−101.1
Mn(NO₃)₂.6H₂O	c	−557.07	−441.2
MnO	c	−92.04	−86.77
MnO₂	c	−124.58	−111.49
Mn₂O₃	c	−229.5	−209.9
Mn₃O₄	c	−331.65	−306.22
MnO.SiO₂	c	−301.3	−282.1
Mn(OH)₂	c	−163.4	−143.1
Mn(OH)₃	c	−221	−190
Mn₃(PO₄)₂	c	−736	
MnSe	c	−26.3	−27.5
MnS	c, green	−47.0	−48.0
MnSO₄	c	−254.18	−228.41
	aq, 400	−265.2	
Mn₂(SO₄)₃	c	−635	
	aq	−657	
Mercury:			
Hg	l	0.00	0.00
HgBr	g	23	18
HgBr₂	c	−40.68	−38.8
	aq	−38.4	−9.74
Hg(C₂H₃O₂)₂	c	−196.3	
	aq	−192.5	−139.2
HgCl₂	c	−53.4	−42.2
	aq	−50.3	−23.25
HgCl	g	19	14
Hg₂Cl₂	c	−63.13	
Hg(CN)₂	c	62.8	
	aq, 1110	66.25	
HgC₂O₄	c	−159.3	
HgH	g	57.1	52.25
HgI₂	c, red	−25.3	−24.0
HgI	g	33	23
Hg₂I₂	c	−28.88	−26.53
Hg(NO₃)₂	aq	−56.8	−13.09
Hg₂(NO₃)₂	aq	−58.5	−15.65
HgO	c, red	−21.6	−13.94
	c, yellow ppt.	−20.8	
Hg₂O	c	−21.6	−12.80
HgS	c, black	−10.7	−8.80
HgSO₄	c	−166.6	
Hg₂SO₄	c	−177.34	−149.12
Molybdenum:			
Mo	c	0.00	0.00
Mo₂C	c	4.36	2.91
Mo₂N	c	−8.3	
MoO₂	c	−130	−118.0
MoO₃	c	−180.39	−162.01
MoS₂	c	−56.27	−54.19
MoS₃	c	−61.48	−57.38
Nickel:			
Ni	c	0.00	0.00
NiBr₂	c	−53.4	
	aq	−72.6	−60.7
Ni₃C	c	9.2	8.88
Ni(C₂H₃O₂)₂	aq	−249.6	−190.1
Ni(CN)₂	aq	230.9	66.3
NiCl₂	c	−75.0	
	aq, 400	−94.34	−74.19
NiF₂	c	−157.5	
	aq	−171.6	−142.9
NiI₂	c	−22.4	
	aq	−42.0	−36.2
Ni(NO₃)₂	c	−101.5	
	aq, 200	−113.5	−64.0
NiO	c	−58.4	−51.7
Ni(OH)₂	c	−129.8	−105.6
Ni(OH)₃	c	−163.2	

Heats and Free Energies of Formation (continued)

Compound	State	Heat of formation‡§ ΔH (formation) at 25°C., kcal./mole	Free energy of formation‖¶ ΔF (formation) at 25°C., kcal./mole
Nickel (Cont.):			
NiS	c	-20.4	
NiSO₄	c	-216	
	aq, 200	-231.3	-187.6
Nitrogen:			
N₂	g	0.00	0.00
NF₃	g	-27	
NH₃	g	-10.96	-3.903
	aq, 200	-19.27	
NH₄Br	c	-64.57	
	aq	-60.27	-43.54
NH₄C₂H₃O₂	c	-148.1	
	aq, 400	-148.58	-108.26
NH₄CN	c	-0.7	
	aq	3.6	20.4
NH₄CNS	c	-17.8	
	aq	-12.3	4.4
(NH₄)₂CO₃	aq	-223.4	-164.1
(NH₄)₂C₂O₄	c	-266.3	
	aq	-260.6	-196.2
NH₄Cl	c	-75.23	-48.59
	aq, 400	-71.20	
NH₄ClO₄	c	-69.4	
	aq	-63.2	-21.1
(NH₄)₂CrO₄	c	-276.9	
	aq	-271.3	-209.3
NH₄F	c	-111.6	
	aq	-110.2	-84.7
NH₄I	c	-48.43	
	aq	-44.97	-31.3
NH₄NO₃	c	-87.40	
	aq, 500	-80.89	
NH₄OH	aq	-87.59	
(NH₄)₂S	aq, 400	-55.21	-14.50
(NH₄)₂SO₄	c	-281.74	-215.06
	aq, 400	-279.33	-214.02
N₂H₄	l	12.06	
N₂H₄.H₂O	l	-57.96	
N₂H₄.H₂SO₄	c	-232.2	
N₂O	g	19.55	24.82
NO	g	21.600	20.719
NO₂	g	7.96	12.26
N₂O₄	g	2.23	23.41
N₂O₅	c	-10.0	
NOBr	l	11.6	19.26
NOCl	g	12.8	16.1
Osmium:			
Os	c	0.00	0.00
OsO₄	c	-93.6	-70.9
	g	-80.1	-68.1
Oxygen:			
O₂	g	0.00	0.00
O₃	g	33.88	38.86
Palladium:			
Pd	c	0.00	0.00
PdO	c	-20.40	
Phosphorus:			
P	c, white ("yellow")	0.00	0.00
	c, red ("violet")	-4.22	-1.80
P	g	150.35	141.88
P₂	g	33.82	24.60
P₄	g	13.12	5.89
PBr₃	l	-45	
PBr₅	c	-60.6	
PCl₃	g	-70.0	-65.2
	l	-76.8	-63.3
PCl₅	g	-91.0	-73.2
PH₃	g	2.21	-1.45
PI₃	c	-10.9	
P₂O₅	c	-360.0	
POCl₃	g	-138.4	-127.2
Platinum:			
Pt	c	0.00	0.00
PtBr₄	c	-40.6	
	aq	-50.7	
PtCl₂	c	-34	
PtCl₄	c	62.6	
	aq	-82.3	
PtI₄	c	-18	
Pt(OH)₂	c	-87.5	-67.9
PtS	c	-20.18	-18.55
PtS₂	c	-26.64	-24.28
Potassium:			
K	c	0.00	0.00
K₃AsO₃	aq	-323.0	
K₃AsO₄	aq	-390.3	-355.7
KH₂AsO₄	c	-271.2	-236.7
KBr	c	-94.06	-90.8
	aq, 400	-89.19	-92.0
Potassium (Cont.):			
KBrO₃	c	-81.58	-60.30
	aq, 1667	-71.68	
KC₂H₃O₂	c	-173.80	
	aq, 400	-177.38	-156.73
KCl	c	-104.348	-97.76
	aq, 400	-100.164	-98.76
KClO₃	c	-93.5	-69.30
	aq, 400	-81.34	
KClO₄	c	-103.8	-72.86
	aq, 400	-101.14	
KCN	c	-28.1	
	aq, 400	-25.3	-28.08
KCNO	c	-99.6	
	aq	-94.5	-90.85
KCNS	c	-47.0	
	aq, 400	-41.07	-44.08
K₂CO₃	c	-274.01	
	aq, 400	-280.90	-264.04
K₂C₂O₄	c	-319.9	
	aq, 400	-315.5	-293.1
K₂CrO₄	c	-333.4	
	aq, 400	-328.2	-306.3
K₂Cr₂O₇	c	-488.5	
	aq, 400	-472.1	-440.9
KF	c	-134.50	
	aq, 180	-138.36	-133.13
K₃Fe(CN)₆	c	-48.4	
	aq	-34.5	
K₄Fe(CN)₆	c	-131.8	
	aq	-119.9	
KH	c	-10	-5.3
KHCO₃	c	-229.8	
	aq, 2000	-224.85	-207.71
KI	c	-78.88	-77.37
	aq, 500	-73.95	-79.76
KIO₃	c	-121.69	-101.87
	aq, 400	-115.18	-99.68
KIO₄	aq	-98.1	
KMnO₄	c	-192.9	-169.1
	aq, 400	-182.5	-168.0
K₂MoO₄	aq, 880	-364.2	-342.9
KNH₂	c	-28.25	
KNO₂	aq	-86.0	-75.9
KNO₃	c	-118.08	-94.29
	aq, 400	-109.79	-93.68
K₂O	c	-86.2	
K₂O.Al₂O₃.4H₂O	c, leucite	-1379.6	
	gls	-1368.2	
K₂O.Al₂O₃.6H₂O	c, adularia	-1810.7	
	c, microcline	-1784.5	
	gls	-1747	
KOH	c	-102.02	
	aq, 400	-114.96	-105.0
K₃PO₄	aq	-397.5	
K₃PO₄	aq	-478.7	-443.3
KH₂PO₄	c	-362.7	-326.1
K₂PtCl₄	c	-254.7	
	aq	-242.6	-226.5
K₂PtCl₆	c	-299.5	-263.6
	aq, 9400	-286.1	
K₂Se	c	-74.4	
	aq	-83.4	-99.10
K₂SeO₄	aq	-267.1	-240.0
K₂S	c	-121.5	
	aq, 400	-110.75	-111.44
K₂SO₃	c	-267.7	
	aq	-269.7	-251.3
K₂SO₄	c	-342.65	-314.62
	aq, 400	-336.48	-310.96
K₂SO₄.Al₂(SO₄)₃	c	-1178.38	-1068.48
K₂SO₄.Al₂(SO₄)₃.24H₂O	c	-2895.44	-2455.68
K₂S₂O₆	c	-418.62	
Rhenium:			
Re	c	0.00	0.00
ReF₆	g	-274	
Rhodium:			
Rh	c	0.00	0.00
RhO	c	-21.7	
Rh₂O	c	-22.7	
Rh₂O₃	c	-68.3	
Rubidium:			
Rb	c	0.00	0.00
RbBr	c	-95.82	
		-45.0	-52.50
	aq, 500	-90.54	-93.38
RbCN	aq	-25.9	
Rb₂CO₃	c	-273.22	
	aq, 220	-282.61	-263.78

Heats and Free Energies of Formation (continued)

Compound	State†	Heat of formation‡§ ΔH (formation) at 25°C., kcal./mole	Free energy of formation‖¶ ΔF (formation) at 25°C., kcal./mole
Rubidium (Cont.):			
RbCl..................	c	−105.06	−98.48
	g	−53.6	−57.9
	aq, ∞	−101.06	−100.13
RbF..................	c	−133.23	
	aq, 400	−139.31	−134.5
RbHCO₃..................	c	−230.01	
	aq, 2000	−225.59	−209.07
RbI..................	c	−81.04	
	g	−31.2	−40.5
	aq, 400	−74.57	−81.13
RbNH₂..................	c	−27.74	
RbNO₃..................	c	−119.22	
	aq, 400	−110.52	−95.05
Rb₂O..................	c	−82.9	
Rb₂O₂..................	c	−107	
RbOH..................	c	−101.3	
	aq, 200	−115.8	−106.39
Ruthenium:			
Ru..................	c	0.00	0.00
RuS₂..................	c	−46.99	−44.11
Selenium:			
Se..................	c, I, hexagonal	0.00	0.00
	c, II, red, monoclinic	0.2	
Se₂Cl₂..................	l	−22.06	−13.73
SeF₆..................	g	−246	−222
SeO₂..................	c	−56.33	
Silicon:			
Si..................	c	0.00	0.00
SiBr₄..................	l	−93.0	
SiC..................	c	−28	−27.4
SiCl₄..................	l	−150.0	−133.9
	g	−142.5	−133.0
SiF₄..................	g	−370	−360
SiH₄..................	g	−14.8	−9.4
SiI₄..................	c	−29.8	
Si₃N₄..................	c	−179.25	−154.74
SiO₂..................	c, cristobalite, 1600° form	−202.62	
	c, cristobalite, 1100° form	−202.46	
	c, quartz	−203.35	−190.4
	c, tridymite	−203.23	
Silver:			
Ag..................	c	0.00	0.00
AgBr..................	c	−23.90	−23.02
Ag₂C₂..................	c	84.5	
AgC₂H₃O₂..................	c	−95.9	
	aq	−91.7	−70.86
AgCN..................	c	33.8	38.70
Ag₂CO₃..................	c	−119.5	−103.0
Ag₂C₂O₄..................	c	−158.7	
AgCl..................	c	−30.11	−25.98
AgF..................	c	−48.7	
	aq, 400	−53.1	−47.26
AgI..................	c	−15.14	−16.17
AgIO₃..................	c	−42.02	−24.08
AgNO₂..................	c	−11.6	3.76
	aq	−2.9	9.99
AgNO₃..................	c	−29.4	−7.66
	aq, 6500	−24.02	−7.81
Ag₂O..................	c	−6.95	−2.23
Ag₂S..................	c	−5.5	−7.6
Ag₂SO₄..................	c	−170.1	−146.8
	aq	−165.8	−139.22
Sodium:			
Na..................	c	0.00	0.00
Na₃AsO₃..................	aq, 500	−314.61	
Na₃AsO₄..................	c	−366	
	aq, 500	−381.97	−341.17
NaBr..................	c	−86.72	
	aq, 400	−86.33	−87.17
NaBrO..................	aq	−78.9	
NaBrO₃..................	aq, 400	−68.89	−57.59
NaC₂H₃O₂..................	c	−170.45	
	aq, 400	−175.450	−152.31
NaCN..................	c	−22.47	
	aq, 200	−22.29	−23.24
NaCNO..................	c	−96.3	
	aq	−91.7	−86.00
NaCNS..................	c	−39.94	
	aq, 400	−38.23	−39.24
Na₂CO₃..................	c	−269.46	−249.55
	aq, 1000	−275.13	−251.36
NaCO₂NH₂..................	c	−142.17	
Na₂C₂O₄..................	c	−313.8	
	aq, 600	−309.92	−283.42
NaCl..................	c	−98.321	−91.894
	aq, 400	−97.324	−93.92
Sodium (Cont.):			
NaClO₃..................	c	−83.59	
	aq, 400	−78.42	−62.84
NaClO₄..................	c	−101.12	
	aq, 476	−97.66	−73.29
Na₂CrO₄..................	c	−319.8	
	aq, 800	−323.0	−296.58
Na₂Cr₂O₇..................	aq, 1200	−465.9	−431.18
NaF..................	c	−135.94	−129.0
	aq, 400	−135.711	−128.29
NaH..................	c	−14	−9.30
NaHCO₃..................	c	−226.0	−202.66
	aq	−222.1	−202.87
NaI..................	c	−69.28	
	aq, ∞	−71.10	−74.92
NaIO₃..................	aq, 400	−112.300	−94.84
Na₂MoO₄..................	c	−364	
	aq	−358.7	−333.18
NaNO₂..................	c	−86.6	
	aq	−83.1	−71.04
NaNO₃..................	c	−111.71	−87.62
	aq, 400	−106.880	−88.84
Na₂O..................	c	−99.45	−90.06
Na₂O₂..................	c	−119.2	−105.0
Na₂O.SiO₂..................	c	−383.91	−361.49
Na₂O.Al₂O₃.3SiO₂..................	c, natrolite	−1180	
Na₂O.Al₂O₃.4SiO₂..................	c	−1366	
NaOH..................	c	−101.96	−90.60
	aq, 400	−112.193	−100.18
Na₃PO₃..................	aq, 1000	−389.1	
Na₃PO₄..................	c	−457	
	aq, 400	−471.9	−428.74
Na₂PtCl₄..................	aq	−237.2	−216.78
Na₂PtCl₆..................	aq	−272.1	
	c	−280.9	
Na₂Se..................	c	−59.1	
	aq, 440	−78.1	−89.42
Na₂SeO₄..................	c	−254	
	aq, 800	−261.5	−230.30
Na₂S..................	c	−89.8	
	aq, 400	−105.17	−101.76
Na₂SO₃..................	c	−261.2	−240.14
	aq, 800	−264.1	−241.58
Na₂SO₄..................	c	−330.50	−302.38
	aq, 1100	−330.82	−301.28
Na₂SO₄.10H₂O..................	c	−1033.85	−870.52
Na₂WO₄..................	c	−391	
	aq	−381.5	−345.18
Strontium:			
Sr..................	c	0.00	0.00
SrBr₂..................	c	−171.0	
	aq, 400	−187.24	−182.36
Sr(C₂H₃O₂)₂..................	c	−358.0	
	aq	−364.4	−311.80
Sr(CN)₂..................	aq	−59.5	−54.50
SrCO₃..................	c	−290.9	−271.9
SrCl₂..................	c	−197.84	
	aq, 400	−209.20	−195.86
SrF₂..................	c	−289.0	
Sr(HCO₃)₂..................	aq	−459.1	−413.76
SrI₂..................	c	−136.1	
	aq, 400	−156.70	−157.87
Sr₃N₂..................	c	−91.4	−76.5
Sr(NO₃)₂..................	c	−233.2	
	aq, 400	−228.73	−185.70
SrO..................	c	−140.8	−133.7
SrO.SiO₂..................	gls	−364	
SrO₂..................	c	−153.3	−139.0
Sr₂O..................	c	−153.6	
Sr(OH)₂..................	c	−228.7	
	aq, 800	−239.4	−208.27
Sr₃(PO₄)₂..................	c	−980	
	aq	−985	−881.54
SrS..................	c	−113.1	
	aq	−120.4	−109.78
SrSO₄..................	c	−345.3	
	aq, 400	−345.0	−309.30
SrWO₄..................	c	−393	
Sulfur:			
S..................	c, rhombic	0.00	0.00
	c, monoclinic	−0.071	−0.023
	l, λ	0.257	0.072
	l, λμ equilibrium	....	0.071
	g	53.25	43.57
S₂..................	g	31.02	19.36
S₆..................	g	27.78	13.97
S₈..................	g	27.090	12.770
S₂Br₂..................	l	−4	
SCl₄..................	l	−13.7	

Heats and Free Energies of Formation (continued)

Compound	State†	Heat of formation‡§ ΔH (formation) at 25°C., kcal./mole	Free energy of formation‖¶ ΔF (formation) at 25°C., kcal./mole
Sulfur (Cont.):			
S_2Cl_2	l	−14.2	−5.90
S_2Cl_4	l	−24.1	
SF_6	g	−262	−237
SO	g	19.02	12.75
SO_2	g	−70.94	−71.68
SO_3	g	−94.39	−88.59
	l	−103.03	−88.28
	c, α	−105.09	−88.22
	c, β	−105.92	−88.34
	c, γ	−109.34	−88.98
SO_2Cl_2	g	−82.04	−74.06
	l	−89.80	−75.06
Tantalum:			
Ta	c	0.00	0.00
TaN	c	−51.2	−45.11
Ta_2O_5	c	−486.0	−453.7
Tellurium:			
Te	c	0.00	0.00
$TeBr_4$	c	−49.3	
$TeCl_4$	c	−77.4	−57.4
TeF_6	g	−315	−292
TeO_2	c	−77.56	−64.66
Thallium:			
Tl	c	0.00	0.00
TlBr	c	−41.5	−39.43
	aq	−28.0	−32.34
TlCl	c	−49.37	−44.46
	aq	−38.4	−39.09
$TlCl_3$	c	−82.4	
	aq	−91.0	−44.25
TlF	aq	−77.6	−73.46
TlI	c	−31.1	−31.3
	aq	−12.7	−20.09
$TlNO_3$	c	−58.2	−36.32
	aq	−48.4	−34.01
Tl_2O	c	−43.18	
Tl_2O_3	c	−120	
TlOH	c	−57.44	−45.54
	aq	−53.9	−45.35
Tl_2S	c	−22	
Tl_2SO_4	c	−222.8	−197.79
	aq, 800	−214.1	−191.62
Thorium:			
Th	c	0.00	0.00
$ThBr_4$	c	−281.5	
	aq	−352.0	−295.31
ThC_2	c	−45.1	
$ThCl_4$	c	−335	
	aq	−392	−322.32
ThI_4	aq	−292	−246.33
Th_3N_4	c	−309.0	−282.3
ThO_2	c	−291.6	−280.1
$Th(OH)_4$	c, "soluble"	−336.1	
$Th(SO_4)_2$	c	−632	
	aq	−668.1	−549.2
Tin:			
Sn	c, II, tetragonal	0.00	0.00
	c, III, "gray," cubic	0.6	1.1
$SnBr_2$	c	−61.4	
	aq	−60.0	−55.43
$SnBr_4$	c	−94.8	
	aq	−110.6	−97.66
$SnCl_2$	c	−83.6	
	aq	−81.7	−68.94
$SnCl_4$	l	−127.3	−110.4
	aq	−157.6	−124.67
SnI_2	c	−38.9	
	aq	−33.3	−30.95

Compound	State†	Heat of formation‡§ ΔH (formation) at 25°C., kcal./mole	Free energy of formation‖¶ ΔF (formation) at 25°C., kcal./mole
Tin (Cont.):			
SnO	c	−67.7	−60.75
SnO_2	c	−138.1	−123.6
$Sn(OH)_2$	c	−136.2	−115.95
$Sn(OH)_4$	c	−268.9	−226.00
SnS	c	−18.61	
Titanium:			
Ti	c	0.00	0.00
TiC	c	−110	−109.2
$TiCl_4$	l	−181.4	−165.5
TiN	c	−80.0	−73.17
TiO_2	c, III, rutil	−225.0	−211.9
	amorphous	−214.1	−201.4
Tungsten:			
W	c	0.00	0.00
WO_2	c	−130.5	−118.3
WO_3	c	−195.7	−177.3
WS_2	c	−84	
Uranium:			
U	c	0.00	0.00
UC_2	c	−29	
UCl_3	c	−213	
UCl_4	c	−251	
U_3N_4	c	−274	−249.6
UO_2	c	−256.6	−242.2
$UO_2(NO_3)_2.6H_2O$	c	−756.8	−617.8
UO_3	c	−291.6	
U_3O_8	c	−845.1	
Vanadium:			
V	c	0.00	0.00
VCl_2	c	−147	
VCl_3	l	−187	
VCl_4	l	−165	
VN	c	−41.43	−35.08
V_2O_2	c	−195	
V_2O_3	c	−296	−277
V_2O_4	c	−342	−316
V_2O_5	c	−373	−342
Zinc:			
Zn	c	0.00	0.00
ZnSb	c	−3.6	−3.88
$ZnBr_2$	c	−77.0	−72.9
	aq, 400	−93.6	
$Zn(C_2H_3O_2)_2$	c	−259.4	
	aq, 400	−269.4	−214.4
$Zn(CN)_2$	c	17.06	
$ZnCO_3$	c	−192.9	−173.5
$ZnCl_2$	c	−99.9	−88.8
	aq, 400	−115.44	
ZnF_2	aq	−192.9	−166.6
ZnI_2	c	−50.50	−49.93
	aq	−61.6	
$Zn(NO_3)_2$	aq, 400	−134.9	−87.7
ZnO	c, hexagonal	−83.36	−76.19
$Zn.SiO_2$	c	−282.6	
$Zn(OH)_2$	c, rhombic	−153.66	
ZnS	c, wurtzite	−45.3	−44.2
$ZnSO_4$	c	−233.4	
	aq, 400	−252.12	−211.28
Zirconium:			
Zr	c	0.00	0.00
ZrC	c	−29.8	−34.6
$ZrCl_4$	c	−268.9	
ZrN	c	−82.5	−75.9
ZrO_2	c, monoclinic	−258.5	−244.6
$Zr(OH)_4$	c	−411.0	
$ZrO(OH)_2$	c	−337	−307.6

† The physical state is indicated as follows: c, crystal (solid); l, liquid; g, gas; pls, glass or solid supercooled liquid; aq, in aqueous solution. A number following the symbol aq applies only to the values of the heats of formation (not to those of free energies of formation); and indicates the number of moles of water per mole of solute; when no number is given, the solution is understood to be dilute. For the free energy of formation of a substance in aqueous solution, the concentration is always that of the hypothetical solution of unit molality.

‡ The increment in heat content, ΔH, in the reaction of forming the given substance from its elements in their standard states. When ΔH is negative, heat is evolved in the process, and, when positive, heat is absorbed.

§ The heat of solution in water of a given solid, liquid, or gaseous compound is given by the difference in the value for the heat of formation of the given compound in the solid, liquid, or gaseous state and its heat of formation in aqueous solution. The following two examples serve as an illustration of the procedure: (1) For NaCl(c) and NaCl(aq, 400H₂O), the values of ΔH(formation) are, respectively, −98.321 and −97.324 kg.-cal. per mole. Subtraction of the first value from the second gives ΔH = 0.998 kg.-cal. per mole for the reaction of dissolving crystalline sodium chloride in 400 moles of water. When this process occurs at a constant pressure of 1 atm., 0.998 kg.-cal. of energy are absorbed. (2) For HCl(g) and HCl(aq, 400H₂O), the values for ΔH(formation) are, respectively, −22.06 and −39.85 kg.-cal. per mole. Subtraction of the first from the second gives ΔH = −17.79 kg.-cal per mole for the reaction of dissolving gaseous hydrogen chloride in 400 moles of water. At a constant pressure of 1 atm. 17.79 kg.-cal. of energy are evolved in this process.

‖ The increment in the free energy, ΔF, in the reaction of forming the given substance in its standard state from its elements in their standard states. The standard states are: for a gas, fugacity (approximately equal to the pressure) of 1 atm.; for a pure liquid or solid, the substance at a pressure of 1 atm.; for a substance in aqueous solution, the hypothetical solution of unit molality, which has all the properties of the infinitely dilute solution except the property of concentration.

¶ The free energy of solution of a given substance from its normal standard state as a solid, liquid, or gas to the hypothetical one molal state in aqueous solution may be calculated in a manner similar to that described in footnote § for calculating the heat of solution.

Appendix: Heat Combustion

| Compound | Formula | State | Heat of combustion, $-\Delta Hc°$, at 25°C. and constant pressure, to form | | | | | |
| | | | H₂O (liq.) and CO₂ (gas) | | | H₂O (gas) and CO₂ (gas) | | |
			Kcal./mole	Cal./g.	B.t.u./lb.	Kcal./mole	Cal./g.	B.t.u./lb.
Hydrogen	H₂	gas	68.3174	33,887.6	60,957.7	57.7979	28,669.6	51,571.4
Carbon	C	solid, graph.	94.0518	7,831.1	14,086.8			
Carbon monoxide	CO	gas	67.6361	2,414.7	4,343.6			
Paraffins								
Methane	CH₄	gas	212.798	13,265.1	23,861	191.759	11,953.6	21,502
Ethane	C₂H₆	gas	372.820	12,399.2	22,304	341.261	11,349.6	20,416
Propane	C₃H₈	gas	530.605	12,033.5	21,646	488.527	11,079.2	19,929
Propane	C₃H₈	liq.*	526.782	11,946.8	21,490	484.704	10,992.5	19,774
n-Butane	C₄H₁₀	gas	687.982	11,837.3	21,293	635.384	10,932.3	19,665
n-Butane	C₄H₁₀	liq.*	682.844	11,748.9	21,134	630.246	10,843.9	19,506
2-Methylpropane (Isobutane)	C₄H₁₀	gas	686.342	11,809.1	21,242	633.744	10,904.1	19,614
2-Methylpropane (Isobutane)	C₄H₁₀	liq.*	681.625	11,727.9	21,096	629.027	10,822.9	19,468
n-Pentane	C₅H₁₂	gas	845.16	11,714.6	21,072	782.04	10,839.7	19,499
n-Pentane	C₅H₁₂	liq.	838.80	11,626.4	20,914	775.68	10,751.5	19,340
2-Methylbutane (Isopentane)	C₅H₁₂	gas	843.24	11,688.0	21,025	780.12	10,813.1	19,451
2-Methylbutane (Isopentane)	C₅H₁₂	liq.	837.31	11,605.8	20,877	774.19	10,730.9	19,303
2,2-Dimethylpropane (Neopentane)	C₅H₁₂	gas	840.49	11,649.8	20,956	777.37	10,775.0	19,382
2,2-Dimethylpropane (Neopentane)	C₅H₁₂	liq.	835.18	11,576.2	20,824	772.06	10,701.4	19,250
n Hexane	C₆H₁₄	gas	1,002.57	11,634.5	20,928	928.93	10,780.0	19,391
n-Hexane	C₆H₁₄	liq.	995.01	11,546.8	20,771	921.37	10,692.2	19,233
2-Methylpentane	C₆H₁₄	gas	1,000.87	11,614.8	20,893	927.23	10,760.2	19,356
2-Methylpentane	C₆H₁₄	liq.	993.71	11,531.7	20,743	920.07	10,677.1	19,206
3-Methylpentane	C₆H₁₄	gas	1,001.51	11,622.2	20,906	927.87	10,767.6	19,369
3-Methylpentane	C₆H₁₄	liq.	994.25	11,538.0	20,755	920.61	10,683.4	19,218
2,2-Dimethylbutane	C₆H₁₄	gas	998.17	11,583.5	20,837	924.53	10,728.9	19,299
2,2-Dimethylbutane	C₆H₁₄	liq.	991.52	11,506.3	20,698	917.88	10,651.7	19,161
2,3-Dimethylbutane	C₆H₁₄	gas	1,000.04	11,605.2	20,876	926.40	10,750.6	19,338
2,3-Dimethylbutane	C₆H₁₄	liq.	993.05	11,524.0	20,730	919.41	10,669.5	19,192
n-Heptane	C₇H₁₆	gas	1,160.01	11,577.2	20,825	1,075.85	10,737.2	19,314
n-Heptane	C₇H₁₆	liq.	1,151.27	11,489.9	20,668	1,067.11	10,650.0	19,157
2-Methylhexane	C₇H₁₆	gas	1,158.30	11,560.1	20,795	1,074.14	10,720.2	19,284
2-Methylhexane	C₇H₁₆	liq.	1,149.97	11,477.0	20,645	1,065.81	10,637.0	19,134
3-Methylhexane	C₇H₁₆	gas	1,158.94	11,566.5	20,806	1,074.78	10,726.6	19,295
3-Methylhexane	C₇H₁₆	liq.	1,150.55	11,482.8	20,655	1,066.39	10,642.8	19,145
3-Ethylpentane	C₇H₁₆	gas	1,159.56	11,572.7	20,817	1,075.40	10,732.7	19,306
3-Ethylpentane	C₇H₁₆	liq.	1,151.13	11,488.6	20,666	1,066.97	10,648.6	19,155
2,2-Dimethylpentane	C₇H₁₆	gas	1,155.61	11,533.3	20,746	1,071.45	10,693.3	19,235
2,2-Dimethylpentane	C₇H₁₆	liq.	1,147.85	11,455.8	20,607	1,063.69	10,615.9	19,096
2,3-Dimethylpentane	C₇H₁₆	gas	1,157.28	11,549.9	20,776	1,073.12	10,710.0	19,265
2,3-Dimethylpentane	C₇H₁₆	liq.	1,149.09	11,468.2	20,629	1,064.93	10,628.3	19,118
2,4-Dimethylpentane	C₇H₁₆	gas	1,156.60	11,543.1	20,764	1,072.44	10,703.2	19,253
2,4-Dimethylpentane	C₇H₁₆	liq.	1,148.73	11,464.6	20,623	1,064.57	10,624.7	19,112
3,3-Dimethylpentane	C₇H₁₆	gas	1,156.73	11,544.4	20,766	1,072.57	10,704.5	19,255
3,3-Dimethylpentane	C₇H₁₆	liq.	1,148.83	11,465.6	20,625	1,064.67	10,625.7	19,114
2,2,3-Trimethylbutane	C₇H₁₆	gas	1,155.94	11,536.6	20,752	1,071.78	10,696.6	19,241
2,2,3-Trimethylbutane	C₇H₁₆	liq.	1,148.27	11,460.0	20,614	1,064.11	10,620.1	19,104
n-Octane	C₈H₁₈	gas	1,317.45	11,533.9	20,747	1,222.77	10,705.0	19,256
n-Octane	C₈H₁₈	liq.	1,307.53	11,447.1	20,591	1,212.85	10,618.2	19,100
2-Methylheptane	C₈H₁₈	gas	1,315.76	11,519.1	20,721	1,221.08	10,690.2	19,230
2-Methylheptane	C₈H₁₈	liq.	1,306.28	11,436.1	20,572	1,211.60	10,607.2	19,080
3-Methylheptane	C₈H₁₈	gas	1,316.44	11,525.1	20,732	1,221.76	10,696.2	19,240
3-Methylheptane	C₈H₁₈	liq.	1,306.92	11,441.7	20,582	1,212.24	10,612.8	19,091
4-Methylheptane	C₈H₁₈	gas	1,316.57	11,526.2	20,734	1,221.89	10,697.3	19,243
4-Methylheptane	C₈H₁₈	liq.	1,307.09	11,443.2	20,584	1,212.41	10,614.3	19,093
3-Ethylhexane	C₈H₁₈	gas	1,316.87	11,528.8	20,738	1,222.19	10,699.9	19,247
3-Ethylhexane	C₈H₁₈	liq.	1,307.39	11,445.8	20,589	1,212.71	10,616.9	19,098
2,2-Dimethylhexane	C₈H₁₈	gas	1,313.56	11,499.9	20,686	1,218.88	10,671.0	19,195
2,2-Dimethylhexane	C₈H₁₈	liq.	1,304.64	11,421.8	20,546	1,209.96	10,592.9	19,055
2,3-Dimethylhexane	C₈H₁₈	gas	1,316.13	11,522.4	20,727	1,221.45	10,693.5	19,236
2,3-Dimethylhexane	C₈H₁₈	liq.	1,306.86	11,441.2	20,581	1,212.18	10,612.3	19,090
2,4-Dimethylhexane	C₈H₁₈	gas	1,314.83	11,511.0	20,706	1,220.15	10,682.1	19,215
2,4-Dimethylhexane	C₈H₁₈	liq.	1,305.80	11,431.9	20,564	1,211.12	10,603.0	19,073
2,5-Dimethylhexane	C₈H₁₈	gas	1,314.05	11,504.2	20,694	1,219.37	10,675.3	19,203
2,5-Dimethylhexane	C₈H₁₈	liq.	1,305.00	11,424.9	20,551	1,210.32	10,596.0	19,060
3,3-Dimethylhexane	C₈H₁₈	gas	1,314.65	11,509.4	20,703	1,219.97	10,680.5	19,212
3,3-Dimethylhexane	C₈H₁₈	liq.	1,305.68	11,430.9	20,562	1,211.00	10,602.0	19,071
3,4-Dimethylhexane	C₈H₁₈	gas	1,316.36	11,524.4	20,730	1,221.68	10,695.5	19,239
3,4-Dimethylhexane	C₈H₁₈	liq.	1,307.04	11,442.8	20,583	1,212.36	10,613.9	19,092
2-Methyl-3-ethylpentane	C₈H₁₈	gas	1,316.79	11,528.1	20,737	1,222.11	10,699.2	19,246
2-Methyl-3-ethylpentane	C₈H₁₈	liq.	1,307.08	11,447.5	20,592	1,212.90	10,618.6	19,101
3-Methyl-3-ethylpentane	C₈H₁₈	gas	1,315.88	11,520.2	20,723	1,221.20	10,691.3	19,232
3-Methyl-3-ethylpentane	C₈H₁₈	liq.	1,306.80	11,440.7	20,580	1,212.12	10,611.8	19,089
2,2,3-Trimethylpentane	C₈H₁₈	gas	1,314.66	11,509.5	20,703	1,219.98	10,680.6	19,213
2,2,3-Trimethylpentane	C₈H₁₈	liq.	1,305.83	11,432.2	20,564	1,211.15	10,603.3	19,073
2,2,4-Trimethylpentane	C₈H₁₈	gas	1,313.69	11,501.0	20,688	1,219.01	10,672.1	19,197
2,2,4-Trimethylpentane	C₈H₁₈	liq.	1,305.29	11,427.5	20,556	1,210.61	10,598.6	19,065
2,3,3-Trimethylpentane	C₈H₁₈	gas	1,315.54	11,517.2	20,717	1,220.86	10,688.3	19,226
2,3,3-Trimethylpentane	C₈H₁₈	liq.	1,306.64	11,439.3	20,577	1,211.96	10,610.4	19,086
2,3,4-Trimethylpentane	C₈H₁₈	gas	1,315.29	11,515.0	20,713	1,220.61	10,686.1	19,222
2,3,4-Trimethylpentane	C₈H₁₈	liq.	1,306.28	11,436.1	20,572	1,211.60	10,607.2	19,080
2,2,3,3-Tetramethylbutane	C₈H₁₈	gas	1,313.27	11,497.3	20,682	1,218.59	10,668.4	19,191
2,2,3,3-Tetramethylbutane	C₈H₁₈	solid	1,303.03	11,407.7	20,520	1,208.35	10,578.8	19,029
n-Nonane	C₉H₂₀	gas	1,474.90	11,500.2	20,687	1,369.70	10,680.0	19,211
n-Nonane	C₉H₂₀	liq.	1,463.80	11,413.6	20,531	1,358.60	10,593.4	19,056
n-Decane	C₁₀H₂₂	gas	1,632.34	11,473.0	20,638	1,516.63	10,659.7	19,175
n-Decane	C₁₀H₂₂	liq.	1,620.06	11,386.7	20,483	1,504.35	10,573.4	19,020
n-Undecane	C₁₁H₂₄	gas	1,789.78	11,450.8	20,598	1,663.55	10,643.2	19,145
n-Undecane	C₁₁H₂₄	liq.	1,776.32	11,364.7	20,443	1,650.09	10,557.0	18,990
n-Dodecane	C₁₂H₂₆	gas	1,947.23	11,432.2	20,564	1,810.48	10,629.4	19,120
n-Dodecane	C₁₂H₂₆	liq.	1,932.59	11,346.3	20,410	1,795.84	10,543.4	18,966
n-Tridecane	C₁₃H₂₈	gas	2,104.67	11,416.5	20,536	1,957.40	10,617.6	19,099

| Compound | Formula | State | Heat of combustion, $-\Delta Hc°$, at 25°C. and constant pressure, to form | | | | | |
| | | | H$_2$O (liq.) and CO$_2$ (gas) | | | H$_2$O (gas) and CO$_2$ (gas) | | |
			Kcal./mole	Cal./g.	B.t.u./lb.	Kcal./mole	Cal./g.	B.t.u./lb.
Tridecane	C$_{13}$H$_{28}$	liq.	2,088.85	11,330.6	20,382	1,941.58	10,531.8	18,945
Tetradecane	C$_{14}$H$_{30}$	gas	2,262.11	11,402.9	20,512	2,104.32	10,607.5	19,081
Tetradecane	C$_{14}$H$_{30}$	liq.	2,245.11	11,317.2	20,358	2,087.32	10,521.8	18,927
Pentadecane	C$_{15}$H$_{32}$	gas	2,419.55	11,391.2	20,491	2,251.24	10,598.7	19,065
Pentadecane	C$_{15}$H$_{32}$	liq.	2,401.37	11,305.6	20,337	2,233.06	10,513.2	18,911
Hexadecane	C$_{16}$H$_{34}$	gas	2,577.00	11,380.9	20,472	2,398.17	10,591.1	19,052
Hexadecane	C$_{16}$H$_{34}$	liq.	2,557.64	11,295.4	20,318	2,378.81	10,505.6	18,898
Heptadecane	C$_{17}$H$_{36}$	gas	2,734.44	11,371.8	20,456	2,545.09	10,584.3	19,039
Heptadecane	C$_{17}$H$_{36}$	liq.	2,713.90	11,286.4	20,302	2,524.55	10,498.9	18,886
Octadecane	C$_{18}$H$_{38}$	gas	2,891.88	11,363.7	20,441	2,692.01	10,578.3	19,028
Octadecane	C$_{18}$H$_{38}$	liq.	2,870.16	11,278.4	20,288	2,670.29	10,493.0	18,875
Nonadecane	C$_{19}$H$_{40}$	gas	3,049.33	11,356.5	20,428	2,838.94	10,572.9	19,019
Nonadecane	C$_{19}$H$_{40}$	liq.	3,026.43	11,271.2	20,275	2,816.04	10,487.7	18,865
Eicosane	C$_{20}$H$_{42}$	gas	3,206.77	11,350.0	20,416	2,985.86	10,568.1	19,010
Eicosane	C$_{20}$H$_{42}$	liq.	3,182.69	11,264.7	20,263	2,961.78	10,482.8	18,857
Alkyl benzenes								
Benzene	C$_6$H$_6$	gas	789.08	10,102.4	18,172	757.52	9,698.4	17,446
Benzene	C$_6$H$_6$	liq.	780.98	9,998.7	17,986	749.42	9,594.7	17,259
Methylbenzene (toluene)	C$_7$H$_8$	gas	943.58	10,241.4	18,422	901.50	9,784.7	17,601
Methylbenzene (toluene)	C$_7$H$_8$	liq.	934.50	10,142.8	18,245	892.42	9,686.1	17,424
Ethylbenzene	C$_8$H$_{10}$	gas	1,101.13	10,372.4	18,658	1,048.53	9,876.9	17,767
Ethylbenzene	C$_8$H$_{10}$	liq.	1,091.03	10,277.2	18,487	1,038.43	9,781.7	17,596
1,2-Dimethylbenzene (o-xylene)	C$_8$H$_{10}$	gas	1,098.54	10,348.0	18,614	1,045.94	9,852.5	17,723
1,2-Dimethylbenzene (o-xylene)	C$_8$H$_{10}$	liq.	1,088.16	10,250.2	18,438	1,035.56	9,754.7	17,547
1,3-Dimethylbenzene (m-xylene)	C$_8$H$_{10}$	gas	1,098.12	10,344.0	18,607	1,045.52	9,848.5	17,716
1,3-Dimethylbenzene (m-xylene)	C$_8$H$_{10}$	liq.	1,087.92	10,247.9	18,434	1,035.32	9,752.4	17,543
1,4-Dimethylbenzene (p-xylene)	C$_8$H$_{10}$	gas	1,098.29	10,345.6	18,610	1,045.69	9,850.1	17,719
1,4-Dimethylbenzene (p-xylene)	C$_8$H$_{10}$	liq.	1,088.16	10,250.2	18,438	1,035.56	9,754.7	17,547
Propylbenzene	C$_9$H$_{12}$	gas	1,258.24	10,469.1	18,832	1,195.12	9,943.9	17,887
Propylbenzene	C$_9$H$_{12}$	liq.	1,247.19	10,377.2	18,667	1,184.07	9,852.0	17,722
Isopropylbenzene (cumene)	C$_9$H$_{12}$	gas	1,257.31	10,461.4	18,818	1,194.19	9,936.2	17,873
Isopropylbenzene (cumene)	C$_9$H$_{12}$	liq.	1,246.52	10,371.6	18,657	1,183.40	9,846.4	17,712
1-Methyl-2-ethylbenzene	C$_9$H$_{12}$	gas	1,256.66	10,456.0	18,808	1,193.54	9,930.8	17,864
1-Methyl-2-ethylbenzene	C$_9$H$_{12}$	liq.	1,245.26	10,361.1	18,638	1,182.14	9,835.9	17,693
1-Methyl-3-ethylbenzene	C$_9$H$_{12}$	gas	1,255.92	10,449.8	18,797	1,192.80	9,924.6	17,853
1-Methyl-3-ethylbenzene	C$_9$H$_{12}$	liq.	1,244.71	10,356.5	18,630	1,181.59	9,831.3	17,685
1-Methyl-4-ethylbenzene	C$_9$H$_{12}$	gas	1,255.59	10,447.1	18,792	1,192.47	9,921.9	17,848
1-Methyl-4-ethylbenzene	C$_9$H$_{12}$	liq.	1,244.45	10,354.4	18,626	1,181.33	9,829.2	17,681
1,2,3-Trimethylbenzene (hemimellitene)	C$_9$H$_{12}$	gas	1,254.08	10,434.5	18,770	1,190.96	9,909.3	17,825
1,2,3-Trimethylbenzene (hemimellitene)	C$_9$H$_{12}$	liq.	1,242.36	10,337.0	18,594	1,179.24	9,811.8	17,650
1,2,4-Trimethylbenzene (pseudocumene)	C$_9$H$_{12}$	gas	1,253.04	10,425.8	18,754	1,189.92	9,900.7	17,809
1,2,4-Trimethylbenzene (pseudocumene)	C$_9$H$_{12}$	liq.	1,241.58	10,330.5	18,583	1,178.46	9,805.3	17,638
1,3,5-Trimethylbenzene (mesitylene)	C$_9$H$_{12}$	gas	1,252.53	10,421.6	18,747	1,189.41	9,896.4	17,802
1,3,5-Trimethylbenzene (mesitylene)	C$_9$H$_{12}$	liq.	1,241.19	10,327.2	18,577	1,178.07	9,802.1	17,632
Butylbenzene	C$_{10}$H$_{14}$	gas	1,415.44	10,546.3	18,971	1,341.80	9,997.6	17,984
Butylbenzene	C$_{10}$H$_{14}$	liq.	1,403.46	10,457.0	18,810	1,329.82	9,908.4	17,823
Alkyl cyclopentanes								
Cyclopentane	C$_5$H$_{10}$	gas	793.39	11,313.1	20,350	740.79	10,563.1	19,001
Cyclopentane	C$_5$H$_{10}$	liq.	786.54	11,215.5	20,175	733.94	10,465.4	18,825
Methylcyclopentane	C$_6$H$_{12}$	gas	948.72	11,273.4	20,279	885.60	10,523.3	18,930
Methylcyclopentane	C$_6$H$_{12}$	liq.	941.14	11,183.5	20,117	878.02	10,433.2	18,768
Ethylcyclopentane	C$_7$H$_{14}$	gas	1,106.21	11,266.9	20,267	1,032.57	10,516.9	18,918
Ethylcyclopentane	C$_7$H$_{14}$	liq.	1,097.50	11,178.2	20,108	1,023.86	10,428.2	18,758
Propylcyclopentane	C$_8$H$_{16}$	gas	1,263.56	11,260.9	20,256	1,179.40	10,510.8	18,907
Propylcyclopentane	C$_8$H$_{16}$	liq.	1,253.74	11,173.4	20,099	1,169.58	10,423.3	18,750
Butylcyclopentane	C$_9$H$_{18}$	gas	1,421.10	11,257.7	20,250	1,326.42	10,507.6	18,901
Butylcyclopentane	C$_9$H$_{18}$	liq.	1,410.10	11,170.5	20,094	1,315.42	10,420.5	18,745
Alkyl cyclohexanes								
Cyclohexane	C$_6$H$_{12}$	gas	944.79	11,226.7	20,195	881.67	10,476.7	18,846
Cyclohexane	C$_6$H$_{12}$	liq.	936.88	11,132.7	20,026	873.76	10,382.7	18,676
Methylcyclohexane	C$_7$H$_{14}$	gas	1,099.59	11,199.5	20,146	1,025.95	10,449.5	18,797
Methylcyclohexane	C$_7$H$_{14}$	liq.	1,091.13	11,113.3	19,991	1,017.49	10,363.3	18,642
Ethylcyclohexane	C$_8$H$_{16}$	gas	1,257.90	11,210.4	20,166	1,173.74	10,460.4	18,816
Ethylcyclohexane	C$_8$H$_{16}$	liq.	1,248.23	11,124.3	20,011	1,164.07	10,374.3	18,661
Propylcyclohexane	C$_9$H$_{18}$	gas	1,415.12	11,210.3	20,165	1,320.44	10,460.3	18,816
Propylcyclohexane	C$_9$H$_{18}$	liq.	1,404.34	11,124.9	20,012	1,309.66	10,374.9	18,663
Butylcyclohexane	C$_{10}$H$_{20}$	gas	1,572.74	11,213.0	20,170	1,467.54	10,463.0	18,821
Butylcyclohexane	C$_{10}$H$_{20}$	liq.	1,560.78	11,127.8	20,017	1,455.58	10,377.8	18,668
Monoolefins								
Ethene (ethylene)	C$_2$H$_4$	gas	337.234	12,021.7	21,625	316.195	11,271.7	20,276
Propene (propylene)	C$_3$H$_6$	gas	491.987	11,692.3	21,032	460.428	10,942.3	19,683
1-Butene	C$_4$H$_8$	gas	649.757	11,581.3	20,833	607.679	10,831.3	19,484
cis-2-Butene	C$_4$H$_8$	gas	648.115	11,552.0	20,780	606.037	10,802.0	19,431
trans-2-Butene	C$_4$H$_8$	gas	647.072	11,533.4	20,747	604.994	10,783.4	19,397
Methylpropene (isobutene)	C$_4$H$_8$	gas	646.134	11,516.7	20,716	604.056	10,766.7	19,367
1-Pentene	C$_5$H$_{10}$	gas	806.85	11,505.1	20,696	754.25	10,755.1	19,346
2-Pentene	C$_5$H$_{10}$	gas	805.34	11,483.5	20,657	752.74	10,733.5	19,308
cis-2-Pentene	C$_5$H$_{10}$	gas	804.26	11,468.1	20,629	751.66	10,718.1	19,280
2-Methyl-1-butene	C$_5$H$_{10}$	gas	803.17	11,452.6	20,601	750.57	10,702.6	19,252
3-Methyl-1-butene	C$_5$H$_{10}$	gas	804.93	11,477.7	20,646	752.33	10,727.7	19,297
2-Methyl-2-butene	C$_5$H$_{10}$	gas	801.68	11,431.3	20,563	749.08	10,681.3	19,214
Acetylenes								
Ethyne (acetylene)	C$_2$H$_2$	gas	310.615	11,930.2	21,460	300.096	11,526.2	20,734
Propyne (methylacetylene)	C$_3$H$_4$	gas	463.109	11,559.8	20,794	442.070	11,034.6	19,849
1-Butyne (ethylacetylene)	C$_4$H$_6$	gas	620.86	11,478.7	20,648	589.302	10,895.2	19,599
2-Butyne (dimethylacetylene)	C$_4$H$_6$	gas	616.533	11,398.7	20,504	584.974	10,815.2	19,455
1-Pentyne	C$_5$H$_8$	gas	778.03	11,422.5	20,547	735.95	10,804.7	19,436
2-Pentyne	C$_5$H$_8$	gas	774.33	11,368.2	20,449	732.25	10,750.4	19,338
3-Methyl-1-butyne	C$_5$H$_8$	gas	776.13	11,394.6	20,497	734.05	10,776.8	19,386

saturation pressure.

Appendix: The Molar Heat Capacities of Gases in the Ideal Gas (Zero Pressure) State

		a	$b \times 10^2$	$c \times 10^5$	$d \times 10^9$	Temperature Range, K
Acetylenes and Diolefins						
Acetylene	C_2H_2	5.21	2.2008	-1.559	4.349	273–1500
Methylacetylene	C_3H_4	4.21	4.073	-2.192	4.713	273–1500
Dimethylacetylene	C_4H_6	3.54	5.838	-2.760	4.974	273–1500
Propadiene	C_3H_4	2.43	4.693	-2.781	6.484	273–1500
1,3-Butadiene	C_4H_6	-1.29	8.350	-5.582	14.24	273–1500
Isoprene	C_5H_8	-0.44	10.418	-6.762	16.93	273–1500
Combustion Gases (Low Range)						
Nitrogen	N_2	6.903	-0.03753	0.1930	-0.6861	273–1800
Oxygen	O_2	6.085	0.3631	-0.1709	0.3133	273–1800
Air		6.713	0.04697	0.1147	-0.4696	273–1800
Hydrogen	H_2	6.952	-0.04576	0.09563	-0.2079	273–1800
Carbon monoxide	CO	6.726	0.04001	0.1283	-0.5307	273–1800
Carbon dioxide	CO_2	5.316	1.4285	-0.8362	1.784	273–1800
Water vapor	H_2O	7.700	0.04594	0.2521	-0.8587	273–1800
Combustion Gases (High Range)†						
Nitrogen	N_2	6.529	0.1488	-0.02271	—	273–3800
Oxygen	O_2	6.732	0.1505	-0.01791	—	273–3800
Air		6.557	0.1477	-0.02148	—	273–3800
Hydrogen	H_2	6.424	0.1039	-0.007804	—	273–3800
Carbon monoxide	CO	6.480	0.1566	-0.02387	—	273–3800
Water vapor	H_2O	6.970	0.3464	-0.04833	—	273–3800

Sulfur Compounds						
Sulfur	S_2	6.499	0.5298	−0.3888	0.9520	273–1800
Sulfur dioxide	SO_2	6.157	1.384	−0.9103	2.057	273–1800
Sulfur drioxide	SO_3	3.918	3.483	−2.675	7.744	273–1300
Hydrogen sulfide	H_2S	7.070	0.3128	0.1364	−0.7867	273–1800
Carbon disulfide	CS_2	7.390	1.489	−1.096	2.760	273–1800
Carbonyl sulfide	COS	6.222	1.536	−1.058	2.560	273–1800
Paraffinic Hydrocarbons						
Methane	CH_4	4.750	1.200	0.3030	−2.630	273–1500
Ethane	C_2H_6	1.648	4.124	−1.530	1.740	273–1500
Propane	C_3H_8	−0.966	7.279	−3.755	7.580	273–1500
n-Butane	C_4H_{10}	0.945	8.873	−4.380	8.360	273–1500
i-Butane	C_4H_{10}	−1.890	9.936	−5.495	11.92	273–1500
n-Pentane	C_5H_{12}	1.618	10.85	−5.365	10.10	273–1500
n-Hexane	C_6H_{14}	1.657	13.19	−6.844	13.78	273–1500
Monoolefinic Hydrocarbons						
Ethylene	C_2H_4	0.944	3.735	−1.993	4.220	273–1500
Propylene	C_3H_6	0.753	5.691	−2.910	5.880	273–1500
1-Butene	C_4H_8	−0.240	8.650	−5.110	12.07	273–1500
i-Butane	C_4H_8	1.650	7.702	−3.981	8.020	273–1500
cis-2-Butene	C_4H_8	−1.778	8.078	−4.074	7.890	273–1500
trans-2-Butene	C_4H_8	2.340	7.220	−3.403	6.070	273–1500
Cycloparaffinic Hydrocarbons						
Cyclopentane	C_5H_{10}	−12.957	13.087	−7.447	16.41	273–1500
Methylcyclopentane	C_6H_{12}	−12.114	15.380	−8.915	20.03	273–1500
Cyclopentane	C_6H_{12}	−15.935	16.454	−9.203	19.27	273–1500
Methylcyclopentane	C_7H_{14}	−15.070	18.972	−10.989	24.09	273–1500

Source: O. Hougen, K. Watson, and R. A. Ragatz, Chemical Process Principles, Part 1, John Wiley & Sons, New York, 1954.

The Molar Heat Capacities of Gases in the Ideal Gas (Zero Pressure) State
(continued)

	a	$b \times 10^2$	$c \times 10^5$	$d \times 10^9$	Temperature Range, K
Aromatic Hydrocarbons					
Benzene C_6H_6	-8.650	11.578	-7.540	18.54	273–1500
Toluene C_7H_8	-8.213	13.357	-8.230	19.20	273–1500
Ethylbenzene C_8H_{10}	-8.398	15.935	-10.003	23.95	273–1500
Styrene C_8H_8	-5.968	14.354	-9.150	22.03	273–1500
Cumene C_9H_{12}	-9.452	18.686	-11.869	28.80	273–1500
Halogens and Halogen Acids					
Fluorine F_2	6.115	0.5864	-0.4186	0.9797	273–2000
Chlorine Cl_2	6.8214	0.57095	-0.5107	1.547	273–1500
Bromine Br_2	8.051	0.2462	-0.2128	0.6406	273–1500
Iodine I_2	8.504	0.13135	-0.10684	0.3125	273–1800
Hydrogen fluoride HF	7.201	-0.1178	0.1576	-0.3760	273–2000
Hydrogen chloride CHl	7.244	-0.1820	0.3170	-1.036	273–1500
Hydrogen bromide HBr	7.169	-0.1604	0.3314	-1.161	273–1500
Hydrogen iodide HI	6.702	0.04546	0.1216	-0.4813	273–1900
Chloromethanes					
Methyl chloride CH_3Cl	3.05	2.596	-1.244	2.300	273–1500
Methylene chloride CH_2CL_2	4.20	3.419	-2.3500	6.068	273–1500
Chloroform $CHCl_3$	7.61	3.461	-2.668	7.344	273–1500
Carbon tetrachloride CCl_4	12.24	3.400	-2.995	8.828	273–1500
Phosgene $COCl_2$	10.35	1.653	-0.8408	—	273–1000
Thiophosgene $CSCl_2$	10.80	1.859	-1.045	—	273–1000

Cyanogens

Cyanogens	$(CN)_2$	9.82	1.4858	-0.6571	—	273–1000
Hydrogen cyanide	HCN	6.34	0.8375	-0.2611	—	273–1500
Cyanogen chloride	CNCl	7.97	1.0745	-0.5265	—	273–1000
Cyanogen bromide	CNBr	8.82	0.9084	-0.4367	—	273–1000
Cyanogen iodide	CNI	9.69	0.7213	-0.3265	—	273–1000
Acetonitrile	CH_3CN	5.09	2.7634	-0.9111	—	273–1200
Acrylic nitrile	CH_2CHCN	4.55	4.1039	-1.6939	—	273–1000

Oxides of Nitrogen

Nitric oxide	NO	6.461	0.2358	-0.07705	0.08729	273–3800
Nitric oxide	NO	7.008	-0.02244	0.2328	-1.000	273–1500
Nitrous oxide	N_2O	5.758	1.4004	-0.8508	2.526	273–1500
Nitorgen dioxide	NO_2	5.48	1.365	-0.841	1.88	273–1500
Nitrogen tetroxide	N_2O_4	7.9	4.46	-2.71	—	273–600

Oxygenated Hydrocarbons

Formaldehyde	CH_2O	5.447	0.9739	0.1703	-2.078	273–1500
Acetaldehyde	C_2H_4O	4.19	3.164	-0.515	-3.800	273–1000
Methanol	CH_4O	4.55	2.186	-0.291	-1.92	273–1000
Ethanol	C_2H_6O	4.75	5.006	-2.479	4.790	273–1500
Ethylene oxide	C_2H_4O	-1.12	4.925	-2.389	3.149	273–1000
Ketene	C_2H_2O	4.11	2.966	-1.793	4.22	273–1500

Miscellaneous Hydrocarbons

Cyclopropane	C_3H_6	-6.481	8.206	-5.577	15.61	273–1000
Isopentane	C_5H_{12}	-2.273	12.434	-7.097	15.86	273–1500
Neopentane	C_5H_{12}	-3.865	13.305	-8.018	18.83	273–1500
o-Xylene	C_8H_{10}	-3.789	14.291	-8.354	18.80	273–1500
m-Xylene	C_8H_{10}	-6.533	14.905	-8.831	20.05	273–1500
p-Xylene	C_8H_{10}	-5.334	14.220	-7.984	17.03	273–1500

The Molar Heat Capacities of Gases in the Ideal Gas (Zero Pressure) State
(continued)

		a	$b \times 10^2$	$c \times 10^5$	$d \times 10^9$	Temperature Range, K
C3 Oxygenated Hydrocarbons						
Carbon suboxide	C_3O_2	8.203	3.073	-2.081	5.182	273–1500
Acetone	C_3H_6O	1.625	6.661	-3.737	8.307	273–1500
1-Propyl alcohol	C_3H_8O	0.7936	8.502	-5.016	11.56	273–1500
n-Propyl alcohol	C_3H_8O	-1.307	9.235	-5.800	14.14	273–1500
Allyl alcohol	C_3H_6O	0.5203	7.122	-4.259	9.948	273–1500
Chloroethenes						
Chloroethene	C_2H_3Cl	2.401	4.270	-2.751	6.797	273–1500
1,1-Dichloroethene	$C_2H_2Cl_2$	5.899	4.383	-3.182	8.516	273–1500
cis-1,2-Dichloroethene	$C_2H_2Cl_2$	4.336	4.691	-3.397	9.010	273–1500
trans-1,2-Dichloroethene	$C_2H_2Cl_2$	5.661	4.295	-3.022	7.891	273–1500
Trichloroethene	C_2HCl_3	9.200	4.517	-3.600	10.10	273–1500
Tetrachloroethene	C_2Cl_4	15.11	3.799	-3.179	9.089	273–1500
Nitrogen Compounds						
Ammonia	NH_3	6.5846	0.61251	0.23663	-1.5981	273–1500
Hydrazine	N_2H_4	3.890	3.554	-2.304	5.990	273–1500
Methylamine	CH_5N	2.9956	3.6101	-1.6446	2.9505	273–1500
Dimethylamine	C_2H_7N	-0.275	6.6152	-3.4826	7.1510	273–1500
Trimethylamine	C_3H_9N	-2.098	9.6187	-5.5488	12.432	273–1500

CHAPTER THREE

Reaction Rate Expression

INTRODUCTION

Chemical reactions are processes in which reactants are transformed into products. In some processes, the change occurs directly and the complete description of the mechanism of the reaction present can be relatively obtained. In complex processes, the reactants undergo a series of step-like changes and each step constitutes a reaction in its own right. The overall mechanism is made up of contributions from all the reactions that are sometimes too complex to determine from the knowledge of the reactants and products alone. Chemical kinetics and reaction mechanism as reviewed in Chapter 1 can often provide a reasonable approach in determining the reaction rate equations. Chemical kinetics is concerned with analyzing the dynamics of chemical reactions. The raw data of chemical kinetics are the measurements of the reactions rates. The end product explains these rates in terms of a complete reaction mechanism. Because a measured rate shows a statistical average state of the molecules taking part in the reaction, chemical kinetics does not provide information on the energetic state of the individual molecules.

Here, we shall examine a series of processes from the viewpoint of their kinetics and develop model reactions for the appropriate rate equations. The equations are used to arrive at an expression that relates measurable parameters of the reactions to constants and to concentration terms. The rate constant or other parameters can then be determined by graphical or numerical solutions from this relationship. If the kinetics of a process are found to fit closely with the model equation that is derived, then the model can be used as a basis for the description of the process. Kinetics is concerned about the quantities of the reactants and the products and their rates of change. Since reactants disappear in reactions, their rate expressions are given a negative sign. The amounts of the products increase and their rates of change are therefore positive.

REACTION RATE EQUATION

The rate of a reaction is the number of units of mass of some participating reactants that is transformed into a product per unit time and per unit volume of the system. The rate of a closed homogeneous reaction (that is, no gain or loss of material during the reaction) is determined by the composition of the reaction mixture, the temperature, and pressure. The pressure from an equation of state can be determined together with the temperature and composition.

Consider a single-phase reaction

$$aA + bB \rightarrow cC + dD \tag{3-1}$$

The reaction rate for reactant A can be expressed as

$$\left(-r_A\right) = -\frac{1}{V}\frac{dN_A}{dt} = \frac{\left(\text{amount of A disappearing}\right)}{\left(\text{volume}\right)\left(\text{time}\right)}$$

$$= \frac{\text{moles}}{m^3 \cdot \text{sec}} \tag{3-2}$$

where the minus sign means "disappearance" and

$(-r_A)$ = rate of disappearance of A
N_A = number of moles
V = system volume
t = time

The reaction rates of the individual components are related by

$$\frac{-r_A}{a} = \frac{-r_B}{b} = \frac{r_C}{c} = \frac{r_D}{d} \tag{3-3}$$

The rate is defined as an intensive variable, and the definition is independent of any particular reactant or product species. Because the reaction rate changes with time, we can use the time derivative to express the instantaneous rate of reaction since it is influenced by the composition and temperature (i.e., the energy of the material). Thus,

$$(-r_A) = f\{\text{temperature, concentration}\} \tag{3-4}$$

The Swedish chemist Arrhenius first suggested that the temperature dependence of the specific reaction rate k could be correlated by an equation of the type $k(T) = k_o e^{-E/RT}$. Therefore,

$$(-r_A) = kC_A^a = k_o e^{-E/RT} C_A^a$$

where E = activation energy (J/mol)
    k_o = frequency factor
    a = reaction order
    C_A = concentration of reactant A
    T = absolute temperature, K
    R = gas constant = 1.987 cal/mol • K = 8.314 J/mol •K

The reaction rate usually rises exponentially with temperature as shown in Figure 3-1. The Arrhenius equation as expressed in Chapter 1 is a good approximation to the temperature dependency. The temperature dependent term fits if plotted as ln (rates) versus 1/T at fixed concentration C_A, C_B (Figure 3-2).

At the same concentration, but two different temperatures,

$$\ln \frac{\text{rate}_1}{\text{rate}_2} = \ln \frac{k_1}{k_2} = \frac{E}{R}\left(\frac{1}{T_2} - \frac{1}{T_1}\right) \tag{3-5}$$

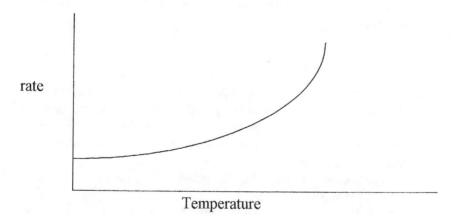

Figure 3-1. Reaction rate as a function of temperature.

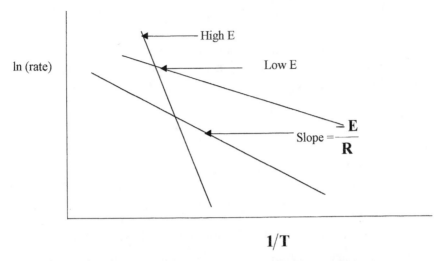

Figure 3-2. Temperature dependency of the reaction rate.

High values of the activation energy E are examples of gas phase reaction, which can only proceed at a high temperature (examples are free radical reactions and combustion). Low values of E can be found in enzymes, cellular and life related reactions, and reactions that occur at room temperature.

Composition affects the rate of reaction by fitting a simple expression to the data as shown in Figure 3-3. The concentration dependent term is found by guessing the rate equation, and seeing whether or not it fits the data.

Boudart [1] expressed the many variables that have influenced reaction rates as:

1. The rate of a chemical reaction depends on temperature, pressure, and composition of the system under investigation.
2. Certain species that do not appear in the stoichiometric equation for the reaction can affect the reaction rate even when they are present in only trace amounts. These materials are known as catalysts or inhibitors, depending on whether they increase or decrease the reaction rate.
3. At a constant temperature, the rate of reaction decreases with time or extent of the reaction.
4. Reactions that occur in systems that are far removed from equilibrium give the rate expressions in the form:

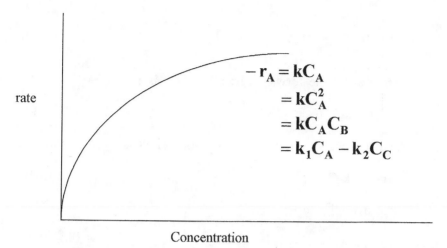

$$-r_A = kC_A$$
$$= kC_A^2$$
$$= kC_A C_B$$
$$= k_1 C_A - k_2 C_C$$

Concentration

Figure 3-3. Reaction rate as a function of concentration.

$$r = kf(C_{Ai}) \tag{3-6}$$

where $f(C_{Ai})$ is a function that depends on the concentration (C_A) of the various species present in the system, such as reactants, products, catalysts, and inhibitors. The function $f(C_{Ai})$ may also depend on the temperature. The coefficient k is the reaction constant. It does not depend on the composition of the system and is also independent of time in an isothermal system.

5. The rate constant k varies with the absolute temperature T of the system according to the Arrhenius law: $k = k_o e^{-E/RT}$.

6. The function $f(C_{Ai})$ in Equation 3-6 is temperature independent and can be approximated as:

$$f(C_{Ai}) = \prod_i C_i^{\beta_i} \tag{3-7}$$

where the product $\prod$ is taken over all components of the system. The exponents β_i are the orders of the reaction with respect to each of the species present in the system. The algebraic sum of the exponents is called the total order or overall order of the reaction.

7. For a system in which both forward and reverse reactions are important, the net rate of reaction can be expressed as the difference between the rate in the forward direction and that in the opposite direction. For example:

$$A + B \underset{k_2}{\overset{k_1}{\rightleftharpoons}} C + D \qquad (3\text{-}8)$$

The rate of the forward direction $(-r_A)$ is :

$$(-r_A) = k_1 C_A C_B \qquad (3\text{-}9)$$

and the rate of the reverse direction $(+r_A)$ is:

$$(+r_A) = k_2 C_C C_D \qquad (3\text{-}10)$$

The overall rate is:

$$(-r_A)_{net} = (-r_A) - (+r_A) \qquad (3\text{-}11)$$

The net rate of disappearance of $A = k_1 C_A C_B - k_2 C_C C_D$ at equilibrium where $(-r_A)_{net} = 0$. Therefore,

$$\frac{\vec{k}_1}{\overleftarrow{k}} = \left(\frac{C_C C_D}{C_A C_B} \right)_{Equilibrium} = K_C \qquad (3\text{-}12)$$

REACTION ORDERS

The manner in which the reaction rate varies with the concentrations of the reactants and products can be shown by the orders of the reaction. These are the powers to which the concentrations are raised and are expressed as:

$$(-r_A) = k C_A^\alpha C^\beta \qquad (3\text{-}13)$$

where α^{th} is the order with respect to component A, and β^{th} is the order with respect to component B. The overall order of the reaction n is:

$$n = \alpha + \beta \qquad (3\text{-}14)$$

The exponents α and β may be integers or fractions and may be both positive and negative values, as well as the value of zero. In some cases, the exponents are independent of temperature. Where the experimental data fit the expressions in the form of Equation 3-13, the exponents will vary slightly with temperature. In these cases, the observed correlation should be applied only in a restricted temperature interval.

Not all reactions can be expressed to have a specific order. A specific example is the synthesis of ammonia expressed as

$$N_2 + 3H_2 \overset{k_1}{\underset{k_2}{\rightleftharpoons}} 2NH_3$$

The Temkin-Pyzhev equation describes the net rate of synthesis over promoted iron catalyst as:

$$\left(+r_{NH_3}\right)_{net} = k_1 C_{N_2}\left(\frac{C_{H_2}^3}{C_{NH_3}^2}\right)^a - k_2\left(\frac{C_{NH_3}^2}{C_{H_2}^3}\right)^b \tag{3-15}$$

where a and b are empirical exponents that vary with reaction conditions such as temperature, pressure, and conversion. Equation 3-15 does not describe the initial rate at zero partial pressure of NH_3.

The units of the reaction rate constant k vary with the overall order of the reaction. These units are those of a rate divided by the nth power of concentration as evident from Equations 3-13 and 3-14.

$$k = \frac{(-r_A)}{C^n} = \frac{moles/(volume - time)}{(moles/volume)^n} \tag{3-16}$$

or

$$k = time^{-1}(moles)^{1-n}\frac{1}{(volume)^{1-n}}$$

$$= time^{-1}\left(\frac{moles}{volume}\right)^{-n+1} \tag{3-17}$$

Consider a reaction involving a reactant such that A → products. The rate equations corresponding to a zero, first, second, and third order reaction together with their corresponding units are:

1. Zero order (n = 0; k = moles/m^3-sec)

$$(-r_A) = kC_A^0 \qquad\qquad (3\text{-}18)$$

2. First order (n = 1; k = 1/sec)

$$(-r_A) = kC_A \qquad\qquad (3\text{-}19)$$

3. Second order (n = 2; k = m^3/moles-sec)

$$(-r_A) = kC_A^2 \qquad\qquad (3\text{-}20)$$

4. Third order (n = 3; k = (m^3/moles)$^2 \cdot$sec^{-1})

$$(-r_A) = kC_A^3 \qquad\qquad (3\text{-}21)$$

DETERMINING THE ORDER OF REACTIONS

ZERO ORDER REACTIONS

The rate of a chemical reaction is of a zero order if it is independent of the concentrations of the participating substances. The rate of reaction is determined by such limiting factors as:

1. In radiation chemistry, the energy, intensity, and nature of radiation.
2. In photochemistry, the intensity and wave length of light.
3. In catalyzed processes, the rates of diffusion of reactants and availability of surface sites.

If the rate of the reaction is independent of the concentration of the reacting substance A, then the amount dC_A by which the concentration of A decreases in any given unit of time dt is constant throughout the course of the reaction. The rate equation for a constant volume batch system (i.e., constant density) can be expressed as:

$$\left(-r_A\right) = -\frac{dC_A}{dt} = k \qquad\qquad (3\text{-}22)$$

The negative sign indicates that the component A is removed from the system, and k is the velocity constant with the units as moles/m^3-sec. Assuming that at time t_1 the concentration of A is C_{AO}, and at time t_2 the concentration is C_{Af}, integrating Equation 3-22 between these limits gives:

$$-\int_{C_{AO}}^{C_{Af}} dC_A = k \int_{t_1}^{t_2} dt \qquad (3\text{-}23)$$

$$-(C_{Af} - C_{AO}) = k(t_2 - t_1)$$

$$k = \frac{(C_{AO} - C_{Af})}{(t_2 - t_1)} \qquad (3\text{-}24)$$

If $t_1 = 0$, Equation 3-24 reduces to

$$C_{Af} = C_{AO} - kt_2 \qquad (3\text{-}25)$$

Plotting the concentration (C_A) versus time t gives a straight line, where C_{AO} is the intercept and k is the slope. The velocity constant k may include arbitrary constants resulting from various limiting factors such as diffusion constants and a fixed intensity of absorbed light. In terms of the fractional conversion X_A

$$C_{AO}X_A = kt \qquad (3\text{-}26)$$

Because C_A cannot be negative, the following is obtained

$$C_{AO} - C_A = C_{AO} X_A = kt \quad \text{for} \quad t < \frac{C_{AO}}{k} \qquad (3\text{-}27)$$

This means that the conversion is proportional to time. Figure 3-4 shows plots of the zero order rate equations. Examples of zero order reactions are the intensity of radiation within the vat for photochemical reactions or the surface available in certain solid catalyzed gas reactions.

FIRST ORDER REACTIONS

Consider the reaction $A \xrightarrow{k} B$. The rate of a first order reaction is proportional to the first power of the concentration of only one component. Assuming that there is no change in volume, temperature,

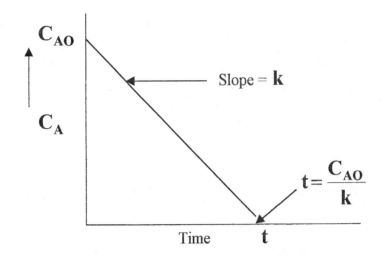

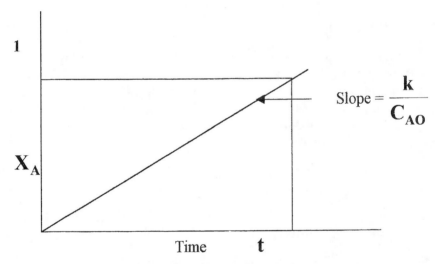

Figure 3-4. Zero order reaction.

or any other factor that may affect the reaction, the amount dC_A that undergoes a chemical change in the short time interval dt depends only on the amount of component A present at that instant. The rate for a first order reaction for a constant volume batch system (i.e., constant density) can be expressed as:

$$(-r_A) = \frac{-dC_A}{dt} = kC_A \qquad (3\text{-}28)$$

Rearranging and integrating Equation 3-28 between the limits gives

$$-\int_{C_{AO}}^{C_{Af}} \frac{dC_A}{C_A} = k\int_{t_1}^{t_2} dt \tag{3-29}$$

$$-\ln\left[\begin{matrix}C_{Af}\\C_A\\C_{AO}\end{matrix}\right] = k(t_2 - t_1) \tag{3-30}$$

$$-\ln\left[C_{Af} - C_{AO}\right] = k(t_2 - t_1) \tag{3-31}$$

$$-\ln\frac{C_{Af}}{C_{AO}} = k(t_2 - t_1) \tag{3-32}$$

At $t_1 = 0$ and $t_2 = t$. Therefore,

$$-\ln\frac{C_{Af}}{C_{AO}} = kt \tag{3-33}$$

The fractional conversion X_A for a given reactant A is defined as the fraction of the reactant converted into product or

$$X_A = \frac{N_{AO} - N_A}{N_{AO}} \tag{3-34}$$

For a constant density system, the volume V remains constant

$$C_A = \frac{N_A}{V} = \frac{N_{AO}(1 - X_A)}{V} = C_{AO}(1 - X_A) \tag{3-35}$$

Differentiating Equation 3-35 gives

$$dC_A = -C_{AO}dX_A \tag{3-36}$$

Substituting Equation 3-36 into Equation 3-28 gives

$$\frac{dX_A}{dt} = k(1 - X_A) \tag{3-37}$$

Rearranging and integrating Equation 3-37 between the limits t = 0, $X_A = 0$ and $t = t_f$ and $X_A = X_{Af}$ gives

$$\int_0^{X_{Af}} \frac{dX_A}{(1-X_A)} = k \int_0^{t_f} dt$$

$$-\ln(1-X_{Af}) = kt_f \qquad (3\text{-}38)$$

A plot of $-\ln(1 - X_A)$ or $-\ln C_{Af}/C_{AO}$ versus t gives a straight line through the origin. The slope of the line is the velocity constant k, as represented by Figure 3-5.

The unit of the velocity constant k is sec^{-1}. Many reactions follow first order kinetics or pseudo-first order kinetics over certain ranges of experimental conditions. Examples are the cracking of butane, many pyrolysis reactions, the decomposition of nitrogen pentoxide (N_2O_5), and the radioactive disintegration of unstable nuclei. Instead of the velocity constant, a quantity referred to as the half-life $t_{1/2}$ is often used. The half-life is the time required for the concentration of the reactant to drop to one-half of its initial value. Substitution of the appropriate numerical values into Equation 3-33 gives

$$k = \frac{1.}{t_{1/2}} \ln 2 = \frac{0.693}{t_{1/2}} \qquad (3\text{-}39)$$

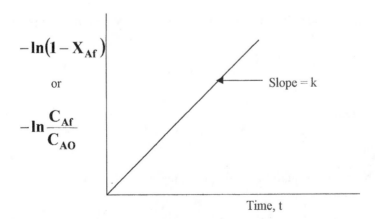

Figure 3-5. First order reaction.

Equation 3-39 shows that in the first order reactions, the half-life is independent of the concentration of the reactant. This basis can be used to test whether a reaction obeys first order kinetics by measuring half-lives of the reaction at various initial concentrations of the reactant.

SECOND ORDER REACTIONS

A second order reaction occurs when two reactants A and B interact in such a way that the rate of reaction is proportional to the first power of the product of their concentrations. Another type of a second order reaction includes systems involving a single reactant. The rate at any instant is proportional to the square of the concentration of a single reacting species.

Case 1

Consider the reaction $A + B \xrightarrow{\text{k}}$ products. The rate equation for a constant volume batch system (i.e., constant density) is:

$$\left(-r_A\right) = \frac{-dC_A}{dt} = \frac{-dC_B}{dt} = kC_A C_B \tag{3-40}$$

The amount of A and B that have reacted at any time t can be described by the following mechanism and set of equations.

From stoichiometry:

	A	**B**
Amount at time t = 0	C_{AO}	C_{BO}
Amount at time t = t	C_A	C_B
Amounts that have reacted	$C_{AO} - C_A$	$C_{BO} - C_B$

$$C_B = C_{BO} - (C_{AO} - C_A) \tag{3-41}$$

Substituting Equation 3-41 into Equation 3-40, rearranging and integrating between the limits gives

$$\int_{C_{AO}}^{C_A} \frac{-dC_A}{C_A\left[C_{BO} - \left(C_{AO} - C_A\right)\right]} = k\int_0^t dt \tag{3-42}$$

Equation 3-42 is resolved into partial fractions as

$$\frac{1}{C_A\left[(C_{BO}-C_{AO})+C_A\right]} \equiv \frac{p}{C_A} + \frac{q}{\left[(C_{BO}-C_{AO})+C_A\right]} \tag{3-43}$$

$$1 \equiv p\left[(C_{BO}-C_{AO})+C_A\right] + qC_A \tag{3-44}$$

Equating the coefficients of C_A and the constant on both the right and left side of Equation 3-44 gives

"Const" $1 = p(C_{BO}-C_{AO})$

"C_A" $0 = p + q$

where

$$p = 1/(C_{BO}-C_{AO}) \tag{3-45}$$

and

$$q = -1/(C_{BO}-C_{AO}) \tag{3-46}$$

Substituting p and q into Equation 3-43 and integrating between the limits gives

$$-\left\{\int_{C_{AO}}^{C_A}\frac{dC_A}{(C_{BO}-C_{AO})C_A} - \int_{C_{AO}}^{C_A}\frac{dC_A}{(C_{BO}-C_{AO})\left[(C_{BO}-C_{AO})+C_A\right]}\right\}$$

$$= k\int_0^t dt \tag{3-47}$$

$$\frac{1}{(C_{BO}-C_{AO})}\left\{\ln\frac{C_{AO}}{C_A} + \ln\left[\left\{\frac{(C_{BO}-C_{AO})+C_A}{(C_{BO}-C_{AO}+C_{AO})}\right\}\right]\right\} = kt \tag{3-48}$$

$$\ln\left\{\left(\frac{C_{AO}}{C_{BO}}\right)\left[\frac{(C_{BO}-C_{AO})+C_A}{C_A}\right]\right\} = k(C_{BO}-C_{AO})t \tag{3-49}$$

Figure 3-6 shows that a plot of

$$\ln\left\{\left(\frac{C_{AO}}{C_{BO}}\right)\left[\frac{(C_{BO}-C_{AO})+C_A}{C_A}\right]\right\}$$

versus time t gives a straight line with the slope $k(C_{BO} - C_{AO})$. If C_{BO} is much larger than C_{AO}, then the concentration of B remains approximately constant, and Equation 3-49 approaches Equations 3-33 or 3-38 for the first order reaction. Therefore, the second order reaction becomes a pseudo-first order reaction.

Case 2

Case 2 involves a system with a single reactant. In this case, the rate at any instant is proportional to the square of the concentration of A. The reaction mechanism is $2A \xrightarrow{k} products$. The rate expression for a constant volume batch system (i.e., constant density) is

$$(-r_A) = \frac{-dC_A}{dt} = kC_A^2 = kC_{AO}^2(1-X_A)^2 \tag{3-50}$$

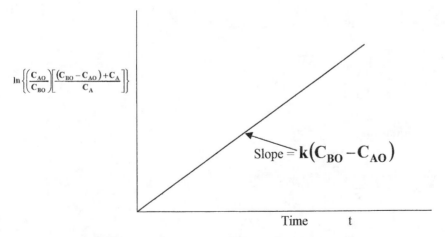

Figure 3-6. Test for the reaction A + B → products.

Rearranging Equation 3-50 and integrating between the limits yields

$$\int_{C_{AO}}^{C_A} \frac{-dC_A}{C_A^2} = k\int_0^t dt \tag{3-51}$$

$$\left[\frac{1}{C_A}\right]_{C_{AO}}^{C_A} = kt \tag{3-52}$$

$$\frac{1}{C_A} - \frac{1}{C_{AO}} = kt \tag{3-53}$$

or

$$\frac{1}{C_A} = \frac{1}{C_{AO}} + kt \tag{3-54}$$

In terms of the fractional conversion, X_A, Equation 3-50 becomes

$$\frac{C_{AO}\, dX_A}{dt} = kC_{AO}^2 (1 - X_A)^2 \tag{3-55}$$

Rearranging Equation 3-55 and integrating gives

$$\int_0^{X_A} \frac{dX_A}{(1-X_A)^2} = kC_{AO}\int_0^t dt \tag{3-56}$$

Equation 3-56 yields

$$\left[\frac{1}{(1-X_A)}\right]_0^{X_A} = kC_{AO}t \tag{3-57}$$

Equation 3-57 gives

$$\frac{X_A}{(1-X_A)} = kC_{AO}t \tag{3-58}$$

Figure 3-7 gives plots of Equations 3-54 and 3-58, respectively.

Consider the second order reaction $2A + B \xrightarrow{\ k\ } products$, which is first order with respect to both A and B, and therefore second order overall. The rate equation is:

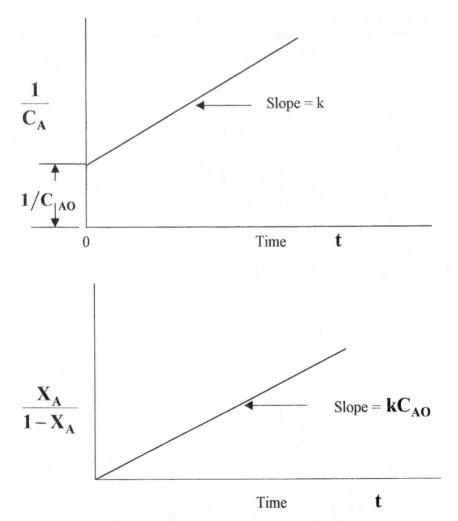

Figure 3-7. Test for the reaction 2A → products.

$$\left(-r_A\right)=\frac{-dC_A}{dt}=kC_AC_B \tag{3-59}$$

From stoichiometry:

	A	B
Amount at time t = 0	C_{AO}	C_{BO}
Amount at time t = t	C_A	C_B
Amounts that have reacted	$(C_{AO}-C_A)$	$C_{BO}-C_B$

and from stoichiometry $(C_{AO} - C_A) = 2(C_{BO} - C_B)$.

Expressing the concentration of B in terms of A and the original amounts C_{AO}, C_{BO} gives

$$C_B=C_{BO}-\frac{1}{2}\left(C_{AO}-C_A\right) \tag{3-60}$$

Substituting Equation 3-60 into Equation 3-59 gives

$$\int_{C_{AO}}^{C_A}\frac{-dC_A}{kC_A\left\{C_{BO}-\frac{1}{2}\left(C_{AO}-C_A\right)\right\}}=\int_0^t dt \tag{3-61}$$

$$-\int_{C_{AO}}^{C_A}\frac{dC_A}{C_A\left\{\left(C_{BO}-\frac{1}{2}C_{AO}\right)+\frac{1}{2}C_A\right\}}=k\int_0^t dt \tag{3-62}$$

Taking the partial fractions of Equation 3-62 with $\alpha = [C_{BO} - (1/2)C_{AO}]$ gives the following:

$$\frac{1}{C_A\left\{\alpha+\frac{1}{2}C_A\right\}}\equiv\frac{p}{C_A}+\frac{q}{\alpha+\frac{1}{2}C_A} \tag{3-63}$$

$$1\equiv p\left(\alpha+\frac{1}{2}C_A\right)+qC_A \tag{3-64}$$

Solving for constants p and q involves equating the "constant" and the coefficient of "C_A" in Equation 3-64:

"Const" $1 = \alpha p$ (3-65)

"C_A" $0 = \dfrac{1}{2} p + q$ (3-66)

$p = \dfrac{1}{\alpha}$ (3-67)

and

$q = \dfrac{-1}{2} p = \dfrac{-1}{2\alpha}$ (3-68)

Substituting the values of p and q into Equation 3-68 and integrating between the limits gives

$$\int_{C_{AO}}^{C_A} \frac{-dC_A}{C_A \left\{ \alpha + \dfrac{1}{2} C_A \right\}} = -\left\{ \int_{C_{AO}}^{C_A} \frac{dC_A}{\alpha C_A} - \int_{C_{AO}}^{C_A} \frac{dC_A}{2\alpha \left(\alpha + \dfrac{1}{2} C_A \right)} \right\}$$

$$= k \int_0^t dt$$ (3-69)

$$-\left\{ \frac{1}{\alpha} \ln \frac{C_A}{C_{AO}} - \frac{1}{\alpha} \ln \left(\frac{2\alpha + C_A}{2\alpha + C_{AO}} \right) \right\} = kt$$ (3-70)

$$\left\{ \ln \frac{C_{AO}}{C_A} + \ln \left(\frac{2\alpha + C_A}{2\alpha + C_{AO}} \right) \right\} = \alpha kt$$ (3-71)

since

$$2\alpha + C_{AO} = 2C_{BO}$$

$$2\alpha + C_A = 2C_B$$

$$\ln\left(\frac{C_{AO}}{C_A} \bullet \frac{C_B}{C_{BO}}\right) = \alpha kt \qquad\qquad (3\text{-}72)$$

A plot of $\ln(C_{AO}/C_A \times C_B/C_{BO})$ versus time t gives a straight line with the slope αk as shown in Figure 3-8. Examples of bimolecular reactions with rate equations similiar to those given in Equation 3-40 are shown in Table 3-1.

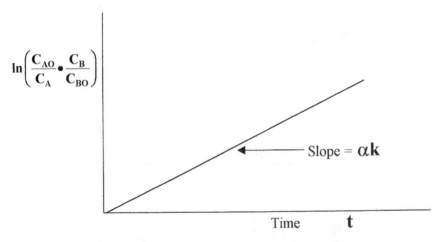

Figure 3-8. Test for the reaction 2A + B → products.

Table 3-1
Second order reaction

1. Alkaline hydrolysis of esters:
   $$CH_3COOC_2H_5 + NaOH \rightarrow CH_3COONa + C_2H_5OH$$

2. Bromination of alkenes:
   $$CH_3CH_2 = CH_2 + Br_2 \rightarrow CH_3CHBrCH_2Br$$

3. Nucleophilic substitution:
   $$Cl^- + CH_3I \rightarrow CH_3Cl + I^-$$

4. Nitroethane iodination:
   $$C_2H_5NO_2 + C_6H_5N \rightarrow C_2H_4NO_2 + C_5H_5NH^+ + I^-$$

EMPIRICAL RATE EQUATIONS OF THE nth ORDER

IRREVERSIBLE REACTIONS

The rate equation for irreversible reactions for a constant volume batch system (i.e., constant density) is

$$(-r_A) = \frac{-dC_A}{dt} = kC_A^n \tag{3-73}$$

Boundary conditions are: at $t = 0$, $C_A = C_{AO}$ and $t = t$, $C_A = C_A$. Integrating Equation 3-73 between the limits gives

$$\int_{C_{AO}}^{C_A} \frac{-dC_A}{C_A^n} = \int_0^t dt \tag{3-74}$$

$$-\left\{ \frac{C_A^{-n+1}}{-n+1} \right\}_{C_{AO}}^{C_A} = kt \tag{3-75}$$

$$\frac{1}{(n-1)} \left\{ C_A^{1-n} - C_{AO}^{1-n} \right\} = kt \tag{3-76}$$

$$C_A^{1-n} - C_{AO}^{1-n} = (n-1)kt \tag{3-77}$$

The value of n must be found by trial-and-error. In terms of the fractional conversion, X_A, the rate equation is

$$(-r_A) = C_{AO} \frac{dX_A}{dt} = kC_{AO}^n (1 - X_A)^n \tag{3-78}$$

Boundary conditions are: at $t = 0$, $X_A = 0$, and $t = t$, $X_A = X_A$. Rearranging and integrating Equation 3-78 between the limits gives

$$\int_0^{X_A} \frac{dX_A}{(1 - X_A)^n} = kC_{AO}^{n-1} \int_0^t dt \tag{3-79}$$

$$\left\{ \frac{(1-X_A)^{-n+1}}{1-n} \right\}_0^{X_A} = kC_{AO}^{n-1}t \qquad (3\text{-}80)$$

$$(1-X_A)^{1-n} - 1 = kC_{AO}^{n-1}(1-n)t \qquad (3\text{-}81)$$

Table 3-2 gives solutions for some chemical reactions using the integration method.

METHOD OF HALF-LIFE $t_{1/2}$

As previously discussed, the half-life of a reaction is defined as the time it takes for the concentration of the reactant to fall to half of its initial value. Determining the half-life of a reaction as a function of the initial concentration makes it possible to calculate the reaction order and its specific reaction rate.

Consider the reaction $A \rightarrow$ products. The rate equation in a constant volume batch reaction system gives

$$(-r_A) = \frac{-dC_A}{dt} = kC_A^n \qquad (3\text{-}82)$$

Rearranging and integrating Equation 3-82 with the boundary conditions $t = 0$, $C_A = C_{AO}$ and $t = t$, $C_A = C_A$ gives

$$-\int_{C_{AO}}^{C_A} \frac{dC_A}{C_A^n} = \int_0^t dt \qquad (3\text{-}83)$$

$$-\left\{ \frac{C_A^{-n+1}}{-n+1} \right\}_{C_{AO}}^{C_A} = kt \qquad (3\text{-}84)$$

$$\frac{1}{(n-1)} \left\{ C_A^{1-n} \right\}_{C_{AO}}^{C_A} = kt \qquad (3\text{-}85)$$

$$\left\{ C_A^{1-n} - C_{AO}^{1-n} \right\} = kt(n-1) \qquad (3\text{-}86)$$

Table 3-2
Rate equations for irreversible processes

Stoichiometry	Order	Rate equation — Differential form	Rate equation — Integrated form	Common unit for rate constant	Relationship between rate constants
a					a
A single reactant, or reactants at equal concentrations					
a	0	$\dfrac{dx}{dt} = k_A$	$k_A = \dfrac{x}{t}$	$mol\ dm^{-3}\ s^{-1}$	a
$A \to Z,\ A + B + \cdots \to Z$	1/2	$\dfrac{dx}{dt} = k_A(a_0 - x)^{1/2}$	$k_A = \dfrac{2}{t}\left[a_0^{1/2} - (a_0 - x)^{1/2}\right]$	$mol^{1/2}\ dm^{-3/2}\ s^{-1}$	$k_A = k_B = k$
$A \to Z,\ A + B + \cdots \to Z$	1	$\dfrac{dx}{dt} = k_A(a_0 - x)$	$k_A = \dfrac{1}{t}\ln\left(\dfrac{a_0}{a_0 - x}\right)$	s^{-1}	$k_A = k_B = k$
$A \to Z,\ A + B + \cdots \to Z$	3/2	$\dfrac{dx}{dt} = k_A(a_0 - x)^{3/2}$	$k_A = \dfrac{2}{t}\left[\dfrac{1}{(a_0 - x)^{1/2}} - \dfrac{1}{a_0^{1/2}}\right]$	$dm^{3/2}\ mol^{-1/2}\ s^{-1}$	$k_A = k_B = k$
$A \to Z,\ A + B + \cdots \to Z$	2	$\dfrac{dx}{dt} = k_A(a_0 - x)^2$	$k_A = \dfrac{x}{t\,a_0(a_0 - x)}$	$dm^3\ mol^{-1}\ s^{-1}$	$k_A = k_B = k$
$A + 2B \to Z$	2	$\dfrac{dx}{dt} = k_A(a_0 - x)(a_0 - 2x)$	$k_A = \dfrac{1}{t\,a_0}\ln\left(\dfrac{a_0 - x}{a_0 - 2x}\right)$	$dm^3\ mol^{-1}\ s^{-1}$	$k_B = 2k_A = 2k$
$A + B + C + \cdots \to Z$	3	$\dfrac{dx}{dt} = k_A(a_0 - x)^3$	$k_A = \dfrac{2a_0 x - x^2}{2t\,a_0^2(a_0 - x)^2}$	$dm^6\ mol^{-2}\ s^{-1}$	$k_A = k_B = k_C = k$
$A \to Z,\ A + B + \cdots \to Z$	n	$\dfrac{dx}{dt} = k_A(a_0 - x)^n$	$k_A = \dfrac{1}{t(n-1)}\left[\dfrac{1}{(a_0 - x)^{n-1}} - \dfrac{1}{a_0^{n-1}}\right]$	$dm^{3n-3}\ mol^{1-n}\ s^{-1}$	$k_A = k_B = k$
Reactants at unequal initial concentrations					
$A + B \to Z$	2	$\dfrac{dx}{dt} = k_A(a_0 - x)(b_0 - x)$	$k_A = \dfrac{1}{t(a_0 - b_0)}\ln\left[\dfrac{b_0(a_0 - x)}{a_0(b_0 - x)}\right]$	$dm^3\ mol^{-1}\ s^{-1}$	$k_A = k_B = k$
$A + 2B \to Z$	2	$\dfrac{dx}{dt} = k_A(a_0 - x)(b_0 - 2x)$	$k_A = \dfrac{1}{t(2a_0 - b_0)}\ln\left[\dfrac{b_0(a_0 - x)}{a_0(b_0 - 2x)}\right]$	$dm^3\ mol^{-1}\ s^{-1}$	$2k_A = k_B = 2k$
$2A + B \to Z$	2	$\dfrac{dx}{dt} = k_A(a_0 - 2x)(b_0 - x)$	$k_A = \dfrac{1}{t(a_0 - 2b_0)}\ln\left[\dfrac{b_0(a_0 - 2x)}{a_0(b_0 - x)}\right]$	$dm^3\ mol^{-1}\ s^{-1}$	$k_A = 2k_B = 2k$
$A \to X$ with autocatalysis[b]	1	$\dfrac{dx}{dt} = k_A(a_0 - x)(x_0 + x)$	$k_A = \dfrac{1}{t(a_0 - x_0)}\ln\left[\dfrac{a_0(x_0 + x)}{x_0(a_0 - x)}\right]$	$dm^3\ mol^{-1}\ s^{-1}$	$k_A = k$

In all cases x is the amount of reactant A consumed per unit volume, and except in the autocatalytic case $x = 0$ when $t = 0$

[a] The equations for the zero order reaction are true for any stoichiometry, but the relationship between the rate constants depends on the stoichiometry.
[b] In the the autocatalytic case some product must be present initially (at concentration x_0) otherwise the reaction will never start.
Source: Laidler, K. J., Chemical Kinetics, 3rd ed. Harper Collins Publishers, 1987.

or

$$t = \frac{1}{k(n-1)}\left\{C_A^{1-n} - C_{AO}^{1-n}\right\}$$

(3-87)

The half-life is defined as the time required for the concentration to drop to half of its initial value, that is, at $t = t_{1/2}$, $C_A = 1/2 C_{AO}$. From Equation 3-86,

$$\left(2^{-1}C_{AO}\right)^{1-n} - C_{AO}^{1-n} = kt_{1/2}(n-1)$$

(3-88)

$$\left(2^{-1} \cdot 2^n - 1\right)C_{AO}^{1-n} = kt_{1/2}(n-1)$$

(3-89)

$$t_{1/2} = \frac{\left(2^{n-1} - 1\right)C_{AO}^{1-n}}{k(n-1)}$$

(3-90)

Similarly, the time required for the concentration to fall to $1/p$ of its initial value gives

$$t_{1/p} = \frac{\left(p^{n-1} - 1\right)C_{AO}^{1-n}}{k(n-1)}$$

(3-91)

Taking the natural logarithm of both sides of Equation 3-90 gives

$$\ln t_{1/2} = \ln\left(2^{n-1} - 1\right) + (1-n)\ln C_{AO} - \ln k - \ln(n-1)$$

(3-92)

$$\ln t_{1/2} = \ln\left\{\frac{2^{n-1} - 1}{k(n-1)}\right\} + (1-n)\ln C_{AO}$$

(3-93)

Plots of $\ln t_{1/2}$ versus $\ln C_{AO}$ from a series of half-life experiments are shown in Figure 3-9. Table 3-3 gives some expressions for reaction half-lives.

The reaction rate constant k is

$$k = \frac{\left(2^{n-1} - 1\right)C_{AO}^{1-n}}{(n-1)\,t_{1/2}}$$

(3-94)

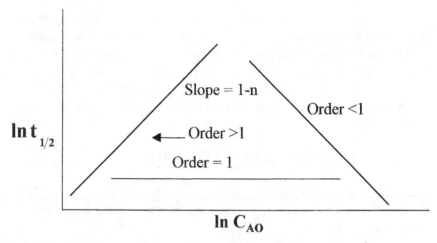

Figure 3-9. Overall order of reaction from a series of half-life experiments each at different initial concentration of reactant.

Table 3-3
Expressions for Reaction Half-Lives

Order	Half-life $t_{1/2}$	Order	Half-life $t_{1/2}$
Single reaction substance A		Reactants in their stoichiometric ratios; $A + \ldots \rightarrow Z$	
0	$\dfrac{a_0}{2k_A}$	0	$\dfrac{a_0}{2k_A}$
1	$\dfrac{\ln 2}{k_A}$	$1(\upsilon_A = k_A[A])$	$\dfrac{\ln 2}{k_A}$
2	$\dfrac{1}{k_A a_0}$	$2(\upsilon_A = k_A[A][B])$	$\dfrac{1}{k_A a_0} = \dfrac{1}{k_A b_0}$
3	$\dfrac{3}{2k_A a_0^2}$	$3(\upsilon_A = k_A[A][B][C])$	$\dfrac{3}{2k_A b_0 c_0}$
n	$\dfrac{2^{n-1}-1}{k_A(n-1)a_0^{n-1}}$		

Source: Laidler, K. J., Chemical Kinetics, *3rd ed., Harper Collins Publishers, 1987.*

The half-life method requires data from several experiments, each at different initial concentration. The method shows that the fractional conversion in a given time rises with increased concentration for orders greater than one, drops with increased concentration for orders less than one, and is independent of the initial concentration for reactions of first order. This also applies to the reaction $A + B \rightarrow$ products when $C_{AO} = C_{BO}$.

PARALLEL REACTIONS

Consider elementary reactions $A \xrightarrow{k_1} B$ and $A \xrightarrow{k_2} C$. The rate equations for these reactions for a constant volume batch system (i.e., constant density) are

$$(-r_A) = \frac{-dC_A}{dt} = k_1 C_A + k_2 C_A = (k_1 + k_2) C_A \tag{3-95}$$

$$(+r_B) = \frac{dC_B}{dt} = k_1 C_A \tag{3-96}$$

$$(+r_C) = \frac{dC_C}{dt} = k_2 C_A \tag{3-97}$$

Rearranging and integrating Equation 3-95 between the limits with the boundary conditions at time $t = 0$, $C_A = C_{AO}$, $C_B = C_{BO}$, $C_C = C_{CO}$, gives:

$$-\int_{C_{AO}}^{C_A} \frac{dC_A}{C_A} = \int_0^t (k_1 + k_2) dt \tag{3-98}$$

$$-\ln \frac{C_A}{C_{AO}} = (k_1 + k_2) t \tag{3-99}$$

In terms of the fractional conversion

$$-\ln(1 - X_A) = (k_1 + k_2) t \tag{3-100}$$

If $-\ln C_A/C_{AO}$ or $-\ln (1 - X_A)$ is plotted against time t, the slope of the line is $(k_1 + k_2)$. Also, dividing Equation 3-96 by Equation 3-97 gives

$$\frac{r_B}{r_C} = \frac{dC_B}{dC_C} = \frac{k_1}{k_2} \tag{3-101}$$

Integrating Equation 3-101 between the limits yields

$$\frac{C_B - C_{BO}}{C_C - C_{CO}} = \frac{k_1}{k_2} \tag{3-102}$$

or $\quad C_B - C_{BO} = \dfrac{k_1}{k_2} C_C - \dfrac{k_1}{k_2} C_{CO}$ $\qquad$ (3-103)

Therefore, the slope of the linear plot C_B versus C_C gives the ratio k_1/k_2. Knowing $k_1 + k_2$ and k_1/k_2, the values of k_1 and k_2 can be determined as shown in Figure 3-10. Concentration profiles of components A, B, and C in a batch system using the differential Equations 3-95, 3-96, 3-97 and the Runge-Kutta fourth order numerical method for the case when $C_{BO} = C_{CO} = 0$ and $k_1 > k_2$ are reviewed in Chapter 5.

An example of parallel reactions involves the two modes of decomposition of an alcohol:

$$C_2H_5OH = C_2H_4 + H_2O \tag{3-104}$$

$$C_2H_5OH = CH_3CHO + H_2 \tag{3-105}$$

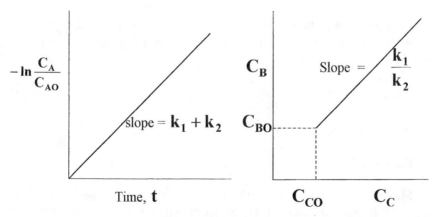

Figure 3-10. Rate constants for two competing elementary first order reactions.

Generally, these occur simultaneously. The relative amount of ethylene and acetaldehyde obtained therefore depends on the relative speed of the two reactions competing with each other for the available alcohol. These speeds, in turn, are determined by the choice of catalyst and temperature.

Parallel reactions often have mechanisms of different reaction order. An example is the solvolysis reactions in the presence of added nucleophiles:

$$C_6H_5SO_2Cl + H_2O \xrightarrow{\ k_1\ } C_6H_5SO_2OH + H^+ + Cl^-$$

$$C_6H_5SO_2Cl + F^- \xrightarrow[H_2O]{\ k_2\ } C_6H_5SO_2F + Cl^- \tag{3-106}$$

In the solvolysis reaction, H_2O is the solvent as well as a reactant, and it is present in large excess. It does not appear in the rate equation, which is expressed by

$$A \xrightarrow{\ k_1\ } C \tag{3-107}$$

$$A + B \xrightarrow{\ k_2\ } D \tag{3-108}$$

The rate equation is:

$$\left(-r_A\right)_{net} = -\frac{dC_A}{dt} = k_1C_A + k_2C_AC_B \tag{3-109}$$

An excess of component B is used, and its concentration remains constant, so that the rate equation reduces to a first order. That is,

$$-\frac{dC_A}{dt} = K_eC_A \tag{3-110}$$

where

$$K_e = k_1 + k_2C_B \tag{3-111}$$

Measurements of K_e at different concentrations of B will enable the rate constants for each of the reactions to be determined. Parallel reactions give a reduction in the yield of the product. This can be improved by incorporating a method that favors the required reaction over the unwanted reaction.

HOMOGENEOUS CATALYZED REACTIONS

A catalyzed reaction can be expressed as

$$A \xrightarrow{\ k_1\ } P \quad \text{(uncatalyzed)} \tag{3-112}$$

$$A + C \xrightarrow{\ k_2\ } P + C \quad \text{(catalyzed)} \tag{3-113}$$

where C and P represent the catalyst and product respectively. The reaction rates for a constant volume batch system (i.e., constant density) are:

$$\left(-r_A\right) = -\left(\frac{dC_A}{dt}\right)_1 = k_1 C_A \tag{3-114}$$

$$\left(-r_A\right) = -\left(\frac{dC_A}{dt}\right)_2 = k_2 C_A C_C \tag{3-115}$$

The overall rate of disappearance of reactant A is:

$$-\frac{dC_A}{dt} = k_1 C_A + k_2 C_A C_C \tag{3-116}$$

The catalyst concentration remains unchanged and integrating Equation 3-116 with the boundary conditions: $t = 0$, $C_A = C_{AO}$ and $t = t$, $C_A = C_A$

$$-\int_{C_{AO}}^{C_A} \frac{dC_A}{C_A} = \int_0^t \left(k_1 + k_2 C_C\right) dt \tag{3-117}$$

$$-\ln \frac{C_A}{C_{AO}} = \left(k_1 + k_2 C_C\right) t = Kt \tag{3-118}$$

where $K = k_1 + k_2 C_C$.

In this case, a series of experiments are performed with varying catalyst concentrations. A plot is made of $K = (k_1 + k_2C_C)$ versus C_C. The slope of the straight line is k_2 and the intercept is k_1, as shown in Figure 3-11.

AUTOCATALYTIC REACTIONS

This type of reaction occurs when one of the products of the reactions acts as a catalyst, and is expressed by

$$A + B \xrightarrow{\ k\ } B + B \tag{3-119}$$

The rate equation for a constant volume batch system is:

$$\left(-r_A\right) = -\frac{dC_A}{dt} = kC_A C_B \tag{3-120}$$

When A is consumed, then the total moles of A and B remain unchanged at any time t and can be expressed as $C_O = C_A + C_B = C_{AO} + C_{BO}$ = constant. Then the rate equation becomes

$$\left(-r_A\right) = -\frac{dC_A}{dt} = kC_A\left(C_O - C_A\right) \tag{3-121}$$

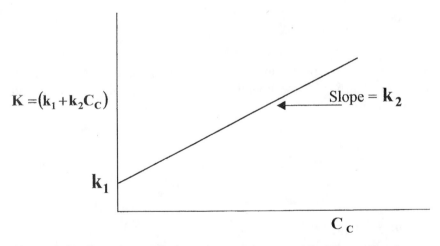

Figure 3-11. Rate constants for a homogeneous catalyzed reaction from a series of experiments with different catalyst concentrations.

Rearranging Equation 3-121 and integrating between the limits gives

$$\int_{C_{AO}}^{C_A} -\frac{dC_A}{C_A(C_O - C_A)} = \int_0^t k\,dt \tag{3-122}$$

Converting the left side of Equation 3-122 into partial fractions gives

$$\frac{1}{C_A(C_O - C_A)} \equiv \frac{p}{C_A} + \frac{q}{C_O - C_A} \tag{3-123}$$

$$1 \equiv p(C_O - C_A) + qC_A$$

Equating the coefficients

Constant $1 = pC_O$

"C_A" $0 = -p + q$

$p = q$

$$p = q = \frac{1}{C_O}$$

Integrating Equation 3-122 between the limits gives

$$-\left\{ \int_{C_{AO}}^{C_A} \frac{1}{C_O} \frac{dC_A}{C_A} + \int_{C_{AO}}^{C_A} \frac{1}{C_O} \frac{dC_A}{(C_O - C_A)} \right\} = k\int_0^t dt \tag{3-124}$$

$$-\left\{ \frac{1}{C_O} \ln\frac{C_A}{C_{AO}} - \frac{1}{C_O} \ln\left(\frac{C_O - C_A}{C_O - C_{AO}}\right) \right\} = kt$$

$$\ln\left\{ \frac{C_{AO}}{C_A}\left(\frac{C_O - C_A}{C_O - C_{AO}}\right) \right\} = \ln\frac{C_B/C_{BO}}{C_A/C_{AO}} = C_O kt \tag{3-125}$$

The fractional conversion, X_A, and the initial reaction ratio, $\theta_B = C_{BO}/C_{AO}$, yields

$$\ln \frac{\theta_B + X_A}{\theta_B(1 - X_A)} = C_{AO}(\theta_B + 1)kt = (C_{AO} + C_{BO})kt \qquad (3\text{-}126)$$

For an autocatalytic reaction some product B must be present if the reaction is to proceed. Starting with a small concentration of B, the rate rises as B is formed, and when A is used up, the rate must drop to zero. A plot of concentration versus time gives a straight line through the origin as shown in Figure 3-12.

IRREVERSIBLE REACTIONS IN SERIES

Consider a first order reaction in series as $A \xrightarrow{k_1} B \xrightarrow{k_2} C$. The rate equations for a constant volume batch system (i.e., constant density) are:

$$(-r_A) = -\frac{dC_A}{dt} = k_1 C_A \qquad (3\text{-}127)$$

$$(-r_B)_{net} = -\frac{dC_B}{dt} = k_2 C_B - k_1 C_A \qquad (3\text{-}128)$$

$$(+r_C) = \frac{dC_C}{dt} = k_2 C_B \qquad (3\text{-}129)$$

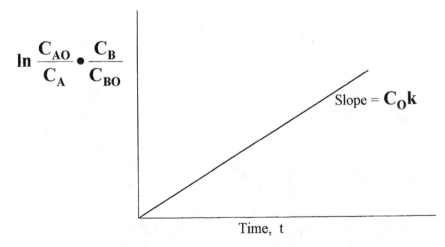

Figure 3-12. Autocatalytic reaction.

The initial conditions are at time $t = 0$, $C_A = C_{A0}$ and $C_B = 0$, $C_C = 0$. From stoichiometry:

	A	**B**	**C**
Amount at time $t = 0$	C_{AO}	0	0
Amount at time $t = t$	C_A	C_B	C_C
Amounts that have reacted	$C_{AO} - C_A$	C_B	C_C

and from stoichiometry $C_{AO} - C_A = C_B = C_C$.
From Equation 3-127, the concentration of A is obtained by integration

$$\frac{dC_A}{dt} = -k_1 C_A \tag{3-130}$$

$$\int_{C_{AO}}^{C_A} \frac{dC_A}{dt} = -k_1 \int_0^t dt \tag{3-131}$$

$$\ln \frac{C_A}{C_{AO}} = -k_1 t$$

Therefore,

$$C_A = C_{AO} e^{-k_1 t} \tag{3-132}$$

The variation in concentration of B is obtained by substituting the concentration of A from Equation 3-132 into Equation 3-128:

$$\frac{dC_B}{dt} + k_2 C_B = k_1 C_{AO} e^{-k_1 t} \tag{3-133}$$

Equation 3-133 is a first order linear differential equation of the form $dy/dx + Py = Q$. The integrating factor is $IF = e^{\int Pdx}$, and the solution is $ye^{\int Pdx} = \int Q e^{\int Pdx} dx + Constant$. Applying this general procedure to Equation 3-133, the integrating factor is $IF = e^{\int k_2 dt}$. Multiplying Equation 3-133 by the integrating factor gives

$$e^{\int k_2 dt} \frac{dC_B}{dt} + k_2 C_B e^{\int k_2 dt} = k_1 C_A e^{\int k_2 dt} \tag{3-134}$$

$$\frac{d}{dt}\left(C_B e^{k_2 t}\right) = k_1 C_A e^{\int k_2 dt} \tag{3-135}$$

$$C_B e^{k_2 t} = \int k_1 C_A e^{\int k_2 dt} dt + \text{Const.} \tag{3-136}$$

where Const. = constant.

$$= k_1 C_{AO} \int e^{-k_1 t} \bullet e^{k_2 t} dt + \text{Const.} \tag{3-137}$$

$$= k_1 C_{AO} \int e^{(k_2 - k_1)t} dt + \text{Const.} \tag{3-138}$$

$$C_B e^{k_2 t} = \frac{k_1 C_{AO} e^{(k_2 - k_1)t}}{\left(k_2 - k_1\right)} + \text{Const.} \tag{3-139}$$

At time t = 0, the concentration of component B is $C_B = 0$. Therefore, the constant Const. becomes

$$\text{Const.} = -\frac{k_1 C_{AO}}{\left(k_2 - k_1\right)} \tag{3-140}$$

Therefore, Equation 3-139 becomes

$$C_B e^{k_2 t} = \frac{k_1 C_{AO} e^{(k_2 - k_1)t}}{k_2 - k_1} - \frac{k_1 C_{AO}}{k_2 - k_1} \tag{3-141}$$

and

$$C_B = \frac{k_1 C_{AO} e^{-k_1 t}}{k_2 - k_1} - \frac{k_1 C_{AO} e^{-k_2 t}}{k_2 - k_1} \tag{3-142}$$

The final concentration of B is:

$$C_B = \frac{k_1 C_{AO}}{k_2 - k_1}\left(e^{-k_1 t} - e^{-k_2 t}\right) \tag{3-143}$$

To obtain the maximum concentration of B, differentiate Equation 3-143 with respect to time t, which gives

$$\frac{dC_B}{dt} = \frac{k_1 C_{AO}}{k_2 - k_1} \left\{ -k_1 e^{-k_1 t} + k_2 e^{-k_2 t} \right\}$$

(3-144)

The values of k_1 and k_2 govern the location and maximum concentration of B, and this occurs at $dC_B/dt = 0$, $t = t_{max}$. Equation 3-144 becomes:

$$0 = -\frac{k_1^2 C_{AO} e^{-k_1 t}}{k_2 - k_1} + \frac{k_1 k_2 C_{AO} e^{-k_2 t}}{k_2 - k_1}$$

$$\frac{k_1^2 C_{AO} e^{-k_1 t}}{k_2 - k_1} = \frac{k_1 k_2 C_{AO} e^{-k_2 t}}{k_2 - k_1}$$

(3-145)

$$e^{(k_2 - k_1) t_{max}} = \frac{k_2}{k_1}$$

(3-146)

$$t_{max}(k_2 - k_1) = \ln \frac{k_2}{k_1}$$

The maximum concentration of B occurs at

$$t_{max} = \frac{\ln\left(\dfrac{k_2}{k_1}\right)}{k_2 - k_1}$$

(3-147)

At t_{max}

$$k_2 e^{-t_{max} k_2} = k_1 e^{-t_{max} k_1}$$

(3-148)

Substituting for $e^{-k_1 t}$ in Equation 3-147 gives

$$\frac{C_{B_{max}}}{C_{AO}} = \frac{1}{(k_2 - k_1)} \left\{ k_2 e^{-k_2 t_{max}} - k_1 e^{-k_2 t_{max}} \right\} = e^{-k_2 t_{max}}$$

(3-149)

Substituting Equation 3-147 into Equation 3-149 gives

$$\frac{C_{B_{max}}}{C_{AO}} = e^{-k_2 \left[\frac{\ln(k_2/k_1)}{k_2 - k_1} \right]}$$

(3-150)

Taking the natural logarithm gives

$$\ln \frac{C_{B_{max}}}{C_{AO}} = \left(-\frac{k_2}{k_2 - k_1} \right) \ln \left(\frac{k_2}{k_1} \right)$$

(3-151)

Treating $k_2/(k_2 - k_1)$ as an exponent and removing the natural logarithm gives

$$\frac{C_{B_{max}}}{C_{AO}} = \left(\frac{k_1}{k_2} \right)^{\left(\frac{k_2}{k_2 - k_1} \right)}$$

(3-152)

This shows that $C_{B_{max}}/C_{AO}$ depends only on k_1 and k_2, and k_2 can be evaluated from $C_{B_{max}}/C_{AO}$ at t_{max}.

From stoichiometry:

$$A \xrightarrow{k_1} B \xrightarrow{k_2} C$$

	A	**B**	**C**
Amount at time t = 0	C_{AO}	0	0
Amount at time t = t	C_A	C_B	C_C

and from stiochiometry, $(C_{AO} - C_A) = (C_B + C_C)$. That is, $C_C = (C_{AO} - C_A) - C_B$. The concentration of C in terms of C_{AO}, k_1, and k_2 are:

$$\frac{C_C}{C_{AO}} = \left\{ 1 - e^{-k_1 t} - \left(\frac{k_1}{k_2 - k_1} \right) \left(e^{-k_1 t} - e^{-k_2 t} \right) \right\}$$

$$= \left\{ 1 + \frac{k_1}{k_2 - k_1} e^{-k_2 t} - \frac{k_2}{k_2 - k_1} e^{-k_1 t} \right\}$$

(3-153)

The concentrations of components A, B, and C vary with time. The concentration profiles of A, B, and C in a batch system using the differential Equations 3-127, 3-128, and 3-129 and velocity constants k_1 and k_2, and employing the Runge-Kutta fourth order numerical method are reviewed in Chapter 5. Important features of consecutive reactions occur in substitution processes. For example $CH_4 + Cl_2 = CH_3Cl + HCl$ and $CH_3Cl + Cl_2 = CH_2Cl_2 + HCl$, and so forth. They also occur frequently in oxidation processes, where the desired product may further oxidize to give an undesired product. An example is the oxidation of methanol, where the desired formaldehyde is readily degraded to carbon dioxide: $CH_3OH \rightarrow HCHO \rightarrow CO_2$.

The formation of resins, tarry matter by consecutive reaction, is prevalent in organic reactions. Figure 3-13a shows the time variations in the concentrations of A, B, and C as given by these equations. The concentration of A falls exponentially, while B goes through a maximum. Since the formation rate of C is proportional to the concentration of B, this rate is initially zero and is a maximum when B reaches its maximum value.

Kinetic Equations 3-143 and 3-153 are obeyed by nucleides undergoing radioactive decay, where the rate constant k_1 is large and k_2 is small. The reactant A is converted rapidly into the intermediate B, which slowly forms C. Figure 3-13b shows plots of the exponentials $e^{-k_1 t}$ and $e^{-k_2 t}$ and of their difference. Since k_2 is small, the exponential $e^{-k_2 t}$ shows a slow decay while $e^{-k_1 t}$ shows a rapid decline. The difference of $e^{-k_2 t} - e^{-k_1 t}$ is shown by the dashed line in Figure 3-13b. The concentration of B is (Equation 3-143) equal to this difference multiplied by C_{AO} (since $k_1 \gg k_2$). Therefore, the concentration of B rapidly rises to the value of C_{AO} and then slowly declines. The rise in concentration C then approximately follows the simple first-order law. Conversely, when k_1 is small and k_2 is large ($k_2 \gg k_1$), the concentration of B is given by Equation 3-143:

$$C_B = \frac{k_1 C_{AO}}{k_2} \left(e^{-k_1 t} - e^{-k_2 t}\right) \tag{3-154}$$

At $t = 0$, $C_B = 0$, but after a short time, relative to the duration of the reaction, the difference $e^{-k_1 t} - e^{-k_2 t}$ reaches the value of unity. The concentration of B is then $C_{AO} k_1 / k_2$, which is much less than C_{AO}. After this short induction period, the concentration of B remains almost

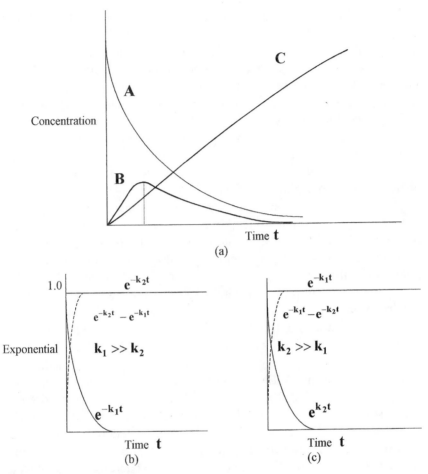

Figure 3-13. (a) Variations in the concentrations of A, B, and C for a reaction occurring by the mechanism $A \xrightarrow{k_1} B \xrightarrow{k_2} C$. (b) Variations with time of the exponentials when $k_1 \gg k_2$. (c) Variations of the exponentials when $k_2 \gg k_1$.

constant, so that generally $dC_B/dt = 0$, the basis of the steady-state treatment.

FIRST ORDER REVERSIBLE REACTIONS

Consider the reaction $A \underset{k_2}{\overset{k_1}{\rightleftharpoons}} B$. The rate equation for a constant volume batch system is:

$$\left(-r_A\right) = -\frac{dC_A}{dt} = k_1 C_A - k_2 C_B \tag{3-155}$$

From stoichiometry:

	A	B
Amount at time t = 0	C_{AO}	C_{BO}
Amount at time t = t	C_A	C_B
Amounts that have reacted	$C_{AO} - C_A$	$C_B - C_{BO}$

and the concentration of B is

$$C_B = (C_{AO} - C_A) + C_{BO} \tag{3-156}$$

Substituting Equation 3-156 into Equation 3-155, rearranging and integrating between the limits gives:

$$\int_{C_{AO}}^{C_A} -\frac{dC_A}{k_1 C_A - k_2 \{(C_{AO} - C_A) + C_{BO}\}} = \int_0^t dt \tag{3-157}$$

$$-\int_{C_{AO}}^{C_A} \frac{dC_A}{(k_1 + k_2)C_A - k_2(C_{AO} - C_{BO})} = \int_0^t dt \tag{3-158}$$

$$-\frac{1}{(k_1+k_2)} \int_{C_{AO}}^{C_A} \frac{dC_A}{C_A - \left(\dfrac{k_2}{k_1+k_2}\right)(C_{AO} - C_{BO})} = \int_0^t dt \tag{3-159}$$

where

$$\alpha = \left(\frac{k_2}{k_1+k_2}\right)(C_{AO} - C_{BO}) \tag{3-160}$$

Substituting Equation 3-160 into Equation 3-159 becomes

$$-\frac{1}{k_1+k_2} \int_{C_{AO}}^{C_A} \frac{dC_A}{C_A - \alpha} = \int_0^t dt \tag{3-161}$$

Integrating Equation 3-161 between the boundary conditions $t = 0$, $C_A = C_{AO}$ and $t = t$, $C_A = C_A$ gives

$$-\frac{1}{k_1 + k_2}\ln\left[C_A - \alpha\right]_{C_{AO}}^{C_A} = t \tag{3-162}$$

That is:

$$-\frac{1}{k_1 + k_2}\left[\ln(C_A - \alpha) - \ln(C_{AO} - \alpha)\right] = t \tag{3-163}$$

Substituting Equation 3-160 into Equation 3-163 gives

$$-\left\{\ln\left(C_A - \frac{k_2(C_{AO} - C_{BO})}{k_1 + k_2}\right)\right.$$
$$\left. - \ln\left(C_{AO} - \frac{k_2(C_{AO} - C_{BO})}{k_1 + k_2}\right)\right\} = (k_1 + k_2)t \tag{3-164}$$

$$-\left\{\ln\left[\frac{(k_1 + k_2)C_A - k_2C_{AO} + k_2C_{BO}}{(k_1 + k_2)}\right]\right.$$
$$\left. - \ln\left[\frac{k_1C_{AO} + k_2C_{AO} - k_2C_{AO} + k_2C_{BO}}{(k_1 + k_2)}\right]\right\} = (k_1 + k_2)t \tag{3-165}$$

$$-\left\{\ln\left[(k_1 + k_2)C_A - k_2(C_{AO} - C_{BO})\right]\right.$$
$$\left. - \ln\left[k_1C_{AO} + k_2C_{BO}\right]\right\} = (k_1 + k_2)t \tag{3-166}$$

or

$$-\ln\left\{\frac{\left[\left(\frac{k_1}{k_2} + 1\right)C_A - (C_{AO} - C_{BO})\right]}{\left(\frac{k_1}{k_2}C_{AO} + C_{BO}\right)}\right\} = (k_1 + k_2)t \tag{3-167}$$

At equilibrium with concentrations of A and B as C_{Ae}, C_{Be}, respectively, then $k_1C_{Ae} = k_2C_{Be}$. Since $dC_A/dt = 0$ at equilibrium

$$\frac{k_1}{k_2} = K = \frac{C_{Be}}{C_{Ae}} \qquad (3\text{-}168)$$

Substituting Equation 3-168 into Equation 3-167 gives

$$-\ln \left\{ \frac{\left(\dfrac{C_{Be}}{C_{Ae}} + 1 \right) \bullet C_A - \left(C_{AO} - C_{BO} \right)}{\left(\dfrac{C_{Be}}{C_{Ae}} \bullet C_{AO} + C_{BO} \right)} \right\} = \left(k_1 + k_2 \right) t \qquad (3\text{-}169)$$

A plot of

$$-\ln \left\{ \frac{\left(C_{Be}/C_{Ae} + 1 \right) \bullet C_A - \left(C_{AO} - C_{BO} \right)}{\left(\left(C_{Be}/C_{Ae} \right) \bullet C_{AO} + C_{BO} \right)} \right\}$$

versus time t gives a straight line with the slope $(k_1 + k_2)$. From the slope and the equilibrium constant K, k_1 and k_2 can be determined. Figure 3-14 shows a plot of the reversible first order reaction.

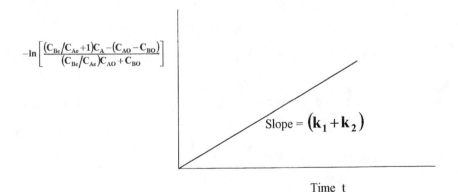

Figure 3-14. First order reversible reaction.

Examples of first-order reversible reactions are gas phase cis-trans isomerization, isomerizations in various types of hydrocarbon systems, and the racemization of α and β glucoses. An example of a catalytic reaction is the ortho-para hydrogen conversion on a nickel catalyst.

SECOND ORDER REVERSIBLE REACTIONS

Bimolecular-type second order reactions are:

$$A + B \underset{k_2}{\overset{k_1}{\rightleftharpoons}} C + D \tag{3-170}$$

$$2\,A \underset{k_2}{\overset{k_1}{\rightleftharpoons}} C + D \tag{3-171}$$

$$2\,A \underset{k_2}{\overset{k_1}{\rightleftharpoons}} 2C \tag{3-172}$$

$$A + B \underset{k_2}{\overset{k_1}{\rightleftharpoons}} 2C \tag{3-173}$$

With the restrictions that $C_{AO} = C_{BO}$ and $C_{CO} = C_{DO} = 0$ in Equation 3-170, the rate equation for a constant volume batch system becomes

$$\left(-r_A\right) = -\frac{dC_A}{dt} = C_{AO}\frac{dX_A}{dt} = k_1 C_A^2 - k_2 C_C^2 \tag{3-174}$$

$$= k_1 C_{AO}^2 \left(1 - X_A\right)^2 - \frac{k_2}{k_1} X_A^2 \tag{3-175}$$

where the fractional conversion $X_A = 0$ at $t = 0$. At equilibrium, $dX_A/dt = 0$ and the final solution is:

$$\ln\left\{\frac{X_{Ae} - \left(2X_{Ae} - 1\right)X_A}{X_{Ae} - X_A}\right\} = 2k_1 C_{AO}\left(\frac{1}{X_{Ae}} - 1\right)t \tag{3-176}$$

All other reversible second-order rate equations have the same solution with the boundary conditions assumed in Equation 3-176. Table 3-4 gives solutions for some reversible reactions.

GENERAL REVERSIBLE REACTIONS

Integrating the rate equation is often difficult for orders greater than 1 or 2. Therefore, the differential method of analysis is used to search the form of the rate equation. If a complex equation of the type below fits the data, the rate equation is:

Table 3-4
Rate equations for opposing reactions

Stoichiometric equation	Rate equation[a]	
	Differential form	**Integrated form**
$A \rightleftharpoons Z$	$\dfrac{dx}{dt} = k_1(a_0 - x) - k_{-1}x$	
$A \rightleftharpoons Z$	$\dfrac{dx}{dt} = k_1(a_0 - x) - k_{-1}(x + x_0)$	$\dfrac{x_e}{a_0}\ln\left(\dfrac{x_e}{x_e - x}\right) = k_1 t$
$2A \rightleftharpoons Z$	$\dfrac{dx}{dt} = k_1(a_0 - x) - k_{-1}\left(\dfrac{x}{2}\right)$	
$A \rightleftharpoons 2Z$	$\dfrac{dx}{dt} = k_1\left(a_0 - \dfrac{x}{2}\right) - k_{-1}x$	
$A \rightleftharpoons Y + Z$	$\dfrac{dx}{dt} = k_1(a_0 - x) - k_{-1}x^2$	$\dfrac{x_e}{2a_0 - x_e}\ln\left[\dfrac{a_0 x_e + x(a_0 - x_e)}{a_0(x_e - x)}\right] = k_1 t$
$A + B \rightleftharpoons Z$	$\dfrac{dx}{dt} = k_1(a_0 - x)^2 - k_{-1}x$	$\dfrac{x_e}{a_0^2 - x_e^2}\ln\left[\dfrac{x_e(a_0^2 - xx_e)}{a_0^2(x_e - x)}\right] = k_1 t$
$A + B \rightleftharpoons Y + Z$	$\dfrac{dx}{dt} = k_1(a_0 - x)^2 - k_{-1}x^2$	$\dfrac{x_e}{2a_0(a_0 - x_e)}\ln\left[\dfrac{x(a_0 - 2x_e) + a_0 x_e}{a_0(x_e - x)}\right] = k_1 t$
$2A \rightleftharpoons Y + Z$	$\dfrac{dx}{dt} = k_1(a_0 - x)^2 - k_1\left(\dfrac{x}{2}\right)^2$	

[a] *In all cases x is the amount of A consumed per unit volume. The concentration of B is taken to be the same as that of A.*

(Source: Laidler, K. J., 1987. Chemical Kinetics, 3rd ed., Harper Collins Publishers.)

$$(-r_A) = -\frac{dC_A}{dt} = k_1 \frac{C_A}{(1 + k_2 C_A)} \tag{3-177}$$

Taking the reciprocals of Equation 3-177 gives

$$\frac{1}{(-r_A)} = \frac{1 + k_2 C_A}{k_1 C_A} = \frac{1}{k_1 C_A} + \frac{k_2}{k_1} \tag{3-178}$$

Plotting $1/(-r_A)$ versus $1/C_A$ would give a straight line with the slope equal to $1/k_1$ and an intercept of k_2/k_1. Another analysis method involves multiplying each side of Equation 3-178 by k_1/k_2 and solving for $(-r_A)$ to give

$$(-r_A) = \frac{k_1}{k_2} - \frac{(-r_A)}{k_2 C_A} \tag{3-179}$$

Plotting $(-r_A)$ versus $(-r_A)/C_A$ gives a straight line with the slope equal to $-1/k_2$ and the intercept equal to k_1/k_2.

SIMULTANEOUS IRREVERSIBLE SIDE REACTION

Consider the chemical reactions

$$A \xrightarrow{k_1} B$$

$$k_2 \downarrow$$

$$C$$

in which the reactions are first order, the initial concentration of A is C_{AO}, and both the concentrations of B and C are zero. The rate equations representing the reactions for a constant volume batch system are:

For component A $\quad (-r_A) = -\dfrac{dC_A}{dt} = (k_1 + k_2)C_A \tag{3-180}$

For component B $\quad (+r_B) = \dfrac{dC_B}{dt} = k_1 C_A \tag{3-181}$

For component C $\left(+r_C\right) = \dfrac{dC_C}{dt} = k_2 C_A$ (3-182)

Rearranging and integrating Equation 3-180 between the limits gives

$$-\int_{C_{AO}}^{C_A} \frac{dC_A}{C_A} = \int_0^t (k_1 + k_2) dt$$ (3-183)

$$-\ln \frac{C_A}{C_{AO}} = (k_1 + k_2) t$$ (3-184)

The solution for the concentration of A is

$$C_A = C_{AO} e^{-(k_1 + k_2)t}$$ (3-185)

Substituting Equation 3-185 into Equation 3-181, rearranging and integrating gives

$$\int_0^{C_B} \frac{dC_B}{C_{AO} e^{-(k_1 + k_2)t}} = k_1 \int_0^t dt$$ (3-186)

Further rearrangement of Equation 3-186 gives

$$\frac{1}{C_{AO}} \int_0^{C_B} dC_B = k_1 \int_0^t e^{-(k_1 + k_2)t} + \text{Constant}$$ (3-187)

$$\frac{C_B}{C_{AO}} = -\frac{k_1}{(k_1 + k_2)} e^{-(k_1 + k_2)t} + \text{Constant}$$ (3-188)

At t = 0, $C_B = 0$

$$\text{Constant} = \frac{k_1}{(k_1 + k_2)}$$

The solution for concentration of B is

$$C_B = \frac{C_{AO}\, k_1}{(k_1 + k_2)}\left\{1 - e^{-(k_1 + k_2)t}\right\} \tag{3-189}$$

Correspondingly, the solution for the concentration of C is

$$C_C = \frac{C_{AO}\, k_2}{(k_1 + k_2)}\left\{1 - e^{-(k_1 + k_2)t}\right\} \tag{3-190}$$

Concentration profiles can be developed with time using the differential Equations 3-180, 3-181, and 3-182, respectively, with the Runge-Kutta fourth order method at known values of k_1 and k_2 for a batch system.

PSEUDO-ORDER REACTION

The results of the types of reaction being considered show that the treatment of kinetic data becomes rapidly more complex as the reaction order increases. In cases where the reaction conditions are such that the concentrations of one or more of the species occurring in the rate equation remain constant, these terms may be included in the rate constant k. The reactions can be attributed to lower order reactions. These types of reactions can be defined as pseudo-nth order, where n is the sum of the exponents of those concentrations that change during the reaction. An example of this type of reaction is in catalytic reaction, where the catalyst concentration remains constant during the reactions.

A kinetic study of a reaction can be simplified by running the reaction with one or more of the components in large excess, so that the concentration remains effectively constant. Mc Tigue and Sime [2] consider the oxidation of aliphatic aldehydes such that ethanal with bromine in aqueous solution follows second-order kinetics:

$$Br_2 + CH_3CHO + H_2O \rightarrow CH_3COOH + 2H^+ + 2Br^- \tag{3-191}$$

The rate equation of the reaction for a constant volume batch system is

$$-\frac{dC_{Br_2}}{dt} = k_2 C_{CH_3CHO} C_{Br_2} \tag{3-192}$$

If there is excessively large aldehyde, its concentration remains constant and the reaction can be referred to as pseudo-first order kinetics. Equation 3-192 can be rewritten as

$$-\frac{dC_{Br_2}}{dt} = k_1 C_{Br_2} \qquad (3\text{-}193)$$

where

$$k_1 = k_2 C_{CH_3CHO} \qquad (3\text{-}194)$$

The second order rate constant k_2 is $k_2 = k_1/C_{CH_3CHO}$, and can be determined by measuring k_1 at known aldehyde concentrations. The use of excess reagent concentrations has important practical consequences where physical methods, such as light absorption or conductance, are used to monitor reactions.

PRACTICAL MEASUREMENTS OF REACTION RATES

Many techniques have been employed in kinetic studies to determine reaction rate constants and reaction orders from either reactant or product concentrations at known times. The most desirable analytical methods allow continuous and rapid measurement of the concentration of a particular component. Any of the methods used to monitor the course of reaction must satisfy the following criteria:

- The method should not interfere with the system by affecting the kinetic processes occurring during the investigation.
- It must give an exact measure of the extent of the reaction.
- The measurement should be representative of the system at the time it is made or at the time the sample analyzed was taken.

Analyses of kinetic data are based on identifying the constants of a rate equation involving the law of mass action and some transfer phenomena. The law of mass action is expressed in terms of concentrations of the species. Therefore, the chemical composition is required as a function of time. Laboratory techniques are used to determine the chemical composition using an instrument that is suitably calibrated to give the required data. The techniques used are classified into two categories, namely chemical and physical methods.

Chemical methods involve removing a portion of the reacting system, quenching of the reaction, inhibition of the reaction that occurs within the sample, and direct determination of concentration using standard analytical techniques—a spectroscopic method. These methods provide absolute values of the concentration of the various species that are present in the reaction mixture. However, it is difficult to automate chemical methods, as the sampling procedure does not provide a continuous record of the reaction progress. They are also not applicable to very fast reaction techniques.

Physical methods involve measuring a physical property of the system as the reaction progresses. It is often possible to obtain a continuous record of the values of the property being measured, which can be transformed into a continuous record of reactant and product concentrations. Examples of physical methods that vary linearly with concentrations include conductance (ionic reagents), absorption of visible or ultraviolet light, optical density, the total pressure of gaseous systems under nearly ideal conditions, and the rotations of polarized light. An essential feature of physical methods involves continuous, rapid response measurement without the need for sampling.

Physical techniques can be used to investigate first order reactions because the absolute concentrations of the reactants or products are not required. Dixon et. al [3] studied the base hydrolysis of cobalt complex, $[Co(NH_3)_5L]^{3+}$, where L = $(CH_3)_2SO$, $(NH_2)_2C = O$, $(CH_3)O_3P = O$ in glycine buffers.

$$\left[Co(NH_3)_5L\right]^{3+} \xrightarrow{\ OH^-\ } \left[Co(NH_3)_5OH\right]^{2+} + L \qquad (3\text{-}195)$$

The rate equation is:

$$-\frac{d\left[Co(NH_3)_5L\right]^{3+}}{dt} = k\left[Co(NH_3)_5L\right]^{3+} \qquad (3\text{-}196)$$

The release of dimethyl sulfoxide is accompanied by an increase in absorbance (D) at 325 nm. The absorbance D is defined as $\log(I_0/I)$, where I_0 and I are the intensities of the incident and transmitted light, respectively. Figure 3-15 illustrates the relationship between concentration and absorbance changes for the hydrolysis of the cobalt

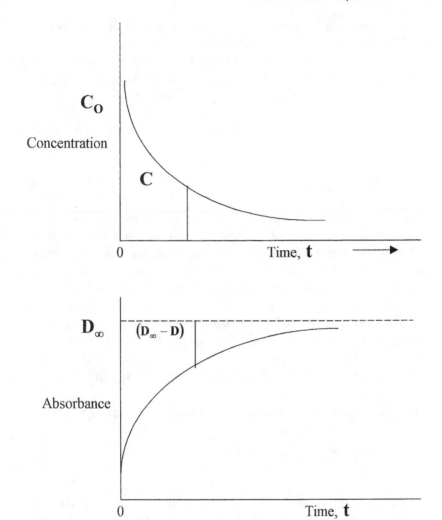

Figure 3-15. Relationship between concentration and absorbance changes for the hydrolysis of $[Co(NH_3)_5L]^{3+}$.

complex. The values of the absorbance during the reaction are related to the concentration of the cobalt complex and is represented by C as follows:

$$C_O \propto (D_\infty - D_O); \qquad C_O = \text{Const.} (D_\infty - D_O)$$
$$C \propto (D_\infty - D); \qquad C = \text{Const.} (D_\infty - D) \qquad (3\text{-}197)$$

where Const. is a proportionality constant related to the absorption coefficients of the reactants and products at 325 nm.

Substituting Equation 3-197 into the integrated first order reaction Equation 3-33 gives the corresponding equations expressed in terms of the solution absorbance:

$$\ln (D_\infty - D) = -kt + \ln (D_\infty - D_O) \tag{3-198}$$

or

$$D_\infty - D = (D_\infty - D_O)e^{-kt} \tag{3-199}$$

Equation 3-199 infers that the absorbance approaches the value at the end of the reaction (infinity value) with the same rate constant k as that for the reaction expressed in terms of the reactant concentration. The required rate constant can be determined from the slope of a plot of $\ln (D_\infty - D_O)$ versus time. The same equations can be written for reactions monitored in terms of optical rotation or conductance.

In the case of a second order reaction, an example is the alkaline hydrolysis of an ester as represented by the following equation:

$$CH_3COOC_2H_5 + OH^- \xrightarrow{\ k_2\ } CH_3COOH + C_2H_5OH \tag{3-200}$$

The rate equation for a constant volume batch system is

$$-\frac{dC_{CH_3COOC_2H_5}}{dt} = k_2 C_{CH_3COOC_2H_5} C_{OH^-} \tag{3-201}$$

Substituting Equation 3-197 into Equation 3-53 gives the following equation for the change in absorbance at the selected wavelength.

$$\frac{1}{(D_\infty - D)} - \frac{1}{(D_\infty - D_O)} = Const.\, k_2 t \tag{3-202}$$

A plot of $1/(D_\infty - D)$ versus time t is linear with the slope = Const.k_2. The rate constant k_2 is determined if the proportionality constant Const. is known, between the absorbance change and the extent of the reaction. The proportionality constant can either be determined by calibrating the system or, more accurately, by studying the reaction under pseudo-first order conditions.

Another useful technique in kinetic studies is the measurement of the total pressure in an isothermal constant volume system. This method is employed to follow the course of homogeneous gas phase reactions that involve a change in the total number of gaseous molecules present in the reaction system. An example is the hydrogenation of an alkene over a catalyst (e.g., platinum, palladium, or nickel catalyst) to yield an alkane:

$$C_nH_{2n} + H_2 \xrightarrow{\text{Pt, Pd, or Ni catalyst}} C_nH_{2n+2} \qquad (3\text{-}203)$$

Nickel is the least active of these catalysts and requires an elevated temperature and pressure, whereas platinum and palladium function adequately at ordinary temperatures and pressures. An example is butylene to butane:

$$C_4H_8 + H_2 \rightarrow C_4H_{10} \qquad (3\text{-}204)$$

In gaseous reactions, the composition term in the rate equation is often expressed as the partial pressure of the reacting species. These pressures are then transformed to concentration.

Consider the reaction $A \rightarrow$ products. The rate equation is:

$$\left(-r_A\right) = k_p p_A^n = kC_A^n \qquad (3\text{-}205)$$

where

$$\left(-r_A\right) = \frac{\text{mol}}{\text{m}^3 \cdot \text{s}} \qquad \left(\frac{\text{mol}}{\text{l} \cdot \text{s}}\right)$$

$$k_p = \frac{\text{mol}}{\text{m}^3 \cdot \text{Pa}^n \cdot \text{s}} \qquad \left(\frac{\text{mol}}{\text{l} \cdot \text{atm}^n \cdot \text{s}}\right)$$

$$p_A^n = \text{Pa}^n \qquad \left(\text{atm}^n\right)$$

$$k = \left(\frac{\text{mol}}{\text{m}^3}\right)^{1-n} \text{s}^{-1} \qquad \left(\frac{\text{mol}}{\text{l} \cdot \text{s}}\right)^{1-n} \text{s}^{-1}$$

For ideal gases, the partial pressure is expressed as:

$$p_A = C_A RT \tag{3-206}$$

where

$$p_A = \text{Pascal, Pa} \qquad (\text{atm})$$

$$C_A = \frac{\text{mol}}{\text{m}^3} \qquad \left(\frac{\text{mol}}{1}\right)$$

$$R = 8.314 \frac{\text{m}^3 \text{Pa}}{\text{mol} \cdot \text{K}} \quad \text{or} \quad 0.082 \cdot \frac{1 \cdot \text{atm}}{\text{mol} \cdot \text{K}}$$

Caution must be exercised when the rate equations with gas phase reactions are expressed in terms of the partial pressure as compared to concentration. This is because each rate gives different activation energy for the same data and for the same reaction. Levenspiel [4] suggests that the difference can be ignored for reactions with reasonably high activation energies as the amount is only a few kJ.

Pressure measurement devices such as a manometer are used without disturbing the system being monitored. Another type of reacting system that can be monitored involves one of the products being quantitatively removed by a solid or liquid reagent that does not affect the reaction. An example is the removal of an acid formed by reactions in the gas phase using hydroxide solutions. From the reaction stoichiometry and measurements of the total pressure as a function of time, it is possible to determine the extent of the reaction and the partial pressure or concentrations of the reactant and product species at the time of measurement.

Consider the following gaseous reaction $aA + bB \rightarrow cC + dD$. Pressure and concentration are related, and for a constant volume reactor with changing number of moles during reaction, the total pressure (π) changes with time, t. For an ideal gas, with any reactant A or B, the partial pressure is expressed as:

$$p_A = C_A RT = p_{AO} + \frac{a}{\Delta n}(\pi_O - \pi) \tag{3-207}$$

where

$$\Delta n = (c + d) - (a + b) \tag{3-208}$$

For any product C or D, the partial pressure is expressed as:

$$p_C = C_C RT = p_{CO} - \frac{c}{\Delta n}\left(\pi_O - \pi\right) \tag{3-209}$$

For an ideal gas at both constant temperature and pressure, but with changing number of moles during reaction, Levenspiel relates that

$$V = V_O(1 + \varepsilon_A X_A) \tag{3-210}$$

where

$$\varepsilon_A = \frac{V_{\text{all A reacted}} - V_{\text{no reaction}}}{V_{\text{no reaction}}} = \frac{V_{X_A=1} - V_{X_A=0}}{V_{X_A=0}} \tag{3-211}$$

Levenspiel considers the cases where the relationship between concentration and conversion of reacting specie is not obvious, but depends on a number of factors.

CASE 1: CONSTANT DENSITY SYSTEMS

This case includes most liquid reactions and also those gas reactions that operate at both constant temperature and pressure with no change in the number of moles during reaction. The relationship between concentration C_A and fractional conversion X_A is as follows:

$$X_A = 1 - \frac{C_A}{C_{AO}} \quad \text{and} \quad dX_A = -\frac{dC_A}{C_{AO}}$$

$$\frac{C_A}{C_{AO}} = 1 - X_A \quad \text{and} \quad dC_A = -C_{AO}dX_A \tag{3-212}$$

where $\varepsilon_A = 0$. The changes in B and C to A are

$$\frac{C_{AO} - C_A}{a} = \frac{C_{BO} - C_B}{b} = \frac{C_C - C_{CO}}{c} \tag{3-213}$$

or

$$\frac{C_{AO} \, X_A}{a} = \frac{C_{BO} X_B}{b} \tag{3-214}$$

CASE 2: CHANGING DENSITY GASES

This case involves constant temperature T and total pressure π. In this case, the density changes since the number of moles change during the reaction, and the volume of a fluid element changes linearly with conversion or $V = V_O(1 + \varepsilon_A X_A)$. The relationship between C_A and X_A is as follows:

$$X_A = \frac{C_{AO} - C_A}{C_{AO} + \varepsilon_A C_A} \quad \text{and} \quad dX_A = \frac{-C_{AO}(1 + \varepsilon_A) dC_A}{(1 + \varepsilon_A C_A)^2}$$

$$\frac{y_A}{y_{AO}} = \frac{C_A}{C_{AO}} = \frac{1 - X_A}{1 + \varepsilon_A X_A} \quad \text{and} \quad \frac{dC_A}{C_{AO}} = \frac{-1 + \varepsilon_A}{(1 + \varepsilon_A X_A)^2} dX_A \tag{3-215}$$

for $\varepsilon_A \neq 0$, where y_A = mole fraction of component A. The changes between the reactants are:

$$\varepsilon_A X_A = \varepsilon_B X_B \tag{3-216}$$

$$\frac{a\varepsilon_A}{C_{AO}} = \frac{b\varepsilon_B}{C_{BO}} \tag{3-217}$$

and for the products and inerts:

$$\frac{y_C}{y_{AO}} = \frac{C_C}{C_{AO}} = \frac{(c/a) X_A + C_{CO}/C_{AO}}{1 + \varepsilon_A X_A} \tag{3-218}$$

$$\frac{y_I}{y_{IO}} = \frac{C_I}{C_{IO}} = \frac{1}{1 + \varepsilon_A X_A} \tag{3-219}$$

CASE 3: GASES WITH VARYING DENSITY, TEMPERATURE, AND TOTAL PRESSURE

Consider the following reaction: $aA + bB \rightarrow cC$; $a + b \neq c$. For an ideal gas behavior with reactant A as the key component, the relationship between concentration C_A, C_{AB}, C_C, and X_A are as follows:

$$X_A = \frac{1 - \dfrac{C_A}{C_{AO}}\left(\dfrac{T\pi_O}{T_O\pi}\right)}{1 + \varepsilon_A \dfrac{C_A}{C_{AO}}\left(\dfrac{T\pi_O}{T_O\pi}\right)} \quad \text{or} \quad \frac{C_A}{C_{AO}} = \frac{1 - X_A}{1 + \varepsilon_A X_A}\left(\frac{T_O\pi}{T\pi_O}\right) \tag{3-220}$$

$$X_A = \frac{\dfrac{C_{BO}}{C_{AO}} - \dfrac{C_B}{C_{AO}}\left(\dfrac{T\pi_O}{T\pi}\right)}{\dfrac{b}{a} + \varepsilon_A \dfrac{C_B}{C_{AO}}\left(\dfrac{T\pi_O}{T_O\pi}\right)} \quad \text{or} \quad \frac{C_B}{C_{AO}} = \frac{\dfrac{C_{BO}}{C_{AO}} - \dfrac{b}{a}X_A}{1 + \varepsilon_A X_A}\left(\frac{T_O\pi}{T\pi_O}\right) \tag{3-221}$$

$$X_A = \frac{\dfrac{C_C}{C_{AO}}\left(\dfrac{T\pi_O}{T_O\pi}\right) - \dfrac{C_{CO}}{C_{AO}}}{\dfrac{c}{a} - \varepsilon_A \dfrac{C_C}{C_{AO}}\left(\dfrac{T\pi_O}{T_O\pi}\right)} \quad \text{or} \quad \frac{C_C}{C_{AO}} = \frac{\dfrac{C_{CO}}{C_{AO}} + \dfrac{c}{a}X_A}{1 + \varepsilon_A X_A}\left(\frac{T_O\pi}{T\pi_O}\right) \tag{3-222}$$

where ε_A is evaluated from stoichiometry at constant temperature T and total pressure π.

For a high-pressure non-ideal gas behavior, the term $(T_O\pi/T\pi_O)$ is replaced by $(Z_O T_O\pi/ZT\pi_O)$, where Z is the compressiblity factor. To change to another key reactant B, then

$$\frac{a\varepsilon_A}{C_{AO}} = \frac{b\varepsilon_B}{C_{BO}} \quad \text{and} \quad \frac{C_{AO}X_A}{a} = \frac{C_{BO}X_B}{b} \tag{3-223}$$

For liquids or isothermal gases with no change in pressure and density, $\varepsilon_A \to 0$ and $(T_O\pi/T\pi_O) \to 1$. Other forms of physical methods include optical measurements that can be used to monitor the course of various reactions such as colorimetry, fluorescence, optical rotation, and refractive indices. Various spectroscopic techniques have been employed in kinetic studies. The absorption of a reacting system of electromagnetic radiation (light, microwaves) is a designated property of the system composition and dimensions. Among the various forms of spectroscopic methods that can be used in kinetic studies are nuclear magnetic resonance, electron spin resonance spectroscopy, visible ultra-violet, and infrared. Figure 3-16 shows a flowchart that will help to

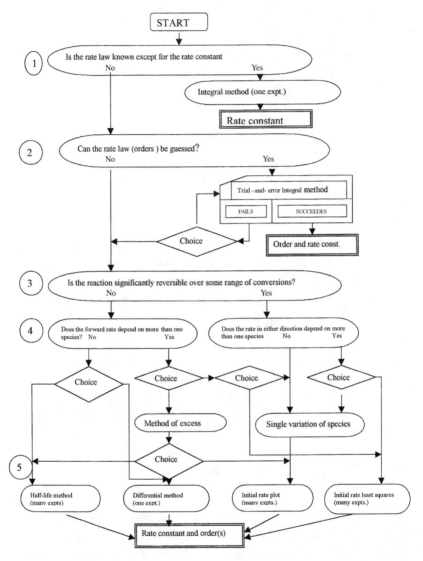

Figure 3-16. Interpretation of rate data. *(Source: Curl, R. L., 1968. Unpublished notes, University of Michigan.)*

determine the rate equation for isothermal batch reaction systems in which only one chemical reaction occurs.

Level 1 in the chart indicates that when the reaction rate law is known, the integral method of analysis may be used after performing an experiment to evaluate the specific reaction rate. This procedure is useful when the reaction order and specific rate at one temperature are already known from previous experiments, and the specific reaction rate at some other temperature is sought. If the reaction rate constant k at two or more different temperatures is known, it is possible to determine the activation energy and the frequency factor.

Level 2 involves the trial-and-error method. After trying zero, first, second, and finally third order, if none of these orders fit the data, another analysis method should be tried. When the reaction order is unknown, cannot be guessed, and the reaction is irreversible, proceed to level 4. If the reaction rate depends on only one species (e.g., isomerization or decomposition reactions), use the differential method that requires only one experiment, or the method of half-lives that requires many experiments, to determine the specific reaction rate and order. However, where certain constraints imposed by a given reaction prevent collating experimental data other than in the initial rate period, neither the differential nor the half-life method may be suitable. Examples include a solid-liquid reactions, where flaking or crumbling of the solid occurs, and certain autocatalytic and simultaneous reactions. In such instances, it is usually best to use the initial rate plot method given in level 5. In this technique, the initial rate is measured at various initial concentrations of the reacting species. The reaction order can be determined from a plot of the logarithm of the initial reaction rate, $\ln (-r_{AO})$, against the logarithm of the initial concentration, $\ln C_{AO}$. The two methods of analysis that require only one experiment to determine the reaction order and the specific rate constant are the integral method (level 2) and the differential method.

If the reaction rate depends on more than one species, use the method of excess coupled either with the half-life method or the differential method. If the method of excess is not suitable, an initial rate plot may be constructed by varying the concentration of one reactant while the concentrations of the others are held constant. This process is repeated until the orders of reaction of each species and the specific reaction rate are evaluated. At level 5, the least-squares analysis can be employed.

Figure 3-16 is helpful in the logical planning of a series of kinetic experiments to determine reaction orders and specific rate constants. However, it is important to remember the main goals and design of the entire experimental analysis. Table 3-5 gives methods used to determine direct or indirect measurements of a species concentration.

(text continued on page 168)

Table 3-5

Methods of analysis	Applications	References
1. Physical methods		
a. Total pressure change	Gas phase reactions performed under constant volume in which there is a change in the total number of moles in the gas phase.	J. Chem. Ed., 46 (1969), 684, A. P. Frost and R. G. Pearson, Kinetics and Mechanism, 2nd ed. (New York: John Wiley and Sons, 1968), Chap. 3.
b. Temperature change	Endothermic or exothermic reactions that are performed adiabatically.	Chem. Engr. Sci., 21 (1966), 397.
c. Volume change	(1) Gas phase reactions performed under constant pressure and involving a change in the total number of moles.	O. Levenspiel, Chemical Reaction Engineering, 2nd ed. p. 91, (New York: John Wiley and Sons, 1972).
	(2) Solid or liquid phase reactions in which there is a density change on reaction.	J. Am. Chem. Soc., 70 (1948), 639. I.E.C. Process Design & Development, 8, No. 1, (1969), 120.
2. Optical methods		
a. Absorption spectrometry	Transitions within molecules, which can be studies by the selective absorption of electro-magnetic radiation. Transitions between electronic levels are found in the UV and visible regions; those between vibrational levels, within the same electronic level are in the IR region	
(1) Visible	Reactions involving one (or at most two) colored compound(s).	J. Chem. Ed. 49, No. 8 (1972), 539. I.E.C. Process Design & Development, 8, No. 1, (1969), 120.
(2) Ultraviolet	The determination of organic compounds, expecially aromatic and heterocyclic substances or compounds with conjugated bonds.	

(3) Infrared	Organic compounds only.	I.E.C. Product Research and Development, 7, No. 1, (1968), 12.
(4) Atomic absorption	Metallic ions.	K. Lund, H. S. Fogler and C. C. McCune, Chem. Engrs. Sci., 28, (1973), 691.
b. Polarimetry	Liquid phase reactions involving optically active species.	Nature 175 (London: 1955), 593.
c. Refractometry	Reaction in which there is a measurable difference between the refractive index of the reactants and that of the products.	

3. Electrochemical methods		
a. Potentiometry	Used in the measurement of the potentials of nonpolarized electrodes under conditions of zero current. Seldom used in organic reactions.	J. Am. Chem. Soc., 68 (1946), Ibid, 69, (1947), 1325.
b. Voltammetry and polarography	Used for dilute electrolytic solutions. Not applicable to reactions that are catalyzed by mercury.	J. Am. Chem. Soc., 71, (1949), 3731.
c. Conductimetry	Reactions involving a change in the number of kind of ions present, thereby changing the electrical conductivity. Suitable for both organic and inorganic reactions.	J. Chem. Soc., 97, (London: 1910), 732 Frost & Pearson, Kinetics and Mechanism, Chap. 3.

4. Nuclear methods		
a. Magnetic resonance spectrometry		
(1) Nuclear magnetic resonance	Used primarily for compounds containing hydrogen.	J. Chem. Ed., 49, No. 8, (1972), 560. Pople, Schneider & Bernstein, High-Resolution Nuclear Magnetic Resonance, (New York: McGraw-Hill, 1959).

Table 3-5
(continued)

Methods of analysis	Applications	References
(2) Electron spin resonance	Used in free radical studies and for estimation of trace amounts of paramagnetic ions.	J. Chem. Phy., 46, No. 2, (1967), 490. Ibid. 48, No. 10, (1968), 4405.
b. Mass spectrometry		Pro. Royal Soc., A199, (London: 1949), 394. I.E.C. Process Design & Development, 8, No. 4, (1969), 450, 456.
c. Nuclear radiation (radioisotopes)	Elucidation of reaction mechanisms.	Trans. Faraday Soc., 30, (1934), 508.
5. Methods of interphase separations		
a. Gas-liquid chromatography	Gas phase reactions; also liquid phase reactions involving only volatile substances.	I.E.C. Product Research and Development, 8, No. 3, (1969), 319. Ibid. 10, No. 2 (1971), 138.

Source: Fogler, H. S., 1974. The Elements of Chemical Kinetics and Reactor Calculations—A Self-Paced Approach, Prentice-Hall, Inc., Englewood Cliffs, NJ

(text continued from page 165)

Most of the references listed refer to a specific chemical kinetics experiment in which the corresponding method analysis was used to obtain the rate data.

DETERMINING REACTION RATE DATA

The principal techniques used to determine reaction rate functions from the experimental data are differential and integral methods.

Differential Method

This method is based on differentiating the concentration versus time data in order to obtain the actual rate of reaction to be tested. All the terms in the equation including the derivative (dC_i/dt) are determined, and the goodness of fit are tested with the experimental data. The following procedures are used to determine the rate constant k and the concentration dependence of the rate equation $f(C_i)$.

1. Set a hypothesis as to the form of the concentration dependent of the rate function $f(C_i)$. This can be of the form

$$(-r_A) = -\frac{dC_A}{dt} = k f(C_i)$$
(3-224)

2. From the experimental data of concentration versus time, determine the reaction rate at various times.
3. Draw a smooth curve through these data.
4. Determine the slope of this curve at selected values of the concentration. The slopes are the rates of reaction at these compositions.
5. Calculate $f(C_i)$ for each composition.
6. Prepare a plot of reaction rate $(-dC_A/dt)$ versus $f(C_i)$. If the plot is linear and passes through the origin, the rate equation is consistent with the data, otherwise another equation should be tested. Figure 3-17 shows a schematic of the differential method.

Integral Method

This method estimates the reaction order based on the reaction stoichiometry and assumptions concerning its mechanism. The assumed rate equation is then integrated to obtain a relation between the composition and time. The following procedures are used for determining the rate equations:

1. Set a hypothesis as to the mathematical form of the reaction rate function. In a constant volume system, the rate equation for the disappearance of reactant A is

$$(-r_A) = -\frac{dC_A}{dt} = k f(C_A)$$
(3-225)

2. Separate the variables and integrate Equation 3-225 to give

$$-\int_{C_{AO}}^{C_A} \frac{dC_A}{f(C_A)} = k \int_0^t dt$$
(3-226)

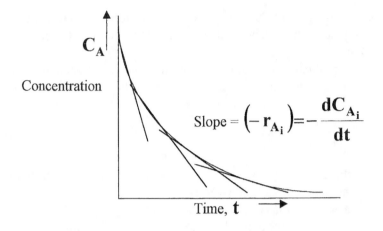

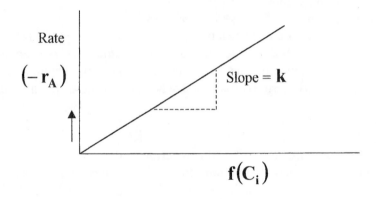

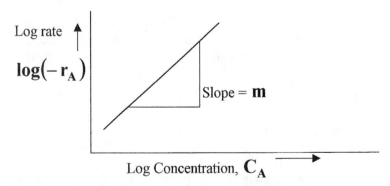

Figure 3-17. Schematics of the differential method for data analysis.

3. Plot the calculated values as shown in Figure 3-18 to give a straight line of slope k.
4. From experimentally determined values of the integral of Equation 3-226, plot these at corresponding times as shown in Figure 3-18.
5. If the data yield a satisfactory straight line passing through the origin, then the reaction rate equation (assumed in step 1) is said to be consistent with the experimental data. The slope of the line is equal to the reaction rate constant k. However, if the data do not fall on a satisfactory straight line, return to step 1 and try another rate equation.

REGRESSION ANALYSES

LINEAR REGRESSION

If the rate law depends on the concentration of more than one component, and it is not possible to use the method of one component being in excess, a linearized least squares method can be used. The purpose of regression analysis is to determine a functional relationship between the dependent variable (e.g., the reaction rate) and the various independent variables (e.g., the concentrations).

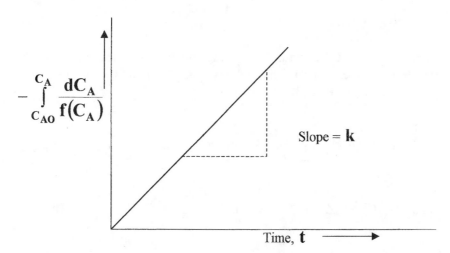

Figure 3-18. Test of reaction rate data using the integral method.

Consider a mole balance on a constant volume batch reactor represented by

$$-\frac{dC_A}{dt} = kC_A^a C_B^b \tag{3-227}$$

Using the method of initial rates gives

$$\left(-\frac{dC_A}{dt}\right)_0 = (-r_A)_0 = kC_{AO}^a C_{BO}^b \tag{3-228}$$

Taking the logarithms of both sides of Equation 3-228 gives

$$\ln\left(-\frac{dC_A}{dt}\right) = \ln k + a \ln C_{AO} + b \ln C_{BO} \tag{3-229}$$

This can be represented in the form

$$Y = C_0 + C_1 X_1 + C_2 X_2 \tag{3-230}$$

where $Y = \ln(-dC_A/dt)$, $C_O = \ln k$, $X_1 = \ln C_{AO}$, $X_2 = \ln C_{BO}$, $C_1 = a$, and $C_2 = b$. If N experimental runs are performed, then for the ith run, Equation 3-230 becomes

$$Y_i = C_0 + C_1 X_{1,i} + C_2 X_{2,i} \tag{3-231}$$

where $X_{1i} = \ln C_{AOi}$, with C_{AOi} being the initial concentration of A for the ith run. Solving for the unknowns C_O, C_1, and C_2 for N experimental runs, i = 1, 2, 3 . . . N, gives

$$\sum_{i=1}^{N} Y_i = NC_0 + C_1 \sum_{i=1}^{N} X_{1i} + C_2 \sum_{i=1}^{N} X_{2i} \tag{3-232}$$

$$\sum_{i=1}^{N} X_{1i} Y_i = C_0 \sum_{i=1}^{N} X_{1i} + C_1 \sum_{i=1}^{N} X_{1i}^2 + C_2 \sum_{i=1}^{N} X_{1i} X_{2i} \tag{3-233}$$

$$\sum_{i=1}^{N} X_{2i} Y_i = C_0 \sum_{i=1}^{N} X_{2i} + C_1 \sum_{i=1}^{N} X_{1i} X_{2i} + C_2 \sum_{i=1}^{N} X_{2i}^2 \tag{3-234}$$

Equations 3-232, 3-233, and 3-234 can be solved simultaneously to determine the unknowns C_O, C_1, and C_2, which are the rate constant k, and the orders (a and b) of the reaction respectively. This type of problem is best expedited using a computer program. Appendix A reviews the multiple regression that is being employed to determine the rate constant k and the orders (a and b) of the reaction. The computer program (PROG3) on the CD-ROM determines the rate constant k and the orders (a and b) of the reaction. Other parameters such as the reaction order, frequency factor k_O, and activation energy E_a can be determined using regression analysis.

NONLINEAR ANALYSIS

Another method for determining rate law parameters is to employ a search for those parameter values that minimize the sum of the squared difference of measured reaction rate and the calculated reaction rate. In performing N experiments, the parameter values can be determined (e.g., E_a, C_O, C_1, and C_2) that minimize the quantity:

$$\sigma^2 = \frac{s^2}{N-K} = \sum_{i=1}^{N} \frac{\left(r_{im} - r_{ic}\right)^2}{N-K-1} \qquad (3\text{-}235)$$

where σ^2 = variance
$\quad s^2 = \Sigma \left(r_{im} - r_{ic}\right)^2$
$\quad N$ = number of runs
$\quad K$ = number of parameters to be determined
$\quad r_{im}$ = measured reaction rate for run i
$\quad r_{ic}$ = calculated reaction rate for run i

Nonlinear least squares curve fitting using the Microsoft Solver is reviewed in Appendix B.

WEIGHTED LEAST SQUARES ANALYSIS

A weighted least-squares analysis is used for a better estimate of rate law parameters where the variance is not constant throughout the range of measured variables. If the error in measurement is corrected, then the relative error in the dependent variable will increase as the independent variable increases or decreases.

Consider a first order reaction with the final concentration expressed by $C_A = C_{AO}e^{-kt}$. If the error in concentration measurement is $0.01C_{AO}$, the relative error in the concentration measurement $[0.01C_{AO}/C_A(t)]$ increases with time. It is possible to minimize the sum of N measurements by:

$$\sigma^2 = \sum_{i=1}^{N} W_i \left[y_i(\exp tl) - y_i(calc) \right]^2 \qquad (3\text{-}236)$$

where W_i is the weighing factor.

The weighted least-squares analysis is important for estimating parameter involving exponents. Examples are the concentration time data for an irreversible first order reaction expressed by $C_A = C_{AO}e^{-kt}$, and the reaction rate-temperature data expressed by $(-r_A) = k_0 C_A e^{-E_a/RT}$. These equations are of the form

$$Y = Ae^{-BX} \qquad (3\text{-}237)$$

where $Y = C_A$ or $(-r_A)$ and $X = t$ or $1/T$, respectively.

Linearizing Equation 3-237 gives

$$\ln Y = \ln A - BX \qquad (3\text{-}238)$$

It is also possible to determine A and B that minimize the weighted sum of squares. The weighting function is the square of the independent variable, and the function to be minimized is

$$\sigma^2 = \sum_{i=1}^{N} y_i^2 \left[\ln y_i(\exp tl) - \ln y_i(calc) \right]^2 \qquad (3\text{-}239)$$

VALIDITY OF LEAST SQUARES FITTING

The validity of least squares model fitting is dependent on four principal assumptions concerning the random error term ε, which is inherent in the use of least squares. The assumptions as illustrated by Bacon and Downie [6] are as follows:

Assumption 1: The values of the operating variables are known exactly. In practice, this is interpreted to mean that any uncertainty associated with a value of an operating variable has much less effect on the response value than the uncertainty associated with a measured value of the response itself.

Assumption 2: The form of the model is appropriate. Statistically, this is expressed as $E(\varepsilon) = 0$ for all data, where ε is the random error.

Assumption 3: The variance of the random error term is constant over the ranges of the operating variables used to collect the data. When the variance of the random error term varies over the operating range, then either weighted least squares must be used or a transformation of the data must be made. However, this may be violated by certain transformations of the model.

Assumption 4: There is no systematic association of the random error for any one data point with the random error for any other data point. Statistically this is expressed as: Correlation $(\varepsilon_u, \varepsilon_v) = 0$. For u, v = 1, 2, . . . n, u ≠ v,

where ε_u = random error for experimental run number u

ε_v = random error for experimental run number v

PROBLEMS AND ERRORS
IN FITTING RATE MODELS

Several methods are used to fit rate models, the two most common of which often give erroneous results. The first is the transformation of a proposed rate model to achieve a model form that is linear in the parameters. An example is the nonlinear model:

$$r = kC_A^a C_B^b \tag{3-240}$$

which can be transformed into linear form by taking logarithm of both sides

$$\ln r = \ln k + a \ln C_A + b \ln C_B \tag{3-241}$$

Equation 3-241 is a linear model of the form

$$Y = C_O + C_1 X_1 + C_2 X_2 \tag{3-242}$$

where $X_1 = \ln C_A$ and $X_2 = \ln C_B$.

Another example of model transformation to achieve linearity is the change from the nonlinear rate equation

$$r = \frac{k_O k_1 C_A}{1 + k_1 C_A + k_2 C_B} \tag{3-243}$$

to the equation

$$\frac{1}{r} = \frac{1}{k_O} + \frac{1}{k_O k_1} \bullet \frac{1}{C_A} + \frac{k_2}{k_O k_1} \bullet \frac{C_B}{C_A} \qquad (3\text{-}244)$$

which is of the linear form

$$Y = C_O + C_1 X_1 + C_2 X_2 \qquad (3\text{-}245)$$

where $X_1 = 1/C_A$ and $X_2 = C_B/C_A$.

The practice of transforming a nonlinear model into linear form can result in invalid estimates of the coefficients and consequent misinterpretation of the fitted model. The problem is associated with the assumption of a constant error variance (Assumption 3). If the measured rates r have a constant error variance, then the errors associated with ln r (Equation 3-241) or 1/r (Equation 3-244) will not have a constant error variance. The degree of violation will depend on the range of the measured reaction rate values. If the data contain both very small rates and very large rates, then the error variance will change appreciably over the data set and ordinary least squares estimates of the coefficients will give poor estimates. However, if the rate values are not spread over a large range, then model transformations of Equations 3-241 and 3-244 will not significantly distort the estimated coefficients. An illustration of the effects of linearizing model transformation of Equation 3-240 is given in Example 3-1.

Generally, a model form is first proposed (i.e., a hypothesis that must be tested) using plots of the data. A simple plot can show obvious inadequacies in the model form and can suggest a better form. Alternatively, it can show excessive scatter in the data and warn against overconfidence in an adequate fit. The least squares estimates of the parameters are determined using a weighted or an optimization (i.e., trial-and-error) search procedure if required. This gives the best estimates if the assumptions are valid. Further testing the adequacy of the fitted model requires using both plots of the residuals and the sum of squares of the residuals. Finally, an estimate should be made of the precision of each parameter estimate by statistical analysis (e.g., 95% confidence intervals for the parameters).

Often, one or more model forms in chemical reaction kinetics may fit the data. Although it is tempting to want to justify a specific model as the mechanism of the reaction, it is preferable to only infer that the model could be the mechanism. It is also desirable that the reaction mechanism taking place be understood in order to solve a problem in reactor design. This is because knowledge of the mechanism will make if possible to fit the experimental data to a theoretical rate expression, which will be more reliable than an empirical fit. Also, the mechanism may require some modifications and optimization for the final design.

Example 3-1

The oxidation of $Fe(CN)_6^{4-}$ to $Fe(CN)_6^{3-}$ by peroxidisulfate, $S_2O_8^{2-}$, can be monitored spectrophotometrically by observing the increase in absorbance at 420 nm, D_{420} in a well-mixed batch system. Assume that the kinetic scheme is:

$$Fe(CN)_6^{4-} + \frac{1}{2}S_2O_8^{2-} \xrightarrow{\ k_2\ } Fe(CN)_6^{3-} + SO_4^{2-}$$

$$-\frac{d}{dt}\left[Fe(CN)_6^{4-}\right] = k_2\left[Fe(CN)_6^{4-}\right]\left[S_2O_8^{2-}\right]$$

Using pseudo-first order conditions with $[S_2O_8^{2-}] = 1.8\times10^{-2}\,M$ and $[Fe(CN)_6^{4-}] = 6.5\times10^{-4}\,M$, the following absorbances were recorded at 25°C:

t/s	0	900	1,800	2,700	3,600	4,500	∞
D_{420}	0.120	0.290	0.420	0.510	0.581	0.632	0.781

Calculate the pseudo-first order rate constant $k_1 = k_2[S_2O_8^{2-}]$ and, hence, k_2.

Solution

Table 3-6 gives $D_\infty - D$ with time t. For a first order rate law, the rate equation is expressed by

$$\ln\frac{(D_\infty - D)}{(D_\infty - D_O)} = -k_1 t \tag{3-246}$$

Equation 3-246 is further expressed by

$$\ln(D_\infty - D) = \ln(D_\infty - D_O) - k_1 t \tag{3-247}$$

Table 3-6

t(sec)	D_{420}	$D_\infty - D$
0	0.120	0.661
900	0.290	0.491
1,800	0.420	0.361
2,700	0.510	0.271
3,600	0.581	0.200
4,500	0.632	0.149
∞	0.781	0.0

that is,

$$D_\infty - D = (D_\infty - D_O)e^{-k_1 t} \qquad (3\text{-}248)$$

Equation 3-248 is of the form

$$Y = Ae^{BX} \qquad (3\text{-}249)$$

Linearizing Equation 3-249 gives

$$\ln Y = \ln A + BX \qquad (3\text{-}250)$$

The computer program PROG1 determines the constants A and B from the regression analysis. Table 3-7 gives the results of the program with the slope –B equal to the reaction rate constant k_1. Figure 3-19 shows a plot of ln $(D_\infty - D)$ against time t.

Table 3-7

t(sec)	Actual ln $(D_\infty - D)$	Estimated ln $(D_\infty - D)$
0	–0.414	–0.416
900	–1.711	–0.713
1,800	–1.019	–1.011
2,700	–1.306	–1.309
3,600	–1.609	–1.609
4,500	–1.904	–1.904

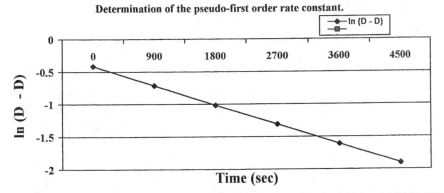

Figure 3-19. Rate constants for the oxidation of $Fe(CN)_6^{4-}$ to $Fe(CN)_6^{3-}$.

The constants for the equation $Y = A^*EXP(B^*X)$ are:

- $A = 0.660$
- $B = -33112 \times 10^{-3}$
- Correlation Coefficient = 0.99997

The results give the pseudo-first order rate constant, and $k_1 = 3.312 \times 10^{-4}$ sec^{-1} and k_2 can be calculated as follows:

$$k_2 = \frac{k_1}{\left[S_2 O_8^{2-} \right]} = \frac{3.312 \times 10^{-4}}{1.8 \times 10^{-2}} \left(\frac{1}{\text{sec}} \cdot \frac{1}{M} \right)$$

$$k_2 = 1.84 \times 10^{-2} \, M^{-1} \cdot \text{sec}^{-1}$$

Example 3-2

First order rate constant k, for the rotation about the C-N bond in N, N-dimethylnicotinamide (3) measured at different temperatures by nuclear magnetic resonance (NMR) are:

T°C	10.0	15.7	21.5	27.5	33.2	38.5	45.7
k sec^{-1}	2.08	4.57	8.24	15.8	28.4	46.1	93.5

Determine the activation energy, E, and the pre-exponential factor, k_O, for the rotation.

Solution

The activation energy E is a measure of the temperature sensitivity of the rate constant. A high E corresponds to a rate constant that increases rapidly with temperature. From the Arrhenius equation

$$k_T = k_O \exp\left(\frac{-E}{RT}\right) \tag{3-251}$$

where k_T = rate constant at known temperature
k_O = pre-exponential factor
E = activation energy
R = gas constant (8.314 J/mol • K)
T = absolute temperature (K = 273.15 + °C)

It is possible to determine the activation energy E and the pre-exponential factor k_0.
Linearizing Equation 3-251 gives

$$\ln k_T = \ln k_O - \frac{E}{RT} \tag{3-252}$$

By plotting $\ln k_T$ against 1/T, the slope equal to $-E/R$ is obtained and the intercept equal to $\ln k_O$. From these known constants, activation energy E and the pre-exponential factor k_O are determined. Equation 3-251 is of the form

Table 3-8

T(°C)	k(sec⁻¹)	T(K)	1/T
10	2.08	283.15	0.00353
15.7	4.57	288.15	0.00346
21.5	8.24	294.65	0.00339
27.5	15.8	300.65	0.00333
33.2	28.4	306.35	0.00326
38.5	46.1	311.65	0.00321
45.7	93.5	318.85	0.00314

$$Y = Ae^{BX} \qquad (3\text{-}253)$$

Linearizing Equation 3-253 gives

$$\ln Y = \ln A + BX \qquad (3\text{-}254)$$

where $\ln Y = \ln k_T$
$\ln A = \ln k_O$
$B = -E/R$
$X = 1/T$

The computer program PROG1 determines the values of constants A and B. Table 3-9 gives the results of the program and the constants A and B
The constants for the equation are:

- $A = 78728 \times 10^{15}$
- $B = -0.94869 \times 10^4$
- Correlation Coefficient = 0.99966

The slope $B = -E/R = -0.9487 \times 10^4$ and $A = k_O = 7.873 \times 10^{14}$. The activation energy $E = R \cdot B = 8.314 \times 10^4$ (J/mol) and $E = 78.9 \times 10^3$ J/mol. The pre-exponential factor is $k_O = 7.87 \times 10^{14}$ sec^{-1}. Figure 3-20 gives a plot of $\ln k_T$ against $1/T$.

Example 3-3

R. T. Dillon (1932) studied the reaction between ethylene bromide and potassium iodide in 99% methanol with the following data:

Table 3-9

1/T	Actual ln k(sec^{-1})	Estimated ln k(sec^{-1})
0.00353	0.732	0.795
0.00346	1.520	1.456
0.00339	2.109	2.103
0.00333	2.760	2.745
0.00326	3.346	3.332
0.00321	3.831	3.859
0.00314	4.538	4.546

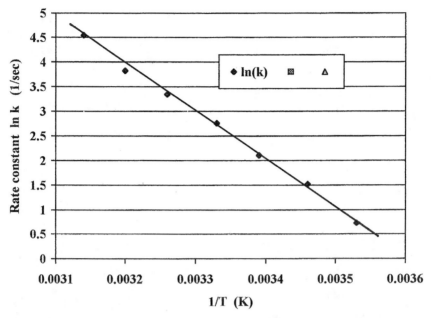

Figure 3-20. Rate constant k, for the rotation about the C-N bond in N,N-dimethylnicotamide (3).

$$C_2H_4Br_2 + 3KI \xrightarrow{\ k\ } C_2H_4 + 2KBr + KI_3$$

Temperature: 59.72°C
Initial KI concentration: 0.1531 kmol/m^3
Initial $C_2H_4Br_2$ concentration: 0.02864 kmol/m^3

Time, ksec	Fraction dibromide reacted
29.7	0.2863
40.5	0.3630
47.7	0.4099
55.8	0.4572
62.1	0.4890
72.9	0.5396
83.7	0.5795

Determine the second order reaction rate constant.

Solution

The reaction in stoichiometry can be represented as

$$A + 3B \xrightarrow{\ k\ } \text{Products} \tag{3-255}$$

Assuming that the reaction is second order in a constant volume batch system, the rate equation is

$$\left(-r_A\right) = \frac{-dC_A}{dt} = kC_A C_B \tag{3-256}$$

From stoichiometry:

	A	B
Amount at t = 0	C_{AO}	C_{BO}
Amount at t = t	C_A	C_B
Amounts that have reacted	$(C_{AO} - C_A)$	$(C_{BO} - C_B)$

and $(C_{BO} - C_B) = 3(C_{AO} - C_A)$. The fractional conversion X_A is defined as

$$X_A = \frac{C_{AO} - C_A}{C_{AO}} \tag{3-257}$$

Therefore,

$$C_A = C_{AO}(1 - X_A) \tag{3-258}$$

Differentiating Equation 3-258 with respect to time t gives

$$-\frac{dC_A}{dt} = C_{AO}\frac{dX_A}{dt} \tag{3-259}$$

The concentration of B is

$$C_B = C_{BO} - 3\left(C_{AO} - C_A\right) \tag{3-260}$$

In terms of the fractional conversion, the concentration of B is

$$C_B = C_{BO} - 3C_{AO}X_A \tag{3-261}$$

Substituting Equations 3-258 and 3-261 into Equation 3-256 gives

$$-\frac{dC_A}{dt} = k\, C_{AO}\left(1 - X_A\right)\left(C_{BO} - 3C_{AO}\, X_A\right) \qquad (3\text{-}262)$$

Substituting Equation 3-259 into Equation 3-262 and rearranging gives

$$\frac{C_{AO}\, dX_A}{C_{AO}\left(1 - X_A\right)\left(C_{BO} - 3C_{AO}\, X_A\right)} = k\, dt \qquad (3\text{-}263)$$

Integrating Equation 3-263 between limits $t = 0$, $X_A = 0$ and $t = t$, $X_A = X_A$ gives

$$\int_0^{X_A} \frac{dX_A}{\left(1 - X_A\right)\left(C_{BO} - 3C_{AO}X_A\right)} = k\int_0^t dt \qquad (3\text{-}264)$$

Equation 3-264 can be further expressed by

$$\int_0^{X_A} \frac{dX_A}{C_{AO}\left(1 - X_A\right)\left(\dfrac{C_{BO}}{C_{AO}} - 3X_A\right)} = k\int_0^t dt \qquad (3\text{-}265)$$

where $\theta_B = C_{BO}/C_{AO}$. Equation 3-265 becomes

$$\int_0^{X_A} \frac{dX_A}{C_{AO}\left(1 - X_A\right)\left(\theta_B - 3X_A\right)} = k\int_0^t dt \qquad (3\text{-}266)$$

Converting Equation 3-266 into partial fraction gives

$$\frac{1}{\left(1 - X_A\right)\left(\theta_B - 3X_A\right)} \equiv \frac{A}{1 - X_A} + \frac{B}{\theta_B - 3X_A} \qquad (3\text{-}267)$$

$$1 = A\left(\theta_B - 3X_A\right) + B\left(1 - X_A\right)$$

Equating the coefficients of the constant and X_A gives

"Const" $1 = A\theta_B + B$ (3-268)

"X_A" $0 = -3A - B$ (3-269)

Adding simultaneous Equations 3-268 and 3-269 gives $1 = A(\theta_B - 3)$. Therefore, $A = 1/(\theta_B - 3)$ and $B = -3A = -3/(\theta_B - 3)$. Equation 3-267 is now expressed as

$$\frac{1}{C_{AO}}\left\{\int_0^{X_A} \frac{dX_A}{(\theta_B - 3)(1 - X_A)} - \int_0^{X_A} \frac{3\,dX_A}{(\theta_B - 3)(\theta_B - 3X_A)}\right\} = k\int_0^t dt$$

$$= \frac{1}{C_{AO}(\theta_B - 3)}\left\{-\ln(1 - X_A) - 3\left(\frac{-1}{3}\right)\ln(\theta_B - 3X_A)\right\} = kt$$

$$= \frac{1}{C_{AO}(\theta_B - 3)}\left\{\ln\frac{(\theta_B - 3X_A)}{(1 - X_A)}\right\} = kt$$ (3-270)

Introducing the initial concentrations of A and B, that is, C_{AO} and C_{BO} into Equation 3-270 gives

$$\frac{1}{C_{AO}\left(\dfrac{C_{BO}}{C_{AO}} - 3\right)}\left\{\ln\left(\frac{C_{BO}/C_{AO} - 3X_A}{1 - X_A}\right)\right\} = kt$$ (3-271)

$$\frac{1}{(C_{BO} - 3C_{AO})}\ln\left\{\frac{C_{BO}/C_{AO} - 3X_A}{(1 - X_A)}\right\} = kt$$ (3-272)

Rearranging Equation 3-272 gives

$$\ln\left\{\frac{C_{BO}/C_{AO} - 3X_A}{1 - X_A}\right\} = kt(C_{BO} - 3C_{AO})$$ (3-273)

Equation 3-273 is of the form $Y = Ae^{BX}$ and plotting $\ln\{[C_{BO}/(C_{AO} - 3X_A)]/1 - X_A\}$ against t gives the slope $B = k(C_{BO} - 3C_{AO})$.

Table 3-10 gives the values of $\ln\{[C_{BO}/(C_{AO} - 3X_A)]/1 - X_A\}$ and t. The computer program PROG1 calculates the slope B from the equation $Y = Ae^{BX}$. From the slope, it is possible to determine the rate constant k.

The constants for the equation are:

- $A = 5.325$
- $B = 0.0056969$
- Correlation Coefficient = 0.99984

The slope $= k(C_{BO} - 3C_{AO}) = 0.0057$

$C_{AO} = 0.02864$, $C_{BO} = 0.1531$

$k(0.1531 - 3 \cdot 0.02864) = 0.0057$

$0.06718 k = 0.0057$

$$k = 0.0848 \ \frac{m^3}{kmol \ k \ sec}$$

$$= 0.0848 \frac{m^3}{kmol - k \ sec} \times 10^3 \frac{1}{m^3} \times \frac{1 \ kmol}{10^3 \ mol}$$

$$= \frac{0.0848}{10^3} \frac{1}{sec \cdot mol} \times 3,600 \frac{sec}{hr}$$

$$= 0.305 \ \frac{1}{mol - hr}$$

Dillon obtained a value for the rate constant $k = 0.300 \ 1/(mol - hr)$. Figure 3-21 shows a plot of $\ln\{[C_{BO}/(C_{AO} - 3X_A)]/1 - X_A\}$ against time t, where $Y = \ln\{[C_{BO}/(C_{AO} - 3X_A)]/1 - X_A\}$.

Table 3-10

Time, t (ksec)	X_A	$1 - X_A$	$C_{BO}/C_{AO} - 3X_A$	$\dfrac{C_{BO}/C_{AO} - 3X_A}{1 - X_A}$	$\ln\left\{\dfrac{C_{BO}/C_{AO} - 3X_A}{1 - X_A}\right\}$
0	0	1	5.3457	5.346	1.676
29.7	0.2863	0.7137	4.4868	6.287	1.838
40.5	0.3630	0.637	4.2567	6.682	1.899
47.7	0.4099	0.5901	4.116	6.975	1.942
55.8	0.4572	0.5428	3.9741	7.321	1.990
62.1	0.4890	0.511	3.8787	7.590	2.027
72.9	0.5396	0.4604	3.7269	8.095	2.091
83.7	0.5795	0.4205	3.6072	8.579	2.149

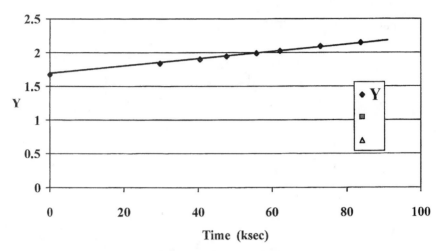

Figure 3-21. Plot for determining the rate constant for the second order reaction.

Table 3-11

Run	Initial pressure (torr) B_2H_6	(torr) Me_2CO	Initial rate $\times 10^3$ (torr/sec)
1	6.0	20.0	0.50
2	8.0	20.0	0.63
3	10.0	20.0	0.83
4	12.0	20.0	1.00
5	16.0	20.0	1.28
6	10.0	10.0	0.33
7	10.0	20.0	0.80
8	10.0	40.0	1.50
9	10.0	60.0	2.21
10	10.0	100.0	3.33

Example 3-4

Table 3-11 gives the initial rate data $[-d(B_2H_6)/dt]$ reported for the gas phase reaction of diborane and acetone at 114°C: $B_2H_6 + 4Me_2CO \rightarrow 2(Me_2CHO)_2BH$. If a rate expression is of the form Rate $= kP_{B_2H_6}^n P_{Me_2CO}^m$, determine n, m, and k.

Solution

$$\text{Rate} = kP_{B_2H_6}^n P_{Me_2CO}^m \tag{3-274}$$

Linearizing the rate expression gives

$$\ln(\text{Rate}) = \ln k + n \ln P_{B_2H_6} + m \ln P_{Me_2CO} \qquad (3\text{-}275)$$

Equation 3-275 is of the form

$$Y = C_0 + C_1 X_1 + C_2 X_2 \qquad (3\text{-}276)$$

where the coefficients $C_0 = k$, $C_1 = n$, and $C_2 = m$. The computer program PROG3 determines the coefficients from the multiple regression of the independent variables $P_{B_2H_6}$ and P_{Me_2CO} and the dependent variables Rate. The results of the program give the coefficients for Equation 3-276 as:

- $C_0 = 4.67755$
- $C_1 = 0.978166$
- $C_2 = 0.954575$
- Correlation Coefficient = 0.9930

Therefore, $C_0 = k = 4.68$, $C_1 = n = 0.98$, and $C_2 = m = 0.95$. The rate equation is expressed as $\text{Rate} = 4.68 P_{B_2H_6}^{0.98} P_{Me_2CO}^{0.95}$, or as a good approximation,

$$\text{Rate} = 4.7 P_{B_2H_6} P_{Me_2CO} \qquad (3\text{-}277)$$

The rate constant k can be determined with values of the exponents, $m = n \cong 1$ as follows:

$$\text{Rate} = k\, P_{B_2H_6}^{1.0} P_{Me_2CO}^{1.0}$$

$$k = \frac{\text{Rate}}{P_{B_2H_6}\, P_{Me_2CO}}$$

The average rate constant $k = 3.85$ and the final rate equation is $\text{Rate} = 3.85 P_{B_2H_6} P_{Me_2CO}$. The computed rate constant from PROG3 is 4.68 and the calculated value from Table 3-12 is 3.85. The percentage deviation between these values is 17.8%. Figure 3-22 shows plots of partial pressures of B_2H_6 and Me_2CO versus the initial rate.

Table 3-12

Run	Initial pressure B_2H_6	(torr) Me_2CO	Initial rate $\times 10^3$ (torr/sec)	Rate constant k
1	6.0	20.0	0.50	4.17
2	8.0	20.0	0.63	3.94
3	10.0	20.0	0.83	4.15
4	12.0	20.0	1.00	4.17
5	16.0	20.0	1.28	4.00
6	10.0	10.0	0.33	3.33
7	10.0	20.0	0.80	4.00
8	10.0	40.0	1.50	3.75
9	10.0	60.0	2.21	3.68
10	10.0	100.0	3.33	3.33

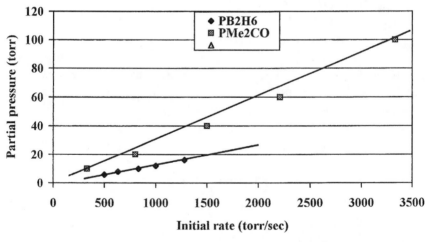

Figure 3-22. Plots of partial pressures of B_2H_6 and Me_2CO versus the initial rate.

Example 3-5

For the irreversible thermal dissociation of paraldehyde at 259°C and constant volume, the following data were obtained:

Time, hrs	0	1	2	3	4	∞
P_{Total}, mm Hg	100	175	220	250	270	300

Determine the order of the reaction and the rate constant.

Solution

Paraldehyde decomposition is represented by $(CH_3CHO)_3 \xrightarrow{k_1}$ $3CH_3CHO$. The stoichiometry of the reaction is of the form $A \xrightarrow{k_1} 3B$. Assuming that the reaction is first order, then the rate equation for a constant volume batch system is:

$$(-r_A) = -\frac{dC_A}{dt} = k_1C_A \tag{3-278}$$

If the reactant and product obey the ideal gas law, the concentration is

$$C_A = \frac{n_A}{V} = \frac{p_A}{RT} \tag{3-279}$$

In terms of the partial pressure p_A,

$$-\frac{dp_A}{dt} = k_1p_A \tag{3-280}$$

Rearranging Equation 3-280 and integrating between the boundary conditions (at $t = 0$, $p_A = p_{AO}$ and at $t = t$, $p_A = p_A$) gives:

$$\int_{p_{AO}}^{p_A} \frac{dp}{p_A} - k_1\int_0^t dt \tag{3-281}$$

Equation 3-281 gives

$$\ln\frac{p_A}{p_{AO}} = -k_1t \tag{3-282}$$

If n_A is the number of moles of paraldehyde at time t and n_{AO}, the moles of paraldehyde at $t = 0$, then at time t, moles of acetaldehyde from stoichiometry are:

	A	**B**
Amount at t = 0	n_{AO}	n_{BO}
Amount at t = t	n_A	n_B
Amounts that have reacted	$(n_{AO} - n_A)$	$n_B - n_{BO}$

and from stoichiometry $3(n_{AO}-n_A) = (n_B-n_{BO})$. Total moles at time $t = n_T = n_A + 3(n_{AO} - n_A) + n_{BO}$, although $n_{BO} = 0$. If the gas law applies, then $n = pV/RT$, and at constant V and T:

$$P_T = p_A + 3\left(p_{AO} - p_A\right)$$

$$= p_A + 3p_{AO} - 3p_A$$

$$P_T = 3p_{AO} - 2p_A$$

or

$$p_A = \frac{1}{2}\left(3p_{AO} - P_T\right)$$

where P_T = total pressure

p_A = partial pressure of A (paraldehyde) at time t

p_{AO} = partial pressure of A at time t = 0

Table 3-13 shows the relationship between the ratio of p_A/p_{AO} versus time t.

A plot of $\ln p_A/p_{AO}$ against time t (Figure 3-23) gives a straight line with the slope equal to the rate constant k_1. Therefore, the assumed first order for the reaction is correct. The relationship between the ratio of p_A/p_{AO} versus time t is represented by the model equation $Y = Ae^{BX}$.

The computer program PROG1 determines the rate constant k_1 from the slope of $Y = Ae^{BX}$. The constants for the equation are:

- A = 1.0082
- B = −0.47105
- Correlation Coefficient = 0.99972

The slope $B = k_1$. Therefore, the model equation $Y = Ae^{BX}$ is $0.471 \ hr^{-1}$.

Table 3-13

Time, hr	P_T, mmHg at 0°C	$p_A = 1/2(3p_{AO} - P_T)$	p_A/p_{AO}
0	$100 = p_{AO}$	100	1.0
1	175	62.5	0.625
2	220	40.0	0.400
3	250	25.0	0.250
4	270	15.0	0.15
∞	300	0	0.0

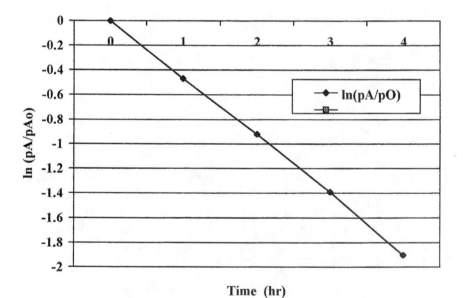

Figure 3-23. Plot of ln p_A/p_{AO} versus time t.

Example 3-6

The reaction $2NOCl \xrightarrow{\ k\ } 2NO + Cl_2$ is studied at 200°C. The concentration of NOCl initially consisting of NOCl only changes as follows:

t, sec	0	200	300	500
C_{NOCl}, gmol/l	0.02	0.016	0.0145	0.012

Determine the order of the reaction and the rate constant.

Solution

If the reaction is second order, the rate equation is:

$$\left(-r_A\right)=-\frac{dC_A}{dt}=kC_A^2 \qquad (3\text{-}283)$$

Rearranging Equation 3-283 and integrating with the boundary conditions, at t = 0, $C_A = C_{AO}$ and at t = t, $C_A = C_A$, gives

$$-\int_{C_{AO}}^{C_A}\frac{dC_A}{C_A^2}=k\int_0^t dt \qquad (3\text{-}284)$$

$$\frac{1}{C_A}-\frac{1}{C_{AO}}=kt \qquad (3\text{-}285)$$

A plot of $1/C_A$ versus t gives a straight line with the slope = k = 0.06661/gmol•s. The computer program PROG1 determines the rate constant for a second order reaction. Equation 3-285 is of the form $1/Y = A + BX$ where the slope B is the rate constant k. The results of the computer program are shown in Table 3-14.

The constants for the equation are:

- $A = 0.496 \times 10^2$
- $B = 0.0666$
- Correlation Coefficient = 0.9992

Table 3-14

Time, sec	Actual $1/C_{NOCl}$, gmol/l	Estimated $1/C_{NOCl}$, gmol/l
0	50	49.6
200	62.5	62.9
300	68.97	69.5
500	83.3	82.8

Figure 3-24 shows the relationship between $1/C_A$ as a function of time t. The graph is a straight line, therefore, the assumed order of the reaction is correct. The slope of the line from the regression analysis is the rate constant k.

The rate constant $k = 0.0666 \dfrac{1}{\text{gmol} \cdot \text{sec}}$

Example 3-7

The gas phase decomposition $A \rightarrow B + 2C$ is conducted in a constant volume reactor. Runs 1 through 5 were conducted at 100°C; run 6 was performed at 110°C (Table 3-15). Determine (1) the reaction order and the rate constant, and (2) the activation energy and frequency factor for this reaction.

Solution

The half-life for the nth order reaction is:

$$t_{1/2} = \frac{2^{n-1} - 1}{k(n-1)} \; \frac{1}{C_{AO}^{n-1}} \qquad (3\text{-}286)$$

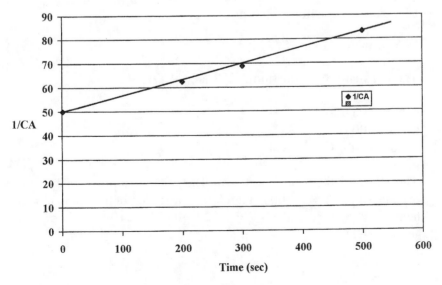

Figure 3-24. Plot of $1/C_A$ against time (sec).

Table 3-15
Half-life $t_{1/2}$ as function of initial concentration C_{AO}

Run number	C_{AO}, gmol/l	Half-life, $t_{1/2}$ min
1	0.025	4.1
2	0.0133	7.7
3	0.01	9.8
4	0.05	1.96
5	0.075	1.30
6	0.025	2.0

Taking the natural logarithm of Equation 3-286 yields

$$\ln\left(t_{1/2}\right) = \ln\frac{2^{n-1} - 1}{k(n-1)} + \left(1 - n\right)\ln C_{AO} \tag{3-287}$$

A plot of C_{AO} versus $t_{1/2}$ on log-log paper should give the slope of the line equal to $(1 - n)$. However, Equation 3-287 can also be represented in the form

$$Y = AX^B \tag{3-288}$$

Linearizing Equation 3-288 gives

$$\ln Y = \ln A + B \ln X \tag{3-289}$$

The constants for Equation 3-289 are:

- A = 0.0878
- B = 1.0032
- Correlation Coefficient = 0.97195

The slope of the line B = 1 – n. The computer program PROG1 gives the slope B = –1.003, therefore, the order of reaction n = 2. When $t_{1/2}$ = 5 min, and C_{AO} = 0.022 gmol/l at 100°C, then from Equation 3-286

$$k_{100} = \frac{2^{n-1} - 1}{(n-1)}\,\frac{1}{C_{AO}^{n-1}\, t_{1/2}}$$

$$= \frac{2-1}{1} \times \frac{1}{0.022} \times \frac{1}{5}$$

$$= 9.09 \ \text{l/gmol} \bullet \text{min}$$

At 110°C, $C_{AO} = 0.025$ gmol/l, $t_{1/2} = 2$ min, and

$$k_{110} = \frac{2-1}{1} \times \frac{1}{0.025} \times \frac{1}{2}$$

$$= 20 \ \text{l/gmol} \bullet \text{min}$$

The activation energy and frequency factor can be determined from the Arrhenius equation

$$k_T = k_O \exp\left(\frac{-E}{RT}\right) \tag{3-290}$$

where k_T = rate constant at known temperature
$\quad\quad\ k_O$ = frequency factor
$\quad\quad\ E$ = activation energy
$\quad\quad\ R$ = ideal gas constant 1.987cal/mol • K

$$k_{100} = k_O \exp\left(\frac{-E}{R(373)}\right) \tag{3-291}$$

$$k_{110} = k_O \exp\left(\frac{-E}{R(383)}\right) \tag{3-292}$$

Dividing Equation 3-291 by Equation 3-292 gives

$$\frac{k_{100}}{k_{110}} = \frac{\exp\left(-E/R(373)\right)}{\exp\left(-E/R(383)\right)} \tag{3-293}$$

Removing the exponential, Equation 3-293 can be expressed as

$$\ln \frac{k_{100}}{k_{110}} = \frac{-E}{R(373)} - \frac{-E}{R(383)} \tag{3-294}$$

$$\ln \frac{9.09}{20.0} = \frac{E}{R}\left(\frac{1}{383} - \frac{1}{373}\right)$$

$$E = \frac{-0.7886 \times 1.987}{-6.999 \times 10^{-5}}$$

$$= 22,388 \frac{cal}{gmol}$$

The frequency factor at 100°C is

$$k_O = \frac{k_{100}}{\exp\left(\dfrac{-E}{RT}\right)}$$

$$= \frac{9.09}{\exp\left(\dfrac{-22,388}{1.987 \times 373}\right)}$$

$$= 1.19 \times 10^{14} \ 1/gmol \bullet min$$

The frequency factor at 110°C is

$$k_O = \frac{k_{110}}{\exp\left(\dfrac{-E}{RT}\right)}$$

$$= \frac{20}{\exp\left(\dfrac{-22,388}{1.987 \times 383}\right)}$$

$$= 1.19 \times 10^{14} \ 1/gmol \bullet min$$

Example 3-8

The hydrolysis of methyl acetate is an autocatalytic reaction and is first order with respect to both methyl acetate and acetic acid. The reaction is elementary, bimolecular and can be considered irreversible at constant volume for design purposes. The following data are given:

- Initial concentration of methyl acetate = 0.45 gmol/l
- Initial concentration of acetic acid = 0.045 gmol/l

The conversion in 1 hr is 65% in a batch reactor. Calculate (1) the rate constant and specify the rate equation, (2) the time at which the rate passes through the maximum, and (3) the type of optimum reactor system needed for the plant to process 200 m³/hr. What would be the reactor volume in this system?

Solution

The hydrolysis of methly acetate can be represented by

$$CH_3COOCH_3 + H_2O \xrightarrow{\ k_1\ } CH_3COOH + CH_3OH$$

$$(A) \qquad\qquad\qquad (B)$$

That is, A → B. For an autocatalytic reaction A + B → B + B and the rate of disappearance of species A is

$$\left(-r_A\right) = -\frac{dC_A}{dt} = k_1 C_A C_B \tag{3-295}$$

The solution for an autocatalytic reaction is

$$\ln\frac{\theta_B + X_A}{\theta_B\left(1-X_A\right)} = \left(C_{AO} + C_{BO}\right)k_1 t \tag{3-296}$$

where

$$\theta_B = \frac{C_{BO}}{C_{AO}} = \frac{0.045}{0.45} = 0.1 \quad \text{at } t = 1 \text{ hr}, \ X_A = 0.65$$

The rate constant k_1 is

$$k_1 = \frac{\ln \dfrac{\theta_B + X_A}{\theta_B(1 - X_A)}}{(C_{AO} + C_{BO}) \, t}$$

$$= \frac{\ln(0.75/0.035)}{(0.495)\,(1)}$$

$$= 6.191 \; 1/\text{gmol} - \text{hr}$$

$$= 6.2 \; 1/\text{gmol} - \text{hr}$$

The rate equation is

$$(-r_A) = -\frac{dC_A}{dt} = 6.2 \, C_A \, C_B \quad \frac{1}{\text{gmol} - \text{hr}}$$

The rate is at maximum when $C_A = C_B$. From stoichiometry:

$$C_{AO} - C_A = C_B - C_{BO}$$

and

$$C_A + C_B = C_{AO} + C_{BO}$$

$$= (0.45 + 0.045)$$

$$= 0.495$$

Therefore,

$$C_A = C_B = \frac{1}{2} \times 0.495$$

$$= 0.2475$$

The corresponding fractional conversion X_A is

$$\frac{C_{AO}-C_A}{C_{AO}} = \frac{0.45-0.2475}{0.45} = 0.45$$

Therefore, the time t at which the rate reaches the maximum is:

$$t = \frac{\ln\left(\dfrac{\theta_B + X_A}{\theta_B(1-X_A)}\right)}{k_1(C_{AO}+C_{BO})}$$

$$= \frac{\ln\left(\dfrac{0.1+0.45}{0.1(1-0.45)}\right)}{(0.62)(0.495)}$$

$$= \frac{\ln\left(\dfrac{0.55}{0.055}\right)}{3.069}$$

$$= 0.75\,hr \ (45\,min)$$

The type of optimum reactor that will process 200 m^3/hr is a continuous flow stirred tank reactor (CFSTR). This configuration operates at the maximum reaction rate. The volume V_R of the reactor can be determined from the design equation:

$$\bar{t} = \frac{V_R}{u}$$

where V_R = volume of the reactor
   u = volumetric flow rate of the fluid
   $\bar{t}$ = mean residence time of the fluid

Therefore, the volume $V_R = \bar{t} \cdot u$

$$= 0.75 \times 200$$

$$= 150 \ m^3$$

Example 3-9

Huang and Dauerman (1969) have studied the acetylation of benzyl chloride in dilute solution at 102°C. Using equimolal concentrations of sodium acetate and benzyl chloride (0.757 kmol/m^3), Table 3-16 lists the reported data on the fraction of benzyl chloride remaining unconverted versus time. Determine the order of the reaction and the reaction rate constant at this temperature.

Solution

The reaction is assumed to be second order as represented by NaAc + $C_6H_5CH_2Cl \rightarrow C_6H_5CH_2Ac + Na^+ + Cl^-$, which can be expressed in terms of components A and B as

$$A + B \xrightarrow{\ \ k\ \ } Products \tag{3-297}$$

The rate equation for component B is

$$\left(-r_B\right) = -\frac{dC_B}{dt} = kC_A C_B \tag{3-298}$$

The stoichiometry for components A and B is:

Table 3-16

Time, t (ksec)	C_B/C_{BO} where B = $C_6H_5CH_2Cl$
10.80	0.945
24.48	0.912
46.08	0.846
54.72	0.809
69.48	0.779
88.56	0.730
109.44	0.678
126.72	0.638
133.74	0.619
140.76	0.590

	A	**B**
Amount at t = 0	C_{AO}	C_{BO}
Amount at t = t	C_A	C_B
Amounts that have reacted	$C_{AO} - C_A =$	$C_{BO} - C_B$

In terms of the fractional conversion of B, X_B

$$X_B = \frac{C_{BO} - C_B}{C_{BO}} \tag{3-299}$$

Equation 3-299 is further expressed as:

$$C_B = C_{BO}(1 - X_B) \tag{3-300}$$

Since the concentration of A and B is the same, then

$$C_A = C_B = C_{BO}(1 - X_B) \tag{3-301}$$

and

$$-\frac{dC_B}{dt} = C_{BO}\frac{dX_B}{dt} \tag{3-302}$$

Substituting Equations 3-300 and 3-302 into Equation 3-298 gives

$$(-r_B) = C_{BO}\frac{dX_B}{dt} = kC_{BO}^2(1 - X_B)^2 \tag{3-303}$$

Rearranging Equation 3-303 and integrating between the limits t = 0, $X_B = 0$ and t = t, $X_B = X_B$ gives

$$\int_0^{X_B} \frac{dX_B}{(1 - X_B)^2} = kC_{BO}\int_0^t dt \tag{3-304}$$

$$\left[\frac{1}{1 - X_B}\right]_0^{X_B} = kC_{BO}t \tag{3-305}$$

$$\left[\frac{1}{1-X_B} - 1\right] = kC_{BO}t \tag{3-306}$$

$$\left[\frac{1}{C_B/C_{BO}} - 1\right] = kC_{BO}t \tag{3-307}$$

The rate constant k can be calculated from Equation 3-307 as:

$$k = \frac{1}{C_{BO}t}\left[\frac{1}{C_B/C_{BO}} - 1\right] \tag{3-308}$$

Table 3-17 shows the calculated values of the rate constant k at varying time t.

Discounting the first and the last two values of the rate constant k, the average value of the rate constant is k = 0.0055 m³/(mol • sec). This shows that the rate of reaction is second order.

Example 3-10

(1) Suppose a first-order reaction (n = 1) is preformed in an iso-thermal batch reactor of constant volume V. Write a material balance

Table 3-17

Time, t (ksec)	C_B/C_{BO}	$\left[\dfrac{1}{C_B/C_{BO}} - 1\right]$	$\dfrac{1}{C_{BO}t}$	$k \dfrac{m^3}{mol \cdot sec}$
10.48	0.945	0.058	0.122	0.0071
24.48	0.912	0.096	0.054	0.0052
46.08	0.846	0.182	0.029	0.0053
54.72	0.809	0.236	0.024	0.0057
69.48	0.779	0.284	0.019	0.0054
88.56	0.730	0.370	0.015	0.0056
109.44	0.678	0.475	0.012	0.0057
126.72	0.638	0.567	0.010	0.0057
133.74	0.619	0.616	0.010	0.0062
140.76	0.590	0.695	0.005	0.0034

on A and integrate it to derive the expression $C_A = C_{AO} \exp(-kt)$, where C_{AO} is the concentration of A in the reactor at $t = 0$.

(2) The gas-phase decomposition of sulfuryl chloride, $SO_2Cl_2 \rightarrow SO_2 + Cl_2$, is thought to follow a first order rate law. The reaction is performed in a constant volume, isothermal batch reactor, and the concentration of SO_2Cl_2 is measured at several reaction times, with the following results.

t(min)	4.0	20.2	40.0	60.0	120.0	180.0
C_A(mol/l)	0.0158	0.0152	0.0144	0.0136	0.0116	0.0099

Use these results to verify the proposed law and determine the rate constant k. Give both the value and units of k.

Solution

(1) Assuming that the reaction is first order in an isothermal batch reactor of constant volume, then the rate equation for the reaction $A \xrightarrow{k_1}$ Products is

$$\left(-r_A\right) = -\frac{1}{V}\frac{dn_A}{dt} = k_1 C_A \tag{3-309}$$

where the number of moles n_A is

$$n_A = C_A V \tag{3-310}$$

where C_A = concentration of species A
     V = volume of the batch reactor

Substituting Equation 3-310 into Equation 3-309 yields

$$\left(-r_A\right) = -\frac{1}{V} d\left(VC_A\right) = k_1 C_A$$

$$= -\frac{dC_A}{dt} = k_1 C_A \tag{3-311}$$

since V is constant.

Rearranging Equation 3-311 and integrating between the limits at $t = 0$, $C_A = C_{AO}$ and $t = t$, $C_A = C_A$ gives

$$\int_{C_{AO}}^{C_A} \frac{dC_A}{C_A} = -k_1 \int_0^t dt \qquad (3\text{-}312)$$

and

$$\ln \frac{C_A}{C_{AO}} = -k_1 t \qquad (3\text{-}313)$$

Therefore,

$$C_A = C_{AO} \exp(-k_1 t) \qquad (3\text{-}314)$$

(2) Using Equation 3-314 for the gas-phase decomposition of sulfuryl chloride, it can represented by

$$Y = Ae^{BX} \qquad (3\text{-}315)$$

Linearizing Equation 3-315 gives

$$\ln Y = \ln A + BX \qquad (3\text{-}316)$$

Table 3-18 shows the concentration of SO_2Cl_2, C_A as a function of time, t.

A plot of $\ln C_A$ versus time t gives a straight line with slope $(-B)$ equal to the rate constant k_1. The constants of the equation $Y = Ae^{BX}$ are:

Table 3-18

Time (min)	C_A, mol/l	ln C_A, mol/l
4.0	0.0158	−4.1477
20.2	0.0152	−4.1865
40.0	0.0144	−4.2405
60.0	0.0136	−4.2977
120.0	0.0166	−4.4568
180.0	0.0099	−4.6152

- A = 0.015999
- B = -0.26711×10^{-2}
- Correlation Coefficient = 0.99991

The rate constant $k_1 = -B = 0.00267 \text{ sec}^{-1}$. Figure 3-25 gives a plot of ln C_A versus time t (min).

Example 3-11

A gas decomposition reaction with stoichiometry 2A $\rightarrow$ 2B + C follows a second order rate law $r_d(\text{mol}/\text{m}^3 \cdot \text{s}) = kC_A^2$, where C_A is the reactant concentration in mol/m^3. The rate constant k varies with the reaction temperature according to the Arrhenius law:

$$k\left(\text{m}^3/\text{mol} \cdot \text{s}\right) = k_O \exp\left(\frac{-E}{RT}\right)$$

where $k_O(\text{m}^3/\text{mol·s})$ = pre-exponential factor
E(J/mol) = reaction activation energy
R = gas constant

(1) Suppose the reaction is performed in a batch reactor of constant volume V(m^3) at a constant temperature T(K), beginning with pure A

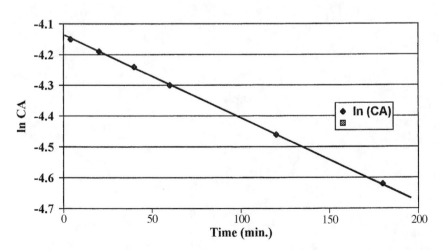

Figure 3-25. Plot of ln C_A versus t.

at a concentration C_{AO}. Write a differential balance on A and integrate it to obtain an expression for $C_A(t)$ in terms of C_{AO} and k.

(2) Let P_O(atm) be the initial reactor pressure. Prove that $t_{1/2}$, the time required to achieve 50% conversion of A in the reactor, equals RT/kp_O. Assume an ideal gas behavior.

(3) The decomposition of nitrous oxide (N_2O) to nitrogen and oxygen is preformed in a 5.0 1 batch reactor at a constant temperature of 1,015 K, beginning with pure N_2O at several initial pressures. The reactor pressure P(t) is monitored, and the times ($t_{1/2}$) required to achieve 50% conversion of N_2O are noted in Table 3-19. Use these results to verify that the N_2O decomposition reaction is second order and determine the value of k at T = 1,015 K.

(4) The same experiment is performed at several other temperatures at a single initial pressure of 1.0 atm. The results are shown in Table 3-20. Determine the Arrhenius law parameters (k_O and E) for the reaction.

Solution

(1) Since the reaction is carried out in a batch system of constant volume, the rate expression for a second order rate law is

$$\left(-r_A\right) = -\frac{1}{V}\frac{dn_A}{dt} = -\frac{dC_A}{dt} = kC_A^2 \tag{3-317}$$

where $k = k_O \exp(-E/RT)$.

Table 3-19

P_O(atm)	0.135	0.286	0.416	0.683
$t_{1/2}$(sec)	1,060	500	344	209

Table 3-20

T(K)	900	950	1,000	1,050
$t_{1/2}$(sec)	5,464	1,004	219	55

Rearranging Equation 3-317 and integrating between the limits $t = 0$, $C_A = C_{AO}$ and $t = t$, $C_A = C_A$ gives

$$-\int_{C_{AO}}^{C_A} \frac{dC_A}{C_A^2} = k\int_0^t dt \qquad (3\text{-}318)$$

Integrating Equation 3-318 yields

$$\left. \int_{C_{AO}}^{C_A} \left[\frac{1}{C_A} \right] \right. = kt \qquad (3\text{-}319)$$

$$\frac{1}{C_A} - \frac{1}{C_{AO}} = k_O \exp\left(-\frac{E}{RT} \right) t \qquad (3\text{-}320)$$

Therefore, the concentration of species A in terms of C_{AO} and k is

$$\frac{1}{C_A} = \frac{1}{C_{AO}} + k_O \exp\left(-\frac{E}{RT} \right) t \qquad (3\text{-}321)$$

(2) The half-life $t_{1/2}$ equation for the general nth-order reaction is

$$t_{1/2} = \frac{\left(2^{n-1} - 1\right) C_{AO}^{1-n}}{k(n-1)} \qquad (3\text{-}322)$$

Since the reaction order is two, that is $n = 2$, this can be substituted in Equation 3-322 to obtain

$$t_{1/2} = \frac{C_{AO}^{-1}}{k} = \frac{1}{kC_{AO}} \qquad (3\text{-}323)$$

where

$$C_{AO} = \frac{n_{AO}}{V} = \frac{p_{AO}V}{RTV} = \frac{p_{AO}}{RT} \qquad (3\text{-}324)$$

Substituting Equation 3-324 into Equation 3-323 gives

$$t_{1/2} = \frac{1}{k p_{AO}/RT}$$

$$= \frac{RT}{k p_{AO}} \tag{3-325}$$

(3) The decomposition of nitrous oxide (N_2O) to nitrogen and oxygen is represented by

$$2N_2O \to 2N_2 + O_2 \tag{3-326}$$

The half-life $t_{1/2}$ of the general nth order reaction is

$$t_{1/2} = \frac{\left(2^{n-1} - 1\right) C_{AO}^{1-n}}{k(n-1)}$$

Therefore, using Equation 3-324 gives

$$t_{1/2} = \frac{2^{n-1} - 1}{k(n-1)} \left(\frac{p_{AO}}{RT}\right)^{1-n} \tag{3-327}$$

Taking the natural logarithm of Equation 3-327 gives

$$\ln\left(t_{1/2}\right) = \ln \frac{2^{n-1} - 1}{k(n-1)} + (1-n) \ln \left(\frac{p_{AO}}{RT}\right) \tag{3-328}$$

Equation 3-328 can be expressed in the form

$$Y = AX^B \tag{3-329}$$

Linearizing Equation 3-329 gives

$$\ln Y = \ln A + B \ln X \tag{3-330}$$

where the slope $B = 1 - n$.

From Table 3-19, we can construct the independent variable p_O/RT at the constant temperature of 1,015 K and the gas constant $R = 0.08206$ (l • atm/mol • k) (Table 3-21). The dependent variable is $t_{1/2}$.

Table 3-21

p_O, atm	$t_{1/2}$ (sec)	$p_O/RT \times 10^{-3}$	$t_{1/2}$-estimated
0.135	1,060	1.6208	1,059.8
0.286	500	3.4337	501.7
0.416	344	4.9945	342.2
0.683	209	8.2001	209.5

The constants for Equation 3-330 from PROG1 are:

• A = 1.7834
• B = –0.992
• Correlation Coefficient = 1.0

The slope B = –0.9922 = 1 – n. Therefore, the value of n = 1.992 $\cong$ 2 and the reaction is second order. The value of the rate constant k at T = 1,015 K is

$$t_{1/2} = \frac{RT}{kp_O}$$

Therefore,

$$k = \frac{RT}{p_O t_{1/2}} = \frac{0.08206 \times 1,015}{0.135 \times 1,060} \left(\frac{1 \cdot atm}{mol - K} \cdot \frac{K}{atm \cdot sec} \right)$$

$$= 0.582 \frac{1}{mol \cdot sec}$$

(**4**) Applying the Arrhenius equation

$$k_T = k_O \exp\left(\frac{-E}{RT} \right) \tag{3-331}$$

Linearizing Equation 3-331 gives

$$\ln k_T = \ln k_O - \frac{E}{RT} \tag{3-332}$$

The rate constant k_T can be determined at varying temperature T from the half-life $t_{1/2}$ as follows:

$$t_{1/2} = \frac{2^{n-1}-1}{k_T(n-1)}\left(\frac{RT}{p_{AO}}\right)^{n-1} \tag{3-327}$$

At the initial pressure of 1 atm, $p_{AO} = 1$ and $n = 2$. Therefore, $t_{1/2}$ in Equation 3-327 is $t_{1/2} = RT/k_T$ and $k_T = RT/t_{1/2}$. Table 3-22 can be constructed from Table 3-20 for values of k_T, the dependent variable, from varying temperature T and $t_{1/2}$. Equation 3-331 is of the form

$$Y = Ae^{BX} \tag{3-333}$$

Linearizing the model Equation 3-333 gives

$$\ln Y = \ln A + BX \tag{3-334}$$

where the constant $A = k_O$, and the slope $B = -E/R$. The computer program PROG1 determines the constants A and B. The constants for Equation 3-330 are:

- $A = 0.3378 \times 10^{13}$
- $B = -0.29875 \times 10^5$
- Correlation Coefficient = 0.9999

The gas constant $R = 0.08206$ l • atm/mol • K.

Table 3-22

T(K)	$t_{1/2}(sec)$	1/T	$k_T\left(\dfrac{RT}{t_{1/2}}\right)$(l/mol • sec)
900	5,464	0.00111	0.0135
950	1,004	0.00105	0.0776
1,000	219	0.001	0.3747
1,050	55	0.00095	1.5666

The activation energy $-\dfrac{E}{R} = B = -0.2988 \times 10^5$

$$= 0.2988 \times 10^5 \times 0.08206$$

$$= 2,452 \ \dfrac{1}{\text{mol}}$$

The value of the pre-exponential factor $k_O = 3.378 \times 10^{12}$ l/mol • sec.

Example 3-12

For the synthesis of ammonia, $N_2 + 3H_2 \rightarrow 2NH_3$, over an iron catalyst, develop the rate expression for the following mechanism

$$H_{2(g)} + 2S \underset{k_{-1}}{\overset{k_1}{\rightleftharpoons}} 2HS \tag{3-335}$$

$$N_{2(g)} + 2S \xrightarrow{k_2} 2NS \tag{3-336}$$

$$NS + 3HS \underset{k_{-3}}{\overset{k_3}{\rightleftharpoons}} NH_3S + 3S \tag{3-337}$$

$$NH_3S \underset{k_{-4}}{\overset{k_4}{\rightleftharpoons}} NH_{(g)} + S \tag{3-338}$$

Solution

Boudart (1972) introduced the assumption of the most abundant surface intermediate (masi). This assumption suggests that the sites occupied by all species except the most abundant surface intermediate is regarded as negligible compared to those filled by the most abundant intermediate and to those which are empty.

The rate expression for the above mechanism is based on the following assumptions:

1. The most abundant surface intermediate is NS

   that $C_T = C_s + C_{NS}$. $\tag{3-339}$

2. Steps 1, 3, and 4 are in equilibrium and step 2 is the rate-determining step.
3. The conditions at which the reaction $N_2 + 3H_2 \rightarrow 2NH_3$ are performed are such that the reverse rate is negligible.

C_T = total concentration of sites and defined as

$$\left(\frac{\text{total moles of sites}}{\text{unit mass of catalyst}} \right)$$

C_S = concentration of vacant sites
C_{HS} = concentration of adsorbed hydrogen

$$\left(\frac{\text{moles of hydrogen adsorbed}}{\text{unit mass of catalyst}} \right)$$

C_{NS} = concentration of the most abundant surface intermediate
C_{NH_3S} = concentration of adsorbed ammonia

$$\left(\frac{\text{moles of ammonia adsorbed}}{\text{unit mass of catalyst}} \right)$$

p_{N_2} = partial pressure of nitrogen

p_{NH_3} = partial pressure of ammonia

$(-r_{N_2})$ = rate of reaction of nitrogen, moles of nitrogen disappearing per unit time per unit mass of catalyst

Since step 2 is the rate-determining step, then the rate expression is:

$$\left(-r_{N_2}\right) = k_2 p_{N_2} C_S^2 \tag{3-340}$$

The net rate of the disappearance of H_2 is given by

$$\left(-r_{H_2}\right)_{net} = k_1 p_{H_2} C_S^2 - k_{-1} C_{HS}^2 \tag{3-341}$$

On the basis of the pseudo-steady state assumption, the net rate of disappearance is zero, therefore

$$k_1 p_{H_2} C_S^2 = k_{-1} C_{HS}^2$$

and

$$K_1 = \frac{k_1}{k_{-1}} = \frac{C_{HS}^2}{p_{H_2} C_S^2} \tag{3-342}$$

$$K_3 = \frac{k_3}{k_{-3}} = \frac{C_S^3 C_{NH_3S}}{C_{NS} C_{HS}^3} \tag{3-343}$$

$$K_4 = \frac{k_4}{k_{-4}} = \frac{C_S p_{NH_3}}{C_{NH_3S}} \tag{3-344}$$

From Equation 3-343, the concentration of the most abundant surface intermediate is

$$C_{NS} = \frac{C_s^3 C_{HN_3S}}{K_3 C_{HS}^3} \tag{3-345}$$

The concentration of adsorbed ammonia from Equation 3-344 is

$$C_{NH_3S} = \frac{C_S p_{NH_3}}{K_4} \tag{3-346}$$

The concentration of adsorbed hydrogen from Equation 3-342 is

$$C_{HS} = \left(K_1 p_{H_2} C_S^2\right)^{1/2} \tag{3-347}$$

Substituting Equations 3-346 and 3-347 into Equation 3-345 gives

$$C_{NS} = \frac{C_S^3 C_S p_{NH_3}}{K_3 K_4 \left(K_1 p_{H_2} C_S^2\right)^{3/2}} \tag{3-348}$$

$$= \frac{p_{NH_3} C_S}{K p_{H_2}^{1.5}} \tag{3-349}$$

where $K = K_1^{1.5} K_3 K_4$. Substituting Equation 3-349 into Equation 3-339 yields

$$C_T = \frac{p_{NH_3} C_S}{K p_{H_2}^{1.5}} + C_S \tag{3-350}$$

or

$$C_S = \frac{C_T}{\dfrac{p_{NH_3}}{K p_{H_2}^{1.5}} + 1} \tag{3-351}$$

Substituting Equation 3-351 into the rate expression of Equation 3-340 gives

$$\left(-r_{N_2}\right) = k_2 p_{N_2} \left[\frac{C_T}{\dfrac{p_{NH_3}}{K p_{H_2}^{1.5}} + 1} \right]^2 \tag{3-352}$$

REFERENCES

1. Boudart, M., *Kinetics of Chemical Processes,* Chapter 1, Prentice Hall, Englewood Cliffs, NJ, 1968.
2. Mc. Tigue, P. T. and Sime, J. M., *J. Chem. Soc.,* 1303, 1963.
3. Dixon, N. E., Jackson, W. G., Marty, W., and Sargeson, A. M., "Base Hydrolysis of Pentaamminecobalt (III) Complexes of Urea, Dimethyl Sulfoxide and Trimethyl Phosphate," *Iorg., Chem.,* 21, 688–697, 1982.
4. Levenspiel, O., *The Chemical Reactor Omnibook,* OSU Book Stores, Inc., 1993.
5. Boudart, M., "Two-Step Catalytic Reactions," *AIChEJ,* Vol. 18, No. 3, pp. 465–478, 1972.
6. Bacon, D. W. and J. Downie, "Evaluation of Rate Data III," AIChE MI Series E: Kinetics Vol. 2, Reactors and Rate Data, AIChE, 1981.

7. Holt, M. J. and Norris, A. C., "A New Approach to the Analysis of First-Order Kinetic Data," *Journal of Chemical Education,* Vol. 54, No. 7, pp. 426–428, 1977.

8. Norris, A. C., *Computational Chemistry—An Introduction to Numerical Methods,* John Wiley & Sons Ltd., 1981.

9. Weller, S., "Analysis of Kinetic Data for Heterogeneous Reactions," *AIChE. J,* Vol. 2, No. 1, pp. 59–61, 1956.

10. Cox, B. G., *Modern Liquid Phase Kinetics,* Oxford Chemistry Primers, Oxford University Press, 1996.

11. Hill, C. G. Jr., *An Introduction to Chemical Kinetics & Reactor Design,* John Wiley & Sons, New York, 1977.

12. Fogler, H. S., *Elements of Chemical Reaction Engineering,* 3rd ed. Prentice-Hall International Series, 1999.

13. Dillon, R. T., *J. Am. Chem. Soc.,* 54 (952), 1932.

14. Huang and Dauerman, *Ind. Eng. Chem. Product Research & Development,* 8, 227, 1969.

CHAPTER FOUR

Industrial and Laboratory Reactors

INTRODUCTION

Chemical reactors are the most important features of a chemical process. A reactor is a piece of equipment in which the feedstock is converted to the desired product. Various factors are considered in selecting chemical reactors for specific tasks. In addition to economic costs, the chemical engineer is required to choose the right reactor that will give the highest yields and purity, minimize pollution, and maximize profit. Generally, reactors are chosen that will meet the requirements imposed by the reaction mechanisms, rate expressions, and the required production capacity. Other pertinent parameters that must be determined to choose the correct type of reactor are reaction heat, reaction rate constant, heat transfer coefficient, and reactor size. Reaction conditions must also be determined including temperature of the heat transfer medium, temperature of the inlet reaction mixture, inlet composition, and instantaneous temperature of the reaction mixture.

An important factor in reactor operation is the outlet degree of conversion. Operating conditions such as temperature, pressure, and degree of agitation, are related for the most economic operation. The optimum reactor that will best meet the process requirements requires a review of whether the process is continuous or batch, and whether a combination of reactor types or multiple reactors in series or parallel would be most adequate. It is also important to determine whether the mode of operation involves either an isothermal (i.e., constant temperature) or an adiabatic (i.e., heat does not exchange with the surroundings) condition, whether a single pass operation is best, or whether recycling is needed to achieve the desired degree of conversion of the raw feedstock. The degree of conversion affects the economics of separating the reaction mixture and the costs of returning the unconverted reactant back into the reaction.

In chemical laboratories, small flasks and beakers are used for liquid phase reactions. Here, a charge of reactants is added and brought to reaction temperature. The reaction may be held at this condition for a predetermined time before the product is discharged. This batch reactor is characterized by the varying extent of reaction and properties of the reaction mixture with time. In contrast to the flasks are large cylindrical tubes used in the petrochemical industry for the cracking of hydrocarbons. This process is continuous with reactants in the tubes and the products obtained from the exit. The extent of reaction and properties, such as composition and temperature, depend on the position along the tube and does not depend on the time.

Another classification refers to the shape of the vessel. In the case of the laboratory vessel installed with a stirrer, the composition and temperature of the reaction is homogeneous in all parts of the vessel. This type of vessel is classified as a stirred tank or well mixed reactor. Where there is no mixing in the direction of flow as in the cylindrical vessel, it is classified as a plug flow or tubular flow reactor.

Knowledge of the composition and temperature at each point of the reactor enables the designer to describe the behavior of a chemical reactor. Concentrations of species at any point may change either because the species are consumed by chemical reaction or they reach this position via mass transfer. Correspondingly, the temperature at any point may change because the heat is being absorbed or released by chemical reaction or heat transfer. The rate of the chemical reaction and the rate of mass and heat transfer affect the concentration and temperature of a given section of the system. Concentration, temperature, and molecular properties determine the reaction rate. This process occurs through microkinetic properties that are identical in any chemical reactor (i.e., if the temperature and composition in two reactors are the same, then the reaction rate is also the same). Macrokinetic properties of the reactor affect the outcome of the process through the kinetics of heat and mass transfer. The rates of mass and heat transfer depend on the properties relative to the reactor, such as size of the reactor, size and speed of the impeller, and the area of heat exchanging surfaces.

Reactors without the effect of macrokinetic properties are composed of elements that are either perfectly insulated from the viewpoint of mass and heat transfer or at equilibrium with the surroundings. Reactors that are free of the effect of macrokinetic properties are classified as:

- Batch isothermal perfectly stirred reactor (the reaction mixture is at equilibrium with the heat transfer medium).
- Batch adiabatic perfectly stirred reactor.
- Semi-batch perfectly stirred reactor.
- Continuous isothermal perfectly stirred flow reactor (the reaction mixture is at equilibrium with the heat transfer medium).
- Continuous adiabatic perfectly stirred flow reactor.
- Continuous isothermal plug flow reactor (the reactor mixture is at thermal equilibrium with the surroundings).
- Continuous adiabatic plug flow reactor.

Knowledge of these types of reactors is important because some industrial reactors approach the idealized types or may be simulated by a number of ideal reactors. In this chapter, we will review the above reactors and their applications in the chemical process industries. Additionally, multiphase reactors such as the fixed and fluidized beds are reviewed. In Chapter 5, the numerical method of analysis will be used to model the concentration-time profiles of various reactions in a batch reactor, and provide sizing of the batch, semi-batch, continuous flow stirred tank, and plug flow reactors for both isothermal and adiabatic conditions.

BATCH ISOTHERMAL PERFECTLY STIRRED REACTOR

The concept of a batch reactor assumes that the reaction is instantaneously charged (i.e., filled) and perfectly homogenized in the reactor. Also, its temperature is immediately adjusted to that of the heat transfer medium. Therefore, the chemical reaction takes place at the temperature of the heat transfer medium under perfect mixing. The process is stopped as soon as the degree of conversion is achieved.

Figure 4-1 shows two nozzles at the top of a batch reactor where charging of the reactants occurs. Batch reactors are used extensively in a final scale-up of an industrial plant. The choice of a batch reaction over a continuous system is often a result of special considerations. The size of batch reactors range from 5 gal (19 l) in small industrial pilot plants to 10,000–20,000 gal (38,000–76,000 l) in large plants.

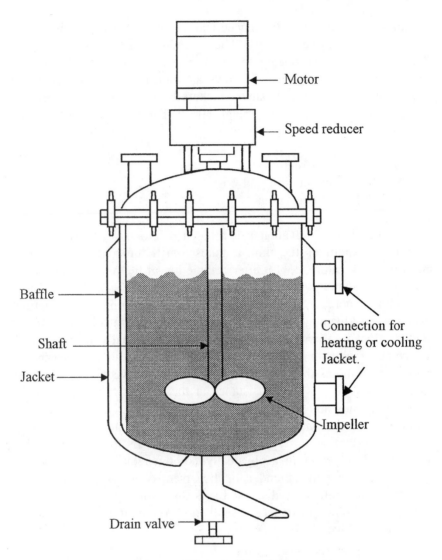

Figure 4-1. A batch homogeneous-type reactor.

When larger sizes are required, the design may include multiple units of batch reactors arranged in parallel.

In small industrial pilot plants, a batch system may be employed for preliminary information. Also, batch reactors can be used in these plants to obtain small quantities of the new product for further evaluations such as purity, yield, and sales. At the industrial level, batch reactors are used in the pharmaceutical, biochemical, or multi-product plants as in the dye industry. These reactors or autoclaves require suitable access for inserting agitators, coils, or other internal devices and for cleaning. Figure 4-2 shows a steam-jacketed autoclave and Figure 4-3 is a 120-gal, steam-jacketed autoclave for processing organic chemicals at 2,000 psi and 300°F. The advantages of a batch reactor are:

- Simple in construction.
- Small instrumentation and cost.
- Flexibility of operation.

The principal disadvantage of a batch reactor is in the labor cost. Labor cost includes the time it takes to fill the reactor, heat it to reaction temperature, cool it after completion of the reaction, discharge the reactor contents, and clean the reactor for the next batch. These procedures increase the overall labor costs per unit of production. Another disadvantage involves the difficulty to control heat transfer and product quality. Chemical reaction rates usually increase with temperature and with more intimate contact between reactants. Mechanical agitation promotes the flow of heat by forcing convection of the mass and by reducing the film resistance at the vessel wall. Additionally, agitation breaks up agglomerated solids thereby increasing the contact surface and the rate at which reacting species come into close proximity.

An important purpose of agitation or mixing is to bring a number of materials together in a physically homogeneous mixture. Two or more fluids are either blended or dispersed as emulsions; fluids and finely divided solids are dispersed as suspensions, gases dispersed as fluids, or soluble substances dissolved. Mixing of process fluids is reviewed in Chapter 7.

SEMI-BATCH REACTORS

Figure 4-4 shows a semi-batch reactor with outside circulation and the addition of one reactant through the pump. Semi-batch reactors have some reactants that are charged into the reactor at time zero, while other reactants are added during the reaction. The reactor has no outlet stream. Some reactions are unsuited to either batch or continuous operation in a stirred vessel because the heat liberated during the reaction may cause dangerous conditions. Under these

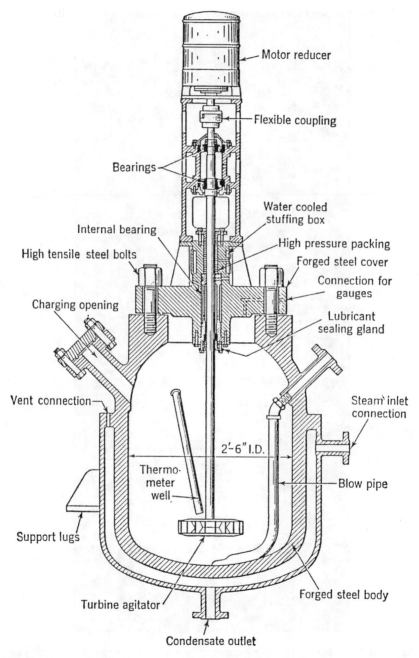

Figure 4-2. Jacketed batch reactor. *(Source: D. B. Gooch, "Autoclaves Pressure—Temperature Reactions," Ind. Eng. Chem., 35, 927–946, 1943. Used with permission from the American Chemical Society.)*

Figure 4-3. A 120-gal steam-jacketed autoclave for processing organic chemicals at 2,000 psi and 300°F. *(Source: D. B. Gooch, "Autoclaves Pressure— Temperature Reactions," Ind. Eng. Chem., 35, 927–946, 1943. Used with permission from the American Chemical Society.)*

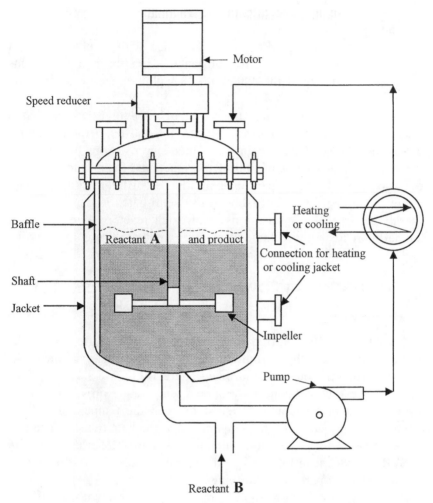

Figure 4-4. A semi-batch reactor with outside circulation.

circumstances, reactions are sometimes performed under semi-batch conditions. This means that one or more reagents that do not react are completely entered in a vessel and the final reagent is then added under controlled conditions. No product is withdrawn until the entire reagent has been added and the reaction has proceeded to the required extent. As an example, consider the reaction $A + B \xrightarrow{k_1} C$, where the reactor is first charged with reagent A and later reagent B is added continuously until the reaction is complete. Examples of this reaction process are:

- Reaction of a gas with a liquid where the gas bubbles through the liquid.
- Reaction of a highly reactive substance with a relatively inert substance where the reactive substance, if present in large amounts at the beginning of the reaction, either polymerize or decompose.

A semi-batch reactor has the same disadvantages as the batch reactor. However, it has the advantages of good temperature control and the capability of minimizing unwanted side reactions by maintaining a low concentration of one of the reactants. Semi-batch reactors are also of value when parallel reactions of different orders occur, where it may be more profitable to use semi-batch rather than batch operations. In many applications semi-batch reactors involve a substantial increase in the volume of reaction mixture during a processing cycle (i.e., emulsion polymerization).

CONTINUOUS FLOW ISOTHERMAL PERFECTLY STIRRED TANK REACTOR

A continuous flow stirred tank reactor (CFSTR) differs from the batch reactor in that the feed mixture continuously enters and the outlet mixture is continuously withdrawn. There is intense mixing in the reactor to destroy any concentration and temperature differences. Heat transfer must be extremely efficient to keep the temperature of the reaction mixture equal to the temperature of the heat transfer medium. The CFSTR can either be used alone or as part of a series of battery CFSTRs as shown in Figure 4-5. If several vessels are used in series, the net effect is partial backmixing.

It is easy to maintain good temperature control with a CFSTR. However, a disadvantage is that the conversion of reactant per volume of reactor is the smallest of the flow reactors. Therefore, very large reactors in series are needed to achieve high conversions. For example, the first reactor could be run to give a 50% conversion, yielding a high rate of reaction and subsequently reducing the total reactor volume. The next reactor might run from 50%–80% conversion and the third from 80%–90% until the desired conversion is reached. The effect of this process is a continuous reaction system that has a much lower volume, but has more equipment items because of the reactor vessels required.

Industrial reactors operate in the steady state with the volume, concentration, and temperature of the reaction mixture being constant

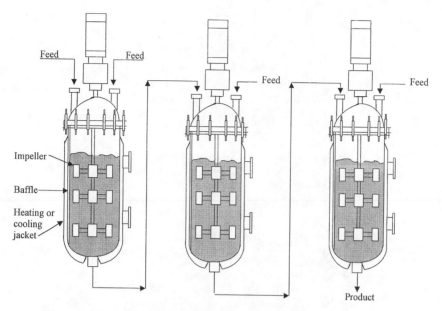

Figure 4-5. A battery of continuous flow stirred tank reactors.

with time. It is approximated that a steady state operation is reached when about five to ten times the reactor volume of the reaction mixture has passed through the reactor. In the continuously flow stirred tank reactor, the reaction takes place at the temperature and the degree of conversion of the outlet stream. This gives the reactor its characteristic features from either the batch or semi-batch reactor.

Figure 4-6 is a variation of the continuous homogeneous reactor (baffled tank) in which backmixing has a considerable effect. The advantage of this type of reactor is its low cost per unit volume. It is often used when a long holding time would require a tubular reactor that is too long or too expensive. Agitation is not required and a high degree of backmixing is not harmful to the yield.

CONTINUOUS ISOTHERMAL PLUG FLOW TUBULAR REACTOR

The plug flow tubular reactor is a heat exchanger where the reaction occurs in the tubes. Construction is often varied. For example, the reactor may consist of a tube placed in a bath, a tube in a jacket, or a number of tubes immensed in a heat transfer medium for the reactor

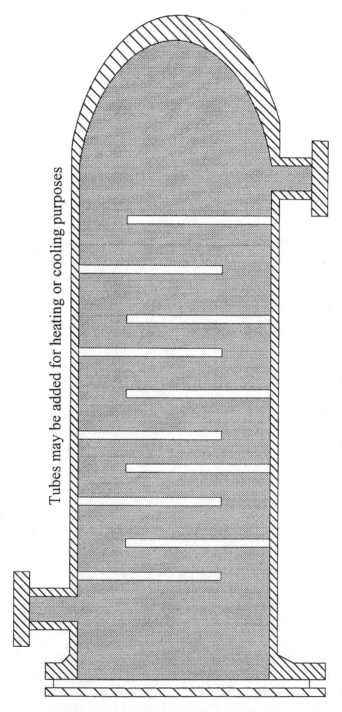

Figure 4-6. Baffled tank reactor.

to operate isothermally. The exchange of heat between the reaction mixture and the medium must be so intense so as to heat the reaction mixture instantaneously and eliminate the effect of the reaction heat. The heat transfer medium must flow through the jacket in excess to keep its temperature unaffected by the reaction heat.

Plug flow is an idealized flow of fluids where all particles in a given cross-section have identical velocity and direction of motion. During plug flow, particles of different age do not mingle and there is no backmixing. All particles that enter the reactor at the same time must leave simultaneously. The essential features of the plug flow reactor require that there be no longitudinal mixing of fluid elements as they move through the reactor, and that all fluid elements take the same length of time to move from the reactor inlet to the outlet. The plug flow can be described as a piston flow model. This is because the reaction occurring within differentially thin slugs of fluid, fill the entire cross-section of the tube and are separated from one another by hypothetical pistons that prevent axial mixing. These plugs of material move as units through the reactor, with the assumption that the velocity profile is flat as the fluid traverses the tube diameter. Each plug of fluid is assumed to be uniform in temperature, composition, and pressure and thus can be assumed that radial mixing is infinitely rapid.

The tubular plug flow reactor is relatively easy to maintain with no moving parts, and it usually produces the highest conversion per reactor volume of any of the flow reactors. Other advantages are:

- High throughput.
- Little or no backmixing.
- Close temperature control.

The disadvantages are:

- Expensive instrumentation.
- High operating cost (maintenance, cleaning).
- Nonuniform heat flux of the radiant section of furnace.

The principal disadvantage of the tubular reactor is the difficulty in controlling the temperature within the reactor. This often results in hot spots especially when the reaction is exothermic. The tubular reactor can be in the form of one long tube or one of a number of shorter reactors arranged in a tube bank (Figure 4-7).

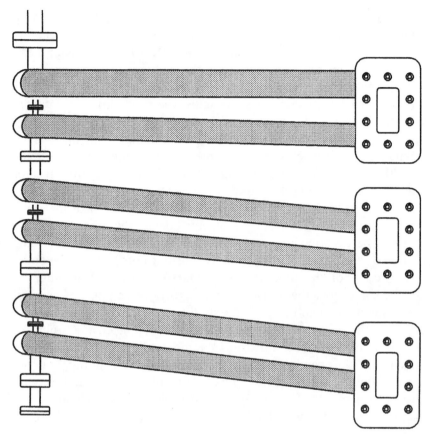

Figure 4-7. Longitudinal tubular reactor.

CONTINUOUS MULTIPHASE REACTORS

Figure 4-8 shows a continuous reactor used for bubbling gaseous reactants through a liquid catalyst. This reactor allows for close temperature control. The fixed-bed (packed-bed) reactor is a tubular reactor that is packed with solid catalyst particles. The catalyst of the reactor may be placed in one or more fixed beds (i.e., layers across the reactor) or may be distributed in a series of parallel long tubes. The latter type of fixed-bed reactor is widely used in industry (e.g., ammonia synthesis) and offers several advantages over other forms of fixed beds.

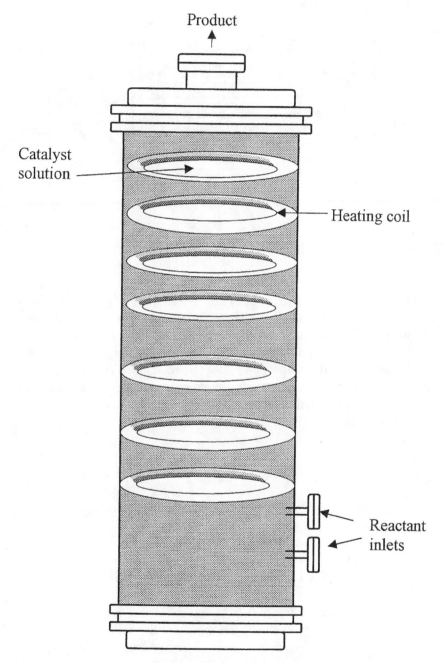

Figure 4-8. Heterogeneous continuous reactor.

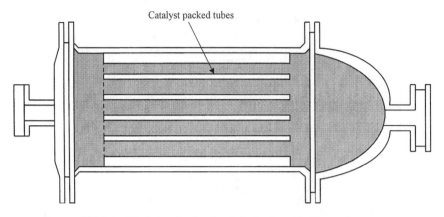

Figure 4-9. Longitudinal catalytic fixed-bed reactor.

The type of reactor shown in Figure 4-9 is used in a heterogeneous reaction system involving catalyzed gas reactions. The fixed-bed reactor gives less chance for backmixing, but channeling of the gas flow through the catalyst bed causes ineffective use of parts of the reactor bed. An advantage of the fixed-bed reactor is that for most reactions, it gives the highest conversion per weight of catalyst of any catalytic reactor. Another advantage is it provides large volumes of processed reactants. The disadvantages are:

- The catalysts are highly prone to deactivation.
- The catalysts often require regeneration after a relatively short period of operation. This may incur additional cost.
- It is difficult to control the heat-transfer in the catalyst bed.
- Some part of the catalyst surface remains unused as a result of the reaction system and the rate-controlling step.

FLUIDIZED BED SYSTEM

The fluidized bed in Figure 4-10 is another common type of catalytic reactor. The fluidized bed is analogous to the CFSTR in that its contents though heterogeneous are well mixed, resulting in an even temperature distribution throughout the bed.

In a fluidized bed reactor, the solid material in the form of fine particles is contained in a vertical cylindrical vessel. The fluid stream

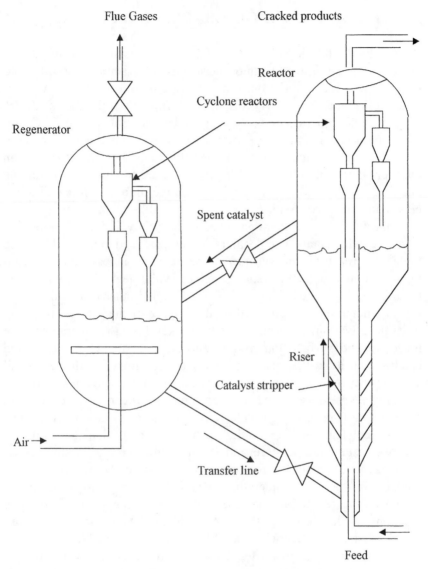

Figure 4-10. Fluid catalytic cracking unit.

is passed up through the particles at a rate strong enough for them to lift and not fall back into the fluidized phase above its free surface by carryover in the fluid stream. The bed of particles in this state shows the appearance of boiling. In heterogeneous catalytic reactions, the catalyst loses its activity with operating time.

FLUID CATALYTIC CRACKING (FCC) UNIT

The fluid catalytic cracking (FCC) process, as shown in Figure 4-10, converts straight-run atmospheric gas oil, vacuum gas oils, and heavy stocks recovered from other operations into high-octane gasoline, light fuel oils, slurry oil, and olefin-rich light gases. These products undergo further processing and separation in the FCC unit main fractionator and other vessels downstream of the FCC reactor. The gasoline produced has good overall octane characteristics and an excellent octane number. The catalysts used are mixtures of crystalline alumina silicates (known as zeolites), active alumina, silica-alumina, clay, and rare earth oxides.

In the FCC, an oil feed composed of heavy hydrocarbon molecules is mixed with catalyst and enters a fluidized bed reactor. The long molecules react on the surface of the catalyst and are cracked into lighter product molecules (e.g., gasoline), which leave the reactor from the top. During the cracking process, carbon and other heavy uncracked organic materials are deposited on the surface of the catalyst resulting in its deactivation. The catalyst is then taken into a regenerator where the deposited surface material is burned with air. The regenerated catalyst returns to the reactor after it has been mixed with fresh feed. The activity of the newer catalysts is so intense that much of the cracking takes place in the line returning the regenerated catalyst to the reactor. This process is referred to as the transfer line cracking.

A salient feature of the fluidized bed reactor is that it operates at nearly constant temperature and is, therefore, easy to control. Also, there is no opportunity for hot spots (a condition where a small increase in the wall temperature causes the temperature in a certain region of the reactor to increase rapidly, resulting in uncontrollable reactions) to develop as in the case of the fixed bed reactor. However, the fluidized bed is not as flexible as the fixed bed in adding or removing heat. The loss of catalyst due to carryover with the gas stream from the reactor and regenerator may cause problems. In this case, particle attrition reduces their size to such an extent where they are no longer fluidized, but instead flow with the gas stream. If this occurs, cyclone separators placed in the effluent lines from the reactor and the regenerator can recover the fine particles. These cyclones remove the majority of the entrained equilibrium size catalyst particles and smaller fines. The catalyst fines are attrition products caused by the collision of catalyst particles with each other and the reactor walls. These fines can escape capture in the cyclones because the removal efficiency of cyclones for particles of uniform density decreases with decreasing particle size. The catalyst fines captured in the reactor cyclones are transferred to the catalyst regenerator. Here, the fines are carried with the exhausted air and combustion products known as fuel gas. Most of the fine catalyst particles that are entrained in the flue gas are first captured in a two-stage cyclone within the regenerator vessel and then returned to the catalyst bed. Coker [1] illustrated the design of a cyclone separator, which is an economical device for removing particulate solids from a fluid system. Cyclone separators have been successfully employed in catalytic cracker operations. Designing cyclones often requires a balance between the desired collection efficiency, pressure drop, space limitations, and installation cost.

The advantages of the fluidized bed are:

- Savings in operating expenses due to heat recovery in the reaction-regeneration steps.
- Rapid mixing of reactants-solids and high heat transfer rates.
- Easy to control both the heat transfer and the fluid flow system.

The disadvantages are:

- Backmixing due to particle distribution in dense and dilute phases.
- Inefficient contacting due to solids movement and the bypassing of solids by bubbles.
- Possible channeling, slugging, and attrition of catalyst.
- Possible agglomeration and sintering of fine particles in the dilute phase under certain conditions (e.g., high temperature).

The advantages of the ease of catalyst replacement or regeneration are offset by the high cost of the reactor and catalyst regeneration equipment.

DEEP CATALYTIC CRACKING UNIT

An improved FCC unit is the deep catalytic cracking unit (DCC) that is designed to produce gasoline from vacuum gas oil (VGO). The

DCC uses heavy VGO as feedstock and has the same features has the FCC but with the following differences: special catalyst, high catalyst-to-oil ratio, higher steam injection rate, operating temperature, residence time, and lower operating pressure.

In the DCC unit, the hydrocarbon feed is dispersed with steam and cracked using a hot solid catalyst in a riser, and enters a fluidized bed reactor. A known injection system is employed to achieve the desired temperature and catalyst-to-oil contacting. This maximizes the selective catalytic reactions. The vaporized oil and catalyst flow up the riser to the reactor where the reaction conditions can be varied to complete the cracking process. The cyclones that are located in the top of the reactor effect the separation of the catalyst and the hydrocarbon vapor products. The steam and reaction products are discharged from the reactor vapor line and enter the main fractionator where further processing ensure the separation of the stream into valuable products.

The formed coke on the catalyst particles during cracking reduces its activity and selectivity. The spent catalyst passes into a stripping zone where steam is used to displace the entrained and adsorbed hydrocarbons, which leave the reactor with the products. Stripped catalyst particles are transported into the regenerator where the particles are contacted with air under controlled conditions. The regeneration process is the same as in the FCC unit. The DCC has been shown to produce polymer grade propylene from heavy gas oils, and it produces three and a half times more propylene and less than half the gasoline than a conventional FCC unit. Figure 4-11 illustrates a DCC unit and Figure 4-12 represents a typical DDC plant that produces propylene, which is integrated to a petrochemical complex. Table 4-1 compares the operating variables of the deep catalytic cracking (DCC), fluidized catalytic cracking (FCC) and steam cracking (SC) units.

Another classification involves the number of phases in the reaction system. This classification influences the number and importance of mass and energy transfer processes in the design. Consider a stirred mixture of two liquid reactants A and B, and a catalyst consisting of small particles of a solid added to increase the reaction rate. A mass transfer resistance occurs between the bulk liquid and the surface of the catalyst particles. This is because the small particles tend to move with the liquid. Consequently, there is a layer of stagnant fluid that surrounds each particle. This results in reactants A and B transferring through this layer by diffusion in order to reach the catalyst surface. The diffusion resistance gives a difference in concentration between

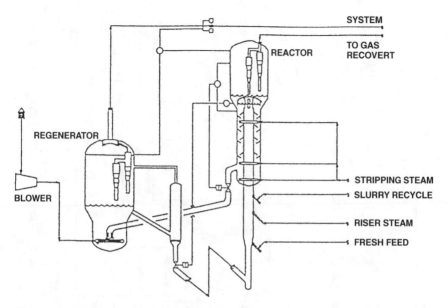

Figure 4-11. Deep catalytic cracking process flow diagram. *(Courtesy of Stone & Webster Engineering Corporation, © 1977 Stone & Webster Engineering Corporation.)*

Table 4-1
Operating parameters

Item	Deep Catalytic Cracking (DCC)	Fluidized Catalytic Cracking (FCC)	Steam Cracking (SC)
Typical Residence Time (sec)	10–60	1–30	0.1–0.2
Catalyst/Oil	9–15	5–10	
Steam, wt% of feed	10–30	1–10	30–80
Reactor Temperature, °F	950–1,070	950–1,020	1,400–1,600
Pressure, psig	10–20	15–30	15

the bulk fluid and the catalyst surface. The developing rate expression is used to account for the intrinsic kinetics at the catalyst surface and the mass transfer process. The coupling of the intrinsic kinetics and mass transfer is due to the heterogeneous nature of the system.

In contrast to reactors involving the use of solid catalyst phases, there are reactors that use two liquid phases. An example is the

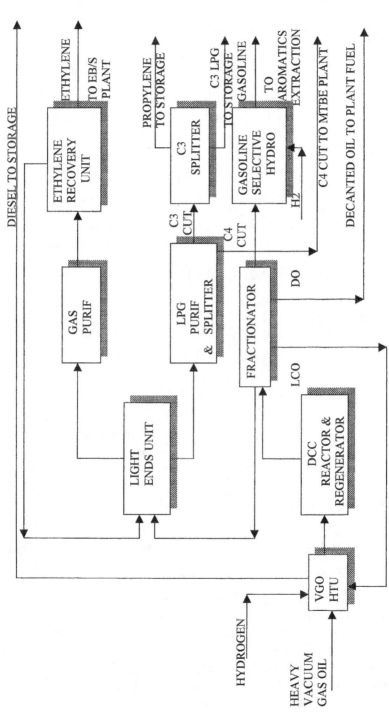

Figure 4-12. DCC plant petrochemicals integration. *(Courtesy of Stone & Webster Engineering Corporation, © 1977 Stone & Webster Engineering Corporation.)*

catalytic alkylation reaction of an olefin (e.g., butylene) with isobutane to give C_8-isomers (alkylate). Here, the two liquid hydrocarbon streams enter the bottom of the tubular flow vessel as shown in Figure 4-13. They are then dispersed as bubbles in a co-current continuous stream of liquid hydrofluoric acid (HF) that acts as a catalyst. Phase separation occurs at the top of the reactor where the lighter alkylate product is removed, and the heavier HF stream is recycled to the bottom of the reactor. In this case, the reaction occurs when both the olefin and isobutane reach the acid. This results in interphase mass transfer and the kinetics of the reaction in the acid phase.

Tubular reactors are used for reactions involving a gas and a liquid. In this arrangement, the gas phase is dispersed as bubbles at the bottom of a tubular vessel. The bubbles then rise through the continuous liquid phase that flows downwards as shown in Figure 4-14. An example of this process is the removal of organic pollutants from water by noncatalytic oxidation with pure oxygen.

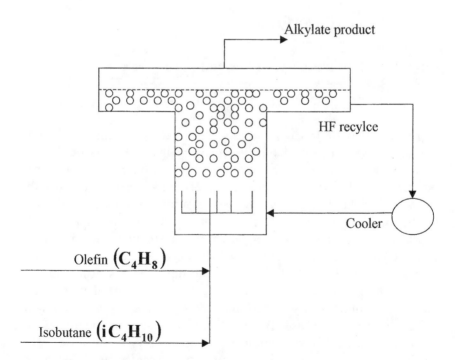

Figure 4-13. Liquid-liquid heterogeneous tubular flow reactor (e.g., alkylation of olefins and isobutane). *(Source: J. M. Smith,* Chemical Engineering Kinetics, *3rd ed., McGraw-Hill, Inc., 1981.)*

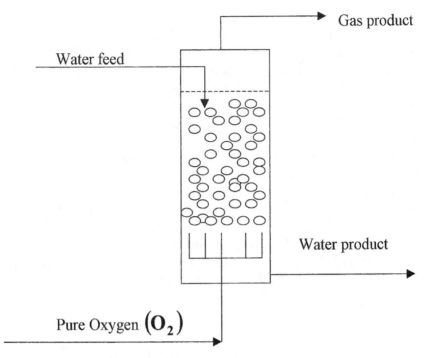

Figure 4-14. A two-phase tubular flow reactor (e.g., gas-liquid bubble reactor for oxidation of pollutants in water). *(Source: J. M. Smith,* Chemical Engineering Kinetics, *3rd ed., McGraw-Hill, Inc. 1981.)*

The effect of physical processes on reactor performance is more complex than for two-phase systems because both gas-liquid and liquid-solid interphase transport effects may be coupled with the intrinsic rate. The most common types of three-phase reactors are the slurry and trickle-bed reactors. These have found wide applications in the petroleum industry. A slurry reactor is a multi-phase flow reactor in which the reactant gas is bubbled through a solution containing solid catalyst particles. The reactor may operate continuously as a steady flow system with respect to both gas and liquid phases. Alternatively, a fixed charge of liquid is initially added to the stirred vessel, and the gas is continuously added such that the reactor is batch with respect to the liquid phase. This method is used in some hydrogenation reactions such as hydrogenation of oils in a slurry of nickel catalyst particles. Figure 4-15 shows a slurry-type reactor used for polymerization of ethylene in a slurry of solid catalyst particles in a solvent of cyclohexane.

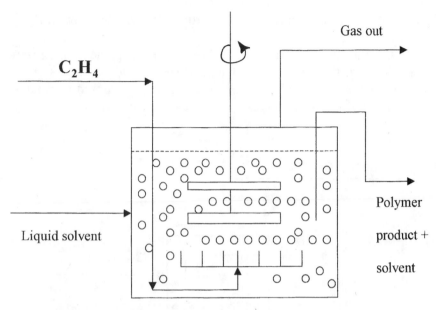

Figure 4-15. A slurry-type reactor (e.g., for C_2H_4 polymerization). *(Source: J. M. Smith,* Chemical Engineering Kinetics, *3rd ed., McGraw-Hill, Inc., 1981.)*

A heterogeneous tubular reactor that incorporates three phases where gas and liquid reactants are contacted with the solid catalyst particles, is classified as a trickle-bed reactor. The liquid is usually allowed to flow down over the bed of catalyst, while the gas flows either up or down through the void spaces between the wetted pellets. Co-current downflow of the gas is generally preferred because it allows for better distribution of liquid over the catalyst bed and higher liquid flow rates are possible without flooding.

In most applications, the reaction occurs between a dissolved gas and a liquid-phase reactant in the presence of a solid catalyst. In some cases, the liquid is an inert medium and the reaction takes place between the dissolved gases at the solid surface. These reactors have many diverse applications in catalytic processes and are used extensively in the chemical industry. Trickle-bed reactors have been developed by the petroleum industry for hydrodesulfurization, hydrocracking, and hydrotreating of various petroleum fractions of relatively high boiling point. Under reaction conditions, the hydrocarbon feed is frequently a vapor-liquid mixture that reacts at liquid hourly space velocities (LHSV in volume of fresh feed, as liquid/volume of bed, hr) in the range of 0.5–4, as in the case of hydrodesulfurization shown in Figure 4-16.

Unlike the slurry reactor, a trickle-bed reactor approaches plug flow behavior and, therefore, the problem of separating the catalyst from the product stream does not exist. The low ratio of liquid to catalyst in the reactor minimizes the extent of homogeneous reaction. However,

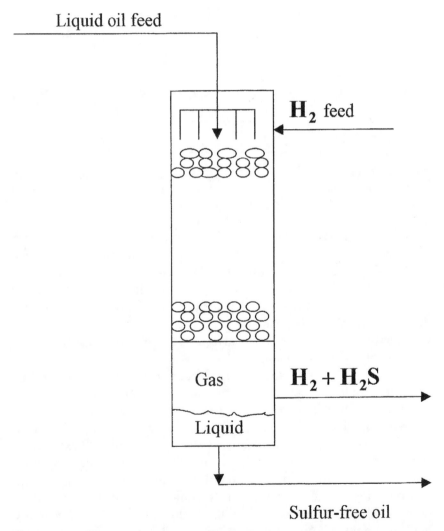

Figure 4-16. Trickle-bed (tubular reactor) for hydrodesulfurization. *(Source: J. M. Smith, Chemical Engineering Kinetics, 3rd ed., McGraw-Hill, Inc., 1981.)*

if the reaction is substantially exothermic, the evolved heat may cause portions of the bed not to be wetted resulting in poor contacting of catalyst and liquid. Satterfield [2], Ramanchandran, and Chaudhari [3] have given detailed design procedures of three-phase catalytic reactors. Table 4-2 shows the various applications of these reactors.

DETERMINING LABORATORY REACTORS

The success of designing industrial reactors greatly depends on accurate and reliable laboratory data. These data are derived from the

Table 4-2
Applications of three-phase reactors

1. Slurry Reactor

(A) Hydrogenation of:
  • fatty acids over a supported nickel catalyst.
  • 2-butyne-1,4-diol over a $Pd-CaCO_3$ catalyst.
  • glucose over a Raney nickel catalyst.
(B) Oxidation of:
  • C_2H_4 in an inert liquid over a $PdCl_2$-carbon catalyst.
  • SO_2 in inert water over an activated carbon catalyst.
(C) Hydroformation of CO with high-molecular weight olefins on either a cobalt or ruthenium complex bound to polymers.
(D) Ethynylation: Reaction of acetylene with formaldehyde over a $CaCl_2$-supported catalyst.

2. Trickle Bed Reactors

(A) Hydrodesulfurization: Removal of sulfur compounds from crude oil by reaction with hydrogen on C0 – Mo on alumina.
(B) Hydrogenation of:
  • aniline over a Ni-clay catalyst.
  • 2-butyne, 1,4-diol over a supported Cu – Ni catalyst.
  • benzene, α – CH_3 styrene, and crotonaldehyde.
  • aromatics in napthenic lube oil distillate.
(C) Hydrodenitrogenation of:
  • lube oil distillate.
  • cracked light furnace oil.
(D) Oxidation of:
  • cumene over activated carbon.
  • SO_2 over carbon.

Source: C. N. Satterfield, AIChE J., 21, 209 (1975); P. A. Ramachandran and R. V. Chaudari, Chem. Eng., 87 (24), 74 (1980); R. V. Chaudari and P. A. Ramachandran, AIChE J., 26, 177 (1980).

reaction kinetic study from laboratory reactors, which are useful in scale-up. The design and construction of laboratory and pilot plant reactors together with experimental programs can be both time consuming and expensive, therefore it is imperative that the correct choice of laboratory reactor be used. Weekman [4] presented an excellent review on the choice of various laboratory reactors. The criteria employed to determine the various types of these reactors are listed in Table 4-3.

Sampling of a two-fluid phase system containing powdered catalyst can be problematic and should be considered in the reactor design. In the case of complex reacting systems with multiple reaction paths, it is important that isothermal data are obtained. Also, different activation energies for the various reaction paths will make it difficult to evaluate the rate constants from non-isothermal data.

Measurements of the true reaction times are sometimes difficult to determine due to the two-phase nature of the fluid reactants in contact with the solid phase. Adsorption of reactants on the catalyst surface can result in catalyst-reactant contact times that are different from the fluid dynamic residence times. Additionally, different velocities between the vapor, liquid, and solid phases must be considered when measuring reaction times. Various laboratory reactors and their limitations for industrial use are reviewed below.

DIFFERENTIAL REACTOR

The differential reactor is used to evaluate the reaction rate as a function of concentration for a heterogeneous system. It consists of a tube that contains a small amount of catalyst as shown schematically in Figure 4-17. The conversion of the reactants in the bed is extremely small due to the small amount of catalyst used, as is the change in reactant concentration through the bed. The result is that the reactant concentration through the reactor is constant and nearly equal to the

Table 4-3
Criteria used to determine laboratory reactors

1. Ease of sampling and product composition analysis.
2. Degree of isothermality.
3. Effectiveness of contact between catalyst and reactant.
4. Handling of catalyst decay.
5. Ease of construction and cost.

Catalyst

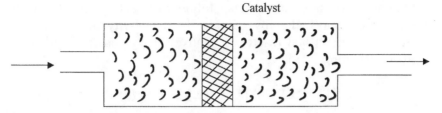

Figure 4-17. Differential reactor. *(Source: V. W. Weekman, "Laboratory Reactors and Their Limitations," AlChEJ, Vol. 20, p. 833, 1974. Used with permission of the AlChEJ.)*

inlet concentration. Therefore, the reactor is considered gradientless, and the reaction rate is spatially uniform in the bed. Because of the small conversion, heat release is also small, and the reactor operates in an isothermal manner.

The differential reactor is simple to construct and inexpensive. However, during operation, care must be taken to ensure that the reactant gas or liquid does not bypass or channel through the packed catalyst, but instead flows uniformly across the catalyst. This reactor is a poor choice if the catalyst decays rapidly, since the rate of reaction parameters at the start of a run will be different from those at the end of the run.

FIXED BED (INTEGRAL) REACTOR

Advantages of the fixed bed reactors are its ease of construction, good sampling and product analysis, higher conversions, and no catalyst or product separation problems. Rapid sampling that provides instantaneous data during catalyst decay may cause channeling or bypassing of some of the catalyst by the reactant stream. This can be minimized by careful attention to the distribution of the liquid and gas. The main problem with the fixed bed design is achieving uniform isothermal temperatures. Significant axial and radial temperature gradients can result with severe exothermic or endothermic reactions. Different products are formed at different reaction paths. This results in changes in the reaction mechanism with changing temperature along the length of the reactor and, consequently, makes it difficult to evaluate the various reaction rate constants.

If the catalyst decays during the experiment, the reaction rates will be significantly different at the end of the experiment than at the beginning. Also, the reaction may follow different reaction paths as the catalyst decays, thereby varying the selectivity of a particular product. The fixed bed reactor is relatively easy and inexpensive to construct (Figure 4-18).

STIRRED BATCH REACTOR (SBR)

The stirred batch reactor contains catalyst dispersed as slurry. There is better contact between the catalyst and the fluid in this reactor than with either the differential or integral reactors. Separation of the product from the catalyst must be accomplished by the sampling system, which can be a problem. Samples of fluids are usually passed through cyclones or withdrawn through filters or screens to separate the catalyst and fluid, consequently stopping the reaction. Due to sufficient mixing, isothermality is good and accurate residence time measurements should be possible. Since all three phases are contained in the reactor, it may provide the most accurate measurement of contact time of all reactors provided the reaction can be rapidly quenched at the end of the experiment. The decaying of the catalyst poses the same problem as with the fixed bed reactor. Consequently, the activity and selectivity will vary during the course of data collection. The stirred batch reactor (Figure 4-19) is fairly simple to construct at a reasonable cost.

STIRRED CONTAINED SOLIDS REACTOR (SCSR)

Figure 4-20 shows a typical design of a stirred contained solids reactor. Here, the catalyst particles are mounted in the paddles that

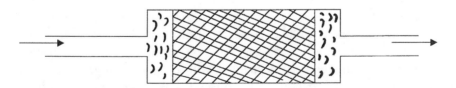

Figure 4-18. Fixed bed (integral) reactor. *(Source: V. W. Weekman, "Laboratory Reactors and Their Limitations," AIChEJ, Vol. 20, p. 833, 1974. Used with permission of the AIChEJ.)*

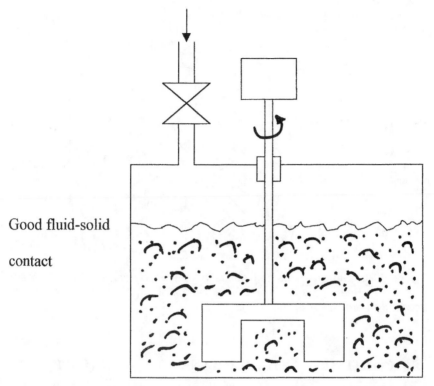

Good fluid-solid

contact

Figure 4-19. Stirred batch reactor. *(Source: V. W. Weekman, "Laboratory Reactors and Their Limitations," AIChEJ, Vol. 20, p. 833, 1974. Used with permission of the AIChEJ.)*

rotate at high speeds to minimize external mass transfer effects, and also maintain well-mixed fluid contents. This type of operation provides good isothermal conditions, which can be maintained, and there is good contact between the catalyst and the fluid. However, if the catalyst particle size is small, there may be difficulties in containing the particles in the paddle screens. This reactor rates well in ease of sampling and analysis of the product composition. The residence time of the solid is accurately known and, with good mixing, the gas-vapor residence times can also be measured quite accurately. The disadvantage of this type of reactor is its unsteady state, which can affect its selectivity. However, it is unable to generate useful data when the catalyst being studied decays.

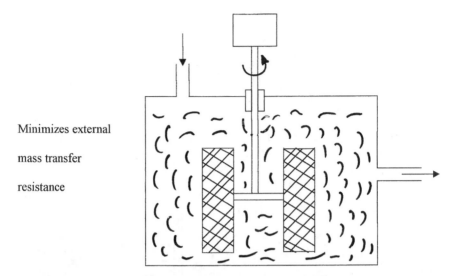

Minimizes external

mass transfer

resistance

Figure 4-20. Stirred contained solids reactor. *(Source: V. W. Weekman, "Laboratory Reactors and Their Limitations," AIChEJ, Vol. 20, p. 833, 1974. Used with permission of the AIChEJ.)*

CONTINUOUS STIRRED TANK REACTOR (CSTR)

In this reactor (Figure 4-21), the catalyst is charged to the reactor together with the fluid feed, and the catalyst leaves the reactor in the product stream at the same rate that it is charged into the reactor. The catalyst in the reactor maintains the same level of catalytic activity at all times. However, as with the stirred batch reactors, the catalyst slurry in the reactor presents some sampling problems. This requires either quenching or rapidly separating the catalyst from the reaction mixture, otherwise the sample may continuously react at different temperatures as it cools, disguising the selectivity behavior. Since the reactor is well mixed, isothermality and fluid solid contact are good. The reactor operates in the steady state because catalyst and reactants are continuously added, which eliminates any possible catalyst decay selectivity disguise.

STRAIGHT-THROUGH TRANSPORT REACTOR (STTR)

The transport reactor (Figure 4-22) is widely used in the production of gasoline from heavier petroleum fractions. In this reactor, either an

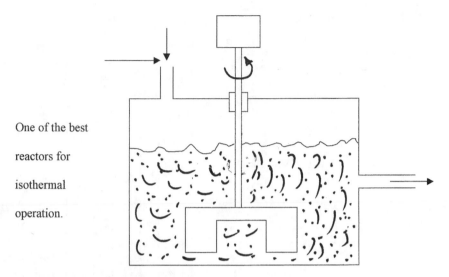

One of the best

reactors for

isothermal

operation.

Figure 4-21. Continuous stirred tank reactor. *(Source: V. W. Weekman, "Laboratory Reactors and Their Limitations," AIChEJ, Vol. 20, p. 833, 1974. Used with permission of the AIChEJ.)*

inert carrier gas or the reactant itself transports the catalyst through the reactor. Since the catalyst passes through with the reactant, it is necessary to achieve either rapid quenching or rapid catalyst reactant separation. The possibility of catalyst decay selectivity disguise is completely eliminated because the catalyst and reactants are continuously fed. For highly endothermic or exothermic reactions, isothermal operation is difficult to achieve, hence, a poor-to-fair rating in this category. For high velocities where there is little slip between the catalyst and reactant phases, there are accurate measurements of residence time. At lower velocities, there may be slip between the phases that can lead to difficulties in accurately determining the contact time. Since the transport reactors are a length of tubing, they are easier to construct, but salt or sand baths may be required in order to maintain isothermal operation. Additionally, the construction rating is fair-to-good because product-catalyst separation facilities are required.

RECIRCULATING TRANSPORT REACTOR (RTR)

Adding a recirculating loop to the transport reactor, a well-mixed condition is achieved provided the recirculation rate is large with

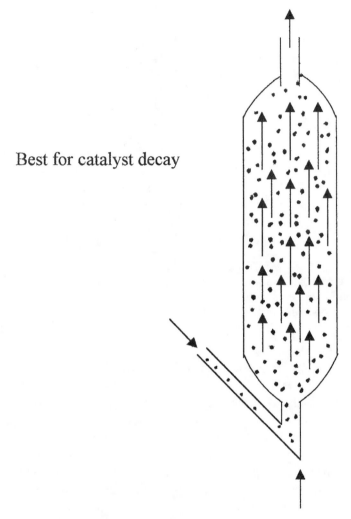

Best for catalyst decay

Figure 4-22. Straight-through transport reactor.

respect to the feed rate (Figure 4-23). Isothermal operation is attained due to this well-mixed condition. Since the reactor is operated at a steady state, the kinetic parameters measured at the beginning of the experiment will be the same as those measured at the end. Because fresh catalyst is mixed with decayed catalyst from the recycle, the product distribution and the kinetic parameters may not be the same as those measured in a straight-through transport reactor where the gas

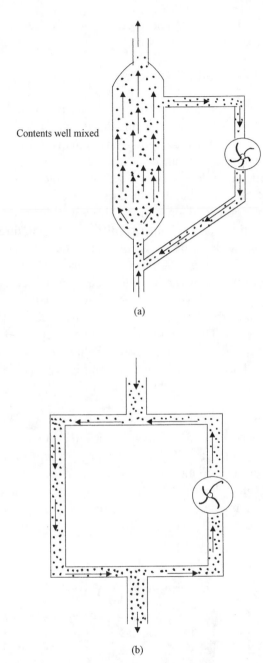

Contents well mixed

(a)

(b)

Figure 4-23. Recirculationg transport reactors. *(Source: V. W. Weekman, "Laboratory Reactors and Their Limitations," AIChEJ, Vol. 20, p. 833, 1974. Used with permission of the AIChEJ.)*

contacts only fresh catalyst. The recirculation further provides more complexity to the construction, which gives it a lower rating.

PULSE REACTOR

In the pulse reactor, a small pulse of reactant is charged with a small amount of catalyst. The product from the reactor is directly fed to a chromatograph to minimize sampling problems (Figure 4-24). A wide range of conversion levels can be achieved so sampling and analysis of product composition do not pose a serious problem. A small amount of catalyst can be surrounded by a large heat sink to minimize its deviation from isothermal operation. However, high exothermic or endothermic reactions can result in significant temperature difference. The difficulty with the pulse reactor is the change in the catalyst surface concentrations during the pulse. Consequently, the adsorbed species change during the course of the reaction, which could lead to selectivity disguise. However, if all the reaction paths are identically altered by these adsorbed species, then the pulse reactor may be useful for selectivity studies. This is an unsteady state reactor and short pulses of reactant can follow the instantaneous behavior, resulting in a fair-to-good rating. The problems in construction are identical to the differential reactor and are slightly compounded by the need to introduce accurate pulses of reactant.

Table 4-4 summarizes the ratings of the various reactors. The CFSTR and the recirculating transport reactor are the best choices because they are satisfactory in every category except for construction. The stirred batch and contained solid reactors are satisfactory if the catalyst under study does not decay. If the system is not limited by internal diffusion in the catalyst pellet, larger pellets could be used and the stirred-contained solids reactor is the better choice. However,

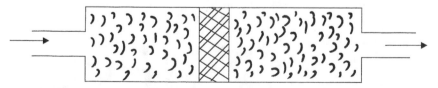

Figure 4-24. Pulse reactors. *(Source: V. W. Weekman, "Laboratory Reactors and Their Limitations," AIChEJ, Vol. 20, p. 833, 1974. Used with permission of the AIChEJ.)*

Table 4-4
Summary of reactor ratings, gas-liquid, powdered catalyst, decaying catalyst system

Reactor type	Sampling and analysis	Isothermality	Solid contact	Fluid-decaying catalyst	Ease of construction
Differential	P-F	F-G	F	P	G
Fixed bed	G	P-F	F	P	G
Stirred batch	F	G	G	P	G
Stirred-contained solids	G	G	F-G	P	F-G
Continuous flow stirred tank	F	G	F-G	F-G	P-F
Straight-through transport	F-G	P-F	F-G	G	F-G
Recirculating transport	F-G	G	G	F-G	P-F
Pulse	G	F-G	P	F-G	G

G = good, F = fair, P = poor.
Source: V. W. Weekman, "Laboratory Reactors and Their Limitations", AIChEJ, Vol. 20, p. 833, 1974. Used with permission of the AIChEJ.

if the catalyst is non-decaying and heat effects are negligible, the fixed bed (integral) reactor is the best choice because of its ease of construction and operation. The pulse reactor is most satisfactory in systems that do not strongly adsorb or where the adsorbed species do not relatively alter the reaction paths. In cases where the reaction system is extremely critical, more than one reactor type is used in determining the reaction rate law parameters.

LOOP REACTORS

A loop reactor is a continuous steel tube or pipe, which connects the outlet of a circulation pump to its inlet. Reactants are fed into the loop, where the reaction occurs, and product is withdrawn from the loop. Loop reactors are used in place of batch stirred tank reactors in a variety of applications including chlorination, ethoxylation, hydrogenation, and polymerization. A loop reactor is typically much smaller than a batch reactor producing the same amount of product. Mass transfer is often the rate-limiting step in gas-liquid reactions, and a loop reactor design increases mass transfer, while reducing reactor size and improving process yields. An example is an organic material that has been chlorinated in a glass-lined batch stirred tank reactor, with chlorine fed through a dip pipe. Replacing the stirred tank reactor with a loop reactor, with chlorine fed to the recirculating liquid stream through an eductor, reduced reactor size, increased productivity, and reduced chlorine usage. Figure 4-25 shows a schematic of a loop reactor system. Table 4-5 compares the advantages of a loop reactor to a batch stirred tank reactor.

GUIDELINES FOR SELECTING BATCH PROCESSES

Douglas [5] gives an excellent review in selecting a batch process in favor of a continuous process. The factors that favor batch operations are summarized as follows:

- *Production rates:*
  - Sometimes batch process, if the plants have production capacity less than 10×10^6 lb/yr (5×10^6 kg/hr).
  - Usually batch process, if the plants have production capacity less than 1×10^6 lb/yr (0.5×10^6 kg/hr).
  - Where multiproduct plants are produced using the same processing equipment.

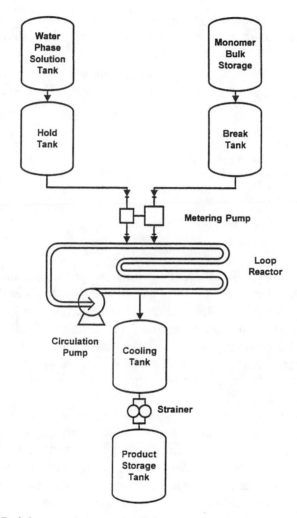

Figure 4-25. A loop reactor production system. *(Source: M. Wilkinson and K. Geddes, "An Award Winning Process." Chemistry in Britain, December, 1050–1052, 1993.)*

- *Market forces:*
  - — Where products are seasonal (e.g., fertilizers).
  - — Short product lifetime (e.g., organic pigments).
- *Operational problems:*
  - — Long reaction times (i.e., when chemical reactions are slow).
  - — Handling slurries at low flowrates.
  - — Rapidly fouling materials (e.g., materials foul equipment so rapidly that shutdown and frequent cleaning are required).

Table 4-5
Effect of reactor design on size and productivity
for a gas-liquid reaction

Reactor type	Batch stirred tank reactor	Loop reactor
Reactor size (l)	8,000	2,500
Chlorination time (hr)	16	4
Productivity (kg/hr)	370	530
Chlorine usage (kg/100 kg product)	33	22
Caustic usage in vent scrubber (kg/100 kg product)	31	5

Source: Center for Chemical Process Safety (CCPS). "Guidelines for Engineering Design for Process Safety." New York: AIChE, 1993a.

Regardless of their inherent advantages and disadvantages, the choice of the proper design from any of the industrial reactors for a certain operation greatly depends on the specific reaction system. This involves an extensive study of many factors that contribute to the optimum. The performance of an industrial reactor also depends on performing the reaction at a number of scales to gain sufficient confidence to make the correct predictions. This may require predicting the effect of concentration and temperature on selectivity using a laboratory or pilot plant reactor, followed by predicting the temperature and mixing patterns on an industrial plant. This approach is much preferred in industry than the empirical scale-up method because only two types of reactors (laboratory and pilot plant) are needed.

Greater options are available in the design of a full-scale reactor once the heat transfer and mixing characteristics are defined. Additionally, the criteria for similarity are essential between the laboratory and full-scale plant. The scale-up of reactors is covered in Chapter 13. The safety of controlling and operating chemical reactions and associated hazards form an essential aspect of chemical manufacture. The manufacture of various chemical products involves the processing of reactive chemicals that are toxic, explosive, or flammable. The principal types of reactors for processing organic reactions such as polymerization, sulphonation, nitration, halogenation, or alkylation are batch and semi-batch reactors. In these operations, the reactions are exothermic and may lead to overheating. As the temperature rises, reaction rates increase exponentially, which is characterized by progressive increases

Table 4-6
Design guidelines for reactors

1. Single irreversible reactions (not autocatalytic)

(A) Isothermal—always use a plug flow reactor
(B) Adiabatic
   1. Plug flow if the reaction rate monotonically decreases with conversion
   2. CFSTR operating at the maximum reaction rate followed by a plug flow section

2. Single reversible reactions—adiabatic

(A) Maximum temperature—adiabatic
(B) A series of adiabatic beds with a decreasing temperature profile if exothermic

3. Parallel reactions-composition effects

(A) For A → R (desired) and A → S (waste), where the ratio of the reaction rates is $r_R/r_S = (k_1/k_2)C_A^{a_1-a_2}$
   1. If $a_1 > a_2$, keep C_A high
      a. Use batch or plug flow
      b. High pressure, eliminate inerts
      c. Avoid recycle of products
      d. Can use a small reactor
   2. If $a_1 < a_2$, keep C_A low
      a. Use a CFSTR with a high conversion
      b. Large recycle of products
      c. Low pressure, add inerts
(B) For A + B → R (desired) and A + B → S (waste), where the ratio of the rates is $r_R/r_S = (k_1/k_2)C_A^{a_1-a_2}C_B^{b_1-b_2}$
   1. If $a_1 > a_2$ and $b_1 > b_2$, both C_A and C_B high
   2. If $a_1 < a_2$ and $b_1 > b_2$, then C_A low, C_B high
   3. If $a_1 > a_2$ and $b_1 < b_2$, then C_A high, C_B low
   4. If $a_1 < a_2$ and $b_1 < b_2$, both C_A and C_B low
   5. See Figure 4-26 for various reactor configuration

4. Consecutive reactions-composition effects.

(A) A → R (desired) and R → S (waste)—minimize the mixing of streams with different compositions.

5. Parallel reactions-temperature effects $r_R/r_S = (k_1/k_2)f(C_A, C_B)$

(A) If $E_1 > E_2$, use a high temperature
(B) If $E_1 < E_2$, use an increasing temperature profile

6. Consecutive reactions-temperature effects $A \xrightarrow{k_1} R \xrightarrow{k_2} S$

(A) If $E_1 > E_2$, use a decreasing temperature profile—not very sensitive
(B) If $E_1 < E_2$, use a low temperature

Source: J. M. Douglas, Conceptual Design of Chemical Processes, *McGraw-Hill, Inc., 1988.*

in the generated heat rate. When the rate of heat generation is greater than the rate of heat absorbed, a thermal runaway will occur with disastrous consequences (as the incidents at Seveso and Bhopal have shown). A review of these incidents in batch reactors reveals the main causes are:

- Inadequate control systems and safety backup systems.
- Inadequate operational procedures, including training.
- Inadequate engineering design for heat transfer.
- Inadequate understanding of the chemistry process and thermochemistry.

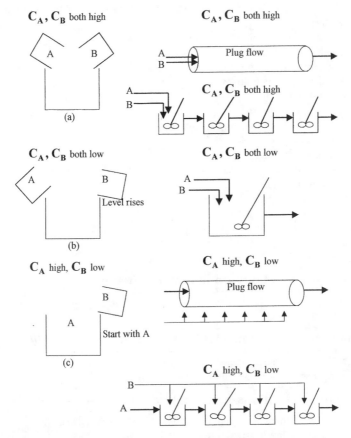

Figure 4-26. Parallel reactions. *(Source: O. Levenspiel,* Chemical Reaction Engineering, *2nd ed., Wiley, New York, 1972.)*

Analysis of thermal runaway reactions is reviewed in Chapter 12. Table 4-6 gives various guidelines for the design of reactors and Figure 4-26 illustrates various reactor configurations.

REFERENCES

1. Coker, A. K., "Understanding cyclone design," *Chem. Eng Progr,* pp. 51–55, December 1993.
2. Satterfield, C. N., *Mass Transfer in Heterogeneous Catalysis,* MIT Press, Cambridge, Mass, 1970.
3. Ramachandran, P. A. and Chaudhari, R. V., *Three Phase Catalytic Reactors,* Gordon and Breach Science Publishers.
4. Weekman, V. W., "Laboratory Reactors and Their Limitations," *AIChEJ,* Vol. 20, No. 5, pp. 833–839, 1974.
5. Douglas, J. M., *Conceptual Design of Chemical Processes,* McGraw-Hill, Inc., 1988.
6. Horak, J. and Pasek, J., *Design of Industrial Chemical Reactors from Laboratory Data,* Heyden & Son Ltd., 1978.
7. Smith, J. M., *Chemical Engineering Kinetics,* 3rd ed., McGraw-Hill Book Company, 1981.

CHAPTER FIVE

Introduction to Reactor Design Fundamentals for Ideal Systems

INTRODUCTION

In Chapter 4, the various types of laboratory and industrial reactors were reviewed. Reactor design is primarily concerned with the type and size of reactor and its method of operation for the required process. Important considerations are given at an early stage in the design of any process where chemical reaction is required. Designers must first consider the thermodynamics involving the equilibrium of reaction before investigating the chemical kinetics in relation to the reaction mechanisms. Figure 5-1 succinctly illustrates the logic of this approach. In this figure, consideration is given to the possibility of using either a gas phase or liquid phase reaction, whether the reaction is endothermic (i.e., absorbing heat) or exothermic (i.e., releasing heat), and whether or not the reaction is reversible.

There are advantages to using the liquid phase rather than the gas phase operation. For example, for the desired product the reactor may be smaller. This is because the physical properties of liquids are greater than those for gases, namely the heat capacities and thermal conductivities, factors which increase the heat transfer. Additionally, the equipment size is small resulting in lower power requirements and capital costs. The main disadvantages are corrosion and catalyst losses. In considering a liquid system, all operating conditions must fall within the two-phase region. If the critical temperature is not significantly above the desired reaction temperature, high operating pressures are potentially hazardous and expensive to contain, especially if one of the reactants is a noncondensable gas, which is required at high partial pressure.

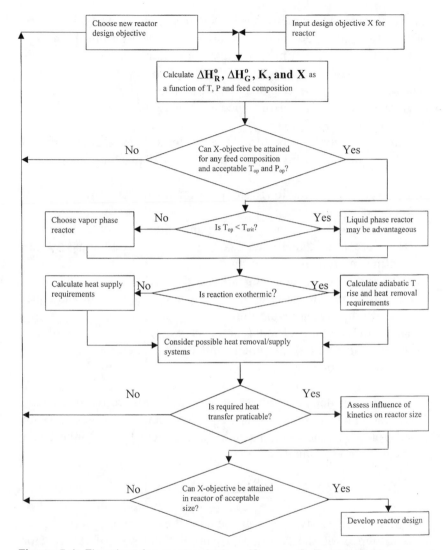

Figure 5-1. Flowchart for reactor design. *(Source: R. Scott with N. Macleod, "Process Design Case Studies." Used with permission of IChemE, 1992.)*

In the case of thermodynamics, the designer can investigate the nature of the reaction heat and whether the reaction is reversible. If these exothermic reactions are irreversible, attention may be focused on the influence of reactor design on conversion and with heat transfer control. An objective of reactor design is to determine the size and type of reactor and mode of operation for the required job. The choice of reactors depends on many factors such as safety, environment, and profit. Optimization techniques are often employed during the design stage to establish the optimum design from the profit viewpoint. This includes factors such as raw materials, initial and operating costs, and the market value of the finished products. The designer also requires knowledge of reactor performance before reviewing an optimization technique.

Chemical reactions are performed in reactor systems that are derived from one of the following basic types of model reactors:

- The well-mixed reaction system with uniform composition that is operated batchwise.
- The semibatch reactor where the incoming and outgoing mass flows are not equal to each other, and the total mass of the reacting mixture is not constant.
- The continuously operated stirred tank reaction in which the composition of the reaction mixture is assumed uniform and equal to the composition at the outlet.
- The tubular (plug flow) reactor in which piston flow of the reacting mixture is assumed, and there is neither mixing nor diffusion in the flow direction.

The following details establish reactor performance, considers the overall fractional yield, and predicts the concentration profiles with time of complex reactions in batch systems using the Runge-Kutta numerical method of analysis.

A GENERAL APPROACH

The rate equation involves a mathematical expression describing the rate of progress of the reaction. To predict the size of the reactor required in achieving a given degree of conversion of reactants and a fixed output of the product, the following information is required:

- Composition changes
- Temperature changes
- Mixing patterns
- Mass transfer
- Heat transfer

Information on the composition and temperature changes is obtained from the rate equation, while the mixing patterns are related to the intensity of mixing and reactor geometry. Heat transfer is referred to as the exothermic or endothermic nature of the reactions and the mass transfer to the heterogeneous systems.

GENERAL MASS BALANCE

For the ideal reactors considered, the design equations are based on the mass conservation equations. With this in mind, a suitable component is chosen (i.e., reactant or product). Consider an element of volume, δV, and the changes occurring between time t and t + δt (Figure 5-2):

$$\begin{pmatrix} \text{Input} \\ + \\ \text{Appearance} \\ \text{by reaction} \end{pmatrix} - \begin{pmatrix} \text{Output} \\ + \\ \text{Disappearance} \\ \text{by reaction} \end{pmatrix} = \text{Accumulation} \qquad (5\text{-}1)$$

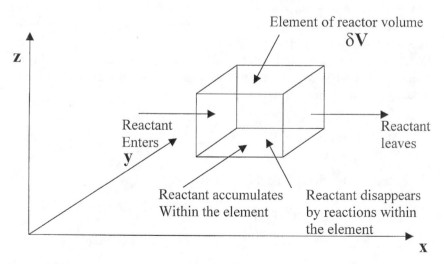

Figure 5-2. Material balance for a volume element of the reactor.

- *Input:* Includes the amount of reactant entering the system δV in δt by flow plus the amount of reactant formed by reaction in δV during δt.
- *Output:* Includes the amount of reactant leaving δV in δt by flow plus the amount of reactant destroyed by reaction in δV during δt.
- *Accumulation:* Is the amount of reactant in δV at $(t + \delta t)$ minus the amount of reactant in δV at t.
- *Ideal reactors:* Refers to the assumed mixing patterns in the reactor.
- *Perfect mixing:* Describes the contents as so well mixed that both composition and temperature are uniform throughout the system. The term "perfect mixing" is unambiguous as it refers to instantaneous and complete mixing on the molecular scale. Realistically, no reactor can attain this ideal behavior just as no tubular reactor can achieve a true piston (plug) flow situation. However, it is possible to design reactors that closely approach these conditions. Fluid mixing, residence time distribution, and micro and macro scale are reviewed in Chapters 7 and 8.

IDEAL ISOTHERMAL REACTORS

BATCH REACTORS

A well-mixed batch reactor has no input or output of mass. The amounts of individual components may change due to reaction, but not because of flow into or out of the system. A typical batch reactor is shown in Figure 5-3.

Consider a well-mixed batch reactor with a key reactant A, during time t to time $t + \delta t$, where δt is very small. For a well-mixed batch system, assume the following:

- Perfect mixing
- An isothermal operation
- Fluid density is constant

Rearranging the mass balance in Equation 5-1, gives

$$\underset{\text{No flow}}{\text{INPUT}} = \underset{\text{No flow}}{\text{OUTPUT}} + \underset{\substack{\text{Concentration of A} \\ \text{changes with time}}}{\text{ACCUMULATION}}$$

$$+ \underset{\substack{\text{Disappearance of A} \\ \text{by reaction}}}{}$$

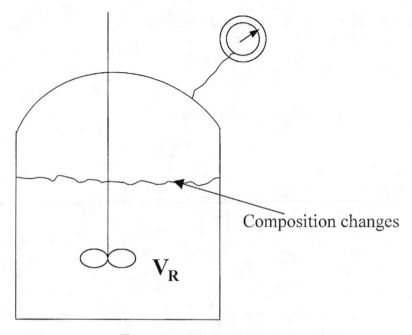

Figure 5-3. The batch reactor.

Assume δV_R to be V_R (i.e., the volume of the fluid in the reactor), the total reactor volume. Suppose that at time $t = 0$, the number of moles of A in $V_R = N_{AO}$, and at time $t = t$, the number of moles of A in $V_R = N_A$.

The rate of reaction $(-r_A)$ also varies with time t, since the concentration C_A changes with time. Using Taylor's series $y = f(t)$, gives

$$f(t+\delta t)=f(t) + \frac{df(t)}{dt}\delta t + \frac{d^2f(t)}{dt^2}\bullet\frac{\delta t^2}{2}+\ldots\ldots \tag{5-2}$$

when δt is small, terms containing δt become negligible.

time	t	$t + \delta t$
Moles of A in V_R	N_A	$N_A + \dfrac{dN_A}{dt}\delta t +\ldots..$
Net rate of reaction of A	$(-r_A)$	$\left(-r_A\right)+\dfrac{d\left(-r_A\right)}{dt}\delta t +\ldots.$

The average rate of reaction during δt is

$$\left(-r_A\right) + \frac{1}{2}\frac{d\left(-r_A\right)}{dt}\delta t + \ldots\ldots \tag{5-3}$$

During δt, OUTPUT = ACCUMULATION

$$\left(t+\delta t\right) - t$$

$$-\left\{\left(-r_A\right) + \frac{1}{2}\frac{d\left(-r_A\right)}{dt}\delta t\right\}V_R \bullet \delta t = \left\{N_A + \frac{dN_A}{dt}\delta t - N_A\right\} \tag{5-4}$$

$$-\left\{\left(-r_A\right) + \frac{1}{2}\frac{d\left(-r_A\right)}{dt}\delta t\right\}V_R = \frac{dN_A}{dt} \tag{5-5}$$

As $\delta t \to 0$, the term $1/2[d(-r_A)/dt]\, \delta t \to 0$, and Equation 5-5 becomes

$$-\left(-r_A\right)V_R = \frac{dN_A}{dt} \tag{5-6}$$

Rearranging Equation 5-6 gives

$$\left(-r_A\right) = -\frac{1}{V_R}\frac{dN_A}{dt} \tag{5-7}$$

In terms of the fractional conversion X_A

$$X_A = \frac{N_{AO} - N_A}{N_{AO}} \tag{5-8}$$

Rearranging Equation 5-8 gives

$$N_A = N_{AO}\left(1 - X_A\right) \tag{5-9}$$

Differentiating Equation 5-9 gives

$$dN_A = -N_{AO}\, dX_A \tag{5-10}$$

Substituting Equation 5-10 into Equation 5-7 gives

$$\left(-r_A\right)V_R = N_{AO}\frac{dX_A}{dt} \tag{5-11}$$

Rearranging and integrating Equation 5-11 between the limits at $t = 0$, $X_A = 0$ and at $t = t$, $X_A = X_{AF}$ results in

$$\int_0^t dt = N_{AO}\int_0^{X_{AF}}\frac{dX_A}{\left(-r_A\right)V_R}$$

$$t = N_{AO}\int_0^{X_{AF}}\frac{dX_A}{\left(-r_A\right)V_R} \tag{5-12}$$

where the concentration $C_A = \dfrac{N_A}{V_R}\left(\dfrac{moles}{volume}\right)$ (5-13)

The fractional conversion X_A at constant volume is

$$X_A = \frac{C_{AO} - C_A}{C_{AO}} \tag{5-14}$$

Substituting Equation 5-13 into Equation 5-11, rearranging and integrating between the limits at $t = 0$, $X_A = 0$ and at $t = t$, $X_A = X_{AF}$ yields

$$\int_0^t dt = C_{AO}\int_0^{X_{AF}}\frac{dX_A}{\left(-r_A\right)} \tag{5-15}$$

$$t = C_{AO}\int_0^{X_{AF}}\frac{dX_A}{\left(-r_A\right)} \tag{5-16}$$

Equation 5-16 is the time required to achieve a conversion X_A for either isothermal or non-isothermal operation. Equation 5-16 can also be expressed in terms of concentration at constant fluid density as

$$t = - \int_{C_{AO}}^{C_{AF}} \frac{dC_A}{(-r_A)} =, \quad \text{for } \varepsilon = 0 \qquad (5\text{-}17)$$

Equations 5-12, 5-16, and 5-17 are the performance equations for a batch reactor.

When the volume of reacting mixture changes proportionately with conversion, as in the case of gas phase reactions, this implies working volume changes with time,

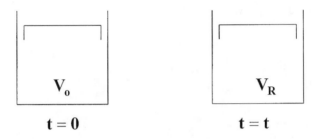

| Initial volume V_0 at $t = 0$ | final volume V_R at $t = t$ |

and Equation 5-12 becomes

$$V_R = V_O(1 + \varepsilon_A X_A) \qquad (5\text{-}18)$$

where V_O = initial volume of the reactor
$\quad\quad V_R$ = volume at time t

$$t = N_{AO} \int_0^{X_{AF}} \frac{dX_A}{(-r_A)V_O(1 + \varepsilon_A X_A)} = \frac{N_{AO}}{V_O} \int_0^{X_{AF}} \frac{dX_A}{(-r_A)(1 + \varepsilon_A X_A)}$$

$$= C_{AO} \int_0^{X_{AF}} \frac{dX_A}{(-r_A)(1 + \varepsilon_A X_A)} \qquad (5\text{-}19)$$

where ε_A is the fractional change in volume of the system between no conversion and complete conversion of reactant A given by

$$\varepsilon_A = \frac{V_{X_A=1} - V_{X_A=0}}{V_{X_A=0}} \qquad (5\text{-}20)$$

Batch reactors are charged with reactants, closed, and heated to the reaction temperature, which may be maintained (i.e., isothermally) for the duration of the reaction. After the reaction is completed, the mixture cooled, and the reactor opened, the product is discharged and the reactor is cleaned for the next batch. In industrial operations, the cycle time is constant from one batch to the next. The time required for filling, discharging, heating, cooling, and cleaning the reactor forms an integral part of the total batch cycle time (t_b) and is referred to as the turnaround time (t_t). The total batch cycle time t_b is the reaction t_r time plus the turnaround time t_t. This is expressed as:

$$t_b = t_r + t_t \tag{5-21}$$

From these various estimates, the total batch cycle time t_b is used in batch reactor design to determine the productivity of the reactor. Batch reactors are used in operations that are small and when multi-products are required. Pilot plant trials for sales samples in a new market development are carried out in batch reactors. Use of batch reactors can be seen in pharmaceutical, fine chemicals, biochemical, and dye industries. This is because multi-product, changeable demand often requires a single unit to be used in various production campaigns. However, batch reactors are seldom employed on an industrial scale for gas phase reactions. This is due to the limited quantity produced, although batch reactors can be readily employed for kinetic studies of gas phase reactions. Figure 5-4 illustrates the performance equations for batch reactors.

Following are examples for finding the time of an isothermal batch reactor for a given conversion of the reactant and other pertinent variables, and for gas phase reaction.

Example 5-1

Consider an isothermal batch reactor for a given conversion of the reactant

$$aA + bB \rightarrow cC + dD \tag{5-22}$$

where component A is the limiting reactant. The stoichiometric coefficients are: a = 1, b = 2, c = 1, and d = 0. The initial moles of components A, B, C, and D are $N_A = 0.001$ gmol, $N_B = 0.003$ gmol,

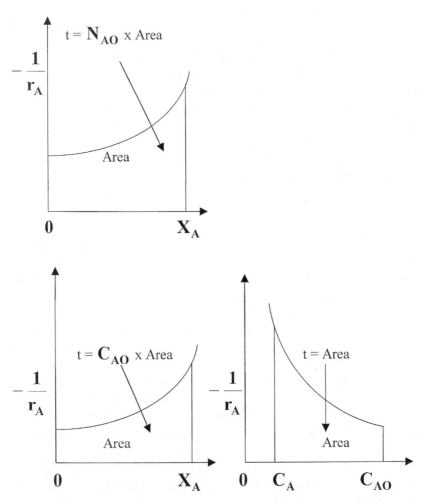

Figure 5-4. The performance equations for batch reactors.

$N_C = 0.0$ gmol, $N_D = 0.0$ gmol, respectively. A mixture of A and B is charged into a 1-liter reactor. Determine the holding time required to achieve 90% fractional conversion of A ($X_A = 0.9$). The rate constant is $k = 1.0 \times 10^5 [(\text{liter})^2/(\text{gmol}^2 \cdot \text{min})]$ and the reaction is first order in A, second order in B and third order overall.

Solution

Assuming a constant volume batch system, from the stoichiometric coefficients of Equation 5-22

$$A + 2B \xrightarrow{\text{k}} C \tag{5-23}$$

where component A is the limiting reactant. The material balance on component A in the batch reactor is

$$(-r_A) = -\frac{1}{V_R}\frac{dN_A}{dt} \tag{5-7}$$

where the concentration $C_A = N_A/V_R$ and the rate Equation 5-7 becomes

$$(-r_A) = -\frac{dC_A}{dt} = kC_A C_B^2 \tag{5-24}$$

The concentrations of A and B can be expressed in terms of the fractional conversion and the stoichiometry of the reaction as:

Component	A	B
At time t = 0	C_{AO}	C_{BO}
At time t = t	C_A	C_B
Amount reacted	$(C_{AO} - C_A)$	$(C_{BO} - C_B)$

From the stoichiometry,

$$2(C_{AO} - C_A) = C_{BO} - C_B \tag{5-25}$$

The concentration of B is expressed as

$$C_B = C_{BO} - 2(C_{AO} - C_A) \tag{5-26}$$

Using the fractional conversion X_A at constant volume

$$X_A = \frac{C_{AO} - C_A}{C_{AO}} \tag{5-13}$$

Therefore, the amount reacted

$$C_{AO} - C_A = C_{AO}X_A \tag{5-27}$$

or

$$C_A = C_{AO}(1 - X_A) \tag{5-28}$$

Differentiation of A in Equation 5-27 gives

$$-dC_A = C_{AO}\,dX_A \tag{5-29}$$

Substituting Equations 5-26, 5-28, and 5-29 into Equation 5-24 gives

$$C_{AO}\frac{dX_A}{dt} = kC_{AO}(1 - X_A)\left\{C_{BO} - 2(C_{AO} - C_A)\right\}^2 \tag{5-30}$$

$$= kC_{AO}(1 - X_A)(C_{BO} - 2C_{AO}X_A)^2 \tag{5-31}$$

$$C_{AO}\frac{dX_A}{dt} = kC_{AO}(1 - X_A)\left\{C_{AO}\left[\frac{C_{BO}}{C_{AO}} - 2X_A\right]\right\}^2$$

$$\frac{dX_A}{dt} = kC_{AO}^2(1 - X_A)(\theta_B - 2X_A)^2 \tag{5-32}$$

where $\theta_B = C_{BO}/C_{AO}$. Rearranging Equation 5-32 and integrating between the limits $t = 0$, $X_A = 0$, $t = t_{\text{holding time}}$, $X_A = X_{AF=0.9}$ gives

$$\int_{X_{A=0}}^{X_{A=0.9}} \frac{dX_A}{(1 - X_A)(\theta_B - 2X_A)^2} = kC_{AO}^2 \int_{t=0}^{t=t_{\text{holding time}}} dt \tag{5-33}$$

The integral on the left side of Equation 5-33 can be expressed by partial fraction to give

$$\frac{1}{(1 - X_A)(\theta_B - 2X_A)^2} \equiv \frac{A}{1 - X_A} + \frac{B}{\theta_B - 2X_A} + \frac{C}{(\theta_B - 2X_A)^2} \tag{5-34}$$

Rearranging Equation 5-34 gives

$$1 \equiv A(\theta_B - 2X_A)^2 + B(1 - X_A)(\theta_B - 2X_A) + C(1 - X_A) \tag{5-35}$$

Putting $X_A = 1$ into Equation 5-35 gives $1 = A(\theta_B - 2)^2$ or

$$A = \frac{1}{(\theta_B - 2)^2} \tag{5-36}$$

Putting $X_A = 0$ into Equation 5-35 gives

$$1 = A\theta_B^2 + B\theta_B + C \tag{5-37}$$

Putting $X_A = \theta_B/2$ into Equation 5-35 yields

$$1 = C\left(1 - \frac{\theta_B}{2}\right) \tag{5-38}$$

or

$$C = \frac{2}{2 - \theta_B} \tag{5-39}$$

Substituting Equations 5-36 and 5-39 into Equation 5-37 gives

$$1 = \frac{\theta_B^2}{(\theta_B - 2)^2} + B\theta_B + \frac{2}{(2 - \theta_B)} \tag{5-40}$$

$$B = -\frac{2}{(\theta_B - 2)^2} \tag{5-41}$$

The integral of Equation 5-34 between the limits becomes

$$I = \int_{X_A = 0}^{X_A = 0.9} \frac{dX_A}{(1 - X_A)(\theta_B - 2X_A)^2}$$

$$= \frac{1}{(\theta_B - 2)^2} \int_0^{X_A} \frac{dX_A}{1 - X_A} - \frac{2}{(\theta_B - 2)^2} \int_0^{X_A} \frac{dX_A}{(\theta_B - 2X_A)}$$

$$+ \frac{2}{(2 - \theta_B)} \int_0^{X_A} \frac{dX_A}{(\theta_B - 2X_A)^2} \tag{5-42}$$

$$I = -\frac{1}{(\theta_B - 2)^2} \left[\ln(1 - X_A) \right]_0^{X_A} + \frac{1}{(\theta_B - 2)^2} \left[\ln(\theta_B - 2X_A) \right]_0^{X_A}$$

$$+ \frac{1}{(2 - \theta_B)} \left[\frac{1}{(\theta_B - 2X_A)} \right]_0^{X_A} \tag{5-43}$$

or

$$\frac{1}{(\theta_B - 2)^2} \left[\ln\left\{ \frac{(\theta_B - 2X_A)}{\theta_B(1 - X_A)} \right\} - \frac{2X_A(\theta_B - 2)}{\theta_B(\theta_B - 2X_A)} \right] = kC_{AO}^2 t_{holding\ time} \tag{5-44}$$

The holding time $t_{holding\ time}$ is given by the expression

$$t_{holding\ time} = \frac{1}{kC_{AO}^2} \frac{1}{(\theta_B - 2)^2} \left[\ln\left\{ \frac{(\theta_B - 2X_A)}{\theta_B(1 - X_A)} \right\} - \frac{2X_A(\theta_B - 2)}{\theta_B(\theta_B - 2X_A)} \right] \tag{5-45}$$

Given the initial concentrations of A and B, the fractional conversion X_A, and the rate constant k as the following: $C_{AO} = 0.001$ gmol/1, $C_{BO} = 0.003$ gmol/1, $X_A = 0.9$, $k = 10^5 [(liter)^2/(gmol^2 \cdot min)]$, and $\theta_B = C_{BO}/C_{AO} = 3$, gives

$$\frac{1}{(3 - 2)^2} \left[\ln\left\{ \frac{(3 - 2 \times 0.9)}{3(1 - 0.9)} \right\} - \frac{2 \times 0.9(3 - 2)}{3(3 - 2 \times 0.9)} \right] = 10^5 \times (0.001)^2 \, t_{holding\ time}$$

$$1[1.386 - 0.5] = 10^{-1} t_{holding\ time}$$

or $t_{holding} = 8.86$ min. In other words, it would take approximately 9 minutes to convert 90% of reactant A charged to the 1-liter reactor. The Microsoft Excel spreadsheet (EXAMPLE5-1.xls) was used to calculate the holding time.

Example 5-2

Determine the conversion for an isothermal batch reactor using the stoichiometry of Example 5-1 and the same values of initial concentrations of A, B, C, and D in a reactor volume of 1 liter operating for 4 minutes. The rate constant is $k = 10^5 [(liter)^2/(gmol^2 \cdot min)]$.

Solution

An initial guess for the fractional conversion $X_{A,guess}$ on the limiting reactant A is substituted into the derived Equation 5-45. The time t_{guess} for the fractional conversion is calculated and compared with the given holding time ($t_{holding}$). If $t_{guess} < t_{holding}$, assume another guess for the fractional conversion until $t_{guess} \cong t_{holding}$. The fractional conversion $X_{A,guess}$ that gives $t_{guess} \cong t_{holding}$ is the actual conversion for the batch operation at that holding time. Table 5-1 gives the results of the simulation exercise for the 4-min batch operation (using Excel spreadsheet in Example 5-1).

In this exercise, the fractional conversion X_A for the 4-min isothermal batch operation is 77.1%. Obviously, the required answer is obtained through trial-and-error. Equation 5-45 can be incorporated into a computer program or mathematical package such as Mathcad, Maple, Polymath, or Mathematica.

Related calculation procedures can be made with different stoichiometry, which strongly influences the final form of the design expression. Also, different rate expressions often result in different relationships between time and conversion. In complex reactions, analytical integration of the design equation may be cumbersome and tedious, and perhaps impossible to solve. In such cases, the designer should resort to numerical integration methods such as the Euler Simpson's method or graphical evaluation of $\int_0^X f(X)dX$ by plotting $f(X)$ versus X and determining the area beneath the curve.

Example 5-3

Gas A decomposes irreversibly to form gas B according to the reaction A → 2B,

$$\left(-r_A\right) = kC_A^2 \tag{5-46}$$

The reaction is second order and is performed in an isothermal constant pressure batch reactor. Determine the conversion with time

Table 5-1

$X_{A,guess}$	0.75	0.76	0.77	0.771	0.772
t_{guess}, min	3.598	3.782	3.979	3.999	4.02

for the reaction assuming that A and B are ideal gases, and starting with pure A at N_{AO}. Derive a general expression for time t in terms of the fractional conversion X_A for an nth order reaction expressed by A → bB.

Solution

Since the volume depends on conversion or time in a constant pressure batch reactor, consider the mole balance in relation to the fractional conversion X_A. From the stoichiometry,

Component	A	B
At time t = 0	N_{AO}	0
At time t = t	N_A	N_B
Amount reacted	$(N_{AO} - N_A)$	N_B

From the stoichiometry,

$$2(N_{AO} - N_A) = N_B \tag{5-47}$$

the fractional conversion X_A is

$$X_A = \frac{N_{AO} - N_A}{N_{AO}} \tag{5-8}$$

From Equation 5-8, the amount reacted in terms of the fractional conversion and the initial moles is

$$N_{AO} - N_A = N_{AO}X_A \tag{5-48}$$

and

$$N_A = N_{AO}(1 - X_A) \tag{5-9}$$

Differentiating Equation 5-9 gives

$$dN_A = -N_{AO}dX_A \tag{5-10}$$

Therefore, the amount of component B is

$$N_B = 2N_{AO}X_A \tag{5-49}$$

ε_A is the fractional change in volume of the system between no conversion and complete conversion of reactant A given by

$$\varepsilon_A = \frac{V_{X_A=1} - V_{X_A=0}}{V_{X_A=0}} \tag{5-20}$$

For the gas phase reaction A $\rightarrow$ 2B, $\varepsilon_A = (2 - 1)/1 = 1$ or

$$V = V_O(1 + \varepsilon_A X_A) \tag{5-18}$$

Since two moles of B are formed for every mole of A that disappears, then the mole balance table becomes

Component	Mole initially at time t = 0	Moles at t = t
A	N_{AO}	$N_A = N_{AO}(1 - X_A)$
B	0	$2N_{AO}X_A$
Total moles	N_{AO}	$N_{AO}(1 + X_A)$
Volume	V_O	$V_O(1 + X_A)$

The batch reactor design equation for component A is

$$(-r_A) = -\frac{1}{V}\frac{dN_A}{dt} = kC_A^2 \tag{5-50}$$

Substituting Equations 5-9, 5-10, and 5-18 into Equation 5-50 yields

$$(-r_A) = \frac{N_{AO}}{V_O(1 + X_A)}\frac{dX_A}{dt} = k\left\{\frac{N_{AO}(1-X_A)}{V_O(1+X_A)}\right\}^2 \tag{5-51}$$

Rearranging Equation 5-51 gives

$$\frac{(1+X_A)dX_A}{(1-X_A)^2} = k\frac{N_{AO}}{V_O}dt \tag{5-52}$$

Integrating Equation 5-52 between the limits of fractional conversion $t = 0$, $X_A = 0$ and $t = t$, $X_A = X_A$ gives

$$\int_0^{X_A} \frac{(1+X_A)}{(1-X_A)^2} dX_A = k \frac{N_{AO}}{V_O} \int_0^t dt \tag{5-53}$$

Expressing the left side of Equation 5-53 into partial fraction gives

$$I = \frac{1+X_A}{(1-X_A)^2} \equiv \frac{A}{1-X_A} + \frac{B}{(1-X_A)^2} \tag{5-54}$$

$$1 + X_A = A(1-X_A) + B \tag{5-55}$$

Putting $X_A = 1$ into Equation 5-55 gives $2 = B$. Putting $X_A = 0$ into Equation 5-55 yields

$$1 = A + B \tag{5-56}$$

and $A = -1$. Substituting the values of A and B into the right side of Equation 5-54 and integrating between the limits results in

$$I = \int_0^{X_A} -\frac{1 \, dX_A}{(1-X_A)} + \int_0^{X_A} \frac{2 \, dX_A}{(1-X_A)^2} \tag{5-57}$$

$$I = \left[\ln(1-X_A)\right]_0^{X_A} + \left[\frac{2}{1-X_A}\right]_0^{X_A} = kC_{AO}t$$

$$\left\{\ln(1-X_A) + \frac{2X_A}{1-X_A}\right\} = kC_{AO}t$$

or

$$t = \frac{1}{kC_{AO}} \left\{\ln(1-X_A) + \frac{2X_A}{1-X_A}\right\} \tag{5-58}$$

Equation 5-58 gives an expression for the reaction time and fractional conversion for an isothermal constant pressure batch reactor. Considering an nth order irreversible reaction represented by $A \rightarrow bB$ with $(-r_A) = kC_A^n$, the design equation is

$$(-r_A) = -\frac{1}{V}\frac{dN_A}{dt} = kC_A^n$$

or in terms of the fractional conversion X_A,

$$\frac{dX_A}{dt} = kC_{AO}^{n-1}\left(1 - X_A\right)^n\left\{1 + (b-1)X_A\right\}^{1-n} \tag{5-59}$$

or

$$t = \frac{1}{kC_{AO}^{n-1}}\int_0^{X_A}\frac{dX_A}{\left(1 - X_A\right)^n\left\{1 + (b-1)X_A\right\}^{1-n}} \tag{5-60}$$

Equation 5-60 is the generalized equation, which can be used to determine the reaction time of an nth order reaction with b, the stoichiometric coefficient of the product component B.

NUMERICAL METHODS FOR REACTOR SYSTEMS DESIGN

In Chapter 3, the analytical method of solving kinetic schemes in a batch system was considered. Generally, industrial realistic schemes are complex and obtaining analytical solutions can be very difficult. Because this is often the case for such systems as isothermal, constant volume batch reactors and semibatch systems, the designer must review an alternative to the analytical technique, namely a numerical method, to obtain a solution. For systems such as the batch, semibatch, and plug flow reactors, sets of simultaneous, first order ordinary differential equations are often necessary to obtain the required solutions. Transient situations often arise in the case of continuous flow stirred tank reactors, and the use of numerical techniques is the most convenient and appropriate method.

Various types of numerical techniques have been developed based on the degree of accuracy and efficiency. One of the popular methods is the Runge-Kutta fourth order method, which is described in Appendix D. It is computationally efficient and can be generally recommended for all but very stiff sets of first order ordinary differential equations. The availability and the low cost of personal computers now allow designers to focus more attention on the physical insights in the design of reactor systems.

Here, the Runge-Kutta fourth order method is employed to solve a range of reaction kinetic schemes in reactor systems. In using this numerical technique, it is assumed that all concentrations are known at the initial time (i.e., t = 0). This allows the initial rate to be computed, one for each component. In choosing a small time increment, Δt, the concentrations will change very little. These small changes in concentrations can be computed assuming the reaction rates are constant. The new concentrations are used to recalculate the reaction rates. The process is repeated until the sum of Δt attains the specified final reaction time. Appendix D reviews other numerical methods for solving first order differential equations.

SERIES REACTIONS

A series of first order irreversible reactions is one in which an intermediate is formed that can then further react. A generalized series reaction is

$$A \xrightarrow{\ k_1\ } B \xrightarrow{\ k_2\ } C$$
$$\text{(Desired) (Unwanted)}$$

$$(5\text{-}61)$$

in a constant volume batch system and constant temperature T. For the batch reactor, the following rate equations can be written.

For component A disappearing in the system,

$$\left(-r_A\right) = -\frac{dC_A}{dt} = k_1 C_A \tag{5-62}$$

The rate equation of component B disappearing is:

$$\left(-r_B\right)_{net} = -\frac{dC_B}{dt} = k_2 C_B - k_1 C_A \tag{5-63}$$

The rate equation of component C formed is

$$\left(+r_C\right)= \frac{dC_C}{dt}=k_2C_B \tag{5-64}$$

Rearranging Equations 5-62 and 5-63 gives

$$\frac{dC_A}{dt}=-k_1C_A \tag{5-65}$$

$$\frac{dC_B}{dt}=k_1C_A-k_2C_B \tag{5-66}$$

Equations 5-64, 5-65, and 5-66 are first order differential equations, which require initial or boundary conditions. For the batch reactor, these are the initial concentrations of A, B, and C. In addition to the initial concentrations, the rate constants k_1 and k_2 are also required to simulate their concentrations. The concentration profiles depend on the values of k_1 and k_2 (i.e, $k_1 = k_2$, $k_1 > k_2$, $k_2 > k_1$). Assume that at the beginning of the batch process, at time t = 0, C_{AO} = 1.0 mol/m^3, and $C_{BO} = C_{CO}$ = 0. For known values of k_1 and k_2, simulate the concentrations of A, B, and C for 10 minutes at a time interval of t = 0.5 min. A computer program has been developed using the Runge-Kutta fourth order method to determine the concentrations of A, B, and C. The differential Equations 5-64, 5-65, and 5-66 are expressed, respectively, in the form of X-arrays and functions in the computer program as

$$C_A = X(1), \quad \frac{dC_A}{dt}=F(1) \tag{5-67}$$

$$C_B = X(2), \quad \frac{dC_B}{dt}=F(2) \tag{5-68}$$

$$C_C = X(3), \quad \frac{dC_C}{dt}=F(3) \tag{5-69}$$

where

$$F(1)=-K1*X(1) \tag{5-70}$$

$$F(2)=K1*X(1)-K2*X(2) \tag{5-71}$$

$$F(3) = K2 * X(2) \qquad\qquad (5\text{-}72)$$

The computer program BATCH51 determines the concentration profiles with time increment h = Δt = 0.5 min for a period of 10 minutes for $k_1 = k_2 = 1.0$ min^{-1}; $k_1 = 1.0$ min^{-1}, $k_2 = 0.1$ min^{-1}; and $k_1 = 0.1$ min^{-1}, $k_2 = 1.0$ min^{-1}, respectively. Tables 5-2, 5-3, and 5-4, respectively, give the results of the computer program and Figures 5-5, 5-6, and 5-7 illustrate the profiles for varying values of the reaction rate constants. The optimum concentration of the desired product B is determined from the plots.

For $k_2 \gg k_1$ in Figure 5-7, the concentration of the intermediate B remains small. The reaction scheme A $\xrightarrow{k_1}$ B $\xrightarrow{k_2}$ C, therefore, resembles the single reaction A $\xrightarrow{k_2}$ C, as B is negligibly small. As $k_2/k_1 > 50$, the numerical technique becomes unstable and the corresponding equations are known as "stiff differential equations." In this case, other numerical techniques may be used such as the one-point implicit method, the two-point trapezoidal rule, or multi-point methods.

An industrial example of series reactions is the substitution process involving methane and chlorine:

Table 5-2
Simulation of a chemical reaction kinetics
A → B → C in a batch reactor (k_1 = 1.0, k_2 = 1.0)

TIME	CONC. CA	CONC. CB	CONC. CC
.00	1.0000	.0000	.0000
.50	.6068	.3021	.0911
1.00	.3682	.3666	.2652
1.50	.2234	.3337	.4429
2.00	.1355	.2699	.5945
2.50	.0822	.2047	.7130
3.00	.0499	.1491	.8010
3.50	.0303	.1055	.8642
4.00	.0184	.0732	.9084
4.50	.0111	.0500	.9389
5.00	.0068	.0337	.9596
5.50	.0041	.0225	.9734
6.00	.0025	.0149	.9826
6.50	.0015	.0098	.9887
7.00	.0009	.0064	.9927
7.50	.0006	.0042	.9953
8.00	.0003	.0027	.9970
8.50	.0002	.0017	.9981
9.00	.0001	.0011	.9988
9.50	.0001	.0007	.9992
10.00	0.0000	.0005	.9995

Table 5-3
Simulation of a chemical reaction kinetics
A → B → C in a batch reactor (k_1 = 1.0, k_2 = 0.1)

TIME	CONC. CA	CONC. CB	CONC. CC
.00	1.0000	.0000	.0000
.50	.6068	.3827	.0105
1.00	.3682	.5963	.0355
1.50	.2234	.7081	.0685
2.00	.1355	.7591	.1054
2.50	.0822	.7739	.1438
3.00	.0499	.7677	.1824
3.50	.0303	.7493	.2204
4.00	.0184	.7244	.2572
4.50	.0111	.6961	.2928
5.00	.0068	.6664	.3268
5.50	.0041	.6365	.3594
6.00	.0025	.6070	.3905
6.50	.0015	.5784	.4201
7.00	.0009	.5507	.4483
7.50	.0006	.5242	.4752
8.00	.0003	.4989	.5008
8.50	.0002	.4747	.5251
9.00	.0001	.4516	.5483
9.50	.0001	.4296	.5703
10.00	0.0000	.4087	.5913

Table 5-4
Simulation of a chemical reaction kinetics
A → B → C in a batch reactor (k_1 = 0.1, k_2 = 1.0)

TIME	CONC. CA	CONC. CB	CONC. CC
.00	1.0000	.0000	.0000
.50	.9512	.0383	.0105
1.00	.9048	.0596	.0355
1.50	.8607	.0708	.0685
2.00	.8187	.0759	.1054
2.50	.7788	.0774	.1438
3.00	.7408	.0768	.1824
3.50	.7047	.0749	.2204
4.00	.6703	.0724	.2572
4.50	.6376	.0696	.2928
5.00	.6065	.0666	.3268
5.50	.5769	.0636	.3594
6.00	.5488	.0607	.3905
6.50	.5220	.0578	.4201
7.00	.4966	.0551	.4483
7.50	.4724	.0524	.4752
8.00	.4493	.0499	.5008
8.50	.4274	.0475	.5251
9.00	.4066	.0452	.5483
9.50	.3867	.0430	.5703
10.00	.3679	.0409	.5913

Simulation of a series reactions A--->B---->C in a batch reactor

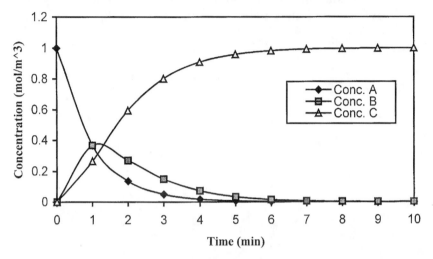

Figure 5-5. Concentrations versus time of A, B, and C in a series reaction $A \xrightarrow{k_1} B \xrightarrow{k_2} C$ for $k_2/k_1 = 1$.

Simulation of a series reactions A--->B---->C in a batch reactor

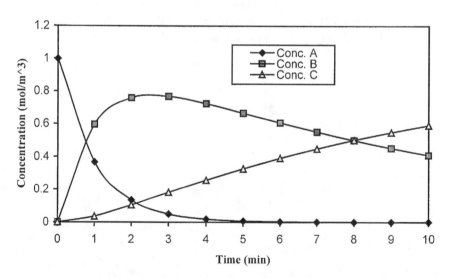

Figure 5-6. Concentrations versus time of A, B, and C in a series reaction $A \xrightarrow{k_1} B \xrightarrow{k_2} C$ for $k_2/k_1 = 0.10$.

Simulation of a series reactions A--->B---->C in a batch reactor

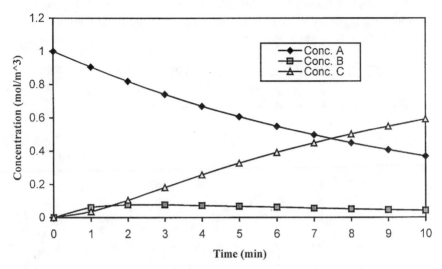

Figure 5-7. Concentrations versus time of A, B, and C in a series reaction $A\xrightarrow{k_1}B\xrightarrow{k_2}C$ for $k_2/k_1 = 10$.

$$CH_4 + Cl_2 \rightarrow CH_3Cl + HCl$$

$$CH_3Cl + Cl_2 \rightarrow CH_2Cl + HCl$$

$$CH_2Cl_2 + Cl_2 \rightarrow CHCl_3 + HCl$$

$$CH_3Cl + Cl_2 \rightarrow CCl_4 + HCl$$

Series reactions also occur in oxidation processes where the required product may oxidize further. An example is the production of maleic anhydride from the oxidation of benzene. In this case, the maleic anhydride can be oxidized further to carbon dioxide and water. Another example of a series reaction occurs in the biochemical reaction in which *Pseudomonas ovalis* is used to convert glucose to gluconic acid via gluconolactone in batch culture. This first order reaction is represented by

$$G\xrightarrow{k_1}L\xrightarrow{k_2}P$$

glucose gluconolactone gluconic acid

The time t_{max} at which the concentration of the intermediate (gluconolactone) is maximized and the maximum concentration gluconolactone $(C_L)_{max}$ have already been reviewed in Chapter 3. The simulation of this reaction is reviewed in Chapter 11.

SIMULTANEOUS IRREVERSIBLE SIDE REACTIONS

Consider the reaction schemes

$$
\begin{array}{l}
A \xrightarrow{\ k_1\ } B \\[1em]
\quad \Big\downarrow {\scriptstyle k_2} \\[1em]
\quad C
\end{array}
\qquad (5\text{-}73)
$$

in a constant volume batch isothermal reactor with known values of the rate constants k_1 and k_2, and initial concentrations of $C_{AO} = 1.0$ mol/m^3 and $C_{BO} = C_{CO} = 0.0$ mol/m^3. The reactions are first order. The concentrations of A, B, and C can be simulated with time using the following rate equations:

$$
\left(-r_A\right) = -\frac{dC_A}{dt} = \left(k_1 + k_2\right)C_A \qquad (5\text{-}74)
$$

$$
\left(+r_B\right) = \frac{dC_B}{dt} = k_1 C_A \qquad (5\text{-}75)
$$

$$
\left(+r_C\right) = \frac{dC_C}{dt} = k_2 C_A \qquad (5\text{-}76)
$$

Rearranging Equation 5-74 gives

$$
\frac{dC_A}{dt} = -\left(k_1 + k_2\right)C_A \qquad (5\text{-}77)
$$

Equations 5-75, 5-76, and 5-77 are first order differential equations. The concentrations of A, B, and C can be simulated for $k_1 = 0.25$ hr^{-1}

and $k_2 = 0.75$ hr^{-1} for a period of 5 hours with time increment h = Δt = 0.5 hr^{-1}. Using the developed Runge-Kutta fourth order computer program BATCH52, the concentrations of A, B, and C can be predicted for any specified time duration. Figure 5-8 shows the plots of these concentrations. The figure shows the exponential decrease with time of component A, while B and C increase asymptotically to their ultimate concentrations.

REVERSIBLE SERIES REACTIONS

Consider chemical reaction schemes of the form

$$A \underset{k_2}{\overset{k_1}{\rightleftharpoons}} B \xrightarrow{k_3} C \qquad\qquad (5\text{-}78)$$

in which A reacts to form B and C in a constant volume batch reactor under isothermal condition. The rate equations for components A, B, and C are

**Simulation of a simultaneous irreversible side reaction
A---->B, A---->C**

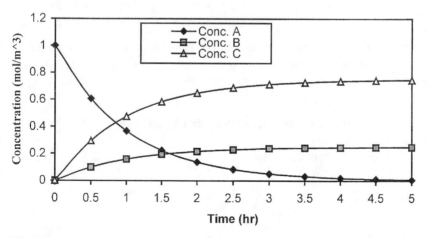

Figure 5-8. Concentrations versus time of A, B, and C in a simultaneous irreversible reaction for $k_1 = 0.25$ hr^{-1} and $k_2 = 0.75$ hr^{-1}.

$$\left(-r_A\right)_{net} = -\frac{dC_A}{dt} = k_1 C_A - k_2 C_B \tag{5-79}$$

$$\left(-r_B\right)_{net} = -\frac{dC_B}{dt} = k_2 C_B + k_3 C_B - k_1 C_A \tag{5-80}$$

$$\left(+r_C\right) = \frac{dC_C}{dt} = k_3 C_B \tag{5-81}$$

Rearranging Equations 5-79 and 5-80 gives

$$\frac{dC_A}{dt} = k_2 C_B - k_1 C_A \tag{5-82}$$

$$\frac{dC_B}{dt} = k_1 C_A - \left(k_2 + k_3\right) C_B \tag{5-83}$$

Equations 5-81, 5-82 and 5-83 are first order differential equations that can be solved simultaneously using the Runge-Kutta fourth order method. Consider two cases:

- *Case I:*   At time t = 0, $C_{AO} = 1.0$, $C_{BO} = C_{CO} = 0$, and $k_1 = k_2 = 1$, $k_3 = 10$
- *Case II:* At time t = 0, $C_{AO} = 1.0$, $C_{BO} = C_{CO} = 0$, and $k_1 = 1$, $k_2 = k_3 = 10$

The developed batch program BATCH53 simulates the concentrations of A, B, and C with time step $\Delta t = 0.05$ hr for 1 hour. Figures 5-9 and 5-10 show plots of the concentrations versus time for both cases.

CONSECUTIVE REVERSIBLE REACTIONS

Consider the chemical reversible reactions

$$A \underset{k_2}{\overset{k_1}{\longleftrightarrow}} B \underset{k_4}{\overset{k_3}{\longleftrightarrow}} C \tag{5-84}$$

in a constant volume batch reactor under isothermal condition. For first order reaction kinetics, the rate equations are:

Simulation of a reversible series reaction
A<===>B---->C

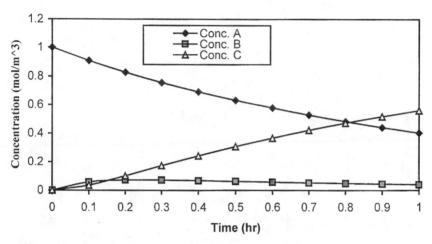

Figure 5-9. Concentrations versus time of A, B, and C in a simultaneous irreversible reaction for $k_1 = k_2 = 1.0$ hr^{-1} and $k_3 = 10$ hr^{-1}.

Simulation of a reversible series reaction
A<===>B---->C

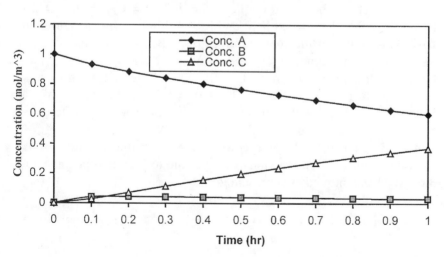

Figure 5-10. Concentrations versus time of A, B, and C in a simultaneous irreversible reaction for $k_1 = 1.0$ hr^{-1} and $k_2 = k_3 = 10$ hr^{-1}.

$$\left(-r_A\right)=-\frac{dC_A}{dt}=k_1C_A-k_2C_B \tag{5-85}$$

$$\left(-r_B\right)_{net}=-\frac{dC_B}{dt}=\left(k_2+k_3\right)C_B-k_1C_A-k_4C_C \tag{5-86}$$

$$\left(-r_C\right)=-\frac{dC_C}{dt}=k_4C_C-k_3C_B \tag{5-87}$$

Rearranging Equations 5-85, 5-86, and 5-87 gives

$$\frac{dC_A}{dt}=k_2C_B-k_1C_A \tag{5-88}$$

$$\frac{dC_B}{dt}=k_1C_A+k_4C_C-\left(k_2+k_3\right)C_B \tag{5-89}$$

$$\frac{dC_C}{dt}=k_3C_B-k_4C_C \tag{5-90}$$

Equations 5-88, 5-89, and 5-90 are first order differential equations and the Runge-Kutta fourth order method with the boundary conditions is used to determine the concentrations versus time of the components.

At the start of the batch, $t = 0$, $C_{AO} = 1$, $C_{BO} = C_{CO} = 0$, $k_1 = 1.0$ hr^{-1}, $k_2 = 2.0$ hr^{-1}, $k_3 = 3.0$ hr^{-1}, and $k_4 = 4.0$ hr^{-1}. Computer program BATCH54 simulates the concentrations of A, B, and C for the duration of 2 hours with a time increment $h = \Delta t = 0.2$ hr. Figure 5-11 shows the concentrations versus time of A, B, and C. Note that when increasing the time increment beyond $h = \Delta t = 0.3$ hr, the values of the concentrations become very unstable and the corresponding differential equations are said to be "stiff."

An industrial example of a consecutive reversible reaction is the catalytic isomerization reactions of n-hexane to 2-methyl pentane and 3-methyl pentane and is represented as:

$$nC_6H_{14} \underset{k_2}{\overset{k_1}{\longleftrightarrow}} 2-\text{methyl pentane} \underset{k_4}{\overset{k_3}{\longleftrightarrow}} 3-\text{methyl pentane}$$

$$A \qquad\qquad\qquad B \qquad\qquad\qquad C$$

where 2-methyl pentane is the most desirable product.

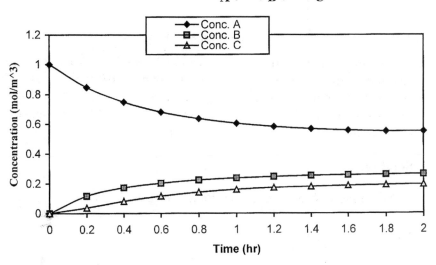

Figure 5-11. Concentrations versus time of A, B, and C in a simultaneous irreversible reaction for k_1 = 1.0 hr^{-1}, k_2 = 2.0 hr^{-1}, k_3 = 3 hr^{-1}, and k_4 = 4 hr^{-1}.

PARALLEL SECOND ORDER REACTIONS

Consider an irreversible chemical reactions scheme of the form

$$A + B \xrightarrow{\ k_1\ } C$$

$$A + C \xrightarrow{\ k_2\ } D \tag{5-91}$$

in which A reacts with B to form C and with further reaction of A with C to form D in a constant volume batch reactor under isothermal conditions. The rate equations for components A, B, C, and D are

$$\left(-r_A\right)_{net} = -\frac{dC_A}{dt} = k_1 C_A C_B + k_2 C_A C_C \tag{5-92}$$

$$\left(-r_B\right) = -\frac{dC_B}{dt} = k_1 C_A C_B \tag{5-93}$$

$$\left(-r_C\right)_{net} = -\frac{dC_C}{dt} = k_2 C_A C_C - k_1 C_A C_B \tag{5-94}$$

$$\left(+r_D\right) = \frac{dC_D}{dt} = k_2 C_A C_C \tag{5-95}$$

Rearranging Equations 5-92, 5-93, and 5-94 gives

$$\frac{dC_A}{dt} = -\left(k_1 C_A C_B + k_2 C_A C_C\right) \tag{5-96}$$

$$\frac{dC_B}{dt} = -k_1 C_A C_B \tag{5-97}$$

$$\frac{dC_C}{dt} = k_1 C_A C_B - k_2 C_A C_C \tag{5-98}$$

Equations 5-95, 5-96, 5-97, and 5-98 are first order differential equations and the Runge-Kutta fourth order method is used to simulate the concentrations, with time, of components A, B, C, and D at varying ratios of k_2/k_1 of 0.5, 1.0, and 2.0 for a time increment $\Delta t = 0.5$ min. At the start of the batch process, the initial concentrations at time t = 0, are $C_{AO} = C_{BO} = 1.0$ mol/m^3, $C_{CO} = C_{DO} = 0.0$ mol/m^3. The computer program BATCH55 simulates the concentrations of A, B, C, and D for a period of 5 minutes. Figures 5-12, 5-13, and 5-14, respectively, show the profiles of the concentrations with time, and the effect of k_2/k_1 on the concentrations of species of A, B, C, and D. The maximum conversion of component C depends on the ratio of k_2/k_1.

COMPLEX REACTIONS IN A BATCH SYSTEM

Industrial chemical reactions are often more complex than the earlier types of reaction kinetics. Complex reactions can be a combination of consecutive and parallel reactions, sometimes with individual steps being reversible. An example is the chlorination of a mixture of benzene and toluene. An example of consecutive reactions is the chlorination of methane to methyl chloride and subsequent chlorination to yield carbon tetrachloride. A further example involves the chlorination of benzene to monochlorobenzene, and subsequent chlorination

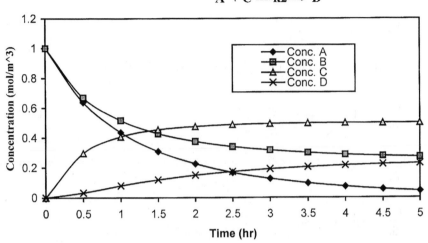

Figure 5-12. Concentrations versus time of A, B, C, and D in parallel second order reactions for k_1 = 1.0 m^3/mol • min and k_2 = 0.5 m^3/mol • min.

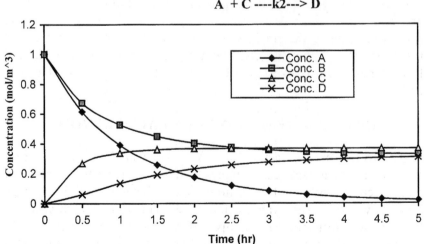

Figure 5-13. Concentrations versus time of A, B, C, and D in parallel second order reactions for k_1 = 1.0 m^3/mol • min and k_2 = 1.0 m^3/mol • min.

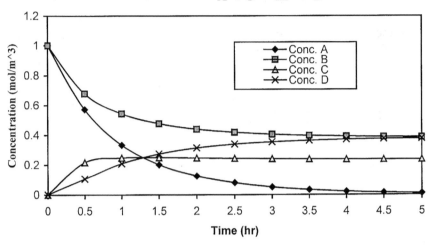

Simulation of parallel second order reactions

A + B ----k1---> C

A + C ----k2---> D

Figure 5-14. Concentrations versus time of A, B, C, and D in parallel second order reactions for k_1 = 1.0 m^3/mol • min and k_2 = 2.0 m^3/mol • min.

of the mono- to yield dichlorobenzene and trichlorobenzene, respectively. These reaction mechanisms are expressed as:

$$C_6H_6 + Cl_2 \xrightarrow{k_1} C_6H_5Cl + HCl$$

$$C_6H_5Cl + Cl_2 \xrightarrow{k_2} C_6H_4Cl_2 + HCl$$

$$C_6H_4Cl_2 + Cl_2 \xrightarrow{k_3} C_6H_3Cl_3 + HCl$$

Analysis in terms of series and parallel steps is:

$$C_6H_6 \rightarrow C_6H_5Cl \rightarrow C_6H_4Cl_2 \rightarrow C_6H_3Cl_3$$

$$\nearrow C_6H_5Cl$$

$$Cl_2 \longrightarrow C_6H_4Cl_2$$

$$\searrow C_6H_3Cl_3$$

When a substance participates in several reactions at the same time as exemplified in the above reaction, its net formation rate or disappearance is the algebraic sum of its rates in the elementary reactions.

The following examples review some complex reactions and determine the concentrations history for a specified period using the Runge-Kutta fourth order numerical method.

Example 5-4

Consider the reaction scheme

$$A \xrightarrow{k_1} B$$

$$B \underset{k_3}{\overset{k_2}{\rightleftharpoons}} C$$

$$B \xrightarrow{k_4} D \tag{5-99}$$

in a constant volume batch reactor under isothermal conditions. The rate constants are $k_1 = 5 \text{ hr}^{-1}$, $k_2 = 4 \text{ hr}^{-1}$, and $k_3 = k_4 = 3 \text{ hr}^{-1}$. The initial concentrations are $C_{AO} = 16$ mol/l and $C_{BO} = C_{CO} = C_{DO} = 0$ mol/l. The concentrations of A, B, C, and D with time for a period of 2 hours will be determined.

Solution

Assuming that the reactions are first order in a constant volume batch reactor, the rate equations for components A, B, C, and D, respectively, are:

$$\left(-r_A\right) = -\frac{dC_A}{dt} = k_1 C_A \tag{5-100}$$

$$\left(-r_B\right)_{net} = -\frac{dC_B}{dt} = \left(k_2 + k_4\right)C_B - k_1 C_A - k_3 C_C \tag{5-101}$$

$$\left(-r_C\right) = -\frac{dC_C}{dt} = k_3 C_C - k_2 C_B \tag{5-102}$$

$$\left(+r_D\right)=\frac{dC_D}{dt}=k_4C_B \tag{5-103}$$

Rearranging Equations 5-100, 5-101, and 5-102 gives

$$\frac{dC_A}{dt}=-k_1C_A \tag{5-104}$$

$$\frac{dC_B}{dt}=k_1C_A+k_3C_C-\left(k_2+k_4\right)C_B \tag{5-105}$$

$$\frac{dC_C}{dt}=k_2C_B-k_3C_C \tag{5-106}$$

Equations 5-103, 5-104, 5-105, and 5-106 are first order differential equations. The computer program BATCH56 simulates the above differential equations with time increment $=\Delta t=0.2$ hr. Table 5-5 gives the results of the simulation and Figure 5-15 shows plots of the concentrations versus time.

Example 5-5

Consider the simulation of a constant volume batch reaction of the form

$$A + D \xrightarrow{\ k_1\ } B$$

$$B + D \xrightarrow{\ k_2\ } C$$

$$D + D \xrightarrow{\ k_3\ } E \tag{5-107}$$

The reactor initially contains 0.12 mol/m^3 of A and 0.4 mol/m^3 of D with the rate constants $k_1=0.4$, $k_2=0.2$, and $k_3=0.05$ m^3/mol • min. What are the compositions of A, B, C, and D after 5 minutes?

Solution

Assume the rates to be second order in a constant volume batch system and determine the rate equations as follows.

Table 5-5
Simulation of complex chemical reactions
A → B, B ↔ C, B → D in a batch reactor
(k_1 = 5.0, k_2 = 4.0, k_3 = 3.0, k_4 = 3.0)

TIME	CONC. CA	CONC. CB	CONC. CC	CONC. CD
.00	16.0000	.0000	.0000	.0000
.20	6.000	3.563	3.541	2.896
.40	2.250	3.536	4.705	5.509
.60	.844	2.923	4.666	7.568
.80	.316	2.345	4.164	9.175
1.00	.119	1.885	3.549	10.447
1.20	.044	1.526	2.962	11.468
1.40	.017	1.241	2.447	12.295
1.60	.006	1.013	2.012	12.969
1.80	.002	.828	1.650	13.520
2.00	.001	.677	1.352	13.970

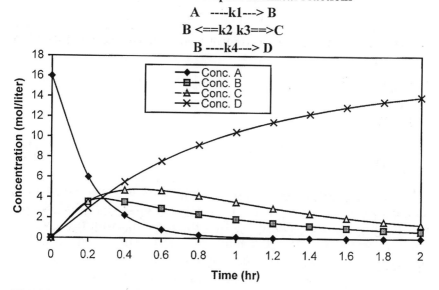

Simulation of complex chemical reactions
A ----k1---> B
B <==k2 k3==>C
B ----k4---> D

Figure 5-15. Concentrations versus time of A, B, C, and D in complex reactions for k_1 = 5.0 hr^{-1}, k_2 = 4 hr^{-1}, and k_3 = k_4 = 3 hr^{-1}.

For component A, the reaction rate is expressed by

$$\left(-r_A\right) = -\frac{dC_A}{dt} = k_1 C_A C_D \tag{5-108}$$

The reaction rate of component B is

$$\left(-r_B\right)_{net} = -\frac{dC_B}{dt} = k_2 C_B C_D - k_1 C_A C_D \tag{5-109}$$

The rate of appearance of component C is

$$\left(+r_C\right) = \frac{dC_C}{dt} = k_2 C_B C_D \tag{5-110}$$

The reaction rate of component D is

$$\left(-r_D\right)_{net} = -\frac{dC_D}{dt} = k_1 C_A C_D + k_2 C_B C_D + k_3 C_D^2 \tag{5-111}$$

Rearranging Equations 5-108, 5-109, and 5-111, respectively, gives

$$\frac{dC_A}{dt} = -k_1 C_A C_D \tag{5-112}$$

$$\frac{dC_B}{dt} = k_1 C_A C_D - k_2 C_B C_D \tag{5-113}$$

$$\frac{dC_D}{dt} = -\left(k_1 C_A C_D + k_2 C_B C_D + k_3 C_D^2\right) \tag{5-114}$$

Equations 5-110, 5-112, 5-113, and 5-114 are first order differential equations and the Runge-Kutta fourth order numerical method is used to determine the concentrations of A, B, C, and D, with time, with a time increment h = Δt = 0.5 min for a period of 10 minutes. The computer program BATCH57 determines the concentration profiles at an interval of 0.5 min for 10 minutes. Table 5-6 gives the results of the computer program and Figure 5-16 shows the concentration profiles of A, B, C, and D from the start of the batch reaction to the final time of 10 minutes.

Table 5-6
Simulation of complex chemical reactions
A + D → B, B + D → C, D + D → E in a batch reactor
(k_1 = 0.4, k_2 = 0.2, k_3 = 0.05)

TIME	CONC. CA	CONC. CB	CONC. CC	CONC. CD
.00	.2000	.0000	.0000	.4000
.50	.1850	.0147	.0003	.3809
1.00	.1717	.0272	.0011	.3634
1.50	.1599	.0378	.0022	.3473
2.00	.1494	.0469	.0037	.3324
2.50	.1400	.0547	.0053	.3187
3.00	.1315	.0613	.0071	.3060
3.50	.1239	.0670	.0091	.2942
4.00	.1169	.0720	.0111	.2831
4.50	.1106	.0762	.0131	.2728
5.00	.1048	.0799	.0152	.2631
5.50	.0996	.0831	.0173	.2541
6.00	.0947	.0858	.0195	.2456
6.50	.0902	.0882	.0216	.2375
7.00	.0861	.0902	.0236	.2300
7.50	.0823	.0920	.0257	.2228
8.00	.0788	.0935	.0277	.2160
8.50	.0755	.0948	.0297	.2096
9.00	.0724	.0959	.0317	.2035
9.50	.0696	.0968	.0336	.1977
10.00	.0669	.0975	.0355	.1922

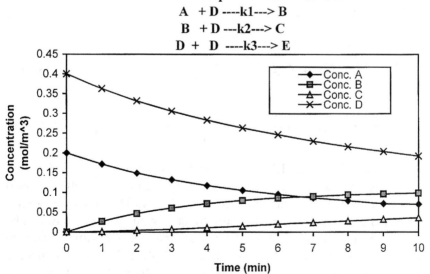

Simulation of complex chemical reactions
A + D ----k1---> B
B + D ---k2---> C
D + D ----k3---> E

Figure 5-16. Concentrations versus time of A, B, C, and D in complex reactions for k_1 = 0.40 m^3/mol•min, k_2 = 0.2 m^3/mol•min, and k_3 = 0.05 m^3/mol•min.

Example 5-6

Consider the reaction scheme

$$A \xrightleftharpoons[k_2]{k_1} B \xrightarrow{k_3} C$$

$$k_5 \Big\uparrow \Big\downarrow k_4 \tag{5-115}$$

$$D$$

in a constant volume batch reactor under isothermal conditions, and with the rate constants $k_1 = 0.4$ hr^{-1}, $k_2 = 0.1$ hr^{-1}, $k_3 = 0.16$ hr^{-1}, $k_4 = 0.08$ hr^{-1}, and $k_5 = 0.13$ hr^{-1}. The original compositions in the batch are:

$C_A = 7$ gmol/l

$C_B = 2.0$ gmol/l

$C_C = 0$ gmol/l

$C_D = 0.5$ gmol/l

Determine the concentrations of each component as a function of time for 2 hours.[1]

Solution

From the reaction scheme, it can be assumed the rate equations are first order. The rate of disappearance of component A is:

$$\left(-r_A\right)_{net} = -\frac{dC_A}{dt} = k_1 C_A - k_2 C_B \tag{5-116}$$

The rate of disappearance of component B is:

---

[1]D. M. Himmelblau and K. B. Bischoff, Process Analysis & Simulation-Deterministic Systems, John Wiley, 1968.

$$\left(-r_B\right)_{net} = -\frac{dC_B}{dt} = \left(k_2 + k_3 + k_4\right)C_B - k_1C_A - k_5C_D \tag{5-117}$$

The rate of formation of component C is:

$$\left(r_C\right) = \frac{dC_C}{dt} = k_3C_B \tag{5-118}$$

The rate of disappearance of component D is:

$$\left(-r_D\right)_{net} = -\frac{dC_D}{dt} = k_5C_D - k_4C_B \tag{5-119}$$

Rearranging Equations 5-116, 5-117, and 5-119, respectively, gives

$$\frac{dC_A}{dt} = k_2C_B - k_1C_A \tag{5-120}$$

$$\frac{dC_B}{dt} = k_1C_A + k_5C_D - \left(k_2 + k_3 + k_4\right)C_B \tag{5-121}$$

$$\frac{dC_D}{dt} = k_4C_B - k_5C_D \tag{5-122}$$

Equations 5-118, 5-120, 5-121, and 5-122 are first order differential equations. A simulation exercise on the above equations using the Runge-Kutta fourth order method, can determine the number of moles with time increment h = Δt = 0.2 hr for 2 hours. Computer program BATCH58 evaluates the number of moles of each component as a function of time. Table 5–7 gives the results of the simulation, and Figure 5-17 shows the plots of the concentrations versus time.

Example 5-7

A pair of reactions

$$A + B \xrightarrow{\ k_1\ } 2C$$

$$A + C \xrightarrow{\ k_2\ } D \tag{5-123}$$

Table 5-7
Simulation of complex chemical reactions in a batch reactor
(k_1 = 0.4, k_2 = 0.1, k_3 = 0.16, k_4 = 0.08, k_5 = 0.13)

TIME	CONC. CA	CONC. CB	CONC. CC	CONC. CD
.00	7.0000	2.0000	.0000	.5000
.20	6.5043	2.4030	.0706	.5220
.40	6.0538	2.7437	.1531	.5494
.60	5.6440	3.0294	.2456	.5809
.80	5.2707	3.2670	.3465	.6158
1.00	4.9302	3.4623	.4543	.6532
1.20	4.6193	3.6206	.5677	.6924
1.40	4.3350	3.7465	.6856	.7329
1.60	4.0747	3.8441	.8072	.7741
1.80	3.8361	3.9170	.9314	.8155
2.00	3.6170	3.9685	1.0576	.8569

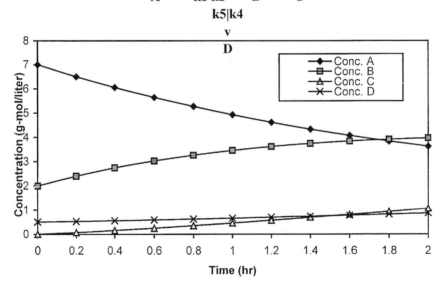

Figure 5-17. Concentrations versus time of A, B, C, and D in complex reactions for k_1 = 0.4 hr^{-1}, k_2 = 0.1 hr^{-1}, k_3 = 0.16 hr^{-1}, k_4 = 0.08 hr^{-1}, and k_5 = 0.13 hr^{-1}.

is carried out in a constant volume batch reactor with $k_1 = 0.1$ ft³/
(lb-mol)(min) and $k_2 = 0.05$ ft³/(lb-mol)(min). The concentrations of
A and B are such that at time t = 0, $C_{AO} = 3C_{BO} = 0.9$ lb-mol/ft³.
The concentrations of C and D at initial time t = 0 are $C_{CO} = C_{DO} =$
0.0 lb-mol/ft³. Determine the concentrations of A, B, C, and D as a
function of time for the duration of 12 minutes.[2]

Solution

For complete

reaction to **D**

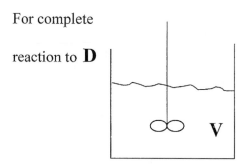

The batch reactor

Assuming that the rate equations are second order, the rate of
disappearance of component A is:

$$\left(-r_A\right) = -\frac{dC_A}{dt} = k_1 C_A C_B + k_2 C_A C_C \qquad (5\text{-}124)$$

The rate of disappearance of component B is

$$\left(-r_B\right) = -\frac{dC_B}{dt} = k_1 C_A C_B \qquad (5\text{-}125)$$

Rearranging Equation 5-124 and 5-125 gives

$$\frac{dC_A}{dt} = -\left(k_1 C_A C_B + k_2 C_A C_C\right) \qquad (5\text{-}126)$$

[2]S. M. Walas, Reaction Kinetics for Chemical Engineers, McGraw-Hill, NY, 1959.

$$\frac{dC_B}{dt} = -k_1 C_A C_B \tag{5-127}$$

The material balance from the stoichiometry is

Component	A	B	C	D
At time t = 0	C_{AO}	C_{BO}	0	0
At time t = t	C_A	C_B	C_C	C_D
Amount reacted	$C_{AO} - C_A$	$C_{BO} - C_B$	C_C	C_D

The reaction steps are parallel with respect to component A and series with respect to component B. That is, $A + B \rightarrow 2C$; $2A + 2C \rightarrow 2D$; $3A + B \rightarrow 2D$:

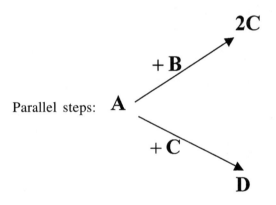

Parallel steps:

Series steps: $B \xrightarrow{+A} 2C \xrightarrow{+2A} 2D$

For an isothermal, well-mixed constant volume batch reactor, at time t = 0: n_{AO} mols of A, n_{BO} mols of B, and $n_{AO} = 3n_{BO}$. At any time, t: n_A mols of A, n_B mols of B, n_C mols of C, and n_D mols of D. n_D mols requires $n_D/2$ mols of B and $3n_D/2$ mols of A. n_C mols requires $n_C/2$ mols of B and $n_C/2$ mols of A. Consequently:

$$\frac{n_D}{2} + \frac{n_C}{2} = n_{BO} - n_B \quad \text{(Amount of B reacted)}$$

$$\frac{3n_D}{2} + \frac{n_C}{2} = n_{AO} - n_A \quad \text{(Amount of A reacted)}$$

Therefore,

$$\frac{3n_D}{2} + \frac{3n_C}{2} = 3n_{BO} - 3n_B$$

and

$$\frac{3n_D}{2} + \frac{n_C}{2} = n_{AO} - n_A$$

Mols of C: $n_C = 3(n_{BO} - n_B) - (n_{AO} - n_A)$ or

Concentration of C: $C_C = 3(C_{BO} - C_B) - (C_{AO} - C_A)$ (5-128)

Mols of D: $n_D = (n_{AO} - n_A) - (n_{BO} - n_B)$ or

Concentration of D: $C_D = (C_{AO} - C_A) - (C_{BO} - C_B)$ (5-129)

but $C_{AO} = 3C_{BO}$. Therefore, the material balance on component C is

$$C_C = C_A - 3C_B \qquad (5\text{-}130)$$

and concentration of component D becomes

$$C_D = (0.9 - C_A) - (0.3 - C_B) \qquad (5\text{-}131)$$

$$C_D = 0.6 - C_A + C_B \qquad (5\text{-}132)$$

This means $C_D = 0.6$ when $C_A = C_B = 0$. Substituting Equation 5-130 into Equation 5-126 gives

$$\frac{dC_A}{dt} = -k_2 C_A (C_A - 3C_B) - k_1 C_A C_B \qquad (5\text{-}133)$$

The differential Equations 5-127 and 5-133 are solved using the Runge-Kutta fourth order while components C and D are calculated from the mass balance of Equations 5-130 and 5-132, respectively, at a time increment of $h = \Delta t = 0.5$ min. The computer program BATCH59 calculates the concentrations of A, B, C, and D as function of time for 12 minutes. Table 5-8 gives the results of the simulation and Figure 5-18 shows the concentration profiles of the components.

THE SEMIBATCH REACTOR

Some chemical reactions are unsuitable to either batch or continuous operation in a stirred vessel because the heat liberated (generated) during the reaction may be highly exothermic and likely to cause dangerous conditions. Under these circumstances, reactions are sometimes carried out under semibatch conditions. This means that one or more reagents, which do not react, are entered completely into a batch container; the final reagent is then added under controlled conditions. Product is not withdrawn until the entire reagent has been added and the reaction has proceeded to the required extent. A semibatch reactor operates under an unsteady state condition. When the reaction heat is high there is better control of the energy, which evolves by regulating

Table 5-8
The concentration-time curve of a complex chemical reaction
of A + B $\xrightarrow{k_1}$ 2C, A + C $\xrightarrow{k_2}$ D
(k_1 = 0.1, k_2 = 0.05)

TIME	CONC. CA	CONC. CB	CONC. CC	CONC. CD
.00	.9000	.3000	.0000	.0000
.50	.8866	.2869	.0259	.0003
1.00	.8734	.2745	.0498	.0011
1.50	.8605	.2629	.0718	.0024
2.00	.8477	.2519	.0920	.0042
2.50	.8352	.2415	.1106	.0063
3.00	.8229	.2317	.1277	.0088
3.50	.8109	.2224	.1435	.0116
4.00	.7991	.2137	.1580	.0146
4.50	.7875	.2054	.1714	.0179
5.00	.7761	.1975	.1837	.0214
5.50	.7650	.1900	.1949	.0250
6.00	.7541	.1829	.2053	.0288
6.50	.7435	.1762	.2148	.0327
7.00	.7331	.1698	.2235	.0368
7.50	.7228	.1638	.2315	.0409
8.00	.7129	.1580	.2389	.0451
8.50	.7031	.1525	.2456	.0494
9.00	.6935	.1473	.2517	.0538
9.50	.6841	.1423	.2573	.0582
10.00	.6750	.1375	.2624	.0626
10.50	.6660	.1330	.2670	.0670
11.00	.6572	.1287	.2712	.0715
11.50	.6486	.1245	.2750	.0759
12.00	.6402	.1206	.2785	.0804

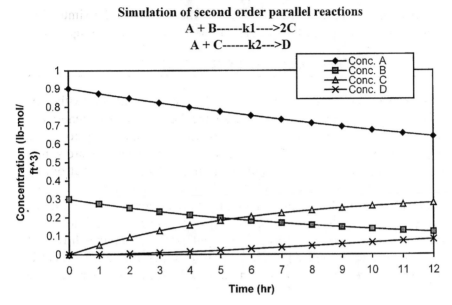

Figure 5-18. Concentrations versus time of A, B, C, and D with second order chemical reactions for k_1 = 0.4 ft^3/lb-mol • min and k_2 = 0.05 ft^3/lb-mol • min.

the addition rate of one of the reactants. The semibatch reactor also allows some degree of control over concentration of the reaction mixture, and therefore the rate of reaction.

Consider the reaction A + B $\xrightarrow{\ k_1\ }$ C that is performed in a semibatch reactor as shown in Figure 5-19, where the reaction is first order and carried out by the controlled addition of reactant B. Assuming that the reaction is first order with respect to both A and B (i.e., second order overall), the rate of disappearance of components A and B is $(-r_A) = (-r_B) = k_1 C_A C_B$.

Following considers the material balances of both components A and B. From the material balance Equation 5-1

$$\begin{array}{ccccc}
\text{Input} & & \text{Output} & \text{Disappearance} & \text{Rate of} \\
\text{Rate} & = & \text{Rate} & + \quad \text{by reaction} & + \quad \text{Accumulation} \\
& & & (-r_A)V & \\
& & & (-r_B)V &
\end{array}$$

$$(5\text{-}1)$$

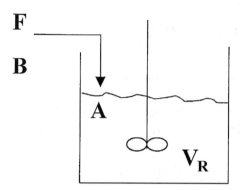

Figure 5-19. The semibatch reactor.

At constant density

$$\frac{dV_R}{dt} = F \tag{5-134}$$

with an initial condition, Equation 5-134 can be integrated to give

$$\int dV_R = F \int dt + \text{Const.} \tag{5-135}$$

$$V_R = Ft + \text{Const.} \tag{5-136}$$

at $t = 0$, $V_R = V_O$ and Const. $= V_O$. Therefore, the volume of the semibatch mixture is

$$V_R = Ft + V_O \tag{5-137}$$

Considering the material balance on reagent B, from

$$\begin{array}{c} \text{Input} \\ \text{Rate} \end{array} = \begin{array}{c} \text{Output} \\ \text{Rate} \end{array} + \begin{array}{c} \text{Disappearance} \\ \text{by reaction} \end{array} + \begin{array}{c} \text{Rate of} \\ \text{Accumulation} \end{array} \tag{5-1}$$

$$FC_{BO} = 0 + k_1 C_A C_B V_R + \frac{d}{dt}(V_R C_B) \tag{5-138}$$

Therefore,

$$\frac{d}{dt}\left(V_R\,C_B\right) = F\,C_{BO} - k_1\,C_A\,C_B\,V_R \tag{5-139}$$

The material balance on reactant A, from

$$\begin{array}{ccccccc} \text{Input} & = & \text{Output} & + & \text{Disappearance} & + & \text{Rate of} \\ \text{Rate} & & \text{Rate} & & \text{by reaction} & & \text{Accumulation} \end{array} \tag{5-1}$$

$$0 = 0 + k_1 C_A C_B V_R + \frac{d}{dt}\left(V_R\,C_A\right) \tag{5-140}$$

Therefore,

$$\frac{d}{dt}\left(V_R C_A\right) = -k_1 C_A C_B V_R \tag{5-141}$$

The material balance on product C is

$$\frac{d}{dt}\left(V_R C_C\right) = k_1 C_A C_B V_R \tag{5-142}$$

Using the differentiation of a product on A gives

$$V_R \frac{dC_A}{dt} + C_A \frac{dV_R}{dt} + k_1 C_A C_B V_R = 0 \tag{5-143}$$

or

$$V_R \frac{dC_A}{dt} = -k_1 C_A C_B V_R - C_A \frac{dV_R}{dt} \tag{5-144}$$

but $\dfrac{dV_R}{dt} = F$ $\tag{5-134}$

Therefore,

$$\frac{dC_A}{dt} = -k_1 C_A C_B - \frac{C_A}{V_R}\frac{dV_R}{dt} \tag{5-145}$$

Substituting Equation 5-134 into Equation 5-145 yields

$$\frac{dC_A}{dt} = -k_1 C_A C_B - \frac{C_A F}{V_R} \tag{5-146}$$

Material balance on B gives

$$V_R \frac{dC_B}{dt} + C_B \frac{dV_R}{dt} = FC_{BO} - k_1 C_A C_B V_R \tag{5-147}$$

Rearranging Equation 5-147 yields

$$\frac{dC_B}{dt} = \frac{FC_{BO}}{V_R} - k_1 C_A C_B - \frac{C_B}{V_R} \frac{dV_R}{dt} \tag{5-148}$$

that is,

$$\frac{dC_B}{dt} = \frac{FC_{BO}}{V_R} - k_1 C_A C_B - \frac{C_B F}{V_R} \tag{5-149}$$

Material balance on C gives

$$V_R \frac{dC_C}{dt} + C_C \frac{dV_R}{dt} = k_1 C_A C_B V_R \tag{5-150}$$

Rearranging Equation 5-150 gives

$$\frac{dC_C}{dt} = k_1 C_A C_B - \frac{C_C}{V_R} \frac{dV_R}{dt} \tag{5-151}$$

Therefore,

$$\frac{dC_C}{dt} = k_1 C_A C_B - \frac{C_C F}{V_R} \tag{5-152}$$

The volume of the batch with respect to time t is

$$V_R = Ft + V_O \tag{5-137}$$

Equations 5-146, 5-149, and 5-152 are first order differential equations. The concentration profiles of A, B, C, and the volume V of the batch using Equation 5-137 is simulated with respect to time using the Runge-Kutta fourth order numerical method.

Equations 5-137, 5-146, 5-149, and 5-152 can be incorporated respectively in the subprogram of a computer code as follows

$$VR = U * T + V \tag{5-153}$$

$$F(1) = -(K1 * X(1) * X(2)) - (X(1) * U/VR) \tag{5-154}$$

$$F(2) = (U^*) - (K1 * X(1) * X(2)) - (X(2) * U/VR) \tag{5-155}$$

$$F(3) = (K1 * X(1) * X(2)) - (X(3) * U/VR) \tag{5-156}$$

where

$$F(1) = \frac{dC_A}{dt}, \quad F(2) = \frac{dC_B}{dt}, \quad F(3) = \frac{dC_C}{dt}$$

and

$$X(1) = C_A, \quad X(2) = C_B, \quad X20 = C_{BO}, \quad X(3) = C_C$$

Semibatch reactor conditions are illustrated in Table 5-9.

The computer program BATCH510 was developed incorporating Equations 5-156, 5-157, 5-158, and 5-159 in the subprogram of the Runge-Kutta fourth order program. The results of the simulation are

Table 5-9

Velocity constant	k_1 = 10 l/hr • mol
Initial concentrations	C_{AO} = 1.0 mol/l
	C_B = 0.0 mol/l
	C_{CO} = 0.0 mol/l
Charge rate	U = 100 l/hr
Charge concentration	C_{BO} = 1.0 mol/l
Initial volume	V = 100 l

shown in Table 5-10. Figure 5-20 shows the concentration profiles of A, B, C, and volume V during a period of 1 hour.

Semibatch reactors can be employed when parallel reactions of different orders occur. In such cases, it may be economical to use a semibatch rather than a batch operation.

CONTINUOUS FLOW STIRRED TANK REACTOR (CFSTR)

The contents of a continuously operated stirred tank are assumed to be perfectly mixed, so that the properties (e.g., concentration, temperature) of the reaction mixture are uniform in all parts of the system. Therefore, the conditions throughout the tank are the same and equal to the conditions at the outlet. This means that the volume element can be taken as the volume, V_R, of the entire contents. Additionally, the composition and temperature at which the reaction occurs are the same as the composition and temperature of any exit stream. A continuous flow stirred tank reactor as shown in Figure 5-21 assumes that the fluid is perfectly well mixed.

Additionally, the following is also assumed:

• Isothermal operation (i.e., constant temperature)
• Steady state condition
• No change in liquid density

Table 5-10
Simulation of the semibatch reator concentrations involving the chemical reaction A + B → C (k_1 = 10.0)

TIME	VOLUME	CONC. CA	CONC. CB	CONC. C
.00	.0000	1.0000	.0000	.0000
.10	110.0000	.8767	.0586	.0324
.20	120.0000	.7480	.0814	.0853
.30	130.0000	.6323	.0938	.1369
.40	140.0000	.5319	.1034	.1823
.50	150.0000	.4457	.1124	.2210
.60	160.0000	.3717	.1217	.2533
.70	170.0000	.3082	.1318	.2800
.80	180.0000	.2538	.1427	.3018
.90	190.0000	.2073	.1546	.3191
1.00	200.0000	.1676	.1676	.3324

Simulation of a semibatach reactor concentrations versus time of the type A + B ---k1----> C

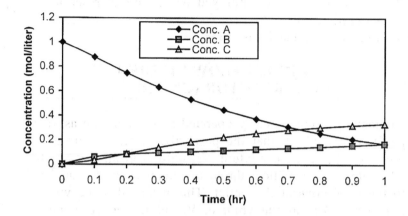

Simulation of a semibatach reactor volume versus time of the type A + B ---k1----> C

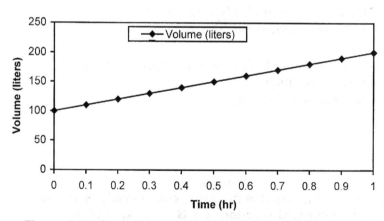

Figure 5-20. Semibatch reactor concentrations, volume versus time.

The following notations are used:

$$u = \text{volumetric flowrate} \left(\frac{\text{volume}}{\text{unit time}} \right)$$

$$(-r_A)_{net} = \frac{\text{moles of A disappearing by reaction}}{(\text{unit volume})(\text{time})}$$

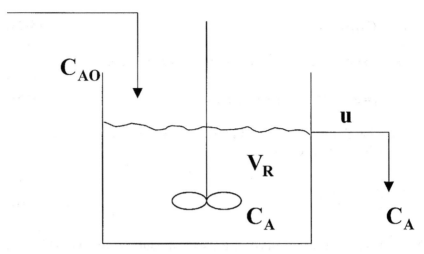

Figure 5-21. The continuous flow stirred tank reactor (CFSTR).

M_A = molecular weight

$$C_A = \text{concentration} \left(\frac{\text{moles}}{\text{volume}} \right)$$

$\delta V = V_R$ = volume of the reactor

Accumulation = 0 (no change with time)

Input during δt = Output during δt

By flow $u \cdot \delta t \cdot C_{AO} \cdot M_A = u \cdot \delta t \cdot C_A \cdot M_A$

By reaction $\qquad\qquad + (-r_A) \cdot V_R \cdot \delta t \cdot M_A$ $\qquad$ (5-157)

$$u C_{AO} = u C_A + (-r_A) V_R \qquad (5-158)$$

In terms of the fractional conversion X_A,

$$X_A = \frac{C_{AO} - C_A}{C_{AO}} \qquad (5-14)$$

$$C_A = C_{AO}(1 - X_A) \tag{5-28}$$

Substituting Equation 5-28 into Equation 5-158 gives

$$uC_{AO} = uC_{AO}(1 - X_A) + (-r_A)V_R \tag{5-159}$$

Equation 5-159 can be rearranged in terms of the residence time $\bar{t}$ of fluid in the reactor as

$$\bar{t} = \frac{V_R}{u} = \frac{C_{AO}X_A}{(-r_A)} \tag{5-160}$$

X_A and $(-r_A)$ are determined at the exit stream conditions, which are the same conditions as in the reactor of volume V_R.

For constant density system, the performance equation in terms of C_A becomes

$$\bar{t} = \frac{V_R}{u} = \frac{C_{AO} - C_A}{(-r_A)} \tag{5-161}$$

These expressions relate the terms X_A, $(-r_A)$, V_R, and u. Therefore, the fourth term can be determined if the remaining terms are known. When designing, the size of the reactor needed for a given duty or the extent of conversion in a reactor of given size can be directly obtained.

IRREVERSIBLE REACTIONS

Consider the nth-order irreversible reaction of the form A $\rightarrow$ products, $(-r_A) = kC_A^n$, in a constant density single-stage CFSTR. If n = 1, Equation 5-158 becomes

$$uC_{AO} = uC_A + kC_A V_R \tag{5-162}$$

where $\bar{t} = V_R/u$, the mean residence time. Rearranging Equation 5-162 yields

$$C_A = \frac{C_{AO}}{1 + k\bar{t}} \tag{5-163}$$

If the reaction is of the form $A \xrightarrow{k} B$, the concentration of component B can be determined from the stoichiometry:

$$C_{AO} - C_A = C_B - C_{BO} \tag{5-164}$$

If $C_{BO} = 0$, Equation 5-164 becomes

$$C_B = C_{AO} - C_A \tag{5-165}$$

where

$$C_A = \frac{C_{AO}}{1 + k\bar{t}} \tag{5-163}$$

The concentration of B then becomes

$$C_B = C_{AO} - \frac{C_{AO}}{1 + k\bar{t}}$$

$$= \frac{C_{AO}k\bar{t}}{1 + k\bar{t}} \tag{5-166}$$

The mean residence time is evaluated in terms of the fractional conversion as defined by $C_A = C_{AO}(1 - X_A)$.

The material balance for the single CFSTR in terms of X_A for the first order irreversible reaction is

$$uC_{AO} = uC_A + kC_A V_R$$

$$uC_{AO} = uC_{AO}(1 - X_A) + kC_{AO}(1 - X_A)V_R$$

$$uC_{AO}X_A = kC_{AO}V_R(1 - X_A) \tag{5-167}$$

The fractional conversion X_A is determined from Equation 5-167 in terms of the mean residence time $\bar{t} = (V_R/u)$ to yield

$$X_A = \frac{k\bar{t}}{1 + k\bar{t}} \tag{5-168}$$

REVERSIBLE REACTION

Consider the first order reversible reaction of the form

$$A \underset{k_2}{\overset{k_1}{\rightleftharpoons}} B, \quad (-r_A) = k_1 C_A - k_2 C_B$$

in a single-stage CFSTR at constant density. The material balance is

$$uC_{AO} = uC_A + (-r_A)V_r$$
$$= uC_A + (k_1 C_A - k_2 C_B)V_R \tag{5-169}$$

From the stoichiometry:

Component	A	B
At time t = 0	C_{AO}	0
At time t = t	C	C_B
Amount reacted	$(C_{AO} - C_A)$	C_B

Therefore,

$$C_{AO} - C_A = C_B \tag{5-170}$$

Substituting Equation 5-170 into Equation 5-169 gives

$$uC_{AO} = uC_A + \left[k_1 C_A - k_2(C_{AO} - C_A)\right] V_R \tag{5-171}$$

Rearranging Equation 5-171 for C_A in terms of the mean residence time $\bar{t}$ and C_{AO} yields

$$C_A = \frac{C_{AO}(1 + k_2 \bar{t})}{\left[1 + (k_1 + k_2)\bar{t}\right]} \tag{5-172}$$

As $\bar{t} \to \infty$, the compositions of A and B will be in equilibrium, that is

$$K = \frac{k_1}{k_2} = \frac{C_{B_{equ}}}{C_{A_{equ}}} = \frac{X_{A_{equ}}}{1 - X_{A_{equ}}}$$

However, if $k_2 = 0$, Equation 5-172 becomes an expression for the irreversible first order. The mean residence time is determined in terms of the fractional conversion X_A from Equation 5-171 as follows

$$\left(C_{AO} - C_A\right) = X_A C_{AO} \tag{5-14}$$

and

$$C_A = C_{AO}\left(1 - X_A\right) \tag{5-28}$$

Substituting Equation 5-14 and 5-28 into Equation 5-171 gives

$$uC_{AO} = uC_{AO}\left(1 - X_A\right) + \left[k_1 C_{AO}\left(1 - X_A\right) - k_2 C_{AO} X_A\right]V_R \tag{5-173}$$

$$uC_{AO}X_A = C_{AO}\left[k_1\left(1 - X_A\right) - k_2 X_A\right]V_R \tag{5-174}$$

The mean residence time $\bar{t}$ in terms of the fractional conversion X_A from Equation 5-174 is

$$\bar{t} = \frac{X_A}{k_1\left(1 - X_A\right) - k_2 X_A} \tag{5-175}$$

Similarly, the mean residence time $\bar{t}$ in terms of C_A is

$$\bar{t} = \frac{C_{AO} - C_A}{k_1 C_A - k_2\left(C_{AO} - C_A\right)} \tag{5-176}$$

SERIES REACTION

Consider the series first order irreversible reaction of the form $A \xrightarrow{\ k_1\ } B \xrightarrow{\ k_2\ } C$ in a constant density CFSTR. The material balance of component A is $uC_{AO} = uC_A + k_1 C_A V_R$. The material balance of component B is $uC_{BO} = uC_B + (k_2 C_B - k_1 C_A)V_R$. The material balance of component C is $uC_{CO} = uC_C - k_2 C_B V_R$. Assuming that $C_{BO} = C_{CO} = 0$, then the exit concentration of components A, B, and C are

$$C_A = \frac{C_{AO}}{1 + k\bar{t}}$$

$$C_B = \frac{k_1 C_{AO} \bar{t}}{(1+k_1\bar{t})(1+k_2\bar{t})}$$

$$C_C = \frac{k_1 k_2 C_{AO} \bar{t}^2}{(1+k_1\bar{t})(1+k_2\bar{t})} \tag{5-177}$$

The maximum concentration of B is determined by $dC_B/d\bar{t}_{max} = 0$:

$$\frac{dC_B}{d\bar{t}_{max}} = k_1 C_{AO} \left\{ \bar{t}_{max} (1+k_1\bar{t}_{max})^{-1} (1+k_2\bar{t}_{max})^{-1} \right\} = 0 \tag{5-178}$$

Using the differentiation of a product as

$$\frac{d}{dt}(uvw) = VW\frac{du}{dt} + UW\frac{dv}{dt} + UV\frac{dw}{dt}$$

on Equation 5-178 gives

$$\frac{dC_B}{d\bar{t}_{max}} = 0 = \frac{k_1 C_{AO}(1 - k_1 k_2 \bar{t}_{max}^2)}{(1+k_1\bar{t}_{max})^2 (1+k_2\bar{t}_{max})^2} \tag{5-179}$$

$$k_1 C_{AO}(1 - k_1 k_2 t_{max}^2) = 0$$

$$k_1 C_{AO} = 0$$

or

$$1 - k_1 k_2 t_{max}^2 = 0$$

$$\bar{t}_{max} = \frac{1}{\sqrt{k_1 k_2}} \tag{5-180}$$

Substituting the maximum time $\bar{t}_{max}$ from Equation 5-180 into Equation 5-177 for the maximum concentration of B followed by further mathematical manipulation gives

$$C_{B_{max}} = \frac{k_1 C_{AO}}{\left(k_1^{1/2} + k_2^{1/2}\right)} \tag{5-181}$$

Equation 5-181 is useful in obtaining the maximum yield of component B in a CFSTR.

SECOND ORDER IRREVERSIBLE REACTION

Consider the liquid phase second order irreversible reaction of the form A + B $\rightarrow$ products in a CFSTR at constant density under stable conditions.

Assume that the reaction is irreversible and first order with respect to each species A and B and second order overall. The reaction rate for component A is represented by $(-r_A) = kC_A C_B$.

Also assume that the tank is well mixed, and the concentrations of A and B are uniform at any time in the tank. Additionally, assume that the total holdup in the reactor is steady, and that flowrate through it is constant at u m^3/min.

Applying the material balance of Equation 5-1 on species A gives

$$uC_{AO} = uC_A + kC_A C_B V_R + V_R \frac{dC_A}{dt} \tag{5-182}$$

At steady state condition, $V_R \dfrac{dC_A}{dt} = 0$

The material balance on component B is

$$uC_{BO} = uC_B + kC_A C_B V_R + V_R \frac{dC_B}{dt} \tag{5-183}$$

At steady state condition, $V_R \dfrac{dC_B}{dt} = 0$

Equations 5-182 and 5-183 become

$$uC_{AO} = uC_A + kC_A C_B V_R \tag{5-184}$$

$$uC_{BO} = uC_B + kC_A C_B V_R \tag{5-185}$$

where u is the volumetric flowrate entering the reactor. Rearranging Equation 5-184 gives

$$u(C_{AO} - C_A) = kC_A C_B V_R \qquad (5\text{-}186)$$

From stoichiometry,

$$C_{AO} - C_A = C_{BO} - C_B$$

or

$$C_B = C_{BO} - (C_{AO} - C_A) \qquad (5\text{-}187)$$

Also, the concentration of A in terms of the initial concentration C_{AO} and fractional conversion X_A is

$$C_A = C_{AO}(1 - X_A) \qquad (5\text{-}28)$$

or

$$C_{AO} - C_A = C_{AO} X_A \qquad (5\text{-}27)$$

Substituting Equations 5-27, 5-28, and 5-187 into Equation 5-186 gives

$$uC_{AO} X_A = kC_{AO}(1 - X_A)(C_{BO} - C_{AO} X_A) V_R \qquad (5\text{-}188)$$

Rearranging Equation 5-188, the residence time $\bar{t}$ can be determined for the conversion of species A in terms of the initial concentrations of species A, B, and the rate constant k as

$$\bar{t} = \frac{X_A}{k(1 - X_A)(C_{BO} - C_{AO} X_A)} \qquad (5\text{-}189)$$

With $\theta_B = (C_{BO}/C_{AO})$ Equation 5-189 becomes

$$\bar{t} = \frac{X_A}{kC_{AO}(1 - X_A)(\theta_B - X_A)} \qquad (5\text{-}190)$$

Equation 5-190 enables the residence time to be determined for a given conversion of A and the initial concentrations of A and B in a

CFSTR. Figure 5-22 illustrates the graphical performance equations for CFSTRs.

Example 5-8

A continuous flow stirred tank reactor is used to decompose a dilute solution of species A. The decomposition is irreversible and first order with velocity constant at 2.5 hr^{-1}. The reactor volume is 15 m^3. What flowrate of feed solution can this reactor treat if 90% decomposition is required?

Solution

Assuming a steady state condition (i.e., accumulation = 0)

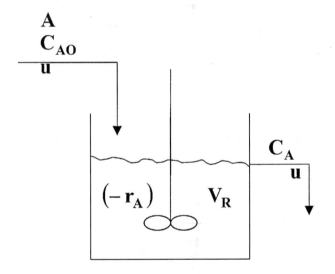

CFSTR with inflow of component A

The reaction scheme for the decomposition of A in a CFSTR is represented by A → products, $(-r_A) = kC_A$. The mass balance can be written as

$$uC_{AO} = uC_A + (-r_A)V_R \qquad (5\text{-}158)$$

or

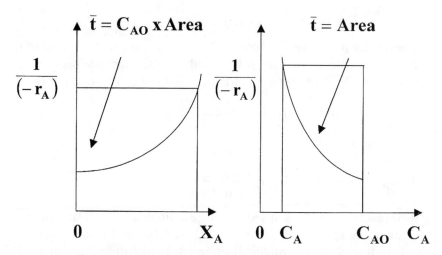

Figure 5-22. Graphical representation of the design equations for a CFSTR.

$$u(C_{AO} - C_A) = kC_A V_R \tag{5-191}$$

The outlet concentration of A in terms of the fractional conversion X_A is

$$C_A = C_{AO}(1 - X_A) \tag{5-28}$$

or

$$(C_{AO} - C_A) = C_{AO} X_A \tag{5-27}$$

Substituting Equation 5-27 and 5-28 into Equation 5-191 yields $uC_{AO}X_A = kC_{AO}(1 - X_A)V_R$. The volumetric rate u is

$$u = \frac{k(1 - X_A)V_R}{X_A}$$

$$= \frac{2.5 \times (1 - 0.9) \times 15}{0.9}$$

$$= 4.17 \frac{m^3}{hr}$$

Example 5-9

The reaction between ammonia (NH_3) and formaldehyde (HCHO) to produce hexamine $N_4(CH_2)_6$ was studied at 36°C (Kermode & Stevens,[*] 1965) in a CFSTR of 490 dm³ stirred at 1,800 rpm. The reaction is

$$4NH_3 + 6HCHO \rightarrow N_4(CH_2)_6 + 6H_2O$$

and the rate equation is

$$\left(-r_{NH_3}\right) = 0.0649 C_{NH_3} C_{HCHO}^2 \ \frac{mol}{dm^3 \bullet sec}$$

The reactants were each fed to the reactor at a rate of 1.5 dm³/sec with an ammonia concentration of 4.02 mol/dm³ and a formaldehyde concentration of 6.32 mol/dm³. Determine the exit concentrations.

Solution

Assuming steady state condition (i.e., accumulation = 0)

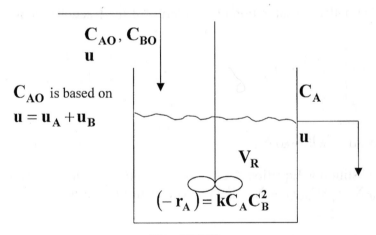

The CFSTR.

The reaction in terms of A and B is 4A + 6B → products. Material balance of species A gives

$$uC_{AO} = uC_A + \left(-r_A\right)V_R \qquad (5\text{-}158)$$

[*]*Canad. J. Chem. Eng., 43, 68 (1965).*

where

$$(-r_A) = 0.0649 C_{NH_3} C_{HCHO}^2 \frac{mol}{dm^3 \cdot sec}$$

That is,

$$uC_{AO} = uC_A + 0.0649 C_A C_B^2 V_R \tag{5-192}$$

From stoichiometry:

Component	A	B
At time t = 0	C_{AO}	C_{BO}
At time t = t	C_A	C_B
Amount reacted	$(C_{AO} - C_A)$	$(C_{BO} - C_B)$

Using the stoichiometry,

$$\frac{1}{4}(C_{AO} - C_A) = \frac{1}{6}(C_{BO} - C_B) \tag{5-193}$$

with C_{AO} = 4.02 mol/dm^3 and C_{BO} = 6.32 mol/dm^3. The exit concentration of B can be expressed by $C_B = C_{BO} - 1.5 (C_{AO} - C_A)$ or

$$C_B = 0.29 + 1.5 C_A \tag{5-194}$$

The residence time $\bar{t}$ = V_R/u = 490/3.0 = 163.3 sec. Equation 5-192 now becomes

$$C_{AO} = C_A + 0.0649 \bar{t} C_A C_B^2 \tag{5-195}$$

Substituting Equation 5-194 and the value of the residence time $\bar{t}$ into Equation 5-192 gives

$$4.02 = C_A + 10.598 C_A (0.29 + 1.5 C_A)^2 \tag{5-196}$$

Further expansion and rearrangement of Equation 5-196 yields

$$23.85 C_A^3 + 9.22 C_A^2 + 1.89 C_A - 4.02 = 0 \tag{5-197}$$

Equation 5-197 is a polynomial of the third degree, and by employing either a numerical method or a spreadsheet package such as Microsoft Excel, the roots (C_A) of the equation can be determined. A developed computer program PROG51 using the Newton-Raphson method to determine C_A was used. The Newton-Raphson method for the roots C_A of Equation 5-197 is

$$C_{A,n+1} = C_{A,n} - \frac{f(C_{A,n})}{f'(C_{A,n})} \tag{5-198}$$

Equation 5-197 can be expressed as a function of concentration of A as

$$f(C_A) = 23.85C_A^3 + 9.22C_A^2 + 1.89C_A - 4.02 \tag{5-199}$$

The derivative of Equation 5-199 is

$$f'(C_A) = 71.55\,C_A^2 + 18.44\,C_A + 1.89 \tag{5-200}$$

Substituting Equations 5-199 and 5-200 into Equation 5-198 and using a starting value of the root as 0.25, the solution for the root C_A after four iterations was 0.4124. The same result was obtained using Microsoft Excel (see Appendix B). Table 5-11 gives the computer results of program PROG51. Appendix D gives the Newton-Raphson method for solving for the root of an equation.

Table 5-11
Newton-Raphson method for a nonlinear equation

```
INITIAL GUESS OF THE ROOT:        .3500

AT ITERATION   1     X=.42052E+00      F(X)=.17874E+00

AT ITERATION   2     X=.41250E+00      F(X)=.25137E-02

AT ITERATION   3     X=.41239E+00      F(X)=.48670E-06

TOLERANCE MET IN  3 ITERATIONS THE FINAL ROOT: X=.41239E+00   F(X)=.48670E-0
```

The exit concentration $C_A = 0.4124$ mol/dm³ and

$$C_B = 0.29 + 1.5 \times 0.4124$$

$$= 0.9086 \, \text{mol/dm}^3$$

MULTI-STAGE CONTINUOUS FLOW STIRRED TANK REACTOR

In a single continuous flow stirred tank reactor, a portion of the fresh feed could exit immediately in the product stream as soon as the reactants enter the reactor. To reduce this bypassing effect, a number of stirred tanks in series is frequently used. This reduces the probability that a reactant molecule entering the reactor will immediately find its way to the exiting product stream. The exit stream from the first stirred tank serves as the feed to the second, the exit stream from the second reactor serves as the feed to the third, and so on. For constant density, the exit concentration or conversion can be solved by consecutively applying Equation 5-158 to each reactor. The following derived equations are for a series of three stirred tanks (Figure 5-23) with constant volume V_R.

Consider the first order reaction $A \xrightarrow{\ k\ } B$ in a battery of three continuous flow stirred tank reactors, where

V_R = volume of fluid in each stage
$(-r_{Ai})$ = rate of reaction per unit volume in stage i. This changes with i.

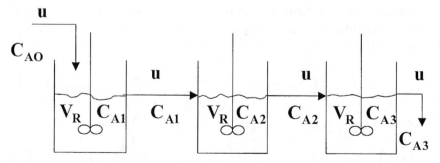

Figure 5-23. Battery of continuous flow stirred tank reactors.

Consider the mass balance on key component A

$$\frac{\text{Input}}{\text{rate}} = \frac{\text{Output}}{\text{rate}} + \frac{\text{Disappearance}}{\text{by reaction}} + \text{Accumulation}$$

Accumulation = 0, therefore for the key component A, the mass balance is:

Stage 1: $\quad u \cdot C_{AO} = u C_{A1} + (-r_{A1}) V_R$

Stage 2: $\quad u \cdot C_{A1} = u C_{A2} + (-r_{A2}) V_R$

Stage 3: $\quad u \cdot C_{A2} = u C_{A3} + (-r_{A3}) V_R$ $\qquad$ (5-201)

For the ith stage, the mass balance is:

$$u \cdot C_{Ai-1} = u \cdot C_{Ai} + (-r_{Ai}) V_R \qquad (5\text{-}202)$$

Dividing Equation 5-202 by u gives

$$C_{Ai-1} = C_{Ai} + (-r_{Ai}) \frac{V_R}{u} \qquad (5\text{-}203)$$

where $\bar{t} = V_R/u$ = mean residence time of fluid in the stage. Equation 5-203 becomes

$$C_{Ai-1} = C_{Ai} + (-r_{Ai})\bar{t} \qquad (5\text{-}204)$$

Consider the first order reaction $A \xrightarrow{k} B$ in a battery of three continuous flow stirred tank reactors with volumes V_1, V_2, and V_3. The material balance for stage 1 CFSTR is

$$u C_{AO} = u C_{A1} + (-r_{A1}) V_1 \qquad (5\text{-}205)$$

where

$$(-r_{A1}) = k C_{A1} \qquad (5\text{-}206)$$

Equation 5-205 becomes

$$uC_{AO} = uC_{A1} + kC_{A1}V_1 \tag{5-207}$$

Rearranging Equation 5-207 with the mean residence time $\bar{t} = V_1/u$ gives

$$C_{A1} = \frac{C_{AO}}{\left(1 + k\bar{t}_1\right)} \tag{5-208}$$

or

$$\frac{C_{AO}}{C_{A1}} = 1 + k\bar{t}_1 \tag{5-209}$$

The material balance for stage 2 CFSTR is

$$uC_{A1} = uC_{A2} + \left(-r_{A2}\right)V_2 \tag{5-210}$$

Equation 5-210 is further expressed by

$$uC_{A1} = uC_{A2} + kC_{A2}V_2 \tag{5-211}$$

where $\bar{t}_2 = V_2/u$. The exit concentration C_{A2} can be determined as

$$C_{A2} = \frac{C_{A1}}{\left(1 + k\bar{t}_2\right)} \tag{5-212}$$

Substituting Equation 5-208 into Equation 5-212 gives

$$C_{A2} = \frac{C_{AO}}{\left(1 + k\bar{t}_1\right)\left(1 + k\bar{t}_2\right)} \tag{5-213}$$

or

$$\frac{C_{AO}}{C_{A2}} = \left(1 + k\bar{t}_1\right)\left(1 + k\bar{t}_2\right) \tag{5-214}$$

Where $\bar{t}_1 = \bar{t}_2 = \bar{t}$, Equation 5-214 becomes

$$\frac{C_{AO}}{C_{A2}} = (1 + k\bar{t})^2 \tag{5-215}$$

The material balance for a stage 3 CFSTR is

$$uC_{A2} = uC_{A3} + (-r_{A3})V_3 \tag{5-216}$$

Equation 5-216 is further expressed by

$$uC_{A2} = uC_{A3} + kC_{A3}V_3 \tag{5-217}$$

where $\bar{t}_3 = V_3/u$. The exit concentration C_{A3} can be determined as

$$C_{A3} = \frac{C_{A2}}{(1 + k\bar{t}_3)} \tag{5-218}$$

Substituting Equation 5-214 into Equation 5-218 gives

$$C_{A3} = \frac{C_{AO}}{(1 + k\bar{t}_1)(1 + k\bar{t}_2)(1 + k\bar{t}_3)} \tag{5-219}$$

Rearranging Equation 5-219 with $\bar{t}_1 = \bar{t}_2 = \bar{t}_3 = \bar{t}$ gives

$$C_{A3} = \frac{C_{AO}}{(1 + k\bar{t})^3} \tag{5-220}$$

or

$$\frac{C_{AO}}{C_{A3}} = (1 + k\bar{t})^3 \tag{5-221}$$

Example 5-10

The pair of reactions $A + B \xrightarrow{k_1} 2C$ and $A + C \xrightarrow{k_2} D$ are conducted in a four-stage CFSTR with $C_{AO} = 0.9$ mol/m^3, $C_{BO} = 0.3$ mol/m^3, and $C_{CO} = C_{DO} = 0$ mol/m^3. The residence time in each stage is 10 minutes. Determine the exit concentrations of A and B in the four CFSTRs. The rate expressions are

$(-r_A) = k_1 C_A C_B + k_2 C_A C_C$ and $(-r_B) = k_1 C_A C_B$, where $k_1 = 0.3$ m^3/mol•min and $k_2 = 0.15$ m^3/mol•min.

Solution

The following is a battery of continuous flow stirred tank reactors with $C_{AO} = 0.9$ mol/m^3 and $C_{BO} = 0.3$ mol/m^3 in the first tank, and where $V_1 = V_2 = V_3 = V_4 = V_R$, u = volumetric flow rate, and the residence time, $\bar{t} = V_R/u = 10$ mins.

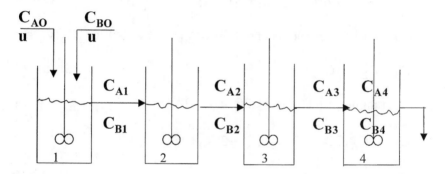

1st CFSTR

The material balance on species A is

$$uC_{AO} = uC_{A1} + \left(k_1 C_{A1} C_{B1} + k_2 C_{A1} C_C\right) V_R \qquad (5\text{-}222)$$

The stoichiometry between species A, B, and C is

$$C_C = 3\left(C_{BO} - C_B\right) - \left(C_{AO} - C_A\right)$$
$$= 0.9 - 3C_B - 0.9 + C_A$$

$$C_C = C_A - 3C_B \qquad (5\text{-}223)$$

Substituting Equation 5-223 into Equation 5-222 gives $C_{AO} = C_{A1} + C_{A1}\bar{t}\,[k_1 C_{B1} + k_2(C_{A1} - 3C_{B1})]$ or

$$0.9 = C_{A1} + 3C_{A1}C_{B1} + 1.5C_{A1}(C_{A1} - 3C_{B1}) \qquad (5\text{-}224)$$

Material balance on species B is $uC_{BO} = uC_{B1} + k_1C_{A1}C_{B1}V_R$ or

$$0.3 = C_{B1} + 3C_{A1}C_{B1} \qquad (5\text{-}225)$$

2nd CFSTR

Material balance on species A is $uC_{A1} = uC_{A2} + [k_1C_{A2}C_{B2} + k_2C_{A2}(C_{A2} - 3C_{B2})]V_R$ or

$$C_{A1} = C_{A2} + 10[0.3C_{A2}C_{B2} + 0.15C_{A2}(C_{A2} - 3C_{B2})] \qquad (5\text{-}226)$$

Material balance on species B is $uC_{B1} = uC_{B2} + k_1C_{A2}C_{B2}V_R$ or

$$C_{B1} = C_{B2} + 3C_{A2}C_{B2} \qquad (5\text{-}227)$$

Rearranging Equations 5-226 and 5-227 yields

$$C_{A2} + 3C_{A2}C_{B2} + 1.5C_{A2}(C_{A2} - 3C_{B2}) = C_{A1}$$

$$C_{B2} + 3C_{A2}C_{B2} = C_{B1} \qquad (5\text{-}228)$$

3rd CFSTR

The material balances on species A and B are

$$C_{A3} + 3C_{A3}C_{B3} + 1.5C_{A3}(C_{A3} - 3C_{B3}) = C_{A2}$$

$$C_{B3} + 3C_{A3}C_{B3} = C_{B2} \qquad (5\text{-}229)$$

4th CFSTR

The material balances on species A and B are

$$C_{A4} + 3C_{A4}C_{B4} + 1.5C_{A4}(C_{A4} - 3C_{B4}) = C_{A3}$$

$$C_{B4} + 3C_{A4}C_{B4} = C_{B3} \qquad (5\text{-}230)$$

There are eight nonlinear equations involving the material balances of species A and B in the four stirred tank reactors. Rearranging these equations yields the following:

$$1.5C_{A1}^2 - 1.5C_{A1}C_{B1} + C_{A1} - 0.9 = 0$$

$$3C_{A1}C_{B1} + C_{B1} - 0.3 = 0$$

$$1.5C_{A2}^2 - 1.5C_{A2}C_{B2} + C_{A2} - C_{A1} = 0$$

$$3C_{A2}C_{B2} + C_{B2} - C_{B1} = 0$$

$$1.5C_{A3}^2 - 1.5C_{A3}C_{B3} + C_{A3} - C_{A2} = 0$$

$$3C_{A3}C_{B3} + C_{B3} - C_{B2} = 0$$

$$1.5C_{A4}^2 - 1.5C_{A4}C_{B4} + C_{A4} - C_{A3} = 0$$

$$3C_{A4}C_{B4} + C_{B4} - C_{B3} = 0 \qquad (5\text{-}231)$$

A computer program PROG52 was developed using the Newton-Raphson's method to determine the outlet concentrations of species A and B from the four stirred tank reactors. The eight Equations 5-231 are supplied as functions in the subroutine of PROG52 as:

$F(1) = 1.5 * X(1) * X(1) - 1.5 * X(1) * X(2) + X(1) - 0.9$

$F(2) = 3.0 * X(1) * X(2) + X(2) - 0.3$

$F(3) = 1.5 * X(3) * X(3) - 1.5 * X(3) * X(4) + X(3) - X(1)$

$F(4) = 3.0 * X(3) * X(4) + X(4) - X(2)$

$F(5) = 1.5 * X(5) * X(5) - 1.5 * X(5) * X(6) + X(5) - X(3)$

$F(6) = 3.0 * X(5) * X(6) + X(6) - X(4)$

$F(7) = 1.5 * X(7) * X(7) - 1.5 * X(7) * X(8) + X(7) - X(5)$

$F(8) = 3.0 * X(7) * X(8) + X(8) - X(6)$

where $C_{A1} = X(1)$, $C_{B1} = X(2)$; $C_{A2} = X(3)$, $C_{B2} = X(4)$; $C_{A3} = X(5)$, $C_{B3} = X(6)$; $C_{A4} = X(7)$, $C_{B4} = X(8)$.

The exit concentrations of species A and B from the computer results and the initial guesses of $C_{A1} = 0.1$, $C_{B1} = 0.1$, $C_{A2} = 0.1$, $C_{B2} = 0.1$, $C_{A3} = 0.1$, $C_{B3} = 0.1$, $C_{A4} = 0.1$, $C_{B4} = 0.1$ are:

$$C_{A1} = X(1) = 0.54599, \quad C_{B1} = X(2) = 0.11372$$
$$C_{A2} = X(3) = 0.3703, \quad C_{B2} = X(4) = 0.05386$$
$$C_{A3} = X(5) = 0.2718, \quad C_{B3} = X(6) = 0.02958$$
$$C_{A4} = X(7) = 0.21107, \quad C_{B4} = X(8) = 0.01792$$

The computer program PROG52 can be used to solve any number of nonlinear equations. The partial derivatives of the functions are estimated by the difference quotients when a variable is perturbed by an amount equal to a small value (Δ) used in the program to perturb the X-values.

EQUAL SIZE CFSTR IN SERIES

Consider a system of N continuous flow stirred tank reactors in series as shown in Figure 5-24. Although the concentration is uniform from one tank to another, there is a change in concentration as the fluid traverses between the CFSTRs. This is illustrated in Figure 5-25. The drop in concentration implies that the larger the number of CFSTRs in series, the closer the system would behave as plug flow.

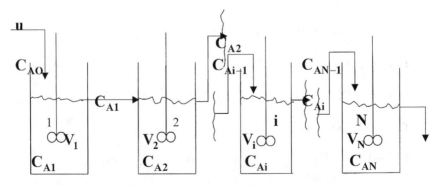

Figure 5-24. System of N equal size CFSTRs in series.

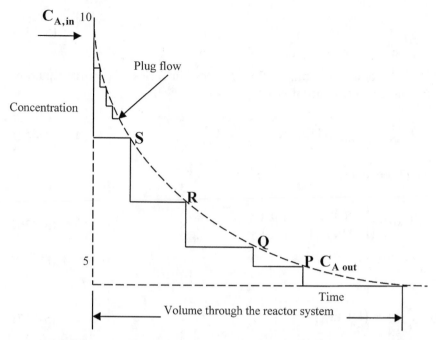

Figure 5-25. Concentration profile through an N stage CFSTR system.

A general equation can be derived for the exit concentration C_{AN} as follows.

The material balance for stage i CFSTR is

$$u C_{Ai-1} = u C_{Ai} + (-r_{Ai}) V_i \tag{5-202}$$

Equation 5-202 is further expressed by

$$u C_{Ai-1} = u C_{Ai} + k C_{Ai} V_i \tag{5-232}$$

where $\bar{t}_i = V_i/u$. Equation 5-232 becomes

$$C_{Ai} = \frac{C_{Ai-1}}{(1 + k\bar{t}_i)} \tag{5-233}$$

or

$$\frac{C_{Ai-1}}{C_{Ai}} = 1 + k\bar{t}_i \qquad\qquad (5\text{-}234)$$

For N equal size mixed flow reactors in series, a general equation for the exit concentration C_{AN} is

$$C_{AN-1} = C_{AN} + kC_{AN}\bar{t}_N \qquad\qquad (5\text{-}235)$$

which becomes

$$C_{AN} = \frac{C_{AN-1}}{\left(1 + k\bar{t}_N\right)} = \frac{C_{AO}}{\left(1 + k\bar{t}_N\right)^N} \qquad\qquad (5\text{-}236)$$

or

$$\frac{C_{AO}}{C_{AN}} = \left(1 + k\bar{t}_N\right)^N \qquad\qquad (5\text{-}237)$$

Rearranging Equation 5-237, the residence time distribution for the system is

$$\bar{t}_{N,\,\text{reactors}} = N\bar{t}_i = \frac{N}{k}\left[\left(\frac{C_{AO}}{C_{AN}}\right)^{1/N} - 1\right] \qquad\qquad (5\text{-}238)$$

Using the limit where $N \to \infty$, Equation 5-238 reduces to the plug flow equation

$$\bar{t}_{\text{plug}} = \frac{1}{k}\ln\left(\frac{C_{AO}}{C_A}\right) \qquad\qquad (5\text{-}239)$$

For most types of kinetics, the stepwise changes in concentration results in a smaller average reaction rate than it would otherwise if the same feed materials were in a batch or plug (tubular) reactor. Therefore, to obtain the same output the volume of the reaction space must be larger, and in some instances much larger as in the case where only a single tank is used. By arranging several tanks in series the volume of the reaction space can be substantially reduced, as illustrated in Figure 5-25. However, the volume per unit of output will still be greater than is required for either a batch or tubular reactor.

Using several CFSTR systems in series, some of the desired characteristics of the CFSTR system can be achieved, such as good mixing and heat transfer and easier and less expensive maintenance, while at the same time approaching the performance of a tubular system. However, this can be offset by the cost of using several smaller reactors, which will be greater than the cost of a single larger reactor. In such cases, there may be some economic tradeoff between reactor size and cost.

SECOND ORDER REACTION IN A CFSTR

Consider the second order reaction of two components A + B $\rightarrow$ products, $\theta_A = C_{BO}/C_{AO}$, where

$$\left(-r_A\right) = \left(-r_B\right) = kC_A C_B = kC_A^2 \tag{5-240}$$

When the reactant ratio is unity (i.e., $\theta_A = 1$), the mass balance for component A in the continuous flow stirred tank reactor is

$$uC_{AO} = uC_A + \left(-r_A\right)V_R + V_R \frac{dC_A}{dt} \tag{5-241}$$

where accumulation is zero $V_R dC_A/dt = 0$ and Equation 5-241 becomes

$$uC_{AO} = uC_A + kC_A^2 V_R \tag{5-242}$$

where $\bar{t} = V_R/u$. Equation 5-242 becomes

$$C_{AO} = C_A + k\bar{t}C_A^2 \tag{5-243}$$

Rearranging Equation 5-243 gives

$$k\bar{t}C_A^2 + C_A + C_{AO} = 0 \tag{5-244}$$

Equation 5-244 is a quadratic equation of the form $ax^2 + bx + c = 0$, where the root of the equation is

$$x = \frac{-b \pm \left(b^2 - 4ac\right)^{0.5}}{2a}$$

The exit concentration C_A for the second order reaction is

$$C_A = \frac{-1 + \left(1 + 4k\bar{t}C_{AO}\right)^{0.5}}{2k\bar{t}} \tag{5-245}$$

nth order kinetics in a CFSTR.

The nth order kinetics in a CFSTR, as represented by $(-r_A) = kC_n^A$, yields

$$u\,C_{AO} = uC_A + (-r_A)\,V_R$$

$$uC_{AO} = uC_A + kC_A^n\,V_R \tag{5-246}$$

Rearranging Equation 5-246 gives

$$k\bar{t}C_A^n + C_A - C_{AO} = 0 \tag{5-247}$$

where $\bar{t} = V_R/u$.

Equation 5-247 is a polynomial, and the roots (C_A) are determined using a numerical method such as the Newton-Raphson as illustrated in Appendix D. For second order kinetics, the positive sign (+) of the quadratic Equation 5-245 is chosen. Otherwise, the other root would give a negative concentration, which is physically impossible. This would also be the case for the nth order kinetics in an isothermal reactor. Therefore, for the nth order reaction in an isothermal CFSTR, there is only one physically significant root $(0 < C_A < C_{AO})$ for a given residence time $\bar{t}$.

Example 5-11

---

(1) For a cascade of N CFSTRs of equal volume, V_R, in which the first order forward reaction $A \xrightarrow{\ k\ } P$ occurs with a throughput u, show that the system fractional conversion is

$$1 - \frac{1}{\left[1 + \dfrac{kV_R}{u}\right]^N}$$

whatever the feed concentration of A.

(2) A sample CFSTR of volume V_R has one inlet stream rate u, containing A at concentration C_{AO} and a second inlet stream of rate αu, containing B at concentration C_{BO}. The reaction is $A + B \xrightarrow{k} 2P$ with the rate first order with respect to both A and B. Obtain an expression for the concentration of A at the reactor exit in terms of V_R and the conditions in the feed streams.

Solution

(1) Considering the material balance on species A in the first CFSTR at constant density gives $uC_{AO} = uC_{A1} + (-r_A)V_R$ where $(-r_A) = kC_A$:

$$uC_{AO} = uC_{A1} + kC_{A1}V_R$$

Rearranging yields

$$C_{A1} = \frac{C_{AO}}{1 + \dfrac{kV_R}{u}}$$

For the second CFSTR, the exit concentration C_{A2} is

$$C_{A2} = \frac{C_{A1}}{1 + \dfrac{kV_R}{u}} = \frac{C_{AO}}{\left(1 + \dfrac{kV_R}{u}\right)^2}$$

For the Nth CFSTR, Equation 5-237 gives

$$C_{AN} = \frac{C_{AN-1}}{1 + \dfrac{kV_R}{u}} = \frac{C_{AO}}{\left[1 + \dfrac{kV_R}{u}\right]^N}$$

where $\dfrac{C_{AN}}{C_{AO}} = \dfrac{1}{\left[1 + \dfrac{kV_R}{u}\right]^N}$

The fractional conversion $X_A = 1 - \dfrac{C_{AN}}{C_{AO}}$ is

$$= 1 - \frac{1}{\left[1 + \dfrac{kV_R}{u}\right]^N}$$

(2) A CFSTR with two inlets A and B is shown below.

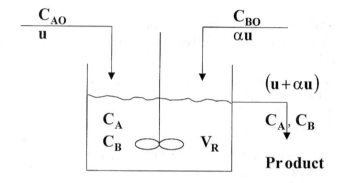

The reaction $A + B \xrightarrow{\ k\ } 2P$ and the rate expression of species A or B is $(-r_A) = (-r_B) = kC_AC_B$. The material balance on species A is

$$uC_{AO} = u(1+\alpha)C_A + (-r_A)V_R$$

$$uC_{AO} = u(1+\alpha)C_A + kC_AC_B V_R$$

or

$$C_A = \frac{uC_{AO}}{u(1+\alpha) + kV_RC_B} \tag{5-248}$$

The material balance on species B is

$$\alpha uC_{BO} = u(1+\alpha)C_B + (-r_B)V_R$$

$$\alpha uC_{BO} = u(1+\alpha)C_B + kC_AC_B V_R$$

or

$$C_B = \frac{u\alpha C_{BO}}{\left[u(1+\alpha) + kC_A V_R\right]} \tag{5-249}$$

Substituting Equation 5-249 into Equation 5-248 gives

$$C_A = \frac{uC_{AO}}{\left\{u(1+\alpha) + kV_R\left[\dfrac{\alpha u C_{BO}}{u(1+\alpha) + kC_A V_R}\right]\right\}} \tag{5-250}$$

Rearranging Equation 5-250 gives

$$(1+\alpha)kV_R C_A^2 + \left\{u(1+\alpha)^2 + (\alpha C_{BO} - C_{AO})kV_R\right\}C_A$$

$$-u(1+\alpha)C_{AO} = 0 \tag{5-251}$$

Equation 5-251 is a quadratic equation of the form $ax^2 + bx + c = 0$, where the root of the equation is

$$x = \frac{-b \pm (b^2 - 4ac)^{0.5}}{2a}$$

Therefore, the exit concentration of species A in terms of V_R and the conditions in the feed streams ($\alpha, u, C_{AO}, C_{BO}$) is

$$C_A = \frac{-\left[u(1+\alpha)^2 + (\alpha C_{BO} - C_{AO})kV_R\right] + \sqrt{\left[u(1+\alpha)^2 + (\alpha C_{BO} - C_{AO})kV_R\right]^2 + 4u(1+\alpha)^2 kC_{AO}V_R}}{2(1+\alpha)kV_R}$$

INTERMEDIATE CONVERSION OF N − 1 CFSTR

Consider a series of continuous flow stirred tank reactors of equal size with inlet and exit conversions as X_0 and X_N. The intermediate optimal conversions $X_1, X_2, X_3 \ldots X_i \ldots X_{N-1}$ can be determined, which will minimize the overall reactor size. Levenspiel [1] has shown that the optimum size ratio of CFSTRs is achieved from the plot of $1/(-r_A)$ versus X_A, when the slope of the rate curve at point X_i equals the diagonal of the rectangle, as this gives both the intermediate conversion as well as the size of the units. Adesina [2] has employed a numerical method to determine the optimum $(N - 1)$ intermediate conversions, which also minimizes the overall reactor size for a series of N equal-sized stirred tanks. Adesina's methodology depends on the rate expression being differentiated twice. The following determines the intermediate conversions $X_1, X_2, X_3 \ldots X_i \ldots X_{N-1}$ of a series of $N - 1$ equal-sized continuous flow stirred tank reactors.

For the first CFSTR the optimal selection of the conversion X_1 on the $1/(-r_A)$ versus X_A plot is such that the diagonal of the rectangle must possess the same slope as the tangent to the curve at point X_1 (Figure 5-26). Therefore, the first CFSTR gives

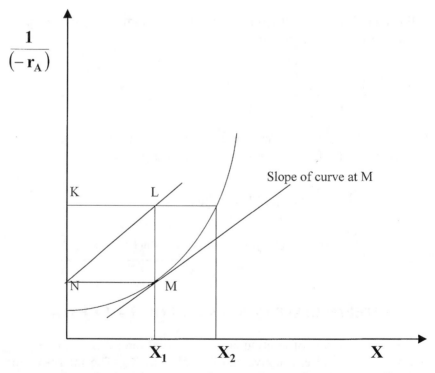

Figure 5-26. Maximization of rectangle applied to find the optimum intermediate conversion and optimum sizes of mixed flow reactors in series.

$$\frac{f(X_2)-f(X_1)}{X_1-X_0}=\frac{df(X)}{dX}\bigg|_{X_1} \tag{5-252}$$

where $f(X) = 1/(-r_A)$ and $\big|_{x_1}$ means that the slope is evaluated at X_1. Rearranging Equation 5-252 gives $w_1 = f(X_2) - f(X_1) - \Psi(X_1)(X_1 - X_0) = 0$.

For the second and third CFSTRs:

$$w_2 = f(X_3)-f(X_2)-\Psi(X_2)(X_2-X_1)=0$$

$$w_3 = f(X_4)-f(X_3)-\Psi(X_3)(X_3-X_2)=0$$

For the ith CFSTR:

$$w_i = f(X_{i+1}) - f(X_i) - \Psi(X_i)(X_i - X_{i-1})=0$$

For the $(N - 1)^{th}$ CFSTR:

$$w_{N-1} = f(X_N)-f(X_{N-1})-\Psi(X_{N-1})(X_{N-1}-X_{N-2}) = 0 \tag{5-253}$$

The simultaneous solution of Equation 5-253 yields the unknown intermediate conversions X_1, X_2, X_3 . . . X_i . . . X_{N-1}, where the optimal size is $(V/F_{AO})_i$ for the ith reactor.

$$\left(\frac{V}{F_{AO}}\right)_i =(X_i -X_{i-1})\Psi(X_i), \quad i=1, 2, ...N \tag{5-254}$$

and $\Psi(X) = [df(X)/dX]$, where

V = reactor volume, liter, m^3
F_{AO} = feed molar rate, mol/min

Using the Newton-Raphson method for solving the nonlinear system of Equations 5-253 gives

$$\gamma^{(m)} =\underline{X}^{(m)} -\underline{X}^{(m-1)} =-\underline{J}^{-1}\left(\underline{X}^{(m-1)}\right)\underline{w}\left(\underline{X}^{(m-1)}\right)$$

$$m = 1,2,........ \tag{5-255}$$

where $\underline{X}^{(m)}$ is the column vector of conversions at the mth iteration. This is noted as

$$\underline{X}^{(m)} = \left[X_1^{(m)} \;\; X_2^{(m)} \;\; X_3^{(m)} \ldots\ldots\ldots X_{N-1}^{(m)} \right]^T \qquad (5\text{-}256)$$

$X^{(0)}$ is the guessed initial conversion vector, and the Jacobian matrix at the $(m - 1)^{th}$ iteration, $J(X^{(m-1)})$, has the dimension of $(N - 1) \times (N - 1)$. Equation 5-255 has a quadratic convergent property, which gives the final solution after few iterations. The convergence criterion is:

$$\sqrt{\dfrac{\displaystyle\sum_{i=1}^{N}\left(\dfrac{X_i^{(m)} - X_i^{(m-1)}}{X_i^{(m)}}\right)^2}{N}} \;\; \le \varepsilon = 10^{-2} \qquad (5\text{-}257)$$

The Jacobian matrix is tri-diagonal since the elements are

$$\frac{\partial w_i}{\partial X_q} = 0 \quad i = 1, 2, \ldots\ldots\ldots(N-1) \quad q = 1, 2, \ldots\ldots(i-2)$$

$$\frac{\partial w_i}{\partial X_i} = \left(X_{i-1} - X_i\right)\frac{\partial \Psi(X_i)}{\partial X_i} - 2\Psi(X_i)$$

$$\frac{\partial w_i}{\partial X_{i-1}} = \Psi(X_{i-1})$$

$$\frac{\partial w_i}{\partial X_{i+1}} = \Psi(X_{i+1})$$

$$\frac{\partial w_i}{\partial X_p} = 0 \quad p = i+2, \, i+3, \ldots\ldots\ldots(N-1) \qquad (5\text{-}258)$$

and is represented by

$$J(X)=\begin{bmatrix} \dfrac{\partial w_1}{\partial X_1} & \dfrac{\partial w_1}{\partial X_2} & 0 & .. & 0 \\[2ex] \dfrac{\partial w_2}{\partial X_1} & \dfrac{\partial w_2}{\partial X_2} & \dfrac{\partial w_2}{\partial X_3} & .. & 0 \\[2ex] 0 & \dfrac{\partial w_3}{\partial X_2} & \dfrac{\partial w_3}{\partial X_3} & \dfrac{\partial w_3}{\partial X_4} & 0 \\[2ex] : & : & : & : & : \\[2ex] 0 & 0 & \dfrac{\partial w_i}{\partial X_{i-1}} & \dfrac{\partial w_i}{\partial X_i} & \dfrac{\partial w_i}{\partial X_{i+1}} \\[2ex] : & : & : & : & : \\[2ex] 0 & .. & 0 & \dfrac{\partial w_{N-1}}{\partial X_{N-2}} & \dfrac{\partial w_{N-1}}{\partial X_{N-1}} \end{bmatrix} \qquad (5\text{-}259)$$

Adesina has shown that it is superfluous to carry out the inversion required by Equation 5-255 at every iteration of the tri-diagonal matrix J. The vector γ^m is readily computed from simple operations between the tri-diagonal elements of the Jacobian matrix and the vector. The methodology can be employed for any reaction kinetics. The only requirement is that the rate expression be twice differentiable with respect to the conversion. The following reviews a second order reaction and determines the intermediate conversions for a series of CFSTRs.

Example 5-12

Styrene (A) and butadiene (B) are to be treated in a series of CFSTRs each possessing 26.5 m^3 capacity. The initial concentration of A is 0.795 and B is 3.55 kg mol/m^3. The flowrate is 20 m^3/hr. The rate equation is $(-r_A) = 0.036C_A C_B$ kg $\bullet$ mol/m^3 – hr. Determine the intermediate conversions, if the eleventh stirred tank is required to effect 80% conversion of the limiting reactant.

Solution

Assume that the reaction between A and B is second order and is represented by A + B $\rightarrow$ products where A is the limiting reactant. The rate expression is

$$\left(-r_A\right) = kC_A C_B \qquad\qquad (5\text{-}240)$$

From the stoichiometry, $C_{AO} - C_A = C_{BO} - C_B$ and $C_B = C_{BO} - (C_{AO} - C_A)$. The fractional conversion X_A of A is $X_A = (C_{AO} - C_A)/C_{AO}$ or $C_{AO} - C_A = C_{AO}X_A$. Therefore, C_A in terms of the fractional conversion X_A and initial concentration C_{AO} is $C_A = C_{AO}(1 - X_A)$ and

$$C_B = C_{BO} - \left(C_{AO} - C_A\right)$$

$$= C_{BO} - C_{AO}X_A$$

where $\theta_B = C_{BO}/C_{AO}$. The concentration of B is $C_B = C_{AO}(\theta_B - X_A)$. Substituting the concentrations of A and B in Equation 5-240 gives

$$\left(-r_A\right) = kC_{AO}^2\left(1 - X_A\right)\left(\theta_B - X_A\right) \qquad\qquad (5\text{-}260)$$

Equation 5-260 is used to determine the intermediate conversions $X_1, X_2, X_3, \ldots X_{10}$. This is accomplished first by determining the inverse of Equation 5-260, and then differentiating the inverse function twice with respect to the fractional conversion X_A. Hence, the inverse of Equation 5-260 is

$$\frac{1}{\left(-r_A\right)} = \frac{1}{kC_{AO}^2\left(1 - X_A\right)\left(\theta_B - X_A\right)} \qquad\qquad (5\text{-}261)$$

Differentiating Equation 5-261 with respect to X_A yields

$$\frac{d}{dX_A}\left(\frac{1}{\left(-r_A\right)}\right) = \frac{\left(1 - X_A\right) + \left(\theta_B - X_A\right)}{kC_{AO}^2\left[\left(1 - X_A\right)\left(\theta_B - X_A\right)\right]^2} \qquad\qquad (5\text{-}262)$$

Further differentiation of Equation 5-262 gives after mathematical manipulations using any of the calculus of either the product or quotient involving two variables

$$\frac{d}{dx}(uv) = u\frac{dv}{dx} + v\frac{du}{dx}$$

or

$$\frac{d}{dx}\left(\frac{u}{v}\right) = \frac{v\dfrac{du}{dx} - u\dfrac{dv}{dx}}{v^2}$$

$$\frac{d}{dX_A}\left[\frac{d}{dX_A}\left(\frac{1}{(-r_A)}\right)\right]$$

$$= \frac{2}{kC_{AO}^2}\left\{\frac{\left[(1-X_A)+(\theta_B-X_A)\right]^2 - \left[(1-X_A)(\theta_B-X_A)\right]}{\left[(1-X_A)(\theta_B-X_A)\right]^3}\right\} \qquad (5\text{-}263)$$

Equations 5-261, 5-262, and 5-263 are incorporated as functions in the computer program CFSTR51. They are represented by

$$\text{RATE(XX)} = \text{CONST1} * \text{CAO} * *2 * (1-\text{XX}) * (\text{THETAB} - \text{XX})$$

$$\text{RATEINV(XX)} = 1/\text{RATE(XX)}$$

$$\text{DER(XX)} = \left((1.0-\text{XX})+(\text{THETAB}-\text{XX})\right) /$$
$$\left(\text{CAO}**2*\text{CONST1}*\left((1.0-\text{XX})*(\text{THETAB}-\text{XX})\right)**2\right)$$

$$\text{DDER(XX)} = 2*\left(\left((1.0-\text{XX})+(\text{THETAB}-\text{XX})\right)\right)**2$$
$$-\left((1.0-\text{XX})*(\text{THETAB}-\text{XX})\right)) /$$
$$\left(\text{CAO}**2*\text{CONST1}*\left((1.0-\text{XX})*(\text{THETAB}-\text{XX})\right)**3\right)$$

Known values of C_{AO}, C_{BO}, and k, and guessed values of the intermediate conversions X_1, X_2, X_3, . . . X_{10} are given as data in the computer program. The results are obtained after a few iterations and are shown in Table 5-12.

GRAPHICAL SOLUTION OF THE CASCADE OF N – CFSTR IN SERIES

The CFSTR equation of the nth stage

$$uC_{A,n-1} = uC_{A,n} + (-r_A)_n V_R \qquad (5\text{-}235)$$

Table 5-12
Results of the intermediate conversions in a reactor train of CFSTRs involving the second order irreversible reaction kinetics A + B → products

NUMBER OF TANKS:	11
CONCENTRATION OF REACTANT A IN FEED kg-mol/m^3:	.7950
CONCENTRATION OF REACTANT B IN FEED kg-mol/m^3:	3.5500
REACTION RATE CONSTANT k (m^3/kg-mol-hr):	.0360
INLET CONVERSION:	0.0000E+00

INTERMEDIATE CONVERSIONS IN THE C.F.S.T.Rs TRAINS

- -

FINAL RESULTS SATISFYING TOLERANCE LIMIT IN THE C.F.S.T.R. TRAINS ARE:	
X(2):	0.1363E+00
X(3):	0.2535E+00
X(4):	0.3546E+00
X(5):	0.4419E+00
X(6):	0.5175E+00
X(7):	0.5829E+00
X(8):	0.6396E+00
X(9):	0.6888E+00
X(10):	0.7313E+00
X(11):	0.7682E+00
EXIT CONVERSION:	0.8000E+00
REQUIRED NUMBER OF TOLERANCE LIMIT:	9

COLUMN VECTOR CONVERSION

- -

```
    -0.6220E-03
    -0.4798E-04
     0.1739E-02
     0.4975E-02
     0.9361E-02
     0.1380E-01
     0.1638E-01
     0.1498E-01
     0.8692E-02
```

can be rearranged in the form

$$\left(-r_A\right)_n = -\frac{1}{\bar{t}_n}C_{A,n} + \frac{1}{\bar{t}_n}C_{A,n-1}$$

or

$$\frac{\left(-r_A\right)_n}{C_{A,n-1} - C_{A,n}} = \frac{1}{\bar{t}_n} \tag{5-264}$$

where $\bar{t}_n$ is the mean residence time.

At a given inlet concentration, $C_{A,n-1}$, Equation 5-264 is linear in exit concentration. A plot of $(-r_A)$ versus C_A for the limiting component A is shown in Figure 5-27. Equation 5-264, on arrangement,

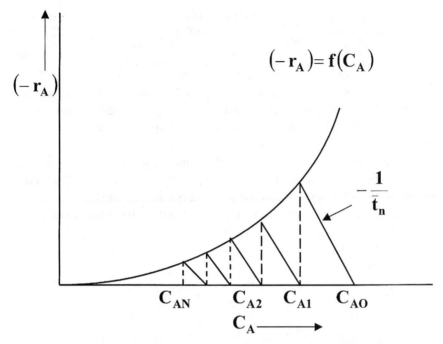

$$(-r_A) = f(C_A)$$

$$-\frac{1}{\bar{t}_n}$$

C_{AN} C_{A2} C_{A1} C_{AO}

$C_A \longrightarrow$

Figure 5-27. Graphical representation of a number of CFSTRs in series.

can be viewed as a slope of $-1/\bar{t}_n$, which intersects the abscissa (X-axis) at the point representing the inlet concentration. The abscissa of the point where this straight line intersects the $(-r_A)$ versus C_A curve gives the exit concentration for each stage in the cascade. This represents the inlet to the next stage. If the size of the cascade reactor is the same (i.e., $V_1 = V_2 = \ldots V_N$), then the residence time is the same throughout the system, and the straight lines are parallel. The procedure is repeated until the desired concentration is obtained. The number of parallel lines drawn until the final concentration is reached gives the number of stages required for the desired conversion. In many cases, $\bar{t}_n$ must be estimated because the purpose is to move from C_{AO} to C_{AN} in about four or five stages.

SPACE TIME (ST) AND SPACE VELOCITY (SV)

Space time and space velocity are used to determine the performance of flow reactors. Space velocity, SV, is defined as F_A/VC_{AO}, that is,

$$SV = \frac{F_A}{VC_{AO}} = \left(\frac{uC_{AO}}{VC_{AO}} \right)$$

The term V/F_A for flow reactors is used to evaluate the size of the reactor required to achieve the conversion X_A of a reactant feed rate F_A. It is the number of reactor volumes of feed at specified conditions that can be treated in unit time, where F_A is the molar flowrate (uC_{AO}) and V is the volume of the reactor.

Space time, ST, is defined as the time required to process one reactor volume of feed measure at specified conditions. The relationship between space velocity SV and ST is as follows:

If C_{AO} is the concentration of the reactant in the feed, then $F_A = uC_{AO}$ and

$$\frac{V}{F_A} = \frac{V}{uC_{AO}}$$

or

$$\frac{V}{u} = C_{AO} \left(\frac{V}{F_A} \right)$$

$$= \frac{\left(\dfrac{\text{moles of A entering}}{\text{volume of feed}} \right) (\text{volume of reactor})}{\left(\dfrac{\text{moles of A entering}}{\text{time}} \right)} \left\{ \frac{\left[\left(\dfrac{mol}{m^3} \times m^3 \right) \right]}{\dfrac{mol}{sec}} \right\}$$

The ratio of V/u has the unit of time and its reciprocal u/V is called the space velocity:

$$ST = \frac{V}{u} \left(\frac{\text{reactor volume}}{\text{volumetric feed rate}} \right)$$

The space velocity SV is often used with conversion to describe the overall reactor performance. It is common in the petroleum and petrochemical industries to plot conversion against space velocity to describe the effect of feed rate on the performance of a flow system.

The reactor feed may be preheated and the feed pressure may alter. The volumetric flowrate of gases depends on the reactor temperature and pressure at fixed mass flowrate. In many cases, the feed is liquid at room temperature, while the reaction mixture is a gas at the higher temperature in the reactor. In these instances, the performance of the reactor is represented as conversion and selectivity against the liquid hourly space velocity (LHSV). This is defined as

$$LHSV = \frac{\text{volume of liquid feed per hour}}{\text{volume of reactor}}$$

Another commonly used term is the gas hourly space velocity (GHSV), which is defined as

$$GHSV = \frac{\text{volume of gaseous feed per hour}}{\text{volume of reactor}}$$

For gaseous feeds, space velocity SV is normally determined in terms of volumes measured at standard conditions T = 25°C and P = 1.013 bar.

Space time ST is equal to the residence time in a plug flow reactor only if the volumetric flowrate remains constant throughout the reactor. The residence time depends on the change in the flowrate through the reactor, as well as V/u. The change in u depends on the variation in temperature, pressure, and the number of moles. The concept of SV with conversions in the design of a plug flow reactor is discussed later in this chapter.

FRACTIONAL CONVERSION, YIELD, AND SELECTIVITY IN REACTORS

Consider the following reactions, namely the second order irreversible reaction aA + bB → rR + sS, the series reaction A → B → C, and the parallel reaction A → B A → C

Fractional conversion, X_A

In the above reactions, the fractional conversion is defined as

$$\text{conversion} = \frac{\text{loss of reactant}}{\text{feed of reactant}}$$

For reactant A, the conversion is

$$X_A = \frac{N_{AO} - N_A}{N_A} = \frac{F_{AO} - F_A}{F_{AO}}$$

At constant density (i.e., the volume is constant), X_A is defined as

$$X_A = \frac{C_{AO} - C_A}{C_{AO}}$$

Selectivity

Selectivity is defined as the ratio of the desired product to the amount of limiting reactant that has undergone chemical change. That is

$$S_R = \frac{\text{amount of desired product}}{\text{amount of limiting reactant that has undergone chemical change}}$$

For the second order irreversible reaction, the selectivity of species R is

$$S_R = \frac{N_R}{N_{AO} - N_A}$$

where N_{AO}, N_A = moles of reactant A

N_R = moles of product R

If the system is at constant density, then the selectivity of species R is

$$S_R = \frac{C_R}{C_{AO} - C_A}$$

For the series $A \rightarrow B \rightarrow C$ and parallel reaction $A \rightarrow B \, A \rightarrow C$, the selectivity of the wanted species B is

$$S_B = \frac{N_B}{N_{AO} - N_A}$$

If the system is at constant density, S_B is defined as

$$S_B = \frac{C_B}{C_{AO} - C_A}$$

or

$$S_B = \frac{C_B}{C_B + C_C}$$

Yield

The yield of the product is the amount of the desired product formed divided by the amount of the reactant fed. That is

$$Y_R = \frac{\text{desired product formed}}{\text{reactant fed}}$$

For the reaction aA $\rightarrow$ rR, the yield is defined by

$$Y_R = \left(\frac{a}{r}\right)\left(\frac{C_R}{C_{AO}}\right)$$

where a and r are the stoichiometric coefficients, C_R is the concentration of the desired product, and C_{AO} is the inlet concentration of the reactant.

For the series reaction A $\rightarrow$ B $\rightarrow$ C and parallel reaction A $\rightarrow$ B A $\rightarrow$ C, the yield of the desired product (B) is

$$Y_B = \frac{C_B}{C_{AO}}$$

RELATIONSHIP BETWEEN CONVERSION, SELECTIVITY, AND YIELD

A relationship between the yield, selectivity and conversion can be obtained on the components as follows. The conversion X_A of component A is

$$X_A = \frac{C_{AO} - C_A}{C_{AO}}$$

The yield of component B is

$$Y_B = \frac{C_B}{C_{AO}}$$

and the selectivity of component B is

$$S_B = \frac{C_B}{C_B + C_C} = \frac{C_B}{C_{AO} - C_A}$$

Therefore,

$$Y_B = \frac{C_B}{C_{AO}} = \left(\frac{C_B}{C_{AO} - C_A} \right)\left(\frac{C_{AO} - C_A}{C_{AO}} \right) = S_B \cdot X_B$$

Therefore,

$$Y_B = S_B \cdot X_B \tag{5-265}$$

For a well-stirred tank reactor involving a second order irreversible reaction of the form $aA + bB \rightarrow rR + sS$, the yield of R is

$$Y_R = \frac{C_R}{C_{AO}}\left(\frac{a}{r} \right) = \left(\frac{C_R}{C_{AO} - C_A} \right) \cdot \left(\frac{a}{r} \right) \cdot \left(\frac{C_{AO} - C_A}{C_{AO}} \right)$$

$$= S_R \cdot \left(\frac{a}{r} \right) X_B \tag{5-266}$$

For a plug flow or batch reactor, the yield of component R is defined as

$$Y_R = - \frac{1}{C_{AO}} \int_{C_{AO}}^{C_A} Y_R \, dC_A = \int_0^{X_A} Y_R \cdot dX_A \tag{5-267}$$

For a cascade of well-stirred reactors, the yield of component R is defined as

$$Y_R = \frac{1}{(A)} \sum_{i=1}^{N} Y_{R,i} \Delta(A)_i = \frac{1}{C_{AO}} \sum_{i=1}^{N} Y_{R,i} \left(C_{Ai} - C_{A,i-1} \right) \qquad (5\text{-}268)$$

SELECTIVITY

The concept of selectivity and its application are essential in the industrial evaluation of complex chemical reactions when the objective is to attain maximum production of the desired product. In industrial operations, the goal is to establish the best operating conditions for the desired product. In complex chemical reactions, two or more products are produced whose formation may interfere with the generation of the desired product. In such cases, production is improved of the desired product by employing the selectivity concept to either accelerate its reaction rate or suppress the formation rates of undesired products. This procedure may be performed by examining a single expression for selectivity instead of examining the effect of appropriate reaction parameters on individual rate equations. This is referred to as selectivity maximization because it is a search for the maximum value of selectivity within the permissible range of process parameters. This value determines the optimum operational conditions for the desired products.

The following details mathematical expressions for instantaneous (point or local) or overall (integral) selectivity in series and parallel reactions at constant density and isothermal conditions. An instantaneous selectivity is defined as the ratio of the rate of formation of one product relative to the rate of formation of another product at any point in the system. The overall selectivity is the ratio of the amount of one product formed to the amount of some other product formed in the same period of time.

Consider a series reaction $A \xrightarrow{k_1} B \xrightarrow{k_2} C$ occurring in a batch system at constant volume with $C_{AO} \neq 0$, and $C_{BO} = C_{CO} = 0$. The rate equations for these reactions are

$$\frac{dC_A}{dt} = -k_1 C_A \qquad (5\text{-}269)$$

$$\frac{dC_B}{dt} = k_1 C_A - k_2 C_B \qquad (5\text{-}270)$$

$$\frac{dC_C}{dt} = k_2 C_B \tag{5-271}$$

The equations of components A, B, and C when integrated are

$$C_A = C_{AO}\, e^{-k_1 t} \tag{5-272}$$

$$C_B = \frac{k_1 C_{AO}}{k_2 - k_1}\left(e^{-k_1 t} - e^{-k_2 t}\right) \tag{5-273}$$

and

$$C_C = C_{AO}\left\{1 - \frac{k_2}{k_2 - k_1}e^{-k_1 t} + \frac{k_1}{k_2 - k_1}e^{-k_2 t}\right\} \tag{5-274}$$

If component B is the desired product and C is unwanted, the instantaneous selectivity of B relative to C is expressed as S_C^B from Equations 5-270 and 5-271 as

$$S_C^B = \frac{dC_B/dt}{dC_C/dt}$$

$$= \frac{k_1 C_A - k_2 C_B}{k_2 C_B}$$

$$= \frac{k_1 C_A}{k_2 C_B} - 1 \tag{5-275}$$

The overall selectivity of B relative to C, $S_C^{\bar{B}}$ is $S_C^{\bar{B}} = C_B/C_C$ from Equations 5-273 and 5-274 is

$$S_C^{\bar{B}} = \frac{k_1}{k_2 - k_1}\frac{\left(e^{-k_1 t} - e^{-k_2 t}\right)}{\left[1 - \dfrac{k_2}{k_2 - k_1}e^{-k_1 t} + \dfrac{k_1}{k_2 - k_1}e^{-k_2 t}\right]} \tag{5-276}$$

In certain instances, the concentrations of reaction participants in the rate and product distribution equation are expressed in terms of concentrations of one selected reactant. In the series reaction $A \xrightarrow{k_1} B \xrightarrow{k_2} C$, dividing Equation 5-270 by 5-271 yields

$$\frac{dC_B}{dC_A} = \frac{k_2 C_B}{k_1 C_A} - 1 \tag{5-277}$$

Integrating Equation 5-277 by substituting C_B/C_A and subsequent manipulation yields

$$C_B = \frac{k_1 C_{AO}}{k_2 - k_1} \left[\frac{C_A}{C_{AO}} - \left(\frac{C_A}{C_{AO}} \right)^{\frac{k_2}{k_1}} \right] \tag{5-278}$$

The concentration of component C by material balance gives

$$C_C = C_{AO} - C_A - C_B$$

$$= C_{AO} \left[1 - \frac{k_2}{k_2 - k_1} \frac{C_A}{C_{AO}} + \frac{k_1}{k_2 - k_1} \left(\frac{C_A}{C_{AO}} \right)^{\frac{k_2}{k_1}} \right] \tag{5-279}$$

Similarly, for a parallel reaction, in which all the steps are of the same order

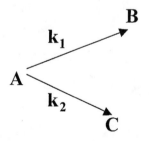

with $C_{AO} \neq 0$, $C_{BO} = C_{CO} = 0$. The rate equations are

$$\frac{dC_A}{dt} = -(k_1 + k_2) C_A \tag{5-280}$$

$$\frac{dC_B}{dt} = k_1 C_A \tag{5-281}$$

$$\frac{dC_C}{dt} = k_2 C_A \tag{5-282}$$

Integrating Equation 5-280 gives

$$C_A = C_{AO} e^{-(k_1 + k_2)t} \tag{5-283}$$

Substituting Equation 5-283 into Equation 5-281 gives

$$\frac{dC_B}{dt} = k_1 C_{AO} e^{-(k_1 + k_2)t} \tag{5-284}$$

Integrating Equation 5-284 with the boundary conditions gives

$$C_B = \frac{k_1 C_{AO}}{k_1 + k_2} \left[1 - e^{-(k_1 + k_2)t} \right] \tag{5-285}$$

Similarly,

$$C_C = \frac{k_2 C_{AO}}{k_1 + k_2} \left[1 - e^{-(k_1 + k_2)t} \right] \tag{5-286}$$

The instantaneous selectivity of B relative to C is S_C^B

$$S_C^B = \frac{dC_B/dt}{dC_C/dt} = \frac{k_1 C_A}{k_2 C_A} = \frac{k_1}{k_2}$$

The overall selectivity of B relative to C is S_C^B

$$S_C^{\bar{B}} = \frac{C_B}{C_C} = \frac{k_1}{k_2} \tag{5-287}$$

This shows that both the instantaneous and the overall selectivity of B relative to C are the same, and depends only on k_1 and k_2. The two selectivities differ for most complex reactions.

Selectivity of a Parallel Reaction in a CFSTR

Consider the first order parallel reaction

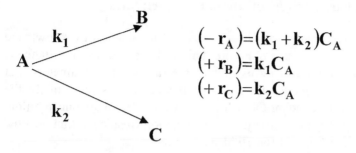

$$(-r_A) = (k_1 + k_2)C_A$$
$$(+r_B) = k_1 C_A$$
$$(+r_C) = k_2 C_A$$

in a well-mixed stirred tank. The mass balance on component A is

$$uC_{AO} = uC_A + (-r_A)V_R$$

$$uC_{AO} = uC_A + (k_1 + k_2)C_A V_R$$

$$C_{AO} = C_A\left[1 + (k_1 + k_2)\bar{t}\right]$$

The residence time $\bar{t}$ is

$$\bar{t} = \frac{C_{AO} - C_A}{(k_1 + k_2)C_A}$$

The exit concentration of A is

$$C_A = \frac{C_{AO}}{\left[1 + (k_1 + k_2)\bar{t}\right]} \tag{5-288}$$

The exit concentrations of B and C in terms of C_{AO} are

$$C_B = k_1 \bar{t} \frac{C_{AO}}{\left[1 + (k_1 + k_2)\bar{t}\right]} \tag{5-289}$$

$$C_C = k_2 \bar{t} \frac{C_{AO}}{\left[1 + (k_1 + k_2)\bar{t}\right]} \tag{5-290}$$

The instantaneous and the overall selectivity of B relative to C are the same and are expressed as

$$S_C^B = S_C^{\bar{B}} = \frac{k_1}{k_2} \tag{5-291}$$

Reaction Characteristics in Defining Selectivities

In developing mathematical expressions for selectivities, knowledge of the rate equations are required. This is because the instantaneous selectivity is defined in terms of the rate ratios. The parameters that affect the instantaneous and the overall selectivities are exactly the same as those influencing the reaction rates, namely, the concentration, temperature, activation energy, time of reaction (residence time in flow reactors), catalysts, and the fluid mechanics.

Effect of Temperature on Selectivity

For the various reaction mechanisms used in determining both instantaneous and the overall selectivities, selectivity depends on the energy of activation obtained from the Arrhenius equation [(k = $k_o exp(-E/RT)$)], the temperature, initial concentration, and the time of reaction. From the Arrhenius equation, the specific reaction rate k is an integral part of the selectivity expressions. Furthermore, analyzing selectivity expressions may indicate an enhanced effect of the temperature on selectivity. Maximizing the expressions for both instantaneous and overall selectivities may depend on the following:

- The time of reaction at constant temperature.
- The temperature at constant time of reaction.
- Both the time and temperature.

Example 5-13

A well-mixed batch reactor is used for performing the isothermal liquid phase reaction

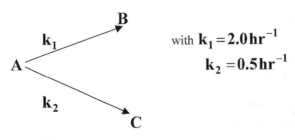

with $k_1 = 2.0\,hr^{-1}$

$k_2 = 0.5\,hr^{-1}$

Each reaction step is first order with respect to reactant A and the initial concentration of A is 1.0 kmol/m^3.

(1) Determine the yield of B with respect to A for any given initial concentration.

(2) Determine the time required to achieve a 60% conversion of A to B or C.

Solution

(1) Assuming the reaction is first order in a constant volume batch system, the rate equations of A, B, and C, respectively, are:

$$\left(-r_A\right)_{net} = \frac{dC_A}{dt} = -\left(k_1 + k_2\right)C_A \tag{5-280}$$

$$\left(+r_B\right) = \frac{dC_B}{dt} = k_1 C_A \tag{5-281}$$

$$\left(+r_C\right) = \frac{dC_C}{dt} = k_2 C_A \tag{5-282}$$

Integrating Equation 5-280 between the boundary conditions t = 0, $C_A = C_{AO}$ and t = t, $C_A = C_A$ gives

$$C_A = C_{AO}e^{-\left(k_1 + k_2\right)t} \tag{5-283}$$

Substituting Equation 5-283 into Equation 5-281 gives

$$\frac{dC_B}{dt} = k_1 C_{AO}e^{-\left(k_1 + k_2\right)t} \tag{5-284}$$

Integrating Equation 5-284 between the boundary conditions gives

$$C_B = \frac{k_1 C_{AO}}{k_1 + k_2}\left\{1 - e^{-\left(k_1 + k_2\right)t}\right\} \tag{5-285}$$

The yield of B with respect to A is Y_B

$$Y_B = \frac{C_B}{C_{AO}}$$

$$= \frac{k_1 C_{AO}}{(k_1 + k_2)} \cdot \frac{\left\{1 - e^{-(k_1 + k_2)t}\right\}}{C_{AO}}$$

$$= \frac{k_1}{(k_1 + k_2)} \cdot \left\{1 - e^{-(k_1 + k_2)t}\right\}$$

Substituting the values of k_1 and k_2 in terms of time t gives the yield of B with respect to A as $Y_B = 0.8(1 - e^{-2.5t})$.

(2) The time required to achieve 60% conversion of A to B or C is

Conversion of A, $X_A = 1 - \dfrac{C_A}{C_{AO}} = 0.6$

$$\frac{C_A}{C_{AO}} = 0.4 = e^{-(k_1 + k_2)t}$$

$$\ln\left(\frac{C_A}{C_{AO}}\right) = \ln(0.4) = -(k_1 + k_2)t$$

That is,

$\ln 0.4 = -2.5t$

$t = 0.366$ hr (≈ 22 min)

PLUG FLOW REACTOR

For a tubular (plug flow) reactor, the conditions at any point in the reactor are independent of time, and the linear velocity v of the reacting mixture is the same at every point in a cross-section S perpendicular to the flow direction and equal to (G/ρS). The composition of the reaction mixture depends on the distance L from the inlet point.

A tubular plug flow (Figure 5-28) reactor assumes that mixing of fluid does not take place, the velocity profile is flat, and both temperature and composition are uniform at any cross-section in the reactor.

At any instant, because gas-phase reactions are often carried out in tubular systems, the mass flowrate G and C_i' the concentration of i in moles per unit mass is used. The mass flowrate G does not change with position when fluid density changes as is the case with u, the volumetric flowrate (Figure 5-29).

Consider the reaction A → Products, where A is the key reactant.

$(-r_A)$ = Net rate of disappearance of A by reaction per unit volume [moles/(unit time, unit volume)]

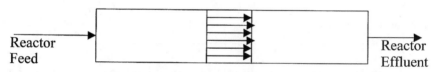

Figure 5-28. A tubular (plug flow) reactor.

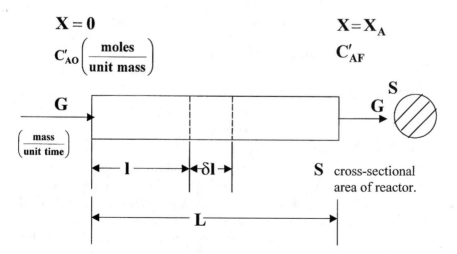

Figure 5-29. Piston flow reactor differential element.

The following assumptions should be made:

- There is a steady state operation (i.e., no change with time in the system)
- Plug flow
- The system operates isothermally (i.e., the rate constant does not change with l)

Now consider the mass balance for reactant A over the control volume $S\delta l$.

Concentration of A	l	$1 + \delta l$
	C'_A	$C'_A + \dfrac{dC'_A \, \delta l}{dl} + \ldots$
Rate of reaction	$\ldots (-r_A)$	$(-r_A) + \dfrac{d(-r_A) \, \delta l}{dl} + \ldots$

The average rate of reaction is given by the arithmetic mean of the rates at l and $1 + \delta l$. That is,

$$(-r_A)_{average} = (-r_A)_l + \frac{1}{2} \frac{d(-r_A)_l}{dl} \delta l + \ldots\ldots \qquad (5\text{-}286)$$

Since there is a steady state operation,

INPUT RATE = OUTPUT RATE

By flow: $GC'_A = G\left\{ C'_A + \dfrac{dC'_A}{dl} \delta l + \ldots \right\}$

$$+ \left\{ (-r_A)_l + \frac{1}{2} \frac{d(-r_A)_l}{dl} \delta l + \ldots \right\} S\delta l$$

$$\left(\frac{mass}{time} \cdot \frac{moles}{mass} \right) \qquad (5\text{-}287)$$

Equation 5-287 reduces to

$$-G\frac{dC_A'}{dl}\delta l = \left\{(-r_A)_l + \frac{1}{2}\frac{d(-r_A)_l}{dl}\delta l\right\}S\delta l \tag{5-288}$$

As $\delta l \to 0$, $(1/2)[d(-r_A)_l/dl]\delta l$ becomes increasingly small and Equation 5-288 then becomes

$$-\frac{G}{S}\frac{dC_A'}{dl} = (-r_A) \tag{5-289}$$

Consider the case where both volumetric flowrate u and density ρ are constant. A relationship between the mass flowrate G and the volumetric flow is:

$$G = u \cdot \rho \tag{5-290}$$

By definition,

$$C_A = C_A' \cdot \rho \tag{5-291}$$

That is,

$$\frac{\text{moles}}{\text{unit volume}} = \frac{\text{moles}}{\text{unit mass}} \times \frac{\text{mass}}{\text{unit volume}}$$

Differentiating Equation 5-291 gives

$$dC_A = \rho \cdot dC_A' \tag{5-292}$$

or

$$dC_A' = \frac{dC_A}{\rho} \tag{5-293}$$

Substituting Equations 5-290 and 5-293 into Equation 5-289 gives

$$-\frac{u \cdot \rho}{S} \cdot \frac{dC_A}{\rho\,dL} = (-r_A) \tag{5-294}$$

or

$$\left(-r_A\right) = -\frac{u}{S} \cdot \frac{dC_A}{dL} \tag{5-295}$$

where the average flow velocity in the reactor is

$$\frac{u}{S} \equiv v_L \left(\frac{length}{time}\right)$$

Equation 5-295 becomes

$$\left(-r_A\right) = -v_L \cdot \frac{dC_A}{dL} \tag{5-296}$$

Separating the variables and integrating Equation 5-296 with the limits at $L = 0$, $C_A = C_{AO}$ and at $L = L$, $C_A = C_{AF}$, Equation 5-296 becomes

$$-\int_{C_{AO}}^{C_{AF}} \frac{dC_A}{\left(-r_A\right)} = \frac{1}{v_L} \int_0^L dL \tag{5-297}$$

or

$$-\int_{C_{AO}}^{C_{AF}} \frac{dC_A}{\left(-r_A\right)} = \frac{L}{v_L} \tag{5-298}$$

where L is the total length of the reactor and the reactor volume, $v_R = S \cdot L$.

$$\frac{L}{v_L} = \frac{length}{length/time} \equiv \bar{t}, \text{ mean residence time}$$

$$\bar{t} \equiv \frac{L}{v_L} = \frac{L \cdot S}{v_L \cdot S} = \frac{V_R}{u}, \text{ plug flow} \tag{5-299}$$

Substituting Equation 5-299 into Equation 5-298 gives

$$\bar{t}_{\text{plug flow}} = \frac{V_R}{u} = -\int_{C_{AO}}^{C_{AF}} \frac{dC_A}{(-r_A)} \tag{5-300}$$

or volume of plug flow reactor is

$$V_{R\,\text{plug flow}} = -u \int_{C_{AO}}^{C_{AF}} \frac{dC_A}{(-r_A)} \tag{5-301}$$

In terms of the fractional conversion X_A, that is

$$X_A = \frac{C_{AO} - C_A}{C_{AO}} \tag{5-13}$$

Rearranging Equation 5-13 gives

$$C_A = C_{AO}(1 - X_A) \tag{5-28}$$

Differentiating Equation 5-28 gives

$$dC_A = -C_{AO}\,dX_A \tag{5-29}$$

Substituting Equation 5-29 into Equation 5-296 gives

$$(-r_A) = v_L\,C_{AO}\frac{dX_A}{dL} \tag{5-302}$$

Integrating Equation 5-302 between the limits at $L = 0$, $X = 0$ and $L = L$, $X = X_A$, and rearranging yields

$$\frac{1}{v_L}\int_0^L dL = C_{AO}\int_0^{X_A} \frac{dX_A}{(-r_A)} \tag{5-303}$$

$$\frac{L}{v_L} = C_{AO}\int_0^{X_A} \frac{dX_A}{(-r_A)} \tag{5-304}$$

Therefore,

$$\bar{t} \equiv \frac{L}{v_L} = \frac{L \cdot S}{v_L \cdot S} = \frac{V_R}{u} \qquad (5\text{-}305)$$

In terms of the fractional conversion X_A, the mean residence time $\bar{t}$ is

$$\bar{t}_{\text{plug flow}} = \frac{V_R}{u} = C_{AO} \int_0^{X_A} \frac{dX_A}{(-r_A)} \qquad (5\text{-}306)$$

The design equations for plug flow in concentration and fractional conversion are:

$$\bar{t}_{\text{plug}} = \frac{V_R}{u} = -\int_{C_{AO}}^{C_{AF}} \frac{dC_A}{(-r_A)} = C_{AO} \int_0^{X_A} \frac{dX_A}{(-r_A)} \qquad (5\text{-}307)$$

Equation 5-307 is the design performance equation for a plug flow reactor at constant density. Figure 5-30 shows the profiles of these equations.

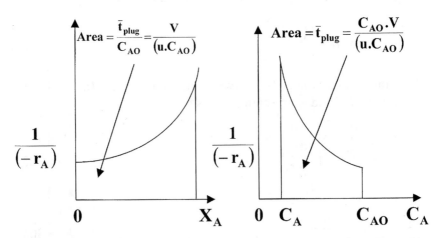

Figure 5-30. Graphical representation of the design equations for a plug flow reactor.

The area of $1/(-r_A)$ versus C_A or X_A plots is determined for all design equations for batch, CFSTR, and plug flow reactors by employing the Simpson's rule.

For zero-order homogeneous reaction

$$\bar{t}_{plug} = \frac{V_R}{u} = C_{AO} \int_0^{X_A} \frac{dX_A}{(-r_A)}$$

where

$$(-r_A) = kC_{AO}^0 = k$$

$$\frac{V_R}{u} = \frac{C_{AO}}{k} \cdot X_A$$

or

$$k\frac{V_R}{u} = k\bar{t}_{plug} = C_{AO}X_A$$

For first order irreversible reaction at constant density $A \rightarrow$ products, $(-r_A) = kC_A$, the plug flow design equation is

$$\bar{t}_{plug} = \frac{V_R}{u} = -\int_{C_{AO}}^{C_{AF}} \frac{dC_A}{(-r_A)}$$

$$= -\int_{C_{AO}}^{C_{AF}} \frac{dC_A}{kC_A}$$

$$\bar{t}_{plug} = \frac{V_R}{u} = \frac{1}{k}\ln\left(\frac{C_{AO}}{C_{AF}}\right)$$

In terms of the fractional conversion X_A

$$\bar{t}_{plug} = \frac{V_R}{u} = C_{AO} \int_0^{X_A} \frac{dX_A}{(-r_A)}$$

where

$$\left(-r_A\right) = kC_A = kC_{AO}\left(1 - X_A\right)$$

$$\bar{t}_{plug} = \frac{V_R}{u} = \frac{C_{AO}}{kC_{AO}} \int_0^{X_A} \frac{dX_A}{\left(1 - X_A\right)}$$

$$\bar{t}_{plug} = \frac{V_R}{u} = -\frac{1}{k}\ln\left(1 - X_A\right) \tag{5-308}$$

When the volume of the reaction mixture changes proportionately with conversion,

$$V = V_0\left(1 + \varepsilon_A X_A\right) \text{ and } \frac{C_A}{C_{AO}} = \frac{1 - X_A}{1 + \varepsilon_A X_A}$$

For first order reaction:

$$\left(-r_A\right) = kC_A$$

$$= kC_{AO}\frac{\left(1 - X_A\right)}{\left(1 + \varepsilon_A X_A\right)} \tag{5-309}$$

Substituting Equation 5-309 into Equation 5-307 gives

$$\bar{t}_{plug} = C_{AO} \int_0^{X_{AF}} \frac{dX_A}{\dfrac{kC_{AO}\left(1 - X_A\right)}{\left(1 + \varepsilon_A X_A\right)}}$$

$$= \frac{1}{k}\left\{ \int_0^{X_{AF}} \frac{1}{\left(1 - X_A\right)}dX_A + \int_0^{X_{AF}} \frac{\varepsilon_A X_A}{\left(1 - X_A\right)}dX_A \right\} \tag{5-310}$$

Integration of Equation 5-310 yields

$$\bar{t}_{plug} = \frac{1}{k}\left\{-\varepsilon_A X_{AF} - \ln(1 - X_{AF})\left[1 + \varepsilon_A\right]\right\}$$ (5-311)

First order reversible reaction $A \overset{k_1}{\underset{}{\rightleftharpoons}} rR$ where $C_{RO}/C_{AO} = \theta_R$, at any constant ε_A and $(-r_A) = k_1 C_A - k_2 C_R^{k_2}$ and at equilibrium conversion X_{Ae}, the design equation is:

$$k_1 \bar{t}_{plug} = \left(\frac{\theta_R + rX_{Ae}}{\theta_R + r}\right)\left[-(1 + \varepsilon_A X_{Ae})\ln\left(\frac{X_{Ae} - X_A}{X_{Ae}}\right) - \varepsilon_A X_A\right]$$ (5-312)

For second order irreversible reaction with equimolar feed $A + B \rightarrow$ products or $2A \rightarrow$ products at any constant ε_A, the design equation is

$$C_{AO}k\bar{t}_{plug} = 2\varepsilon_A(1 + \varepsilon_A)\ln(1 - X_A) + \varepsilon_A^2 X_A + (\varepsilon_A + 1)^2 \frac{X_A}{(1 - X_A)}$$ (5-313)

The design equations for both batch and plug flow systems show that at constant density, $\varepsilon_A = 0$, $\bar{t}_{plug}$ and $\bar{t}_{batch}$ are identical, and either of these equations can be used interchangeably. Where systems change in density, there is no direct relationship between the batch and plug flow equations. In such cases, the correct equation must be applied for the particular problem.

HETEROGENEOUS TUBULAR REACTOR

If the reaction in a tubular reactor proceeds in a solid catalyst (e.g., in the form of a packed bed of catalyst pellets), the conversion rate is often given per unit mass of solids $(-r_A')$. In this case, the total mass of solids M_L required for a certain degree of conversion is

$$M_L = u_m \int_0^{X_A} \frac{dX_A}{M_j(-r_A')}$$ (5-314)

where M_j = molar mass of species, kg/kmol
   $(-r_A')$ = molar rate of conversion per unit mass of the solid kmol/kg • s
   u_m = mass flowrate, kg/s
   M_L = mass of solid or catalyst, kg

The reactor volume is calculated from M_L and the bulk density of the catalyst material. $(-r'_A)$ depends not only on composition and temperature, but also on the nature and size of the catalyst pellets and the flow velocity of the mixture. In a heterogeneous reaction where a solid catalyst is used, the reactor load is often determined by the term space velocity, SV. This is defined as the volumetric flow at the inlet of the reactor divided by the reaction volume (or the total mass of catalyst), that is

$$SV \equiv \frac{u_m}{\rho_i V_R} \left(or \ \frac{u_m}{\rho_i M_L} \right) \tag{5-315}$$

where ρ_i is the density at the inlet conditions.

DESIGN EQUATION FOR SYSTEMS OF VARIABLE DENSITY

The general design Equation 5-307 was derived with no assumptions made concerning the variation of density. It is valid for constant density and variable density systems. To use the design equation for variable density reacting mixtures, the rate $(-r)$ must be expressed as a function of the fractional conversion X that properly accounts for the changes in volume as the reaction proceeds. An approximate relation for V(X) for gases is derived by using the ideal gas law. Moles N is accounted for in a general gas-phase reaction including inerts. Consider the following reaction:

$aA + bB \rightarrow rR + sS + (Inerts)$

At X = 0

$N_{AO} \quad N_{BO} \quad N_{RO} \quad N_{SO} \quad N_I$

For X > 0

$$N_{AO}(1-X_A) \quad N_{BO}-\frac{b}{a}N_{AO}X_A \quad N_{RO}+\frac{r}{a}N_{AO}X_A \quad N_{SO}+\frac{s}{a}N_{AO}X_A \quad N_I$$

Then the total mole for X > 0 is

$$N_T = N_{AO} - N_{AO} X_A + N_{BO} - \frac{b}{a} N_{AO} X_A + N_{RO} + \frac{r}{a} N_{AO} X_A$$

$$+ N_{SO} + \frac{s}{a} N_{AO} X_A + \dots\dots + N_I \qquad (5\text{-}316)$$

or

$$N_T = N_O + \left(\frac{r + s + \ \dots\dots - a - b - \dots}{a} \right) N_{AO} X_A \qquad (5\text{-}317)$$

where

N_O = total initial mole

$$= N_{AO} + N_{BO} + \dots\dots + N_{RO} + N_{SO} + \dots\dots N_I \qquad (5\text{-}318)$$

Substituting Equation 5-317 into the ideal gas law gives the total volume V:

$$V = \frac{N_T RT}{P}$$

$$= \frac{N_O RT}{P} + \left(\frac{r + s + \ \dots - a - b - \dots}{a} \right) \frac{N_{AO} RT X_A}{P} \qquad (5\text{-}319)$$

or

$$V = V_O + \left(\frac{r + s + \dots\dots - a - b -}{a} \right) V_{AO} X_A \qquad (5\text{-}320)$$

Assuming that the total pressure is constant (i.e., when the pressure drop in the reactor is neglected), Equation 5-320 becomes

$$V = V_O \left(1 + \varepsilon_A X_A \right) \qquad (5\text{-}18)$$

where Levenspiel refers to ε_A as the "expansion factor." This is identical to the change in the total volume per unit change in conversion as a fraction of the initial volume, V_O at X = 0.

Comparing Equations 5-320 and 5-18

$$\varepsilon_A = \left(\frac{r+s+\dots\dots-a-b}{a}\right)\left(\frac{V_{AO}}{V_O}\right) \tag{5-320a}$$

However, in practice, ε_A is much more easily calculated for a given reaction by directly using Equation 5-18

$$\varepsilon_A = \frac{V-V_O}{V_O X_A} \tag{5-20}$$

The expansion factor ε_A can be determined if V is known at some nonzero value of X. The most likely condition at which V can be determined is at total conversion, where the volume is readily computed from the stoichiometry. Then

$$\varepsilon_A = \frac{V_1 - V_O}{V_O}$$

where V_1 is the total volume at $X = 1$. Consider the reaction

$2NH_3 \rightleftharpoons 3H_2 + N_2$ for a feed of pure ammonia.

V at X = 0

$2V_A$ 0   0 $(V_O = 2V_A)$

V at X = 1

0            $3V_A$ V_A $(V_1 = 4V_A)$

and

$$\varepsilon_A = \frac{4V_A - 2V_A}{2V_A} = 1$$

The same reaction with a feed of 25% NH_3 and 75% inerts yields

$2NH_3 \rightleftharpoons 3H_2 + N_2 + (Inerts)$

V at X = 0

$2Va$ 0   0 $6V_A$ $(V_O = 8V_A)$

V at X = 1

0            $3V_A$ V_A $6V_A$ $(V_1 = 10V_A)$

and

$$\varepsilon_A = \frac{10V_A - 8V_A}{8V_A} = 0.25$$

These calculations show that the volume dependency of a gas-phase reaction is a function not only of the stoichiometry, but also of the inerts content of the reacting mixture. The sensitivity of volume to conversion is lowered as the inerts increase. The expansion factor, ε_A, is positive for reactions producing a net increase in moles, negative for a decrease in moles, and $\varepsilon_A = 0$ for reactions producing no net changes and at constant volume.

DESIGN EQUATIONS FOR HETEROGENEOUS REACTIONS

Heterogeneous reactions involve two or more phases. Examples are gas-liquid reactions, solid catalyst-gas phase reactions and products, and reactions between two immiscible liquids. Catalytic reactions as illustrated in Chapter 1 involve a component or species that participates in various elementary reaction steps, but does not appear in the overall reaction. In heterogeneous systems, mass is transferred across the phase.

There are several ways of defining the rate for heterogeneous systems. In solid catalyzed gaseous reactions, the rate can be defined in several ways:

• A catalyst volume basis
• A void volume (gas volume) basis
• A catalyst mass basis
• A catalyst surface area basis
• A reactor volume basis

These rates are related to the material balance by considering steady state plug flow systems. The differential material balance on component A in a heterogeneous reaction is

$$F_{AO}\, dX_A = (-r_A)\, dV = (-r_A')\, dW = (-r_A'')\, dS$$

$$= (-r_A''')\, dV_p = (-r_A^{iv})\, dV_r \qquad (5\text{-}321)$$

where F_{AO} = inlet molar flowrate of A feed (mol/s) $(u_0 C_{AO})$

C_{AO} = initial molar concentration of the key component A at feed (mol/m^3)

u_0 = volumetric flowrate of feed (m^3/sec)

X_A = fractional conversion of A

V = void volume or gas phase volume, m^3

W = mass of catalyst, kg

S = catalyst surface area, m^2

V_p = volume of catalytic solid (pellet volume), m^3

V_r = reactor volume, m^3

$(-r_A)$ = rate of disappearance of A based on volume of fluid, mol/m$^3 \cdot$ sec

$(-r_A')$ = rate of disappearance of A based on unit mass of catalyst, mol/kg $\cdot$ sec

$(-r_A'')$ = rate of disappearance of A based on unit surface, mol/m$^2 \cdot$ sec

$(-r''')$ = rate of disappearance of A based on unit volume of solid, mol/m$^3 \cdot$ sec

$(-r_A^{iv})$ = rate of disappearance of A based on unit volume of reactor, mol/m$^3 \cdot$ sec

For constant volume systems, Equation 5-321 becomes

$$\frac{V}{F_{AO}} = \frac{t}{C_{AO}} = \int_0^{X_A} \frac{dX_A}{(-r_A)} = \frac{V}{W} \int_0^{X_A} \frac{dX_A}{(-r_A')} = \frac{V}{S} \int_0^{X_A} \frac{dX_A}{(-r_A'')}$$

$$= \frac{V}{V_p} \int_0^{X_A} \frac{dX_A}{(-r_A''')} = \frac{V}{V_r} \int_0^{X_A} \frac{dX_A}{(-r_A^{iv})} \qquad (5\text{-}322)$$

The experimental study of solid catalyzed gaseous reactions can be performed in batch, continuous flow stirred tank, or tubular flow reactors. This involves a stirred tank reactor with a recycle system flowing through a catalyzed bed (Figure 5-31). For integral analysis, a rate equation is selected for testing and the batch reactor performance equation is integrated. An example is the rate on a catalyst mass basis in Equation 5-322.

$$\frac{t\,W}{C_{AO}\,V} = \int_0^{X_A} \frac{dX}{(-r_A')} \qquad (5\text{-}323)$$

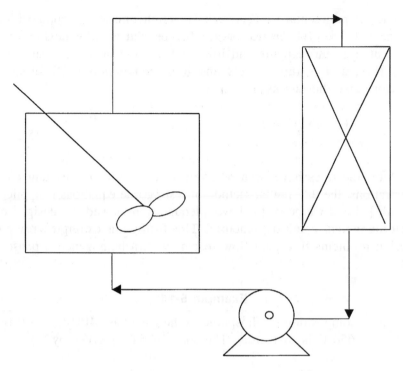

Figure 5-31. Recycle reactor for catalyst evaluation.

For each experiment, the left and right sides of Equation 5-323 are determined. The values of both sides are plotted and tested for linearity.

There are two types of continuous flow stirred tank catalytic reactors. One has the catalyst in a spinning basket, while the movement of the basket provides gas-catalyst contact, coupled with gaseous mixing. The other has a fixed catalyst bed and the gas is circulated at high rates through the packed catalyst. Most of the gas exiting from the catalyst bed is recycled back through it, and the net throughput is small. The high flowrates ensure good heat transfer to the catalyst and eliminate gas phase resistance to mass transfer. The high level of recirculation results in low conversion per pass, and the system can be analyzed as if it were a perfect mixer. The specific rate is determined as:

$$\left(-r_A'\right) = \frac{F_{AO}\, X_A}{W} \tag{5-324}$$

Each experimental run gives the reaction rate at the composition of the exit fluid. Tubular reactors can be operated as differential reactors (i.e., at high throughputs and low conversions) or as integral reactors (i.e., at low throughputs and high conversions). Differential reactors give the rate as:

$$\left(-r_A'\right) = \frac{F_{AO}\, \Delta X_A}{\Delta W} \tag{5-325}$$

After the rates have been determined at a series of reactant concentrations, the differential method of testing rate equations is applied. Smith [3] and Carberry [4] have adequately reviewed the designs of heterogeneous catalytic reactors. The following examples review design problems in a plug flow reactor with a homogeneous phase.

Example 5-14

The decomposition of phosphine in the gas phase $4PH_3(g) \rightarrow P_4(g) + 6H_2$ at 650°C is first order. The specific rate is given by

$$\log k = -\frac{18,963}{T} + 2\log T + 12.3 \quad \left(\sec^{-1}\right)$$

Determine what size plug flow reactor, operating at 650°C and 5.0 atm, produces 85% conversion of the feed consisting of 4 kg/mols of pure phosphine per hour.

Solution

The rate constant k at 650°C (923 K) is

$$\log_{10} k = -\frac{18,963}{923} + 2\log_{10}(923) + 12.3$$

$$k \approx 0.0033\, \sec^{-1} \left(12.0\, hr^{-1}\right)$$

Consider the reaction $4PH_3(g) \rightarrow P_4(g) + 6H_2$ or $4_A \rightarrow C + 6D$, where $A = PH_3$, $C = P_4$, and $D = H_2$. Because the reaction is first order, then

$$(-r_A) = kC_A$$

$$= 12C_A \left(\frac{kg \cdot mol}{m^3 \cdot hr} \right)$$

The volume of plug flow reactor in Equation 5-307 is given by

$$V_R = uC_{AO} \int_0^{X_{AF}} \frac{dX_A}{(-r_A)}$$

$$= uC_{AO} \int_0^{X_{AF}=0.85} \frac{dX_A}{kC_A}$$

At constant pressure of 5.0 atm, the concentration C_A is

$$C_A = C_{AO} \left(\frac{1 - X_A}{1 + \varepsilon_A X_A} \right)$$

Therefore, the volume of the plug flow reactor is

$$V_R = uC_{AO} \int_0^{0.85} \frac{dX_A}{kC_{AO} \left(\dfrac{1 - X_A}{1 + \varepsilon_A X_A} \right)}$$

Integration of the above equation yields Equation 5-311, which is

$$V_R = \frac{u}{k} \left\{ -\varepsilon_A X_A - \ln(1 - X_A)[1 + \varepsilon_A] \right\}_{0.0}^{0.85}$$

where u = volumetric flowrate m^3/hr

$\varepsilon_A = (7 - 4)/4 = 0.75$

$X_A = 0.85$

$F_{AO} = 4[(kg \cdot mol)/hr] = uC_{AO}$

Assuming an ideal gas

$$pV = nRT$$

$$\frac{n}{V} = C_A = \frac{p_{AO}}{RT} \quad \left(\frac{kg \bullet mol}{m^3} \right)$$

Gas constant $R = 0.08206 \quad \left(\dfrac{l \bullet atm}{gmol \bullet K} \right)$

$p_{AO} = 5.0$ atm

$T = 923$ K

$$C_{AO} = \frac{5.0}{(0.08206)(923)} \left(\frac{\cancel{atm}}{\dfrac{m^3 \bullet \cancel{atm}}{kg\,mol \bullet \cancel{K}} \bullet \cancel{K}} \right)$$

$$= 0.066 \quad \frac{kg \bullet mol}{m^3}$$

$$u = \frac{F_{AO}}{C_{AO}}$$

$$= \frac{4.0}{0.066} \left(\frac{\cancel{kg \bullet mol}}{hr} \times \frac{m^3}{\cancel{kg \bullet mol}} \right)$$

$$= 60.59 \quad \left(\frac{m^3}{hr} \right)$$

$$V_R = \frac{60.59}{12.0} \{ -(0.75 \times 0.85) - \ln(1.0 - 0.85)(1 + 0.85) \} \quad \left\{ \frac{m^3}{\cancel{hr}} \cancel{hr} \right\}$$

$$V_R = 14.5 \, m^3$$

Example 5-15

Consider an aqueous feed of A and B (600 l/min, 150 mmol of A/l, 300 mmol of B/l) that is being converted to product in a plug flow reactor. The stoichiometry and rate equation are:

$$A + B \rightarrow C \quad (-r_A) = 300 C_A C_B \quad \frac{mol}{l \cdot min}$$

Find the volume of the reactor needed for 95% conversion of A to product.

Solution

Assuming that the reaction is second order, the design Equation 5-307 for a plug flow reactor is

$$\bar{t}_{plug} = \frac{V_R}{u} = - \int_{C_{AO}}^{C_{AF}} \frac{dC_A}{(-r_A)} = C_{AO} \int_0^{X_A} \frac{dX_A}{(-r_A)}$$

From the stoichiometry $C_B = C_{BO} - (C_{AO} - C_A)$, the conversion of X_A is $(C_{AO} - C_A)/C_{AO}$. Therefore, $C_A = C_{AO}(1 - X_A)$ and $C_{AO} - C_A = C_{AO}X_A$. The rate expression for the second order is

$$(-r_A) = k C_A C_B \quad \left(\frac{mol}{l \cdot min} \right)$$

$$k = 300 \quad \left(\frac{1}{mol \cdot min} \right)$$

The design equation in terms of the fractional conversion X_A becomes

$$\frac{V_R}{u} = C_{AO} \int_0^{X_{AF}=0.95} \frac{dX_A}{k C_A C_B}$$

$$= C_{AO} \int_0^{X_{AF}=0.95} \frac{dX_A}{k C_{AO}(1 - X_A)(C_{BO} - C_{AO}X_A)}$$

where $C_{BO}/C_{AO} = \theta_B$

$$\frac{V_R}{u} = C_{AO} \int_0^{0.95} \frac{dX_A}{kC_{AO}^2(1-X_A)(\theta_B - X_A)}$$

Integration of the above equation involving partial fractions and further manipulation yields

$$V_R = \frac{u}{kC_{AO}(\theta_B - 1)} \ln \left\{ \left(\frac{\theta_B - X_A}{1 - X_A} \right) \right\}_{X_A=0}^{X_{AF}=0.95}$$

$$u = 600 \frac{1}{\min} \quad C_{AO} = 150 \frac{mmol}{l} \quad C_{BO} = 300 \frac{mmol}{l}$$

$$k = 300 \frac{1}{mol \cdot min} \quad \theta_B = \frac{C_{BO}}{C_{AO}} = 2.0$$

$$V_R = \frac{600}{300 \times 150 \times 10^{-3}(2-1)} \left\{ \ln \left(\frac{2-0.95}{1-0.95} \right) - \ln \left(\frac{2-0}{1-0} \right) \right\}$$

$$V_R = 31.34 \ l$$

Example 5-16

Hougen and Watson in an analysis of Kassel's data for the homogeneous, vapor-phase dehydrogenation of benzene in a tubular-flow reactor considered two reactions:

1. $2C_6H_6(g) \rightarrow C_{12}H_{10}(g) + H_2(g)$
2. $C_6H_6(g) + C_{12}H_{10}(g) \rightarrow C_{18}H_{14}(g) + H_2(g)$

The rate equations are:

$$r_1 = 14.94 \times 10^6 \, e^{\frac{-15,200}{T}} \left(p_B^2 - \frac{p_D p_H}{K_1} \right) \quad \left(\frac{lb \ moles \ benzene \ reacted}{hr \cdot ft^3} \right)$$

$$r_2 = 8.67 \times 10^6 \, e^{\frac{-15,200}{T}} \left(p_B p_D - \frac{p_T \, p_H}{K_2} \right)$$

$$\left(\frac{\text{lb moles triphenyl produced or diphenyl reacted}}{\text{hr} \cdot \text{ft}^3} \right)$$

where p_B = partial pressure of benzene, atm

p_D = partial pressure of diphenyl, atm

p_T = partial pressure of triphenyl, atm

p_H = partial pressure of hydrogen, atm

K_1, K_2 = equilibrium constants for the two reactions in terms of partial pressures

The data on which the rate equations are based were obtained at a total pressure of 1 atm and temperatures of 1,265°F and 1,400°F in a 0.5-in. tube, 3 ft long.

To design a tubular reactor that will operate at 1 atm pressure and 1,400°F:

1. Determine the total conversion of benzene to di- and triphenyl as a function of space velocity.

2. Determine the reactor volume required to process 10,000 lb/hr of benzene (the feed is pure benzene) as a function of the total conversion.

3. Determine the space velocity for the maximum concentration of diphenyl. Assume that the reaction is operated isothermally and that no other reactions are significant.[3]

Solution

Since the reactor is isothermal, the equilibrium constants K_1 and K_2 are estimated at 1,400°F from equations developed by Hougen and Watson. The results are:

$K_1 = 0.312$

$K_2 = 0.480$

---

[3]Source: J. M. Smith, Chemical Engineering Kinetics, 3rd ed. McGraw-Hill, 1981.

At 1,400°F (1,033 K) the two rates in terms of the disappearance rate of benzene are

$$r_1 = \left(14.96 \times 10^6\right) e^{-14.7} \left(p_B^2 - \frac{p_D \cdot p_H}{0.312} \right)$$

$$= 6.23\left(p_B^2 - \frac{p_D \cdot p_H}{0.312} \right) \tag{5-326}$$

$$r_2 = \left(8.67 \times 10^6\right) e^{-14.7} \left(p_B \cdot p_D - \frac{p_T \cdot p_H}{0.480} \right)$$

$$= 3.61\left(p_B \cdot p_D - \frac{p_T \cdot p_H}{0.480} \right) \tag{5-327}$$

The design equation for a plug flow is

$$\frac{V}{F_A} = \int_{X_A=0}^{X_A=X_{AF}} \frac{dX_A}{(-r_A)}$$

If the mass balances are based on benzene, then the two equations are

$$\frac{V}{F} = \int \frac{dX_1}{r_1} \tag{5-328}$$

$$\frac{V}{F} = \int \frac{dX_2}{r_2} \tag{5-329}$$

where the conversion X_1 is the pound moles of benzene disappearing by reaction 1 per pound mole of the feed, and the conversion X_2 is the pound moles of benzene disappearing by reaction 2 per pound mole of the feed.

Based on 1.0 mole of entering benzene, the moles of each component at conversions X_1 and X_2 are

Component	Mole at conversions X_1 and X_2
Hydrogen (H_2)	$(1/2)X_1 + X_2$
Diphenyl ($C_{12}H_{10}$)	$(1/2)X_1 - X_2$
Benzene (C_6H_6)	$1 - X_1 - X_2$
Triphenyl ($C_{18}H_{14}$)	X_2
Total moles	1.0

Since the total mole equals 1.0, the mole fractions of each component are also given by these quantities. If the components are assumed to behave as ideal gases, then the partial pressures are:

Component	Partial pressure p_i
Hydrogen (H_2), P_H	$1.0[(1/2)X_1 + X_2] = (1/2)X_1 + X_2$
Diphenyl ($C_{12}H_{10}$), P_D	$(1/2)X_1 - X_2$
Benzene (C_6H_6), p_B	$1 - X_1 - X_2$
Triphenyl ($C_{18}H_{14}$), p_T	X_2

Substituting the partial pressures of the components in Equations 5-326 and 5-327 gives

$$r_1 = 6.23 \left\{ \left(1 - X_1 - X_2\right)^2 - \frac{\left(\frac{1}{2}X_1 - X_2\right)\left(\frac{1}{2}X_1 + X_2\right)}{0.312} \right\} \tag{5-330}$$

$$r_2 = 3.61 \left\{ \left(1 - X_1 - X_2\right)\left(\frac{1}{2}X_1 - X_2\right) - \frac{X_2\left(\frac{1}{2}X_1 + X_2\right)}{0.480} \right\} \tag{5-331}$$

Substituting Equations 5-330 and 5-331 in the design Equations 5-328 and 5-329, values of exit conversions X_1, X_2, and the composition of the components are computed for various values of V/F.

The computer program PLUG51 employing the Runge-Kutta fourth order numerical method was used to determine the conversions and the compositions of the components. Applying the Runge-Kutta method, Equations 5-328 and 5-329 in differential forms are

$$\frac{dX_1}{d(V/F)} = r_1(X_1, X_2)$$

$$= 6.23 \left\{ (1 - X_1 - X_2)^2 - \frac{\left(\frac{1}{2}X_1 - X_2\right)\left(\frac{1}{2}X_1 + X_2\right)}{0.312} \right\} \quad (5\text{-}332)$$

$$\frac{dX_2}{d(V/F)} = r_2(X_1, X_2)$$

$$= 3.61 \left\{ (1 - X_1 - X_2)\left(\frac{1}{2}X_1 - X_2\right) - \frac{X_2\left(\frac{1}{2}X_1 + X_2\right)}{0.480} \right\} \quad (5\text{-}333)$$

Writing Equations 5-332 and 5-333 as function programs to determine V/F, X_1 (the conversion of benzene in reaction 1), X_2 (the conversion of benzene in reaction 2), XT (the total conversion of X_1 and X_2). The procedure calculates X_1 and X_2 for the first increment of $\Delta(V/F)$, beginning at the entrance to the reactor. The rates at the entrance are given by Equations 5-330 and 5-331 with $X_1 = X_2 = 0$.

Choose an increment of $\Delta(V/F) = 0.005$ ft/(lb • mol feed – hr) to begin the calculation.

$$Y = 1.0 - X(1) - X(2)$$

$$Z = 0.5 * X(1) - X(2)$$

$$W = 0.5 * X(1) + X(2)$$

$$F(1) = 6.23 * \left(Y**2 - \left(\frac{Z*W}{0.312}\right) \right)$$

$$F(2) = 3.61 * \left((Y*Z) - \left(\frac{X(2)*W}{0.480}\right) \right)$$

$$XT = XEND(1) + XEND(2) \quad (5\text{-}334)$$

The computer program PLUG51 used Equation 5-334 to determine the conversions and the compositions of the components. Table 5-13 illustrates the results of the computer program, and Figure 5-32 shows the plots of the rates of each reaction as a function of V/F. In both instances, the rates decrease toward zero as V/F increases. Figure 5-33 shows the plot of the total conversion X_T versus V/F.

The reactor volume required to process 10,000 lb/hr of benzene is estimated from Table 5-13. For a total conversion of 50.8%,

$$V/F = 0.2$$

$$V = 0.2\,F$$

$$= 0.2 \times \frac{10,000}{78} \left(\frac{lb}{hr} \bullet \frac{ft^3 \bullet hr}{lb\,mol} \bullet \frac{lb\,mol}{lb} \right)$$

$$= 25.64\,ft^3$$

The computer results from Table 5-13 show the calculated compositions of benzene, diphenyl, triphenyl, and hydrogen. At a fixed feedrate, increasing V/F values correspond to movement through the plug flow reactor (i.e., increasing reactor volume). Thus, these results illustrate how the composition varies with position in the reactor. Here, the mole fraction of benzene decreases steadily as the reaction mixture progresses in the reactor, while the composition of diphenyl increases and reaches a maximum between 1,684 and 1,723 hr^{-1} and thereafter decreases. This is often typical of an intermediate in consecutive reactions.

Details of the Runge-Kutta fourth order numerical method and other improved methods are illustrated in Appendix D.

COMPARISON OF IDEAL REACTORS

By comparing the design equations of batch, CFSTR, and plug flow reactors, it is possible to establish their performances. Consider a single stage CFSTR.

(text continued on page 390)

Table 5-13
Conversion versus V/F for the dehydrogenation of benzene using the Runge-Kutta fourth order method

V/F	SV	X1	X2	XT	C6H6	C12H10	C18H14	H2
.0050	75800.0	.0302	.0001	.0303	.9697	.0150	.00013	.0152
.0100	37900.0	.0586	.0005	.0591	.9409	.0288	.00051	.0298
.0150	25266.7	.0852	.0011	.0863	.9137	.0415	.00110	.0437
.0200	18950.0	.1102	.0019	.1121	.8879	.0533	.00187	.0570
.0250	15160.0	.1338	.0028	.1365	.8635	.0641	.00279	.0697
.0300	12633.3	.1558	.0038	.1597	.8403	.0741	.00384	.0818
.0350	10828.6	.1766	.0050	.1816	.8184	.0833	.00501	.0933
.0400	9475.0	.1960	.0063	.2023	.7977	.0918	.00626	.1043
.0450	8422.2	.2143	.0076	.2219	.7781	.0996	.00760	.1148
.0500	7580.0	.2315	.0090	.2405	.7595	.1068	.00899	.1247
.0550	6890.9	.2476	.0104	.2581	.7419	.1134	.01043	.1342
.0600	6316.7	.2628	.0119	.2747	.7253	.1195	.01192	.1433
.0650	5830.8	.2770	.0134	.2904	.7096	.1251	.01343	.1519
.0700	5414.3	.2904	.0150	.3053	.6947	.1302	.01497	.1601
.0750	5053.3	.3029	.0165	.3194	.6806	.1349	.01651	.1680
.0800	4737.5	.3147	.0181	.3328	.6672	.1393	.01807	.1754
.0850	4458.8	.3258	.0196	.3454	.6546	.1433	.01963	.1825
.0900	4211.1	.3362	.0212	.3574	.6426	.1469	.02119	.1893
.0950	3989.5	.3460	.0227	.3687	.6313	.1503	.02273	.1957
.1000	3790.0	.3552	.0243	.3794	.6206	.1533	.02427	.2019
.1050	3609.5	.3638	.0258	.3896	.6104	.1561	.02580	.2077
.1100	3445.5	.3719	.0273	.3992	.6008	.1586	.02731	.2133
.1150	3295.7	.3795	.0288	.4083	.5917	.1610	.02880	.2186
.1200	3158.3	.3867	.0303	.4170	.5830	.1631	.03027	.2236
.1250	3032.0	.3934	.0317	.4252	.5748	.1650	.03172	.2284
.1300	2915.4	.3998	.0332	.4329	.5671	.1667	.03315	.2330
.1350	2807.4	.4057	.0346	.4403	.5597	.1683	.03456	.2374
.1400	2707.1	.4113	.0359	.4473	.5527	.1697	.03594	.2416
.1450	2613.8	.4166	.0373	.4539	.5461	.1710	.03729	.2456
.1500	2526.7	.4215	.0386	.4601	.5399	.1721	.03862	.2494
.1550	2445.2	.4262	.0399	.4661	.5339	.1732	.03992	.2530
.1600	2368.8	.4305	.0412	.4717	.5283	.1741	.04120	.2565
.1650	2297.0	.4346	.0424	.4771	.5229	.1749	.04245	.2598
.1700	2229.4	.4385	.0437	.4822	.5178	.1756	.04367	.2629
.1750	2165.7	.4421	.0449	.4870	.5130	.1762	.04486	.2659
.1800	2105.6	.4456	.0460	.4916	.5084	.1767	.04603	.2688
.1850	2048.6	.4488	.0472	.4959	.5041	.1772	.04718	.2716
.1900	1994.7	.4518	.0483	.5001	.4999	.1776	.04829	.2742
.1950	1943.6	.4546	.0494	.5040	.4960	.1779	.04939	.2767
.2000	1895.0	.4573	.0505	.5077	.4923	.1782	.05045	.2791
.2050	1848.8	.4598	.0515	.5113	.4887	.1784	.05149	.2814
.2100	1804.8	.4621	.0525	.5146	.4854	.1786	.05251	.2836
.2150	1762.8	.4644	.0535	.5179	.4821	.1787	.05350	.2857
.2200	1722.7	.4664	.0545	.5209	.4791	.1787	.05447	.2877
.2250	1684.4	.4684	.0554	.5238	.4762	.1788	.05542	.2896
.2300	1647.8	.4702	.0563	.5266	.4734	.1788	.05634	.2914
.2350	1612.8	.4719	.0572	.5292	.4708	.1787	.05724	.2932
.2400	1579.2	.4736	.0581	.5317	.4683	.1787	.05811	.2949
.2450	1546.9	.4751	.0590	.5341	.4659	.1786	.05897	.2965
.2500	1516.0	.4765	.0598	.5363	.4637	.1785	.05980	.2981
.2550	1486.3	.4779	.0606	.5385	.4615	.1783	.06062	.2995
.2600	1457.7	.4791	.0614	.5405	.4595	.1782	.06141	.3010
.2650	1430.2	.4803	.0622	.5425	.4575	.1780	.06218	.3023
.2700	1403.7	.4814	.0629	.5444	.4556	.1778	.06293	.3036
.2750	1378.2	.4825	.0637	.5461	.4539	.1776	.06367	.3049
.2800	1353.6	.4834	.0644	.5478	.4522	.1773	.06438	.3061
.2850	1329.8	.4844	.0651	.5494	.4506	.1771	.06508	.3073
.2900	1306.9	.4852	.0658	.5510	.4490	.1769	.06576	.3084
.2950	1284.7	.4860	.0664	.5525	.4475	.1766	.06642	.3094
.3000	1263.3	.4868	.0671	.5539	.4461	.1763	.06707	.3105
.3050	1242.6	.4875	.0677	.5552	.4448	.1761	.06770	.3115
.3100	1222.6	.4882	.0683	.5565	.4435	.1758	.06831	.3124
.3150	1203.2	.4888	.0689	.5577	.4423	.1755	.06891	.3133
.3200	1184.4	.4894	.0695	.5589	.4411	.1752	.06949	.3142
.3250	1166.2	.4899	.0701	.5600	.4400	.1749	.07005	.3150
.3300	1148.5	.4904	.0706	.5611	.4389	.1746	.07061	.3158

Table 5-13
(*continued*)

V/F	SV	X1	X2	XT	C6H6	C12H10	C18H14	H2
.3350	1131.3	.4909	.0711	.5621	.4379	.1743	.07114	.3166
.3400	1114.7	.4914	.0717	.5630	.4370	.1740	.07167	.3174
.3450	1098.6	.4918	.0722	.5640	.4360	.1737	.07218	.3181
.3500	1082.9	.4922	.0727	.5649	.4351	.1734	.07268	.3188
.3550	1067.6	.4926	.0732	.5657	.4343	.1731	.07316	.3194
.3600	1052.8	.4929	.0736	.5665	.4335	.1728	.07363	.3201
.3650	1038.4	.4932	.0741	.5673	.4327	.1725	.07409	.3207
.3700	1024.3	.4935	.0745	.5681	.4319	.1722	.07454	.3213
.3750	1010.7	.4938	.0750	.5688	.4312	.1719	.07498	.3219
.3800	997.4	.4940	.0754	.5694	.4306	.1716	.07540	.3224
.3850	984.4	.4943	.0758	.5701	.4299	.1713	.07582	.3230
.3900	971.8	.4945	.0762	.5707	.4293	.1710	.07622	.3235
.3950	959.5	.4947	.0766	.5713	.4287	.1707	.07661	.3240
.4000	947.5	.4949	.0770	.5719	.4281	.1705	.07699	.3245
.4050	935.8	.4951	.0774	.5725	.4275	.1702	.07737	.3249

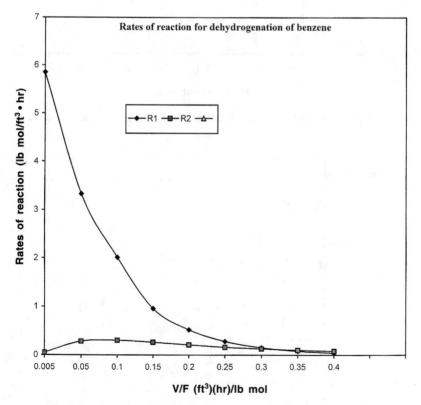

Figure 5-32. Rates of reactions for dehydrogenation of benzene.

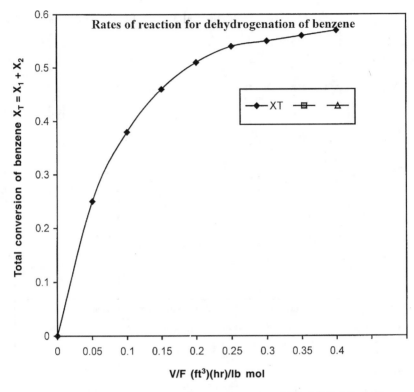

Figure 5-33. Plot of total conversion versus V/F (ft³)(hr)/lb mol.

(text continued from page 387)

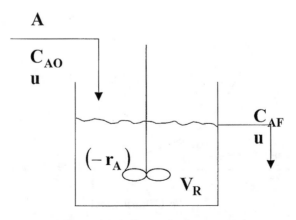

CFSTR with inflow of component A

At constant density, the material balance is $uC_{AO} = uC_A + (-r_A)V_r$ and the volume of the reactor is

$$V_{R_{CFSTR}} = u\left(\frac{C_{AO} - C_{AF}}{(-r_A)}\right)$$

For a well-mixed batch reactor, the design equation is

$$-\frac{1}{V_{R_{batch}}} \cdot \frac{dN_A}{dt} = (-r_A)$$

For constant density (i.e., ρ is constant),

$$C_A = \frac{N_A}{V_{R_{batch}}}$$

and

$$dC_A = \frac{1}{V_{R_{batch}}} \cdot dN_A$$

The above equation becomes

$$-\frac{dC_A}{dt} = (-r_A)$$

Integrating the equation between the limits at $t = 0$, $C_A = C_{AO}$ and $t = t$, $C_A = C_{AF}$ gives

$$-\int_{C_{AO}}^{C_{AF}} \frac{dC_A}{(-r_A)} = \int_0^t dt = t_{batch}$$

Time must be included for emptying, cleaning, and filling the batch reactor (≈ 30 min).

Consider a plug flow reactor:

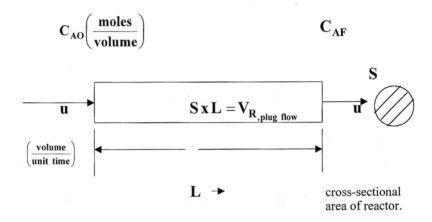

where the design equation is

$$-\int_{C_{AO}}^{C_{AF}} \frac{dC_A}{(-r_A)} = \frac{V_{R_{plug\,flow}}}{u} = \bar{t}_{plug\,flow}$$

Assuming a first order reaction A → Products, the rate expression is $(-r_A) = kC_A$ where $C_{AO} = 10$ kg mol/m^3 and $C_{AF} = 1.0$ kg mol/m^3. The volumetric flowrate of the mixture is 1.0 m^3/hr and the rate constant is k = 1.0 hr^{-1}. Therefore, the volume of the reactor for a single stage CFSTR is

$$V_R = \frac{u(C_{AO} - C_{AF})}{(-r_A) = k_1 C_{AF}}$$

$$= \frac{1.0\,(10 - 1.0)}{1.0 \times 1.0} = 9\,m^3$$

For a batch reactor, the time of the batch is

$$-\int_{C_{AO}}^{C_{AF}} \frac{dC_A}{(-r_A)} = t_{batch}$$

For the first order reaction, the design equation is

$$-\int_{C_{AO}}^{C_{AF}} \frac{dC_A}{k_1 C_A} = t_{\text{batch or plug flow}}$$

$$t = \frac{1}{k_1} \ln\left(\frac{C_{AO}}{C_{AF}}\right)$$

$$= \frac{1}{1.0} \ln\left(\frac{10}{1}\right)$$

$$= 2.303 \text{ hr}$$

$$t_{\text{plug flow}} = \frac{V_{R\,\text{plug flow}}}{u}$$

Therefore, the volume of the plug flow reactor is

$$V_R = u\,t$$

$$= 1.0 \times 2.303$$

$$= 2.303 \,\text{m}^3$$

$$t_{\text{batch}} = 2.303 \,\text{hr}$$

Time added for emptying, cleaning, and filling the batch is 30 min. The total batch processing time is 2.303 + 0.5 = 2.803 hr. Therefore, the volume of batch is $V_{R,\text{batch}} = 2.803 \text{ m}^3$.

The following considers the volume of the CFSTR and plug flow reactors at varying conversion levels of 10%, 20%, 30%, 40%, 50%, 60%, 70%, 80%, 90%, and 95% for a first order reaction.

At 10% conversion $X_A = 0.1$,

$$X_A = \frac{C_{AO} - C_{AF}}{C_{AO}} = 0.1$$

$$C_{AO} - C_{AF} = 0.1 C_{AO}$$

$$C_{AF} = C_{AO}(1.0 - 0.1)$$

$$= 0.9 C_{AO}$$

where C_{AO} = 10 kg mol/m^3 and C_{AF} = 9.0 kg mol/m^3 and

$$V_R = \frac{u(C_{AO} - C_{AF})}{k_1 C_{AF}}$$

$$= \frac{1.0(10.0 - 9.0)}{1 \times 9}$$

$$= 0.111 \text{ m}^3$$

For the plug flow reactor, the design equation at 10% conversion X_A = 0.1 is

$$\bar{t} = \frac{V_R}{u} = \frac{1}{k_1} \ln\left(\frac{C_{AO}}{C_{AF}}\right)$$

$$V_{R, \text{ plug flow}} = \frac{u}{k_1} \ln\left(\frac{C_{AO}}{C_{AF}}\right)$$

$$= \frac{1.0}{1.0} \ln\left(\frac{10}{9}\right)$$

$$= 0.105 \text{ m}^3$$

The Microsoft Excel spreadsheet program REACTOR.xls was used to compute the volumes of the CFSTR, plug flow reactors, and the ratio of the volumes for fractional conversions of 20%, 30%, 40%, 50%, 60%, 70%, 80%, 90%, and 95% for a first order reaction. Table 5-14 shows the volumes of the CFSTR, plug flow reactors, and the ratio of the volumes at given conversion levels.

Table 5-14
Volume of CFSTR and plug flow reactors with respect to conversion level for a first order reaction

Conversion X_A, %	10	20	30	40	50
Volume of CFSTR, m^3	0.111	0.25	0.429	0.667	1.0
Volume of plug flow, m^3	0.105	0.223	0.357	0.511	0.693
Ratio of $\dfrac{V_{R,CFST}}{V_{R,plug\ flow}}$	1.057	1.121	1.202	1.305	1.443
Conversion X_A, %	60	70	80	90	95
Volume of CFSTR, m^3	1.5	2.33	4.0	9.0	19.0
Volume of plug flow, m^3	0.916	1.204	1.609	2.303	2.996
Ratio of $\dfrac{V_{R,CFST}}{V_{R,plug\ flow}}$	1.638	1.935	2.486	3.908	6.342

Table 5-14 shows that there is no significant difference between the volume of the CFSTR and plug flow reactors at low conversions. However, at 70% and higher, more than twice as much volume is required for the CFSTR. Additionally, the ratio of the volume increases with level of conversions and reaction order. Figure 5-34 illustrates the ratio of the volumes of CFSTR and plug flow reactors with conversion.

$1/(-r_A)$ VERSUS C_A PLOT

If $1/(-r_A)$ versus C_A is the plot for the limiting component A in a CFSTR and plug flow system, the mean residence time $\bar{t}_{CFSTR}$ is expressed as

$$\bar{t}_{CFSTR} = \frac{C_{AO} - C_A}{(-r_A)} = \frac{1}{(-r_A)} \bullet (C_{AO} - C_A)$$

For a plug flow reactor, the mean residence time $\bar{t}_{plug}$ is

$$\bar{t}_{plug} = -\int_{C_{AO}}^{C_A} \frac{dC_A}{(-r_A)}$$

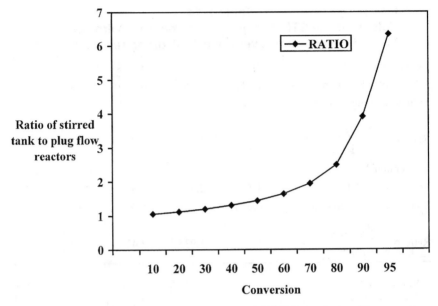

Figure 5-34. Plot of the ratio of volumes of CFSTR and plug flow reactors versus conversion for a first order reaction.

The residence time for the CFSTR is the area under the rectangle of width $(C_{AO} - C_A)$ and height $1/(-r_A)$. The residence time $\bar{t}_{plug}$ is the area under the curve $1/(-r_A)$ from C_{AO} to C_A. Figure 5-35 shows that $1/(-r_A)$ is a monotonically increasing function of $(C_{AO} - C_A)$ if the order of reaction, $n > 0$, and a horizontal line if $n = 0$ and decreases with $(C_{AO} - C_A)$ if $n < 0$.

In Figure 5-36, the area of the CFSTR is larger while $1/(-r_A)$ is a monotonically increasing function of $(C_{AO} - C_A)$ for $n > 1$. Both areas are equal for $n = 0$, and the rectangle has a smaller area for $n < 0$.

At high conversions, the CFSTR requires a larger volume, and as $C_A \to 0$, the $1/(-r_A)$ versus $C_{AO} - C_A$ curve rapidly increases to ∞. In the case where $\bar{t}_{CFSTR}$ becomes very large relative to $\bar{t}_{plug}$, the isothermal plug flow reactor is much preferred to the single-stage CFSTR at high conversions when $n > 0$.

CFSTR AND PLUG FLOW SYSTEMS

Consider a combination of CFSTR and plug flow systems as shown below for a first order irreversible reaction.

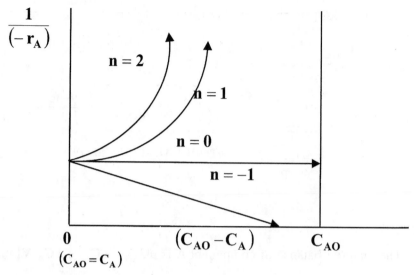

Figure 5-35. Plot of $1/(-r_A)$ versus $(C_{AO} - C_A)$ for the nth order irreversible reactions.

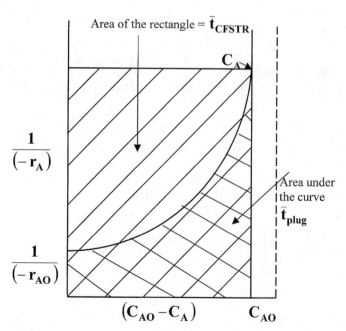

Figure 5-36. Residence time in CFSTR and plug flow reactors from $1/(-r_A)$ versus $C_{AO} - C_A$ plot.

$$(-r_A) = kC_A$$

CFSTR   TUBULAR (PLUG FLOW)

CFSTR

The material balance of component A is $uC_{AO} = uC_{A1} + kC_{A1}V_1$ or

$$C_{A1} = \frac{C_{AO}}{1 + k\bar{t}_1}$$

where $\bar{t}_1 = V_1/u$.

Plug Flow Reactor

The concentration leaving the CFSTR, C_{A1}, is the inlet concentration to the plug flow reactor. The material balance on component A is

$$\bar{t}_{plug} = \bar{t}_2 = \frac{V_2}{u} = -\int_{C_{A1}}^{C_{A2}} \frac{dC_A}{kC_A}$$

The residence time $\bar{t}_2$ in the plug flow is

$$\bar{t}_2 = -\frac{1}{k}\ln\left(\frac{C_{A2}}{C_{A1}}\right)$$

or

$$C_{A2} = C_{A1}e^{-k\bar{t}_2}$$

$$= \left(\frac{C_{AO}}{1 + k\bar{t}_1}\right)e^{-k\bar{t}_2}$$

Consider reversing the reactors as shown below

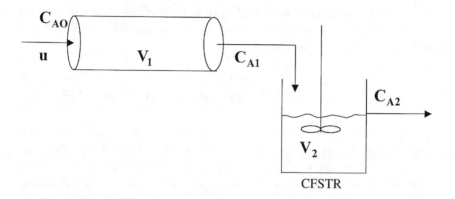

Plug Flow Reactor

$$\bar{t}_1 = \frac{V_1}{u} = -\int_{C_{AO}}^{C_{A1}} \frac{dC_A}{kC_A}$$

$$-k\bar{t}_1 = \ln\left(\frac{C_{A1}}{C_{AO}}\right)$$

or

$$C_{A1} = C_{AO}\, e^{-k\bar{t}_1}$$

CFSTR

$$uC_{A1} = uC_{A2} + kC_{A2}V_2$$

or

$$C_{A2} = \frac{C_{A1}}{1 + k\bar{t}_2} = \left(\frac{C_{AO}}{1 + k\bar{t}_2}\right) e^{-k\bar{t}_1}$$

When $\bar{t}_1 = \bar{t}_2$, that is, the volumes of CFSTR and plug flow are the same, then for the first order irreversible reaction, the expressions are identical for both combinations.

When the reactors are of equal volume, the overall conversion is independent regardless of which reactor precedes. However, generally, the CFSTR should precede the plug flow to minimize the total reactor volume. Figure 5-37 shows the total residence time of the CFSTR and plug flow reactors in series.

DYNAMIC BEHAVIOR OF IDEAL SYSTEMS

The review of the performance equations for the ideal system has been for the steady state situation. This occurs when the process has begun and all transient conditions have died out (that is, no parameters vary with time). In all flow reactors, parameters such as the flowrate, temperature, and feed composition can vary with time at the beginning of the process. It is important for designers to review this situation with respect to fluctuating conditions and the overall control and

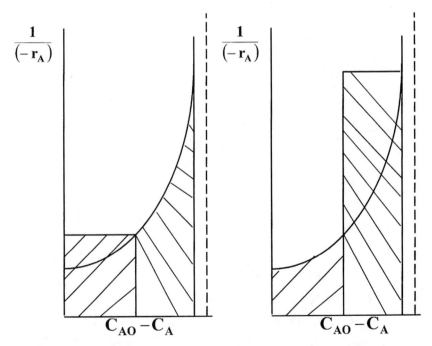

Figure 5-37. Residence time from $1/(-r_A)$ versus $C_{AO} - C_A$ plot for plug flow reactors and CFSTRs in series.

integrity of the plant. Many industrial accidents and explosions occur during startup and shut-down, and by improperly charging the feed to the reactor.

The following considers the unsteady state of a CFSTR:

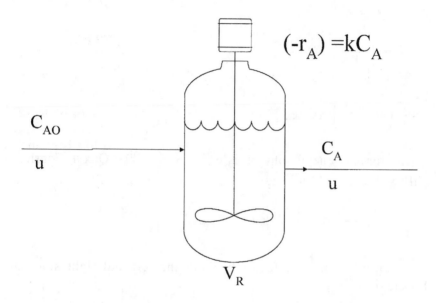

$$(-r_A) = kC_A$$

Consider a CFSTR at constant density. If the flowrate and the reactor volume are constant, the material balance, Equation 5-1, on component A yields

$$uC_{AO} = uC_A + (-r_A)V_R + V_R \frac{dC_A}{dt} \qquad (5\text{-}335)$$

where $V_R \dfrac{dC_A}{dt}$ = accumulation in the reactor.

Rearranging Equation 5-335 gives

$$\frac{dC_A}{dt} + \left(\frac{1}{\bar{t}} + k\right)C_A = \frac{1}{\bar{t}}C_{AO} \qquad (5\text{-}336)$$

where $\bar{t} = V_R/u$ = the mean residence time.

Equation 5-336 is a first order differential equation. The behavior of component A can be predicted with time from the boundary conditions, the flowrate of the feed, composition, and the volume of the reactor.

Equation 5-336 is of the form $dy/dx + Py = Q$ where the integrating factor $I.F. = e^{\int Pdx}$.

$$e^{\int Pdx}\frac{dy}{dx} + e^{\int Pdx}Py = Q \cdot e^{\int Pdx}$$

$$\frac{d}{dx}\left(e^{\int Pdx} \cdot y\right) = Q \cdot e^{\int Pdx}$$

The formal general solution is $e^{\int Pdx} \cdot y = \int e^{\int Pdx} \cdot Qdx + Const.$ In Equation 5-336, the I.F. is

$$I.F. = e^{\int\left(\frac{1+k\bar{t}}{\bar{t}}\right)dt} = e^{\left(\frac{1+k\bar{t}}{\bar{t}}\right)t}$$

Multiplying the I.F. factor by both the left and right sides of Equation 5-336 gives

$$e^{\left(\frac{1+k\bar{t}}{\bar{t}}\right)t}\frac{dC_A}{dt} + e^{\left(\frac{1+k\bar{t}}{\bar{t}}\right)} \cdot \left(\frac{1+k\bar{t}}{\bar{t}}\right)C_A = \frac{1}{\bar{t}} \cdot C_{AO} \cdot e^{\left(\frac{1+k\bar{t}}{\bar{t}}\right)}$$

or

$$\frac{d}{dt}\left(e^{\left(\frac{1+k\bar{t}}{\bar{t}}\right)} \cdot C_A\right) = \frac{1}{\bar{t}} \cdot C_{AO}\, e^{\left(\frac{1+k\bar{t}}{\bar{t}}\right)t} \tag{5-337}$$

Integrating Equation 5-337 yields

$$e^{\left(\frac{1+k\bar{t}}{\bar{t}}\right)t} \cdot C_A = \frac{C_{AO}}{\bar{t}}\int e^{\left(\frac{1+k\bar{t}}{\bar{t}}\right)t} \cdot dt + Const.$$

A general solution is

$$C_A(t)=\frac{C_{AO}}{(1+k\bar{t})}+\text{Const.e}^{-\left(\frac{1+k\bar{t}}{\bar{t}}\right)t}$$ (5-338)

At the boundary condition $t = 0$, $C_A(t) = C_{A\,init}$ and the Const. $= C_{A_{init}} - C_{AO}/(1 + k\bar{t})$, which gives

$$C_A(t)=\frac{C_{AO}}{1+k\bar{t}}+\left\{C_{A_{init}}-\frac{C_{AO}}{1+k\bar{t}}\right\}e^{-\left(\frac{1+k\bar{t}}{\bar{t}}\right)t}$$

or

$$C_A(t)=\frac{C_{AO}+\left\{C_{A_{init}}\left(1+k\bar{t}\right)-C_{AO}\right\}e^{-\left(\frac{1+k\bar{t}}{\bar{t}}\right)t}}{1+k\bar{t}}$$ (5-339)

Figure 5-38 shows plots of the dynamic response to changes in the inlet concentration of component A. The figure represents possible responses to an abrupt change in inlet concentration of an isothermal CFSTR with first order irreversible reaction. The first plot illustrates the situation where the reactor initially contains reactant at $C_{A_{init}}$ and at $t = 0$. The reactant flow at C_{AO} begins until component A reaches a steady state condition. The second plot starts with $C_{A_{init}}$ in the reactor at $t = 0$. Tables 5-15, 5-16, and 5-17, respectively, show common equations for all reactor models involving a first order reversible reaction.

Example 5-17

Acetic anhydride is hydrolyzed at 40°C in a CFSTR. The reactor is initially charged with 0.57 m^3 of an aqueous solution containing 0.487 kmol/m^3 of anhydride. The reactor is heated quickly to 350 K, and at that time, a feed solution containing 0.985 kmol/m^3 of anhydride is run into the reactor at the rate of 9.55×10^{-4} m^3/sec. At the instant the feed stream is introduced, the product pump is started and the product is withdrawn at 9.55×10^{-4} m^3/sec. The reaction is first order with a rate constant of 6.35×10^3 sec^{-1}.

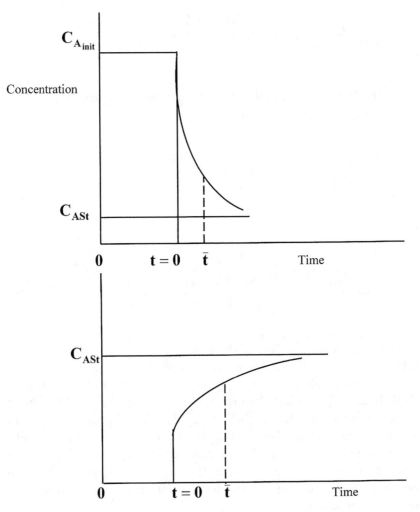

Figure 5-38. Dynamic response in a CFSTR system.

1. Perform the unsteady state balance and solve to obtain the startup transient in the product stream concentrations of anhydride and acid.
2. Determine the concentrations of anhydride and acid after 120 sec.

Solution

The reaction involving acetic anhydride and water in the hydrolysis is

Table 5-15
Reversible Reaction A B R

Total Concentration	$C_t = C_A + C_R$
Arrhenius Law	$k = A\ exp(E_A/RT)$
van't Hoff Equation	$K_{CT_2} = K_{CT_1} exp\left\{\left(\dfrac{-\Delta H_R}{R}\right)\left(\dfrac{1}{T_1} - \dfrac{1}{T_2}\right)\right\}$
Equilibrium Constant	$K_C = \dfrac{k}{k'} = \dfrac{C_{Re}}{C_{Ae}} = \dfrac{C_{Ai} + C_{Ri} - C_{Ae}}{C_{Ae}}$
Conversion	$X_A = 1 - \dfrac{C_A}{C_{Ai}}$
Rate Law	$-\dfrac{dC_A}{dt} = k\left(C_A - \dfrac{C_R}{K_C}\right)$

Source: J. D. Seader, Computer Modeling of Chemical Processes, AIChE Monograph Series, 15, Vol. 81. Reproduced with permission of the American Institute of Chemical Engineers, Copyright © 1985 AIChE. All rights reserved.

$$CH_3CO \cdot O \cdot CO \cdot CH_3 + H_2O \xrightarrow{\ k\ } 2CH_3COOH$$

Acetic anhydride   Water   Acetic acid

or $A + B \xrightarrow{\ k\ } 2C$. Assuming that the reaction is first order with respect to the anhydride, component A, $(-r_A) = kC_A$.

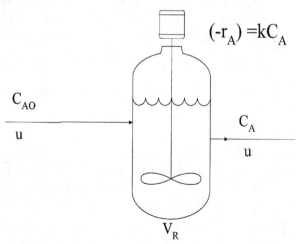

The unsteady state balance of acetic anhydride in a CFSTR.

(text continued on page 408)

Table 5-16
Steady-state reactor models for A $\rightleftharpoons$ R

Case	Type Model	Mass Balances	Energy Balance	Chemical Equilibrium	Kinetics/Transport
1	Overall Conversion	$n_{Af} = n_{A_i}(1 - X_{Af})$ $n_{Rf} = n_{R_i} + n_{A_i}X_{Af}$ $n_t = n_{A_i} + n_{R_i}$	$n_t C_p(T_f - T_i) = n_{A_i}X_{Af}(-\Delta H_r) + Q$	—	—
2	Chemical Equilibrium in Reactor Effluent	Same as Case 1 with $X_{Af} = X_A$	Same as Case 1	$X_{A_e} = 1 - \dfrac{n_t}{(K_c + 1)n_{A_i}}$	—
3	Kinetic for a Single Ideal Stirred-Tank Flow Reactor	Same as Case 1	Same as Case 1	—	$\dfrac{n_{A_i} - n_{Af}}{V_t} = \dfrac{k\rho}{n_t}\left(n_{Af} - \dfrac{n_{Rf}}{K_{cf}}\right)$ $\tau = \dfrac{V\rho}{n_t}$
4	Kinetic for Isothermal Plug-Flow Reactor	$n_A = n_{A_i}(1 - X_A)$ $n_R = n_{R_i} + n_{A_i}X_A$ or $n_R = n_{R_i} + n_{A_i} - n_A$	Same as Case 1 with $T = T_i = T_f$ $(-\Delta H_r)\dfrac{dn_A}{dV} = \dfrac{2q}{R}$ $V = Sz$	—	$-\dfrac{dn_A}{dV} = \dfrac{k\rho}{n_t}\left(n_A - \dfrac{n_R}{K_c}\right)$ $n_A(0) = n_{A_i}$ $\tau = \dfrac{V\rho}{n_t}$
5	Kinetic for Adiabatic or Non-adiabatic Plug-Flow Reactor	Same as Case 4	$n_T C_p \dfrac{dT}{dV} = -(-\Delta H_r)\dfrac{dn_A}{dV} + \dfrac{2q}{R}$ $V = Sz$ $T(0) = T_i$ $n_A(0) = n_{A_i}$ $q = q(z)$ known	—	Same as Case 4 with $k(T)$ and $K_c(T)$
6	Kinetic for Isothermal Plug-Flow Recycle Reactor	Same as Case 4	Same as Case 4	—	$-\dfrac{dn_A}{dV} = \dfrac{k\rho}{(1 + R_c)n_t}\left(n_A - \dfrac{n_R}{K_c}\right)$ $n_A(0) = \dfrac{n_{A_i} + R_c n_{Af}}{1 + R_c}$ $\tau = \dfrac{V\rho}{(1 + R)}$
7	Kinetic/transport for isothermal Plug-Flow Reactor with Axial Dispersion	Same as Case 4	Same as Case 4	$\tau = \dfrac{V\rho}{n_t}$	$D_A \dfrac{d^2 n_A}{dZ^2} - \dfrac{n_t}{\rho S}\dfrac{dn_A}{dz} = k\left(n_A - \dfrac{n_R}{K_c}\right)$ $-\dfrac{D_A \rho S}{n_t}\left(\dfrac{dn_A}{dz}\right)_{z=0^+} + (n_A)_{z=0^+} = n_A$ $\left(\dfrac{dn_A}{dz}\right)_{z=L} = 0$
8	Kinetic/transport for Isothermal Laminar-Flow Reactor with no Axial Dispersion [See Shinohara and Christiansen (1974) for the non-isothermal case]	$C_R = C_{R_i} + C_{A_i} - C_A$	Same as Case 4	—	$2u_{ave}\left[1 - \left(\dfrac{r}{R}\right)^2\right]\dfrac{\partial C_A}{\partial z} =$ $D_A\left[\dfrac{\partial^2 C_A}{\partial r^2} + \dfrac{1}{r}\dfrac{\partial C_A}{2r}\right] - k\left(C_A - \dfrac{C_R}{K_C}\right)$ $C_A = C_A(z,r)$ $C_A(0,r) = C_{Ai} = \dfrac{N_{Ai}\rho}{n_t}$ $\dfrac{\partial C_A(z,R)}{\partial r} = 0; \dfrac{\partial C_A(z,0)}{\partial r} = 0$ $n_{A_i} = (Z)\dfrac{2n_t}{\rho R^2}\int_0^R rC_A(z,r)dr$ $\tau = \dfrac{V\rho}{n_t}$

Source: J. D. Seader, Computer Modeling of Chemical Processes, *Monograph Series 15 Vol. 18. Reproduced with permission of the American Institute of Chemical Engineers. Copyright © 1985, AIChe. All rights reserved.*

Table 5-17
Dynamic reactor models for A $\rightleftharpoons$ R

Case	Type Model	Mass Balances	Energy Balance	Kinetics/Transport
1	Kinetic for Isothermal Uniform-State, Constant-Volume Batch Reactor	$C_A = C_{A_i}(1 - X_A)$ $C_R = C_{R_i} + C_{A_i} - C_A$	$V(-\Delta H_r)\left(\dfrac{dC_A}{dt}\right) = Q$ $T = T_i$	$-\dfrac{dC_A}{dt} = k\left(C_A - \dfrac{C_R}{K_c}\right)$ $C_A(0) = C_{A_i}$
2	Kinetic for Adiabatic or Non-Adiabatic, Uniform-State, Constant-Volume Batch Reactor	Same as Case 1	$V\rho C_p \dfrac{dT}{dt} = -V(-\Delta H_r)\dfrac{dC_A}{dt} + Q$ with $T(0) = T_i$	$-\dfrac{dC_A}{dt} = k\left(C_A - \dfrac{C_R}{K_c}\right)$ with $k = k(T)$ $K_c = K(T)$ $C_A(0) = C_{A_i}$
3	Kinetic for a Single Ideal Stirred-Tank Flow Reactor with Constant-Volume Holdup under Transient Open-Loop Operation	Same as Case 1	Same as Case 2	$-\dfrac{dC_A}{dt} + \dfrac{n_t}{V\rho}(C_{A_i} - C_{A_f}) = k\left(C_A - \dfrac{C_R}{K_c}\right)$ $n_t = n_t(t)$ $k = k(T)$ $K_c = K_c(T)$
4	Kinetic for a Single Ideal Stirred-Tank Flow Reactor under Transient Closed loop Liquid-Level PI Control	$\dfrac{dV}{dt} = \dfrac{n_{ti} - n_{tf}}{\rho}$ $n_{ti} = n_{ti}(t)$ $n_{tf} = \bar{n}_{tf} + K_i\left(E_1 + \dfrac{1}{\tau_1}\int E_1 dt\right)$ $\bar{n}_{tf} = n_{tf(0)}$ $E_1 = H_{set} - H$ $H = \dfrac{V}{S}$	Same as Case 2 with $V = V(t)$	Same as Case 2 with $V = V(t)$
5	Kinetic for an Adiabatic or Non-Adiabatic Plug-Flow Reactor under Transient Open-Loop Operation	$C_R + C_A = C_t$	$n_t C_p \dfrac{\partial T}{\partial V} = -\dfrac{(-\Delta H_r)n_t}{\rho}\dfrac{\partial C_A}{\partial V}$ $+ \dfrac{2q}{R} + C\rho\rho\dfrac{\partial T}{\partial t}$ $V = Sz$ $T = T(V,t)$ $T = T(V,0)$ known $T = T(0,t)$ known $q = q(Z,t)$ known	$-\dfrac{n_t}{\rho}\dfrac{\partial C_A}{\partial V} = k\left(C_A - \dfrac{C_R}{K_c}\right) + \dfrac{\partial C_A}{\partial t}$ $k = k(T),\ K_c = K_c(T)$ $n_t = n_t(t)$ $C_A = C_A(V,t)$ $C_A(0,t)$ known $n_t(0)$ known
6	Kinetic/transport for Isothermal Laminar-Flow Reactor with Axial Dispersion under Transient Open-Loop Operation	$C_R + C_A = C_t$ $C_A = C_A(z,r,t)$	$T = T_i = T_f$ $(-\Delta H_i)\dfrac{dn_A}{dV} = \dfrac{2q}{R}$	$2u_{ave}\left[1 - \left(\dfrac{r}{R}\right)^2\right] + \dfrac{\partial C_A}{\partial t} =$ $D_A\left[\dfrac{\partial^2 C_A}{\partial r^2} + \dfrac{1}{r}\dfrac{\partial C_A}{\partial r} + \dfrac{\partial^2 C_A}{\partial z^2}\right] - k\left(C_A - \dfrac{C_R}{K_c}\right)$ $C_A(z,r,0)$ known $C_A(0,r,t)$ known $u_{ave}(t)$ known $\dfrac{\partial C_A(z,r,t)}{\partial r} = 0;\ \dfrac{\partial C_A(z,0,t)}{\partial r} = 0$ $\dfrac{\partial C_A(L^-,r,t)}{\partial z} = 0$ $C_A(0,r,t) = C_A(0^-,r,t) - \dfrac{D_A}{u}\dfrac{\partial C_A(0^+,r,t)}{\partial z}$

(text continued from page 405)

The unsteady state material balance of component A is

$$uC_{AO} = uC_A + (-r_A)V_R + V_R \frac{dC_A}{dt} \qquad (5\text{-}335)$$

Rearranging Equation 5-335 gives

$$\frac{dC_A}{dt} + \left(\frac{1}{\bar{t}} + k\right)C_A = \frac{1}{\bar{t}}C_{AO} \qquad (5\text{-}336)$$

or

$$\frac{dC_A}{dt} = \frac{1}{\bar{t}}(C_{AO} - C_A) - kC_A \qquad (5\text{-}340)$$

Both Equations 5-336 and 5-340 are first order differential equations, which can be solved to determine the transient concentration of the anhydride.

The transient concentration of acetic anhydride from Equation 5-336 with a first order irreversible reaction is obtained from Equation 5-338

$$C_A(t) = \frac{C_{AO} + \left\{C_{A_{init}}(1 + k\bar{t}) - C_{AO}\right\}e^{-\left(\frac{1 + k\bar{t}}{\bar{t}}\right)t}}{1 + k\bar{t}} \qquad (5\text{-}341)$$

where C_{AO} = feed concentration, kmol/m^3
$C_{A_{init}}$ = initial concentration at time t = 0, kmol/m^3
$\bar{t}$ = V_R/u = mean residence time, sec
V_R = volume of the reactor, m^3
u = volumetric flowrate, m^3/sec
k = rate constant, sec^{-1}
C_{AO} = 0.985 kmol/m^3
$C_{A_{init}}$ = 0.485 kmol/m^3
V_R = 0.55 m^3
u = 9.55 × 10^{-4} m^3/sec
k = 6.35 × 10^3 sec^{-1}

$$\bar{t} = \frac{V_R}{u} = \frac{0.55}{9.55 \times 10^{-4}} = 575.916 \, sec$$

$$1 + k\bar{t} = 1 + 6.35 \times 10^{-3} \times 575.916 = 4.657$$

$$C_A(t) = \frac{0.985}{4.657} + \left\{ \frac{0.485 \times 4.657 - 0.985}{4.657} \right\} e^{-\left(\frac{4.657}{575.916}\right)t}$$

Therefore, the startup anhydride concentration is

$$C_A(t) = 0.212 + 0.273 \, e^{-0.008t} \quad kmol/m^3 \tag{5-342}$$

with time t in sec. The material balance for the acetic acid is

$$(+r_C)V_R = uC_C + V_R \frac{dC_C}{dt} \tag{5-343}$$

or

$$0 = uC_C - kC_A V_R + V_R \frac{dC_C}{dt} \tag{5-344}$$

Rearranging Equation 5-344 gives

$$\frac{dC_C}{dt} + \frac{1}{\bar{t}} C_C = kC_A \tag{5-345}$$

where Equation 5-345 is a first order differential equation. The integrating factor is

$$I.F. = e^{\int Pdx} = e^{\int \frac{1}{\bar{t}} dt} = e^{\frac{t}{\bar{t}}}$$

Multiplying both sides of Equation 5-345 by the I.F. gives

$$e^{\frac{t}{\bar{t}}} \frac{dC_C}{dt} + e^{\frac{t}{\bar{t}}} \cdot \frac{1}{\bar{t}} \cdot C_C = kC_A e^{\frac{t}{\bar{t}}}$$

or

$$\frac{d}{dt}\left(e^{\frac{t}{\bar{t}}} \cdot C_C\right) = kC_A e^{\frac{t}{\bar{t}}}$$ (5-346)

Integrating Equation 5-346 yields

$$e^{\frac{t}{\bar{t}}} \cdot C_C = kC_A \int e^{\frac{t}{\bar{t}}} dt + \text{Const.}$$ (5-347)

$$e^{\frac{t}{\bar{t}}} \cdot C_C = k\bar{t}C_A e^{\frac{t}{\bar{t}}} + \text{Const.}$$ (5-348)

The boundary condition at t = 0, C_C = 0 and $\text{Const.} = -k\bar{t}C_A$. Equation 5-348 becomes

$$C_C = k\bar{t}C_A - k\bar{t}C_A e^{\frac{-t}{\bar{t}}}$$ (5-349)

Substituting Equation 5-342 into Equation 5-349 gives the transient composition of acetic acid as

$$C_C = k\bar{t}C_A\left(1 - e^{\frac{-t}{\bar{t}}}\right)$$ (5-350)

which is

$$C_C = k\bar{t}\left(0.212 + 0.273e^{-0.008t}\right)\left(1 - e^{-0.0017t}\right)$$

$$= \left(0.775 + 0.998e^{-0.008t}\right)\left(1 - e^{-0.0017t}\right)$$

Equations 5-342 and 5-350 are the concentrations of C_A and C_C after 120 sec.

FLOW RECYCLE REACTOR

There are situations where a portion of the product stream in a plug flow reactor is returned to the entrance of the feed stream by means of a pump. The pump can be varied to increase the amount of recycle R from zero to a very large value, thus creating some degree of backmixing to the reactor. Consider the plug flow reactor as shown in Figure 5-39. Component A is fed to the plug flow reactor at constant density.

The volumetric flowrate into the plug flow is u_O and the feed concentration of A is C_{AO}. A portion of A exiting from the reactor is fed back through a pump and mixed with the feed stream, referred to as R (i.e., the recycle ratio). The volumetric flowrate at the entrance of the reactor is $u = u_O(1 + R)$. A balance at the mixing point M gives $uC_{Ai} = u_O C_{AO} + Ru_O C_{Af}$, which is $u_O(1 + R)C_{Ai} = u_O C_{AO} + Ru_O C_{Af}$. The inlet concentration of component A at the entrance of the plug flow reactor is

$$C_{Ai} = \frac{C_{AO} + RC_{Af}}{1 + R} \tag{5-351}$$

The residence time in the plug flow is

$$\bar{t}_{plug} = \frac{V_R}{u} = \frac{V_R}{u_O(1+R)} = -\int_{C_{Ai}}^{C_{Af}} \frac{dC_A}{(-r_A)}$$

or

$$\bar{t}_{plug} = \frac{V_R}{u_O} = -(1+R)\int_{C_{Ai}}^{C_{Af}} \frac{dC_A}{(-r_A)}$$

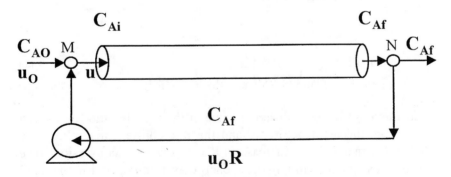

Figure 5-39. The recycle tubular reactor.

For a first order irreversible reaction $(-r_A) = kC_A$, which gives

$$\bar{t}_{plug} = \frac{V_R}{u_O} = -(1+R)\int_{C_{Ai}}^{C_{Af}} \frac{dC_A}{kC_A} = \frac{(1+R)}{k}\ln\left(\frac{C_{Ai}}{C_{Af}}\right) \tag{5-352}$$

Substituting Equation 5-351 into Equation 5-352 yields

$$\bar{t}_{plug} = \frac{V_R}{u_O} = \frac{1+R}{k}\ln\left\{\frac{C_{AO}+RC_{Af}}{C_{Af}(1+R)}\right\} \tag{5-353}$$

Rearranging Equation 5-353 gives

$$\exp\left(\frac{k\bar{t}_{plug}}{1+R}\right) = \frac{1+RC_{Af}/C_{AO}}{\dfrac{C_{Af}}{C_{AO}}(1+R)}$$

or

$$\frac{C_{Af}}{C_{AO}} = \frac{1}{(1+R)\exp\left(\dfrac{k\bar{t}_{plug}}{1+R}\right)-R}$$

The recycle ratio in terms of the fractional conversion X_A for the first order irreversible reaction is

$$X_A = 1 - \frac{1}{(1+R)\exp\left(\dfrac{k\bar{t}_{plug}}{R+1}\right)-R} \tag{5-354}$$

Equation 5-354 determines the value of the recycle ratio for a given conversion, the residence time, and the rate constant in a plug flow reactor. Alternatively, an increase in R will lower the conversion since it produces backmixing in the reactor as it mixes with the feed entrance in the plug flow reactor. As $R \rightarrow \infty$, it yields

$$\bar{t}_{plug} = \frac{V_R}{u_O} = \frac{1+R}{k} \ln\left\{\frac{C_{AO} + RC_{Af}}{C_{Af}(1+R)}\right\}$$

which is

$$\frac{k\bar{t}_{plug}}{1+R} = \ln\left\{\frac{C_{AO} + RC_{Af}}{C_{Af}(1+R)}\right\}$$

or

$$\frac{C_{AO} + RC_{Af}}{C_{Af}(1+R)} = \exp\left(\frac{k\bar{t}_{plug}}{1+R}\right) \tag{5-355}$$

The expansion of e^x is

$$e^x = 1 + x + \frac{1}{2!}x^2 + \frac{1}{3!}x^3 + \ldots$$

Expanding Equation 5-355 yields

$$\frac{C_{AO} + RC_{Af}}{C_{Af}(1+R)} = \left\{1 + \frac{k\bar{t}_{plug}}{1+R} + \frac{1}{2}\left(\frac{k\bar{t}_{plug}}{1+R}\right)^2 + \ldots\right\}$$

as $R \to \infty$, it yields

$$\frac{C_{AO}}{C_{Af}} + R = (1+R)\left\{1 + \frac{k\bar{t}_{plug}}{1+R}\right\}$$

or

$$\frac{C_{AO} - C_{Af}}{C_{Af}} = k\bar{t}_{plug} \tag{5-356}$$

Equation 5-356 is the familiar CFSTR for a first order reaction. As the recycle of the fluid increases, the reactor becomes completely backmixed and eventually approaches the performance of a CFSTR. However, the process increases pumping costs and is, therefore, an inefficient method of creating a CFSTR system. The recycle reactor can also act as an efficient mode in creating a small amount of backmixing into a reactor as in the autocatalytic reaction, where the rate is proportional to the product concentration.

PRESSURE DROP (ΔP) IN PLUG FLOW REACTORS

In a homogeneous plug flow reactor, the pressure drop ΔP corresponding to the desired flowrate is small and has little effect on the operating process. However, it is necessary to determine ΔP as part of the design procedure so that ancillary equipment such as pumps may be adequately specified. If high-viscosity liquids (e.g., polymers) or low-pressure gases are used, ΔP will have a major influence on the design.

In heterogeneous systems ΔP must be critically reviewed, especially if the reaction involves a two-phase mixture of liquid and gas, or if the gas flows through a deep bed of catalyst particles as in the FCC systems. ΔP should be checked early in the design process to assess its influence on the overall plant integrity.

Example 5-18

Consider the application of chemical reactor technology to a commercial reactor.[8] Figure 5-40 illustrates a simplified flow diagram of ethylene glycol production. Consider the reactions in the formation of glycol:

- Hydration of ethylene oxides to ethylene glycol.
- Reactions of part of the glycol with ethylene oxide to form diethylene glycol.
- Further reaction of the ethylene glycol with ethylene oxide.

In the following example, ethylene glycol product was manufactured to a specification that limited the diethylene glycol and higher glycol content; formation of these materials caused a loss in yield. Reactions

---

[8]Source: T. E. Corrigan, G. A. Lessells, M. J. D. Dean, Ind. & Eng. Chem. *Vol. 60, No. 4, April 1968.*

FLOW DIAGRAM

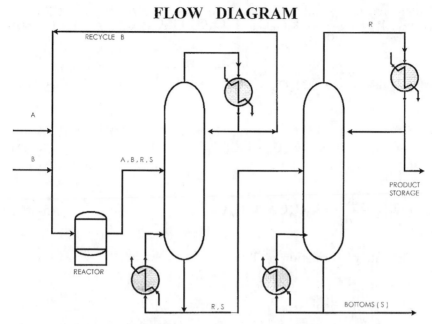

Figure 5-40. A simplified flow diagram of ethylene glycol.

5-357, 5-358, and 5-359, respectively, were all series-parallel reactions represented by

$$A + B \xrightarrow{k_1} R \tag{5-357}$$

$$A + R \xrightarrow{k_2} S \tag{5-358}$$

$$A + S \xrightarrow{k_3} T \tag{5-359}$$

where A = ethylene oxide
     B = water
     R = ethylene glycol
     S = di-ethylene glycol
     T = tri-ethylene glycol

Considering the above reactions at constant volume (i.e., constant density) in a batch system, rates of disappearance for A, B, R, and S are:

$$\left(-r_A\right)_{net} = -\frac{dC_A}{dt} = k_1 C_A C_B + k_2 C_A C_R + k_3 C_A C_S \tag{5-360}$$

$$\left(-r_B\right) = -\frac{dC_B}{dt} = k_1 C_A C_B \tag{5-361}$$

$$\left(-r_R\right) = -\frac{dC_R}{dt} = k_2 C_A C_R - k_1 C_A C_B \tag{5-362}$$

$$\left(-r_S\right) = -\frac{dC_S}{dt} = k_3 C_A C_S - k_2 C_A C_R \tag{5-363}$$

The rate of formation of T is

$$\left(+r_T\right) = \frac{dC_T}{dt} = k_3 C_A C_S \tag{5-364}$$

Rearranging Equations 5-360, 5-361, 5-362, 5-363, and 5-364 gives

$$\frac{dC_A}{dt} = -\left(k_1 C_A C_B + k_2 C_A C_R + k_3 C_A C_S\right) \tag{5-365}$$

$$\frac{dC_B}{dt} = -k_1 C_A C_B \tag{5-366}$$

$$\frac{dC_R}{dt} = k_1 C_A C_B - k_2 C_A C_R \tag{5-367}$$

$$\frac{dC_S}{dt} = k_2 C_A C_R - k_3 C_A C_S \tag{5-368}$$

$$\frac{dC_T}{dt} = k_3 C_A C_S \tag{5-369}$$

For batch and plug flow reactors:

$$\frac{dC_R}{dC_B} = \frac{k_1 C_A C_B - k_2 C_A C_R}{-k_1 C_A C_B}$$

$$= -1 + \frac{k_2}{k_1}\frac{C_R}{C_B} \tag{5-370}$$

where $R = C_R$ and $B = C_B$:

$$\frac{dR}{dB} = -1 + \frac{k_2}{k_1}\frac{R}{B} \tag{5-371}$$

Rearranging Equation 5-371 yields

$$\frac{dR}{dB} - \frac{k_2}{k_1}\frac{R}{B} = -1 \tag{5-372}$$

Equation 5-372 is a first order differential equation of the form

$$\frac{dy}{dx} + P(y) = Q \quad \text{I.F.} = e^{\int P dx}$$

$$e^{\int P dx}\frac{dy}{dx} + e^{\int P dx}P(y) = e^{\int P dx}Q$$

That is,

$$\frac{d}{dx}\left(e^{\int P dx}y\right) = e^{\int P dx}Q$$

The general solution is $e^{\int P dx}y = \int e^{\int P dx}Q dx + C$. The integrating factor of Equation 5-372 is

$$\text{I.F.} = e^{\int P dx} = e^{-\int \frac{K}{B}dB}, \quad \text{where } P = \frac{K}{B} = \frac{k_1}{k_2 B}$$

$$= e^{-K \ln B} = e^{\ln(B)^{-K}} = B^{-K}$$

Multiplying the integrating factor by both the left and right sides of Equation 5-372 gives

$$\frac{dR}{dB}\left(B^{-K}\right) - \frac{K}{B}R \cdot B^{-K} = -1 \cdot B^{-K} \tag{5-373}$$

or

$$\frac{d}{dB}\left(RB^{-K}\right) - KRB^{-K-1} = -B^{-K} \tag{5-374}$$

$$\frac{d}{dB}\left(RB^{-K}\right) = -B^{-K}$$

$$RB^{-K} = \int -B^{-K}dB + \text{Const.} \tag{5-375}$$

$$RB^{-K} = -\left[\frac{B^{-K+1}}{1-K}\right] + \text{Const.}$$

The boundary conditions, at $t = 0$, $B = 1$, $R = R_O$

$$\text{Const} = R_O + \frac{1}{1-K} \tag{5-376}$$

$$RB^{-K} = \frac{-B^{-K+1}}{1-K} + R_O + \frac{1}{1-K} \tag{5-377}$$

and

$$R = \frac{B^K - B}{1-K} + R_O B^K$$

The concentration of R is

$$C_R = \frac{C_B^K - C_B}{1-K} + C_{RO}C_B^K \tag{5-378}$$

The design equations for a CFSTR with perfect mixing, constant fluid density, and steady state operation are as follows. If u is the volumetric flowrate and $K = k_1/k_2$, relative reaction rate constant, where k_1, k_2, and k_3 are the specific reaction rate constants for reactions 5-357, 5-358, and 5-359. The rate expressions of A, B, R, S, and T are

$$\left(-r_A\right)_{net} = k_1 C_A C_B + k_2 C_A C_R + k_3 C_A C_S$$

$$\left(-r_B\right) = k_1 C_A C_B$$

$$\left(-r_R\right) = k_2 C_A C_R - k_1 C_A C_B$$

$$\left(-r_S\right) = k_3 C_A C_S - k_2 C_A C_R$$

$$\left(+r_T\right) = k_3 C_A C_S$$

The material balances of components A, B, R, S, and T are:

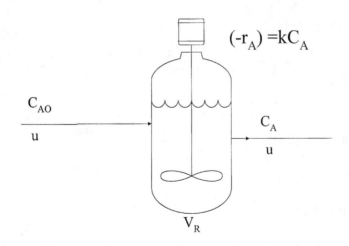

$$(-r_A) = kC_A$$

$$\text{Input rate} = \text{Ouput rate} + \text{Disappearance by reaction} + \text{Accumulation}$$

Mass balance on A is

$$uC_{AO} = uC_A + \left(-r_A\right)V_R \tag{5-379}$$

$$uC_{AO} = uC_A + \left(k_1 C_A C_B + k_2 C_A C_R + k_3 C_A C_S\right)V_R \qquad (5\text{-}380)$$

Mass balance on B is

$$uC_{BO} = uC_B + \left(-r_B\right)V_R \qquad (5\text{-}381)$$

$$uC_{BO} = uC_B + k_1 C_A C_B \, V_R \qquad (5\text{-}382)$$

Mass balance on R is

$$uC_{RO} = uC_R + \left(-r_R\right)V_R \qquad (5\text{-}383)$$

$$uC_{RO} = uC_R + \left(k_2 C_A C_R - k_1 C_A C_B\right)V_R \qquad (5\text{-}384)$$

Mass balance on S is

$$uC_{SO} = uC_S + \left(-r_S\right)V_R \qquad (5\text{-}385)$$

$$uC_{SO} = uC_S + \left(k_3 C_A C_S - k_2 C_A C_R\right)V_R \qquad (5\text{-}386)$$

Mass balance on T is

$$uC_{TO} + \left(+r_C\right)V_R = uC_T \qquad (5\text{-}387)$$

$$0 = uC_T - k_3 C_A C_S V_R \qquad (5\text{-}388)$$

Rearranging Equations 5-382 and 5-384 gives

$$u\left(C_{BO} - C_B\right) = k_1 C_A C_B V_R \qquad (5\text{-}389)$$

$$u\left(C_R - C_{RO}\right) = \left(k_1 C_A C_B - k_2 C_A C_R\right)V_R \qquad (5\text{-}390)$$

Dividing Equation 5-390 by Equation 5-389 yields

$$\frac{C_R - C_{RO}}{C_{BO} - C_B} = \frac{\left(k_1 C_A C_B - k_2 C_A C_R\right)V_R}{k_1 C_A C_B V_R} \qquad (5\text{-}391)$$

That is,

$$\frac{C_R - C_{RO}}{C_{BO} - C_B} = 1 - \frac{k_2}{k_1}\frac{C_R}{C_B}$$

or

$$\frac{R - R_O}{B_O - B} = 1 - \frac{k_2}{k_1}\frac{R}{B} \tag{5-392}$$

Rearranging this equation for R in terms of B_O, B, R_O, and $K = k_1/k_2$ gives

$$R = \frac{B\{(B_O - B) + R_O\}}{B + K(B_O - B)} \tag{5-393}$$

At all t = 0, $B_O = 1$, therefore,

$$R = \frac{B\{(1 - B) + R_O\}}{B + K(1 - B)} \tag{5-394}$$

or

$$C_R = \frac{C_B\{(1 - C_B) + C_{RO}\}}{C_B + K(1 - C_B)} \tag{5-395}$$

Davis et al. [9] have performed studies on the batch hydration of ethylene oxide. Their work determined the value of the product distribution constant K. This value is used in Equation 5-378 to determine the expected performance in a plug flow reactor. This value is also used in Equation 5-394 to illustrate the poor performance that would be obtained with complete backmixing.

The results show that the plant reactor falls between the plug flow and backmixing reactor lines. The distance between lines (2) and (3) in Figure 5-41 measures the degree of backmixing. There is considerable departure from the theoretical plug flow curve. This shows

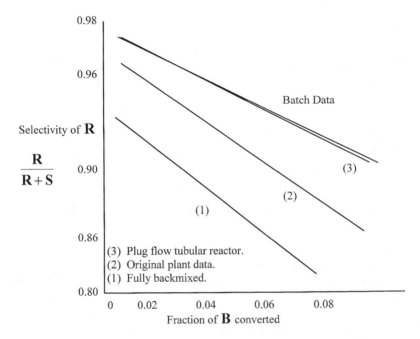

Figure 5-41. Selectivity and conversion for plant data and well-defined residence time distributions.

that there is room for improvement in the process of eliminating or reducing backmixing.

The wide deviation from batch performance is possibly caused by:

- Case 1: Backmixing in the reactor.
- Case 2: Buildup of ethylene glycol in the recycle to the reactor.
- Case 3: Buildup of other reaction products in the recycle.

Since the reactors were a series of baffled tanks, Case 1 is most likely. Case 2 can be evaluated quantitatively by using the kinetics applicable to the reaction and making suitable changes to allow for the recycling of ethylene glycol. Case 3 is extremely unlikely because the impurities formed are water-soluble and should, therefore, be removed by the water purged from the system.

A longitudinal tubular reactor in place of a high backmixing reactor results in substantial savings in processing costs because selectivity and productivity are increased.

REFERENCES

1. Levenspiel, O., *Chemical Reaction Engineering*, 3rd ed., John Wiley, 1998.
3. Adesina, A. A., "Design of CSTRs In Tandem Revisited," *Chemical Engineering Education*, pp. 164–168, Summer 1992.
3. Smith, J. M., *Chemical Engineering Kinetics*, 3rd ed., McGraw-Hill Book Company, New York, 1981.
4. Carberry, J. J., *Chemical and Catalytic Reaction Engineering*, McGraw-Hill Book Company, New York, 1976.
5. Schmidt, L. D., *The Engineering of Chemical Reactions*, Oxford University Press, New York, 1998.
6. Walas, S. M., *Chemical Reaction Engineering Handbook of Solved Problems*, Gordon and Breach Publishers, 1996.
7. Walas, S. M., *Reaction Kinetics for Chemical Engineers*, McGraw-Hill, New York, 1959.
8. Corrigan, T. E., Lessells, A., and Dean, M. J., *Ind. & Eng. Chem.*, Vol. 60, No. 4, April 1968.
9. Davis, et al., *Chem. Eng. Prog. Symp. Ser.* 48, 4, 1952.
10. Smith, E. L., private communications.

CHAPTER SIX

Non-Isothermal Reactors

INTRODUCTION

Most chemical reactions are greatly affected by temperature. The previous chapters discussed reactions at isothermal condition, however, industrial reactors often operate under non-isothermal condition. This is because chemical reactions strongly depend on temperature, either absorbing (i.e., endothermic) or generating (i.e., exothermic) a large amount of heat.

In non-isothermal conditions, the temperature varies from point to point, as does the composition. In an ideal gas mixture, the concentration of a component depends on the temperature (i.e., $C_i = p_i/RT$), and the reaction rate is greatly affected by the temperature through the rate coefficient ($k = k_o e^{-E/RT}$). The reactors described in this chapter operate under both isothermal and non-isothermal conditions. Either a heating/cooling coil or an external heat exchanger can affect the temperature. Some reaction rates double for a rise in temperature of 10–15°C, and temperature affects the properties of the reaction mixture such as density, specific heat, thermal conductivity and viscosity, enthalpy, as well as mixing patterns and the energy for efficient mixing. Temperature coefficients are not as great as for the rate constant, therefore, averaged values of the physical properties may often be used for design purposes.

Some reactors operate under adiabatic conditions, where there is negligible heat input to the reactor. The design of non-isothermal reactors involves the simultaneous solution of both mass and energy balances. In some reactors, such as the packed bed or fluidized catalytic cracking unit (FCCU), momentum balance is considered with both the mass and energy balances. In this section, an energy balance is considered for the CFSTR, batch, and plug flow reactors, and design equations are derived under adiabatic conditions.

There are reactions where the heat of reaction can be employed to preheat the feed when an exothermic reaction is operated at a high temperature (e.g., ammonia $N_2 + 3H_2 \leftrightarrow 2NH_3$ or methanol $CO + 2H_2 \leftrightarrow CH_3OH$ synthesis, water-gas shift reaction $CO + H_2O \leftrightarrow H_2 + CO_2$). These processes may be performed in fixed-bed reactors with an external heat exchanger. The exchanger is primarily used to transfer the heat of reaction from the effluent to the feed stream. The combination of the heat transfer-reaction system is classified as autothermal. These reactors are self-sufficient in energy; however, a high temperature is required for the reaction to proceed at a reasonable rate.

The following also discusses autothermal reactors, the conversion of ammonia, and finally, the optimum-temperature progression, with derived equations for reversible exothermic reactions for which profiles are not readily available.

OPERATING TEMPERATURE, REACTION TYPES, AND TEMPERATURE

The main factors affecting operating temperature are:

• Cost of maintaining the operating temperature (energy).
• Decomposition temperature of reactants and products.
• Possibility of operating at high pressures in the case of gas-phase reactions.
• Effect of temperature and pressure on equilibrium in the case of reversible reactions.
• Effect of temperature on the rate constants associated with each elementary step in a multiple reaction.
• Properties of materials of construction.

In irreversible reactions and in accordance with the Arrhenius equation $[k = k_o\exp(-E/RT)]$, it is possible to operate at the highest possible temperature. In reversible reactions, where

$$aA + bB \underset{k_2}{\overset{k_1}{\rightleftharpoons}} rR + sS \tag{6-1}$$

The equilibrium constant K is related to the Gibbs standard free energy change ΔG_T^o by

$$\Delta G_T^o = \left(rG_R^o + sG_S^o - aG_A^o - bG_B^o \right)$$

$$= -RT \ln K \qquad\qquad (6\text{-}2)$$

where R = universal gas constant

K = k_1/k_2, equilibrium constant

T = absolute temperature (Kelvin)

The van't Hoff equation is expressed as

$$\frac{d \ln K}{dT} = \frac{\Delta H_R}{RT^2}$$

where ΔH_R is the heat of reaction, which is a function of temperature. Considering the case where ΔH_R is constant and integrating the van't Hoff equation between the boundary limits gives

$$\int_{k_1}^{k_2} d \ln K = \frac{\Delta H_R}{R} \int_{T_1}^{T_2} \frac{dT}{T^2}$$

$$\ln \frac{k_2}{k_1} = \frac{-\Delta H_R}{R} \left\{ \frac{1}{T_2} - \frac{1}{T_1} \right\}$$

For endothermic reactions, ΔH_R has a positive numerical value if $T_2 > T_1$ and $k_2 > k_1$. The equilibrium constant K, therefore, increases as the temperature increases. In this case, the reactor is operated at the highest possible temperature.

For exothermic reactions, ΔH_R has a negative numerical value if $T_2 > T_1$ and $k_2 < k_1$. The equilibrium constant K decreases as the temperature increases. Here, the temperature is kept high so that the rate of reaction is high, and then decreased as equilibrium is approached.

PRESSURE AND GAS PHASE REVERSIBLE REACTIONS

Consider the reversible reaction of the form aA + bB $\underset{k_2}{\overset{k_1}{\rightleftharpoons}}$ rR + sS. The activity a_i of component is

$$a_i \text{ (Activity of ith component)} = \frac{\bar{f_i}}{\bar{f_o}}$$

where $\bar{f}_i$ = fugacity of ith component in the mixture

$\bar{f}_o$ = reference fugacity

The equilibrium constant K in terms of the fugacity is expressed as

$$K = \frac{\left(\dfrac{\bar{f}_R}{f_R^o}\right)^r \left(\dfrac{\bar{f}_S}{f_S^o}\right)^s}{\left(\dfrac{\bar{f}_A}{f_A^o}\right)^a \left(\dfrac{\bar{f}_B}{f_B^o}\right)^b} \qquad (6\text{-}3)$$

where f_i^o is usually taken as that of the pure gas at 1 atm, and so $\bar{f}_o = 1$ atm.

Assume that the gas mixture behaves ideally, but the individual gases do not. This yields

$$\bar{f}_i = \frac{x_i\, f_i P}{P}$$

$$= x_i \gamma_i P \qquad (6\text{-}4)$$

where x_i = mole fraction of i in the mixture

f_i = fugacity of pure i

P = total pressure

$\gamma_i = f_i/P$ = fugacity coefficient

Substituting for $\bar{f}_i$ and noting that $f_i^o = 1$ atm, the equilibrium constant K in terms of the fugacity and total pressure becomes

$$K = \frac{\bar{f}_R^r \cdot \bar{f}_S^s}{\bar{f}_A^a \cdot \bar{f}_B^b}$$

$$= \left(\frac{x_R^r \cdot x_S^s}{x_A^a \cdot x_B^b}\right) \left(\frac{\gamma_R^r \cdot \gamma_S^s}{\gamma_A^a \cdot \gamma_B^b}\right) \cdot P^{(r+s-a-b)} \qquad (6\text{-}5)$$

The equilibrium constant K depends only on the temperature and not on the pressure.

Since the product materials are calculated in the number of moles that can be obtained at equilibrium under given conditions of temperature, pressure, and feed composition, Equation 6-5 is modified to give

$$K = \left(\frac{n_R^r \bullet n_S^s}{n_A^a \bullet n_B^b}\right) K_\gamma \left(\frac{P}{\sum n_j}\right)^{(r+s-a-b)} \tag{6-6}$$

where n_j is the number of moles of component j in the reacting system, which also contains n_I moles of inert gas. Therefore,

$$x_j = \frac{n_j}{n_A + n_B + n_R + n_S + n_I}$$

and

$$K_y = \frac{\gamma_R^r \bullet \gamma_S^s}{\gamma_A^a \bullet \gamma_B^b} \tag{6-7}$$

where a, b, r, and s are stoichiometric coefficients

From Equation 6-6, the larger the value for K, the higher the yield of the product species at equilibrium. Also, if ΔG^o is negative, the values of K will be positive, and results in a large equilibrium conversion of reactants into products. Dodge [1] states the following rules to determine whether a reaction is thermodynamically promising at a given temperature:

- $\Delta G < 0$ Reaction is promising
- $0 < \Delta G^o < +10,000$ Reaction of doubtful promise, but requires further study.
- $\Delta G^o > + 10,000$ Very unfavorable; would be possible only under unusual circumstances.

For ideal gases,

$$K = \left(\frac{x_R^r \bullet x_S^s}{x_A^a \bullet x_B^b}\right) \bullet P^{(r+s-a-b)}$$

$$= \frac{p_R^r \bullet p_S^s}{p_A^a \bullet p_B^b} \tag{6-8}$$

where p_i is the partial pressure of component i.

EFFECT OF OPERATING PARAMETERS ON EQUILIBRIUM CONVERSION

TEMPERATURE

The effect of temperature on equilibrium composition can be calculated using the van't Hoff equation. Since the standard heat of reaction is negative $(-\Delta H_R)$ for an exothermic reaction, an increase in temperature results in a decrease in K and a subsequent decrease in conversion. Therefore, an exothermic reaction must be performed at as low a temperature as possible. An endothermic reaction $(+\Delta H_R)$ is positive and K increases with an increase in temperature, as does the equilibrium conversion. Therefore, an endothermic reaction must be performed at an elevated temperature.

PRESSURE

The equilibrium constant K is independent of pressure with standard states. The effect of the pressure is shown in Equation 6-6. K_γ is usually insensitive and may either increase or decrease slightly with pressure. When $(r + s) > (a + b)$, the stoichiometric coefficients, an increase in pressure P results in a decrease in conversion of the reactants to the products (i.e., $A + B \leftrightarrow R + S$). Alternatively, when $(r + s) < (a + b)$, an increase in pressure P results in an increase in the equilibrium conversion. In ammonia synthesis $(N_2 + 3H_2 \leftrightarrow 2NH_3)$, the reaction results in a decrease in the number of moles. Therefore, an increase in pressure causes an increase in equilibrium conversion due to this factor.

ENERGY BALANCE AND HEAT OF REACTION

From the conservation of energy and assuming a steady-state condition:

$$\frac{\text{ENERGY}}{\text{INPUT}} - \frac{\text{ENERGY}}{\text{OUTPUT}} = \frac{\text{ENERGY}}{\text{ACCUMULATION}}$$

This is represented in Figure 6-1.

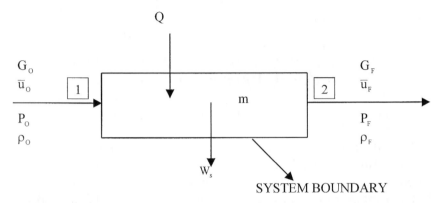

Figure 6-1. Energy balance in chemical reactors, where G = mass flow-rate; $\bar{u}$ = internal energy per unit mass; P = pressure; ρ = fluid density; m = mass of fluid in the system; Q = rate at which heat is transferred to the system; and W_s = rate at which work is done on the surroundings.

Considering the energy balance in this figure, the heat flows into the reactor with the reactant and flows out with the products. The reaction will either generate or absorb heat depending on whether it is exothermic or endothermic. Additionally, stirring the reactants and friction in the CFSTR will generate heat. The heat is finally transferred through the walls. For a CFSTR, assume that the temperature in the system is equal to the temperature at the exit. In other reactor types, the temperature and conversions depend on the position in the reactor.

Consider a constant pressure, steady-state flow process with an inflow of reactant in section (1) and the outflow of the product in section (2) leaving the system boundary (i.e., a chemical reactor).

During time δt from t to t + δt, the energy balance is expressed as

$$G_O\left(\bar{u}_O + \frac{P_O}{\rho_O}\right)\delta t \ - \ G_F\left(\bar{u}_F + \frac{P_F}{\rho_F}\right)\delta t + Q \bullet \delta t - W_s \bullet \delta t$$

$$= \text{Accumulation}\left(m\bar{u} \ + \ \frac{d}{dt}(m\bar{u}) \bullet \delta t\right) - m\bar{u} \qquad (6\text{-}9)$$

Equation 6-9 gives

$$-\left\{G_F\left(\bar{u}_F+\frac{P_F}{\rho_F}\right) - G_O\left(\bar{u}_O+\frac{P_O}{\rho_O}\right)\right\} + Q - W_s = \frac{d}{dt}(m\bar{u}) \qquad (6\text{-}10)$$

Equation 6-10 is the macroscopic energy balance equation, in which potential and kinetic energy terms are neglected. From thermodynamics, the enthalpy per unit mass is expressed as

$$h = \bar{u} + \frac{P}{\rho} \qquad (6\text{-}11)$$

and $\quad \bar{u} = \sum_i \bar{u}_i m_i$

where $\bar{u}_i$ = internal energy per unit mass of component i
$\quad\quad m_i$ = mass fraction of i

Substituting Equation 6-11 into Equation 6-10 yields

$$-\left(G_F h_F - G_O h_O\right) + Q - W_s = \frac{d}{dt}(m\bar{u}) \qquad (6\text{-}12)$$

To use Equation 6-12, a relationship is required between either the internal energy per unit mass $\bar{u}$ or the enthalpy per unit mass h and the state variables such as the temperature T, pressure P, and composition $\overline{m}_i$.

$$(m\bar{u}) = \int_{V_R} \rho\bar{u}\cdot dV_R \quad \text{and} \quad m = \int_{V_R} \rho\cdot dV_R$$

Since mixtures are entering and leaving the system, then

$$\bar{u}_i = \sum_i \overline{m}_i\bar{u}_i$$

where $\overline{m}_i$ = mass fraction of component i
$\quad\quad \bar{u}$ = internal energy per unit mass of mixture

This shows that from thermodynamics the enthalpy is a function of the pressure, temperature, and mass fraction of the component, namely

$$h = f(p, T, \overline{m}_i) \qquad (6\text{-}13)$$

Differentiating Equation 6-13 with respect to T, p, and $\overline{m}_i$ gives

$$\delta h = \sum_i \overline{m}_i \frac{\partial h_i}{\partial p} \cdot \delta p + \sum_i \overline{m}_i \frac{\partial h_i}{\partial T} \cdot \delta T + \sum_i \delta(\overline{m}_i h_i) \qquad (6\text{-}14)$$

From thermodynamics

$$\sum_i \overline{m}_i \frac{\partial h_i}{\partial p} \cdot \delta p = \frac{1}{\rho} \cdot \delta p \qquad (6\text{-}15)$$

where

$$\left(\frac{\partial h}{\partial p} \right)_T = \left[v - T \left(\frac{\partial v}{\partial T} \right)_P \right]$$

$$= \left[\frac{1}{\rho} - T \left(\frac{\partial \frac{1}{\rho}}{\partial T} \right)_P \right]$$

and

$$\sum_i \overline{m}_i \frac{\partial h_i}{\partial T} \cdot \delta T = C_p \cdot \delta T \qquad (6\text{-}16)$$

where C_p is the specific heat per unit mass of mixture.

The last term of Equation 6-14 represents the enthalpy change at constant temperature and pressure due to a change in composition or the degree of conversion. Consider an irreversible reaction of the form aA + bB $\rightarrow$ cC + dD. The change in enthalpy ΔH_R at fixed temperature and pressure is

$$\Delta H_R = (cH_C + dH_D - aH_A - bH_B) \qquad (6\text{-}17)$$

where H_i is the enthalpy of 1 mole of component i.

From the stoichiometry,

$$\frac{-\delta \overline{m}_A}{aM_A} = \frac{-\delta \overline{m}_B}{bM_B} = \frac{+\delta \overline{m}_C}{cM_C} = \frac{+\delta \overline{m}_D}{dM_D} \tag{6-18}$$

where M_i is the molecular weight of component i.

Equation 6-16 makes it possible to express changes in composition in terms of the change in mass fraction of a single component. Thus,

$$\sum_i \delta(\overline{m}_i h_i)\bigg|_{P,T} \equiv \sum_i h_i \delta \overline{m}_i \tag{6-19}$$

Assuming that h_i's are independent of composition and

$$\sum_i h_i \delta \overline{m}_i = \left(cM_C h_C + dM_D h_D - aM_A h_A - bM_B h_B\right)\left(\frac{-\delta \overline{m}_A}{aM_A}\right)$$

$$= \left(cH_C + dH_D - aH_A - bH_B\right)\bullet\left(\frac{-\delta \overline{m}_A}{aM_A}\right)$$

$$= \Delta H_R \left(\frac{-\delta \overline{m}_A}{aM_A}\right) \tag{6-20}$$

Substituting Equations 6-15, 6-16, and 6-20 into Equation 6-14 gives

$$\delta h = \frac{1}{\rho}\bullet\delta p + C_p\bullet\delta T + \left(\frac{-\Delta H_R}{a}\right)\bullet\left(\frac{\delta \overline{m}_A}{M_A}\right) \tag{6-21}$$

For internal energy per unit mass of mixture

$$\delta \overline{u} = -p\bullet\delta\left(\frac{1}{\rho}\right) + C_v\bullet\delta T + \left(\frac{-\Delta \overline{U}_R}{a}\right)\bullet\left(\frac{\delta \overline{m}_A}{M_A}\right) \tag{6-22}$$

ENERGY TRANSFERRED BETWEEN
THE SYSTEM AND SURROUNDINGS

The transfer of energy between the system and its surroundings involves

- Net work done by the system including both mechanical and electrical work, W_s.
- Net heat added to the system from all sources, Q.

From the energy balance Equation 6-12, the rate at which heat is absorbed by the surrounding into system Q is defined as:

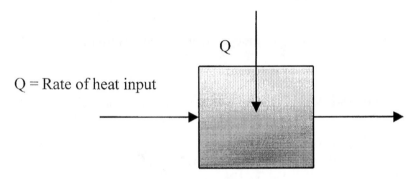

$$Q = \text{Rate of heat input}$$

For an endothermic reaction, Q is a positive value, and is defined as

$$Q = U \cdot A \cdot \Delta T \tag{6-23}$$

where A = heat transfer area of the heating or cooling coil, m^2

    U = overall heat transfer coefficient, $J/m^2 \cdot s \cdot K$ or W/m^2K

    ΔT = temperature difference between the heat transfer fluid and reaction mixture, K

For an exothermic reaction Q is a negative value.

ADDING HEAT TO THE REACTOR Q

In many cases, the heat flow (Q) to the reactor is given in terms of the overall heat transfer coefficient U, the heat exchange area A, and the difference between the ambient temperature, T_a, and the reaction temperature, T. For a continuous flow stirred tank reactor (CFSTR) in which both fluid temperatures (i.e., inside and outside the exchanger) are constant (e.g., condensing steam), Q is expressed as

$$Q = UA (T_a - T) \tag{6-24}$$

If the fluid inside the heat exchanger tubes enters the CFSTR at a temperature T_{a1} and leaves at a temperature T_{a2}, and the temperature inside the CFSTR is spatially uniform at temperature T, then the heat flow Q is expressed to the CFSTR by

$$Q = U \cdot A \cdot \Delta T_{LMTD} = \frac{UA(T_{a1} - T_{a2})}{\ln\left(\dfrac{T_{a1} - T}{T_{a2} - T}\right)} \tag{6-25}$$

When the heat flow varies along the length of the reactor as in a tubular reactor, Equation 6-25 can be integrated along the length of the reactor to obtain the total heat added to the reactor:

$$Q = \int^{A} U(T_a - T)dA = \int^{V} Ua(T_a - T)dV \tag{6-26}$$

where a is the heat exchanger area per unit volume of the reactor.

SHAFT WORK

Work done W_s by the system on the surrounding generally involves the work done by the stirrer. Any stirring within the reactor will evolve heat. W_s is positive when work is done on the system.

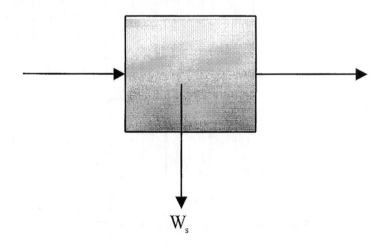

$$W_s$$

If V_R increases at the rate dV_R/dt during the chemical reaction, then for a batch system, the work done is defined by

$$W_s = p \cdot \frac{dV_R}{dt} \qquad\qquad (6\text{-}27)$$

ENERGY BALANCE IN A CFSTR

The following derives an energy balance for a CFSTR based on the following assumptions (Figure 6-2):

- Steady state operation
- Perfect mixing
- Constant coolant temperature, T_c
- Stirrer input energy is neglected W_s
- Constant density of fluid mixture
- No work done by the system on the surrounding

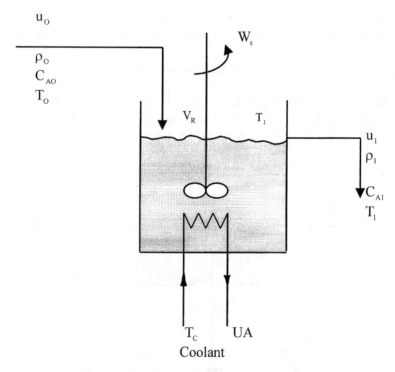

Figure 6-2. Energy balance in a CFSTR.

Since $W_s \approx 0$ and $(d/dt)(m\bar{u}) = 0$, Equation 6-12 becomes

$$-(u_1\rho_1h_1 - u_O\rho_Oh_O) + Q = 0 \tag{6-28}$$

where

$$G_1 = u_1\rho_1,\ G_O = u_O\rho_O \tag{6-29}$$

and $h_i = (\bar{u} + P_i/\rho_i)$.

Since the condition is at steady state, then

$$u_1\rho_1 = u_O\rho_O = u\rho \tag{6-30}$$

In many practical situations, $\rho_1 = \rho_O$ and, therefore, $u_1 = u_O$, and Equation 6-28 becomes $-u\rho\delta h + Q = 0$

$$\text{or } -G\delta h + Q = 0 \tag{6-31}$$

The heat input Q is

$$Q = U \cdot A \cdot (T_C - T_1) \tag{6-32}$$

Substituting Equations 6-30 and 6-32 into Equation 6-28 yields

$$-u\rho(h_1 - h_O) + UA(T_C - T_1) = 0 \tag{6-33}$$

From Equation 6-21

$$\int_{h_O}^{h_1} dh = \int_{T_O}^{T_1} C_p dT + \int_{m_{AO}}^{m_{A1}} \left(\frac{-\Delta H_R}{a}\right)\frac{d\bar{m}_A}{M_A} \tag{6-21}$$

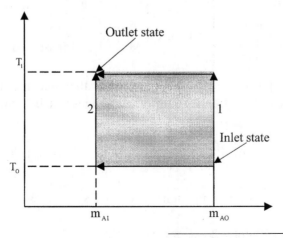

Considering the thermodynamic paths

$$\text{Path 1:} \quad \int_{h_O}^{h_1} dh = \int_{T_O}^{T_1} C_{P_{Feed}} dT + \int_{m_{AO}}^{m_{A1}} \left(\frac{-\Delta H_R}{a}\right)_{T_1} \frac{d\overline{m}_A}{M_A} \qquad (6\text{-}34)$$

$$\text{Path 2:} \quad \int_{h_O}^{h_1} dh = \int_{T_O}^{T_1} C_{P_{Product}} dT + \int_{m_{AO}}^{m_{A1}} \left(\frac{-\Delta H_R}{a}\right)_{T_O} \frac{d\overline{m}_A}{M_A} \qquad (6\text{-}35)$$

Both path 1 or path 2 can be considered for the heat balance equation for a CFSTR. In the case of path 1,

$$-u\rho \int_{T_O}^{T_1} C_p dT - u\rho \left(\frac{-\Delta H_R}{a M_A}\right)_{T_1} \left(\overline{m}_{A1} - \overline{m}_{AO}\right) + UA\left(T_C - T_1\right) = 0 \quad (6\text{-}36)$$

In many practical cases, C_p does not significantly change with composition. Equation 6-36 becomes

$$-u\rho C_p\left(T_1 - T_O\right) - u\rho \left(\frac{-\Delta H_R}{a M_A}\right)\left(\overline{m}_{A1} - \overline{m}_{AO}\right) + UA\left(T_C - T_1\right) = 0 \quad (6\text{-}37)$$

By definition,

$$C_A' \equiv \frac{\overline{m}_A}{M_A}\left(\frac{\text{moles}}{\text{unit mass}}\right) \text{ and } C_A = \frac{\overline{m}_A}{M_A} \cdot \rho \left(\frac{\text{moles}}{\text{unit mass}} \cdot \frac{\text{mass}}{\text{volume}}\right)$$

The heat balance of a CFSTR in terms of the concentration C_A for component A in Equation 6-37 after rearrangement becomes

$$u\rho C_p\left(T_O - T_1\right) + u\left(\frac{-\Delta H_R}{a}\right)\left(C_{AO} - C_{A1}\right) + UA\left(T_C - T_1\right) = 0 \qquad (6\text{-}38)$$

A mass balance on component A in a CFSTR gives

$$uC_{AO} = uC_{A1} + (-r_{A1})V_R \qquad (5\text{-}158)$$

or $u(C_{AO} - C_{A1}) = (-r_{A1})V_R$ (5-158a)

Substituting Equation 5-158a into Equation 6-38 gives

$$u\rho C_p(T_0 - T_1) + \left(\frac{-\Delta H_R}{a}\right)(-r_{A1})V_R + UA(T_C - T_1) = 0 \quad (6\text{-}39)$$

For multiple reaction systems, Equation 6-39 becomes

$$u\rho C_p(T_0 - T_1) + \sum_{i=1}^{R}\left(\frac{-\Delta H_{Ri}}{a}\right)(-r_{Ai})V_R + UA(T_C - T_1) = 0 \quad (6\text{-}40)$$

For a first order irreversible reaction, the energy balance is expressed in terms of the fractional conversion X_A, where

$$X_A = \frac{C_{AO} - C_A}{C_{AO}}$$

$$C_{AO} - C_A = C_{AO}X_A$$

Equation 6-38 now becomes

$$u\rho C_p(T_0 - T_1) + uC_{AO}X_A\left(\frac{-\Delta H_R}{a}\right) + UA(T_C - T_1) = 0 \quad (6\text{-}41)$$

If the temperature in the product is the same as the inlet temperature of the reactant (i.e., $T_0 = T_1$), in which isothermal condition prevails, then Equation 6-38 becomes

$$u\left(\frac{-\Delta H_R}{a}\right)(C_{AO} - C_{A1}) = UA(T_1 - T_C) \quad (6\text{-}42)$$

Or, in terms of the fractional conversion,

$$uC_{AO}X_A\left(\frac{-\Delta H_R}{a}\right) = UA(T_1 - T_C) \quad (6\text{-}43)$$

In an isothermal (constant temperature) reactor as much heat is needed to be supplied from or rejected into the surroundings, as is absorbed or given up by the reaction. In this type of operation, high costs of the equipment are required and, therefore, reactors under isothermal conditions are seldom employed in large-scale manufacturing. Generally, commercial reactors are either adiabatic or polytropic types. In an adiabatic reactor, no heat is added or taken away during the process, and that given up by the reaction is solely used within the system to change its enthalpy. In a polytropic reactor, the heat necessary to sustain the process is either added or taken away from the system. The following example reviews some design problems involving the simultaneous solution of both the mass and energy balances in a CFSTR.

Example 6-1

An endothermic reaction A $\rightarrow$ R is performed in three-stage, continuous flow stirred tank reactors (CFSTRs). An overall conversion of 95% of A is required, and the desired production rate is 0.95×10^{-3} kmol/sec of R. All three reactors, which must be of equal volume, are operated at 50°C. The reaction is first order, and the value of the rate constant at 50°C is 4×10^{-3} sec^{-1}. The concentration of A in the feed is 1 kmol/m^3 and the feed is available at 75°C. The contents of all three reactors are heated by steam condensing at 100°C inside the coils. The overall heat transfer coefficient for the heat-exchange system is 1,500 J/m$^2 \bullet$ sec $\bullet$ °C, and the heat of reaction is $+1.5 \times 10^8$ J/kmol of A reacted.

Calculate (1) the size of the tanks in the reactor system, and (2) the areas of the heating coil in each tank.

Data:

Reaction rate constant = 4×10^{-3} sec^{-1}
Liquid density = 1,000 kg/m^3
Specific heat = 4×10^3 J/kg $\bullet$ °C
Molecular weight of A = 100 kg/kmol

Solution

Assume the following:

- Perfect mixing
- Steady-state operation (i.e., accumulation in the reactor is zero)

- Constant fluid mixture density
- Stirrer input energy is neglected W_s
- No work done by the system on the surroundings

For $A \xrightarrow{k_1} R$, first order endothermic reaction, the rate of reaction is $(-r_A) = K_1 C_A$.

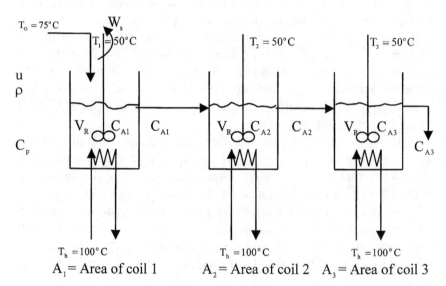

Battery of continuous flow stirred tank reactors (CFSTRs)

The notations are:

A_1, A_2, A_3 = areas of the heating coil
C_{AO} = initial concentration to CFSTR 1
C_{A1}, C_{A2}, C_{A3} = concentrations in CFSTR 1, 2, and 3, respectively
F_{AO} = desired production rate
u = fluid volumetric flowrate
ρ = liquid density
C_p = liquid specific heat
U = overall heat transfer coefficient
M_A = molecular weight of A
V_R = volume of the reactor
$\bar{t}$ = mean residence time (V_R/u)
$(+\Delta H_R)$ = enthalpy of reaction (endothermic)

T_0 = inlet temperature to the first CFSTR
T_1, T_2, T_3 = temperatures in the CFSTRs
T_h = temperature of the heating coils
X_A = overall fractional conversion of A

For a CFSTR, mass balance for the Nth stage is $C_{AO}/C_{AN} = (1 + k\bar{t})^N$. The mass balance for the first stage of the CFSTR on component A is $uC_{AO} = uC_{A1} + (-r_A)V_R + V_R(dC_{A1}/dt)$. Accumulation is zero, which yields $V_R(dC_{A1}/dt) = 0$, and the rate of disappearance of component A is $(-r_A) = k_1C_{A1}$. Hence, $uC_{AO} = uC_{A1} + k_1C_{A1}V_R$.

Rearranging, the exit concentration from the first stage of the CFSTR is $C_{A1} = C_{AO}/(1 + k_1\bar{t})$. The exit concentration from the first stage is now the input concentration to the second stage of the CFSTR. Material balance on the second stage is $uC_{A1} = uC_{A2} + k_1C_{A2}V_R$. Rearranging for the exit concentration of the second stage CFSTR yields $C_{A2} = C_{A1}/(1 + k_1\bar{t})$.

In terms of the inlet concentration, the exit concentration of the second stage becomes $C_{A2} = C_{AO}/(1 + k_1\bar{t})^2$. The exit concentration from the second stage is now the input concentration to the third stage of the CFSTR. Material balance on the third stage CFSTR is $uC_{A2} = uC_{A3} + k_1C_{A3}V_R$. Further rearranging gives the exit concentration from the third stage CFSTR as $C_{A3} = C_{A2}/(1 + k_1\bar{t})$. In terms of the inlet concentration, the exit concentration of the third stage becomes $C_{A3} = C_{AO}/(1 + k_1\bar{t})^3$. The overall conversion of 95% of A indicates that

$$X_A = \frac{C_{AO} - C_{AF}}{C_{AO}} = \frac{C_{AO} - C_{A3}}{C_{AO}} = \frac{95}{100}$$

That is

$$1 - \frac{C_{A3}}{C_{AO}} = 0.95$$

$$C_{A3} = 0.05 C_{AO}$$

where

$$C_{AO} = 1\frac{kmol}{m^3}$$

$$C_{A3} = 0.05 \frac{kmol}{m^3}$$

$$\frac{C_{AO}}{C_{A3}} = \left(1 + k_1 \bar{t}\right)^3$$

That is, $(1 + k_1 \bar{t})^3 = 20$, or

$$1 + k_1 \bar{t} = 2.714$$

$$k_1 \bar{t} = 1.714$$

$$k_1 \frac{V_R}{u} = 1.714$$

$$\frac{V_R}{u} = \frac{1.714}{4 \times 10^{-3}} = 428.5 \text{ sec}$$

u = volumetric flowrate of A in m^3/sec,

$$= \frac{\text{Desired production rate}}{\text{An overall conversion}} \times \frac{1}{\text{Concentration of A in the feed}}$$

$$= \frac{0.95 \times 10^{-3}}{0.95} \times \frac{1}{1} \left\{ \frac{kmol}{sec} \times \frac{1}{\frac{kmol}{m^3}} \right\}$$

$$= 1.0 \times 10^{-3} \frac{m^3}{sec}$$

The size of the reactor V_R is

$$V_R = 428.5 \times u$$

$$= 428.5 \times 10^{-3} m^3$$

$$= 0.429 \ m^3$$

The heat balance operation at constant pressure on the first stage CFSTR is $-G\Delta h + Q = 0$, where $G = \rho u$ and Q = heat absorbed to the reactor tanks.

$$Q = +UA(T_h - T_1)$$

where

$$\Delta h = C_p \Delta T + \left(\frac{-\Delta H_R}{a} \right) \frac{\Delta \overline{m}_A}{M_A}$$

The heat balance equation is expressed as:

$$-u\rho C_p \Delta T - u\rho \left(\frac{-\Delta H_R}{a} \right) \left(\frac{\Delta \overline{m}_A}{M_A} \right) + UA_1(T_h - T_1) = 0$$

$$C_A = \frac{\overline{m}_A}{M_A} = C'_A \cdot \rho \quad \left(\frac{\text{moles}}{\text{unit mass}} \times \frac{\text{mass}}{\text{volume}} \right)$$

$$-u\rho C_p(T_1 - T_O) + u\left(\frac{+\Delta H_R}{a} \right)(C_{A1} - C_{AO}) + UA_1(T_h - T_1) = 0$$

The mass balance on the first stage CFSTR.

$$C_{A1} = \frac{C_{AO}}{1 + k_1 \overline{t}}$$

$$C_{AO} = 1 \frac{\text{kmol}}{\text{m}^3}$$

$$\overline{t} = 428.5 \text{ sec}$$

$$k_1 = 4 \times 10^{-3} \text{ sec}^{-1}$$

$$C_{A1} = \frac{1}{\left(1 + 4 \times 10^{-3} \times 428.5 \right)}$$

$$= 0.3685 \frac{\text{kmol}}{\text{m}^3}$$

$$C_{A1} = 0.369 \frac{\text{kmol}}{\text{m}^3}$$

$$C_{A2} = \frac{C_{AO}}{\left(1 + k_1 \bar{t}\right)^2}$$

$$= \frac{1}{\left(1 + 4 \times 10^{-3} \times 428.5\right)^2}$$

$$C_{A2} = 0.136 \frac{\text{kmol}}{\text{m}^3}$$

Since the reaction is endothermic $(+\Delta H_R/a)$, then the heat balance for the first stage is:

$$-u\rho C_p(T_1 - T_O) + u\left(\frac{+\Delta H_R}{a}\right)(C_{A1} - C_{AO}) + UA_1(T_h - T_1) = 0$$

$$-1 \times 10^{-3} \times 10^3 \times 4 \times 10^3 (50 - 75) + 10^{-3} \times 1.5 \times 10^8$$
$$\times (0.369 - 1.0) + 1,500 \times A_1(100 - 50) = 0$$

$$100,000 - 94,650 + 75,000 A_1 = 0$$

$$A_1 = \frac{-5,350}{75,000}$$

$$= -0.0713 \text{ m}^2$$

The negative value in the area indicates that the first stage of the CFSTR requires a cooling coil.

The second stage CFSTR gives

$$-u\rho C_p(T_2 - T_1) + u\left(\frac{+\Delta H_R}{a}\right)(C_{A2} - C_{A1}) + UA_2(T_h - T_2) = 0$$

Since $T_1 = T_2$ (i.e., isothermal condition), then

$$u\left(\frac{+\Delta H_R}{a}\right)(C_{A2} - C_{A1}) = -UA_2(T_h - T_2)$$

$$10^{-3} \times 1.5 \times 10^8(0.136 - 0.369) = -1,500 \times A_2(100 - 50)$$

$$A_2 = \frac{-0.3495 \times 10^5}{-75,000} = 0.466 \text{ m}^2$$

$$C_{A3} = \frac{C_{AO}}{(1 + k_1\bar{t})^3}$$

$$= \frac{1}{(1 + 4.0 \times 10^{-3} \times 428.5)^3}$$

$$= 0.05 \frac{\text{kmol}}{\text{m}^3}$$

Heat balance on the third stage CFSTR is

$$-u\rho C_p(T_3 - T_2) + u\left(\frac{+\Delta H_R}{a}\right)(C_{A3} - C_{A2}) + UA_3(T_h - T_3) = 0$$

Since $T_2 = T_3$ (i.e., isothermal condition), then

$$u\left(\frac{+\Delta H_R}{a}\right)(C_{A3} - C_{A2}) = -UA_3(T_h - T_3)$$

$$1 \times 10^{-3} \times 1.5 \times 10^8(0.05 - 0.136) = -1,500 \times A_3(100 - 50)$$

$$A_3 = \frac{-10^{-3} \times 1.5 \times 10^8 \times 0.086}{-75,000} = 0.172 \text{ m}^2$$

The calculations show that the volume of each reactor is 0.429 m^3 and a heating coil is not required in the first CFSTR, whose area is 0.0713 m^2. The areas of the subsequent heating coils are 0.466 m^2 and

0.172 m^2, respectively. An Excel spreadsheet program (Example6-1.xls) was developed for this example.

Example 6-2

A second order, liquid phase reaction, $A + B \rightarrow C$, is to be performed in a single-stage CFSTR. Given the following data:

1. Determine the size of the reactor required to achieve 85% conversion of A.
2. Estimate the heat-transfer area needed to maintain the reactor temperature at 27°C.
3. Calculate what the feed temperature must be if the reactor is to be operated adiabatically at 27°C.

Volumetric flowrate	$0.002 \text{ m}^3/\text{sec}$
Feed compositions:	
A	1.0 kmol/m^3
B	1.0 kmol/m^3
Rate constant	$0.01 \text{ m}^3/\text{kmol} \cdot \text{sec}$
Feed temperature	27°C, except for Part 3
Heat of reaction	$-6 \times 10^4 \text{ kJ/kmol of A}$
Mixture density (constant)	800 kg/m^3
Mixture specific heat (constant)	$3.5 \text{ kJ/kg} \cdot \text{K}$
Overall heat-transfer coefficient	$2.0 \text{ kJ/m}^2\text{sec} \cdot \text{K}$
Coolant temperature	10°C
Fractional conversion of A, X_A	0.85

Solution

Assuming the following:

• Perfect mixing
• Steady-state operation (i.e., accumulation in the reactor is zero)
• Constant fluid mixture density
• Stirrer input energy is neglected W_s
• No work done by the system on the surroundings

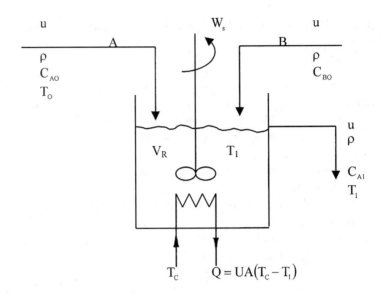

The rate of reaction is $(-r_A) = kC_AC_B$. The material balance on component A is $uC_{AO} = uC_A + (-r_A)V_R$ or $uC_{AO} = uC_A + kC_AC_BV_R$. The fractional conversion is $X_A = 0.85 = (C_{AO} - C_A)/C_{AO}$ or

$$C_A = C_{AO}(1-X_A) \qquad\qquad C_{AO} - C_A = C_{AO}X_A$$

$$= 1.0(1-0.85) \qquad\qquad\quad = 1.0 \times 0.85$$

$$= 0.15 \qquad\qquad\qquad\qquad = 0.85$$

From stoichiometry, $C_{AO} - C_A = C_{BO} - C_B$, or

$$C_B = C_{BO} - (C_{AO} - C_A)$$

$$= C_{BO} - C_{AO}X_A$$

The material balance on component A in terms of the original concentrations of A and B is $uC_{AO} = uC_{AO}(1 - X_A) + kC_{AO}(1 - X_A)(C_{BO} - C_{AO}X_A)V_R$. That is,

$$0.002 \times 1.0 = 0.002 \times 1(1-0.85) + 0.01 \times 1$$

$$\times (1.0 - 0.85)(1.0 - 1.0 \times 0.85)V_R$$

$0.002 = 0.0003 + 0.000225V_R$

$V_R = 7.56 \ m^3$

The heat balance is $-G\Delta h + Q = 0$ or $-\rho u\Delta h + UA(T_C + T_1) = 0$, where

$$\Delta h = C_p\Delta T + \left(\frac{-\Delta H_R}{a}\right)\left(\frac{\Delta \overline{m}_A}{M_A}\right)$$

$$-\rho u\left\{C_p\Delta T + \left(\frac{-\Delta H_R}{a}\right)\left(\frac{\Delta \overline{m}_A}{M_A}\right)\right\} + UA(T_C - T_1) = 0$$

$$-\rho u C_p(T_1 - T_O) - u\left(\frac{-\Delta H_R}{a}\right)(C_A - C_{AO}) + UA(T_C - T_1) = 0$$

Rearranging the above equation gives

$$\rho u C_p(T_O - T_1) + u\left(\frac{-\Delta H_R}{a}\right)(C_{AO} - C_A) = UA(T_1 - T_C)$$

where $T_O = T_1 = 27°C$ (isothermal condition). Therefore, $\rho u C_P (T_O - T_1) = 0$, which gives

$$UA(T_1 - T_C) = u\left(\frac{-\Delta H_R}{a}\right)(C_{AO} - C_A)$$

The heat transfer area A is

$$A = \frac{u\left(\frac{-\Delta H_R}{a}\right)(C_{AO} - C_A)}{U(T_1 - T_C)}$$

$$= \frac{0.002(6\times10^4)(1.0 - 0.15)}{2.0(27 - 10)}\left\{\frac{m^3}{sec}\bullet\frac{kJ}{kmol}\bullet\frac{kmol}{m^3}\bullet\frac{m^2 \ °C\bullet sec}{kJ}\bullet\frac{1}{°C}\right\}$$

$$= 3\,m^2$$

At the adiabatic condition, $Q = 0$, as no heat is added to or removed from the system. The heat balance becomes

$$-\rho u C_p (T_1 - T_O) - u\left(\frac{-\Delta H_R}{a}\right)(C_A - C_{AO}) = 0$$

or $\rho u C_p (T_O - T_1) = u\left(\frac{-\Delta H_R}{a}\right)(C_A - C_{AO})$

or $T_O = T_1 + \dfrac{1}{\rho C_p}\left(\dfrac{-\Delta H_R}{a}\right)(C_A - C_{AO})$

$$= 27 + \frac{1}{(800)(3.5)}\left(6 \times 10^4\right)(0.15 - 1.0)\left\{\frac{m^3}{kg} \bullet \frac{kJ}{kmol} \bullet \frac{kmol}{m^3} \bullet \frac{kg \bullet {}^\circ C}{kJ}\right\}$$

$$= 27 - 18.2$$

$T_O = 8.8°C$

An Excel spreadsheet program (Example6-2.xls) was developed for this example.

Example 6-3

The unsteady state equations for a CFSTR involving a second order reaction with constant heat input are:

Mass balance: $V_R \dfrac{dC_A}{dt} = u(C_{AO} - C_A) - k_O V_R C_A^2 e^{\frac{-E}{RT}}$

Heat balance: $V_R \rho C_P \dfrac{dT}{dt} = u C_p (T_O - T) + \left(\dfrac{-\Delta H_R}{a}\right) k_O V_R C_A^2\ e^{\frac{-E}{RT}} + Q_S$

where $Q_S = m_S \Delta H_V$.

Calculate the outlet concentration and temperature of the reactor with the following data.

Reactor feed rate, u	0.1 ft³/sec
Reactor inlet concentration, C_{AO}	1 lb-mol/ft³
Reactor volume, V_R	100 ft³
Arrhenius pre-exponential factor, k_O	1.907×10^{13} ft/lb-mol • sec
Activation energy, E	44,700 Btu/lb-mol
Gas constant, R	1.987 Btu/lb-mol °R
Specific heat capacity, C_P	0.75 Btu/lb °R
Reactor inlet temperature, T_O	600°R
Heat of reaction, ΔH_R	10,000 Btu/lb-mol
Steam flowrate, m_S	0.711 lb/sec.
Heat of vaporization of steam, ΔH_V	960 Btu/lb
Solution density, ρ	80.0 lb/ft³

Solution

The mass and energy balance equations are expressed respectively as

$$\text{Mass balance: } \frac{dC_A}{dt} = \frac{u}{V_R}\left(C_{AO} - C_A\right) - k_O C_A^2\, e^{\frac{-E}{RT}}$$

$$\text{Energy balance: } \frac{dT}{dt} = \frac{u}{V_R}\left(T_O - T\right) + \frac{\left(\dfrac{-\Delta H_R}{a}\right) k_O C_A^2 e^{\frac{-E}{RT}}}{\rho C_P} + \frac{Q_S}{V_R \rho C_P}$$

where $Q_S = m_S \Delta H_V$.

The two equations are coupled with the temperature and concentration in both mass and heat balances. A computer program CFSTR61 using the Runge-Kutta fourth order method with a step increment for the time $\Delta t = 10.0$ sec was employed to determine both the outlet concentration and temperature for a period of 2,000 sec. Table 6-1 gives typical results from the program and Figure 6-3 shows both concentration and temperature profiles with time. The profiles show that at constant heat input, there is a gradual decrease in the outlet concentration until about 100 sec, which is then followed by a steep decrease until the effluent concentration reaches a steady state at about 2,000 sec. There is a corresponding gradual rise in temperature until about 100 sec, which is followed by a steep increase in the outlet temperature until it reaches a steady value at about 2,000 sec.

Table 6-1
The reactor outlet concentration and temperature
of a CFSTR with constant heat input

Time sec	Outlet concentration lb-mols/ft^3	Outlet temperature °R
0.0	0.5230	609.2
10.0	0.5229	609.4
20.0	0.5228	609.7
30.0	0.5227	609.9
40.0	0.5224	610.1
50.0	0.5221	610.3
100.0	0.5200	611.1
200.0	0.5138	612.2
400.0	0.4998	613.6
600.0	0.4870	614.5
800.0	0.4765	615.3
1,000.0	0.4679	615.9
1,200.0	0.4609	616.4
1,400.0	0.4552	616.8
1,600.0	0.4505	617.1
1,800.0	0.4467	617.4
2,000.0	0.4436	617.6

Example 6-4

A solution containing 0.25 kmol/m^3 of acetic anhydride is to be hydrolyzed in a single CFSTR to give an effluent containing an anhydride concentration of 0.05 kmol/m^3. The volumetric flowrate is 0.05 m^3/min and the working volume of the reactor is 0.75 m^3.

(1) Show that the reactor temperature must be maintained at 34°C given that the reaction is first order with respect to anhydride concentration and the rate constant k can be determined from the relationship

$$k = 0.158 \exp\left(18.55 - \frac{5,529}{T}\right) \text{min}^{-1}$$

where T is the temperature (K).

(2) Given the following design data, determine (a) under what conditions adiabatic operation is feasible, and (b) what cooling area is required if the feed temperature is 30°C?

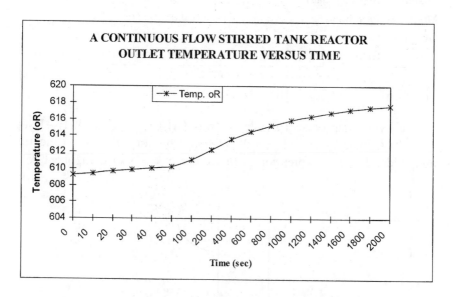

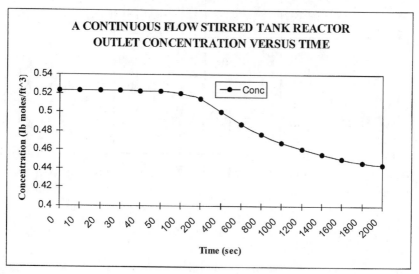

Figure 6-3. Profiles of outlet temperature and concentration with time in a CFSTR.

Heat of reaction	-5×10^4 kcal/kmol of anhydride
Mixture density (constant)	950 kg/m^3
Mixture specific heat	0.95 kcal/kg • °C
Overall heat-transfer coefficient	8 kcal/m^2 min • °C
Cooling water is available at either	15°C or 25°C

Solution

The hydrolysis of acetic anhydride is $CH_3CO • O • COCH_3 + H_2O$ $\rightarrow 2CH_3COOH$ or $A + B \rightarrow 2C$, where component A is the acetic anhydride. The rate expression is first order in terms of component A.

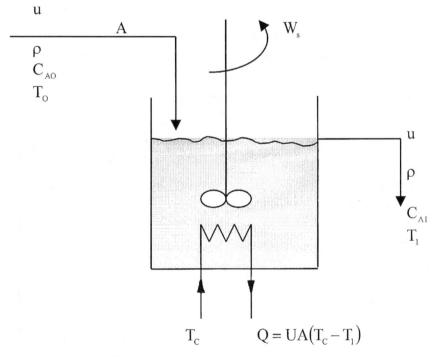

Assume the following:

• Perfect mixing
• Steady-state operation (i.e., accumulation in the reactor is zero)
• Constant fluid mixture density
• Stirrer input energy is neglected W_s
• No work done by the system on the surroundings

The rate expression is $(-r_A) = kC_A$, where $k = 0.158 \exp(18.55 - 5,529/T)$. The material balance on component A gives $uC_A = uC_A + (-r_A)V_R$. That is,

$$uC_{AO} = uC_A + 0.158 e^{\left(18.55 - \frac{5,529}{T}\right)} C_A V_R$$

where $C_{AO} = 0.25$ kmol/m³, $C_A = 0.05$ kmol/m³, $V_R = 0.75$ m³, and $u = 0.05$ m³/min. This yields

$$0.05 \times 0.25 = 0.05 \times 0.05 + 0.158 e^{\left(18.55 - \frac{5,529}{T}\right)} (0.05)(0.75)$$

$$0.0125 = 0.0025 + 0.005925 e^{\left(18.55 - \frac{5,529}{T}\right)}$$

Further rearrangement gives the temperature $T = 306.7$ K, which is

$$T = 306.7 - 273.15$$

$$= 33.55°C \ (\approx 34°C)$$

The operation might be feasible under adiabatic conditions when no heat is added or removed from the reactor (i.e., $Q = 0$), and the heat given up by the reaction is entirely used within the system to change its enthalpy. The inlet temperature of feed reactant is determined as follows.

Consider the energy balance:

$$-G\Delta h + Q = 0 \tag{6-31}$$

where $G = \rho u$ and

$$\Delta h = C_p \Delta T + \left(\frac{-\Delta H_R}{a}\right)\left(\frac{\Delta \overline{m}_A}{M_A}\right)$$

This gives

$$-\rho u \left\{ C_p \Delta T + \left(\frac{-\Delta H_R}{a}\right)\left(\frac{\Delta \overline{m}_A}{M_A}\right) \right\} + Q = 0$$

$$-\rho u C_p (T_1 - T_0) - u\left(\frac{-\Delta H_R}{a}\right)(C_A - C_{AO}) = 0$$

Rearranging the above equation gives

$$\rho u C_p (T_0 - T_1) = u\left(\frac{-\Delta H_R}{a}\right)(C_A - C_{AO})$$

$$T_0 - T_1 = \frac{1}{\rho C_p}\left(\frac{-\Delta H_R}{a}\right)(C_A - C_{AO})$$

$$T_0 - 34.0 = \frac{1}{(950)(0.95)}(5 \times 10^4)(0.05 - 0.25)$$

$$\times \left\{\frac{kcal}{kmol\,A} \cdot \frac{m^3}{kg} \cdot \frac{kg \cdot {}^\circ C}{kcal} \cdot \frac{kmol}{m^3}\right\}$$

$T_0 - 34.0 = -11.08$

$T_0 = 22.92°C$

From the heat balance

$$-\rho u C_p (T_1 - T_0) - u\left(\frac{-\Delta H_R}{a}\right)(C_A - C_{AO}) + UA(T_C - T_1) = 0$$

Rearranging to determine the area of the heating coil, gives

$$A = \frac{\rho u C_p (T_0 - T_1) + u\left(\frac{-\Delta H_R}{a}\right)(C_{AO} - C_A)}{U(T_1 - T_C)}$$

where $T_0 = 30°C$, $T_C = 15°C$, and knowing the other parameters

$$Area = \frac{950 \times 0.05 \times 0.95(30.0 - 34.0) + 0.05(5 \times 10^4)(0.25 - 0.05)}{8(34.0 - 15.0)}$$

Area = 2.10 m²

Therefore, the cooling coil area for the first stage CFSTR is 2.1 m².
An Excel spreadsheet program (Example6-4.xls) was developed for this example.

BATCH REACTOR

Consider a batch reactor as shown in Figure 6-4 with a heat exchange system. There is no flow into or out of the reactor. It can be assumed that the total mass of the mixture, m, is constant and allowing for a change in volume, V_R, from Equation 6-12 gives (no flow implies that $G_O = G_F = 0$)

$$-\left(G_F h_F - G_O h_O\right) + Q - W_S = \frac{d}{dt}(m\bar{u}) \tag{6-12}$$

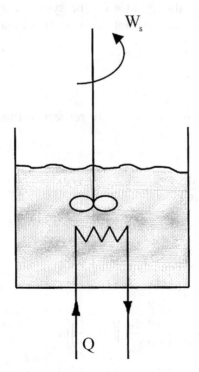

Figure 6-4. Batch reactor with heating coil.

where $Q = UA(T_h - T)$

Q = heat transfer rate, J/s

U = overall heat transfer coefficient, $J/m^2 \cdot s \cdot K$, $W/m^2 \cdot K$

A = area of the heating/cooling coil, m^2

T_h = temperature of the heating coil, K (assumed constant)

T = temperature of the reacting system, K

$$Q - W_s = m \frac{d\overline{u}}{dt} \tag{6-44}$$

The work done by the system is represented by

$$W_s = P \frac{dV_R}{dt}$$

where P = total pressure

dV_R/dt = rate of volume expansion

The work done by the agitator on the system may also have to be considered. However, at constant pressure (e.g., open system)

$$\overline{u} = h - \frac{P}{\rho} \tag{6-11}$$

Differentiating Equation 6-11 with respect to time and multiplying by m gives

$$m \frac{d\overline{u}}{dt} = m \frac{dh}{dt} - m \frac{d}{dt}\left(\frac{P}{\rho}\right)$$

$$= m \frac{dh}{dt} - P \frac{dV_R}{dt} \tag{6-45}$$

From Equation 6-21

$$m \frac{dh}{dt} = m C_p \frac{dT}{dt} + \left(\frac{-\Delta H_R}{aM_A}\right)\left(\frac{d\overline{m}_A}{dt}\right)m \tag{6-46}$$

Using these relationships in Equation 6-44 gives

$$Q - P\frac{dV_R}{dt} = m\frac{d\overline{u}}{dt}$$

$$= m\frac{dh}{dt} - P\frac{dV_R}{dt}$$

$$Q = m\frac{dh}{dt}$$

$$Q = mC_p\frac{dT}{dt} + m\left(\frac{-\Delta H_R}{aM_A}\right)\left(\frac{d\overline{m}_A}{dt}\right) \tag{6-47}$$

At constant volume (e.g., autoclave), using Equations 6-22 and 6-44,

$$Q = mC_v\frac{dT}{dt} + m\left(\frac{-\Delta \overline{U}_R}{aM_A}\right)\left(\frac{d\overline{m}_A}{dt}\right) \tag{6-48}$$

METHODS FOR SOLVING DESIGN EQUATIONS

From the mass balance equation for a batch reactor

$$(-r_A)V_R = \frac{-dN_A}{dt} \tag{5-6}$$

and at constant volume and density

$$\frac{N_A}{V_R} \equiv \frac{\overline{m}_A}{M_A} \cdot \rho \tag{6-49}$$

Substituting Equation 6-49 into Equation 5-6 gives

$$(-r_A)V_R = \frac{-\rho V_R}{M_A} \cdot \frac{d\overline{m}_A}{dt} \tag{6-50}$$

These relationships enable the combination of the mass and the energy balances. For constant volume and density, with $m = \rho V_R$,

$$\frac{-d\overline{m}_A}{dt} = \frac{-M_A}{m} \cdot \frac{dN_A}{dt} = \frac{M_A}{m}(-r_A)V_R \tag{6-51}$$

The mass and energy balances usually must be solved simultaneously by using numerical methods. Equation 6-47 becomes

$$Q = mC_p \frac{dT}{dt} + \frac{dN_A}{dt}\left(\frac{-\Delta H_R}{a}\right) \qquad (6\text{-}52)$$

$$\text{or } Q = mC_p \frac{dT}{dt} - (-r_A)V_R\left(\frac{-\Delta H_R}{a}\right) \qquad (6\text{-}53)$$

Rearranging Equation 6-53 and using $m = \rho V_R$ yields

$$\frac{dT}{dt} = \frac{(-r_A)}{\rho C_p}\left(\frac{-\Delta H_R}{a}\right) + \frac{UA}{\rho V_R C_p}(T_h - T) \qquad (6\text{-}54)$$

For adiabatic conditions, Equation 6-54 becomes

$$\frac{dT}{dt} = \frac{(-r_A)}{\rho C_p}\left(\frac{-\Delta H_R}{a}\right) \qquad (6\text{-}55)$$

In terms of the fractional conversion X_A, Equation 6-52 becomes

$$Q = mC_p \frac{dT}{dt} - N_{AO}\left(\frac{-\Delta H_R}{a}\right)\frac{dX_A}{dt} \qquad (6\text{-}56)$$

For adiabatic condition (i.e., $Q = 0$), Equation 6-56 gives

$$mC_p \frac{dT}{dt} = N_{AO}\left(\frac{-\Delta H_R}{a}\right)\frac{dX_A}{dt} \qquad (6\text{-}57)$$

Since the relationship between dT/dt and dX_A/dt is implicit with respect to time t, Equation 6-57 is expressed as

$$mC_p dT = N_{AO}\left(\frac{-\Delta H_R}{a}\right)dX_A \qquad (6\text{-}58)$$

$$\text{or } T - T_O = \frac{N_{AO}\left(\dfrac{-\Delta H_R}{a}\right)(X_A - X_{AO})}{mC_p} \qquad (6\text{-}59)$$

where T_O and X_{AO} = initial temperature and fractional conversion (i.e., initial conditions in the reactor)

N_{AO} = initial moles

Rearranging Equation 6-59 gives

$$T = T_O + \frac{N_{AO}\left(\frac{-\Delta H_R}{a}\right)(X_A - X_{AO})}{mC_p} \tag{6-60}$$

In an adiabatic operation, $Q = 0$, and as such there is no attempt to heat or cool the contents in the reactor. The temperature T in the reactor rises in an exothermic reaction and falls in an endothermic reaction. It is essential to control T so that it is neither too high nor too low. To assess the design of both the reactor and the heat exchanger required to control T, the material and energy balance equations must be used together with information on rate of reaction and rate of heat transfer because there is an interaction between T and X_A.

The time required to achieve fractional conversion X_A from the mass balance can be determined as

$$t = N_{AO} \int_{X_{AO}}^{X_{AF}} \frac{dX_A}{(-r_A)V_R} \tag{6-61}$$

Example 6-5

Consider the reaction $A \rightarrow B$, $(-r_A) = kC_A$, $k_{350} = 0.65$ min^{-1}, $(-\Delta H_R/a) = 30$(kcal/mol) in a 10-l batch reactor with initial concentration of $A = 2$ mol/l and temperature $T_o = 350$ K. Determine the heat transfer rate required to maintain the temperature at 350 K at 95% conversion.

Solution

For the batch reactor with a first order reaction, the rate expression is

$$(-r_A) = kC_A = -\frac{dC_A}{dt}$$

The concentration of A after integration and mathematical manipulation is $C_A = C_{AO}e^{-kt}$. The rate expression now becomes $(-r_A) = kC_{AO}e^{-kt}$, where k is constant.

The rate of heat transfer Q can be determined from Equation 6-53 as

$$Q = mC_p \frac{dT}{dt} - (-r_A)V_R\left(\frac{-\Delta H_R}{a}\right)$$ (6-53)

Under isothermal conditions, $dT/dt = 0$, and Equation 6-53 becomes

$$Q = -(-r_A)V_R\left(\frac{-\Delta H_R}{a}\right)$$ (6-62)

Substituting the rate expression into Equation 6-62 gives

$$Q = -kC_{AO}\exp(-kt)V_R\left(\frac{-\Delta H_R}{a}\right)\left\{\frac{1}{min}\bullet\frac{mol}{l}\bullet l\bullet\frac{kcal}{mol}\right\}$$

$$Q = -0.65 \times 2.0\ e^{-0.65t} \times 10 \times 30$$

$Q = -390e^{-0.65t}(kcal/min)$. Since $Q < 0$, it shows that heat is removed from the system, which is undergoing exothermic reaction.

The time corresponding to a 95% conversion for the first order reaction is determined by

$$t = N_{AO}\int_{X_{AO}}^{X_{AF}}\frac{dX_A}{(-r_A)V_R}$$ (6-61)

$$\text{or}\ \ t = \frac{1}{k}\int_{X_{AO=0.0}}^{X_{AF}=0.95}\frac{dX_A}{(1-X_A)}$$

$$= -\frac{1}{k}\ln[1-X_A]_{X_{AO}=0.0}^{X_{AF}=0.95}$$

$$t = -\frac{1}{0.65}[\ln(1-0.95) - \ln(1-0)]$$

$$t = 4.6\ min$$

The heat transfer rate $(Q_{t=95\%})$ at 95% conversion is

$$Q = -390e^{-0.65 \times 4.608}$$

$$= -19.51 \text{ kca/min}$$

An Excel spreadsheet program (Example6-5.xls) was developed for this exercise.

Example 6-6

Consider a non-isothermal batch reactor that is operated adiabatically. The reactor contains a liquid reaction mixture in which the reaction $A \xrightarrow{\ k\ } $ Products occurs, where $(-r_A) = kC_A$ and $k = k_O e^{-E/RT}$,

where C_A = concentration of A
   E = activation energy of the reaction
   R = gas constant
   T = absolute temperature
   k_O = pre-exponential factor

Determine the concentration and temperature after 180 sec if the initial concentration and temperature of A are $C_A = 1.0(\text{gmol/l})$, $T = 300$ K, and $E/R = 300$ K.

Solution

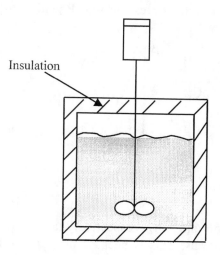

Adiabatic batch reactor.

Assume perfect mixing in the reactor, and because it is operated adiabatically (i.e., no exchange of heat between the reactor and its environment), $Q = 0$ and equation 6-53 becomes

$$mC_p \frac{dT}{dt} = (-r_A) V_R \left(\frac{-\Delta H_R}{a} \right) \tag{6-63}$$

$$\text{or } \frac{dT}{dt} = \frac{(-r_A) V_R \left(\dfrac{-\Delta H_R}{a} \right)}{mC_p} \tag{6-64}$$

Assuming the density of the mixture is constant, then $\rho = m/V_R$, and with C_p taken as the average heat capacity of the reacting mixture, Equation 6-64 becomes

$$\frac{dT}{dt} = \frac{(-r_A) \left(\dfrac{-\Delta H_R}{a} \right)}{\rho C_p} \tag{6-65}$$

$(-\Delta H_R/a)$ is the heat of reaction, which is a function of temperature, but it is assumed constant. The rate expression for a first order reaction is $(-r_A) = k_0 C_A e^{-E/RT}$.

Substituting the rate expression into Equation 6-65 gives

$$\frac{dT}{dt} = \frac{k_0 C_A e^{-E/RT} \left(\dfrac{-\Delta H_R}{a} \right)}{\rho C_p} \tag{6-66}$$

The unsteady state mass balance is $(-r_A) = -(dC_A/dt) = kC_A$

$$\text{or } \frac{dC_A}{dt} = -k_0 C_A e^{-E/RT} \tag{6-67}$$

Equations 6-66 and 6-67, respectively, are two coupled first order differential equations. This is because dC_A/dt is a function of T and C_A while dT/dt is also a function of C_A and T. The Runge-Kutta fourth order numerical method can be applied with initial boundary conditions to simulate both concentration and temperature with time.

In Equation 6-67, $K_1 = -k_O = -0.1$ sec^{-1}, and in Equation 6-66,

$$K_2 = \frac{k_O\left(\dfrac{-\Delta H_R}{a}\right)}{\rho C_p} = 1.0 \quad \left(\frac{1-K}{gmol - sec}\right)$$

By expressing Equations 6-66 and 6-67 in the form

$$\frac{dC_A}{dt} = K_1 C_A e^{-E/RT} \tag{6-68}$$

and

$$\frac{dT}{dt} = K_2 C_A e^{-E/RT} \tag{6-69}$$

respectively, a computer program BATCH61 was developed to simulate the concentration and temperature with time of the adiabatic batch system. Table 6-2 gives the results of the program and Figure 6-5 shows both concentration and temperature profiles with time A in the system.

Example 6-7

Acetylated castor oil is hydrolyzed for the manufacture of drying oils in kettles operated batchwise. Grummit and Fleming (1945)[*] correlated the decomposition data on the basis of a first order reaction as:

Acetylated castor oil (1) $\rightarrow$ CH$_3$COOH (g) + drying oil (1)

$r = kC$

where r = the rate of decomposition (kg of acetic acid produced per m^3/min)

C = concentration of acetic acid, kg/m^3

The batch reactor initially contains 227 kg of acetylated castor and the initial temperature is 613 K. Complete hydrolysis yields 0.156 kg acetic acid per kg of ester. For this reaction, the specific reaction rate constant k is

---

[*]*Ind. Eng. Chem., 37, 485 (1945).*

Table 6-2
Simulation of a non-isothermal batch reactor

Time (sec)	Conc. (gmoles/L)	Temp. K
.000	1.0000	300.0
5.000	.8315	301.7
10.000	.6908	303.1
15.000	.5735	304.3
20.000	.4757	305.2
25.000	.3944	306.1
30.000	.3269	306.7
35.000	.2708	307.3
40.000	.2243	307.8
45.000	.1857	308.1
50.000	.1537	308.5
55.000	.1272	308.7
60.000	.1053	308.9
65.000	.0871	309.1
70.000	.0721	309.3
75.000	.0596	309.4
80.000	.0493	309.5
85.000	.0408	309.6
90.000	.0338	309.7
95.000	.0279	309.7
100.000	.0231	309.8
105.000	.0191	309.8
110.000	.0158	309.8
115.000	.0131	309.9
120.000	.0108	309.9
125.000	.0089	309.9
130.000	.0074	309.9
135.000	.0061	309.9
140.000	.0051	309.9
145.000	.0042	310.0
150.000	.0035	310.0
155.000	.0029	310.0
160.000	.0024	310.0
165.000	.0020	310.0
170.000	.0016	310.0
175.000	.0013	310.0
180.000	.0011	310.0

$$k = \exp\left(35.2 - \frac{22,450}{T}\right)\frac{1}{min} \qquad (6\text{-}70)$$

where T is the temperature, K. The specific heat of the liquid reaction mixture is constant at 0.6 cal/gm-K, molecular weight of acetic acid (M_A) is 60 g/gmol, and the endothermic heat of reaction is $(+\Delta H_R/a)$ = 15,000 cal/gmol of acid produced.

1. Find the relationship between T and the fraction converted for several values of heat input rate, Q = kcal/min.
2. If the batch operation is adiabatic, find the relationship between the conversion (fraction of the acetylated oil that is decomposed) and temperature versus time.

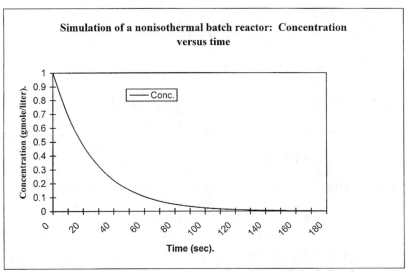

Figure 6-5a: Concentration versus time in an adiabatic batch reactor.

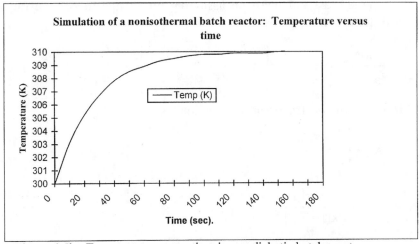

Figure 6-5b: Temperature versus time in an adiabatic batch reactor.

Figure 6-5. Simulation of a non-isothermal batch reactor.

Solution

(1) Rearranging Equation 6-56 yields

$$mC_p \frac{dT}{dt} = Q + N_{AO}\left(\frac{-\Delta H_R}{a}\right)\frac{dX_A}{dt} \tag{6-71}$$

Moles of acetic acid N_{AO} is

$$N_{AO} = \frac{227,000}{60} \times 0.156\left\{g \cdot \frac{gmol}{g} \cdot \frac{kg\ acetic\ acid}{kg\ ester}\right\}$$

$$= 590.2\ gmol\ of\ acetic\ acid$$

Rearranging Equation 6-71, the energy balance equation becomes

$$\frac{dT}{dX_A} = \frac{Q\dfrac{dt}{dX_A} - N_{AO}\left(\dfrac{+\Delta H_R}{a}\right)}{mC_p} \tag{6-72}$$

From the mass balance, and assuming no change in volume for a first order reaction, gives

$$-\frac{1}{V_R}\frac{dN_A}{dt} = (-r_A) = kC_A$$

In terms of the fractional conversion, X_A

$$\frac{dX_A}{dt} = k(1 - X_A) \tag{6-73}$$

or $$\frac{dt}{dX_A} = \frac{1}{k(1 - X_A)} \tag{6-74}$$

Substituting Equation 6-74 into Equation 6-72 gives

$$\frac{dT}{dX_A} = \frac{\dfrac{Q}{k(1 - X_A)} - N_{AO}\left(\dfrac{+\Delta H_R}{a}\right)}{mC_p} \tag{6-75}$$

where m = 227 kg

$\qquad C_p$ = 0.6 cal/g • K

$\qquad N_{AO}$ = 590.2 gmol

$(+\Delta H_R/a)$ = 15.0 kcal/gmol

Equation 6-75 becomes

$$\frac{dT}{dX_A} = \frac{\dfrac{Q}{k(1-X_A)} - (590.2)(15)}{(227)(0.6)}$$

or $\quad \dfrac{dT}{dX_A} = \dfrac{\dfrac{Q}{k(1-X_A)} - 8,853}{136.2}$ $\qquad\qquad$ (6-76)

Equation 6-76 is a first order differential equation. Using the Runge-Kutta fourth order method, it is possible to simulate the temperature at different conversion for varying value of the heat transfer rate Q (kcal/min). A simulation exercise was performed for Q = 0, 500, 1,000, 1,250 and 1,500 kcal/min, respectively, using a developed computer program BATCH62. Table 6-3 gives the computer results and Figure 6-6 shows profiles of the simultaneous solutions of Equations 6-70 and 6-76 for several values of Q. It is seen that the temperature drops drastically for the endothermic reaction with adiabatic operation (i.e., Q = 0), implying that heating should be considered.

(2) The mass balance expression in terms of the fractional conversion assuming no change in volume is

$$\frac{dt}{dX_A} = \frac{1}{k(1-X_A)}$$ $\qquad\qquad$ (6-74)

where the specific reaction rate constant k depends on the temperature and is expressed by

$$k = \exp\left(35.2 - \frac{22,450}{T}\right)\frac{1}{min}$$ $\qquad\qquad$ (6-70)

The reactor temperature T is determined by Equation 6-60 as

Table 6-3
Simulation of a batch reactor for the hydrolysis
of acetylated castor oil

HEAT INPUT Q=   0.   kcal/min

CONVERSION	TEMPERATURE (K)
.00	613.0000
.10	606.5000
.20	600.0000
.30	593.5000
.40	587.0000
.50	580.5000
.60	574.0000
.70	567.5000

HEAT INPUT Q=   500.   kcal/min

CONVERSION	TEMPERATURE (K)
.00	613.0000
.10	608.2667
.20	604.0816
.30	600.5181
.40	597.6075
.50	595.3293
.60	593.6158
.70	592.3709

HEAT INPUT Q=   1000.   kcal/min

CONVERSION	TEMPERATURE (K)
.00	613.0000
.10	609.8633
.20	607.3491
.30	605.4097
.40	603.9646
.50	602.9185
.60	602.1785
.70	601.6639

HEAT INPUT Q=   1250.   kcal/min

CONVERSION	TEMPERATURE (K)
.00	613.0000
.10	610.6076
.20	608.7696
.30	607.4031
.40	606.4144
.50	605.7142
.60	605.2263
.70	604.8903

HEAT INPUT Q=   1500.   kcal/min

CONVERSION	TEMPERATURE (K)
.00	613.0000
.10	611.3203
.20	610.0789
.30	609.1842
.40	608.5518
.50	608.1111
.60	607.8073
.70	607.5994

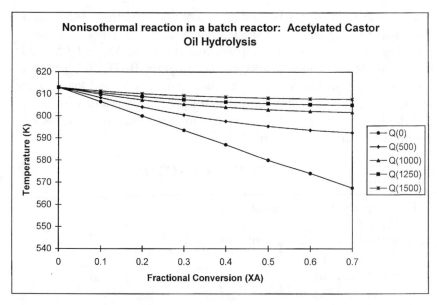

Figure 6-6. Temperature versus fractional conversion profiles for various rates of heat input in a batch reactor.

$$T = T_O + \frac{N_{AO}\left(\dfrac{-\Delta H_R}{a}\right)(X_A - X_{AO})}{mC_p} \qquad (6\text{-}60)$$

$$\text{or} \quad T = T_O - \frac{N_{AO}\left(\dfrac{+\Delta H_R}{a}\right)(X_A - X_{AO})}{mC_p} \qquad (6\text{-}77)$$

where $(+\Delta H_R/a)$ = endothermic enthalpy of reaction
$$T_O = 613 \text{ K}$$
$$X_{AO} = 0$$

Substituting numbers in Equation 6-77 gives

$$T = 613 - \frac{(590)(15.0)}{(227)(0.6)} X_A$$

$$T = 613 - 65X_A \qquad (6\text{-}78)$$

Equation 6-74 is a first order differential equation; substituting Equations 6-70 and 6-78 for the temperature, it is possible to simulate the temperature and time for various conversions at $\Delta X_A = 0.05$. Table 6-4 gives the computer results of the program BATCH63, and Figure 6-7 shows profiles of both fractional conversion and temperature against time. The results show that for the endothermic reaction of $(+\Delta H_R/a) = 15.0$ kcal/gmol, the reactor temperature decreases as conversion increases with time.

PLUG FLOW REACTOR

For Figure 6-8 the notations are:

C_A' = moles per unit mass of species A
G = mass flowrate
ρ = fluid density
C_P = fluid heat capacity
T = fluid temperature
ΔH_R = enthalpy of reaction
d_t = tube diameter
T_C = temperature of coolant fluid is constant
U = overall heat transfer coefficient
S = cross-sectional area of the tubular reactor
  = ($\pi d_t^2/4$ for a cylindrical tube)

Table 6-4
Simulation of a non-isothermal reaction in a batch reactor involving the hydrolysis of acetylated castor oil

CONVERSION	TIME (min)	TEMPERATURE (K)
.0000	.00	613.0
.0500	.24	609.8
.1000	.54	606.5
.1500	.93	603.2
.2000	1.43	600.0
.2500	2.09	596.8
.3000	2.95	593.5
.3500	4.09	590.2
.4000	5.61	587.0
.4500	7.65	583.8
.5000	10.42	580.5

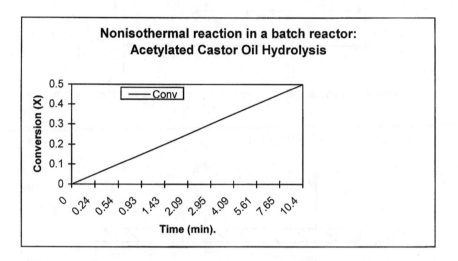

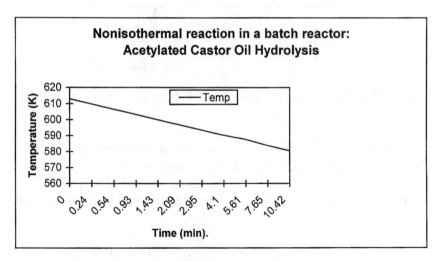

Figure 6-7. Conversion and temperature versus time.

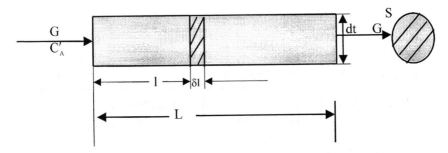

Figure 6-8. Tubular (plug) flow reactor with heat transfer system.

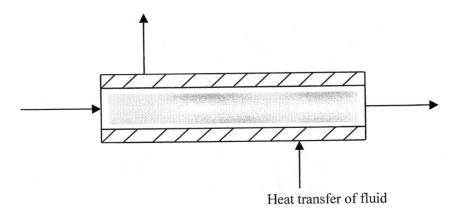

Heat transfer of fluid

To derive the enthalpy balance for a plug flow reactor, assume the following:

- For plug flow, C_A' and T are constant at any cross-section.
- The coolant fluid temperature T_C is constant.
- Steady-state condition (i.e., no accumulation in the system).
- Pressure drop over the reactor length is assumed to be very small (at constant density).

Consider the elemental volume $S\delta l$ at length 1 of the plug flow. Applying the general energy balance in differential form gives

$$-G\delta h + \delta Q = 0 \tag{6-31}$$

From Equation 6-21

$$\delta h = C_p \delta T + \left(\frac{-\Delta H_R}{a} \right) \frac{\delta \overline{m}_A}{M_A} \tag{6-79}$$

and $\delta Q = U \cdot \pi \cdot d_t \cdot \delta l(T_C - T)$ (6-80)

Substituting Equations 6-79 and 6-80 into Equation 6-31 yields

$$-GC_p \delta T - G\left(\frac{-\Delta H_R}{a} \right) \frac{\delta \overline{m}_A}{M_A} + U \cdot \pi \cdot d_t \cdot \delta l(T_C - T) = 0 \tag{6-81}$$

Dividing Equation 6-81 by δl, as $\delta l \rightarrow 0$ gives

$$-GC_p \frac{dT}{dl} - G\left(\frac{-\Delta H_R}{a} \right) \frac{1}{M_A} \frac{d\overline{m}_A}{dl} + U \cdot \pi \cdot d_t(T_C - T) = 0 \tag{6-82}$$

A mass balance on A gives

$$-G \frac{dC'_A}{dl} = (-r_A)S$$

It is known that $C'_A = \overline{m}_A / M_A \,(\text{mols/unit mass})$
If $G/M_A \cdot d\overline{m}_A / dl$ is repalced by $-(-r_A)S$, then

$$GC_p \frac{dT}{dl} + \left(\frac{-\Delta H_R}{a} \right)(-r_A)S + U \cdot \pi \cdot d_t(T - T_C) = 0 \tag{6-83}$$

FRACTIONAL CONVERSION X_A
FOR TUBULAR (PLUG) FLOW

Beginning with $-G\delta h + \delta Q = 0$, the energy balance equation in terms of the fractional conversion X_A gives

$$-G\left\{ C_p \Delta T + \left(\frac{-\Delta H_R}{a} \right) \frac{\Delta \overline{m}_A}{M_A} \right\} + U \cdot \pi \cdot d_t(T_C - T)\delta l = 0$$

$$-\rho u C_p \Delta T - \rho u \left(\frac{-\Delta H_R}{a} \right) \frac{\Delta \overline{m}_A}{M_A} + U \cdot \pi \cdot d_t(T_C - T)\delta l = 0 \tag{6-84}$$

Equation 6-84 in terms of concentration gives

$$-\rho u C_p dT - u\left(\frac{-\Delta H_R}{a}\right) dC_A + U \bullet \pi \bullet d_t \left(T_C - T\right) \delta l = 0$$

or $\rho u C_p dT + u\left(\dfrac{-\Delta H_R}{a}\right) dC_A = U \bullet d_t A\left(T_C - T\right)$

Fractional conversion $X_A = (C_{AO} - C_A)/C_{AO}$ and $-C_{AO}dX_A = dC_A$. Hence,

$$\rho u C_p dT - u\, C_{AO}\left(\frac{-\Delta H_R}{a}\right) dX_A = U \bullet d_t A\left(T_C - T\right) \qquad (6\text{-}85)$$

ADIABATIC TUBULAR (PLUG) FLOW REACTOR

Adiabatic plug flow reactors operate under the condition that there is no heat input to the reactor (i.e., $Q = 0$). The heat released in the reaction is retained in the reaction mixture so that the temperature rise along the reactor parallels the extent of the conversion. Adiabatic operation is important in heterogeneous tubular reactors.

The equation $-G\Delta h + Q = 0$, where

$$\Delta h = C_p \Delta T + \left(\frac{-\Delta H_R}{a}\right)\left(\frac{\Delta \overline{m}_A}{M_A}\right) \text{ gives}$$

$$-G\left\{C_p \Delta T + \left(\frac{-\Delta H_R}{a}\right)\left(\frac{\Delta \overline{m}_A}{M_A}\right)\right\} = 0 \qquad (6\text{-}86)$$

or

$$-\rho u C_p \Delta T - u\left(\frac{-\Delta H_R}{a}\right)\Delta C_A = 0$$

where $C'_A = \overline{m}_A/M_A$ and $C_A = \rho \bullet C'_A$.

Rearranging and integrating Equation 6-86 between the boundary conditions of the tubular reactor, that is $l = 0$, $C_A = C_{AO}$, $T = T_O$ and $l = L$, $C_A = C_A$, $T = T$, gives

$$-\rho u C_p \left(T - T_O \right) = u \left(\frac{-\Delta H_R}{a} \right) \left(C_A - C_{AO} \right)$$

$$\text{or } T - T_O = \left(\frac{1}{\rho C_p} \right) \left(\frac{-\Delta H_R}{a} \right) \left(C_{AO} - C_A \right) \tag{6-87}$$

In terms of the fractional conversion X_A,

$$T - T_O = \left(\frac{1}{\rho C_p} \right) \left(\frac{-\Delta H_R}{a} \right) C_{AO} X_A$$

or

$$T - T_O = \left(\frac{1}{\rho C_p} \right) \left(\frac{-\Delta H_R}{a} \right) \left(C_{AO} - C_A \right) = \left(\frac{1}{\rho C_p} \right) \left(\frac{-\Delta H_R}{a} \right) C_{AO} X_A \tag{6-88}$$

AUTOTHERMAL REACTORS

If a reaction requires a high temperature before proceeding at a normal rate, the products of the reaction will leave the reactor at a high temperature. The heat recovered can be used and the heat of reaction can be used to preheat the feed. There are several methods of achieving this type of operation in a reactor. An external heat exchanger can be employed to transfer the heat of reaction from the effluent to the feeds. Alternatively, the exchanger can be used as an integral part of the reactor or part of the high temperature effluent can be recycled. These processes involving heat transfer and reaction systems are referred to as autothermal. The essential feature of an autothermal reactor system is the feedback of reaction heat that is employed to raise the temperature and, therefore, the reaction rate of the incoming reactant stream. Its advantage is the self-sufficiency in energy even though high temperature is required for the reaction to commence. Another feature of the autothermal system is that an external heat source is required during the startup period while the system is working toward thermal equilibrium. The operation mode for these configurations is adiabatical or with heat transfer to the surrounding. Figures 6-9, 6-10, and 6-11, respectively, show different arrangements of an autothermal reactor

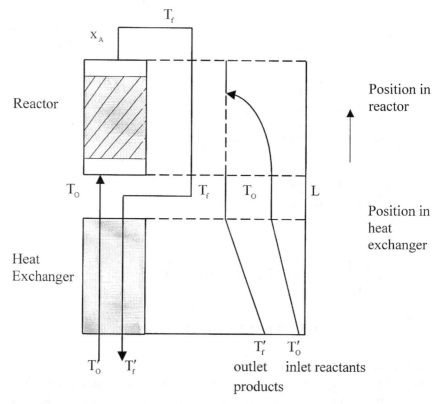

Figure 6-9. An autothermal reactor with an external heat exchanger.

CONVERSION IN AMMONIA SYNTHESIS

In reviewing thermodynamics, the nature of the heat of reaction and whether the reaction is reversible can be investigated. In the synthesis of ammonia, reactors operate at about 200 atm and 350°C and approach the equilibrium conversion of about 70% in each pass.

$$N_2 + 3H_2 \underset{k_2}{\overset{k_1}{\rightleftharpoons}} 2NH_3$$

The ammonia is separated from unreacted H_2 and N_2, and these are recycled back to the reactor. The overall process of a tubular reactor plus separation and recycle produces essentially 100% NH_3 conversion.

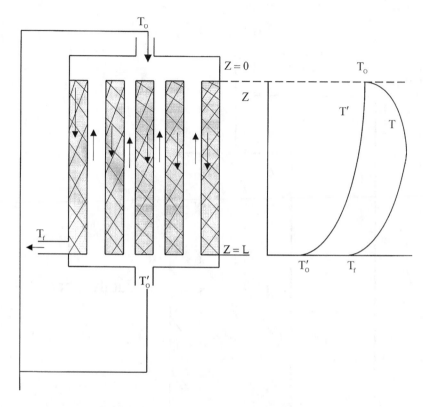

Figure 6-10. An autothermal multi-tubular reactor with an internal heat exchanger.

Water-gas shift: $CO + H_2O \rightleftharpoons CO_2 + H_2$

Ammonia synthesis: $N_2 + 3H_2 \underset{k_2}{\overset{k_1}{\rightleftharpoons}} 2NH_3$

In ammonia synthesis, high temperatures correspond to small reactor volumes. For exothermic reactions, the equilibrium conversion X_e decreases as the temperature increases. Therefore, these reactions are often carried out in a series of adiabatic beds with either intermediate heat exchangers to cool the gases or bypass the cold feed to decrease the temperatures between the beds. Some compromise can be achieved between high temperatures involving small reactor volumes and high equilibrium conversions.

The equilibrium constant K_{eq} for ammonia synthesis is expressed as a function of the partial pressure as

Product

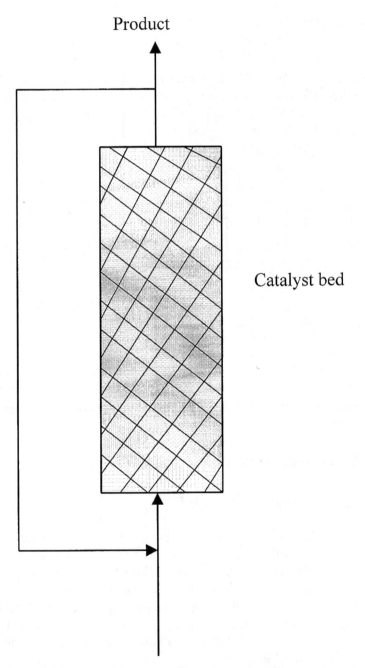

Catalyst bed

Figure 6-11. An autothermal reactor with a recycle arrangement.

$$K_{eq} = \frac{p_{NH_3}^2}{p_{N_2} \cdot p_{H_2}^3}$$

Since the partial pressure is the mole fraction in the vapor phase multiplied by the total pressure, (i.e., $p_i = y_i \cdot P$), the equilibrium constant K_{eq} is expressed as $K_{eq} = K_y \cdot P^{\Delta n}$, where $\Delta n = (2 - 1 - 3)$, the difference between the gaseous moles of the products and the reactants in the ammonia synthesis reaction.

$$K_{eq} = \frac{p_{NH_3}^2}{p_{N_2} \cdot p_{H_2}^3} = \frac{y_{NH_3}^2}{y_{N_2} y_{H_2}^3} \cdot \frac{1}{P^2} = \exp\left(-\frac{\Delta G_R^o}{RT}\right)$$

where y_i = mole fraction in the vapor phase of component i
$\quad$ p_i = partial pressure of component i
$\quad$ P = total pressure
$\quad$ R = gas constant (1.987 cal/gmol•K, 8.314 J/mol•K)
$\quad$ T = absolute temperature, K
$\quad$ ΔG_R^o = standard Gibbs free energy

The equilibrium constant K_{eq} can be determined at any temperature from standard state information on reactants and product. Considering the synthesis of NH_3, the equilibrium conversion X_e can be determined for a stoichiometric feed of H_2 and N_2, at the total pressure. These conversions are determined by the number of moles of each species against conversion X by taking as a basis, 1 mole of N_2.

Component	Number of moles N_i	Mole fraction y_i
N_2	$1 - X$	$\dfrac{1-X}{4-2X} = \dfrac{1-X}{2(2-X)}$
H_2	$3(1 - X)$	$\dfrac{3(1-X)}{4-2X} = \dfrac{3(1-X)}{2(2-X)}$
NH_3	$2X$	$\dfrac{2X}{4-2X} = \dfrac{2X}{2(2-X)}$
Total	$\Sigma N_i = 4 - 2X$	$\Sigma y_i = 1.0$

Substituting the expressions for the mole fractions of H_2, N_2, and NH_3, respectively, for the equilibrium constant K_{eq} gives

$$K_{eq} = \frac{p_{NH_3}^2}{p_{N_2} \cdot p_{H_2}^3} = \frac{y_{NH_3}^2}{y_{N_2} y_{H_2}^3} \cdot \frac{1}{P^2} = \frac{\left[\dfrac{2X}{2(2-X)}\right]^2 P^2}{\left[\dfrac{(1-X)}{2(2-X)}\right] P \left[\dfrac{3(1-X)}{2(2-X)}\right]^3 P^3}$$

$$K_{eq} = \frac{16X^2(2-X)^2}{27(1-X)^4 P^2}$$

Table 6-5 shows the equilibrium constant K_{eq} with the equilibrium partial pressure of NH_3 starting with a stoichiometric mixture of H_2 and N_2 at pressures of 1, 10, and 100 atm. Figure 6-12 shows the relationship between equilibrium conversion X_e versus temperature and pressure for stoichiometric feed.

Table 6-5 shows the conditions for which NH_3 production is possible. Both low temperatures or very high pressures achieve favorable equilibrium. At 25°C, the equilibrium constant is very high, while at higher temperatures, both K_{eq} and P_{NH_3} decrease rapidly. Generally, ammonia synthesis reactors operate at about 350°C and 200 atm with an equilibrium conversion of about 70% in each pass. The NH_3 is separated from unreacted H_2 and N_2, which are recycled back to the reactor. For the overall process involving the tubular reactor, separation and recycle produce about 100% ammonia conversion.

Table 6-5
Equilibrium constant and equilibrium P_{NH_3} versus T and P

T °C	K_{eq}	1 atm	P_{NH_3} 10 atm	100 atm
25	5.27×10^5	0.937	9.80	99.3
100	0.275×10^3	0.660	8.76	95.9
200	0.382	0.172	5.37	81.9
300	0.452×10^{-2}	0.031	2.07	58.0
400	0.182×10^{-3}	0.00781	0.682	34.1
500	0.160×10^{-4}	0.00271	0.259	18.2

Source: Schmidt, L. D, The Engineering of Chemical Reactions, New York, Oxford University Press, 1998.

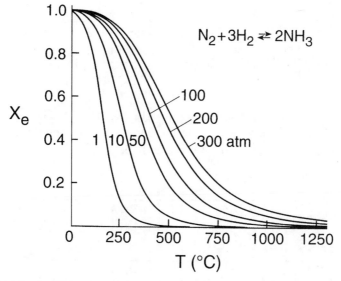

Figure 6-12. Profiles of equilibrium conversion X_e versus temperature T for ammonia synthesis. *(Source: Schmidt, L. D., The Engineering of Chemical Reactions, Oxford University Press, New York, 1998.)*

Example 6-8

A tubular reactor is to be designed for the synthesis of methanol from a stoichiometric mixture of CO and H_2. The reaction occurs in the vapor phase using a solid catalyst in the form of porous spheres: $CO + 2H_2 = CH_3OH$. The average mixture physical and thermo-dynamic data at 500 K and 10 Mpa are

Density	$= 30 \text{ kg/m}^3$
Viscosity	$= 2 \times 10^{-5} \text{ Pa} \cdot \text{s}$
Heat of reaction	$= -100 \text{ kJ/mol } CH_3OH$
Equilibrium constant	$= 0.3 \times 10^4$
Feed	$=$ Stoichiometric mixture at 473 K and 10 Mpa, gas velocity at reactor entrance is 0.15 m/sec
Catalyst	$=$ Porous spheres 3 mm in diameter, packed in tubes 40 mm internal diameter

Comment, from the reactor design and operation viewpoint, on the data.

Solution

$$CO + 2H_2 \underset{k_2}{\overset{k_1}{\rightleftharpoons}} CH_3OH \left(\Delta H_R = -100 \frac{kJ}{mol \cdot CH_3OH} \right)$$

The reaction is reversible and strongly exothermic. The equilibrium yield of CH_3OH decreases as the temperature increases. Hence, a low temperature and increased pressure will be kept.

The equilibrium constant K_{eq} is expressed as

$$K_{eq} = \frac{k_1}{k_2} = \frac{p_{CH_3OH}}{p_{CO} \cdot p_{H_2}^2}$$

Since the partial pressure is the mole fraction in the vapor phase multiplied by the total pressure (i.e., $p_i = y_i P$), the equilibrium constant K_{eq} is expressed as $K_{eq} = K_y \cdot P^{\Delta n}$, where $\Delta n = (1 - 1 - 2)$, the difference between the gaseous moles of the products and the reactants as in the ammonia synthesis reaction.

$$K_{eq} = \frac{p_{CH_3OH}}{p_{CO} \cdot p_{H_2}^2} = \frac{y_{CH_3OH}}{y_{CO} y_{H_2}^2} \cdot \frac{1}{P^2} = exp\left(-\frac{\Delta G_R^O}{RT} \right)$$

The equilibrium constant K_{eq} is determined at any temperature from standard state information on reactants and product. Considering the synthesis of CH_3OH, the equilibrium conversion X_e is determined for a stoichiometric feed of CO and H_2 at the total pressure. These conversions are determined by the number of moles of each species against conversion X by taking as a basis, 1 mole of CO.

Component	Number of moles N_i	Mole fraction y_i
CO	$1 - X$	$\dfrac{1-X}{3-2X}$
H_2	$2(1 - X)$	$\dfrac{2(1-X)}{3-2X}$
CH_3OH	X	$\dfrac{X}{3-2X}$
Total	$\Sigma N_i = 3 - 2X$	$\Sigma y_i = 1.0$

Substituting the expressions for the mole fractions of CO, H_2, and CH_3OH, respectively, for the equilibrium constant K_{eq} yields

$$K_{eq} = \frac{p_{CH_3OH}}{p_{CO} \bullet p_{H_2}^2} = \frac{y_{CH_3OH}}{y_{CO}y_{H_2}^2} \bullet \frac{1}{P^2} = \frac{\dfrac{X}{3-2X}P}{\left(\dfrac{1-X}{3-2X}\right)P\left(\dfrac{2-2X}{3-2X}\right)^2 P^2}$$

$$K_{eq} = \frac{X(3-2X)^2}{4(1-X)^3 P^2}$$

Figure 6-13 shows plots of equilibrium conversion versus temperature. The plots indicate the conversion is low at operating temperature T = 473 K (200°C), but ensures rapid reaction. The conversion per pass is low, therefore, it is important to maintain a high pressure to achieve a high conversion. Modern methanol plants operate at about 250°C and 30–100 atm and give nearly equilibrium conversions using Cu/ZnO catalysts. The unreacted CO and H_2 are recycled back into the reactor.

Reynolds Number

Assuming a reasonable approach to plug flow in the reactor, and assuming the Reynolds number of the fluid in the reactor is

$$Re_p = \frac{d_p \rho v}{\mu}$$

where Re_p = Reynolds number of the fluid in the reactor
d_P = catalyst diameter

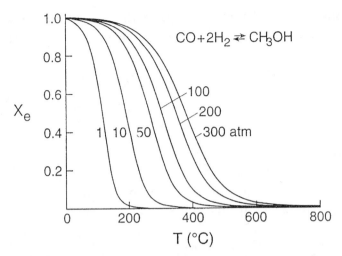

Figure 6-12. Profiles of equilibrium conversion X_e versus temperature T for methanol synthesis. *(Source: Schmidt, L. D., The Engineering of Chemical Reactions, Oxford University Press, New York, 1998.)*

$$\rho = \text{fluid density}$$
$$\mu = \text{fluid viscosity}$$
$$v = \text{fluid velocity}$$

The Reynolds number Re_p is

$$Re_p = \frac{\left(3\times10^{-3}\right)(30)(0.15)}{2\times10^{-5}} \left\{ m \cdot \frac{kg}{m^3} \cdot \frac{m}{sec} \cdot \frac{m \cdot sec}{kg} \right\}$$

$$= 675$$

The Reynolds number is 675, indicating that the fluid flow through the reactor is turbulent.

Example 6-9
Design of Heterogeneous Catalytic Reactors

A bench-scale study of the hydrogenation of nitrobenzene was investigated by Wilson [2]. In this study, nitrobenzene and hydrogen were fed at a rate of 65.9 gmol/hr to a 30 cm internal diameter (ID) reactor containing the granular catalyst. A thermocouple sheath, 0.9

cm in diameter, extended down the center of the tube. The void fraction was 0.424 and the pressure atmospheric. The feed entered the reactor at 427.5 K, and the tube was immersed in an oil batch maintained at the same temperature. The heat transfer coefficient from the mean reaction temperature to the oil bath was determined experimentally to be 8.67 cal/hr • cm^2 • °C. A large excess of hydrogen was used so that the specific heat of the reaction mixture was equal to that of hydrogen. The change in total moles of the reaction was neglected, and the heat of reaction was approximately constant and equal to −152,100 cal/gmol. The feed concentration of nitrobenzene was 5×10^{-7} gmol/cm^3. The global rate of reaction was represented by the expression

$$r_p = 5.79 \times 10^4 \, C^{0.578} \, e^{-\frac{2,958}{T}}$$

where r_P = gmol nitrobenzene reacting/cm^{3}hr, expressed in terms of void volume in the reactor

      C = concentration of nitrobenzene, gmol/cm^3

      T = temperature, K

Calculate the reactor temperature and conversion as a function of reactor length and comment on the results.

Solution

Concentration of nitrobenzene depends on both temperature and conversion. If u is the volumetric flowrate at a point in the reactor where the concentration is C, and u_O is the value at the entrance, the conversion of nitrobenzene is

$$X = \frac{u_O C_O - uC}{u_O C_O} = 1 - \frac{uC}{u_O C_O} \tag{6-89}$$

The volumetric flowrate depends on the temperature and changes as the temperature changes. Assuming a perfect gas behavior,

$$X = 1 - \frac{CT}{C_O T_O} \tag{6-90}$$

Rearranging Equation 6-90 in terms of concentration gives

$$C = \frac{C_0 T_0}{T}(1 - X) \qquad (6\text{-}91)$$

$$= (5 \times 10^{-7})\left(\frac{427.5}{T}\right)(1 - X) \qquad (6\text{-}92)$$

Substituting Equation 6-92 into the rate equation yields

$$r_p = 5.79 \times 10^4 \left\{ 5 \times 10^{-7} \times \frac{427.5}{T}(1 - X) \right\}^{0.578} e^{-\frac{2,958}{T}}$$

$$= 438\left(\frac{1 - X}{T}\right)^{0.578} e^{-\frac{2,958}{T}} \qquad (6\text{-}93)$$

From the differential mass balance (Equation 5-321) in terms of the void volume,

$$F_{AO} dX_A = (-r_A''') dV_p \qquad (5\text{-}321)$$

The feed rate of nitrobenzene is

$$F_A = 65.9(22,400)\left(\frac{427.5}{273}\right)(5.0 \times 10^{-7})$$

$$= 1.16 \text{ gmol/hr}$$

Hence,

$$1.16 \, dX_A = r_p(0.424)\frac{\pi}{4}(9 - 0.81) dl$$

$$1.16 \, dX_A = 2.727(r_p) dl$$

$$\frac{dX_A}{dl} = \frac{2.727}{1.16}(r_p)$$

$$= \frac{2.727}{1.16} \times 438\left(\frac{1 - X}{T}\right)^{0.578} e^{-\frac{2,958}{T}}$$

$$\frac{dX_A}{dl} = 1,038 \left(\frac{1-X}{T} \right)^{0.578} e^{-\frac{2,958}{T}} \tag{6-94}$$

The energy balance for a tubular reactor is:

$$\rho u C_p dT - u\, C_{AO} \left(\frac{-\Delta H_R}{a} \right) dX_A = U\, \pi d_t \left(T_E - T \right) dl \tag{6-95}$$

In terms of the heat transfer coefficient, h_O,

$$\rho u C_p dT - u\, C_{AO} \left(\frac{-\Delta H_R}{a} \right) dX_A = h_O\, \pi d_t \left(T_E - T \right) dl$$

or

$$\rho u C_p dT = u\, C_{AO} \left(\frac{-\Delta H_R}{a} \right) dX_A - h_O\, \pi d_t \left(T - T_E \right) dl$$

$$65.9 \times 6.9 dT = 1.16(152,00) dX_A - 8.67(\pi)(3)(T - 427.5) dl$$

$$\Delta H_R = -152,100 \text{ cal/gmol}$$

$$\frac{dT}{dl} = 385.48 \left(\frac{dX_A}{dl} \right) - 0.178(T - 427.5) \tag{6-96}$$

Substituting Equation 6-94 into Equation 6-96 gives

$$\frac{dT}{dl} = 385.48 \left\{ 1,030 \left(\frac{1-X}{T} \right)^{0.578} e^{-\frac{2,958}{T}} \right\} - 0.178(T - 427.5)$$

$$\frac{dT}{dl} = 397,044.4 \left(\frac{1-X}{T} \right)^{0.578} e^{-\frac{2,958}{T}} - 0.178(T - 427.5) \tag{6-97}$$

Equations 6-94 and 6-97 are first order differential equations, and it is possible to solve for both the conversion and temperature of hydrogenation of nitrobenzene relative to the reactor length of 25 cm. A computer program PLUG61 has been developed employing the Runge-Kutta fourth order method to determine the temperature and conversion using a catalyst bed step size of 0.5 cm. Table 6-6 shows

Table 6-6
Longitudinal temperature profile and conversion in
a reactor for the hydrogenation of nitrobenzene

CATALYST BED DEPTH cm	CONVERSION X	TEMPERATURE K
0.00	0.0000	427.5
0.50	0.0161	433.4
1.00	0.0334	439.2
1.50	0.0520	445.0
2.00	0.0719	450.8
2.50	0.0931	456.6
3.00	0.1157	462.4
3.50	0.1396	468.2
4.00	0.1648	473.9
4.50	0.1913	479.7
5.00	0.2191	485.4
5.50	0.2482	491.1
6.00	0.2785	496.8
6.50	0.3098	502.4
7.00	0.3422	507.9
7.50	0.3755	513.2
8.00	0.4095	518.4
8.50	0.4441	523.3
9.00	0.4790	527.9
9.50	0.5140	532.2
10.00	0.5490	536.0
10.50	0.5835	539.4
11.00	0.6174	542.3
11.50	0.6504	544.6
12.00	0.6822	546.3
12.50	0.7126	547.3
13.00	0.7414	547.6
13.50	0.7685	547.3
14.00	0.7938	546.3
14.50	0.8171	544.7
15.00	0.8384	542.6
15.50	0.8579	539.9
16.00	0.8754	536.7
16.50	0.8912	533.2
17.00	0.9052	529.3
17.50	0.9177	525.2
18.00	0.9288	520.9
18.50	0.9385	516.5
19.00	0.9471	512.1
19.50	0.9545	507.6
20.00	0.9611	503.2
20.50	0.9668	498.8
21.00	0.9717	494.6
21.50	0.9761	490.4
22.00	0.9798	486.4
22.50	0.9830	482.6
23.00	0.9858	478.9
23.50	0.9883	475.4
24.00	0.9903	472.1
24.50	0.9921	469.0
25.00	0.9936	466.0
25.50	0.9949	463.2
26.00	0.9960	460.5
26.50	0.9969	458.1
27.00	0.9977	455.7
27.50	0.9983	453.6
28.00	0.9988	451.5
28.50	0.9992	449.6
29.00	0.9995	447.8
29.50	0.9997	446.2
30.00	0.9999	444.7

the results of the program and Figure 6-14 illustrates the profile of the temperature against the catalyst-bed depth. The results show that as the bed depth increases, the mean bulk temperature steadily increases and reaches a maximum at about 13.0cm from the entrance to the reactor. This maximum temperature is referred to as the "hot spot" in

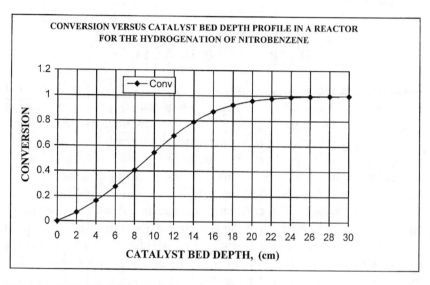

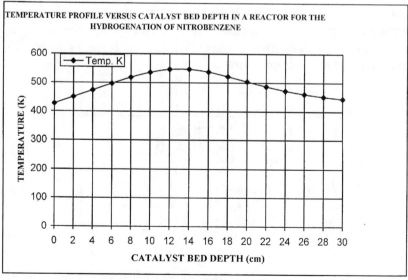

Figure 6-14. Conversion and temperature profiles for the hydrogenation of nitrobenzene.

the reactor. The conversion also increases and attains the maximum at about 24.0 cm from the entrance to the reactor. It is assumed that the radial temperature gradients in the plug flow reactor are negligible. Finally, these results are in agreement with the experimental results of Wilson, where the measured temperature was at the center of the reactor.

TWO-DIMENSIONAL TUBULAR
(PLUG FLOW) REACTOR

Consider a reaction A → products in a tubular reactor with heat exchanged between the reactor and the surroundings. If the reaction is exothermic and heat is removed at the walls, a radial temperature gradient occurs because the temperature at this point is greater than at any other radial position. Reactants are readily consumed at the center resulting in a steep transverse concentration gradient. The reactant diffuses toward the tube axis with a corresponding outward flow of the products. The concentration and radial temperature gradients in this system now make the one-dimensional tubular (plug flow) reactor inadequate. It is necessary to consider both the mass and energy balance equations for the two dimensions l and r. With reference to Figure 6-15, consider the effect of longitudinal dispersion and heat conduction. The following develops both material and energy balance equations for component A in an elementary annulus radius δr and length δl. It is also assumed that equimolecular counter diffusion occurs. Other assumptions are:

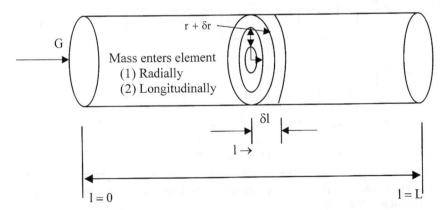

Figure 6-15. Differential section of two-dimensional tubular reactor.

- Laminar flow
- Constant dispersion and thermal conductivity coefficients
- Instantaneous heat transfer between solid catalyst and the reacting ideal gas mixture
- Neglect potential energy
- The reaction rate is independent of total pressure

The component mass balance of species A entering the element longitudinally and radially per unit time is

Mass entering by longitudinal bulk flow
$$= 2\pi r \delta r G(N_A)_l \qquad\qquad (6\text{-}98)$$

Mass entering radially by diffusion
$$= -D_r 2\pi r \delta l \left(\frac{\partial N_A}{\partial r}\right)_r \qquad\qquad (6\text{-}99)$$

Mass entering by longitudinal diffusion
$$= -D_l 2\pi r \delta r \left(\frac{\partial N_A}{\partial l}\right)_l \qquad\qquad (6\text{-}100)$$

Mass leaving by longitudinal bulk flow
$$= 2\pi r \delta r G \left(N_A + \frac{\partial N_A}{\partial l}\delta l\right)_{l+\delta l} \qquad\qquad (6\text{-}101)$$

Mass leaving radially by diffusion
$$= -D_r 2\pi r (r+\delta r)\delta l \left(\frac{\partial N_A}{\partial r} + \frac{\partial^2 N_A}{\partial r^2}\delta r\right)_{r+\delta r} \qquad\qquad (6\text{-}102)$$

Mass leaving by longitudinal diffusion
$$= -D_l 2\pi r \delta r \left(\frac{\partial N_A}{\partial l} + \frac{\partial^2 N_A}{\partial l^2}\delta l\right)_{l+\delta l} \qquad\qquad (6\text{-}103)$$

Mass of component produced by chemical reaction
$$= 2\pi r \delta r \delta l \rho_b(-r_A) \qquad\qquad (6\text{-}104)$$

where D_l = axial diffusivity
$\quad\ \ D_r$ = radial diffusivity

In assuming a steady state, the algebraic sum of the components of mass entering (Equations 6-98, 6-99, and 6-100) and leaving (Equations 6-101, 6-102, 6-103, and 6-104) the element are zero. Expanding the terms evaluated at $(1 + \delta 1)$ and $(r + \delta r)$ in Taylor series about the points 1 and r, and neglecting the second order differences, gives

$$D_r \left\{ \frac{\partial^2 N_A}{\partial r^2} + \frac{1}{r} \frac{\partial N_A}{\partial r} \right\} + D_1 \frac{\partial^2 N_A}{\partial l^2} - G \frac{\partial N_A}{\partial l} = \rho_b \left(-r_A \right) \qquad (6\text{-}105)$$

The corresponding heat balance is

$$k_r \left\{ \frac{\partial^2 T}{\partial r^2} + \frac{1}{r} \frac{\partial T}{\partial r} \right\} + k_1 \frac{\partial^2 T}{\partial l^2} - GC_p \frac{\partial T}{\partial l} = \rho_b \left(\frac{-\Delta H_R}{a} \right) \left(-r_A \right) \qquad (6\text{-}106)$$

where k_1 = axial thermal conductivity
$\quad\;\; k_r$ = radial thermal conductivity

If the reaction rate is a function of pressure, then the momentum balance is considered along with the mass and energy balance equations. Both Equations 6-105 and 6-106 are coupled and highly non-linear because of the effect of temperature on the reaction rate. Numerical methods of solution involving the use of finite difference are generally adopted. A review of the partial differential equation employing the finite difference method is illustrated in Appendix D. Figures 6-16 and 6-17, respectively, show typical profiles of an exothermic catalytic reaction.

Figures 6-16 and 6-17, respectively, show the conversion is higher along the tube axis than at other radial positions and extremely high temperatures can be reached at the tube axis. These temperatures can be sufficiently high to damage the catalyst by overheating. A maximum temperature is also obtained along the reactor length, and the location of this hot spot can alter with changes in catalyst activity.

PRESSURE DROP (ΔP) IN TUBULAR (PLUG FLOW) REACTORS

It is important to determine pressure drop of fluid through tubular reactors, such as packed, fixed, and fluidized bed reactors where, catalysts are employed. ΔP is an important factor that influences the design and operation of such reactors. Ergun [3] developed a useful

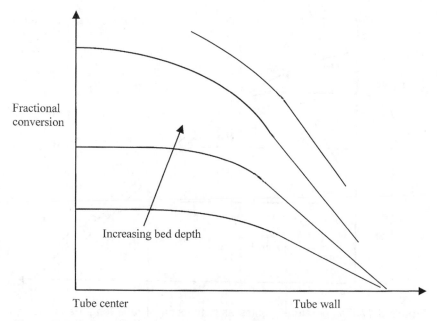

Figure 6-16. Conversion profiles as a function of tube length and radius.

ΔP equation arising from simultaneous turbulent kinetic and viscous energy losses that is applicable to all flow types. Ergun's equation relates the pressure drop per unit depth of packed bed to characteristics such as velocity, fluid density, viscosity, size, shape, surface of the granular solids, and void fraction. The original Ergun equation is:

$$\frac{\Delta P}{L}g_c = \frac{1.75(1-\varepsilon)}{\varepsilon^3} \cdot \frac{G\,v}{D_p} + \frac{150(1-\varepsilon)^2}{\varepsilon^3} \cdot \frac{\mu v}{D_p^2} \qquad (6\text{-}107)$$

where ΔP = pressure drop lb/in.^{2}, N/m^2

       L = unit depth of packed bed, ft, m

      g_c = dimensional constant 32.174(lb$_m$/lb$_f$)(ft/sec^2), 1kg • m/N • sec^2

      μ = viscosity of fluid, lb/ft • hr, kg/m • sec

      v = superficial fluid velocity, ft/sec, m/sec

     D_p = effective particle diameter, ft, m

     G = superficial mass velocity, lb/hr • ft^2, kg/m^{2}s

      ε = void fraction of bed

      ρ = fluid density, lb/ft^3, kg/m^3

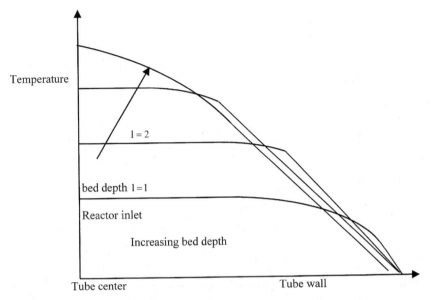

Figure 6-17. Temperature profiles as a function of tube length and radius.

Equation 6-107 gives the total energy loss in fixed beds as the sum of viscous energy loss (the first term on the right side of the equation) and the kinetic or turbulent energy loss (the second term on the right side of the equation). For gas systems, approximately 80% of the energy loss depends on turbulence and can be represented by the second term of Equation 6-107. In liquid systems, the viscous term is the major factor.

For packed beds in either laminar or turbulent flow the Ergun equation is

$$\frac{dP}{dl} = -\frac{\rho v_s^2}{D_p} \frac{(1-\varepsilon)}{\varepsilon^3} \left\{ 1.75 + \frac{150(1-\varepsilon)\mu}{D_p \rho v_s} \right\} \tag{6-108}$$

where D_p = diameter of the packing. For non-spherical packing, $D_p = 6(1 - \varepsilon)/\phi_s S$ where $S = S_o(1 - \varepsilon)$ and $S_o = 6/D_p \phi_s$ where S_o is the surface area per unit volume of solid material in the bed and $\phi_s = 1$ for a spherical particle. Use six times the ratio of volume to surface area of the packing as an effective D_p.

v_s = superficial velocity when the tube is empty

ρ = fluid density

ϕ_s = shape factor of the solid, defined as the quotient of the area of a sphere equivalent to the volume of the particle divided by the actual surface of the particle

μ = fluid viscosity

Equation 6-108 can be expressed in the form

$$-\frac{dP}{dl} = f\frac{\rho v_s^2}{g_c D_p} \tag{6-109}$$

where the friction factor f is

$$f = \frac{1-\varepsilon}{\varepsilon^3}\left[a + \frac{b(1-\varepsilon)}{Re}\right] \tag{6-110}$$

with a = 1.75, b = 150, and Re = $\rho v_s D_p / \mu$

Hanley and Heggs [4] derived a value with a = 1.24 and b = 368. McDonald et al. [5] proposed that a = 1.8 for smooth particles and 4.0 for rough particles and b = 180. Hicks [6] reviewed various ΔP equations and inferred that the Ergun equation is limited to Re/(1 − ε) < 500 and Handley and Hegg's equation to 1,000 < Re/(1 − ε) < 5,000. He developed an equation for the friction factor for spheres, which is of the form

$$f = \frac{6.8(1-\varepsilon)^{1.2}}{\varepsilon^3} \cdot Re^{-0.2} \tag{6-111}$$

which fits Ergun's and Handley and Hegg's data together with the results of Wentz and Thodos [7] at high Reynolds number. This shows that a and b are not true constants as stated by Tallmadge [8], who suggested that a = 1.75 and b = 4.2 $Re^{5/6}$.

Equation 6-108 is also a good approximation for a fluidized bed reactor up to the minimum fluidizing condition. However, beyond this range, fluid dynamic factors are more complex than for the packed bed reactor. Among the parameters that influence the ΔP in a fluidized bed reactor are the different types of two-phase flow, smooth fluidization, slugging or channeling, the particle size distribution, and the gas flowrate. After reaching a peak ΔP at the point of minimum fluidization, the ΔP of a smoothly fluidizing bed drops to a value that approximately corresponds to the static pressure of the bed and remains nearly constant with an increase in the gas flowrate until the entrainment point is reached. A slugging bed displays a wide fluctuation in the ΔP beyond the point of minimum fluidization and a channeling bed exhibits a ΔP far below the bed static pressure. Figure 6-18 shows the behavior of fluidized beds in these three operation modes.

ΔP for laminar flow in a circular tube is given by the Hagen-Poiseuille equation:

$$\frac{dP}{dl} = -8 \frac{v\mu}{R} \qquad (6\text{-}112)$$

For turbulent flow, ΔP is calculated from

$$\frac{dP}{dl} = -4f_F \frac{1}{D} \frac{\rho v^2}{2}$$

$$\text{or } \frac{dP}{dl} = -f_F \left(\frac{\rho v^2}{R} \right) \qquad (6\text{-}113)$$

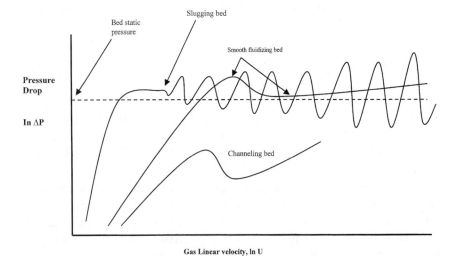

Gas Linear velocity, ln U

Figure 6-18. Relationship between ΔP and gas linear velocity.

or where the Darcy friction factor is four times the Fanning friction factor (i.e., $f_D = 4f_F$)

$$\frac{dP}{dl} = -f_D \left(\frac{\rho V^2}{4R} \right)$$ (6-114)

The Fanning friction factor is given by $f_F = 0.079\ Re^{-0.25}$ and the Reynolds number is $Re = \rho v D_p / 6\mu(1 - \varepsilon)$ for a packed bed consisting of spherical particles and $Re = \rho v D_p \phi_s / 6\mu(1 - \varepsilon)$ for nonspherical particles. The values of ϕ_s for other materials are provided by Perry [9] and listed in the spreadsheet (PACKED.xls) package.

Industrial design problems often occur in tubular reactors that involve the simultaneous solution of ΔP, energy, and mass balances.

Example 6-10

A bed with a height 2.5 m and an internal diameter of 0.035 m is packed with cylindrical particles of diameter $D_P = 0.003$ m. The void fraction of the bed is $\varepsilon = .33$. A gas of density $\rho = 0.9487$ kg/m^3 and dynamic viscosity $\mu = 3.1 \times 10^{-5}$ Pa • s flows through the bed with a superficial mass flow velocity of 1.4 kg/m^{2}s. Determine the pressure drop in the bed.

Solution

From Ergun's Equation 6-107

$$-\frac{dP}{dl} = f \frac{\rho v_s^2}{g_c D_p}$$ (6-109)

where the friction factor f is

$$f = \frac{1-\varepsilon}{\varepsilon^3} \left[a + \frac{b(1-\varepsilon)}{Re} \right]$$ (6-110)

a = 1.75 and b = 150

$$v_s = \frac{1.4}{0.9487} \left(\frac{kg}{m^2 \cdot sec} \cdot \frac{m^3}{kg} \right)$$

$$= 1.475 \text{ m/sec}$$

The Reynolds number, Re

$$Re = \frac{\rho v_s D_p}{\mu}$$

$$= \left(\frac{0.9487 \times 1.475 \times 0.003}{3.1 \times 10^{-5}} \right) \left(\frac{kg}{m^3} \cdot \frac{m}{sec} \cdot \frac{m}{\dfrac{kg}{m \cdot sec}} \right)$$

$$= 135$$

The friction factor f from Equation 6-110 is

$$f = \frac{(1 - 0.33)}{0.33^3} \left[1.75 + \frac{150(1 - 0.33)}{135} \right]$$

$$f = 46.4$$

The pressure drop $-\Delta P_t$ is

$$-\Delta P_t = f \frac{\rho v_s^2 L}{g_c D_p}, \quad \frac{N}{m^2}$$

$$-\Delta P_t = 46.4 \frac{0.9487 \times 1.475^2 \times 2.5}{1 \times 0.003} = 79,809 \ N/m^2 = 0.798 \ bar$$

McDonald et al. equation: a = 1.8, b = 180, f = 50.2, and $-\Delta P_t$ = 86,368 N/m^2 = 0.864 bar.

Hicks equation: f = 43.9 and $-\Delta P_t$ = 75,509 N/m^2 = 0.755 bar.

Tallmadge equation: a = 1.75, b = 4.2 $Re^{5/6}$, f = 55.76, and $-\Delta P_t$ = 95,908 N/m^2 = 0.959 bar.

An Excel spreadsheet program (PACKED.xls) was developed to determine ΔPs of packed beds with known parameters.

THERMAL BEHAVIORS IN FLOW SYSTEMS

The effect of temperature on chemical reaction influences the nature and size of the reactor, and the selectivity of the reaction. Additionally, the temperature effects may primarily determine both the stability and controllability of the reactor. Interesting and important temperature effects can occur in continuous reactor systems (e.g., CFSTR and plug flow reactors) with exothermic reactions. The rate of heat generation increases as the rate of reaction increases, and the reaction rate usually increases with the temperature in accordance with Arrhenius law.

In a CFSTR, it is necessary to perform a material and an energy balance on an element of the system to understand its behavior.

Consider a first order, exothermic reaction ($aA \rightarrow$ products) in a CFSTR having a constant supply of new reagents, and maintained at a steady state temperature T that is uniform throughout the system volume. Assuming perfect mixing and no density change, the material balance equation based on reactants is expressed as $uC_{AO} = uC_A + (-r_A)V_R$, where $(-r_A) = kC_A$.

Substituting the rate expression into the material balance equation gives $uC_{AO} = uC_A + kV_RC_A$ and the concentration of A at any time will be

$$C_A = \frac{C_{AO}}{\left(1+\dfrac{kV_R}{u}\right)} \tag{5-163}$$

The energy balance equation is:

$$-u\rho\Delta h + Q = 0 \tag{6-31}$$

where $\Delta h = C_p\Delta T + (-\Delta H_R/a)(\Delta\overline{m}_A/M_A)$, neglecting the pressure effects on h and the heat transfer rate Q is

$$Q = UA(T_h - T)$$

where A = heat transfer area
  U = overall heat transfer coefficient
  T_h = temperature of the heating medium
  T = temperature of the fluid in the system

The energy balance from Equation 6-39 yields a heating system

$$-\rho uC_p\left(T-T_O\right) + \left(\frac{-\Delta H_R}{a}\right)\left(-r_A\right)V_R + UA\left(T_h - T\right) = 0$$

or $\left(\dfrac{-\Delta H_R}{a}\right)(-r_A)V_R = u\rho C_p(T - T_O) - UA(T_h - T)$ (6-115)

where the left side of Equation 6-115 is Q_g and is expressed as

$$Q_g = \left(\dfrac{-\Delta H_R}{a}\right)(-r_A)V_R$$ (6-116)

that is, the rate at which heat will be generated due to the reaction. The right side of Equation 6-115 is Q_r, referred to as the heat removed by flow and heat exchange:

$$Q_r = -(u\rho C_p T_O + UAT_h) + (u\rho C_p + UA)T$$ (6-117)

Substituting the Arrhenius equation ($k = k_o e^{-E/RT}$) into the material balance for the exit concentration of A gives

$$C_A = \dfrac{C_{AO}}{1 + k_o e^{-E/RT}\dfrac{V_R}{u}}$$ (6-118)

Also, for the first order reaction, $A \rightarrow$ products, the rate expression is $(-r_A) = kC_A$. Substituting the rate expression, the Arrhenius equation, and Equation 6-118 into Equation 6-116 yields

$$Q_g = \dfrac{\left(\dfrac{-\Delta H_R}{a}\right)\bullet V_R \bullet C_{AO} \bullet k_o e^{\left(\frac{-E}{RT}\right)}}{1 + k_o e^{\left(\frac{-E}{RT}\right)}\bullet\dfrac{V_R}{u}}$$ (6-119)

Assuming that the CFSTR is operating at steady state, the rate of heat generation must equal the rate of heat removal from the reactor. Plotting Q_g as a function of T for fixed values of the other variables, an S-shaped curve is obtained as shown in Figure 6-19.

At low values of T, the reaction rates are low and therefore little or no heat is generated. At high values of T, the reaction goes to completion and the entire exotherm is released. The right side of Equation 6-117 gives a linear expression in T as illustrated in Figure 6-20.

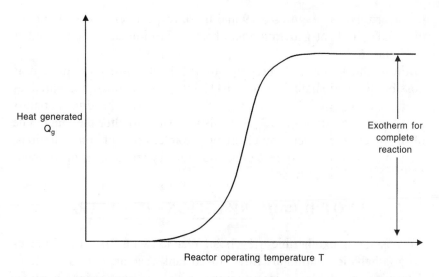

Figure 6-19. Heat generated by reaction versus temperature in a CFSTR.

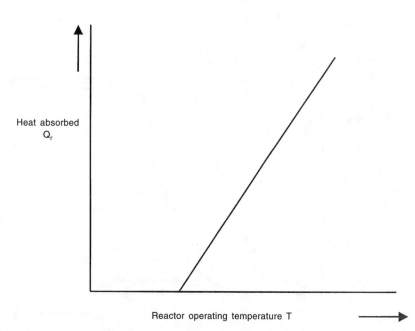

Figure 6-20. Heat absorbed by enthalpy change and loss to environment.

Superimposing Figures 6-19 and 6-20, respectively, possible points of equality in heat generation and heat absorption can be observed as in Figure 6-21.

From the slope and intercept of the heat absorption line, it is possible to manipulate Equation 6-117 by changing operating variables such as u, T_O, and T or design variables such as the dimensionless heat transfer group $UA/\rho u C_p$. It is also possible to alter the magnitude of the reaction exotherm by changing the inlet reactant concentrations. Any of these manipulations can be used to vary the number of locations of the possible steady states.

EXOTHERMIC REACTIONS IN CFSTRs

Consider an exothermic irreversible reaction with first order kinetics in an adiabatic continuous flow stirred tank reactor. It is possible to determine the stable operating temperatures and conversions by combining both the mass and energy balance equations. For the mass balance equation at constant density and steady state condition,

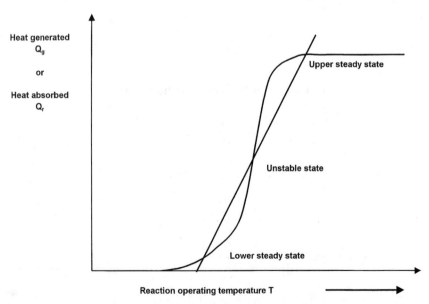

Figure 6-21. Multiple steady states for exothermic reaction in a CFSTR.

$$C_A = \frac{C_{AO}}{(1+k\bar{t})}$$ (5-163)

The conversion $X_A = \frac{C_{AO} - C_A}{C_{AO}} = 1 - \frac{C_A}{C_{AO}}$

Hence, the conversion X_A for the CFSTR becomes

$$X_A = \frac{k\bar{t}}{1+k\bar{t}}$$

In terms of the temperature, the Arrhenius equation $k = k_o \exp(-E/RT)$, the conversion is expressed as

$$X_A = \frac{k_o\bar{t}e^{-E/RT}}{1+k_o\bar{t}e^{-E/RT}}$$ (6-120)

The steady state energy balance for adiabatic operation is determined as follows:

From $-G\Delta h + Q = 0$

where $\Delta h = C_p\Delta T + \left(\frac{-\Delta H_R}{a}\right)\frac{\Delta\overline{m}_A}{M_A}$

and with $G = \rho u$ and $Q = 0$ (adiabatic condition):

$$-\rho u C_p(T-T_o) - \rho u\left(\frac{-\Delta H_R}{a}\right)\frac{\Delta\overline{m}_A}{M_A} = 0$$

In terms of the concentration C_A, where

$$C_A = \rho \bullet \frac{\overline{m}_A}{M_A} \text{ and } \Delta C_A = \rho \bullet \frac{\Delta\overline{m}_A}{M_A}$$

the energy balance becomes

$$-\rho u C_p(T-T_o) - u\left(\frac{-\Delta H_R}{a}\right)(C_A - C_{AO}) = 0$$

or $\left(\dfrac{-\Delta H_R}{a}\right)\left(C_{AO} - C_A\right) = \rho C_p\left(T - T_o\right)$

Introducing the fractional conversion, X_A yields $C_{AO} - C_A = C_{AO}X_A$ and the energy balance becomes

$$\left(\dfrac{-\Delta H_R}{a}\right)C_{AO}X_A = \rho C_p\left(T - T_o\right)$$

The fractional conversion X_A is

$$X_A = \dfrac{\rho C_p}{C_{AO}\left(\dfrac{-\Delta H_R}{a}\right)}\left(T - T_o\right) \qquad (6\text{-}121)$$

Since the heat of reaction usually varies little with temperature, the conversion X_A gives a linear relationship between $T - T_o$ and conversion, and is represented by a straight line as shown in Figure 6-22. For a known reactor and kinetics, and at a given feed temperature T_o, the intersection of the energy balance line with the S-shaped mass

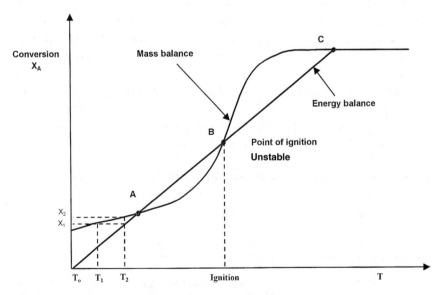

Figure 6-22. Temperature versus conversion for a first order irreversible reaction in an adiabatic continuous flow stirred tank reactor.

balance curve indicates that solutions are possible at three points A, B, and C, as shown in Figure 6-22. If the initial temperature is below the temperature at point A, say T_1, Figure 6-22 shows that the conversion required by the mass balance is greater than the conversion that corresponds to the energy balance. Therefore, the energy evolved will raise the temperature to T_2 and the corresponding conversion X_2. The temperature T_2 is X_2 from Equation 6-120. The reaction rate is too low to justify steady state operation at initial temperature between points A and B, and as such the reaction mixture cools to point A. Correspondingly, the initial temperature between B and C is similar to temperatures below point A. However, in this case, heating will occur until point C is reached. At temperatures above point C, cooling occurs until the temperature drops to C. Point B is unstable because with a small increase in temperature, the heat produced (rapidly rising mass balance curve) is greater than the heat consumed by the reacting mixture (energy balance curve). The excess heat produced will make the temperature rise until point C is reached. Similarly, if the temperature drops slightly below point B, it will continue to drop until point A is reached. Therefore, point B is referred to as the ignition point. If the mixture is raised above this temperature, then the reaction will be self-sustaining.

The relative position of the mass balance curve and the energy balance line in Figure 6-22 depends on various parameters, namely, the chemical properties (k_o, E, ΔH_R) and physical properties (ρ, C_P) of the system and the operating conditions ($\bar{t}$, C_{AO}) in Equation 6-120 and 6-121, respectively. These properties and conditions make it possible to determine whether or not stable operating conditions are permissible and how many stable operating points exist. Figure 6-23 illustrates fractional conversion X_A versus temperature T for reversible exothermic reactions. There is an optimum operating temperature for the given space time value where conversion is maximized. X_A decreases when T is above or below this optimum and, thus, adequate control of heat removal is essential.

THERMAL BEHAVIOR OF A TUBULAR FLOW REACTOR

In a tubular reactor system, the temperature rises along the reactor length for an exothermic reaction unless effective cooling is maintained. For multiple steady states to appear, it is necessary that a

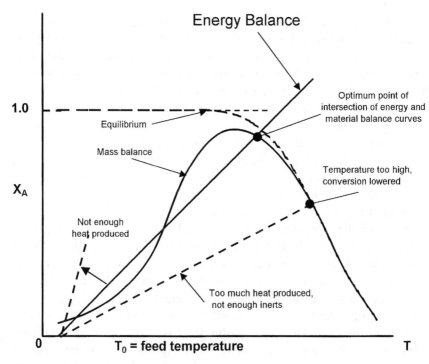

Figure 6-23. Solution of mass and energy balances for reversible exothermic reaction. *(Source: O. Levenspiel 3rd ed.* Chemical Reaction Engineering, *John Wiley and Sons, 1999.)*

feedback mechanism be provided so that the heat generated at one section of the reactor can pass back to an earlier section. In the CFSTR, the mixing provides the feedback. In a tubular flow reactor, the feedback is provided if axial heat conduction is significant or if there is a transfer of heat from the product stream to the feed stream. Figure 6-9 shows a plan for providing autothermal operation in tubular flow with temperature profiles.

If an appropriate thermal feedback mechanism is not provided, the reaction occurs at the lower stationary state where the reaction rate may be negligible. The reaction could be extinguished, if the temperature of the feed entering the reactor drops below some critical value due to fouling of the heat exchange surface.

An important effect in the design of a tubular flow reactor is the development of a radial temperature gradient in a highly exothermic reaction with wall cooling. The temperatures near the tube axis are much higher than those near the tube wall. Consequently, the temperature has an important effect on the reactor design because of the sensitivity of the reaction rate to temperature. The actual reaction rate can be much higher than the reaction rate as calculated on the basis of a constant temperature across the tube.

Example 6-11

Consider a first order adiabatic reaction in a CFSTR with the following characteristics:

Feed temperature, T_o, K	298
Initial concentration, C_{AO}, gmol/l	2
Volumetric flowrate, u, l/sec	0.05
Volume of the reactor, V_R, l	15
Specific heat capacity, C_P, cal/gm – K	1
Density of fluid, ρ, g/l	1,000
Heat of reaction, $(-\Delta H_R/a)$, cal/gmol	60,000

The rate constant k is expressed as $k = \exp(15.32 - 7{,}550/T)$, $\sec^{-1}$.

Determine the operating points for both the mass and heat balance equations.

Solution

The first order reaction is represented by $(-r_A) = kC_A$, and applying the mass balance Equation 6-120 and the heat balance Equation 6-121, respectively, gives the fractional conversion X_A in terms of the mass balance equation:

$$X_A = \frac{k\bar{t}}{1 + k\bar{t}} \qquad (6\text{-}120)$$

and for the heat balance equation:

$$X_A = \frac{\rho C_p}{C_{AO}\left(\dfrac{-\Delta H_R}{a}\right)}(T - T_o) \qquad (6\text{-}121)$$

The residence time $\bar{t} = V_R/u = 15.0/0.05 = 300.0$ secs $(1 \cdot \text{sec/l})$. The mass balance equation now becomes

$$X_A = \frac{300\,k}{1 + 300k}$$

and the heat balance equation is:

$$X_A = \frac{\rho C_p}{C_{AO}\left(\dfrac{-\Delta H_R}{a}\right)}(T - T_o) = \frac{T - 298}{120}$$

The fractional conversions X_A in terms of both the mass balance and heat balance equations were calculated at effluent temperatures of 300, 325, 350, 375, 400, 425, 450, and 475 K, respectively. A Microsoft Excel Spreadsheet (Example6-11.xls) was used to calculate the fractional conversions X_A at varying temperature. Table 6-7 gives the results of the spreadsheet calculation and Figure 6-24 shows profiles of the conversions at varying effluent temperature. The figure shows that the steady state values are $(X_A, T) = (0.02,300), (0.5,362)$, and $(0.95,410)$. The middle point is unstable and the last point is the most desirable because of the high conversion.

Example 6-12

A second order reaction is performed adiabatically in a CFSTR. Use the data in Example 6-11 to plot both conversions for the mass and heat balance equations. The second order rate constant k is

Table 6-7
Fractional conversions X_A (mass and heat balances) at effluent temperatures

Temperature (K)	X_A (Mass)	X_A (Heat)
300	0.016	0.017
325	0.099	0.225
350	0.366	0.433
375	0.709	0.642
400	0.896	0.850
425	0.963	1.058
450	0.986	1.267
475	0.994	1.475

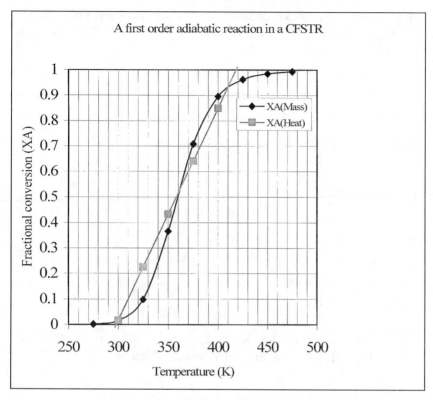

Figure 6-24. Fractional conversions (X_A) versus effluent temperature T (K).

$$k = \exp\left(20 - \frac{9,000}{T}\right)\left(\frac{1}{mol \bullet sec}\right)$$

Solution

The material balance for a second order reaction for species A in a CFSTR is $uC_{AO} = uC_A + (-r_A)V_R$, where

$$\left(-r_A\right) = kC_A^2$$

$$uC_{AO} = uC_A + kC_A^2 V_R$$

Rearranging the above equation in a quadratic form gives $k\bar{t}C_A^2 + C_A - C_{AO} = 0$, which is of the form $ax^2 + bx + c = 0$. The root of the equation is

$$x = \frac{-b \pm \sqrt{b^2 - 4ac}}{2a}$$

For a second order reaction the outlet concentration C_A is

$$C_A = \frac{-1 + \sqrt{1 + 4k\bar{t}C_{AO}}}{2k\bar{t}}$$

where $\bar{t} = V_R/u$

$$\frac{C_A}{C_{AO}} = \frac{-1 + \sqrt{1 + 4k\bar{t}C_{AO}}}{2k\bar{t}C_{AO}}$$

The fractional conversion $X_A = (C_{AO} - C_{AO})/C_{AO} = 1 - C_A/C_{AO}$. Since $X_A < 1$, the smaller root is used, which is

$$X_A = 1 - \left[\frac{-1 + \sqrt{1 + 4k\bar{t}C_{AO}}}{2k\bar{t}C_{AO}} \right]$$

The fractional conversion X_A in terms of the heat balance equation is:

$$X_A = \frac{\rho C_p}{C_{AO}\left(\dfrac{-\Delta H_R}{a} \right)}(T - T_0) = \frac{T - 298}{120}$$

The heat removal line is unchanged irrespective of the kinetics, and the fractional conversion X_A has a qualitatively similar sigmoidal shape for second-order kinetics. Numerical values as in Example 6-11 have been put into both the mass balance and heat balance equations. Fractional conversion X_A from both the mass balance and heat balance equations at effluent temperatures of 300, 325, 350, 375, 400, 425, 450, and 475 K, respectively, were determined using the Microsoft Excel Spreadsheet (Example6-12.xls). Table 6-8 gives the results of

Table 6-8
Fractional conversions X_A (mass and heat balances) at effluent temperatures

Temperature (K)	X_A (Mass)	X_A (Heat)
300	0.026	0.017
325	0.183	0.225
350	0.498	0.433
375	0.740	0.642
400	0.867	0.850
425	0.929	1.058
450	0.960	1.267
475	0.976	1.475

the spreadsheet calculation and Figure 6-25 shows profiles of the conversion at varying effluent temperature. The steady states values in Figure 6-25 are $(X_A, T) = (0.02, 300), (0.32, 335)$, and $(0.88, 405)$. The middle point is unstable and the last point is the most desirable because of the high conversion.

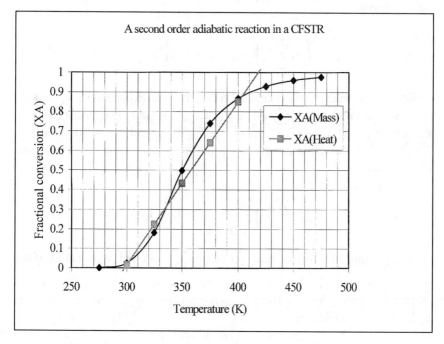

Figure 6-25. Fractional conversions (X_A) versus effluent temperature T (K).

Example 6-13

Determine the operating conditions in a CFSTR for a first order exothermic reaction with heat transfer under the following conditions:

Feed temperature, T_O, K	350
Initial concentration, C_{AO}, gmol/l	1
Temperature of the cooling medium, T_C, K	350
Mean residence time, $\bar{t}$, min	1
$(-\Delta H_R/a)(1/\rho C_p)$, K/gmol–l	200
$UA/u\rho C_p$	1
Reaction rate constant, k, min^{-1}	$\exp(25 - 10,000/T)$

Solution

The heat balance, $-G\Delta h + Q = 0$, gives

$$-\rho u C_p(T - T_O) - \rho u\left(\frac{-\Delta H_R}{a}\right)\frac{\Delta \overline{m}_A}{M_A} + UA(T_C - T) = 0$$

which gives

$$-\rho u C_p(T - T_O) + \left(\frac{-\Delta H_R}{a}\right)(-r_A)V_R + UA(T_C - T) = 0$$

For a first order reaction in a CFSTR, the rate expression $(-r_A) = kC_A$ and the exit concentration C_A is $C_A = C_{AO}/(1 + k\bar{t})$. The fractional conversion $X_A = (C_{AO} - C_A)/C_{AO} = 1 - C_A/C_{AO}$, and $X_A = k\bar{t}/(1 + k\bar{t})$. The heat balance equation in terms of the initial concentration of the feed becomes

$$-\rho u C_p(T - T_O) + \left(\frac{-\Delta H_R}{a}\right)\frac{kC_{AO}}{1 + k\bar{t}}V_R + UA(T_C - T) = 0$$

Rearranging in terms of heat generation Q_g and heat removal Q_r gives

$$\left(\frac{-\Delta H_R}{a}\right)\frac{kC_{AO}V_R}{1 + k\bar{t}} = \rho u C_p(T - T_O) + UA(T - T_C)$$

Further rearrangement in terms of the data in the table gives

$$\left(\frac{-\Delta H_R}{a} \cdot \frac{1}{\rho C_p}\right)\frac{kC_{AO}\bar{t}}{1+k\bar{t}} = \left(T-T_O\right) + \frac{UA}{u\rho C_p}\left(T-T_C\right)$$

where the mean residence time $\bar{t} = V_R/u$.
The heat generation Q_g is

$$Q_g = \left(\frac{-\Delta H_R}{a} \cdot \frac{1}{\rho C_p}\right)\frac{kC_{AO}\bar{t}}{1+k\bar{t}}$$

and the heat removal Q_r is

$$Q_r = \left(T-T_O\right) + \frac{UA}{u\rho C_p}\left(T-T_C\right)$$

Substituting the values in the table for both Q_g and Q_r gives

$$Q_g = \frac{(200)(1)(1)\exp(25-10,000/T)}{1+\exp(25-10,000/T)} \quad \text{and}$$

$$Q_r = (T-350)+(1)(T-350)$$

$$= 2(T-350)$$

A Microsoft Excel Spreadsheet (Example6-13.xls) was used to determine both Q_g and Q_r at varying effluent temperature. Figure 6-26 shows the profiles of the heat generation and heat removal terms, and the intersections give the operating conditions. The values of the fractional conversion X_A corresponding to the temperatures at these points are given in Table 6-9.

VARIABLE COOLANT TEMPERATURE IN A CFSTR

Consider a jacketed CFSTR with an external jacket as shown in Figure 6-27 in which the coolant fluid temperature T_C is not maintained constant. This situation requires solving an energy balance equation on the coolant along with the mass and energy balance

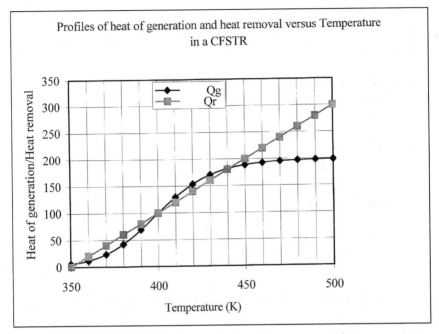

Figure 6-26. Heat of generation/removal versus effluent temperature T(K).

equations on the reactor. If the coolant fluid inlet and outlet temperatures are T_{CO} and T_C, respectively, the coolant volumetric flow rate is u_C, and the contact area with the reactor is A, then the mass and energy balance equations for the jacketed CFSTR with a single reaction A → products become

Mass balance: $uC_{AO} = uC_A + (-r_A)V_R$, which is $C_{AO} - C_A = C_{AO}X_A$ = $k\bar{t}$, where $\bar{t} = V_R/u$.

The heat balance on the CFSTR is

$$-\rho u C_p (T - T_O) + \left(\frac{-\Delta H_R}{a}\right)(-r_A)V_R + UA(T_C - T) = 0$$

and the heat balance on the coolant is

$$-\rho_c u_c C_{pc}(T_C - T_{CO}) + UA(T - T_C) = 0$$

Table 6-9
Q_g and Q_r at effluent temperature

Temperature (K)	Q_g	Q_r
350	5.469	0
360	11.707	20
370	23.279	40
380	42.304	60
390	69.003	80
400	100.000	100
410	129.577	120
420	153.365	140
430	170.244	160
440	181.319	180
450	188.293	200
460	192.612	220
470	195.284	240
480	196.947	260
490	197.993	280
500	198.661	300

Points of intersections between Q_g and Q_r

Temperature (K)	Rate constant	Conversion X_A
354	0.039	0.037
400	1.000	0.500
440	9.706	0.907

where C_{pc} = specific heat capacity of the coolant
u_C = coolant volumetric flowrate
T_C = coolant fluid temperature
T_{CO} = coolant fluid inlet temperature
ρ_c = density of cooling fluid

The residence time of the coolant in the jacket is $\bar{t}_C = V_C/u_C$. Given the three simultaneous equations and with known parameters, it is now possible to solve for C_A, T, and T_C, respectively.

The CFSTR is readily adaptable to both heating and cooling because the fluid can be stirred to promote heat transfer, ensuring better heat management with minimum heat transfer area. A combination of both jacket and internal coil may be necessary to provide the required heat exchange area. Chapter 7 reviews the heat transfer coefficient and the

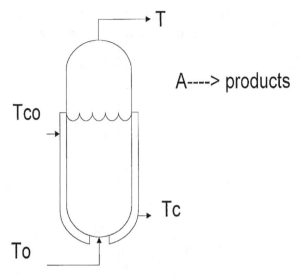

Figure 6-27. A jacketed CFSTR.

time it takes for heating or cooling of process fluids in jacketed agitated vessels.

OPTIMAL DESIGN OF NON-ISOTHERMAL REACTORS

In many cases, the design of reactors has appreciable significant heat changes. The accompanying temperature change affects the size of the reactors. Levenspiel [11,12] has shown that the optimum reactor size is determined by operating isothermally at the maximum allowable temperature. In the case of a reversible exothermic reaction, the ideal optimum reactor size is obtained by following the locus of the maximum rates known as the optimum temperature progression (OTP) on the rate-temperature conversion $(r - T - X_A)$ chart. Levenspiel has provided the $(r - T - X_A)$ chart for a reversible first order reaction $A \leftrightarrow R$. Omoleye et al. [13] have further provided the $(r - T - X_A)$ charts for the generalized types of reversible reactions, and developed mathematical expressions for the optimum temperature progression for these types of reactions at constant density.

The following reviews the analytical treatment of $(r - T - X_A)$ charts and develops computer programs for this class of reversible reactions.

THEORY AND DERIVATIONS FOR THE CONVERSION OF REVERSIBLE REACTIONS

Most chemical reactions of industrial significance fall within the general class of reversible reactions given by

Type 1: $aA \leftrightarrow rR$ $\hspace{3cm}$ (6-122)

Examples: polymerization, isomerization reactions

Type 2: $aA + bB \leftrightarrow rR$ $\hspace{3cm}$ (6-123)

Examples: oxidation, hydrogenation, and halogenation reactions

Type 3: $aA + bB \leftrightarrow rR + sS$ $\hspace{2.5cm}$ (6-124)

Examples: esterification, saponification, and neutralization reactions

Type 4: $aA \leftrightarrow rR + sS$ $\hspace{3cm}$ (6-125)

Examples: decomposition, dehydrogenation, and sublimation reactions

If k_1 and k_2 denote the forward and reverse rate constants respectively, the general rate of disappearance $(-r_A)$ for any type of reaction is defined as:

$$(-r_A) = k_1 \prod_{i=1}^{M} C_i^{\alpha_i} - k_2 \prod_{j=1}^{N} C_j^{\beta_i} \hspace{2cm} (6\text{-}126)$$

where C_i and C_j are the concentrations of reactant i and product j with orders α_i and β_j, respectively. The forward and reverse rate constants k_1 and k_2 have the Arrhenius form

$$k_1 = A_1 \exp\left(\frac{-E_1}{RT}\right) \hspace{2cm} (6\text{-}127)$$

and

$$k_2 = A_2 \exp\left(\frac{-E_2}{RT}\right) \hspace{2cm} (6\text{-}128)$$

For a constant density system, the concentration of any species, K, C_K, during the course of reaction is given by

$$C_K = C_{AO} (\theta_K + \upsilon_K X_A) \tag{6-129}$$

where θ_K = feed concentration ratio of species K to that of A
υ_K = stoichiometric coefficient ratio of species K to that of A; by convection, it is negative for reactants and positive for products

Hence, $\upsilon_A = -1$, $\upsilon_R = \dfrac{r}{a}$, $\upsilon_B = \dfrac{-b}{a}$

Type 1 Reaction: aA $\overset{k_1}{\underset{k_2}{\rightleftharpoons}}$ rR

The rate of disappearance of A may be written as

$$\left(-r_A\right) = k_1 C_A^{\alpha_1} - k_2 C_R^{\beta_1} \tag{6-130}$$

where α_1 and β_1 are the orders of reaction with respect to species A and R, respectively.

From stoichiometry:

	A	**R**
At time t = 0	C_{AO}	0
At time t = t	C_A	C_R
Amount reacted	$C_{AO} - C_A$	C_R

From conversion $X_A = \dfrac{C_{AO} - C_A}{C_{AO}}$ $\tag{5-13}$

$$C_{AO} X_A = C_{AO} - C_A \tag{5-27}$$

and $C_A = C_{AO} (1 - X_A)$ $\tag{5-28}$

The concentration of R is expressed by $C_R = C_{AO} X_A$ or

$$C_R = C_{AO} (\theta_R + \upsilon_R X_A) \tag{6-131}$$

Substituting Equations 5-28 and 6-131 into Equation 6-130 gives

$$(-r_A) = k_1 C_{AO}^{\alpha_1} (1 - X_A)^{\alpha_1} - k_2 C_{AO}^{\beta_1} (\theta_R + \upsilon_R X_A)^{\beta_1} \qquad (6\text{-}132)$$

Rearranging Equation 6-132 yields

$$k_1 C_{AO}^{\alpha_1} (1 - X_A)^{\alpha_1} - k_2 C_{AO}^{\beta_1} (\theta_R + \upsilon_R X_A)^{\beta_1} - (-r_A) = 0 \qquad (6\text{-}133)$$

Equation 6-133 is of the form $f(X_A) = 0$.

It is possible to solve Equations 6-127, 6-128, and 6-133 simultaneously to give the temperature – conversion $(T - X_A)$ data for a specified value of $(-r_A)$ if the values for C_{AO}, θ_R, α_1, β_1, and υ_R are known.

Consider the first order rate of reaction with respect to both the reactants and products. That is, $\alpha_1 = \beta_1 = 1$. Equation 6-133 becomes

$$k_1 C_{AO} (1 - X_A) - k_2 C_{AO} (\theta_R + \upsilon_R X_A) - (-r_A) = 0 \qquad (6\text{-}134)$$

Rearranging Equation 6-134 gives the fractional conversion X_A as

$$X_A = \frac{k_1 C_{AO} - k_2 C_{AO} \theta_R - (-r_A)}{(k_1 C_{AO} + k_2 C_{AO} \upsilon_R)} \qquad (6\text{-}135)$$

$$\text{or } X_A = \frac{k_1 - k_2 \theta_R - (-r_A)/C_{AO}}{k_1 + k_2 \upsilon_R} \qquad (6\text{-}136)$$

Equation 6-136 is the basis for the $r - T - X_A$ chart as developed by Levenspiel [11]. Assuming that there is no product recycle ($\theta_R = 0$) and that the reaction is unimolecular reversible reaction ($\upsilon_R = 1$), Equation 6-136 becomes

$$X_A = \frac{k_1 - (-r_A)/C_{AO}}{k_1 + k_2} \qquad (6\text{-}137)$$

For values of $E_1 = 11{,}600$ cal/gmol, $E_2 = 29{,}600$ cal/gmol, $A_1 = \exp(17.2)$, $A_2 = \exp(41.9)$, and $C_{AO} = 1$ mol/l. It is possible to solve Equation 6-137 with Equations 6-127 and 6-128 simultaneously for the temperature range of 260–400 K. A computer program was developed to determine the fractional conversion for different values of $(-r_A)$ and a temperature range from 260–400 K. Figure 6-28 shows the rate of reaction profile from the computer results. The plot shows the same profile given by Levenspiel [11].

Type 2 Reaction: $aA + bB \underset{k_2}{\overset{k_1}{\rightleftharpoons}} rR$

The general kinetic rate of disappearance of A is:

$$\left(-r_A\right) = k_1 C_A^{\alpha_1} C_B^{\alpha_2} - k_2 C_R^{\beta_1} \tag{6-138}$$

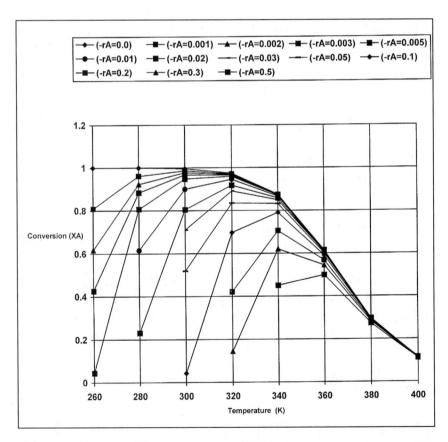

Figure 6-28. Rate of reaction profile for a first order reversible reaction A ↔ R.

From stoichiometry:

	A	**B**	**R**
At time t = 0	C_{AO}	C_{BO}	0
At time t = t	C_A	C_B	C_R
Amount reacted	$C_{AO} - C_A$	$C_{BO} - C_B$	C_R

$$\frac{C_{AO} - C_A}{C_{AO}} = \frac{C_{BO} - C_B}{C_{AO}} = \frac{C_R}{C_{AO}} \qquad (6\text{-}139)$$

$$C_{BO} - C_B = C_{AO}X_A \qquad (6\text{-}140)$$

Therefore,

$$C_B = C_{BO} - C_{AO}X_A$$

$$= C_{AO}\left(\frac{C_{BO}}{C_{AO}} - X_A\right)$$

Hence, $C_B = C_{AO}(\theta_B - X_A)$ or

$$C_B = C_{AO}(\theta_B + \upsilon_B X_A) \qquad (6\text{-}141)$$

where υ_b is the stoichiometric coefficient ratio of species B to that of A, which equals $-b/a$.

The fractional conversion X_A is given by $X_A = (C_{AO} - C_A)/C_{AO}$ and the concentration of A is expressed by

$$C_A = C_{AO}(1 - X_A) \qquad (5\text{-}28)$$

The concentration of R is expressed by

$$C_R = C_{AO}(\theta_R + \upsilon_R X_A) \qquad (6\text{-}131)$$

Substituting Equations 5-28, 6-131, and 6-141 into Equation 6-138 gives

$$(-r_A) = k_1 C_{AO}^{\alpha_1}(1 - X_A)^{\alpha_1} C_{AO}^{\alpha_2}(\theta_B + \upsilon_B X_A)^{\alpha_2}$$

$$- k_2 C_{AO}^{\beta_1}(\theta_R + \upsilon_R X_A)^{\beta_1} \qquad (6\text{-}142)$$

Rearranging Equation 6-142 gives

$$k_1 C_{AO}^{\alpha_1+\alpha_2}\left(1-X_A\right)^{\alpha_1}\left(\theta_B+\upsilon_B X_A\right)^{\alpha_2}$$

$$-k_2 C_{AO}^{\beta_1}\left(\theta_R+\upsilon_R X_A\right)^{\beta_1}-\left(-r_A\right)=0 \tag{6-143}$$

Equation 6-143 represents the $r - T - X_A$ profile equation if $\alpha_1 = \alpha_2 = \beta_1 = 1$ and $-\upsilon_A = -\upsilon_B = \upsilon_R = 1$.

For a product-free feed ($\theta_R = 0$) containing equimolecular concentrations of reactant A and B ($\theta_B = 1$), Equation 6-143 becomes

$$k_1 C_{AO}^2\left(1-X_A\right)^2-k_2 C_{AO}X_A-\left(-r_A\right)=0 \tag{6-144}$$

Equation 6-144 is further expressed to give

$$k_1 C_{AO}^2 X_A^2-\left(k_2 C_{AO}+2k_1 C_{AO}^2\right)X_A+k_1 C_{AO}^2-\left(-r_A\right)=0 \tag{6-145}$$

Equation 6-145 is a quadratic equation and the fractional conversion X_A is given by

$$X_A=\frac{\left(k_2 C_{AO}+2k_1 C_{AO}^2\right)}{\pm\left\{\left(k_2 C_{AO}+2k_1 C_{AO}^2\right)^2-4k_1 C_{AO}^2\left[k_1 C_{AO}^2-\left(-r_A\right)\right]\right\}^{0.5}}{2k_1 C_{AO}^2} \tag{6-146}$$

Using the values of $C_{AO} = 0.1$ mol/l, A_1, A_2, E_1, and E_2 as given in Type 1 reaction, it is possible to solve Equation 6-146 with Equations 6-127 and 6-128 simultaneously for the temperature range of 260–400 K. A computer program was developed to determine the fractional conversion for different values of $(-r_A)$ over the temperature range of 260–400 K. Figure 6-29 shows the reaction profile of the computer results.

Type 3 Reaction: $aA + bB \underset{k_2}{\overset{k_1}{\rightleftharpoons}} rR + sS$

The net rate of disappearance of A is

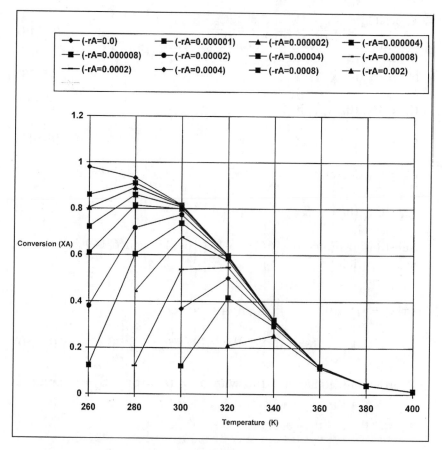

Figure 6-29. Rate of reaction profile for reaction type A + B ↔ R.

$$\left(-r_A\right)= k_1 C_A^{\alpha_1} C_B^{\alpha_2} - k_2 C_R^{\beta_1} C_S^{\beta_2} \tag{6-147}$$

From stoichiometry:

	A	**B**	**R**	**S**
At time t = 0	C_{AO}	C_{BO}	0	0
At time t = t	C_A	C_B	C_R	C_S
Amount reacted	$C_{AO} - C_A$	$C_{BO} - C_B$	C_R	C_S

Diving by the initial concentration of A gives

$$\frac{C_{AO} - C_A}{C_{AO}} = \frac{C_{BO} - C_B}{C_{AO}} = \frac{C_R}{C_{AO}} = \frac{C_S}{C_{AO}} \qquad (6\text{-}148)$$

$$C_A = C_{AO}(1 - X_A) \qquad (5\text{-}28)$$

$$C_R = C_{AO}(\theta_R + \upsilon_R X_A) \qquad (6\text{-}131)$$

$$C_B = C_{AO}(\theta_B + \upsilon_B X_A) \qquad (6\text{-}141)$$

$$C_S = C_{AO} X_A \text{ or}$$

$$C_S = C_{AO}(\theta_S + \upsilon_S X_A) \qquad (6\text{-}149)$$

Substituting Equations 5-28, 6-131, 6-141, and 6-149 into Equation 6-147 yields

$$(-r_A) = k_1 C_{AO}^{\alpha_1}(1 - X_A)^{\alpha_1} C_{AO}^{\alpha_2}(\theta_B + \upsilon_B X_A)^{\alpha_2}$$
$$- k_2 C_{AO}^{\beta_1}(\theta_R + \upsilon_R X_A)^{\beta_1} C_{AO}^{\beta_2}(\theta_S + \upsilon_S X_A)^{\beta_2} \qquad (6\text{-}150)$$

Rearranging Equation 6-150 with $\alpha_1 = \alpha_2 = \beta_1 = \beta_2 = 1$ gives

$$k_1 C_{AO}^2(1 - X_A)(\theta_B + \upsilon_B X_A)$$
$$- k_2 C_{AO}^2(\theta_R + \upsilon_R X_A)(\theta_S + \upsilon_S X_A) - (-r_A) = 0 \qquad (6\text{-}151)$$

If the feed contains equimolar concentrations of reactants A and B, then $\theta_B = 1$. If $-\upsilon_A = -\upsilon_B = \upsilon_R = \upsilon_S = 1$, and there is no product recycle ($\theta_R = \theta_S = 0$), Equation 6-151 becomes

$$k_1 C_{AO}^2(1 - X_A)^2 - k_2 C_{AO}^2 X_A^2 - (-r_A) = 0 \qquad (6\text{-}152)$$

Further expansion and rearrangement of Equation 6-152 gives

$$k_1 C_{AO}^2 - 2k_1 C_{AO}^2 X_A + k_1 C_{AO}^2 X_A^2 - k_2 C_{AO}^2 X_A^2 - (-r_A) = 0 \qquad (6\text{-}153)$$

Equation 6-153 can be expressed in the form of a quadratic equation in terms of the fractional conversion X_A as

$$(k_1 - k_2)X_A^2 - 2k_1 X_A + k_1 - \frac{(-r_A)}{C_{AO}^2} = 0 \qquad (6\text{-}154)$$

The fractional conversion X_A from Equation 6-154 becomes

$$X_A = \frac{2k_1 \pm \left\{ 4k_1^2 - 4(k_1 - k_2)\left[k_1 - (-r_A)/C_{AO}^2\right] \right\}^{0.5}}{2(k_1 - k_2)} \qquad (6\text{-}155)$$

Using the same values of the kinetic parameters as in Type 1, and given $C_{AO} = 0.1$ mol/l, it is possible to solve Equation 6-155 with Equations 6-127 and 6-128 simultaneously to determine the fractional conversion X_A. A computer program was developed to determine the fractional conversion for different values of $(-r_A)$ and a temperature range of 260–500 K. Figure 6-30 shows the reaction profile from the computer results.

If $k_1 = k_2 = 1$ (i.e., the equilibrium constant is unity), the value of X_A in Equation 6-155 becomes infinite and the equation is inapplicable. However, the value of X_A in Equation 6-154 becomes

$$-2k_1 X_A + k_1 - \frac{(-r_A)}{C_{AO}^2} = 0 \qquad (6\text{-}156)$$

$$\text{and } X_A = \frac{k_1 - (-r_A)/C_{AO}^2}{2k_1} \qquad (6\text{-}157)$$

$$\text{or } X_A = \frac{1 - (-r_A)/k_1 C_{AO}^2}{2} \qquad (6\text{-}158)$$

The corresponding $r - T - X_A$ data can be determined from Equation 6-158.

Type 4 Reaction: $aA \underset{k_2}{\overset{k_1}{\rightleftharpoons}} rR + sS$

The net rate of disappearance for this case is:

$$(-r_A) = k_1 C_A^{\alpha_1} - k_2 C_R^{\beta_1} C_S^{\beta_2} \qquad (6\text{-}159)$$

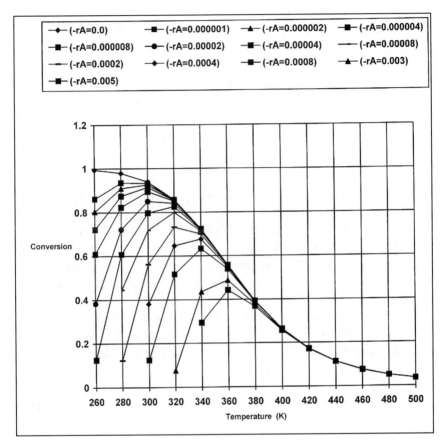

Figure 6-30. Rate of reaction profile for reaction type A + B ↔ R + S.

From stoichiometry:

	A	**R**	**S**
At time t = 0	C_{AO}	0	0
At time t = t	C_A	C_R	C_S
Amount reacted	$C_{AO} - C_A$	C_R	C_S

$$C_A - C_{AO} (1 - X_A) \tag{5-28}$$

$$C_R = C_{AO} (\theta_R + \upsilon_R X_A) \tag{6-131}$$

$$C_S = C_{AO} (\theta_S + \upsilon_S X_A) \tag{6-149}$$

Substituting Equations 5-28, 6-131, and 6-149 into Equation 6-159 gives

$$(-r_A) = k_1 C_{AO}^{\alpha_1}(1 - X_A)^{\alpha_1}$$

$$- k_2 C_{AO}^{\beta_1}(\theta_R + \upsilon_R X_A)^{\beta_1} C_{AO}^{\beta_2}(\theta_S + \upsilon_S X_A)^{\beta_2} \qquad (6\text{-}160)$$

Rearranging Equation 6-160 with $\alpha_1 = \beta_1 = \beta_2 = 1$ gives

$$k_1 C_{AO}(1 - X_A) - k_2 C_{AO}^2(\theta_R + \upsilon_R X_A)(\theta_S + \upsilon_S X_A) - (-r_A) = 0 \quad (6\text{-}161)$$

Again, if the feed contains equimolar concentrations of pure A, with $-\upsilon_A = \upsilon_R = \upsilon_S = 1$, and there is no product recycle ($\theta_R = \theta_S = 0$), Equation 6-161 becomes

$$k_1 C_{AO}(1 - X_A) - k_2 C_{AO}^2 X_A^2 - (-r_A) = 0 \qquad (6\text{-}162)$$

or $\quad k_2 C_{AO}^2 X_A^2 + k_1 C_{AO} X_A - k_1 C_{AO} + (-r_A) = 0 \qquad (6\text{-}163)$

Equation 6-163 is a quadratic equation in terms of the fractional conversion X_A. The roots of X_A become

$$X_A = \frac{-k_1 C_{AO} \pm \left\{ (k_1 C_{AO})^2 - 4 k_2 C_{AO}^2 \left[-k_1 C_{AO} + (-r_A) \right] \right\}^{0.5}}{2 k_2 C_{AO}^2} \qquad (6\text{-}164)$$

or

$$X_A = \frac{-k_1 C_{AO} \pm \left\{ (k_1 C_{AO})^2 + 4 k_2 C_{AO}^2 \left[k_1 C_{AO} - (-r_A) \right] \right\}^{0.5}}{2 k_2 C_{AO}^2} \qquad (6\text{-}165)$$

Using the same values of the kinetic parameters as in Type 3, with $C_{AO} = 1$ mol/l and solving Equations 6-165, 6-127, and 6-128 simultaneously, it is possible to determine the fractional conversion X_A. A computer program was developed to determine the fractional conversion for different values of $(-r_A)$ and a temperature range of 260–500 K. Figure 6-31 illustrates the reaction profile.

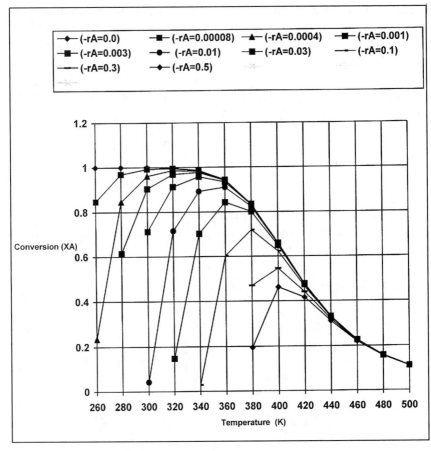

Figure 6-31. Rate of reaction profile for reaction type A ↔ R + S.

OPTIMUM TEMPERATURE PROGRESSION

Following the path of the locus of the maximum rates, referred to as the optimum temperature progression (OTP), determines the ideal optimum reactor size for an exothermic reaction (Figures 6-28, 6-29, 6-30, and 6-31). It can also establish a functional relationship between the optimum temperature, T_{opt}, and the conversion X_A for any type of reaction and kinetics by developing a computer program. This alleviates the cumbersome graphical reactor design [11,12], and the mathematical relation can be incorporated into the reactor design equations.

The following derives equations for the optimum temperature progression for the four types of reactions previously described at constant density. Figures 6-28, 6-29, 6-30, and 6-31 show that the $r - T - X_A$ profile for an exothermic reaction is unimodal. Therefore, the optimum temperature T_{opt} is obtained by performing a partial differentiation of $(-r_A)$ with respect to temperature T at a constant conversion X_A and equating this to zero. That is,

$$\left[\frac{\partial(-r_A)}{\partial T}\right]_{X_A} = 0 \tag{6-166}$$

Therefore, using the general kinetic Equation 6-126 gives

$$\left[\frac{\partial(-r_A)}{\partial T}\right]_{X_A} = \frac{\partial k_1}{\partial T}\prod_{i=1}^{M}C_i^{\alpha_i} - \frac{\partial k_2}{\partial T}\prod_{j=1}^{N}C_j^{\beta_j} = 0 \tag{6-167}$$

Both C_i and C_j are only functions of X_A where

$$k_1 = A_1 \exp\left(\frac{-E_1}{RT}\right) \tag{6-127}$$

and $k_2 = A_2 \exp\left(\frac{-E_2}{RT}\right)$ \hspace{1cm} (6-128)

Differentiating k_1 with respect to T in Equation 6-127 gives

$$\frac{\partial k_1}{\partial T} = \frac{A_1 E_1 \exp(-E_1/RT)}{RT^2} \tag{6-168}$$

and differentiating k_2 with respect to T in Equation 6-128 gives

$$\frac{\partial k_2}{\partial T} = \frac{A_2 E_2 \exp(-E_2/RT)}{RT^2} \tag{6-169}$$

Substituting Equations 6-168 and 6-169 into Equation 6-167 gives

$$\left[\frac{\partial(-r_A)}{\partial T}\right]_{X_A} = \frac{A_1 E_1}{RT^2} \exp\left(\frac{-E_1}{RT}\right) \prod_{i=1}^{M} C_i^{\alpha_i}$$

$$-\frac{A_2 E_2}{RT^2} \exp\left(\frac{-E_2}{RT}\right) \prod_{j=1}^{N} C_j^{\beta_j} = 0 \quad \text{at } T = T_{opt} \qquad (6\text{-}170)$$

Consider the solution of Equation 6-170 for each of the four types of rate expressions to determine the optimum temperature progression at any given fractional conversion X_A.

Type 1 Reaction: $aA \overset{k_1}{\underset{k_2}{\rightleftharpoons}} rR$

The net rate of disappearance of A is

$$(-r_A) = k_1 C_A^{\alpha_1} - k_2 C_R^{\beta_1} \qquad (6\text{-}130)$$

Substituting Equation 6-130 into Equation 6-170 yields

$$\left[\frac{\partial(-r_A)}{\partial T}\right]_{X_A} = \frac{A_1 E_1}{RT^2} \exp\left(\frac{-E_1}{RT}\right) C_A^{\alpha_1}$$

$$-\frac{A_2 E_2}{RT^2} \exp\left(\frac{-E_2}{RT}\right) C_R^{\beta_1} = 0 \quad \text{at } T = T_{opt} \qquad (6\text{-}171)$$

where

$$C_A = C_{AO} (1 - X_A) \qquad (5\text{-}28)$$

$$C_R = C_{AO} (\theta_R + \upsilon_R X_A) \qquad (6\text{-}131)$$

Substituting Equations 6-28 and 6-131 into Equation 6-171 yields

$$\frac{A_1 E_1}{RT^2} \exp\left(\frac{-E_1}{RT}\right) C_{AO}^{\alpha_1} (1 - X_A)^{\alpha_1}$$

$$-\frac{A_2 E_2}{RT^2} \exp\left(\frac{-E_2}{RT}\right) C_{AO}^{\beta_1} (\theta_R + \upsilon_R X_A)^{\beta_1} = 0 \qquad (6\text{-}172)$$

Rearrangement of Equation 6-172 gives

$$A_1E_1 \exp\left(\frac{E_2}{RT} - \frac{E_1}{RT}\right) C_{AO}^{\alpha_1} (1-X_A)^{\alpha_1} = A_2 E_2 C_{AO}^{\beta_1} (\theta_R + \upsilon_R X_A)^{\beta_1}$$

and the optimum temperature progression T_{opt} is

$$T_{opt} = \frac{E_2 - E_1}{R \ln\left\{\dfrac{A_2 E_2 C_{AO}^{\beta_1 - \alpha_1} (\theta_R + \upsilon_R X_A)^{\beta_1}}{A_1 E_1 (1-X_A)^{\alpha_1}}\right\}} \qquad (6\text{-}173)$$

Equation 6-173 suggests that for every value of X_A, there is a corresponding temperature T_{opt} on the optimum temperature progression OTP.

Consider a first order reversible reaction $A \underset{k_2}{\overset{k_1}{\rightleftharpoons}} R$, where the rate expression is $(-r_A) = k_1 C_A - k_2 C_R$. Therefore, $\alpha_1 = \beta_1 = \upsilon_R = 1$. For a feed concentration, $C_{AO} = 1.0$ mol/l with no product ($\theta_R = 0$), using $X_A = 0.485$ with the following kinetic parameters gives

$E_1 = 11,600$ cal/mol
$E_2 = 29,600$ cal/mol
$R = 1.987$ cal/mol $\bullet$ K
$A_1 = \exp(17.2)$
$A_2 = \exp(41.9)$

The optimum temperature progression T_{opt} is

$$T_{opt} = \frac{E_2 - E_1}{R \ln\left\{\dfrac{A_2 E_2 X_A}{A_1 E_1 (1-X_A)}\right\}} \qquad (6\text{-}174)$$

$$T_{opt} = \frac{(29,600 - 11,600)}{1.987 \ln\left\{\dfrac{\exp(41.9)(29,600)(0.485)}{\exp(17.2)(11,600)(0.515)}\right\}}$$

$$= 354.18 \text{ K}$$

This agrees with the value of 354 K at $X_A = 0.485$ obtained from geometrical construction of the optimum temperature progression given by Levenspiel [11]. A Microsoft Excel Spreadsheet (OPTIMUM61.xls) was used to determine the optimum temperature progression (T_{opt}) for varying values of fractional conversion (X_A). Table 6-10 gives the results of the spreadsheet calculation, and Figure 6-32 shows that the optimum temperature progression decreases with increasing fractional conversion.

Table 6-10
Optimum temperature progression for a
first order reversible reaction A $\leftrightarrow$ R

Kinetic parameters	
E_1	11,600
E_2	29,600
R	1.987
A_1	17.2
A_2	41.9
C_{AO}	1

Conv. X_A	Temp (K)
0.1	386.5
0.15	378.0
0.2	373.6
0.25	369.2
0.3	365.4
0.35	362.1
0.4	359.0
0.45	356.1
0.485	354.2
0.5	353.4
0.55	350.6
0.6	347.9
0.65	345.0
0.7	342.1
0.75	338.8
0.8	335.2
0.85	331.0
0.9	325.5
0.95	317.0
0.99	299.6

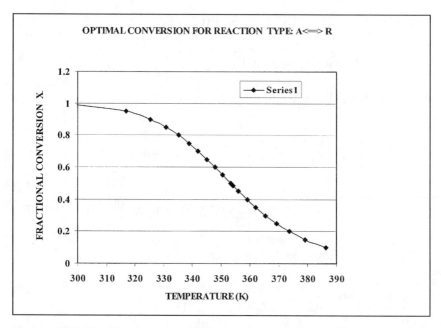

Figure 6-32. Fractional conversion X_A versus T_{opt} for a first order reversible reaction A ↔ R.

Type 2 Reaction: $aA + bB \underset{k_2}{\overset{k_1}{\rightleftharpoons}} rR$

The rate expression is

$$\left(-r_A\right) = k_1 C_A^{\alpha_1} C_B^{\alpha_2} - k_2 C_R^{\beta_1} \tag{6-138}$$

with

$$C_A = C_{AO}(1 - X_A) \tag{5-28}$$

$$C_R = C_{AO}(\theta_R + \upsilon_R X_A) \tag{6-131}$$

$$C_B = C_{AO}(\theta_B + \upsilon_B X_A) \tag{6-141}$$

Substituting Equations 5-28, 6-131, and 6-141 into Equation 6-138 gives

$$\left(-r_A\right) = k_1 C_{AO}^{\alpha_1}\left(1-X_A\right)^{\alpha_1} C_{AO}^{\alpha_2}\left(\theta_B + \upsilon_B X_A\right)^{\alpha_2}$$

$$- k_2 C_{AO}^{\beta_1}\left(\theta_R + \upsilon_R X_A\right)^{\beta_1} \qquad (6\text{-}175)$$

when $[\partial(-r_A)/\partial T]_{X_A} = 0$, then

$$\left[\frac{\partial\left(-r_A\right)}{\partial T}\right]_{X_A} = \frac{A_1 E_1}{RT^2}\exp\left(\frac{-E_1}{RT}\right)C_{AO}^{\alpha_1+\alpha_2}\left(\theta_B + \upsilon_B X_A\right)^{\alpha_2}\left(1-X_A\right)^{\alpha_1}$$

$$- \frac{A_2 E_2}{RT^2}\exp\left(\frac{-E_2}{RT}\right)C_{AO}^{\beta_1}\left(\theta_R + \upsilon_R X_A\right)^{\beta_1} = 0 \qquad (6\text{-}176)$$

Rearranging Equation 6-176 gives

$$\frac{A_1 E_1}{RT^2}\exp\left(\frac{-E_1}{RT}\right)C_{AO}^{\alpha_1+\alpha_2}\left(\theta_B + \upsilon_B X_A\right)^{\alpha_2}\left(1-X_A\right)^{\alpha_1}$$

$$= \frac{A_2 E_2}{RT^2}\exp\left(\frac{-E_2}{RT}\right)C_{AO}^{\beta_1}\left(\theta_R + \upsilon_R X_A\right)^{\beta_1} \qquad (6\text{-}177)$$

The optimum temperature progression at $T = T_{opt}$ from Equation 6-177 is

$$T_{opt} = \frac{E_2 - E_1}{R\ln\left\{\dfrac{A_2 E_2 C_{AO}^{\beta_1-(\alpha_1+\alpha_2)}\left(\theta_R + \upsilon_R X_A\right)^{\beta_1}}{A_1 E_1\left(1-X_A\right)^{\alpha_1}\left(\theta_B + \upsilon_B X_A\right)^{\alpha_2}}\right\}} \qquad (6\text{-}178)$$

For a reaction $A + B \underset{k_2}{\overset{k_1}{\rightleftharpoons}} R$, whose kinetics is first order with respect to the reactants A and B (that is, $\alpha_1 = \alpha_2 = 1$), and the product R (i.e., $\beta_1 = 1$) occurs in a reactor fed with pure equimolar reactant concentration, then $\theta_B = 1$, $\theta_R = 0$, $\upsilon_R = -\upsilon_B = 1$. Putting these values into Equation 6-178 yields

$$T_{opt} = \frac{E_2 - E_1}{R\ln\left\{\dfrac{A_2 E_2 C_{AO}^{-1} X_A}{A_1 E_1\left(1-X_A\right)^2}\right\}} \qquad (6\text{-}179)$$

Using the kinetic parameters in Type 1 reaction at $X_A = 0.5$, the optimum temperature progression T_{opt} is

$$T_{opt} = \frac{(29,600 - 11,600)}{1.987 \ln\left\{\dfrac{\left[(e^{24.7})(29,600)(0.5)\right]}{(11,600)(0.5^2)}\right\}}$$

$$= 344.05 \text{ K}$$

The Microsoft Excel Spreadsheet (OPTIMUM62.xls) was used to determine T_{opt} for varying values of X_A ($0.1 < X_A < 0.99$). Table 6-11 gives the results of the spreadsheet calculation, and Figure 6-33 shows the profile of T_{opt} against X_A. Figure 6-33 gives the same profile as Type 1.

Type 3 Reaction: aA + bB $\underset{k_2}{\overset{k_1}{\rightleftharpoons}}$ rR + sS

The rate expression is

$$(-r_A) = k_1 C_A^{\alpha_1} C_B^{\alpha_2} - k_2 C_R^{\beta_1} C_S^{\beta_2} \qquad (6\text{-}147)$$

with

$$C_A = C_{AO}(1 - X_A) \qquad (5\text{-}28)$$

$$C_R = C_{AO}(\theta_R + \upsilon_R X_A) \qquad (6\text{-}131)$$

$$C_B = C_{AO}(\theta_B + \upsilon_B X_A) \qquad (6\text{-}141)$$

$$C_S = C_{AO}(\theta_S + \upsilon_S X_A) \qquad (6\text{-}149)$$

Substituting Equations 5-28, 6-131, 6-141, and 6-149 into Equation 6-147 gives

$$(-r_A) = k_1 C_{AO}^{\alpha_1}(1 - X_A)^{\alpha_1} C_{AO}^{\alpha_2}(\theta_B + \upsilon_B X_A)^{\alpha_2}$$

$$- k_2 C_{AO}^{\beta_1}(\theta_R + \upsilon_R X_A)^{\beta_1} C_{AO}^{\beta_2}(\theta_S + \upsilon_S X_A)^{\beta_2} \qquad (6\text{-}180)$$

At $[\partial(-r_A)/\partial T]_{X_A} = 0$, Equation 6-180 becomes

Table 6-11
Optimum temperature progression
for reaction type A + B ↔ R

Kinetic parameters	
E_1	11,600
E_2	29,600
R	1.987
A_1	17.2
A_2	41.9
C_{AO}	1

Conv. X_A	Temp (K)
0.1	384.7
0.15	376.4
0.2	370.1
0.25	364.9
0.3	360.2
0.35	356.0
0.4	351.9
0.45	348.0
0.485	345.2
0.5	344.1
0.55	340.1
0.6	336.0
0.65	331.8
0.7	327.2
0.75	322.1
0.8	316.4
0.85	309.5
0.9	300.6
0.95	286.9
0.99	260.0

$$\left[\frac{\partial(-r_A)}{\partial T}\right]_{X_A} = \frac{A_1 E_1}{RT^2}\exp\left(\frac{-E_1}{RT}\right)C_{AO}^{\alpha_1}(1-X_A)^{\alpha_1}C_{AO}^{\alpha_2}$$

$$\times(\theta_B + \upsilon_B X_A)^{\alpha_2} - \frac{A_2 E_2}{RT^2}\exp\left(\frac{-E_2}{RT}\right)$$

$$\times C_{AO}^{\beta_1}(\theta_R + \upsilon_R X_A)^{\beta_1}C_{AO}^{\beta_2}(\theta_S + \upsilon_S X_A)^{\beta_2} = 0 \qquad (6\text{-}181)$$

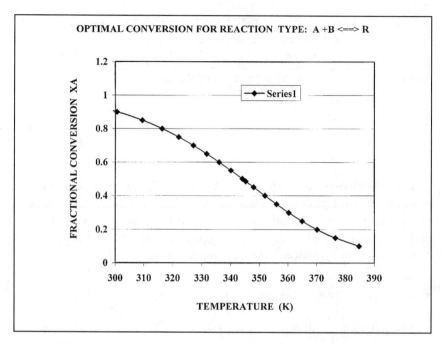

Figure 6-33. Fractional conversion X_A versus T_{opt} for reaction type A + B $\leftrightarrow$ R.

At $T = T_{opt}$ and rearranging Equation 6-181, the optimum temperature progression T_{opt} is

$$T_{opt} = \frac{E_2 - E_1}{R \ln\left\{\dfrac{A_2 E_2 C_{AO}^{\beta_1 + \beta_2 - (\alpha_1 + \alpha_2)} \left(\theta_R + \upsilon_R X_A\right)^{\beta_1} \left(\theta_S + \upsilon_S X_A\right)^{\beta_2}}{A_1 E_1 \left(1 - X_A\right)^{\alpha_1} \left(\theta_B + \upsilon_B X_A\right)^{\alpha_2}}\right\}} \quad (6\text{-}182)$$

Consider the reaction A + B $\underset{k_2}{\overset{k_1}{\rightleftharpoons}}$ R + S, which has a first order kinetics with respect to each participating species, that is, $\alpha_1 = \alpha_2 = \beta_1 = \beta_2 = 1$. Also, from the stoichiometry, $-\upsilon_B = \upsilon_R = \upsilon_S = 1$ and the feed conditions are such that $\theta_R = \theta_S = 0$, $\theta_B = 1$. Equation 6-182 becomes

$$T_{opt} = \frac{E_2 - E_1}{R \ln\left\{\dfrac{A_2 E_2 X_A^2}{A_1 E_1 \left(1 - X_A\right)^2}\right\}} \quad (6\text{-}183)$$

Using the kinetic parameters in Type 1 reaction with $X_A = 0.5$, T_{opt} is

$$T_{opt} = \frac{(29,600 - 11,600)}{1.987 \ln\left\{\dfrac{\left[(e^{24.7})(29,600)(0.5^2)\right]}{(11,600)(0.5^2)}\right\}}$$

$$= 353.36\,K$$

The Microsoft Excel Spreadsheet (OPTIMUM63.xls) was used to evaluate T_{opt} for varying values of X_A. Table 6-12 gives the results of the spreadsheet calculation, and Figure 6-34 illustrates the profile of T_{opt} against X_A, showing that the optimum temperature progression decreases as the fractional conversion increases.

Type 4 Reaction: $aA \underset{k_2}{\overset{k_1}{\rightleftharpoons}} rR + sS$

The net rate of disappearance of A is

$$\left(-r_A\right) = k_1 C_A^{\alpha_1} - k_2 C_R^{\beta_1} C_S^{\beta_2} \tag{6-159}$$

with

$$C_A = C_{AO}(1 - X_A) \tag{5-28}$$

$$C_R = C_{AO}(\theta_R + \upsilon_R X_A) \tag{6-131}$$

$$C_S = C_{AO}(\theta_S + \upsilon_S X_A) \tag{6-149}$$

Substituting Equations 5-28, 6-131 and 6-149 into Equation 6-159 gives

$$\left(-r_A\right) = k_1 C_{AO}^{\alpha_1}\left(1 - X_A\right)^{\alpha_1}$$

$$- k_2 C_{AO}^{\beta_1}\left(\theta_R + \upsilon_R X_A\right)^{\beta_1} C_{AO}^{\beta_2}\left(\theta_S + \upsilon_S X_A\right)^{\beta_2} \tag{6-184}$$

At $[\partial(-r_A)/\partial T]_{X_A} = 0$, then

Table 6-12
Optimum temperature progression
for reaction type A + B ↔ R + S

Kinetic parameters	
E_1	11,600
E2	29,600
R	1.987
A_1	17.2
A_2	41.9
C_{AO}	1

Conv. X_A	Temp (K)
0.1	426.5
0.15	408.7
0.2	396.2
0.25	386.5
0.3	378.4
0.35	371.3
0.4	364.9
0.45	359.0
0.485	355.0
0.5	353.4
0.55	347.9
0.6	342.5
0.65	337.1
0.7	331.4
0.75	325.5
0.8	318.9
0.85	311.2
0.9	301.6
0.95	287.3
0.99	260.1

$$\frac{A_1 E_1}{RT^2} \exp\left(\frac{-E_1}{RT}\right) C_{AO}^{\alpha_1} \left(1 - X_A\right)^{\alpha_1}$$

$$-\frac{A_2 E_2}{RT^2} \exp\left(\frac{-E_2}{RT}\right) C_{AO}^{\beta_1} \left(\theta_R + \upsilon_R X_A\right)^{\beta_1} C_{AO}^{\beta_2} \left(\theta_S + \upsilon_S X_A\right)^{\beta_2} = 0 \quad (6\text{-}185)$$

Rearranging Equation 6-185 yields

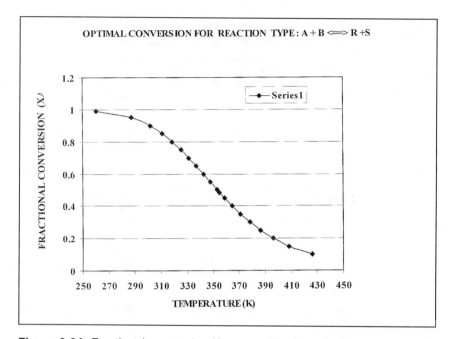

Figure 6-34. Fractional conversion X_A versus T_{opt} for reaction type A + B $\leftrightarrow$ R + S.

$$\frac{A_1 E_1}{RT^2} \exp\left(\frac{-E_1}{RT}\right) C_{AO}^{\alpha_1} (1 - X_A)^{\alpha_1}$$

$$= \frac{A_2 E_2}{RT^2} \exp\left(\frac{-E_2}{RT}\right) C_{AO}^{\beta_1 + \beta_2} (\theta_R + \upsilon_R X_A)^{\beta_1} (\theta_S + \upsilon_S X_A)^{\beta_2}$$

or

$$\exp\left(\frac{E_2 - E_1}{RT}\right) = \frac{A_2 E_2 C_{AO}^{\beta_1 + \beta_2} (\theta_R + \upsilon_R X_A)^{\beta_1} (\theta_S + \upsilon_S X_A)^{\beta_2}}{A_1 E_1 C_{AO}^{\alpha_1} (1 - X_A)^{\alpha_1}} \qquad (6\text{-}186)$$

The optimum temperature progression at $T = T_{opt}$ from Equation 6-186 is

$$T_{opt} = \frac{E_2 - E_1}{R \ln\left\{\dfrac{A_2 E_2 C_{AO}^{\beta_1 + \beta_2 - \alpha_1} (\theta_R + \upsilon_R X_A)^{\beta_1} (\theta_S + \upsilon_S X_A)^{\beta_2}}{A_1 E_1 (1 - X_A)^{\alpha_1}}\right\}} \qquad (6\text{-}187)$$

If $\upsilon_R = \upsilon_S = 1$, $\theta_R = \theta_S = 0$, and $\alpha_1 = \beta_1 = \beta_2 = 1$, then Equation 6-187 becomes

$$T_{opt} = \frac{E_2 - E_1}{R \ln\left\{\dfrac{A_2 E_2 C_{AO} X_A^2}{A_1 E_1 (1 - X_A)}\right\}} \tag{6-188}$$

Using the kinetic parameters in Type 1 reaction with $X_A = 0.5$, T_{opt} is

$$T_{opt} = \frac{(29,600 - 11,600)}{1.987 \ln\left\{\dfrac{\left(e^{24.7}\right)(29,600)(1.0)\left(0.5^2\right)}{(11,600)(0.5)}\right\}}$$

$$= 363.17 \text{ K}$$

The Microsoft Excel Spreadsheet (OPTIMUM64.xls) was used to calculate T_{opt} for values of X_A $(0.1 < X_A < 0.99)$. Table 6-13 gives the results of the spreadsheet calculation, and Figure 6-35 shows the profile of the T_{opt} versus X_A.

Adesina [14] considered the four main types of reactions for variable density conditions. It was shown that if the sums of the orders of the reactants and products are the same, then the OTP path is independent of the density parameter, implying that the ideal reactor size would be the same as no change in density. The optimal rate behavior with respect to T and the optimal temperature progression (T_{opt}) have important roles in the design and operation of reactors performing reversible, exothermic reactions. Examples include the oxidation of SO_2 to SO_3 and the synthesis of NH_3 and methanol CH_3OH.

MINIMUM REACTOR VOLUME AT THE OTP OF A SINGLE CFSTR WITH A REVERSIBLE EXOTHERMIC REACTION

Consider the reversible first order reaction $A \underset{k_2}{\overset{k_1}{\rightleftharpoons}} R$. It is possible to determine the minimum reactor volume at the optimum temperature T_{opt} that is required to obtain a fractional conversion X_A, if the feed is pure A with a volumetric flowrate of u. A material balance for a CFSTR is

Table 6-13
Optimum temperature progression
for reaction type A ↔ R + S

Kinetic parameters	
E_1	11,600
E_2	29,600
R	1.987
A_1	17.2
A_2	41.9
C_{AO}	1

Conv. X_A	Temp (K)
0.1	428.6
0.15	411.7
0.2	400.1
0.25	391.3
0.3	384.1
0.35	378.0
0.4	372.6
0.45	367.7
0.485	364.5
0.5	363.2
0.55	358.9
0.6	354.8
0.65	350.8
0.7	346.7
0.75	342.5
0.8	338.0
0.85	332.9
0.9	326.7
0.95	317.5
0.99	299.7

$$uC_{AO} = uC_A + (-r_A)V_R + V_R \frac{dC_A}{dt} \qquad (5\text{-}335)$$

Assuming there is no accumulation, then

$$V_R \frac{dC_A}{dt} = 0$$

$$uC_{AO} = uC_A + (-r_A)V_R \qquad (5\text{-}158)$$

The rate expression for the first order reversible reaction is

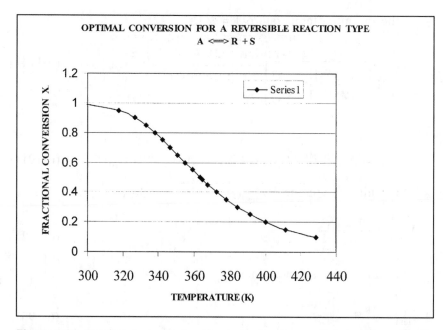

Figure 6-35. Fractional conversion X_A versus T_{opt} for reaction type A $\leftrightarrow$ R + S.

$$(-r_A) = k_1 C_A - k_2 C_R \qquad (6\text{-}189)$$

where $k_1 = A_1 e^{-E_1/RT}$ and $k_2 = A_2 e^{-E_2/RT}$. From stoichiometry, $C_{AO} - C_A = C_R$ and $C_A = C_{AO}(1 - X_A)$, $C_{AO} - C_A = C_{AO}X_A$. The volume V_R of the reactor is

$$V_R = \frac{u(C_{AO} - C_A)}{(-r_A)}$$

$$\text{or } V_R = \frac{u C_{AO} X_A}{(-r_A)} \qquad (5\text{-}160)$$

Substituting the constants k_1 and k_2 into the Equation 6-189 gives

$$(-r_A) = A_1 e^{-E_1/RT} C_A - A_2 e^{-E_2/RT} C_R \qquad (6\text{-}190)$$

or

$$(-r_A) = A_1 e^{-E_1/RT} C_{AO}(1 - X_A) - A_2 e^{-E_2/RT} C_{AO} X_A$$

At the optimum temperature $T = T_{opt}$

$$T_{opt} = \frac{E_2 - E_1}{R \ln\left\{\dfrac{A_2 E_2 X_A}{A_1 E_1(1 - X_A)}\right\}} \tag{6-174}$$

and

$$(-r_A)_{opt} = A_1 e^{-E_1/RT_{opt}} C_{AO}(1 - X_A) - A_2 e^{-E_2/RT_{opt}} C_{AO} X_A \tag{6-191}$$

The minimum volume $V_{R_{min}}$ is

$$V_{R_{min}} = \frac{u C_{AO} X_A}{(-r_A)_{opt}} \tag{6-192}$$

The minimum residence time $\bar{t}_{min}$ is

$$\bar{t}_{min} = \frac{V_{R_{min}}}{u} \tag{6-193}$$

For the minimum volume in a CFSTR under adiabatic condition, Equation 6-192 requires only that the operating temperature within the reactor be T_{opt}. In this case, there is nothing about the mode of operation, such as the feed temperature T_O or heat transfer, either within the reactor or upstream of it. However, if the CFSTR is operated adiabatically for specified conversion (X_A) and molar flowrate $(F_{AO} = u C_{AO})$ without internal heat transfer, T_O must be adjusted to a value obtained from the energy balance as determined by

$$u \rho C_P \left(T_{opt} - T_O\right) = \left(\frac{-\Delta H_R}{a}\right)(u C_{AO})(X_A) \tag{6-194}$$

$$\text{or } G C_P \left(T_{opt} - T_O\right) = \left(\frac{-\Delta H_R}{a}\right)(F_{AO})(X_A) \tag{6-195}$$

OPTIMUM REACTOR SIZE

The purpose of the reactor design is to acquire the optimum reactor size. However, the ideal optimum reactor size $(V/F_{AO})_{opt}$, is obtained by integrating the optimum temperature profile. That is,

$$\left(\frac{V}{F_{AO} = uC_{AO}}\right)_{opt} = \int_{X_{A_i}}^{X_{Af}} \frac{dX_A}{(-r_A)_{opt}}$$ (6-196)

Equation 6-196 may be impossible to obtain because the expression for $(-r_A)_{opt}$ is nonlinear in T_{opt}. The numerical integration method is therefore employed. If T_{opt}, corresponding to the inlet conversion, X_A, is greater than the maximum allowable temperature T_{max} at which the reactor can operate, the integration process must be adjusted so that the reactor operates isothermally at T_{max} until the conversion falls on the optimum temperature progression profile. Then the integration is made to follow the OTP. The following algorithm is used for the modified integration [13].

Step 1: Using the inlet conversion, X_{A_i}, calculate T_{opt}:
    (a) If $T_{opt} \geq T_{max}$, use T_{max} to calculate k_1 and k_2, hence $(-r_A)_{opt}$.
    (b) If $T_{opt} < T_{max}$, use T_{opt} to calculate k_1 and k_2, hence $(-r_A)_{opt}$.
Step 2: Increase X_{A_i}, from X_{A_i} to $X_{A_i} + h$, where h is the step size for X_{A_i} in the numerical integration process. Again, evaluate T_{opt}.
Step 3: Repeat the check in step 1 followed by step 2 until (1b) is satisfied. At this point, the OTP is now being followed.
Step 4: Increase X_{A_i} to $X_{A_i} + h$; estimate T_{opt}. Use this value to calculate k_1 and k_2, and hence $(-r_A)_{opt}$.
Step 5: Repeat step 4 until $X_A = X_{AF}$.
Step 6: Use any numerical method of integration to determine $(V/F_{AO})_{opt}$ in Equation 6-196.

Example 6-14

For an exothermic, liquid-phase, reversible reaction $A \underset{k_2}{\overset{k_1}{\rightleftharpoons}} R$ occurring in a CFSTR, with a feedrate of 600 mol/min^{-1}, a feed concentration (C_{AO}) of 6 mol/L^{-1} ($C_{RO} = 0$) and a fractional conversion (X_A) of 0.65, calculate

1. The optimal temperature (T/K) in the reactor for minimum volume.

2. The minimum volume and residence time required for the reactor.
3. The feed temperature (T_O) for the adiabatic operation at the optimal temperature T_{opt}.

Additional data (rate constants taken from Levenspiel [11]):

$$(-r_A) = r_R = k_1C_A - k_2C_R \text{ mol } L^{-1}\text{min}^{-1}$$

$$k_1 = \exp\left(17.2 - \frac{48,900}{RT}\right), \text{ min}^{-1}$$

$$k_2 = \exp\left(41.9 - \frac{123,800}{RT}\right), \text{ min}^{-1}$$

$$(-\Delta H_R/a) = 70,000 \text{ Jmol}^{-1}; \ C_P(\text{stream}) = 4.2 \text{ Jg}^{-1}K^{-1}$$

$$\rho(\text{stream}) = 1,000 \text{ gL}^{-1}; \ R = 8.314 \text{ Jmol}^{-1}K^{-1}$$

$$C_P(\text{stream}) = 4.2 \text{ J/g} \cdot K$$

Solution

The optimum temperature T_{opt} is determined by Equation 6-174:

$$T_{opt} = \frac{E_2 - E_1}{R \ln\left\{\dfrac{A_2E_2X_A}{A_1E_1(1-X_A)}\right\}} \tag{6-174}$$

$$E_2 = 123,800$$

$$E_1 = 48,900$$

$$A_2 = \exp(41.9) = 15.74 \times 10^{17}$$

$$A_1 = \exp(17.2) = 29.5 \times 10^6$$

$$R = 8.314 \text{ J/mol} \cdot K$$

(1) The optimum temperature T_{opt} is:

$$T_{opt} = \frac{(123,800 - 48,900)}{8.314 \ln \left\{ \dfrac{15.74 \times 10^{17} \times 123,800 \times 0.65}{29.5 \times 10^6 \times 48,900 \times 0.35} \right\}}$$

T_{opt} = 343.2 K (70.1°C)

The minimum volume V_R is determined by Equation 6-192 as

$$V_R = \frac{uC_{AO}X_A}{(-r_A)_{opt}} \tag{6-192}$$

The rate expression at the optimum temperature $T = T_{opt}$ is

$$(-r_A)_{opt} = A_1 e^{-E_1/RT_{opt}} C_{AO}(1 - X_A) - A_2 e^{-E_2/RT_{opt}} C_{AO}X_A$$

$$(-r_A)_{opt} = 29.5 \times 10^6 \exp\left\{ \frac{-48,900}{(8.314 \times 343.2)} \right\} \times 6.0 \times 0.35$$

$$- 15.74 \times 10^{17} \exp\left\{ \frac{-123,800}{(8.314 \times 343.2)} \right\} \times 6.0 \times 0.65$$

$(-r_A)_{opt}$ = 1.353 mol L^{-1} min^{-1}

$F_{AO} = uC_{AO}$

Therefore, the volumetric flowrate

$$u = \frac{F_{AO}}{C_{AO}} = \frac{600}{6} \left(\frac{mol}{min} \cdot \frac{L}{mol} \right) = 100 \frac{L}{min}$$

(2) The minimum volume is determined by

$$V_{R\,min} = \frac{uC_{AO}X_A}{(-r_A)_{opt}} = \frac{100 \times 6.0 \times 0.65}{1.353} \qquad \left\{ \frac{L}{min} \times \frac{mol}{L} \times \frac{L \cdot min}{mol} \right\}$$

$$V_{R_{min}} = 288 \, l$$

The minimum residence time $\bar{t}_{min} = V_{R_{min}}/u = 288/100$

$$\bar{t}_{min} = 2.9 \, min$$

(3) The feed temperature T_O for the adiabatic operation at the optimal temperature T_{opt} is determined from the heat balance, $-G\Delta h + Q = 0$, where $G = \rho u$ and

$$\Delta h = C_p dT + \left(\frac{-\Delta H_R}{a}\right)\left(\frac{\Delta \bar{m}_a}{M_A}\right).$$

Therefore,

$$-\rho u \left\{ C_p \Delta T + \left(\frac{-\Delta H_R}{a}\right)\left(\frac{\Delta \bar{m}_A}{M_A}\right) \right\} + Q = 0$$

At adiabatic condition $Q = 0$

$$-\rho u C_p \left(T_{opt} - T_O\right) - u\left(\frac{-\Delta H_R}{a}\right)\left(C_A - C_{AO}\right) = 0$$

Rearranging the above equation gives

$$\rho u C_p \left(T_O - T_{opt}\right) = -u\left(\frac{-\Delta H_R}{a}\right)\left(C_{AO} - C_A\right)$$

or $\rho u C_p \left(T_O - T_{opt}\right) = -\left(\frac{-\Delta H_R}{a}\right)\left(-r_A\right)_{opt} V_R$

$$T_O = T_{opt} - \frac{1}{u\rho C_p}\left(\frac{-\Delta H_R}{a}\right)\left(-r_A\right)_{opt} V_R$$

$$T_O = 343.2 - \frac{(70,000)(1.353)(288)}{(100)(1,000)(4.2)} \left\{ \frac{\dfrac{J}{mol} \cdot \dfrac{mol}{L \cdot min} \cdot L}{\dfrac{g}{L} \cdot \dfrac{L}{min} \cdot \dfrac{J}{g \cdot K}} \right\}$$

$T_0 = 343.2 - 64.94$

$T_0 = 278.3 \text{ K } (5.1°C)$.

An Excel spreadsheet program (Example6-14.xls) was developed for this example.

REFERENCES

1. Dodge, B. F., *Chemical Engineering Thermodynamics,* McGraw-Hill, New York, 1944.
2. Wilson, K. B., *Trans. Inst. Chem. Engrs*, 24, 77, 1946.
3. Ergun, S., "Fluid flow through packed columns," *Chem. Eng. Prog.* 48, No. 2, 89, 1952.
4. Handley, D. and Heggs, P. J., *Trans. Instn. Chem. Engrs.,* 46-T251, 1968.
5. McDonald, I. F., El-Sayed, M. S., Mow, K., and Dullien, F. A. L., *Ind. Eng. Chem. Fund.,* 18, 199, 1979.
6. Hicks, R. E., *Ind. Eng. Chem. Fund.,* 9, 500, 1970.
7. Wentz, G. A. and Thodos, G., *AIChEJ*, 9, 81, 1963.
8. Tallmadge, J. A., *AIChEJ*, 16, 1092, 1970.
9. Perry, R. and Green, D., *Perry's Chemical Engineer's Handbook,* 6th ed., pp. 5–54, McGraw-Hill International edition, 1984.
10. Smith, J. M., *Chemical Engineering Kinetics,* 3rd ed., McGraw-Hill Book Co., 1981.
11. Levenspiel, O., *Chemical Reaction Engineering,* 3rd ed., John Wiley, 1999.
12. Levenspiel, O., *Chemical Reactor Omnibook,* Oregon State University Press, 1979.
13. Omoleye, J. A., Adesina, A. A., and Udegbunam, E. O., "Optimal design of nonisothermal reactors: Derivation of equations for the rate-temperature conversion profile and the optimum temperature progression for a general class of reversible reactions," *Chem. Eng. Comm.,* Vol. 79, pp. 95–107, 1989.
14. Adesina, A. A., "Non-isothermal reactor design for variable density systems," *The Chemical Eng. J,* 47, pp. 17–24, 1991.
15. Levenspiel, O., private communication.
16. Smith, E. L., private communication.
17. Adesina, A. A., private communication.

CHAPTER SEVEN

Fluid Mixing
in Reactors

INTRODUCTION

The various types of reactors employed in the processing of fluids in the chemical process industries (CPI) were reviewed in Chapter 4. Design equations were also derived (Chapters 5 and 6) for ideal reactors, namely the continuous flow stirred tank reactor (CFSTR), batch, and plug flow under isothermal and non-isothermal conditions, which established equilibrium conversions for reversible reactions and optimum temperature progressions of industrial reactions.

Various types of reactors are increasingly employed in the mixing of reactants to achieve the required yield at a given rate. For an example, in the development of reactor design equations for a CFSTR, an assumption was made that all the molecules move freely in the fluid with no interference from surrounding molecules. The fluid mixing, in this case, was assumed to be instantaneous. Molecules added to the reactors were indistinguishable from those already present. For real fluids, the time required to attain uniform concentration in a CFSTR depends on the intensity of turbulence, which may not be instantaneous. Molecular diffusion allows the movement of different molecules across the boundaries of the liquid elements, thereby reducing the difference between elements. Molecular diffusion is essential to achieve mixing between two liquids and complete the mixing process. If the fluids being mixed are gases, the rate of molecular diffusion is very high so the time required is extremely small. In the case of liquids where the rate of molecular diffusion is slow, knowledge is required of the turbulence so that an estimate of the size of the smallest eddy and time for molecular diffusion can be established.

Danckwerts [1] gave a set of criteria to provide a measure of the level of mixing. The scale of segregation measures the size of the unmixed clumps of pure components; the intensity of segregation describes the effect of molecular diffusion on the mixing process. It is defined as the difference in concentration between the neighboring clumps of fluid.

Levenspiel [2] considered when two fluids are mixed together, the molecular behavior of the dispersed fluid falls between two extremes. If molecules are completely free to move about, the dispersed fluid behaves as a microfluid and exhibits no fluid segregation. At the opposite extreme, the dispersed fluid remains as clumps containing a large number of molecules and is termed a macrofluid. Furthermore, as the macrofluid is transformed to a microfluid by physical mixing processes (e.g., turbulence or molecular diffusion), the degree and scale of segregation (i.e., the average of the segregated clumps) decrease.

An important mixing operation involves bringing different molecular species together to obtain a chemical reaction. The components may be miscible liquids, immiscible liquids, solid particles and a liquid, a gas and a liquid, a gas and solid particles, or two gases. In some cases, temperature differences exist between an equipment surface and the bulk fluid, or between the suspended particles and the continuous phase fluid. The same mechanisms that enhance mass transfer by reducing the film thickness are used to promote heat transfer by increasing the temperature gradient in the film. These mechanisms are bulk flow, eddy diffusion, and molecular diffusion. The performance of equipment in which heat transfer occurs is expressed in terms of forced convective heat transfer coefficients.

This chapter reviews the various types of impellers, the flow patterns generated by these agitators, correlation of the dimensionless parameters (i.e., Reynolds number, Froude number, and Power number), scale-up of mixers, heat transfer coefficients of jacketed agitated vessels, and the time required for heating or cooling these vessels.

MIXING AND AGITATION OF FLUIDS

Many operations depend to a great extent on effective mixing of fluids. Mixing refers to any operation used to change a non-uniform system into a uniform one (i.e., the random distribution of two or more initially separated phases); agitation implies forcing a fluid by mechanical means to flow in a circulatory or other pattern inside a vessel. Mixing is an integral part of chemical or physical processes such as blending, dissolving, dispersion, suspension, emulsification, heat transfer, and chemical reactions.

Dispersion characteristics can be considered as the mixing of two or more immiscible liquids, solids and liquids, or liquids and gases, into a pseudo-homogeneous mass. Small drops are created to provide contact between immiscible liquids. These liquids are mixed for specific purposes, namely solvent extraction, removal or addition of heat, and to affect mass transfer rates in reactors. The terms dispersion and emulsion are often used interchangeably. Dispersion is a general term that implies distribution, whereas emulsion is a special case of dispersion. Dispersion is a two-phase mixture in which drops may coalesce. The material present in a larger quantity is referred to as the continuous phase and the material present in a smaller quantity is called the dispersed phase. An emulsion is a two-phase mixture of very fine drops in which little or no coalescence occurs. The stability of an emulsion depends on surface ion activity, which is a function of particle size. Common dispersions are water and hydrocarbons, and acidic or alkaline solutions combined with organic liquids. Table 7-1 summarizes the principal purposes for agitating fluids.

AGITATION EQUIPMENT

Various types of vessels and tanks of differing geometrical shapes and sizes are used for mixing fluids. The top of the vessel may be open or sealed. A typical batch reactor, as discussed in Chapter 4, is applicable in many operations. The vessel bottom is normally not flat,

Table 7-1
Characteristics for agitating fluids

1. Blending of two miscible or immiscible liquids.
2. Dissolving solids in liquids.
3. Dispersing a gas in a liquid as fine bubbles (e.g., oxygen from air in a suspension of microorganism for fermentation or for activated sludge treatment).
4. Agitation of the fluid to increase heat transfer between the fluid and a coil or jacket.
5. Suspension of fine solid particles in a liquid, such as in the catalytic hydrogenation of a liquid where solid catalyst and hydrogen bubbles are dispersed in the liquid.
6. Dispersion of droplets of one immiscible liquid in another (e.g., in some heterogeneous reaction process or liquid-liquid extraction).

but rounded to eliminate sharp corners or regions into which the fluid currents would not penetrate; dished ends are most common. The liquid depth is approximately equal to the diameter of the tank. An impeller is mounted on an overhung shaft, (i.e., a shaft supported from above). The shaft is motor driven; this is sometimes directly connected to the shaft, but is more often connected through a speed-reducing gearbox. Other attachments include inlet and outlet lines, coils, jackets, and wells for thermometers. Figure 7-1 shows a typical standard tank

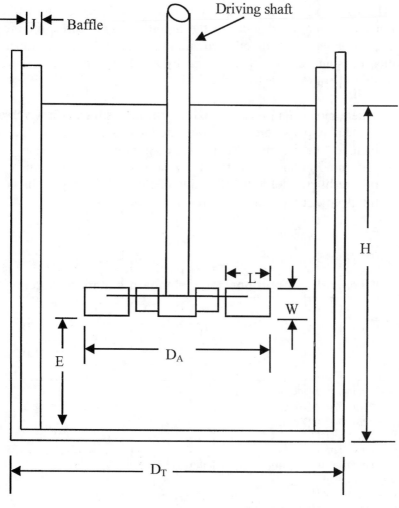

Figure 7-1. Standard tank configuration.

configuration. The geometric proportions of the agitation system, which are considered a typical standard design are given in Table 7-2. These relative proportions form the basis of the major correlations of agitation performance from various studies.

There are cases where $W/D_A = 1/8$ and $J/D_T = 1/10$ for some agitator correlations. Usually, 4 baffles are used and the clearance between the baffles and the wall is about 0.1–0.15 J. This ensures that the liquid does not form stagnant pockets between the baffle and the wall. The number of impeller blades varies from 4 to 16, but is generally between 6 and 8.

Mixing by agitation of liquids normally involves the transfer of momentum from an impeller to the liquid. In some cases, mixing is achieved by gas injection or circulation via a pump loop. An impeller, which is mounted on a shaft driven by an electric motor, is divided into two operation categories:

- Where momentum is transferred by shearing stresses, in which the transfer is perpendicular to the direction of flow. This category includes the rotating disc and cone agitators.
- The momentum is transferred by normal stresses, in which the transfer is parallel to the direction of flow. This category includes the paddle, propeller, and turbo mixer agitators.

Table 7-2
Geometric proportions for a standard agitation system

$\dfrac{D_A}{D_T} = \dfrac{1}{3}$	$\dfrac{H}{D_T} = 1$	$\dfrac{J}{D_T} = \dfrac{1}{12}$
$\dfrac{E}{D_A} = 1$	$\dfrac{W}{D_A} = \dfrac{1}{5}$	$\dfrac{L}{D_A} = \dfrac{1}{4}$

B = number of blades on impeller
R = number of baffles
D_A = agitator diameter
H = liquid height
D_T = tank diameter
E = height of the agitator from the bottom of the tank
J = baffle width
L = agitator blade length
W = agitator blade width

Agitation plays an essential role in the success of many chemical processes, and there is a wide range of commercially available impellers that can provide the optimum degree of agitation for any process. The problem arises in selecting the best impeller for the required process. Equipment manufacturers often provide expert guidance, but it is beneficial for designers and engineers to acquire fundamental knowledge of various types of impellers. The process objective of an impeller is the primary factor that determines its selection. These objectives, summarized in Table 7-1, together with physical properties such as viscosity play an important role in the selection of impellers in laminar, transitional, and turbulent operations. In general, impellers can be classified into two main groups.

- Impellers with a small blade area, which rotate at high speeds. These include turbines and marine propellers.
- Impellers with a large blade area, which rotate at low speeds. These include anchors, paddles, and helical screws.

The latter impellers are very effective for high-viscosity liquids and depend on a large blade area to produce liquid movement throughout the vessel. Since they are low-shear impellers, they are useful for mixing shear-thickening liquids. Figure 7-2 shows a typical gate anchor agitator. Anchor agitators operate very close to the vessel wall with a radial clearance equal to $0.0275 \, D_A$. The shearing action of the anchor blades past the vessel wall produces a continual interchange of liquid between the bulk liquid and the liquid film between the blades and the wall. For heat transfer applications, anchors are fitted with wall scrapers to prevent the buildup of a stagnant film between the anchor and the vessel wall. The anchor impeller is a good blending and heat transfer device when the fluid viscosity is between 5,000 and 50,000 cP (5 and 50 Pas). Below 5,000 cP, there is not enough viscous drag at the tank wall to promote pumping, resulting in a swirling condition. At viscosities greater than 50,000 cP (50 Pas), blending and heat transfer capabilities decrease as pumping capacity declines and the impeller "slips" in the fluid.

Helical screws operate in the laminar range at normally high impeller to vessel diameter ratio (D_A/D_T) with a radial clearance equal to $0.0375 \, D_A$. The impeller usually occupies one-third to one-half of the vessel diameter. They function by pumping liquid from the bottom of a tank to the liquid surface. The liquid returns to the bottom of the

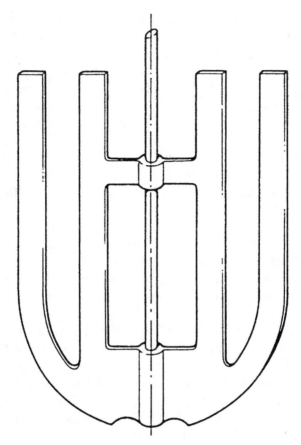

Figure 7-2. Gate anchor agitator. *(Source: Holland, F. A. and Bragg, R. Fluid Flow for Chemical Engineers, 2nd ed., Edward Arnold, 1995.)*

tank to fill the space created when fresh liquid is pumped to the surface. Figure 7-3 shows the flow pattern in a baffled helical screw tank. Baffles set away from the tank wall create turbulence and, thus, enhance the entrainment of liquid in contact with the tank wall. These are not required if the helical screw is placed in an off-centered position because the system becomes self-baffling. These impellers are useful in heat transfer application when it is essential that the fluid closest to the wall moves at high velocities.

Turbulent impellers are classified as axial or radial flow impellers. Axial flow impellers cause the tank fluid to flow parallel to the impeller's rotation axis. Radial flow impellers cause the tank fluid to

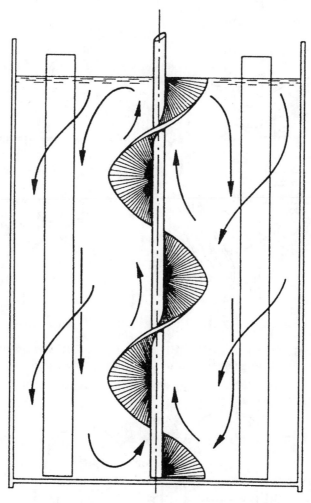

Figure 7-3. Flow pattern in a baffled helical screw system. *(Source: Holland, F. A. and Bragg, R. Fluid Flow for Chemical Engineers, 2nd ed., Edward Arnold, 1995.)*

flow perpendicular to the impeller's rotation axis. Small blade, high-speed impellers are used to mix low to medium viscosity liquids. Figures 7-4 and 7-5, respectively, show the six-flat blade turbine and marine propeller-type agitators. Figure 7-6 shows flat blade turbines used to produce radial flow patterns perpendicular to the vessel wall. In contrast, Figure 7-7 depicts marine-type propellers with axial flow

(text continued on page 562)

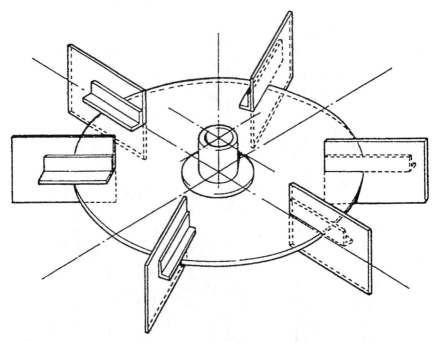

Figure 7-4. Six flat blade turbine. *(Source: Holland, F. A. and Bragg, R.* Fluid Flow for Chemical Engineers, *2nd ed., Edward Arnold, 1995.)*

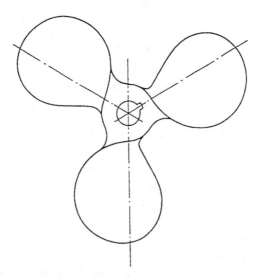

Figure 7-5. Marine propeller. *(Source: Holland, F. A. and Bragg, R.* Fluid Flow for Chemical Engineers, *2nd ed., Edward Arnold, 1995.)*

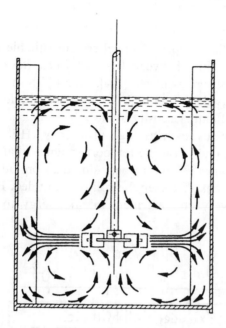

Figure 7-6. Radial flow pattern produced by a flat blade turbine. *(Source: Holland, F. A. and Bragg, R.* Fluid Flow for Chemical Engineers, *2nd ed., Edward Arnold, 1995.)*

Figure 7-7. Axial flow pattern produced by a marine propeller. *(Source: Holland, F. A. and Bragg, R.* Fluid Flow for Chemical Engineers, *2nd ed., Edward Arnold, 1995.)*

(text continued from page 559)

patterns. Both of these types of impellers are suitable to mix liquids with dynamic viscosities between 10 and 50 Pas. Several methods of selecting an impeller are available [3,4]. Figure 7-8 shows one method based on liquid viscosity and tank volume, and Table 7-3 illustrates another based on liquid viscosity alone.

Axial flow devices such as high-efficiency (HE) impellers and pitched blade turbines give better performance than conventional pitched blade turbines. They are best suited to provide the essential flow patterns in a tank that keep the solids suspended. High-efficiency impellers effectively convert mechanical energy to vertical flow

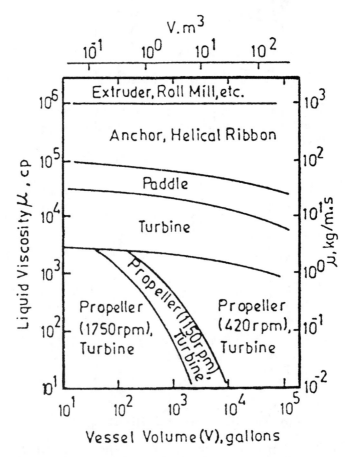

Figure 7-8. Impeller selection. *(Source: Penny, W. R. "Guide to trouble free mixers," Chem. Eng., 77(12), 171, 1970.)*

Table 7-3
Impeller selection guide

Type of impeller	Range of liquid, cP	Viscosity, kg/m – sec
Anchor	$10^2 - 2 \times 10^3$	$10^{-1} - 2$
Propeller	$10^0 - 10^4$	$10^{-3} - 10^1$
Flat-blade turbine	$10^0 - 3 \times 10^4$	$10^{-3} - 3 \times 10^1$
Paddle	$10^2 - 3 \times 10^1$	$10^{-1} - 3 \times 10^1$
Gate	$10^3 - 10^5$	$10^0 - 10^2$
Helical screw	$3 \times 10^3 - 3 \times 10^5$	$3 - 3 \times 10^2$
Helical ribbon	$10^4 - 2 \times 10^6$	$10^1 - 2 \times 10^3$
Extruders	$>10^6$	$>10^3$

Source: Holland, F. A., and Chapman, F. S. Liquid Mixing and Processing in Stirred Tanks, Reinhold, New York, 1966.

required to overcome the effects of gravity on solids in suspension. They also provide the same levels of solids suspension at reduced capital and operating costs.

FLOW PATTERN

In fluid agitation, the direction as well as the magnitude of the velocity is critical. The directions of the velocity vectors throughout an agitated vessel are referred to as the flow pattern. Since the velocity distribution is constant in the viscous and turbulent ranges, the flow pattern in an agitated vessel is fixed.

During the mixing of fluids, it is essential to avoid solid body rotation and a large central surface vortex. When solid body rotation occurs, adequate mixing is not achieved because the fluid rotates as if it were a single mass as shown in Figure 7-9a. Centrifugal force of the fluid causes a central surface vortex to be thrown outward by the impeller. Entrainment of air results if the vortex reaches an impeller, resulting in reduced mixing of the fluids. This situation can be averted by installing baffles on the vessel walls, which impede rotational flow without interfering with radial or longitudinal flow. Effective baffling is attained by installing vertical strips perpendicular to the wall of the tank. With the exception of large tanks, four baffles are adequate to prevent swirling and vortex formation. For propellers, the width of the baffle should be less one-eighteenth the diameter of the tank; for

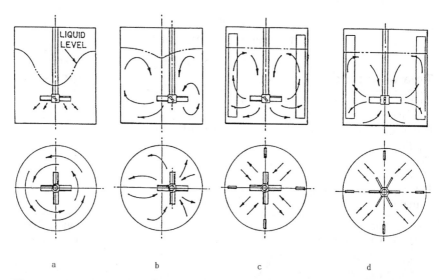

Figure 7-9. Agitator flow patterns. (a) Axial or radial impellers without baffles produce vortex. (b) Off-center location reduces the vortex. (c) Axial impeller with baffles. (d) Radial impeller with baffles. *(Source: Walas, S. M., Chemical Process Equipment—Selection and Design, Butterworths Series in Chemical Engineering, 1988.)*

turbines, less one-twelfth the tank diameter. Figure 7-9 shows the various flow patterns of radial and axial impellers.

Reducing vortex formation may also be achieved by placing an impeller in an off-center position. This creates an unbalanced flow pattern, reducing or eliminating the swirl and thereby increasing or maximizing the power consumption. The exact position is critical, since too far or too little off-center in one direction or the other will cause greater swirling, erratic vortexing, and dangerously high shaft stresses. Changes in viscosity and tank size also affect the flow pattern in such vessels. Off-center mounting of radial or axial flow impellers is readily employed as a substitute for baffled tank installations. It is common practice with propellers, but less with turbine agitators. Off-center mounting can also be useful for a turbine operated in the medium viscosity range and with non-Newtonian fluids where baffles cause stagnation with little swirl of the fluid. Off-center mountings have been quite effective in the suspension of paper pulp. Figure 7-10 illustrates an angular off-center position for propellers, which is effective without using baffles.

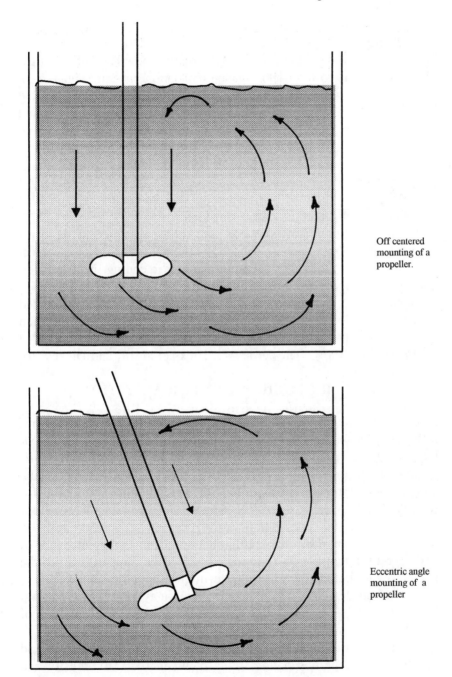

Off centered
mounting of a
propeller.

Eccentric angle
mounting of a
propeller

Figure 7-10. Flow pattern of propellers in an eccentric angle and off-centered
position.

Once swirling stops, the specific flow pattern in the tank depends on the type of impeller. Paddle agitators and flat-blade turbines promote good radial flow in the plane of the impeller with the flow dividing the wall to form two separate circulation patterns (Figure 7-6). One portion flows down along the wall and back to the center of the impeller from below, and the other flows up toward the surface and back to the impeller from above. Propeller agitators drive the liquid down to the bottom of the tank, where the stream spreads radially in all directions toward the wall, flows upward along the wall, and returns to the suction of the propeller from the top. The earlier Figure 7-7 shows the flow pattern of a propeller agitator. Propellers are employed when heavy solid particles are suspended.

Table 7-4 shows flow patterns and applications of some commercially available impellers. Generally, the axial flow pattern is most suitable for flow sensitive operation such as blending, heat transfer, and solids suspension, while the radial flow pattern is ideal for dispersion operations that require higher shear levels than are provided by axial flow impellers. Myers et al. [5] have described a selection of impellers with applications. Further details on selection are provided by Uhl and Gray [6], Gates et al. [7], Hicks et al. [8] and Dickey [9].

POWER REQUIREMENT FOR AGITATION

The flow mechanism in a mixing tank is very complex. Various techniques, including computational fluid dynamics (CFD) and computational fluid mixing (CFM) tools, are employed together with experimental data to establish improvements in mixing with increased yield. Estimating the power consumption for any agitator is essential for design. Generally, the desired requirements for the system to be mixed will categorize the type of impeller to be used. Laboratory tests on the system can establish the appropriate speed for the maintenance of isotropic turbulence in the mixing vessel. Therefore, estimating the power consumption for a large-scale mixing operation must include scale-up considerations. These requirements may be determined from the Navier-Stokes equation or analyzed by a dimensional analysis of the mixing operation.

The power consumed by an agitator depends on its dimensions and the physical properties of the fluids being mixed (i.e., density and viscosity). Since there is a possibility of a gas-liquid surface being

Table 7-4
Impellers and flow patterns

Impeller	Flow Pattern	Name and description	Applications	
		HE-3 Narrow-blade, high-efficiency impeller	Blending, Turbulent heat transfer, Solid suspension,	Upper impeller for gas dispersion, $N_p \sim 0.27$, $N_q \sim 0.5$ (turbulent)
		P-4 Pitched-blade turbine	Blending, Dispersion, Solid suspension,	Heat transfer, Surface motion, $N_p \sim 1.25$, $N_q \sim 0.7$ (turbulent)
		S-4 Straight-blade turbine	Local liquid motion for blending, Dispersion, keeping outlets clear from solids,	$N_p = 3.0$
		Maxflo T Wide-blade, high-efficiency impeller	Blending, Transitional flow, Simultaneous gas dispersion and solid suspension (like mining),	N_p and N_q vary with tip angle and number of blades
		ChemShear Narrow-blade turbine	Liquid-liquid dispersion, Solid-liquid dispersion, Local shear	
		D-6 Flat-blade disc turbine (Rushton turbine)	Gas dispersion, low and intermediate gas flows, Liquid-liquid dispersion, $N_p \sim 5.5$, $N_q \sim 0.75$	
		CD-6 Concave-blade disc turbine (Smith turbine)	Gas dispersion, intermediate and high gas flows	
		Helical ribbon (Double flight shown)	Blending and heat transfer in viscous media ($\mu > 50$ Pa-s or $N_{Re} < 100$) $-N_p \sim 350 / N_{Re}$, $N_{Re} < 100$)	
		Anchor	Heat transfer in viscous media $N_p \sim 400 / N_{Re}$, $N_{Re} < 10$	
		CD-6 / HE-3 / P-4	Gas dispersion and blending for tall reactors Fermentations (food products, pharmaceuticals)	
		CD-6 / HE-3	Combined gas-dispersion, blend- ing, and material drawdown (corn wet milling)	
		Side-entering wide blade impeller (HE3-S or Mark II)	Oil storage, Paper pulp, Wastewater circulation, Flue gas desulphurisation	

Source: Myers, K., et al., Agitation for Success, The Chemical Engineer, *Oct. 10, 1996.*
Reproduced with permission of IChemE.

distorted, as in the formation of a vortex, gravity forces must also be considered.

Consider a stirred tank vessel having a Newtonian liquid of density ρ and viscosity μ is agitated by an impeller of diameter D_A, rotating at a rotational speed N. Let the tank diameter be D_T, the impeller width W, and the liquid depth H. The power P required for agitation of a single-phase liquid can be expressed as:

$$P = f(\rho^a, \mu^b, N^c, g^d, D_A^e, D_T^f, W^g, H^h) \tag{7-1}$$

There are nine variables and three primary dimensions, and therefore by Buckingham's theorem, Equation 7-1 can be expressed by (9-3) dimensionless groups. Employing dimensional analysis, Equation 7-1 in terms of the three basic dimensions (mass M, length L, and time T) yields: Power $= ML^2T^{-3}$.

Substitution of the dimensions into Equation 7-1 gives,

$$ML^2T^{-3} = f\{(ML^{-3})^a, (ML^{-1}T^{-1})^b, T^{-c}, (LT^{-2})^d, L^e, L^f, L^g, L^h\} \tag{7-2}$$

Equating the exponents of M, L, and T on both sides of Equation 7-2 gives

$$\text{M:} \quad 1 = a + b \tag{7-3}$$

$$\text{L:} \quad 2 = -3a - b + d + e + f + g + h \tag{7-4}$$

$$\text{T:} \quad -3 = -b - c - 2d \tag{7-5}$$

From Equation 7-3

$$a = 1 - b \tag{7-6}$$

Substituting Equation 7-6 into Equation 7-4 gives

$$2 = -3(1 - b) - b + d + e + f + g + h$$

$$5 = 2b + d + e + f + g + h \tag{7-7}$$

From Equation 7-5

$$-3 = -b - c - 2d$$

$$b = 3 - c - 2d, \text{ or}$$

$$c = 3 - b - 2d \tag{7-8}$$

From Equation 7-7

$$e = 5 - 2b - d - f - g - h \tag{7-9}$$

Substituting a, c, and e on the right side of Equation 7-1 yields

$$P = K\left(\rho^{1-b},\ \mu^{b},\ N^{3-b-2d},\ g^{d},\ D_A^{5-2b-d-f-g-h},\ D_T^{f},\ W^{g},\ H^{h}\right) \tag{7-10}$$

Rearranging and grouping the exponents yields,

$$P = K\left\{\rho N^3 D_A^5 \left(\frac{\mu}{\rho N D_A^2}\right)^b \left(\frac{g}{N^2 D_A}\right)^d \left(\frac{D_T}{D_A}\right)^f \left(\frac{W}{D_A}\right)^g \left(\frac{H}{D_A}\right)^h\right\} \tag{7-11}$$

or

$$\frac{P}{\rho N^3 D_A^5} = K\left\{\left(\frac{\mu}{\rho N D_A^2}\right)^b \left(\frac{g}{N^2 D_A}\right)^d \left(\frac{D_T}{D_A}\right)^f \left(\frac{W}{D_A}\right)^g \left(\frac{H}{D_A}\right)^h\right\} \tag{7-12}$$

The dimensionless parameters are:

The Power number, $N_p = \dfrac{P g_C}{\rho N^3 D_A^5}$ g_C = dimensional gravitational constant

$$32.174\ \frac{lbm}{lbf} \bullet \frac{ft}{sec^2}$$

$$1\ kg \bullet m/N \bullet sec^2$$

The Reynolds number, $N_{Re} = \dfrac{\rho N D_A^2}{\mu}$

The Froude number, $N_{Fr} = \dfrac{N^2 D_A}{g}$

Substituting these dimensionless numbers into Equation 7-12 yields,

$$N_p = K\left\{N_{Re}^{-b} \, N_{Fr}^{-d} \left(\frac{D_T}{D_A}\right)^f \left(\frac{W}{D_A}\right)^g \left(\frac{H}{D_A}\right)^h\right\} \qquad (7\text{-}13)$$

SIMILARITY

Equality of all groups in Equation 7-13 assures similarity between systems of different sizes. The types of similarity are geometric, kinematic, and dynamic. The last three terms of Equation 7-13 represent the conditions for *geometric similarity,* which require that all corresponding dimensions in systems of different sizes have the same ratio to each other. For geometric similarity, Equation 7-13 becomes

$$N_P = K\,N_{Re}^{-b}\,N_{Fr}^{-d} \qquad (7\text{-}14)$$

The constant K and the exponents b and d must be determined for the particular type of agitator, its size and location in the tank, the dimensions of the tank, and the depth of the liquid.

Kinematic similarity exists between two systems of different sizes when they are geometrically similar and when the ratios of velocities between corresponding points in one system are equal to those in the other.

Dynamic similarity exists between two systems when, in addition to being geometrically and kinematically similar, the ratios of forces between corresponding points in one system are equal to those in the other.

The value of N_{Re} determines whether the flow is laminar or turbulent and is a significant group affecting the power consumption. The Froude number N_{Fr}, representing the ratio of inertial to gravitational forces, is only significant when the liquid in the tank swirls to such an extent that a deep vortex is formed and the wave or surface effects become important. In an unbaffled vessel, a balance between the inertial and gravitational forces determines the shape of any vortex. The Power number N_P may be considered as a drag coefficient or friction factor.

Experimental data on power consumption are generally plotted as a function of the Power number N_P versus Reynolds number N_{Re}, that is by rearranging Equation 7-14.

$$\Phi = \frac{N_P}{N_{Fr}^{-d}} = K N_{Re}^{-b} \tag{7-15}$$

For a fully baffled tank, $b = 0$ and $\Phi = N_P$. A generalized plot of Equation 7-15 is shown in Figure 7-11. The power correlation indicates three ranges of liquid motion: laminar (viscous), transition, and turbulent. The laminar or viscous range occurs below a Reynolds number of 10. The expected result of the Power number being inversely proportional to the Reynolds number is also confirmed by experimental data. The Froude effects are unimportant and a logarithmic plot of the relation between Power number and Reynolds number gives a slope of –1 in this range. Fully turbulent agitation occurs above a Reynolds number of 10,000. The range between these limits can be described as the transition flow because flow patterns change depending on the Reynolds number. Figure 7-12 shows the Power number versus Reynolds number plot for the unbaffled system. Both Figures 7-11 and 7-12 are identical to point C where $N_{Re} \cong 300$. As the Reynolds

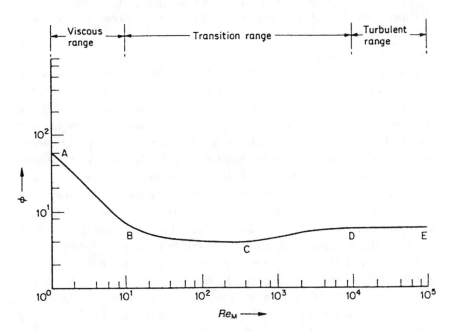

Figure 7-11. Power curve for the standard tank configuration. *(Source: Holland, F. A. and Bragg, R. Fluid Flow for Chemical Engineers, 2nd ed., Edward Arnold, 1995.)*

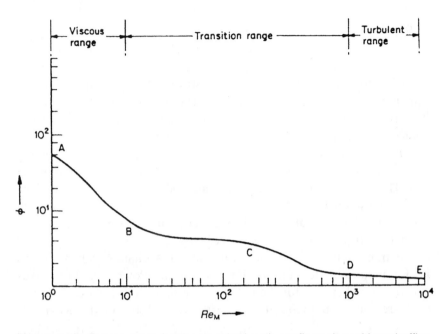

Figure 7-12. Power curve for the standard tank configuration without baffles. *(Source: Holland, F. A. and Bragg, R. Fluid Flow for Chemical Engineers, 2nd ed., Edward Arnold, 1995.)*

number for mixing increases beyond point C in the unbaffled system, vortexing increases and the Power number falls sharply. Figure 7-13 shows the Power number as a function of the Reynolds number for shear thinning fluids. The full line gives the Newtonian Power number obtained by Rushton et al. [10] for a flat-blade turbine system, while the dashed line shows Metzner and Otto's [11] plot for shear thinning liquids. Figure 7-13 illustrates that at no point is the shear thinning power curve higher than the Newtonian power curve. Therefore, the use of the Newtonian power curve to determine the power will give a conservative value when used for shear thinning liquids. Figure 7-14 shows Power number correlations for various types of agitators. In the fully turbulent flow, the curve becomes horizontal and the Power number N_P is independent of the Reynolds number.

Rushton et al. [10] performed extensive measurements of the power requirements for geometrically similar systems and found that for baffled tanks, the Froude number plays no part in determining the power requirements, as vortices do not form in such systems. For unbaffled systems, the Froude number plays a part above N_{Re} of about

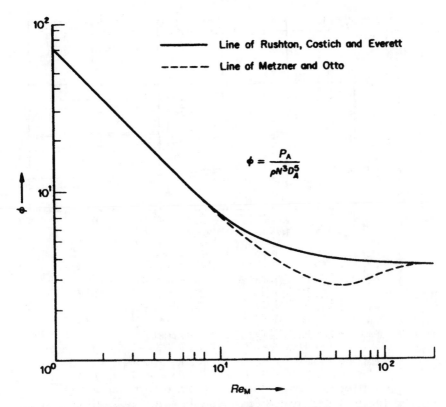

Figure 7-13. Deviation from Newtonian power curve for shear thinning liquids. *(Source: Holland, F. A. and Bragg, R. Fluid Flow for Chemical Engineers, 2nd ed., Edward Arnold, 1995.)*

300. They reported $N_P = 6.3$ in the turbulent range of 10,000. After extensive curve fitting of their experimental data, a single curve was obtained for any particular unbaffled configuration. If Φ is plotted as a function of N_{Re} where Φ is defined as

$$\Phi = N_P \qquad\qquad \text{for } N_{Re} < 300$$

$$\Phi = \frac{N_P}{N_{Fr}^{[(a - \log N_{Re})/b]}} \qquad \text{for } N_{Re} > 300 \qquad (7\text{-}16)$$

where a and b are constants for any configuration (Figure 7-14). Dickey and Fenic [12] observed that the impeller characteristics have significant influence on the Power number correlation.

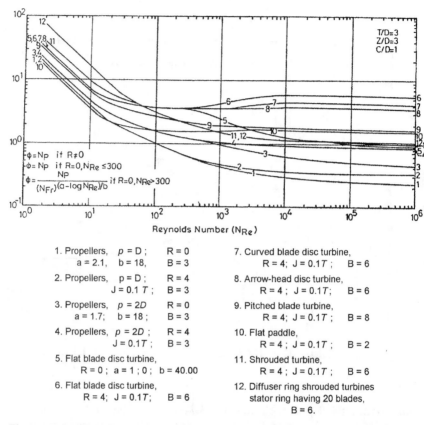

Reynolds Number (N_{Re})

1. Propellers, $p = D$; $R = 0$ $a = 2.1$, $b = 18$, $B = 3$	7. Curved blade disc turbine, $R = 4$; $J = 0.1T$; $B = 6$
2. Propellers, $p = D$; $R = 4$ $J = 0.1\ T$; $B = 3$	8. Arrow-head disc turbine, $R = 4$; $J = 0.1T$; $B = 6$
3. Propellers, $p = 2D$ $R = 0$ $a = 1.7$; $b = 18$; $B = 3$	9. Pitched blade turbine, $R = 4$; $J = 0.1T$; $B = 8$
4. Propellers, $p = 2D$; $R = 4$ $J = 0.1T$; $B = 3$	10. Flat paddle, $R = 4$; $J = 0.1T$; $B = 2$
5. Flat blade disc turbine, $R = 0$; $a = 1$; 0; $b = 40.00$	11. Shrouded turbine, $R = 4$; $J = 0.1T$; $B = 6$
6. Flat blade disc turbine, $R = 4$; $J = 0.1T$; $B = 6$	12. Diffuser ring shrouded turbines stator ring having 20 blades, $B = 6$.

Figure 7-14. Power number versus Reynolds number correlation for common impellers. *(Source: Ruchton et al., Chem. Eng. Prog., 46, No. 8, 495, 1950. Reprinted with permission of AIChE. Copyright © 1950. All rights reserved.)*

Two characteristics of Figure 7-14 are:

1. At low $N_{Re} < 1.0$, $N_P \propto 1/N_{Re}$, independent of the presence of baffles.
2. At high Reynolds, at which most mixing operations are performed, the Power number is constant, that is, $N_P \propto P/\rho N^3 D_A^5$ = constant.

Rushton et al. [10] investigated the effect of varying the tank geometrical ratios and the correlation of the Power number with Reynolds number. At high Reynolds number, it was inferred that,

- Φ is relatively unchanged when D_T/D_A is varied from 2 to 7 for turbine- and propeller-agitated baffled systems.
- Φ is unchanged when H/D_A is varied from 2 to 4.
- Φ is unaltered when E/D_A is changed from 0.7 to 1.6.
- Φ changes to $\Phi \propto (J/D_A)^{0.3}$ when J/D_T is changed from 0.05 to 0.17.
- Φ depends on the number of blades in the turbine impeller as: $\Phi \propto (B/6)^{0.8}$ if $B < 6$ and $\Phi \propto (B/6)^{0.7}$ if $B > 6$ (where B = number of impeller blades).
- If off-centered and inclined propellers without baffles or side-entering propellers without baffles are used, no vortex forms and the Φ versus N_{Re} curve for the corresponding baffled tank can be used to estimate the power requirements.

These conclusions are speculative and experimental curves must be generated if more than one geometrical ratio differs from the standard value.

The power consumed by an agitator at various rotational speeds and physical properties (e.g., viscosity and density) for a system's geometry can be determined from the Power number correlation. The procedure involves:

- Calculating the Reynolds number N_{Re} for mixing.
- Reading the Power number N_P from the appropriate curve, and calculating the power P given by

$$P = N_P \cdot \rho N^3 D_A^5 \tag{7-17}$$

$$\text{or } P = \Phi \rho N^3 D_A^5 \cdot N_{Fr}^{\left[(a - \log N_{Re})/b \right]} \tag{7-18}$$

Equations 7-17 and 7-18 are the power consumed by the agitator. Additional power is required to overcome electrical and mechanical losses. A contingency of motor loading as a percentage (e.g., 85%) is added when selecting the motor. Equation 7-17 can also be rearranged to determine impeller diameter when it is desired to load an agitator impeller to a given power level. The torque delivered to the fluid by an impeller from its speed and power draw is determined by:

$$\tau = \frac{P}{2\pi N} = \frac{N_P \rho N^2 D_A^5}{2\pi} N - m \tag{7-19}$$

The primary pumping capacity of an impeller is determined by the impeller diameter, the Pumping number, and the rotational speed. The Pumping number N_Q is defined by [13]

$$N_Q = \frac{Q_P}{ND_A^3} \qquad (7\text{-}20)$$

The Pumping number is used to determine the pumping rate Q_P of an impeller,

where Q_P = effective pumping capacity, m^3/sec
   N = impeller rotational speed, sec^{-1}
   D_A = impeller diameter, m

Hicks et al. [8] developed a correlation involving the Pumping number and impeller Reynolds number for several ratios of impeller diameter to tank diameter (D_A/D_T) for pitched-blade turbines. From this correlation, Q_P can be determined, and thus the bulk fluid velocity from the cross-sectional area of the tank. The procedure for determining the parameters is iterative because the impeller diameter D_A and rotational speed N appear in both dimensionless parameters (i.e., N_{Re} and N_Q).

Figure 7-15 shows plots of Pumping number N_Q and Power number N_P as functions of Reynolds number N_{Re} for a pitched-blade turbine and high-efficiency impeller. Hicks et al. [8] further introduced the scale of agitation, S_A, as a measure for determining agitation intensity in pitched-blade impellers. The scale of agitation is based on a characteristic velocity, v, defined by

$$v = \frac{Q_P}{A_V} \qquad (7\text{-}21)$$

where v = characteristic velocity, m/sec
   A_V = cross-sectional area of the tank, m^2

The characteristic velocity can be expressed as:

$$v \propto N_Q N D_A \left(\frac{D_A^2}{D_T^2} \right) \qquad (7\text{-}22)$$

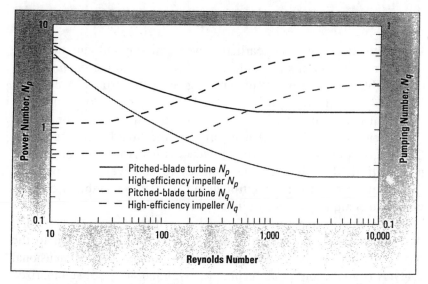

Figure 7-15. Power number and Pumping number as functions of Reynolds number for a pitched-blade turbine and high-efficiency impeller. *(Source: Bakker, A., and Gates L. E., "Properly Choose Mechanical Agitators for Viscous Liquids," Chem. Eng. Prog., pp. 25–34, 1995.)*

In geometrically similar systems, the characteristic velocity becomes

$$v \propto N_Q N D_A \tag{7-23}$$

Thus, during geometric scale-up, the characteristic velocity is held constant by holding $N_Q N D_A$ constant. Q_P is determined from the Pumping number and Figure 7-15. S_A is a linear function of the characteristic velocity and is determined by

$$S_A = 32.8 \ v \tag{7-24}$$

Accordingly, a value of S_A equal to 1 represents a low level and 10 shows a high level of agitation intensity. The 1–10 range of agitation intensity accounts for about 95% or more of all turbine-agitation applications, enabling it to be suited for a wide range of process operations. Gates et al. [14] gave guidelines on how to relate S_A to specific process applications.

MIXING TIME CORRELATION

A distinction was made earlier between mixing and agitation. The third term in liquid mixing is blending. This refers to the intermingling of miscible fluids to produce some degree of uniformity. A criterion for good mixing may only be visual. For example, it could be a particular color from two different color liquids, or the color change of an acid-base indicator that determines the liquid blending times. Characterization of blending in agitated vessels is usually in terms of mixing time. This is the time required to achieve some specified degree of uniformity after introduction of a tracer. Table 7-5 gives various techniques for determining blending time.

Each technique measures a different degree of uniformity, therefore, the time required for blending may differ from one method to the other. The correlation of blending time as derived from dimensional analysis is applicable to all techniques. Uhl and Gray [6] summarized many of the experiments and correlations on blending and mixing times. For a given tank and impeller or geometrically similar systems, the mixing time is predicted to vary inversely with the stirrer speed, as confirmed in various studies [15,16,17,18]. Figure 7-16 shows plots of mixing time (tN) against the Reynolds number N_{Re} for several systems. As an example, a turbine with $D_A/D_T = 1/3$ and $D_T/H = 1$, the value of Nt is 36, for $N_{Re} > 10^3$, compared with a predicted value of 38.

Table 7-5
Methods for determining blending time

Technique	Tracer	Blend time reached when
Grab sample	Any material that can be analyzed.	Samples do not vary more than $\pm X\%$ from final concentration.
Dye introduction	Dyed fluid.	Uniform color is attained.
Conductivity cell	Concentration of salt solution.	Measured conductivity that represents concentration is within $\pm X\%$ of final concentration.
Acid-base indicator	Acid (or base).	Neutralization is complete as determined by color change of indicator.

Source: Dickey, D. S., "Dimensional analysis for fluid agitation system," Chem. Eng., January 5, 1976.

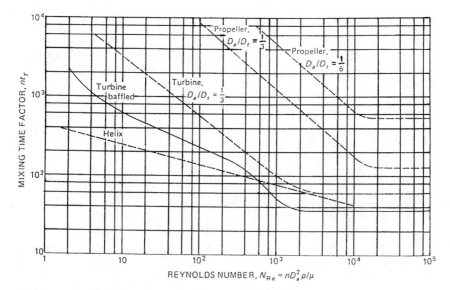

Figure 7-16. Mixing times in agitated vessels. Dashed lines represent unbaffled tanks; solid lines represents a baffled tank. *(Source: McCabe, W. L., et al., Unit Operations of Chemical Engineering, 4th ed., McGraw-Hill Book Company, New York, 1985.)*

Prochazka and Landau [19] developed a mixing time correlation for a single Rushton turbine impeller in a baffled tank in the standard configuration for $N_{Re} > 10^4$:

$$Nt = 0.905 \left(\frac{D_T}{D_A} \right)^{2.57} \log \left(\frac{X_o}{X_c} \right) \tag{7-25}$$

For a propeller, the mixing time is given by:

$$Nt = 3.48 \left(\frac{D_T}{D_A} \right)^{2.05} \log \left(\frac{X_o}{X_c} \right) \tag{7-26}$$

For a pitched-blade turbine, the mixing time is:

$$Nt = 2.02 \left(\frac{D_T}{D_A} \right)^{2.20} \log \left(\frac{X_o}{X_c} \right) \tag{7-27}$$

where X_o = initial value of the degree of inhomogeneity, which varies between 1 and 3; a value of 2 is recommended

X_c = final integral mean value of the local degree of inhomogeneity and is defined as:

$$X_c = \left[\frac{C(t) - C_x}{C_x - C_i} \right] \tag{7-28}$$

where $C(t)$ = instantaneous concentration

C_i = initial concentration

C_x = final concentration

$X_c = 0.05$ for most configurations. Moo-Young et al. [20] correlated their mixing results from

$$Nt = KN_{Re}^a \tag{7-29}$$

where $K = 36$ and $a = 0$ for turbines in baffled tanks for $1,000 < N_{Re} < 10^5$. Sano and Usui [21] developed an expression for mixing times by tracer injection for turbines as:

$$Nt = 3.8 \left(\frac{D_A}{D_T} \right)^{-1.80} \left(\frac{W}{D_T} \right)^{-0.51} n_P^{-0.47} \tag{7-30}$$

where n_P is the number of blades. Gray [22] found the mixing times of helical ribbon impellers to be of the form

$$Nt = 30 \tag{7-31}$$

where N is the rotational speed of the helical ribbon impeller, and t is the batch mixing time. Fasano et al. [23] expressed the blend time for turbulence conditions in a standard baffled tank (i.e., $N_{Re} > 10,000$) as:

$$t_{99} = \frac{4.065}{aN \left(\frac{D_A}{D_T} \right)^b \left(\frac{D_T}{H} \right)^{0.5}} \tag{7-32}$$

where a and b are the mixing rate constants. Table 7-6 shows values of a and b for different impeller types. The constants are for surface addition, however, blend times for similar fluids are relatively insensitive to addition location. Equation 7-32 is limited to the following:

- Newtonian fluids of nearly the same viscosity and density as the bulk fluid.
- Additions of 5% or less, of the fluid volume.
- Additions made to a vessel already undergoing agitation (blend times of stratified fluids can be considerably longer).

The estimated blend time for 95% uniformity ($t_{b,95\%}$) using a standard double flight helical ribbon impeller with ($P_i/D_A = 1$, $W/D_A = 0.1$, and $D_A/D_T = 0.96$) is given by

$$\text{For } N_{Re} \leq 100: \; t_{b,95\%} = \frac{75}{N} \tag{7-33}$$

For anchor impellers of standard geometry ($W/D_A = 0.1$, $D_A/D_T = 0.98$, and $H/D_T = 1.0$), the estimated t_b for $100 < N_{Re} < 10,000$ is given by

$$t_{b,95\%} = \exp\left(12.9 \, N_{Re}^{-0.135}\right) \tag{7-34}$$

where D_A = impeller diameter, m
$\quad\quad D_T$ = tank diameter, m
$\quad\quad H$ = impeller or helix height, m
$\quad\quad N$ = impeller rotational speed, sec^{-1}

Table 7-6
Mixing rate constants for fully turbulent flow regimes ($N_{Re} > 10,000$)

Impeller type	a	b
Six-bladed disc	1.06	2.17
Four-bladed flat	1.01	2.30
Four-bladed 45° pitched	0.641	2.19
Three-bladed high-efficiency	0.272	1.67

Source: Fasano et al. [13], "Advanced Impeller Geometry Boosts Liquid Agitation," Chem. Eng., 102 (8), August 1994.

P_i = pitch of a helical ribbon impeller, m
W = blade width, m

Bakker and Gates [23] compared both Equations 7-33 and 7-34 and inferred that at a Reynolds number of 100, it will take an anchor impeller more than 13 times as long to achieve 95% uniformity as a helical ribbon impeller operating at the same speed. These impellers require cooling to remove the excess heat due to their high power input. The mixing time that was considered relates to tanks operating in closed systems (e.g., batch reactors). In a continuous feed tank, the mixing time is generally shorter than in a closed tank.

Example 7-1

Calculate the power for agitation of a liquid of density 950 kg/m³ and viscosity 250 cP given the following configuration: number of blades B = 6, agitator diameter 0.61 m, and speed at 90 rpm. Other geometrical ratios are shown in Figure 7-1. A disc-mounted flat turbine is used.

Solution

The Reynolds number for mixing is

$$N_{Re} = \frac{\rho N D_A^2}{\mu}$$

N = the number of revolutions per sec is (90/60) = 1.5 rev/sec.

$$N_{Re} = \frac{(950)(1.5)(0.61^2)}{250 \times 10^{-3}} \left\{ \frac{kg}{m^3} \cdot \frac{1}{sec} \cdot \frac{m^2}{\dfrac{kg}{m \cdot sec}} \right\}$$

N_{Re} = 2,121

Using curve 6 in Figure 7-14, the Power number is N_P = 5.0. The theoretical power for mixing is

$$P = N_P \rho N^3 D_A^5 = 5.0 \times 950 \times 1.5^3 \times 0.61^5 \quad \left\{ \frac{kg}{m^3} \cdot \frac{rev^3}{sec^3} \cdot m^5 \right\}$$

$$= 1,353.99 \text{ W}$$

$$= 1.35 \text{ kW} \ (1.82 \text{ hp})$$

(NB: 1 kW = 1.341 hp)

An Excel spreadsheet program (Example 7-1.xls) was developed to determine the theoretical power of any agitator type with given fluid physical properties and tank geometry.

Example 7-2

Calculate the theoretical power for a six-blade, flat-blade turbine without baffles, but with the standard tank configuration shown in Table 7-2. Use the same data as in Example 7-1.

Solution

Since the tank is unbaffled, the Froude number is a factor and its effect is calculated from

$$N_{Fr} = \frac{N^2 D_A}{g} = \frac{\left(1.5^2\right)(0.61)}{9.81} \quad \left\{ \frac{rev}{sec^2} \cdot \frac{m}{\dfrac{m}{sec^2}} \right\}$$

$$= 0.14$$

$$N_{Re} = 2,121$$

The constants a and b for an unbaffled tank R = 0, are a = 1.0 and b = 40. Using curve 5 in Figure 7-14, the Power number is $N_P = 2.0$

$$N_{Fr}^m \quad \text{where} \quad m = \frac{a - \log_{10} N_{Re}}{b}$$

$$m = \frac{1.0 - \log_{10} 2,121}{40} = -0.0582$$

$$N_{Fr}^m = 0.14^{-0.0582} = 1.1212$$

Therefore, power $P = N_P \rho N^3 D_A^5 N_{Fr}^m$

$$= 2.0 \times 950 \times 1.5^3 \times 0.61^5 \times 1.1212 \left(\frac{kg}{m^3} \bullet \frac{rev^3}{sec^3} \bullet m^5 \right)$$

$$= 607.24 \text{ W}$$

$$= 0.61 \text{ kW } (0.81 \text{ hp})$$

A Microsoft Excel spreadsheet (Example7-2.xls) was developed for an unbaffled tank.

Studies on various turbine agitators have shown that geometric ratios that vary from the standard design can cause different effects on the Power number N_P in the turbulent regions [24].

- For the flat, six-blade open turbine, $N_P \propto (W/D_A)^{1.0}$.
- For the flat, six-blade open turbine, varying D_A/D_T from 0.25 to 0.5 has no effect on N_P.
- When two six-blade open turbines are installed on the same shaft and the spacing between the two impellers (vertical distance between the bottom edges of the two turbines) is at least equal to D_A, the total power is 1.9 times a single flat-blade impeller. For two six-blade pitched-blade (45°) turbines, the power is about 1.9 times that of a single pitched-blade impeller.
- A baffled, vertical square tank or a horizontal cylindrical tank has the same Power number as a vertical cylindrical tank.

SCALE-UP OF MIXING SYSTEMS

The calculation of power requirements for agitation is only a part of the mixer design. In any mixing problem, there are several defined objectives such as the time required for blending two immiscible liquids, rates of heat transfer from a heated jacket per unit volume of the agitated liquid, and mass transfer rate from gas bubbles dispersed by agitation in a liquid. For all these objectives, the process results are to achieve the optimum mixing and uniform blending.

The process results are related to variables characterizing mixing, namely geometric dimensions, stirrer speed (rpm), agitator power, and physical properties of the fluid (e.g., density, viscosity, and surface tension) or their dimensionless combinations (e.g., the Reynolds number, Froude number, and Weber number, $\rho N^2 D_A^3 / \sigma$). Sometimes, empirical relationships are established to relate process results and agitation parameters. Often, however, such relationships are non-existent. Laboratory scales of equipment using the same materials as on a large scale are then experimented with, and the desired process result is obtained. The laboratory system can then be scaled-up to predict the conditions on the larger system.

For some scale-up problems, generalized correlations as shown in Figures 7-11, 7-12, 7-13, and 7-14 are available for scale-up. However, there is much diversity in the process to be scaled-up, and as such no single method can successfully handle all types of scale-up problems.

Various methods of scale-up have been proposed; all based on geometric similarity between the laboratory equipment and the full-scale plant. It is not always possible to have the large and small vessels geometrically similar, although it is perhaps the simplest to attain. If geometric similarity is achievable, dynamic and kinematic similarity cannot often be predicted at the same time. For these reasons, experience and judgment are relied on with aspects to scale-up.

The main objectives in a fluid agitation process are [25]:

- Equivalent liquid motion (e.g., liquid blending where the liquid motion or corresponding velocities are approximately the same in both cases).
- Equivalent suspension of solids, where the levels of suspension are identical.
- Equivalent rates of mass transfer, where mass transfer is occurring between a liquid and a solid phase, between liquid-liquid phases, or between gas and liquid phases, and the rates are identical.

A scale ratio R is used for scale-up from the standard configuration as shown in Table 7-2. The procedure is:

1. Determine the scale-up ratio R, assuming that the original vessel is a standard cylinder with $D_{T1} = H_1$. The volume V_1 is

$$V_1 = \frac{\pi D_{T1}^2}{4} \cdot H_1 = \frac{\pi D_{T1}^3}{4} \tag{7-35}$$

The ratio of the volumes is then

$$\frac{V_2}{V_1} = \frac{\pi D_{T2}^3/4}{\pi D_{T1}^3/4} = \frac{D_{T2}^3}{D_{T1}^3} \tag{7-36}$$

The scale-up ratio R is

$$R = \frac{D_{T2}}{D_{T1}} = \left(\frac{V_{T2}}{V_{T1}}\right)^{\frac{1}{3}} \tag{7-37}$$

Using the value of R, calculate the new dimensions for all geometric sizes. That is,

$$D_{A2} = RD_{A1}, \; J_2 = RJ_1, \; W_2 = RW_1$$

$$E_2 = RE_1, \; L_2 = RL_1, \; H_2 = RH_1$$

or

$$R = \frac{D_{A2}}{D_{A1}} = \frac{D_{T2}}{D_{T1}} = \frac{W_2}{W_1} = \frac{H_2}{H_1} = \frac{J_2}{J_1} = \frac{E_2}{E_1}$$

2. The selected scale-up rule is applied to determine the agitator speed N_2 from the equation:

$$N_2 = N_1\left(\frac{1}{R}\right)^n = N_1\left(\frac{D_{T1}}{D_{T2}}\right)^n \tag{7-38}$$

- where $n = 1$ for equal liquid motion
- $n = 3/4$ for equal suspension of solids
- $n = 2/3$ for equal rates of mass transfer (corresponding equivalent power per unit volume, which results in equivalent interfacial area per unit volume)

The value of n is based on theoretical and empirical considerations and depends on the type of agitation problem.
3. Knowing the value of N_2, the required power can be determined using Equation 7-17 and the generalized Power number correlation.

Other possible ways of scaling up are constant tip speed $u_T(\pi N D_A)$, and a constant ratio of circulating capacity to head Q/h.

Since $P \propto N^3 D_A^5$ and $V \propto D_A^3$ then

$$\frac{P}{V} \propto N^3 D_A^2 \tag{7-39}$$

For scale-up from system 1 to system 2 involving geometrically similar tanks and same liquid properties, the following equations can be applied:

$$N_1 D_{A1} = N_2 D_{A2}$$

For a constant tip speed,

$$\frac{N_2}{N_1} = \frac{D_{A1}}{D_{A2}} \tag{7-40}$$

For a constant ratio of circulating capacity to head, Q/h,

$$N_1^3 D_{A1}^2 = N_2^3 D_{A2}^2 \tag{7-41}$$

Example 7-3

Scraper blades set to rotate at 35 rpm are used for a pilot plant addition of liquid ingredients into a body-wash product. What should the speed of the blades be in a full-scale plant, if the pilot and the full-scale plants are geometrically similar in design? Assume scale-up is based on constant tip speed, diameter of the pilot plant scraper blades is 0.6 m, and diameter of the full-scale plant scraper blades is 8 ft.

Solution

The diameter of the full scale plant scraper blades = 8.0×0.3048 = 2.4384 m (2.4 m).

Assuming constant tip speed,

$$\frac{N_2}{N_1} = \frac{D_{A1}}{D_{A2}} \tag{7-42}$$

where N_1 = scraper speed of pilot plant

N_2 = scraper speed of full-scale plant

D_{A1} = diameter of pilot plant scraper blades

D_{A2} = diameter of full-scale plant scraper blades

$$N_2 = \frac{N_1 \, D_{A1}}{D_{A2}}$$

$$= \frac{(35)(0.6)}{(2.4)}$$

$$= 8.75 \text{ rpm}$$

Example 7-4

During liquid makeup production, color pigments (i.e., solid having identical particle size) are added to the product via a mixer. In the pilot plant, this mixer runs at 6,700 rpm and has a diameter head of 0.035 m. Full-scale production is geometrically similar and has a mixer head diameter of 0.12 m. Determine the speed of the full-scale production mixer head. What additional information is required for the motor to drive this mixer? Assume that power curves are available for this mixer design, and the scale-up basis is constant power/unit volume.

Solution

For constant power per unit volume, Equation 7-39 is applied: P/V $\propto N^3D_A^2$ or $N_1^3D_{A1}^2 = N_2^3D_{A2}^2$. Therefore,

$$N_2 = N_1 \left(\frac{D_{A1}}{D_{A2}} \right)^{2/3}$$

where N_1 = 6,700 rpm

D_{A1} = 0.035 m

D_{A2} = 0.12 m

$$N_2 = 6,700 \left(\frac{0.035}{0.12} \right)^{2/3}$$

$N_2 = 2,946.7$ rpm

$N_2 \approx 2,950$ rpm

The power required for mixing is $P = N_p \rho N^3 D_A^5$, where the Power number (N_P) is a function of the Reynolds number [i.e., $N_P = f(N_{Re})$]:

$$N_{Re} = \frac{\rho N D_A^2}{\mu}$$

The plant must be provided with the viscosity of the product and its density after addition of the pigments.

Example 7-5

A turbine agitator with six flat blades and a disk has a diameter of 0.203 m. It is used in a tank with a diameter of 0.61 m and height of 0.61 m. The width is $W = 0.0405$ m. Four baffles are used with a width of 0.051 m. The turbine operates at 275 rpm in a liquid having a density of 909 kg/m^3 and viscosity of 0.02 Pas.

Calculate the kW power of the turbine and kW/m^3 of volume. Scale up this system to a vessel whose volume is four times as large, for the case of equal mass transfer rate.

Solution

The Reynolds number for mixing is N_{Re}. The number of revolutions per sec, $N = 275/60 = 4.58$ rev/sec.

$$N_{Re} = \frac{\rho N D_A^2}{\mu}$$

$$= \frac{(909)(4.58)(0.203)^2}{0.02} \left\{ \frac{kg}{m^3} \cdot \frac{rev}{sec} \cdot m^2 \cdot \frac{m \cdot sec}{kg} \right\}$$

$$= 8,578.1$$

$N_{Re} \approx 8,600$

Using curve 6 in Figure 7-14, the Power number $N_P = 6.0$. The power of the turbine $P = N_p \rho N^3 D_A^5$:

$$P = (6.0)(909)(4.58^3)(0.203^5) \left\{ \frac{kg}{m^3} \cdot \frac{rev^3}{sec^3} \cdot m^5 \right\}$$

$$= 0.1806 \text{ kW } (0.24 \text{ hp})$$

The original tank volume $V_1 = \pi D_{T1}^3/4$. The tank diameter $D_{T1} = 0.61$:

$$V_1 = \frac{(\pi)(0.61)^3}{4}$$

$$V_1 = 0.178 m^3$$

The power per unit volume is P/V

$$\frac{P}{V} = \frac{0.1806}{0.178}$$

$$= 1.014 \text{ kW/m}^3$$

For the scale-up of the system, the scale-up ratio R is

$$R = \frac{V_2}{V_1} = \frac{\pi D_{T2}^3/4}{\pi D_{T1}^3/4} = \frac{D_{T2}^3}{D_{T1}^3}$$

$$R = \left(\frac{V_2}{V_1} \right)^{\frac{1}{3}} = \frac{D_{T2}}{D_{T1}} \tag{7-37}$$

where $V_2 = 4V_1$

$$V_2 = 4(0.178)$$

$$= 0.712 \text{ m}^3$$

$$R = (4)^{\frac{1}{3}} = 1.587$$

The dimensions of the larger agitator and tank are:

$$D_{A2} = RD_{A1} = 1.587 \times 0.203 = 0.322 \text{ m}$$

$$D_{T2} = RD_{T1} = 1.587 \times 0.61 = 0.968 \text{ m}$$

For equal mass transfer rate n = 2/3.

$$N_2 = N_1 \left(\frac{1}{R}\right)^{\frac{2}{3}} \qquad\qquad (7\text{-}38)$$

$$= 4.58 \left(\frac{1}{1.587}\right)^{\frac{2}{3}}$$

$$= 3.37 \text{ rev/sec}$$

The Reynolds number N_{Re} is

$$N_{Re} = \frac{\rho N_2 D_{A2}^2}{\mu}$$

$$= \frac{(909)(3.37)(0.322)^2}{0.02} \left\{\frac{\text{kg}}{\text{m}^3} \cdot \frac{\text{rev}}{\text{sec}} \cdot \text{m}^2 \cdot \frac{\text{m} \cdot \text{sec}}{\text{kg}}\right\}$$

$$= 15,880.9$$

$$N_{Re} \approx 16,000$$

Using curve 6 in Figure 7-14, $N_P = 6.0$. Power required by the agitator is $P_2 = N_p \rho N_2^3 D_{A2}^5$

$$P_2 = (6.0)(909)(3.37)^3(0.322)^5 \left\{\frac{\text{kg}}{\text{m}^3} \cdot \frac{\text{rev}^3}{\text{sec}^3} \cdot \text{m}^5\right\}$$

$$P_2 = 722.57 \text{ W}$$

$$= 0.723 \text{ kW} \ (0.97 \text{ hp})$$

The power per unit volume P/V is:

$$\frac{P_2}{V_2} = \frac{0.723}{0.712}$$

$$= 1.015 \text{ kW/m}^3$$

A Microsoft Excel spreadsheet (Example 7-5.xls) was developed for this example.

MIXING TIME SCALE-UP

Predicting the time for obtaining concentration uniformity in a batch mixing operation can be based on model theory. Using the appropriate dimensionless groups of the pertinent variables, a relationship can be developed between mixing times in the model and large-scale systems for geometrically similar equipment.

Consider the mixing in both small and large-scale systems to occur in the turbulent region, designated as S and L respectively. Using the Norwood and Metzner's correlation [26], the mixing time for both systems is

$$\frac{t_S\left(N_S D_{AS}^2\right)^{2/3} g^{1/6} D_{AS}^{1/2}}{H_S^{1/2} \cdot D_{TS}^{3/2}} = \frac{t_L\left(N_L D_{AL}^2\right)^{2/3} g^{1/6} D_{AL}^{1/2}}{H_L^{1/2} \cdot D_{TL}^{3/2}} \tag{7-43}$$

Applying the scale-up rule of equal mixing times, and rearranging Equation 7-43, yields

$$\left(\frac{N_L}{N_S}\right)^{\frac{2}{3}} = \left(\frac{D_{TL}}{D_{TS}}\right)^{\frac{3}{2}} \left(\frac{D_{AS}}{D_{AL}}\right)^{\frac{4}{3}} \left(\frac{D_{AS}}{D_{AL}}\right)^{\frac{1}{2}} \left(\frac{H_L}{H_S}\right)^{\frac{1}{2}} \tag{7-44}$$

Assuming geometric similarity,

$$\frac{H_L}{H_S} = \frac{D_{AL}}{D_{AS}} \tag{7-45}$$

$$\frac{D_{TL}}{D_{TS}} = \frac{D_{AL}}{D_{AS}} \tag{7-46}$$

Substituting Equations 7-45 and 7-46 into Equation 7-44 gives

$$\left(\frac{N_L}{N_S}\right)^{\frac{2}{3}} = \left(\frac{D_{AL}}{D_{AS}}\right)^{\frac{3}{2}} \left(\frac{D_{AS}}{D_{AL}}\right)^{\frac{4}{3}} \left(\frac{D_{AS}}{D_{AL}}\right)^{\frac{1}{2}} \left(\frac{D_{AL}}{D_{AS}}\right)^{\frac{1}{2}} \tag{7-47}$$

$$\left(\frac{N_L}{N_S}\right)^{\frac{2}{3}} = \left(\frac{D_{AL}}{D_{AS}}\right)^{\frac{1}{6}}$$

or

$$\left(\frac{N_L}{N_S}\right) = \left(\frac{D_{AL}}{D_{AS}}\right)^{\frac{1}{4}} \tag{7-48}$$

The exponent n for the mixing time scale-up rule is 0.25.
The power P of the agitator for both large and small systems is

$$\frac{P_L}{\rho N_L^3 D_{AL}^5} = \frac{P_S}{\rho N_S^3 D_{AS}^5} \tag{7-49}$$

where

$$\frac{P_L}{P_S} = \left(\frac{N_L}{N_S}\right)^3 \left(\frac{D_{AL}}{D_{AS}}\right)^5 \tag{7-50}$$

Substituting Equation 7-48 into Equation 7-50 yields

$$\frac{P_L}{P_S} = \left(\frac{D_{AL}}{D_{AS}}\right)^{0.75} \left(\frac{D_{AL}}{D_{AS}}\right)^5 \tag{7-51}$$

or

$$\frac{P_L}{P_S} = \left(\frac{D_{AL}}{D_{AS}}\right)^{5.75} \tag{7-52}$$

The power per unit volume P/V for both large- and small-scale systems is:

$$\frac{P_L/V_L}{P_S/V_S} = \frac{P_L \Big/ \dfrac{\pi D_{TL}^3}{4}}{P_S \Big/ \dfrac{\pi D_{TS}^3}{4}}$$

$$= \frac{P_L}{P_S} \cdot \left(\frac{D_{TS}}{D_{TL}}\right)^3 \tag{7-53}$$

Substituting Equations 7-46 and 7-52 into Equation 7-53 gives

$$\frac{(P/V)_L}{(P/V)_S} = \left(\frac{D_{AL}}{D_{AS}}\right)^{5.75} \left(\frac{D_{AS}}{D_{AL}}\right)^3$$

$$= \left(\frac{D_{AL}}{D_{AS}}\right)^{2.75} \tag{7-54}$$

Table 7-7 summarizes correlations for the effects of equipment size on the rotational speed needed for the same mixing time by various investigators.

The relationships in Table 7-7 show that the rotational speed to obtain the same batch mixing time is changed by a small power of the increase in linear equipment dimension as equipment size is changed. Equation 7-49 shows that greater power is required for a large-scale system compared to a smaller system. Often, the power required for a larger system may be prohibitive, thus modification of the scale-up rule is needed (e.g., $t_L = 10t_S$ or $t_L = 100t_S$) to obtain a lower power requirement. It should be noted that relaxation of mixing time requirements may not pose other problems. For example, if the mixing is accompanied by a chemical reaction in a CSTR, assuming that the Norwood-Metzner [26] correlation for mixing time (t) is still applicable, it must be ensured that the mixing time in the larger scale ($t_L = 10t_S$ or $t_L = 100t_S$) is less than 5% of the average residence time of the liquids in the reactor, otherwise the conversion

Table 7-7

Effect of equipment size on rotational speed needed for the same mixing time

Relationship between N and D	Equipment	$\Delta\rho$	Equation	Investigator
$N \propto D^{-1/6}$	Propellers, no baffles	Not zero	$\left(\dfrac{\theta ND^2 p}{v}\right)\left(\dfrac{\rho D^2 N^2}{g Z_L \Delta\rho}\right)^{0.25} = 9$	van de Vusse [17]
$N \propto D^{-0.1 \text{ to } -0.2}$	Paddles, turbines	Not zero	$\left(\dfrac{\theta Q}{v}\right) \propto \left(\dfrac{\rho D^2 N^2}{g Z_L \Delta\rho}\right)^{-0.3}$	van de Vusse [17]
$N = \text{constant}$	Propellers, paddles, turbines	Zero		van de Vusse[17]
$N \propto D^{-1/5}$	Propellers	Zero	$\theta = \dfrac{C_1 Z_L^{1/2} T}{N_{Re}^{1/6} (ND^2)^{4/6} g^{1/6}}$	Fox and Gex [18]
$N \propto D^{1/4}$	Turbines	Zero		Norwood and Metzner [26]

Source: Gray, J. B., Mixing I Theory and Practice, V. W. Uhl and J. B. Gray, Eds. Academic Press Inc., 1966.

and product distribution will be affected [27]. Figure 7-17 summarizes the scale-up relationships for many of the important and controlling functions, depending on the nature of the process equipment. The figure identifies which curves apply to laminar (L) and turbulent (T) flow patterns in the fluid being subjected to the mixing operation. The

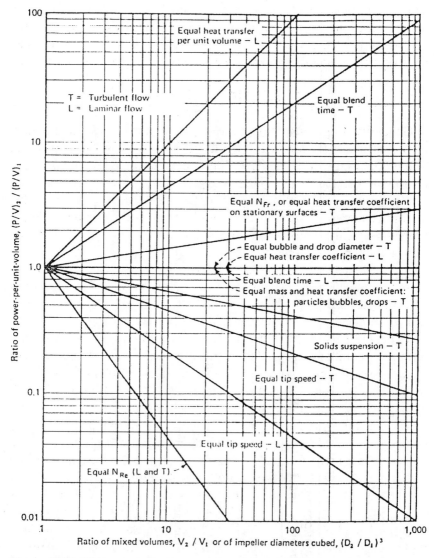

Figure 7-17. Mixing scale-up factors referenced to experienced ratios of power per unit volume. *(Source: Penny, W. R., Chem. Engr., p. 88, March 22, 1971.)*

scale-up chart only applies to systems of similar geometry. When the geometry is different, special and specific analyses of the system are required.

Samant and Ng [28] compared various scale-up rules for agitated reactors. They suggested that a scale-up rule of power per unit volume and constant average residence time (where the power per unit volume and average residence time cannot be increased) is the most suited in many operations. However, this still may not improve or preserve the performance of the systems. Therefore, adequate consideration must be given to a tradeoff between performance and operating constraints.

STATIC MIXERS

Generally, static or motionless mixers are tubular devices in which mixing elements are inserted in the line through which fluid is pumped. The mixing units or elements are often installed in a circular arrangement; however, they can be adapted to rectangular or other arrangements. The pressure drop incurred as the process stream flows through the mixing elements provides the energy required for the mixing. Pressure drop through the units varies depending on design and whether the flow is laminar or turbulent.

A typical static mixer consists of stationary mixing elements, aligned in series and turned on their main axis by 90°. They split the fluid flow into several partial flows. These partial flows are then redirected and recombined to form a mixture, which is homogeneous with respect to concentration and temperature. Mixing in the elements occurs through a combination of flow-splitting and shearing at the junction of successive elements employing a stretching and folding mechanism within the elements. The degree of homogeneity as well as the pressure drop increases with the number of mixing elements.

Various types of static mixers are used in the chemical process industry [29]. Table 7-8 lists their wide range of applications. The type of static mixer used often plays an essential role in the mixing mechanism, which greatly affects the flow pattern and pressure drop (ΔP) of fluids. The most common mixers are Kenics, Koch, Sulzer SMX, and Ross LPD. The Kenics and Ross LPD have a light structure. They can be employed for low-viscosity fluids in the transient and turbulent flow regimes with relatively low ΔP. The Sulzer SMX is suitable for viscous and non-Newtonian fluids in the laminar flow regime. However, a high ΔP is required to overcome the fluid drag.

Table 7-8
Application of static mixers in the chemical process industry

Industry	Applications
Chemicals	Chlorination and oxidation Steam injection Acid and base dilution Fast reactions
Food Processing	Acid washing of fats and oils Constituent blending Starch slurry and cooking
Mineral Processing	Slurry dilution Metal recovery by solvent extraction
Paints and Resins	Coloring and tinting Solvent blending Emissions monitoring and control
Petrochemicals and Refining	Gaseous reactant blending Gasoline blending Emissions monitoring and control
Pharmaceuticals	Nutrient blending Sterilization pH control
Polymers and Plastics	Reactant/catalyst blending Thermal homogenization Plug-flow finishing reactors
Pulp and Paper	Chemical and coatings preparation Stock dilution and consistency control Addition of bleaching chemicals
Water and Waste Treatment	Polymer dilution Disinfection, aeration, and dechlorination

Source: Myers, K. J., et al., Chem. Eng. Prog., *June 1997. Reproduced with permission of AIChEJ. © Copyright 1997. All rights reserved.*

The Kenics static mixer provides continuous inline mixing and processing, has no moving parts, and requires no external power or regular maintenance. Static mixers have proven to be effective for many process applications. Although useful for a wide range of viscosity fluids, some of the units have given excellent performance in the mixing of molten polymers. This type of unit is particularly useful in liquid-liquid mixing; some units can be designed for solid-liquid dispersion, gas-gas, and gas-liquid mixing/dispersion. Static mixers are used for laminar, transitional, and turbulent blending of miscible liquids, as tubular reactors in laminar flow heat exchangers, for laminar and turbulent homogenization, for dispersion of immiscible phases, and for interphase mass transfer between immiscible phases. When two immiscible fluids are subjected to the shear forces within a specific number of Kenics mixer units, drops of one phase are produced within the other. Intimate contact between the two fluids is achieved because of the mixing characteristics. An average drop size is established for a given flowrate and depends on the diameter of the mixer, the average linear velocity, the geometry of the mixer elements, and the physical properties of the two fluids. Drop size significantly effects the stability of a dispersed system, and dimensional analysis shows that the drop size of the dispersed phase is controlled by the Weber number (N_{We}), defined as:

$$N_{We} = \frac{\rho v^2 D}{\sigma}$$

where ρ = density of fluid
   D = inside tube diameter
   v = average axial velocity in the mixer
   σ = interfacial surface tension

Some units can be adapted for solid-solid blending. The residence time in the units can be varied or adjusted, which makes them suitable for certain types of reactions. Figure 7-18 shows a cross-section of a flanged Kenics static mixer with four helical mixing elements, and Table 7-9 lists the characteristics of static mixers.

PRINCIPLES OF OPERATION

The Kenics static mixer unit is a series of fixed, helical elements enclosed within a tubular housing. The fixed geometric design of the unit produces the following unique patterns of flow division and radial mixing simultaneously. In laminar flow, a processed material divides at the leading edge of each element and follows the channels created by the element shape. At each succeeding element, the two channels are further divided, resulting in an exponential increase in stratification.

Figure 7-18. Cross-section of flanged Kenics Static Mixer with four helical mixing elements. *(Source: Chemineer-Liquid-Agitation.)*

Table 7-9
Advantages of static mixers

- No moving parts
- Little or no maintenance requirements
- Small space requirements
- Available in many construction materials
- No power requirements other than pumping
- Mixing achieved in short conduit lengths
- Minimal chance of material hangup or plugging
- Short residence times
- Narrow residence time distribution
- Enhanced heat transfer
- Cost effective

Source: Myers, K. J., et al., Chem. Eng. Prog., *June 1997. Reproduced with permission of AIChEJ. © Copyright 1997. All rights reserved.*

The number of striations produced is 2^n, where n is the number of elements (Figure 7-19).

RADIAL MIXING

In either laminar or turbulent flow, rotational circulation of a processed material around its own hydraulic center in each channel of the mixer causes radial mixing of the material. All processed material is continuously and completely intermixed, virtually eliminating radial gradients in temperature, velocity, and material composition.

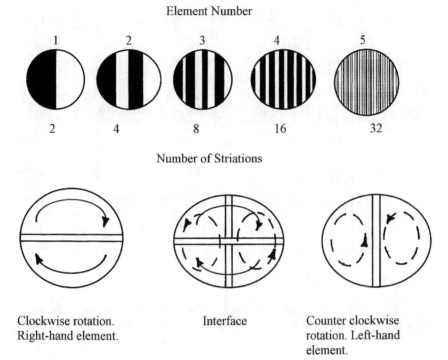

Figure 7-19. Principles of operation of static mixer modules.

The HEV static mixer produces complete stream uniformity through controlled vortex structures generated by the mixing elements. The element geometry takes advantage of the naturally occurring vortices induced by the element edges. These principles result in a reliable scale-up. Generally, the Kenics static mixer maximizes efficiency without the wasted energy and material blockage that occur in other restrictive motionless mixers.

The pressure drop in the Kenics mixer of the same length and diameter as an empty pipe can be determined from

$$\Delta P_{SM} = K \cdot \Delta P \tag{7-55}$$

where ΔP_{SM} is the pressure drop in the Kenics mixer and K is a function of the geometry of the mixer elements and the Reynolds number. The value for K is determined from Figures 7-20 and 7-21 depending on the type of flow regimes. Table 7-10 shows the Kenics mixer specification parameters for varying pipe sizes.

(text continued on page 604)

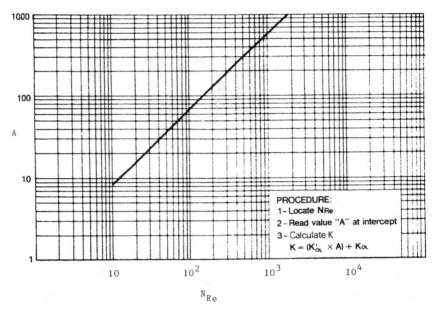

Figure 7-20. A-factor versus Reynolds number in the laminar flow region $10 < N_{Re} < 2,000$. *(Source: Chen, S. J., Kenics technical data KTEK-2, 1978.)*

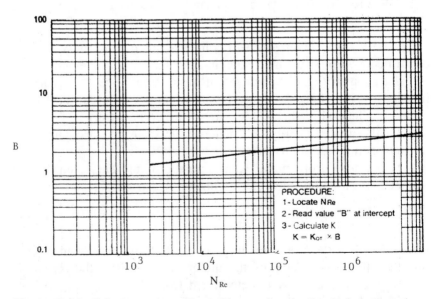

Figure 7-21. B-factor versus Reynolds number in the turbulent regions. *(Source: Chen, S. J., Kenics technical data KTEK-2, 1978.)*

Table 7-10
Kenics mixer specification

NOM. PIPE SIZE	HOUSING SCHEDULE	OUTSIDE DIA. Inch.	mm	INSIDE DIA. Inch.	mm	*MOD. LENGTH^f Feet	m	K_{OL}	K'_{OL}	K_{OT}
½	40	0.84	21.34	0.62	15.75	0.51	0.16	6.00	0.075	40.7
¾	40	1.06	26.92	0.82	20.83	0.65	0.20	5.23	0.050	23.5
1	40	1.32	33.53	1.05	26.67	0.90	0.27	5.79	0.069	36.3
1	80	1.32	33.53	0.96	24.38	0.85	0.26	5.57	0.062	31.4
1½	40	1.90	48.26	1.61	40.89	1.27	0.39	5.72	0.071	36.8
1½	80	1.90	48.26	1.50	38.10	1.27	0.39	5.53	0.065	32.6
2	40	2.38	60.45	2.07	52.58	1.71	0.52	5.70	0.068	35.1
2	80	2.38	60.45	1.94	49.28	1.67	0.51	5.54	0.062	31.6
2½	40	2.88	73.15	2.47	62.74	2.30	0.70	5.04	0.053	24.3
2½	80	2.88	73.15	2.32	58.93	1.92	0.59	5.58	0.066	33.8
3	40	3.50	88.90	3.07	77.98	2.82	0.86	4.94	0.052	23.6
3	80	3.50	88.90	2.90	73.66	2.82	0.86	4.82	0.049	21.4
4	40	4.50	114.3	4.03	102.36	3.37	1.03	5.08	0.058	26.9
4	80	4.50	114.3	3.83	97.28	3.18	0.97	5.16	0.060	28.2
6	40	6.63	168.4	6.07	154.18	4.88	1.49	5.19	0.060	28.6
6	80	6.63	168.4	5.76	146.30	4.88	1.49	5.08	0.057	26.2
8	40	8.63	219.2	7.98	202.69	6.26	1.91	5.14	0.061	28.4
10	40	10.75	273.05	10.02	254.51	7.79	2.37	5.07	0.060	27.8
12	40	12.75	323.85	11.94	303.28	9.66	2.94	4.88	0.056	24.8

^f Add (T+1/8) inch per flange up to a maximum of 3/8 inch per flange when ordering with flanges

(T= wall thickness of housing) (1/8 inch = 3.18 mm, 3/8 inch = 9.53 mm). * Mod. Length is based on Model 6.

(Source: Chen, S.J. Pressure Drop in the Kenics Mixer, Kenics Technical data KTEK-2, 1978).

Because there are no moving parts in the Kenics mixer, only the processed materials are in motion. Therefore, the only energy required for the mixer is the energy required to overcome the pressure drop (ΔP). The general equation for calculating the pressure drop in an empty pipe for isothermal incompressible fluids is given by

$$\Delta P = 4f_F \left(\frac{L}{D_{SM}} \right) \frac{\rho v^2}{2g_c} \qquad (7\text{-}56)$$

For laminar flow (i.e., $N_{Re} < 2,000$), ΔP is given by

$$\Delta P = 3.4 \times 10^{-5} \left(\frac{L}{D_{SM}^4} \right) \left(\frac{\mu W}{\rho} \right) = 2.73 \times 10^{-4} \left(\frac{L}{D_{SM}^4} \right) (\mu Q) \qquad (7\text{-}57)$$

ΔP in S.I. unit is:

$$\Delta P = 113.2 \left(\frac{L}{D_{SM}^4} \right) \left(\frac{\mu W}{\rho} \right) = 6.79 \left(\frac{L}{D_{SM}^4} \right) (\mu \rho) \qquad (7\text{-}58)$$

For turbulent flow, ΔP is given by

$$\Delta P = 2.16 \times 10^{-4} f_D \left(\frac{L}{D_{SM}^5} \right) \rho Q^2 = 3.36 \times 10^{-6} f_D \left(\frac{L}{D_{SM}^5} \right) \left(\frac{W^2}{\rho} \right) \qquad (7\text{-}59)$$

ΔP in S.I. unit is:

$$\Delta P = 2.252 f_D \left(\frac{L}{D_{SM}^5} \right) \rho Q^2 = 625.3 f_D \left(\frac{L}{D_{SM}^5} \right) \left(\frac{W^2}{\rho} \right) \qquad (7\text{-}60)$$

where ΔP = pressure drop in a pipe (psi, bar)
  ρ = fluid density (lb/ft^3, kg/m^3)
  f_D = Darcy friction factor ($f_D = 4f_F$)
  L = length of a static mixer (ft, m)
  v = fluid velocity (ft/sec, m/sec)
  g_C = dimensional gravitational constant (32.714 lb$_m$/lb$_f$ • ft/sec^2, 1 kg • m/N • sec^2)

D_{SM} = inside diameter of a static mixer (in, mm)
Q = volumetric flowrate (U.S. gal/min, l/min)
W = mass flowrate (lb/hr, kg/hr)

The Reynolds number N_{Re} is determined by

$$N_{Re} = 50.6 \frac{Q \cdot \rho}{\mu \cdot D_{SM}} = 6.31 \frac{W}{\mu \cdot D_{SM}} \qquad (7\text{-}61)$$

For S.I. unit, N_{Re} is determined by

$$N_{Re} = 21.22 \frac{Q \cdot \rho}{\mu \cdot D_{SM}} = 354 \frac{W}{\mu \cdot D_{SM}} \qquad (7\text{-}62)$$

where μ is the fluid viscosity, cP. For $0 < N_{Re} < 2,000$, the friction factor is

$$f_D = \frac{64}{N_{Re}} \qquad (7\text{-}63)$$

At higher Reynolds numbers, the friction factor is affected by the roughness of the surface, measured as the ratio ε/D of projections on the surface to the diameter of the pipe. Glass and plastic pipe essentially have $\varepsilon = 0$. Table 7-11 gives the pipe roughness of various materials. Figure 7-22 shows plots of Darcy friction factor versus Reynolds number for various pipe sizes. Alternatively, an explicit equation for the friction factor is given by [30]:

Table 7-11
Pipe roughness of materials

Material	Pipe roughness ε(ft)	Pipe roughness ε(mm)
Riveted steel	0.003–0.03	0.9–9.0
Cast iron	0.00085	0.25
Galvanized iron	0.0005	0.15
Asphalted cast iron	0.0004	0.12
Commercial steel or wrought iron	0.00015	0.046
Drawn tubing	0.000005	0.0015

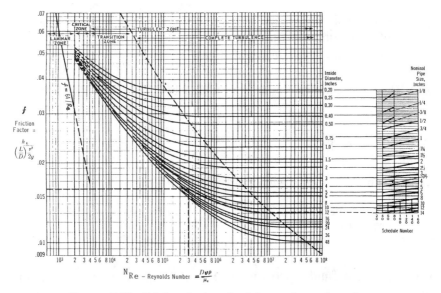

Figure 7-22. Friction factor chart for various pipe sizes.

$$\frac{1}{\sqrt{f_C}} = -4\log\left\{\frac{\varepsilon/D_{SM}}{3.7} - \frac{5.02}{N_{Re}}\log A\right\} \tag{7-64}$$

where $A = (\varepsilon/D_{SM})/3.7 + (6.7/N_{Re})^{0.9}$
f_C = Chen friction factor
ε = pipe roughness

A more explicit equation is given by [31]:

$$f = 1.6364\left[\ln\left(\frac{0.135\varepsilon}{D_{SM}} + \frac{6.5}{N_{Re}}\right)\right]^{-2} \tag{7-65}$$

The Darcy friction factor f_D is four times the Fanning (f_F) friction factor (i.e., $f_D = 4f_F = f_C$).

TWO-PHASE FLOW PRESSURE DROP (GAS-LIQUID) ESTIMATION

The Lockhart and Martinelli [32] correlation is employed to estimate the two-phase pressure drop in the Kenics mixer. The pressure drop is calculated for each phase, assuming that each phase is flowing alone in the unit. The pressure drop for each phase is related to the X-factor defined by

$$X = \left(\frac{\Delta P_L}{\Delta P_G}\right)^{0.5}$$

(7-66)

where X = a factor (dimensionless)
ΔP_L = pressure drop of liquid phase only (psi/element)
ΔP_G = pressure drop of gas phase only (psi/element)

The two-phase pressure drop is obtained by multiplying either the liquid-phase drop by ϕ_L^2 or the gas-phase pressure drop by ϕ_G^2. Figure 7-23 gives the Lockhart-Martinelli correlation between X and ϕ's

where $\phi_{L_{TT}}$ = liquid-phase pressure drop correction factor with both phases in turbulent flow region
$\phi_{L_{VV}}$ = liquid-phase pressure drop correction factor with both phases in viscous (or laminar) flow region

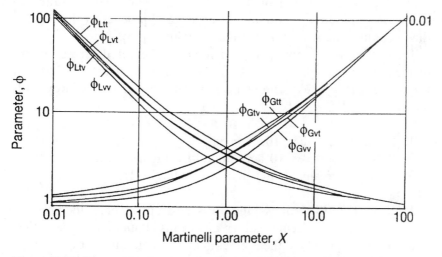

Figure 7-23. Parameters for pressure drop in liquid/gas flow through horizontal pipes. (*Source: Lockhead and Martinelli, Chem. Engr. Prog., 45, 39, 1949.*)

$\phi_{G_{TT}}$ = gas-phase pressure drop correction factor with both phases in turbulent flow region

$\phi_{G_{VV}}$ = gas-phase pressure drop correction factor with both phases in viscous (or laminar) flow region

The values midway between the TT and VV curves in Figure 7-23 are used for flow systems with either turbulent gas phase and viscous liquid phase or vice versa.

The total pressure drop is:

$$\Delta P = \Delta P_L \cdot \phi_{L_{TT}}^2$$

$$\text{or } \Delta P = \Delta P_G \cdot \phi_{G_{TT}}^2 \qquad (7\text{-}67)$$

Coker [33] has developed a computer program to apply the Lockhart-Martinelli correlation to determine the total pressure drop of the two-phase flow based on the vapor phase pressure drop. The program also determines the gas-liquid phase regime employing a modified Baker's map used to determine the flow regimes for horizontal gas-liquid flow. Static mixers generate smaller gas bubbles than stirred tank reactors (STR) and bubble columns, resulting in improved mass transfer rates 10 to 100 times those of a STR. Recently, a static mixer was successfully employed to enhance gas-liquid mixing and mass transfer in a process intensification of a packed-bed reactor. This resulted in increased productivity, elimination of a gel byproduct in the reactor, thus avoiding the need for frequent shutdown [34].

In most process piping applications, the required Kenics mixer diameter is the same as the existing process line diameter. The K-factor for a specific process application is determined by the Reynolds number as follows [35]:

1. If N_{Re} is less than 10, $K = K_{OL}$.
2. If N_{Re} is between 10 and 2,000, use Figure 7-20.
3. If N_{Re} is greater than 2,000, use Figure 7-21.

Multiply K by the empty-pipe pressure drop to obtain the pressure drop caused by the Kenics mixer model installation. The theoretical horsepower required by the Kenics mixer is determined by

$$hp = (0.262)(\Delta P_{SM})(q) \qquad (7\text{-}68)$$

where ΔP_{SM} = pressure drop in the Kenics Mixer (psi)
     q = volumetric flowrate (ft^3/sec)

To design the Kenics Mixer for a blending application, the procedure is as follows:

1. Calculate the Reynolds number (N_{Re}) for the existing process line.
2. Choose the appropriate Standard Model from Table 7-12.
3. Determine the pressure drop (ΔP) in the existing process line as a result of Standard Model installation.

If ΔP is excessive, repeat steps 1 through 3 using a larger mixer diameter.

Li et al. [36] performed an extensive study on ΔP in a Sulzer SMX static mixer with both Newtonian and non-Newtonian fluids. They showed that ΔP increased by a factor of 23 in a SMX static mixer in the laminar flow regime. Figure 7-24 shows their correlation between the Fanning friction factor and the Reynolds number for experimental points under various operating conditions.

Computational fluid dynamics (CFD) simulations are a valuable tool in studying the flow and mixing in static mixers [37]. Static mixer technology has been employed to solve many process mixing problems. Additionally, static mixers are used in tubular laminar flow reactors. Laminar open-pipe chemical reactors give a broad residence time distribution (RTD). The residence time distribution (Chapter 8) can be narrowed by the addition of helical elements, thus approaching the ideal plug flow system. The design data and calculations of static

Table 7-12
The correct number of Kenics Mixer
models for blending applications

N_{Re} (based on empty pipe)	Standard model required
<10	18
10–100	12
100–1,000	6
1,000–5,000	4
>5,000	2

Source: Chen, S. J., Pressure drop in the Kenics mixer: Kenics Technical data KTEK-2, 1988.

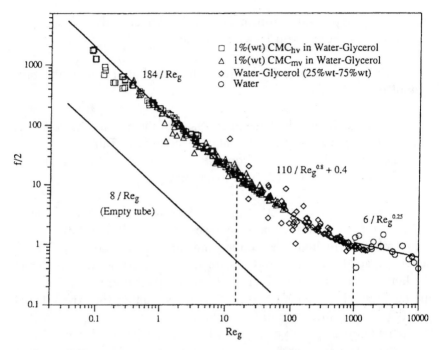

Figure 7-24. Friction factor versus Reynolds number in Sulzer SMX Static Mixer. *(Source: Li et al.,* Trans. IChemE, *Vol. 75, Part A, pp. 792–798, 1997. Reproduced with permission of the Trans. IChemE.)*

mixers can be incorporated into knowledge-based computer programs that produce optimal and cost effective designs for a wide range of applications [38]. Figure 7-25 shows a flowsheet that illustrates the pertinent issues to consider when selecting the proper mixer.

Example 7-6

What Kenics Mixer model of 3-in. Schedule 40 is required to process a Newtonian fluid with a viscosity of 150,000 cP, a density of 60 lb/ft^3, and a flowrate of 650 lb/hr? What is the pressure drop (ΔP) and theoretical horsepower?

Solution

From Table 7-10, $D_{SM} = 3.07$ in., $K_{OL} = 4.94$, $K'_{OL} = 0.052$, and $K_{OT} = 23.6$.

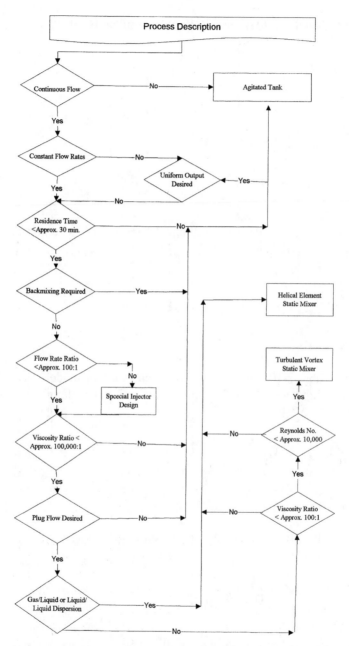

Figure 7-25. Flowchart shows the main issues in deciding the use of either a static mixer or an agitated tank. *(Source: Myers, K., et al.,* Chem. Eng. Prog., *June 1997. Used with permission of AIChE. Copyright © 1997. All rights reserved.)*

From Equation 7-61, the Reynolds number is

$$N_{Re} = 6.31 \cdot \frac{W}{\mu D_{SM}} \tag{7-61}$$

$$= 6.31 \cdot \frac{650}{(1.5 \times 10^5)(3.07)}$$

$$= 0.0089(8.9 \times 10^{-3})$$

Because $N_{Re} < 10$, from Table 7-12 model 18 is required. The Darcy friction factor f_D is:

$$f_D = \frac{64}{N_{Re}} = \frac{64}{(8.9 \times 10^{-3})} \tag{7-63}$$

$$= 7,185$$

From Table 7-10, the length of the Kenics mixer model is L = 2.82 ft. Therefore, the length of four modules is (i.e., the length of model 6 mixer = 2.82 ft):

$$L = \left(\frac{18}{6} \times 2.82 \right) ft$$

L = 8.46 ft

The pressure drop ΔP in the empty tube is:

$$\Delta P = 3.4 \times 10^{-5} \left(\frac{L}{D_{SM}^4} \right) \left(\frac{\mu W}{\rho} \right) \tag{7-57}$$

$$= (3.4 \times 10^{-5}) \left(\frac{8.46}{3.07^4} \right) \left(\frac{1.5 \times 10^5 \times 650}{60} \right)$$

$$= 5.26 \text{ psi}$$

Because $N_{Re} < 10$, Figure 7-20 gives $K = K_{OL} = 4.94$. The pressure drop in a model 18 mixer is:

$$\Delta P_{SM} = K \cdot \Delta P \qquad (7\text{-}55)$$

$$= 4.94 \times 5.26$$

$$= 25.98 \text{ psi}$$

The required theoretical horsepower from Equation 7-68 is:

$$hp = (0.262)(\Delta P_{SM})(q) \qquad (7\text{-}68)$$

$$hp = (0.262)(25.98)\left(\frac{650}{60 \times 3,600}\right)$$

$$= 0.02 \text{ hp}$$

Example 7-7

Determine which 2-in. Schedule 40 mixer model is required to process water-like fluids at a flowrate of 10 gpm. What is the required theoretical horsepower? (Data: viscosity = 1 cP; Density = 62.4 lb_m/ft^3; assume $\varepsilon = 0.00015$ ft)

Solution

From Table 7-10,

$$D_{SM} = 2.07 \text{ in.}$$
$$K_{OL} = 5.70$$
$$K'_{OL} = 0.068$$
$$K_{OT} = 35.1$$

Using Equation 7-61, the Reynolds number is:

$$N_{Re} = 50.6 \frac{Q \cdot \rho}{\mu \cdot D_{SM}} \qquad (7\text{-}61)$$

$$= 50.6 \frac{(10)(62.4)}{(1)(2.07)}$$

$$= 15{,}253.3 \ (1.5 \times 10^4)$$

Because $N_{Re} > 2000$, model 2 is required (Table 7-12). The Darcy friction factor f_D from Equation 7-64 is $f_D = 0.0293$. The length for model 2 is $L = 1.71/3 = 0.57$ ft (i.e., the length for model 6 mixer = 1.71 ft). The pressure drop (ΔP) in the empty pipe is:

$$\Delta P = 2.16 \times 10^{-4} \ f_D \left(\frac{L}{D_{SM}^5} \right) \rho Q^2 \qquad (7\text{-}59)$$

$$\Delta P = 2.16 \times 10^{-4} \ (0.0293) \left(\frac{0.57}{2.07^5} \right) (62.4)(10^2)$$

$$= 5.92 \times 10^{-4} \ \text{psi}$$

Because $N_{Re} > 2000$, from Figure 7-21, B = 1.85

B = 1.85 at $N_{Re} = 1.5 \times 10^4$

$K = K_{OT} \times B$

$\quad = 35.1 \times 1.85$

$\quad = 64.94$

The pressure drop in the Kenics Mixer is:

$$\Delta P_{SM} = K \cdot \Delta P \qquad (7\text{-}55)$$

$$= 64.94 \times 5.92 \times 10^{-4}$$

$$= 0.038 \ \text{psi}$$

The required theoretical horsepower is:

$$hp = (0.262)(\Delta P_{SM})(q) \qquad (7\text{-}68)$$

$$= (0.262)(0.038)\left(\frac{10 \times 0.134}{60}\right)$$

$$= 2.25 \times 10^{-4} \text{ hp}$$

Example 7-8

A liquid-gas mixture is to flow in a 3-in. Schedule 40 Kenics mixer. Estimate the pressure drop of the unit. The system conditions and physical properties are:

	Liquid	Gas
Flowrate, lb/hr	1,000	3,000
Density, lb/ft^3	62.4	0.077
Viscosity, cP	1.0	0.00127
Surface tension, dyne/cm	15.0	

The length of 3-in. Schedule 40 pipe from Table 7-8 is 2.82 ft. Assume $\varepsilon = 0.00015$ ft.

Solution

A computer program (TWOPHASE) was developed that uses the Lockhart-Matinelli correlation and determines the total pressure drop based on the vapor phase pressure drop. The total length of the unit depends on the nature of the Reynolds number. The program also calculates the gas-liquid phase regime employing a modified Baker's map [33]. Table 7-13 gives the results of the two-phase pressure drop.

HEAT TRANSFER IN AGITATED VESSELS

Agitated vessels with an external jacket or an internal coil are increasingly employed in biotechnology and other process applications. The most common type of jackets consists of an outer cylinder that surrounds part of the vessel. The heating or cooling medium circulates in the annular space between the jacket and vessel walls. Alternatively, condensation of vapor (e.g., steam or a proprietary heat transfer medium) may serve for heating and vaporization of liquid (e.g., a refrigerant) may serve for cooling. The heat is transferred through the wall of the vessel. Circulation baffles can be used in the annular space

Table 7-13
Two-phase pressure drop calculation in a pipeline

```
TWO-PHASE PRESSURE DROP CALCULATION IN A PIPE LINE
*********************************************************************
     PIPE INTERNAL DIAMETER, inch:                         3.070
     EQUIVALENT LENGTH OF PIPE, ft.:                         .000
     ACTUAL LENGTH OP PIPE, ft:                            2.820
     TOTAL LENGTH OF PIPE, ft.:                            2.820
     LIQUID DENSITY, lb/ft^3.:                            62.400
     LIQUID VISCOSITY, cP:                                 1.000
     SURFACE TENSION, dyne/cm.:                           15.000
     LIQUID FLOWRATE, lb/hr.:                           1000.00
     LIQUID REYNOLDS NUMBER:                              2055.
     LIQUID FRICTION FACTOR:                               .0311
     PRESSURE DROP OF LIQUID PER 100ft., psi/100ft:        .0007
     GAS FLOW RATE, lb/hr.:                             3000.00
     GAS DENSITY, lb/ft^3.:                                .077
     GAS VISCOSITY, cP:                                    .001
     GAS REYNOLDS NUMBER:                             4855216.
     GAS FRICTION FACTOR:                                  .0174
     PRESSURE DROP OF GAS PER 100ft, psi/100ft:           2.7309
     FLOW REGIME IS:                                     ANNULAR
     LOCKHART-MARTINELLI TWO PHASE FLOW MODULUS:           .0157
     VELOCITY OF FLUID IN PIPE, ft./sec.:               210.4991
     BAKER PARAMETER IN THE LIQUID PHASE:                 1.644
     BAKER PARAMETER IN THE GAS PHASE:                57508.660
     TWO-PHASE FLOW MODULUS:                              1.4561
     PRESSURE DROP OF TWO-PHASE MIXTURE, psi/100ft.:      3.9765
     OVERALL PRESSURE DROP OF THE TWO-PHASE, psi:          .1121
     INDEX  4553. IS LESS THAN 10000.0 PIPE EROSION IS UNLIKELY
```

to increase the velocity of the liquid flowing through the jacket, thus enhancing the heat transfer coefficient. An alternative is to introduce the fluid via a series of nozzles spaced down the jacket. In this case, the momentum of the jets issuing from the nozzles develops a swirling motion in the jacket liquid. The spacing between the jacket and vessel wall depends on the size of the vessel, however, it ranges from 50 mm for small vessels to 300 mm for larger vessels. Figure 7-26 shows different configurations of jacketed vessels.

The pitch of the coils and the area covered can be selected to provide the heat transfer area required. Standard pipe sizes from 60 mm to 120 mm outside diameter area are often used. Half-pipe construction can produce a jacket capable of withstanding a higher pressure than conventional jacket design.

The rate of heat transfer to or from an agitated liquid mass in a vessel depends on the physical properties of the liquid (e.g., density, viscosity, and specific heat) and of the heating or cooling medium, the vessel geometry, and the degree of agitation. The type and size of the agitator and its location also influence the rate. An agitator is selected on the basis of material properties and the processing required. The heat transfer forms part of a process operation such as suspended

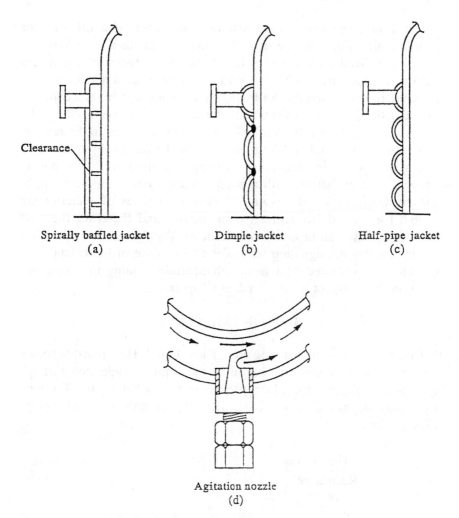

Spirally baffled jacket   Dimple jacket   Half-pipe jacket
(a)   (b)   (c)

Agitation nozzle
(d)

Figure 7-26. Types of jacket construction.

or dissolving solids, dispersing a gas in a liquid, emulsifying immiscible liquids, or regulating chemical reactions.

When processing is controlled by heat transfer variables, a log mean temperature difference (ΔT_{LMTD}) and heat transfer surface area will predominate over the agitation variables. Provided it is sufficient to give a homogeneous process fluid temperature, increased agitation can only reduce the inside film resistance, which is one of a number of resistances that determines the overall heat transfer coefficient.

Many jacketed vessels are reactors; the types of exothermic or endothermic effects reviewed in Chapter 6 must then be taken into account. Stirred tank reactors in which an exothermic reaction is performed may involve the removal of substantial amounts of heat from the reacting mixture. Refluxing of a boiling solvent is a common method; the heat of vaporization of the solvent is removed by the reflux condenser, and the condensed solvent is returned to the reactor. Other methods include cooling the walls of the reactor by means of a jacket with a cooling medium, inserting a cooling coil, or using an external heat exchanger with a pump around system. In many applications using jacketed vessels, successive batches of material are heated (or cooled) to a given temperature, and therefore the heat transfer involves an unsteady state process. Proper care is essential in terms of charging, agitation, and adequate cooling of the reactants to prevent the generated heat from subsequently leading to a runaway reaction. This aspect is reviewed in Chapter 12.

DESIGN EQUATION

Consider a vessel containing an agitated liquid. Heat transfer occurs mainly through forced convection in the liquid, conduction through the vessel wall, and forced convection in the jacket media. The heat flow may be based on the basic film theory equation and can be expressed by

$$\text{Rate} = \frac{\text{Driving force}}{\text{Resistance}}$$

$$\text{or} \quad \frac{Q}{A} = \frac{\Delta T}{1/U} \tag{7-69}$$

In an idealized situation, the vessel and its jacket each operate continuously under isothermal conditions. Rearranging Equation 7-69 becomes:

$$Q = UA\Delta T \tag{7-70}$$

In a realistic continuous situation, where the vessel contents are at constant temperature, but with different jacket inlet and outlet temperatures, Equation 7-70 is expressed as:

$$Q = UA\Delta T_{LMTD} \tag{7-71}$$

where ΔT_{LMTD} is the log mean temperature difference between the bulk temperature of the vessel contents, t, and the temperature in the jacket, T. ΔT_{LMTD} is expressed as

$$\Delta T_{LMTD} = \frac{(t_2 - T_2) - (t_1 - T_1)}{\ln\left[(t_2 - T_2)/(t_1 - T_1)\right]} \tag{7-72}$$

where t_1 = entering fluid temperature in the vessel
t_2 = leaving fluid temperature in the vessel
T_1 = entering fluid temperature in the jacket
T_2 = leaving fluid temperature in the jacket

The overall heat transfer coefficient U is determined from a series of resistances to the transfer of heat, namely

$$\frac{1}{U} = \frac{1}{h_i} + FF_i + \frac{x_w}{k} + FF_j + \frac{1}{h_j} \tag{7-73}$$

where h_i = coefficient on process side of heat transfer area, i.e., inside surface of jacketed vessel or outside surface of internal coil, $W/m^2{}^\circ C$
FF_i = fouling factor, inside vessel, $m^2{}^\circ C/W$
x_w = wall thickness of vessel or coil, mm
k = thermal conductivity, $W/m^\circ C$
FF_j = fouling factor, inside jacket, $m^2{}^\circ C/W$
h_j = coefficient on inside surface of jacket, $W/m^2{}^\circ C$

When the heat transfer is through internal coils or tubular baffles, the difference between the inner and outer heat transfer surfaces may be significant.

Inside Film (h_i) Coefficients

When applying the following equations for calculating film coefficients in jacketed vessels, the physical property data should be accurate. This is especially important for thermal conductivity k, as its value can have a major impact on the calculated film coefficient and vary widely.

The inside film heat transfer coefficient (h_i) can be calculated from the following Nusselt number correlation:

$$Nu = C N_{Re}^a N_{Pr}^b \left(\frac{\mu_b}{\mu_w}\right)^c f\left(\frac{D_T}{H}, \frac{W}{D_A}\right) \tag{7-74}$$

where $f(D_T/H, W/D_A)$ represents various geometric correction factors. For a geometrically similar system, Equation 7-74 becomes:

$$Nu = C \bullet N_{Re}^a \bullet N_{Pr}^b \left(\frac{\mu_b}{\mu_w}\right)^c \tag{7-75}$$

For agitated vessels,

$$\frac{h_i D_T}{k_f} = C \left(\frac{\rho N D_A^2}{\mu}\right)^a \left(\frac{C_p \mu}{k}\right)^b \left(\frac{\mu_b}{\mu_w}\right)^c \tag{7-76}$$

where h_i = heat transfer coefficient to vessel wall or coil, W/m²°C
$\quad D_A$ = agitator diameter, m
$\quad D_T$ = tank diameter, m
$\quad N$ = agitator speed (rev/sec)
$\quad \rho$ = density, kg/m³
$\quad k_f$ = thermal conductivity, W/m°C
$\quad C_P$ = specific heat capacity, J/kg°C
$\quad \mu_b$ = viscosity at bulk fluid temperature, [(N•s)/m²][kg/(m•sec)]
$\quad \mu_w$ = viscosity at the wall temperature, [(N•s)/m²][kg/(m•sec)]

The values of constant C and the exponents a, b, and c depend on the type of agitator, whether baffles are used and their type, and whether the transfer is via the vessel wall or to coils. Baffles are normally used in most applications, and the values of a, b and c in the literature are 2/3, 1/3, and 0.14 respectively. Tables 7-14 and 7-15 give typical correlations for various agitator types.

Fouling Factors and Wall Resistances

Experience and judgment as to fouling severity are required to estimate fouling factors (FF_i, FF_j) to determine the overall heat transfer coefficient. These will vary with time and depend on the frequency and efficiency of vessel cleaning. Wall resistances can be significant and should be calculated from accurate thermal conductivity data.

Outside Coefficients (h_o) Jacketed Vessels

Annular Jacket with Spiral Baffling

In heat transfer applications, this jacket is considered a helical coil if certain factors are used for calculating outside film coefficients. The equivalent heat transfer diameter, D_e, for a rectangular cross-section is equal to 4 w (w being the width of the annular space). Velocities are calculated from the actual cross-section of the flow area, pw (p being the pitch of the spiral baffle), and the effective mass flowrate W' through the passage. The effective mass flowrate is approximately 60% of the total mass flowrate of the jacket.

$$W' \approx 0.6 \, W \tag{7-77}$$

At a given Reynolds number, heat transfer coefficients of coils, particularly with turbulent flow, are higher than those of long, straight pipes, due to friction. This also applies to flow through an annular jacket with spiral baffling.

At $N_{Re} > 10,000$ the Sieder-Tate equation for straight pipe, $1 + 3.5(D_e/D_c)$ can be used to calculate the outside film coefficient.

$$\frac{h_o D_e}{k} = 0.027 \left(N_{Re}\right)^{0.8} \left(N_{Pr}\right)^{0.33} \left(\frac{\mu_b}{\mu_w}\right)^{0.14} \left\{1 + 3.5\left(\frac{D_e}{D_c}\right)\right\} \tag{7-78}$$

where D_e = equivalent diameter for heat transfer, mm (ft)
$\quad D_c$ = Mean or centerline diameter of internal coil helix, mm (ft)
$\quad h_j$ = heat transfer coefficient on inside surface of jacket
$\quad \mu_b$ = viscosity at bulk fluid temperature, [(N•s)/m^2][kg/(m•sec)]
$\quad \mu_w$ = viscosity at the wall temperature, [(N•s)/m^2][kg/(m•sec)]

$$N_{Re} = \frac{\rho \cdot v \cdot D_e}{\mu}$$

(text continued on page 628)

Table 7-14
Equations for calculating inside film coefficients (h_i) of jacketed agitated vessels

Agitator type	Baffled?	Reynolds number (N_{RE})	Nusselt number (N_{Nu})	Remarks
Paddle	Yes/No	$20 < N_{RE} < 4,000$	$0.415 \ (N_{RE})^{0.67} \ (N_{Pr})^{0.33} \ (\mu_b/\mu_w)^{0.14}$	Vessel geometry is discussed by Holland and Chapman [4]
Paddle	Yes/No	$>4,000$	$0.36 \ (N_{RE})^{0.67} \ (N_{Pr})^{0.33} \ (\mu_b/\mu_w)^{0.14}$	Vessel geometry is discussed by Holland and Chapman [4]
Flat blade turbine	Yes/No	<400	$0.54 \ (N_{RE})^{0.67} \ (N_{Pr})^{0.33} \ (\mu_b/\mu_w)^{0.14}$	$D_A/D_T = 1/3$, $H/D_T = 1.0$. Six-bladed turbine. Standard geometry.
Flat blade turbine	Yes	>400	$0.74 \ (N_{RE})^{0.67} \ (N_{Pr})^{0.33} \ (\mu_b/\mu_w)^{0.14}$	$D_A/D_T = 1/3$, $H/D_T = 1.0$. Six-bladed turbine. Standard geometry.
Flat blade turbine	Yes	$2,000 < N_{RE} < 700,000$	$1.10 \ (N_{RE})^{0.62} \ (N_{Pr})^{0.33} \ (\mu_b/\mu_w)^{0.14}$	
Propeller	Yes	$>5,000$	$0.64 \ (N_{RE})^{0.67} \ (N_{Pr})^{0.33} \ (\mu_b/\mu_w)^{0.14}$	Three blades
Propeller	Yes	No limitation	$0.54 \ (N_{RE})^{0.67} \ (N_{Pr})^{0.25} \ (\mu_b/\mu_w)^{0.14}$	45° pitched, four-blade impeller. Equation is based on limited data with regard to propeller pitch and vessel baffling. Divide h_i obtained with this equation by a factor of about 1.3.
Retreating blade turbine	Yes	No limitation	$0.33 \ (N_{RE})^{0.67} \ (N_{Pr})^{0.33} \ (\mu_b/\mu_w)^{0.14}$	Glassed-steel impeller. Three retreating blades. The lower constant (0.33) for the glassed-steel impeller is attributed to greater slippage around its curved surfaces than around the sharp corners of the alloy-steel impeller.

Impeller type	Baffled	Reynolds range	Equation	Remarks
Retreating blade turbine	Yes	No limitation	$0.37 (N_{RE})^{0.67} (N_{Pr})^{0.33} (\mu_b/\mu_w)^{0.14}$	Alloy-steel impeller. Three retreating blades.
Retreating blade turbine	No	No limitation	$0.68 (N_{RE})^{0.67} (N_{Pr})^{0.33} (\mu_b/\mu_w)^{0.14}$	Six retreating blades.
Propeller	Yes	No limitation	$0.54 (N_{RE})^{0.67} (N_{Pr})^{0.25} (\mu_b/\mu_w)^{0.14}$	45° pitched four bladed impeller. Equation is based on limited data with regard to propeller pitch and vessel baffling. Divide hi obtained with this equation by a factor of about 1.3.
Anchor	No	$30 < N_{RE} < 300$	$1.0 (N_{RE})^{0.67} (N_{Pr})^{0.33} (\mu_b/\mu_w)^{0.18}$	The overall heat transfer coefficient U, varies inversely with the anchor-to-wall clearance. Anchor to wall clearance is less than 1 in.
Anchor	No	$300 < N_{RE} < 4,000$	$0.38 (N_{RE})^{0.67} (N_{Pr})^{0.33} (\mu_b/\mu_w)^{0.18}$	Similar condition as before.
Anchor	No	$4,000 < N_{RE} < 37,000$	$0.55 (N_{RE})^{0.67} (N_{Pr})^{0.25}(\mu_b/\mu_w)^{0.14}$	Anchor to wall clearance of 1 to 5.125 in. Vessel geometry is illustrated by Holland and Chapman [4]
Helical ribbon	No	<130	$0.248 (N_{RE})^{0.5} (N_{Pr})^{0.33} (\mu_b/\mu_w)^{0.14}$ $\times (e/DA)^{-0.22} (i/D)^{-0.28}$	e = clearance, $(D_T - D_A)/2$, ft; D_A = impeller diameter, ft; i = agitator-ribbon pitch, ft
Helical ribbon	No	>130	$0.248 (N_{RE})^{0.67} (NPr)^{0.33} (\mu_b/\mu_w)^{0.14}$ $\times (i/D)^{-0.25}$	Same as above.

Table 7-15
Equations for calculating outside film coefficients (h_O) of jacketed agitated vessels

Jacket type	Reynolds number (N_{Re})	Nusselt number (N_{Nu})	Remarks
Annular jacket with spiral baffling	>10,000	$0.027\ (N_{Re})^{0.8}(N_{Pr})^{0.33}(\mu_b/\mu_w)^{0.14}(1 + 3.5D_e/D_c)$	This jacket is considered a special case of a helical coil if certain factors are incorporated into equations for calculating outside-film coefficients. In the equations at left and below, the equivalent heat transfer diameter D_e, for a rectangular cross-section is equal to four times the width of the annular space, w and D_c is the mean or centerline diameter of the coil helix. Velocities are calculated from the actual cross-section of the flow area, pw, where p is the pitch of the spiral baffle, and from the effective mass flowrate, W', through the passage. The leakage around spiral baffles is considerable, amounting to 35–50% of the total mass flowrate. The effective mass flowrate is about 60% of the total mass flowrate to the jacket: W' = 0.6 W. The N_{Nu} for this equation should be expressed in terms of $D_e(N_{Nu} = h_j D_e/k)$ as should the Reynolds number ($N_{Re} = D_e v\rho/\mu$), k being thermal conductivity, v being velocity, and ρ being density.

Annular jacket with spiral baffling	$<2{,}100$	$1.86\,(N_{Re})^{0.33}(N_{Pr})^{0.33}(D_e/L)^{0.33}\,(\mu_b/\mu_w)^{0.14}$	Same as the above. L is length of coil or jacket passage, ft
Annular jacket with spiral baffling	$2{,}100<N_{Re}<10{,}000$		Use the above equations depending on the value of N_{Re}
Annular jacket, no baffles	Laminar flow	$1.02(N_{Re})^{0.45}(N_{Pr})^{0.33}(\mu_b/\mu_w)^{0.14}(D_e/L)^{0.4}$ $\times (D_{jo}/D_{ji})^{0.8}\,(N_{Gr})^{0.05}$	D_{ji} and D_{jo} are the inside and outside diameters of the jacket respectively. For this equation, $D_e = D_{jo} - D_{ji}$. The Grashof number $N_{Gr} = D_e^3 \rho g \beta \Delta t_G/\mu^2$ were D_e is equivalent diameter, g is acceleration due to gravity, β is coefficient of volumetric expansion, μ is viscosity, ρ is density, and Δt_G is the difference between the temperature at the wall and that in the bulk fluid. N_{Gr} must be calculated from fluid properties at the bulk temperature.
Annular jacket with spiral baffling	$<2{,}100$	$1.86(N_{Re})^{0.33}(N_{Pr})^{0.33}(D_e/L)^{0.33}(\mu_b/\mu_w)^{0.14}$	The Nusselt and Reynolds numbers must be calculated with D_e as the diameter term.
Annular jacket, no baffles	Turbulent	$0.027(N_{Re})^{0.8}(N_{Pr})^{0.33}(\mu_b/\mu_w)^{0.14}$ $\times (1 + 3.5D_e/D_c)$	For the equivalent heat transfer diameter for turbulent flow, use $D_e = [(D_{jo})^2 - (D_{ji})^2]/D_{ji}$, where D_{ji} and D_{jo} are the inside and outside diameters of the jacket respectively. The cross-sectional flow area, $A_x = \pi[(D_{jo})^2 - (D_{ji})^2]/4$

Table 7-15
(continued)

Jacket type	Reynolds number (N_{Re})	Nusselt number (N_{Nu})	Remarks
Annular jacket with spiral baffling	$210 < N_{Re} < 10{,}000$		Use the above equations depending on the value of N_{Re}.
Half-pipe coil jacket	Laminar flow	$18.6(N_{Re})^{0.33}(N_{Pr})^{0.33}(D_e/L)^{0.33}(\mu_b/\mu_w)^{0.14}$	When pipe coils are made with a semicircular cross-section, $D_e = \pi d_{ci}/2$, where d_{ci} is the inner diameter of the pipe, in feet. For calculating the velocity, the cross-sectional flow area equals $\pi d_{ci}^2/8$. When pipe coils are made with a 120° central angle, $D_e = 0.0708\, d_{ci}$ and the cross-sectional area equals $0.154(d_{ci})^2$.
Half-pipe coil jacket	Turbulent flow	$0.027(N_{Re})^{0.2}(N_{Pr})^{0.33}(\mu_b/\mu_w)^{0.14} \times (1 + 3.5D_e/D_c)$	D_c is the mean diameter of the coil.
Half-pipe coil jacket	Transition flow		Use the above equations depending of the value of N_{Re}.
Dimple jacket	Laminar flow	$1.86(N_{Re})^{0.33}(N_{Pr})^{0.33}(D_e/L)^{0.33}(\mu_b/\mu_w)^{0.14}$	The equivalent diameter D_e, in a dimpled jacket equals 0.66 in. the cross-sectional flow area equals 1.98 in.^{2} per foot of vessel circumference.

| Dimple jacket | Turbulent flow | $0.27(N_{Re})^{0.8}(N_{Pr})^{0.33}(\mu_b/\mu_w)^{0.14}$ | The coefficients are not very accurate due to turbulence created by the dimples in the flow steam. |
| Dimple jacket | Transition flow | | Determine N_{Nu} from the above equations depending on the value of N_{Re}. |

The dimensionless parameters:

Reynolds number, $N_{Re} = \dfrac{\rho VD}{\mu} = \dfrac{\rho N D_A^2}{\mu}$

where ρ = density
V = velocity
D = diameter
D_A = impeller diameter
N = rotational speed of the agitator
μ = viscosity

Prandtl number, $N_{Pr} = \dfrac{C_p \mu}{k}$

where C_p = specific heat
μ = viscosity
k = thermal conductivity

Nusselt number, $N_{Nu} = \dfrac{hD}{k}$

where h = heat transfer coefficient
D = diameter
k = thermal conductivity

Viscosity number, μ_b/μ_w

where μ_b = viscosity at the bulk fluid temperature
μ_w = viscosity at the wall surface temperature

$(\mu_b/\mu_w)^{0.14} \cong 1.0$ for water

Grashof number, $N_{Gr} = \dfrac{D_e^3 \rho^2 g \beta \Delta t_G}{\mu^2}$

where D_e = equivalent diameter
ρ = density
g = acceleration due to gravity
β = coefficient of volumetric expansion
Δt_G = difference between the temperature at the wall and that in the bulk fluid

(text continued from page 621)

$$N_{Pr} = \frac{C_p \bullet \mu}{k}$$

At $N_{Re} < 2,100$

$$\frac{h_o D_e}{k} = 1.86 (N_{Re})^{0.33} (N_{Pr})^{0.33} \left(\frac{D_e}{L}\right)^{0.33} \left(\frac{\mu_b}{\mu_w}\right)^{0.14} \tag{7-79}$$

where L is the length of the coil or jacket passage, mm (ft).

Annular Jacket with No Baffles

In the case of steam condensation, a film heat transfer coefficient h_j is used. In the case of liquid circulation, velocities will be very low because of the large cross-sectional area.

Outside heat transfer coefficients for unbaffled jackets under laminar flow conditions can be calculated from,

$$\frac{h_j D_e}{k} = 1.02 (N_{Re})^{0.45} (N_{Pr})^{0.33} \left(\frac{D_e}{L}\right)^{0.4} \left(\frac{\mu_b}{\mu_w}\right)^{0.14}$$

$$\times \left(\frac{D_{jo}}{D_{ji}}\right)^{0.8} (N_{Gr})^{0.05} \tag{7-80}$$

where D_{ji} = inside diameter of the jacket
D_{jo} = outside diameter of the jacket
$D_e = D_{jo} - D_{ji}$

The Grashof number N_{Gr} is expressed by

$$N_{Gr} = \frac{D_e^2 \rho^2 g \beta \Delta t_G}{\mu_b^2}$$

where g = acceleration due to gravity
ρ = fluid density

β = coefficient of volumetric expansion

Δt_G = the difference between the temperature at the wall and that in the bulk fluid

μ_b = viscosity at bulk fluid temperature

Evidently, from the low value of the exponent in Equation 7-80, the contribution from natural convection and, hence, its practical significance is small.

The following equation can be used to predict heat transfer coefficients from coils to tank walls in agitated tanks.

$$\frac{hD_T}{k_f} = C\left(\frac{\rho ND_A^2}{\mu}\right)^{2/3}\left(\frac{C_p\mu}{k}\right)^{1/3}\left(\frac{\mu_b}{\mu_w}\right)^{1/4} \tag{7-81}$$

where C is a constant. Table 7-16 gives values of C for various agitator types and surface [39].

A software package (MIXER) was developed to determine the heat transfer coefficient for any type of agitator and surface using the value in Table 7-16, fluid physical properties, agitator speed, and diameter.

HEAT TRANSFER AREA

Surface area for heating or cooling agitated vessels can be provided by either external jacketing or internal coils (or tubular baffles). Jacketing is usually preferred because of:

Table 7-16
Constant (C) for various impellers

Agitator	Surface	C
Turbine	Jacket	0.62
Turbine	Coil	1.50
Paddle	Jacket	0.36
Paddle	Coil	0.87
Anchor	Jacket	0.46
Propeller	Jacket	0.54
Propeller	Coil	0.83

Source: Chopey, N. P. and Hicks, T. G., Handbook of Chemical Engineering Calculations, *McGraw-Hill Book Co., 1984.*

- Cheaper construction materials because the jacket is not in contact with process fluid.
- Less tendency to foul.
- Easier cleaning and maintenance.
- Fewer problems in circulating catalysts and viscous fluids.
- Larger heat-transfer surface, unless significant reactor volume is taken up by the coils.
- Helical jackets may allow thinner walls to be used for pressure vessels.
- No restriction is placed on agitator type, whereas if a coil is installed it restricts agitator dimensions.

Coils should be considered only if jacketing alone does not provide a sufficient heat transfer area, if jacket pressure exceeds 150 psig, or if high-temperature vacuum processing is required. The coil offers the advantage of a higher overall film coefficient because of thinner walls with the latter conditions, but the wall resistance may not be significant compared to that on the process side (e.g., with a viscous liquid).

Example 7-9

Determine the heat transfer coefficient from a coil immersed in an agitated vessel with a diameter of 10 ft (3.048 m). The agitator is a paddle measuring 3.5 ft (1.01 m) in diameter and revolving at 200 rev/min. The fluid properties are:

ρ = density = 720 kg/m^3
μ_b = viscosity = 4.13 cP = 4.13 × 10^{-3} (Pa • s)
C_p = specific heat = 2.9 kJ/kg • K
k = thermal conductivity = 0.17 W/mK

Assume $(\mu_b/\mu_w)^{0.14}$ = 1.0.

Solution

From Table 7-16, for a paddle type agitator, C = 0.87. The heat transfer coefficient from Equation 7-76 becomes

$$\frac{hD_T}{k_f} = 0.87 \left(\frac{\rho ND_A^2}{\mu}\right)^{2/3} \left(\frac{C_p \mu}{k}\right)^{1/3} \left(\frac{\mu_b}{\mu_w}\right)^{1/4} \tag{7-81}$$

N = number of revolution per sec is 200/60 = 3.3333 rev/sec.
The Reynolds number, $N_{Re} = \rho N D^2_A/\mu$ is:

$$N_{Re} = \frac{(720)(3.3333)(1.01)^2}{4.13 \times 10^{-3}} \left(\frac{kg}{m^3} \cdot \frac{rev}{sec} \cdot \frac{m^2}{\frac{kg}{m \cdot sec}} \right)$$

$$= 592,794$$

The Prandtl number, $N_{Pr} = C_p\mu/k$ is

$$N_{Pr} = \frac{(2.9 \times 10^3)(4.13 \times 10^{-3})}{0.17} \left(\frac{J}{kg \cdot K} \cdot \frac{kg}{m \cdot sec} \cdot \frac{sec \cdot m \cdot K}{J} \right)$$

$$= 70.45$$

The heat transfer coefficient is:

$$h_v = 0.87 \left(\frac{0.17}{3.048} \right) (592,794)^{2/3} (70.45)^{1/3}$$

$$h_v = 1,414 \ W/m^2 \ K$$

Table 7-17 shows the computer results from the MIXER software.

Scale-Up with Heat Transfer

The scale-up criterion of constant heat transfer coefficient is suitable when the predominant problem of the reactor involves the removal of heat. The magnitude of the heat transfer coefficient is governed by the intensity of stirring within the reactor, and is represented by:

$$\frac{hD_T}{k} = C \left(\frac{\rho N D_A^2}{\mu} \right)^{0.65} \left(\frac{C_p\mu}{k} \right)^{0.33} \left(\frac{\mu_b}{\mu_w} \right)^{0.24} \tag{7-82}$$

where C is a constant that depends on the agitator design and h is the required inside film heat transfer coefficient.

Table 7-17
Heat transfer coefficient to fluids in a vessel
using mechanical agitated coils or jacket

AGITATOR:	PADDLE
SURFACE:	COIL
VALUE OF A:	0.870
DIAMETER OF VESSEL, m.:	3.048
THERMAL CONDUCTIVITY, W/m.K.:	0.170
DIAMETER OF AGITATOR, m.:	1.010
SPEED OF AGITATOR, rev/min.:	200.000
DENSITY OF FLUID, kg/m^3:	720.000
VISCOSITY OF FLUID, 10^(-3) Pa.s:	4.130
SPECIFIC HEAT CAPACITY, kJ/kg.K:	2.900
VISCOSITY AT BULK FLUID TEMPERATURE, 10^(-3) Pa.s:	1.000
VISCOSITY AT SURFACE TEMPERATURE, 10^(-3) Pa.s:	1.000
REYNOLDS NUMBER:	592794.
PRANDTL NUMBER:	70.5
HEAT TRANSFER COEFFICIENT, W/m^2.K.:	1414.2

To scale-up a reactor from V_1 to V_2 with geometrically similar systems having similar bulk average temperatures (i.e., the physical properties of the fluids are identical), Equation 7-82 becomes

$$\frac{h_2 D_{T2}}{h_1 D_{T1}} = \left(\frac{N_2 D_{A2}^2}{N_1 D_{A1}^2}\right)^{0.65} \tag{7-83}$$

$$\text{or} \quad \frac{h_2 D_{T2}}{h_1 D_{T1}} = \left(\frac{N_2}{N_1}\right)^{0.65}\left(\frac{D_{A2}}{D_{A1}}\right)^{1.30} \tag{7-84}$$

$$\left(\frac{N_2}{N_1}\right)^{0.65} = \frac{h_2 D_{T2}}{h_1 D_{T1}}\left(\frac{D_{A1}}{D_{A2}}\right)^{1.30} \tag{7-85}$$

where $D_{T2}/D_{T1} = D_{A2}/D_{A1}$.
Equation 7-85 becomes

$$\left(\frac{N_2}{N_1}\right)^{0.65} = \frac{h_2 D_{A2}}{h_1 D_{A1}}\left(\frac{D_{A1}}{D_{A2}}\right)^{1.30} \tag{7-86}$$

At equal heat transfer coefficients, $h_1 = h_2$

$$\left(\frac{N_2}{N_1}\right)^{0.65} = \left(\frac{D_{A2}}{D_{A1}}\right)^{-0.30} \tag{7-87}$$

$$\text{or } \frac{N_2}{N_1} = \left(\frac{D_{A2}}{D_{A1}}\right)^{-0.46} \tag{7-88}$$

Assuming that the equation is in the turbulent range, the Power numbers will be equal. The ratio of the power per unit volume (P/V) for large and small scales can be expressed by

$$\frac{(P/V)_2}{(P/V)_1} = \frac{\rho N_2^3 D_{A2}^5 / D_{A2}^3}{\rho N_1^3 D_{A1}^5 / D_{A1}^3} = \frac{N_2^3 \, D_{A2}^2}{N_1^3 \, D_{A1}^2} \tag{7-89}$$

Substituting of Equation 7-88 into Equation 7-89 gives

$$\frac{(P/V)_2}{(P/V)_1} = \left(\frac{D_{A2}}{D_{A1}}\right)^2 \left(\frac{D_{A2}}{D_{A1}}\right)^{-1.38}$$

$$= \left(\frac{D_{A2}}{D_{A1}}\right)^{0.62} \tag{7-90}$$

The power per unit volume thus increases slightly. For equal heat transfer coefficients on small and large scales, the larger tank will use an impeller at a lower speed.

$$\text{Where } \frac{V_2}{V_1} = \left(\frac{D_{T2}}{D_{T1}}\right)^3 \tag{7-36}$$

$$\frac{N_2}{N_1} = \left(\frac{V_2}{V_1}\right)^{-0.15} \tag{7-91}$$

Having achieved the same heat transfer coefficient on the larger scale, the heat removal facilities must be increased because the heat generation is proportional to V_2/V_1, but the surface area of the vessel has increased by $(V_2/V_1)^{2/3}$. This can be done by adding coils in the reactor. Larger areas can be added by using an external heat exchanger and a pump around system. In some cases, it may be possible to lower the coolant temperature and thereby increase the rate of heat flow through the existing surface. However, this is usually fixed by stability considerations, which require that the coolant temperature be within a few degrees of the reaction temperature.

LIQUID–SOLID AGITATION

In certain cases, the primary process objective is to keep solid particles in suspension. Areas of application involve catalytic reactions, crystallization, precipitation, ion exchange, and adsorption. Axial flow and pitched-blade turbines are best suited in providing the essential flow patterns in a tank to keep the solids in suspension. The suspended solid is characterized by two parameters:

- Particle density, ρ_P.
- Particle size: the mean diameter, $\overline{d}_P$ or the particle size distribution.

Various correlations are provided for calculating the minimum speed of the agitator N_{min} to keep a given solid in suspension. Zwietering [40] developed the following equation:

$$N_{min} = \Psi \left(\frac{D_T}{D_A} \right)^\alpha \frac{g^{0.45} \left(\rho_P - \rho_L \right)^{0.45} \mu_L^{0.1} \overline{d}_P^{0.2} W_S^{0.13}}{D_A^{0.85} \rho_L^{0.55}} \qquad (7\text{-}92)$$

where D_A = impeller diameter, m
$\quad D_T$ = tank diameter, m
$\quad W_S$ = weight ratio of solid to liquid in percentage
$\quad \rho_L$ = liquid density
$\quad \mu_L$ = liquid viscosity

Ψ and α depend on the characteristics of the stirrer.

For agitator types of propellers, turbines with flat blades and paddles, Ψ and α are 1.5 and 1.4, respectively. The criterion for Equation 7-92 is the absence of any immobile solid on the bottom of the tank.

Weisman and Efferding [41] in constrast, related the degree of agitation to the height H_S occupied by the solid suspension, which is expressed as:

$$\frac{H_S - \left[E + (n_T - 1)D_A\right]}{D_T} = 0.23 \ln\left[\frac{P(\epsilon_S)^{-0.66}}{n_T g \rho_M V u_S}\left(\frac{D_A}{D_T}\right)^{0.5}\right] + 0.1 \quad (7\text{-}93)$$

where E = distance from reaction bottom to agitation system
$\quad\quad H_S$ = height occupied by the solid suspension
$\quad\quad n_T$ = number of stirring components (1 to 3)
$\quad\quad V$ = volume to be stirred
$\quad\quad \epsilon_S$ = volumetric fraction of solid
$\quad\quad \rho_M$ = density of the suspension
$\quad\quad u_S$ = particle sedimentation rate

Equation 7-93 was established for turbine agitators with flat blades and $Z/D_A = 0.5$. The criteria for Equation 7-93 relate to a specific type of suspension. The distribution of the solid as a function of the height in the liquid is not uniform in every case. Therefore, the uniformity can only be approximated by obtaining a circulation rate Q as high as possible. Nienow [42] found that H/E = 7 in tanks for which $H/D_T = 1.0$.

Recently, Corpstein et al. [43] found that high-efficiency impellers provide the same levels of solids suspension at reduced capital and operating costs. They introduced the term "just-suspended" for the most commonly encountered level of liquid solid agitation. This occurs when none of the solid particles remains stationary on the bottom of the vessel for longer than 1–2 sec. They developed a correlation of the speed required to achieve just-suspended conditions as:

$$N_{js} = k\left[\left(\frac{\rho_S - \rho_L}{\rho_L}\right)u_t\right]^{0.28} f(X)\; f\left(\frac{D_A}{D_T}\right)\left(\frac{D_T}{D_o}\right)^{-n} \quad (7\text{-}94)$$

where k = applicable coefficient, k = 15.0 for a pitched-blade turbine;
$\quad\quad\quad$ k = 23.0 for a high-efficiency impeller
$\quad\quad D_o$ = a reference scale (0.29 m)
$\quad\quad D_A$ = impeller diameter, m
$\quad\quad D_T$ = tank diameter, m

$f(X)$ = solids-loading factor, $f(X)$, is a non-linear function for up to 5% solids loading

n = scale-up exponent

N_{js} = just-suspended speed, s^{-1}

u_t = terminal settling velocity for a particle, m/sec

ρ_l = density of liquid

ρ_S = density of solids

X = solids loading (solids mass/slurry mass)

The required power can be determined by:

$$P = N_P \, \rho_{sl} N^3 D_A^5 \qquad (7\text{-}95)$$

where N = impeller rotation speed, s^{-1}

N_P = impeller Power number

P = Impeller power draw, W

ρ_{sl} = density of slurry, kg/m^3

Computation fluid mixing and computational fluid dynamic techniques have increasingly been used to elucidate solids distribution in agitated vessels [44].

BATCH HEATING AND COOLING OF FLUIDS

Heating or cooling of process fluids in a batch-operated vessel is common in the chemical process industries. The process is unsteady state in nature because the heat flow and/or the temperature vary with time at a fixed point. The time required for the heat transfer can be modified, by increasing the agitation of the batch fluid, the rate of circulation of the heat transfer medium in a jacket and/or coil, or the heat transfer area. Bondy and Lippa [45] and Dream [46] have compiled a collection of correlations of heat transfer coefficients in agitated vessels. Batch processes are sometimes disadvantageous because:

- Use of heating or cooling medium is intermittent.
- The liquid being processed is not readily available.
- The requirements for treating time require holdup.
- Cleaning or regeneration is an integral part of the total operating period.

The variables in batch heating or cooling processes are surface requirement, time, and temperature. Heating a batch may be by external means (e.g., a jacket or coil) or by withdrawing and recirculating process liquid through an external heat exchanger. In either case, assumptions are made to facilitate calculation, namely,

- The overall heat transfer coefficient U is constant for the process and over the entire surface.
- Liquid flowrates are at steady state.
- Specific heats are constant for the process.
- The heating or cooling medium has a constant inlet temperature.
- Agitation gives a uniform batch fluid temperature.
- There is no phase change.
- Heat losses are negligible.

The following discusses various heating or cooling process conditions in a batch vessel and the processing time relationships.

BATCH HEATING: INTERNAL COIL, ISOTHERMAL HEATING MEDIUM

When an agitated batch containing M of fluid with specific heat c and initial temperature t is heated using an isothermal condensing heating medium T_1, the batch temperature t_2 at any time θ can be derived by the differential heat balance. For an unsteady state operation as shown in Figure 7-27, the total number of heat transferred is q', and per unit time θ is:

$$
\begin{array}{cccc}
\text{I} & \text{II} & \text{III} & \text{IV}
\end{array}
$$

$$
dq = \frac{dq'}{d\theta} = Mc\frac{dt}{d\theta} = UA\Delta t \tag{7-96}
$$

$$
\begin{array}{cc}
\text{Accumulation} & \text{Transfer} \\
\text{in the batch} & \text{rate}
\end{array}
$$

where

$$
\Delta t = T_1 - t \tag{7-97}
$$

Equating III and IV gives

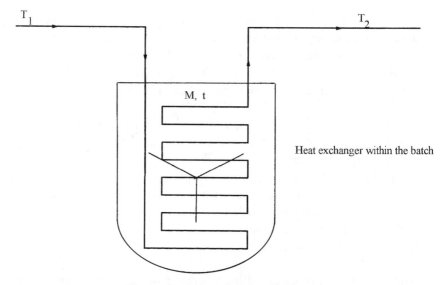

Figure 7-27. Agitated batch vessel.

$$Mc\frac{dt}{d\theta}=UA\Delta t \tag{7-98}$$

Rearranging Equation 7-98 gives

$$\frac{dt}{\Delta t}=\frac{UA}{Mc}\bullet d\theta \tag{7-99}$$

Integration of Equation 7-99 between the limits gives

$$\int_{t_1}^{t_2}\frac{dt}{T_1-t}=\frac{UA}{Mc}\int_0^\theta d\theta \tag{7-100}$$

Integration of Equation 7-100 from t_1 to t_2 while the batch processing time passes from 0 to θ yields:

$$\ln\left(\frac{T_1-t_1}{T_1-t_2}\right)=\frac{UA}{Mc}\bullet\theta$$

$$\text{or} \quad \theta = \frac{Mc}{UA} \ln\left(\frac{T_1 - t_1}{T_1 - t_2}\right) \tag{7-101}$$

where A = heat transfer surface area
  c = specific heat of batch liquid
  M = weight of batch liquid
  T_1 = heating medium temperature
  t_1 = initial batch temperature
  t_2 = final batch temperature
  U = overall heat transfer coefficient
  θ = time

Example 7-10

A tank containing 50,000 lb (22,679.5 kg) material with a specific heat of 0.5 Btu/lb • °F (2.1 kJ/kg • K) is to be heated from 68°F (293 K) to 257°F (398 K). The tank contains a heating coil with a heat transfer surface of 100 ft² (9.29 m²), and the overall heat transfer coefficient from the coil to the tank contents of 150 Btu/hr • ft²°F (850 W/m² • K). Calculate the time required to heat the tank contents with steam condensing at 320°F (433 K).

Solution

Select and apply the appropriate heat transfer formula.
When heating a batch with an internal coil with an isothermal heating medium, the following equation applies:

$$\ln\left(\frac{T_1 - t_1}{T_1 - t_2}\right) = \frac{UA}{Mc} \cdot \theta \tag{7-101}$$

$$\ln\left(\frac{433 - 293}{433 - 398}\right) = \frac{(850)\,(9.29)}{(22{,}679.5)(2.1)\,(10^3)}\theta \left\{\frac{W}{m^2 K} \cdot \frac{m^2}{kg} \cdot \frac{kg \cdot K}{J}\right\}$$

$$\theta = \frac{(1.386)(22,679.5)(2.1)(10^3)}{(850)(9.29)(3,600)} \ \text{hr}$$

$$= 2.32 \ \text{hr}$$

Computer software (BATCH) was developed to determine the time required for heating or cooling process fluids in a batch system. Table 7-18 gives the computer results of Example 7-10.

BATCH REACTOR HEATING AND COOLING TEMPERATURE PREDICTION

Start up of a jacketed batch reactor requires control of the heat-up and cool-down rates. This involves determining and setting the jacket heat transfer fluid temperatures. An alternative is to make a trial heat-up and incorporate the results into a time-dependent heat transfer equation:

$$\theta = \ln\left(\frac{T_1 - t_1}{T_1 - t_2}\right) \bullet \left(\frac{Mc}{UA}\right) \tag{7-101}$$

Equation 7-101 can also be used to calculate the heat-up time for non-isothermal heating (e.g., by hot-water jacketing), provided that the difference between the outlet and inlet jacket temperatures is not greater than 10% of the difference between the batch and average water temperature [47].

Table 7-18
Batch heating: internal coil isothermal heating

HEAT TRANSFER SURFACE AREA, m^2:	9.290
SPECIFIC HEAT OF LIQUID, kJ/kg.K:	2.100
WEIGHT OF BATCH LIQUID, kg.:	22679.500
HEATING MEDIUM TEMPERATURE, K:	433.000
INITIAL BATCH TEMPERATURE, K:	293.000
FINAL BATCH TEMPERATURE, K:	398.000
OVERALL HEAT TRANSFER COEFFICIENT, W/m^2.K:	850.000
TIME, hr.:	2.323

Assuming that M, c, U, and A are constants, where

$$K = \frac{UA}{Mc} \qquad (7\text{-}102)$$

Equation 7-101 becomes

$$\theta = \ln\left(\frac{T_1 - t_1}{T_1 - t_2}\right) \cdot \frac{1}{K} \qquad (7\text{-}103)$$

Rearranging Equation 7-103 gives the jacket temperature as a function of time as:

$$T_1 = \frac{t_1 - t_2 e^{K\theta}}{1 - e^{K\theta}} \qquad (7\text{-}104)$$

Thus, by taking a series of readings during a trial heat-up, K can be determined. The heat-up and cool-down times for varying jacket temperatures can then be predicted.

Example 7-11

Assume that in Example 7-10, the overall cycle time for a batch reaction is 8 hrs. The cycle time will include 2 hrs for heat-up and 3 hrs for cool-down. The batch will be heated from 20°C to the reaction temperature of 60°C, then cooled to 35°C. Using a hot-water jacket temperature of 80°C, it took 15 min to heat the batch from 20°C and 30°C. Calculate the jacket temperatures required for heat-up and cool-down.

Solution

From Equation 7-102,

$$K = \frac{UA}{Mc} = \frac{(850)(9.29)}{(22,679.5)(2.1)(1,000)} \quad \left\{\frac{J}{\sec \bullet m^2 K} \bullet m^2 \bullet \frac{1}{kg} \bullet \frac{kg \bullet K}{J}\right\}$$

$$K = 0.00017 \ \sec^{-1}$$

The jacket temperature required for a 2 hr heat-up can be obtained from Equation 7-104 as:

$$T_1 = \frac{t_1 - t_2 e^{K\theta}}{1 - e^{K\theta}} \tag{7-104}$$

$$= \frac{20 - 60 e^{\left(0.00017\frac{1}{sec} \times 2 \times 3,600\,sec\right)}}{1 - e^{\left(0.00071\frac{1}{sec} \times 2 \times 3,600\,sec\right)}}$$

$$= 77°C$$

The jacket temperature required for a 3 hr cool-down is:

$$T_1 = \frac{t_1 - t_2 e^{K\theta}}{1 - e^{K\theta}}$$

$$= \frac{60 - 35 e^{\left(0.00017\frac{1}{sec} \times 3 \times 3,600\,sec\right)}}{1 - e^{\left(0.00017\frac{1}{sec} \times 3 \times 3,600\,sec\right)}}$$

$$= 30.3°C$$

A Microsoft Excel spreadsheet (Example 7-11.xls) was developed for predicting the jacket temperature required for either heating up or cooling down reactants in a batch reactor.

BATCH COOLING: INTERNAL COIL ISOTHERMAL COOLING MEDIUM

Consider the same arrangement as before containing M of liquid with specific heat c and initial temperature T_1 cooled by an isothermal-vaporizing medium of temperature t_1. If T is the batch temperature at any time θ, then

$$\frac{dq'}{d\theta} = -Mc\frac{dT}{d\theta} = UA\Delta t \tag{7-105}$$

where

$$\Delta t = T - t_1 \tag{7-106}$$

Then

$$-Mc\frac{dT}{d\theta} = UA\Delta t \tag{7-107}$$

Substituting Equation 7-106 into Equation 7-107 and rearranging gives:

$$-\int_{T_1}^{T_2} \frac{dT}{T - t_1} = \int_0^\theta \frac{UA}{Mc} \cdot d\theta \tag{7-108}$$

Integration from T_1 to T_2 while the time passes from 0 to θ gives

$$\ln\left(\frac{T_1 - t_1}{T_2 - t_1}\right) = \frac{UA}{Mc} \cdot \theta$$

$$\text{or} \quad \theta = \frac{Mc}{UA} \ln\left(\frac{T_1 - t_1}{T_2 - t_1}\right) \tag{7-109}$$

where A = heat transfer surface area
    c = specific heat of batch liquid
    M = weight of batch liquid
    T_1 = initial batch temperature
    T_2 = final batch temperature
    t_1 = cooling medium temperature
    U = overall heating transfer coefficient
    θ = time

BATCH HEATING: NON-ISOTHERMAL HEATING MEDIUM

The non-isothermal heating medium has a constant flowrate W_h, specific heat C_h, and inlet temperature T_1, but a variable outlet temperature. For an unsteady state operation:

$$\begin{array}{cccc} \text{I} & \text{II} & \text{III} & \text{IV} \end{array}$$

$$\frac{dq'}{d\theta} = Mc\frac{dt}{d\theta} = W_h C_h (T_1 - T_2) = UA\Delta t_{LMTD} \tag{7-110}$$

The log mean temperature difference Δt_{LMTD} is:

$$\Delta t_{LMTD} = \frac{T_1 - T_2}{\ln\left(\dfrac{T_1 - t}{T_2 - t}\right)} \tag{7-111}$$

Equating III and IV in Equation 7-111 and rearranging gives:

$$\frac{W_h C_h (T_1 - T_2)}{UA} = \frac{T_1 - T_2}{\ln\left(\dfrac{T_1 - t}{T_2 - t}\right)} \tag{7-112}$$

Equation 7-112 becomes:

$$\ln\left(\frac{T_1 - t}{T_2 - t}\right) = \frac{UA}{W_h C_h} \tag{7-113}$$

$$\frac{T_1 - t}{T_2 - t} = e^{\frac{UA}{W_h C_h}} \tag{7-114}$$

Rearranging Equation 7-114 gives

$$T_2 = t + \frac{T_1 - t}{e^{\frac{UA}{W_h C_h}}} \tag{7-115}$$

$$\text{where} \quad K_1 = e^{\frac{UA}{W_h C_h}} \tag{7-116}$$

Equating II and III in Equation 7-110 and substituting Equation 7-115 into Equation 7-110 gives:

$$Mc\frac{dt}{d\theta} = W_hC_h\left\{T_1 - \left(t + \frac{T_1 - t}{K_1}\right)\right\}$$

$$= W_hC_h\left(\frac{K_1 - 1}{K_1}\right)(T_1 - t) \qquad (7\text{-}117)$$

Rearranging Equation 7-117 and integrating from t_1 to t_2 while the processing time passes from 0 to θ gives:

$$\int_{t_1}^{t_2}\frac{dt}{T_1 - t} = \int_0^{\theta}\frac{W_hC_h}{Mc}\left(\frac{K_1 - 1}{K_1}\right)d\theta \qquad (7\text{-}118)$$

Integrating Equation 7-118 gives

$$\ln\left(\frac{T_1 - t_1}{T_1 - t_2}\right) = \left(\frac{W_hC_h}{Mc}\right)\left(\frac{K_1 - 1}{K_1}\right)\theta$$

$$\text{or}\quad \theta = \left(\frac{K_1}{K_1 - 1}\right)\left(\frac{Mc}{W_hC_h}\right)\ln\left(\frac{T_1 - t_1}{T_1 - t_2}\right) \qquad (7\text{-}119)$$

where A = heat transfer surface area
    c = specific heat of batch liquid
    C_h = heating medium specific heat
    M = weight of batch liquid
    T_1 = heating medium temperature
    t_1 = initial batch temperature
    t_2 = final batch temperature
    U = overall heat transfer coefficient
    W_h = heating medium flowrate
    θ = time

BATCH COOLING: NON-ISOTHERMAL COOLING MEDIUM

When cooling a batch with internal coil and a non-isothermal cooling medium, the following equation can be applied.

$$\frac{dq'}{d\theta} = -Mc\frac{dT}{d\theta} = W_cC_c(t_2 - t_1) = UA\Delta t_{LMTD} \qquad (7\text{-}120)$$

where $K_2 = e^{\frac{UA}{W_cC_c}}$

and $\ln\left(\dfrac{T_1 - t_1}{T_2 - t_1}\right) = \left(\dfrac{W_cC_c}{Mc}\right)\left(\dfrac{K_2 - 1}{K_2}\right)\theta$

or $\theta = \left(\dfrac{K_2}{K_2 - 1}\right)\left(\dfrac{Mc}{W_cC_c}\right)\ln\left(\dfrac{T_1 - t_1}{T_2 - t_1}\right) \qquad (7\text{-}121)$

where A = heat transfer surface area
c = specific heat of batch liquid
C_c = coolant specific heat
M = weight of batch liquid
T_1 = initial batch temperature
T_2 = final batch temperature
t_1 = initial coolant temperature
U = overall heat transfer coefficient
W_c = coolant flowrate
θ = time

Example 7-12

For the tank described in Example 7-10, calculate the time required to cool the batch from 257°F (398 K) to 104°F (313 K) if cooling water is available at a temperature of 86°F (303 K) with a flowrate of 10,000 lb/hr (4535.9 kg/hr).

Solution

Select and apply the appropriate heat transfer formula. When cooling a batch with internal coil and a non-isothermal cooling medium, the following equation can be applied.

$$\frac{dq'}{d\theta} = -Mc\frac{dT}{d\theta} = W_cC_c(t_2 - t_1) = UA\Delta t_{LMTD} \qquad (7\text{-}120)$$

where $K_2 = e^{UA/W_cC_c}$

and $\ln\left(\dfrac{T_1 - t_1}{T_2 - t_1}\right) = \left(\dfrac{W_cC_c}{Mc}\right)\left(\dfrac{K_2 - 1}{K_2}\right)\theta$ (7-121)

$K_2 = e^{UA/W_cC_c}$

$$= e^{\left\{\dfrac{(850)(9.29)(3,600)}{(4,535.9)(4.12)\left(10^3\right)}\right\}} \left\{\dfrac{J}{sec \bullet m^2 K} \bullet \dfrac{m^2}{\dfrac{kg}{hr} \bullet \dfrac{hr}{3,600\ sec}} \bullet \dfrac{kg \bullet K}{10^3 J}\right\}$$

$= 4.58$

$\ln\left(\dfrac{T_1 - t_1}{T_2 - t_1}\right) = \left(\dfrac{W_cC_c}{Mc}\right)\left(\dfrac{K_2 - 1}{K_2}\right)\theta$

$$\ln\left(\dfrac{398 - 303}{313 - 303}\right) = \left\{\dfrac{(4,535.9)(4.12)}{(22,679.5)(2.1)}\right\}\left\{\dfrac{4.58 - 1}{4.58}\right\}$$

$$\times\ \theta\left\{\dfrac{kg}{hr} \bullet \dfrac{kJ}{kg.K} \bullet \dfrac{1}{kg} \bullet \dfrac{kg\ K}{kJ}\right\}\theta$$

$\theta = 7.34$ hr

Table 7-19 gives the computer results from the software (BATCH) for batch cooling, non-isothermal cooling medium.

BATCH HEATING: EXTERNAL HEAT EXCHANGER, ISOTHERMAL HEATING MEDIUM

Figure 7-28 illustrates the arrangement in which the fluid in the tank is heated by an external heat exchanger. The heating medium is isothermal; therefore any type of exchanger with steam in the shell

Table 7-19
Batch cooling: non-isothermal cooling medium

HEAT TRANSFER SURFACE AREA, m^2:	9.290
SPECIFIC HEAT OF LIQUID, kJ/kg.K:	2.100
COOLING MEDIUM SPECIFIC HEAT, kJ/kg.K:	4.120
WEIGHT OF BATCH LIQUID, kg.:	22679.500
COOLING MEDIUM TEMPERATURE, K:	303.000
INITIAL BATCH TEMPERATURE, K:	398.000
FINAL BATCH TEMPERATURE, K:	313.000
OVERALL HEAT TRANSFER COEFFICIENT, W/m^2.K:	850.000
COOLING MEDIUM FLOW RATE, kg/hr.:	4535.900
TIME, hr.:	7.34

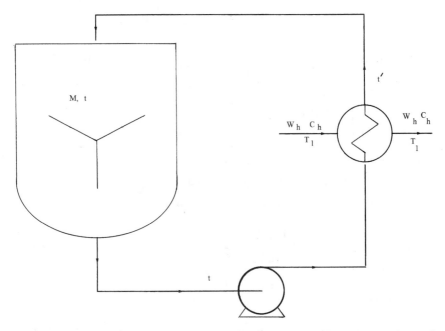

Figure 7-28. Batch heating through an external heat exchanger, isothermal heating medium.

or tubes can be used (i.e., no advantage in the magnitude of Δt can be observed by using a multi-pass design, such as a 2:4 type). The variable temperature from the exchanger t' will defer from the variable tank temperature t. An energy balance around the tank and the heat exchanger gives:

$$\frac{dq'}{dt} \overset{\text{I}}{=} \quad Mc\frac{dt}{d\theta} \overset{\text{II}}{=} \quad W_hC_h(t'-t) \overset{\text{III}}{=} \quad UA\Delta t_{LMTD} \qquad (7\text{-}122)$$

Heat   Heat entering Transfer rate
accumulation the batch by in the external
in the batch recirculation   exchanger

The log mean temperature difference Δt_{LMTD} is:

$$\Delta t_{LMTD} = \frac{(T_1-t)-(T_1-t')}{\ln\left(\dfrac{T_1-t}{T_1-t'}\right)}$$

$$= \frac{t'-t}{\ln\left(\dfrac{T_1-t}{T_1-t'}\right)} \qquad (7\text{-}123)$$

Equating II and III in Equation 7-122 gives:

$$W_hC_h(t'-t) = UA\,\Delta t_{LMTD} \qquad (7\text{-}124)$$

That is:

$$W_hC_h(t'-t) = UA\,\frac{(t'-t)}{\ln\left\{\dfrac{T_1-t}{T_1-t'}\right\}} \qquad (7\text{-}125)$$

Rearranging Equation 7-125 gives

$$\ln\left(\frac{T_1-t}{T_1-t'}\right) = \frac{UA}{W_hC_h} \qquad (7\text{-}126)$$

Equation 7-126 can be expressed as:

$$T_1-t = e^{\frac{UA}{W_hC_h}}\left(T_1-t'\right) \qquad (7\text{-}127)$$

where $K_3 = e^{\frac{UA}{W_h C_h}}$

$$T_1 - t = K_3(T_1 - t') \qquad (7\text{-}128)$$

Therefore,

$$t' = T_1 - \left(\frac{T_1 - t}{K_3}\right) \qquad (7\text{-}129)$$

Equating I and II in Equation 7-122 gives

$$Mc\frac{dt}{d\theta} = W_h C_h(t' - t) \qquad (7\text{-}130)$$

Substituting Equation 7-129 into Equation 7-130 and rearranging yields

$$\frac{Mc}{W_h C_h} \cdot \frac{dt}{d\theta} = \left\{ T_1 - \left(\frac{T_1 - t}{K_3}\right) \right\} - t$$

$$= \frac{(K_3 - 1)(T_1 - t)}{K_3} \qquad (7\text{-}131)$$

Rearranging Equation 7-131 and integrating from t_1 to t_2 while the time passes from 0 to θ gives:

$$\int_{t_1}^{t_2} \frac{dt}{T_1 - t} = \left(\frac{K_3 - 1}{K_3}\right)\left(\frac{W_h C_h}{Mc}\right)\int_0^\theta d\theta \qquad (7\text{-}132)$$

which yields $\ln\left(\frac{T_1 - t_1}{T_1 - t_2}\right) = \left(\frac{K_3 - 1}{K_3}\right)\left(\frac{W_h C_h}{Mc}\right)\theta$

$$\text{or } \theta = \left(\frac{K_3}{K_3 - 1}\right)\left(\frac{Mc}{W_h C_h}\right)\ln\left(\frac{T_1 - t_1}{T_1 - t_2}\right) \tag{7-133}$$

where A = heat transfer surface area
    c = specific heat of batch liquid
    C_h = heating medium specific heat
    M = weight of batch liquid
    T_1 = heating medium temperature
    t_1 = initial batch temperature
    t_2 = final batch temperature
    U = overall heat transfer coefficient
    W_h = heating medium flowrate
    θ = time

BATCH COOLING: EXTERNAL HEAT EXCHANGER, ISOTHERMAL COOLING MEDIUM

When cooling a batch with an external heat exchanger and an isothermal cooling medium, the equation is:

$$\ln\left(\frac{T_1 - t_1}{T_2 - t_1}\right) = \left(\frac{W_c C_c}{Mc}\right)\left(\frac{K_4 - 1}{K_4}\right)\theta$$

$$\text{or } \theta = \left(\frac{K_4}{K_4 - 1}\right)\left(\frac{Mc}{W_c C_c}\right)\ln\left(\frac{T_1 - t_1}{T_2 - t_1}\right) \tag{7-134}$$

where $K_4 = e^{UA/W_c C_c}$
    A = heat transfer surface area
    c = specific heat of batch liquid
    C_c = coolant specific heat
    M = weight of batch liquid
    T_1 = initial batch temperature
    T_2 = final batch temperature
    t_1 = initial coolant temperature
    U = overall heat transfer coefficient
    W_c = coolant flowrate
    θ = time

BATCH COOLING: EXTERNAL HEAT EXCHANGER (COUNTER-CURRENT FLOW), NON-ISOTHERMAL COOLING MEDIUM

When cooling a batch with an external heat exchanger and a non-isothermal cooling medium, the following equation can be used:

$$\ln\left(\frac{T_1 - t_1}{T_2 - t_1}\right) = \left(\frac{K_5 - 1}{M}\right)\left(\frac{W_b W_c C_c}{K_5 W_c C_c - W_b c}\right)\theta$$

$$\text{or } \theta = \left(\frac{K_5 W_c C_c - W_b c}{W_b W_c C_c}\right)\left(\frac{M}{K_5 - 1}\right)\ln\left(\frac{T_1 - t_1}{T_2 - t_1}\right) \tag{7-135}$$

where $K_5 = \exp[UA(1/W_b c - 1/W_c C_c)]$
      A = heat transfer surface area
      c = specific heat of batch liquid
      C_c = coolant specific heat
      M = weight of batch liquid
      T_1 = initial batch temperature
      T_2 = final batch temperature
      t_1 = initial coolant temperature
      U = overall heat transfer coefficient
     W_b = batch flowrate
     W_c = coolant flowrate
      θ = time

Example 7-13

Calculate the time required to cool the batch described in Example 7-12 if an external heat exchanger with a heat transfer surface of 200 ft^2 (18.58 m^2) is available. The batch material is circulated through the exchanger at the rate of 25,000 lb/hr (11,339.8 kg/hr). The overall heat transfer coefficient in the heat exchanger is 200 Btu/hr • ft^2 • °F (1,134 W/m^{2}K).

Solution

Select and apply the appropriate heat transfer formula. When cooling a batch with an external heat exchanger and a non-isothermal cooling medium, the following equation can be used:

$$\ln\left(\frac{T_1-t_1}{T_2-t_1}\right)=\left(\frac{K_5-1}{M}\right)\left(\frac{W_bW_cC_c}{K_5W_cC_c-W_bc}\right)\theta \qquad (7\text{-}135)$$

where

$$K_5=\exp\left\{UA\left(\frac{1}{W_bc}-\frac{1}{W_cC_c}\right)\right\}$$

$$=\exp\left\{\frac{(1,134)(18.58)(3,600)}{1,000}\left(\frac{1}{(11,339.8)(2.1)}-\frac{1}{(4,535.9)(4.2)}\right)\right\}$$

$$\times\left\{\frac{J}{\sec\bullet K}\bullet m^2\bullet\frac{hr}{kg}\bullet\frac{3,600\sec}{hr}\bullet\frac{kg\bullet K}{10^3 J}\right\}$$

$$=0.4509$$

$$\ln\left(\frac{T_1-t_1}{T_2-t_1}\right)=\left(\frac{K_5-1}{M}\right)\left(\frac{W_bW_cC_c}{K_5W_cC_c-W_bc}\right)\theta$$

$$\ln\left(\frac{398-303}{313-303}\right)=\left(\frac{0.4509-1}{22,679.5}\right)$$

$$\times\left(\frac{(11,339.8)(4,535.9)(4.2)}{(0.4509)(4,535.9)(4.2)-(11,339.8)(2.1)}\right)\theta$$

$$\theta=6.55\,hr$$

Table 7-20 shows the computer results using an external heat exchanger involving a non-isothermal cooling medium.

BATCH HEATING: EXTERNAL HEAT EXCHANGER AND NON-ISOTHERMAL HEATING MEDIUM

When heating a batch reactor with an external heat exchanger and non-isothermal heating, the following equation applies:

Table 7-20
Batch heating/cooling of fluids external heat exchanger:
(counter-current flow) non-isothermal cooling medium

HEAT TRANSFER SURFACE AREA, m^2:	18.580
SPECIFIC HEAT OF LIQUID, kJ/kg.K:	4.200
COOLING MEDIUM SPECIFIC HEAT, kJ/kg.K:	2.100
WEIGHT OF BATCH LIQUID, kg.:	22679.500
COOLING MEDIUM TEMPERATURE, K:	303.000
INITIAL BATCH TEMPERATURE, K:	398.000
FINAL BATCH TEMPERATURE, K:	313.000
OVERALL HEAT TRANSFER COEFFICIENT, W/m^2.k:	1134.000
BATCH FLOW RATE, kg/hr.:	11339.800
COOLING MEDIUM FLOW RATE, kg/hr.:	4535.900
TIME, hr.:	6.55

$$\ln\left(\frac{T_1 - t_1}{T_1 - t_2}\right) = \left(\frac{K_6 - 1}{M}\right)\left(\frac{W_b W_h C_h}{K_6 W_h C_h - W_b c}\right)\theta$$

$$\text{or }\ \theta = \left(\frac{M}{K_6 - 1}\right)\left(\frac{K_6 W_h C_h - W_b c}{W_b W_h C_h}\right)\ln\left(\frac{T_1 - t_1}{T_1 - t_2}\right) \tag{7-136}$$

where $K_6 = \exp[UA(1/W_b c - 1/W_h C_h)]$
A = heat transfer surface area
c = specific heat of batch liquid
C_h = heating medium specific heat
M = weight of batch liquid
T_1 = heating medium temperature
t_1 = initial batch temperature
t_2 = final batch temperature
U = overall heat transfer coefficient
W_b = batch flowrate
W_h = heating medium flowrate
θ = time

Tables 7-21 and 7-22 summarize the rules-of-thumb involving mixing, agitation, and reactors, respectively [48]. The following considerations are essential during mixing of fluids in a reactor [49]:

1. Whenever reaction rates are of the same magnitude as, or faster than, the mixing rate in a stirred reactor, mixing will have a

Table 7-21
Rules-of-thumb: mixing and agitation

1. Mild agitation is obtained by circulating the liquid with an impeller at superficial velocities of 0.1–0.2 ft/sec, and intense agitation at 0.7–1.0 ft/sec.

2. Intensities of agitation with impeller in baffled tanks are measured by power input, hp/1,000 gal, and impeller tip speeds:

Operation	hp/1,000 gal	Tip speed (ft/min)
Blending	0.2–0.5	
Homogeneous reaction	0.5–1.5	7.5–10
Reaction with heat transfer	1.5–5.0	10–15
Liquid-liquid mixtures	5	15–20
Liquid-gas mixtures	5.0–10	15–20
Slurries	10	

3. Proportions of a stirred tank relative to the diameter D: liquid level = D; turbine impeller diameter = D/3; impeller level above bottom = D/3; impeller blade with = D/15; four vertical baffles with width = D/10.

4. Propellers are a maximum of 18 in.; turbine impellers to 9 ft.

5. Gas bubbles sparged at the bottom of the vessel will result in mild agitation at a superficial gas velocity of 1 ft/min and severe agitation at 4 ft/min.

6. Suspension of solids with a settling velocity of 0.03 ft/sec is accomplished with either turbine or propeller impellers, but when the settling velocity is above 0.15 ft/sec intense agitation with a propeller is needed.

7. Power to drive a mixture of a gas and a liquid can be 25%–50% less than the power to drive the liquid alone.

8. In–line blenders are adequate when a 1–2 sec contact time is sufficient, with power inputs of 0.1–0.2 hp/gal.

Source: Walas, S. M., Chemical Process Equipment Selection and Design, *Butterworths Series in Chemical Engineering, 1988.*

serious impact on results. Poor mixing is a primary source of variability in products made in batch reactors. The results for a reaction run in a poorly mixed CFSTR may deviate strongly from those expected.

2. There is no single "correct" agitator type. Different agitator designs may perform equally well, or equally poorly, for a given application. Although some detailed design calculations can be performed, workable designs are often developed by trial and error.

Table 7-22
Rules-of-thumb: Reactors

1. The rate of reaction in every instance must be established in the laboratory, and the residence time or space velocity and product distribution eventually must be found in a pilot plant.

2. Dimensions of catalyst particles are 0.1 mm in fluidized beds, 1 mm in slurry beds, and 2–5 mm in fixed beds.

3. The optimum proportions for stirred tank reactors are when the liquid level is equal to the tank diameter, but at high pressures slimmer proportions are economical.

4. Power input to a homogeneous reaction stirred tank is 0.5–1.5 hp/1,000 gal, but three times this amount when heat is to be transferred.

5. Ideal continuous stirred tank reactor (CSTR) behavior is approached when the mean residence time is 5–10 times the length needed to achieve homogeneity, which is accomplished with 500–2,000 revolutions of a properly designed stirrer.

6. Batch reactions are conducted in stirred tanks for small daily production rates or when the reaction times are long, or when some condition such as feed rate or temperature must be programmed in some way.

7. Relatively slow reactions of liquids and slurries are conducted in continuous stirred tanks. A battery of four or five in series is most economical.

8. Tubular flow reactors are suited to high production rates at short residence times (sec or min) and when substantial heat transfer is needed. Embedded tubes or shell-and-tube construction are then used.

9. In granular catalyst packed reactors, the residence time distribution often is no better than that of a five-stage CSTR battery.

10. For conversions under about 95% of equilibrium, the performance of a five-stage CSTR battery approaches plug flow.

Source: Walas, S. M., Chemical Process Equipment Selection and Design, Butterworths Series in Chemical Engineering, 1988.

3. Many reactions involve shear-sensitive materials, which severely limit the maximum mixing rate and make impeller and reactor design important. Mixing becomes the limiting factor.

DESIGN OF MIXING SYSTEMS

The following procedure may be adopted in the design of a mixing vessel for a given process.

Table 7-23
Mixing equipment specifications

Mixer data sheet	(PROCEDE)	Equipment No. (Tag)	
		Function	
		Sheet No.	

Operating Data

					1
					2
No. OF MACHINES		WORKING		STANDBY	3
SIZE OF CHARGE					4
RATE OF CHARGING					5
TIME ACTUALLY MIXING		CONTIN. DUTY		INTERMIT. DUTY	6
TYPE OF MIXING (turbulent / moderate / light)					7
SOLIDS CONTENT		SOLIDS S.G.			8
LIQUID VISCOSITY		LIQUIDS S.G.			9
SLURRY VISCOSITY (APPARENT)					10
PARTICLE SIZE ANALYSIS					11
SOLIDS SETTLING VELOCITY					12

Vessel Data

						13
						14
DEPTH OF VESSEL						15
DEPTH OF LIQUID		MAX	NORMAL	MIN		16
ANGLE OF AGITATOR						17
SIZE OF APERTURE FOR IMPELLER						18
WORKING PRESSURE						19
WORKING TEMPERATURE						20
DEPTH OF VESSEL						21

Technical Data

						22
						23
TYPE OF MIXER						24
No. OF BLADES		DRAWING No.				25
No. OF SETS OF BLADES		ELECTRICITY SUPPLY	Volts	phase	Hz	26
SPEED		ABSORBED POWER (hp/kW)				27
SHAFT DIAMETER		TYPE OF MOTOR				28
CRITICAL SPEED		RECOMMENDED MOTOR POWER (hp/kW)				29
TYPE OF SEAL OR GLAND		RECOMMENDED MOTOR SPEED (rpm)				30
METHOD OF SUPPORT		INERTIA				31
TOTAL LOAD		STARTING TORQUE				32
WITHDRAWAL HEIGHT REQUIRED		OPERATING TORQUE				33
TYPE OF BEARINGS		TYPE OF GEAR BOX				34
ANGLE OF BLADES		VEE BELT/DIRECT DRIVE				35

Design Standards and Inspection

				36
				37
DESIGN CODE		MAX. ERECTION WEIGHT		38
HYDROSTATIC TEST PRESSURE		SHIPPING WEIGHT		39
DRGS and DATA REQ.		SHIPPING VOLUME		40
INSPECTION		TOTAL WEIGHT		41

Materials of construction

				42
				43
SHAFT		IMPELLER		44
SUPPORTS				45
VESSEL		SEAL OR GLANDS		46
BEARINGS				47
				48
DATE OF ENQUIRY		DATE OF ORDER		49
DRG. No.		ORDER No.		50
MANUFACTURER				51
REMARKS				
				53
				54
				55
				56
				57

	Date	Engineering	Process	REV	By	Appr.	Date	REV	By	Appr.	Date	
Prepared				3				6				58
Checked				2				5				59
Approved				1				4				60
												61

Service		Company		Address		62
Equipment No.						63
Project No.						64

Reproduced with permission of PROCEDE.

Table 7-24
Reactor vessel specifications

Vessel data sheet (PROCEDE)			
Equipment No. (Tag)			
Function			
Sheet No.			

	SHELL	JACKET FULL/HALF COIL	INTERNAL COIL	
Operating Data				1
				2
No. REQUIRED		CAPACITY		3
SPECIFIC GRAVITY OF CONTENTS		COMPUTED (yes or no)		4
	SHELL	JACKET FULL/HALF COIL	INTERNAL COIL	5
CONTENTS				6
DIAMETER				7
LENGTH				8
DESIGN CODE				9
MAX. WORKING PRESSURE				10
DESIGN PRESSURE				11
MAX. WORKING TEMP				12
DESIGN TEMP				13
TEST PRESSURE (HYDROSTATIC)				14
TEST PRESSURE (AIR)				15
MATERIALS				16
JOINT FACTOR				17
CORROSION ALLOWANCE				18
THICKNESS				19
END TYPE	THICKNESS	JOINT FACTOR		20
END TYPE	THICKNESS	JOINT FACTOR		21
TYPE OF SUPPORT	THICKNESS	MATERIAL		22
WIND LOAD DESIGN	RADIOGRAPHY %	STRESS RELIEF		23
INTERNAL BOLTS MATERIAL	TYPE	NUTS		24
EXTERNAL BOLTS MATERIAL	TYPE	NUTS		25
INSULATION (SEP. ORDER)	INSULATION FITTING ATTACHMENT BY			26
GASKET MATERIAL	INSPECTION BY			27
PAINTING				28
WEIGHT	EMPTY			29
FULL OF LIQUID	OPERATING			30
INTERNALS and EXTERNALS	DATE OF ENQUIRY	DATE OF ORDER		31
ORDER No.	DRG. No.			32
MANUFACTURER				33
REMARKS AND NOTES:- UNLESS OTHERWISE STATED ALL FLANGE BOLT HOLES TO BE				34
OFF-CENTRE OF VESSEL CENTRE LINES N/S and E/W (NOT RADIALLY)				35
				36
				37
				38
				39
				40
A				41
B				42
C				43
D				44
E				45
F				46
G				47
H				48
H				49
K				50
K				51
M				52
N				53
P				54

REF	No.	DUTY	NOM BORE mm/ins	PIPE WALL THICKNESS	TYPE	CLASS FLANGE SPEC	MATERIAL	BRANCH COMPEN'N	REMARKS	
	BRANCH									55
										56
										57

Prepared			3			6				58
Checked			2			5				59
Approved			1			4				60
	Date	Engineering	Process	REV	By	Appr.	Date	REV	By	Appr.	Date	61

Service	Company	Address	62
Equipment No.			63
Project No.			64

Reproduced with permission of PROCEDE.

1. Study the properties of the liquid and physical requirements, and choose the type of impeller.
2. Select size ratios preferably the same as the standard values (Table 7-2) to avoid experimentation or based on small-scale studies.
3. Select impeller diameter D_A for the larger system to accommodate the system being mixed. This leaves the impeller speed N as the only independent variable.
4. Choose N based on scale-up studies or commonly used rules.
5. Calculate the power P, mixing time θ, etc. Account for mechanical losses and the like when selecting a motor.
6. Change N and P to standard values.
7. Iterate and see if alternative designs requiring lower power exists.
8. Perform mechanical designs (e.g., to obtain shaft diameter, supports, bearing designs, etc.).

Table 7-23 shows a mixing equipment specification sheet, which can be helpful as a general checklist. Generally, the specification sheet should not be completely relied on for a mixing problem, unless the problem is known or data are known that can be given to the manufacturer (e.g., blending, dispersing, or dissolving crystals). For unique problems, laboratory data should be carried out under the guidance of technical advice from the manufacturer or other qualified authority, in order that adequate scale-up data are taken and evaluated. It is essential that both a description and dimensions are given for the vessel to be used. Otherwise, request the manufacturer to recommend the type best suited to the service. Table 7-24 provides a reactor vessel specification data sheet.

REFERENCES

1. Danckwerts, P. V., *Appl. Sci. Research,* A3, 279, 1953.
2. Levenspiel, O., *Chemical Reaction Engineering,* 3rd ed., John Wiley & Sons, New York 1999.
3. Penny, W. R., "Guide to trouble free mixers," *Chem. Eng.,* 77 (12), 171, 1970.
4. Holland, F. A. and Chapman, F. S., *Liquid Mixing and Processing in Stirred Tanks,* Reinhold, New York, 1966.
5. Myers, K. J., Reeder, M., and Bakker, A., "Agitating for success," *The Chemical Engineer,* pp. 39–42, 1996.

6. Uhl, V. W. and Gray, J. B., Eds. *Mixing Theory and Practice,* Volume 1, Academic Press Inc., New York, 1966.

7. Gates, L. E., Henley, T. L., and Fenic, J. G., "How to select the optimum turbine agitator," *Chem. Eng.,* p. 110, Dec. 8, 1975.

8. Hicks, R. W., Morton, J. R., and Fenic, J. G., "How to design agitators for desired process response," *Chem. Eng.,* p. 102, April 26, 1976.

9. Dickey, D. S., "Succeed at stirred tank reactor design," *Chem. Eng.,* pp. 22–31, Dec. 1991.

10. Rushton, J. H., Costich, E. W., and Everett, H. J., "Power characteristics of mixing impellers," *Chem. Eng. Prog,* 46, 395, 1950.

11. Metzner, A. B. and Otto, R. E., "Agitation of non-Newtonian fluids," *AIChEJ,* 3, pp. 3–10, 1957.

12. Dickey, S. D. and Fenic, J. G., "Dimensional analysis for fluid agitation systems," *Chem. Eng.,* Jan 5, 1976.

13. Fasano, J. B., Bakker, A., and Penny, W. R., "Advanced impeller geometry boosts liquid agitation," *Chem. Eng.,* 10(8), pp. 110–116, August 1994.

14. Gates, L. E., Hicks, R. W., and Dickey, D. S., "Application guidelines for turbine agitators," *Chem. Eng.,* 83, pp. 165–170, Dec. 6, 1976.

15. Cutter, L. A., *AIChEJ,* 12, 35, 1966.

16. Moo-Young, M., Tichar, K., and Dullien, F. A. L., *AIChEJ,* 18, 178, 1972.

17. van de Vusse, J. G., "Mixing by agitation of miscible liquids," *Chem. Eng. Sci.,* 4, 178, 1955.

18. Fox, E. A., and Gex, V. E., "Single phase blending of liquids," *AIChEJ,* 2, 539, 1956.

19. Prochazka, J. and Landau, J., Coll Czech Chem Cummun, 26: 2961, 1961.

20. Moo-Young, M., Tichar, K., and Takahashi, F. A. L., "The blending efficiencies of some impellers in batch mixing," *AIChEJ,* 18(1), pp. 178–182, 1972.

21. Sano, Y. and Usui, H., "Interrelations among mixing time, power number and discharge flow rate number in baffled mixing vessels," *J. Chem. Eng.,* Japan, 18:47–52, 1985.

22. Gray, J. B., "Batch mixing of viscous liquids," *Chem. Eng. Progr.,* 59, No. 3, p. 59, 1963.

23. Bakker, A. and Gates, L. E., "Properly choose mechanical Agitators for viscous liquids," *Chem. Eng. Prog.,* Dec. 1995.

24. Bates, R. L., Fondy, P. L., and Corpstein, R. R., *I.E.C. Proc. Des.,* 2, 310, 1963.
25. Rautzen, R. R., Corpstein, R. R., and Dickey, D.S., "How to use scale-up methods for turbine agitators," *Chem. Eng.,* Oct. 25, 1976.
26. Norwood, K. W. and Metzner, A. B., "Flow patterns and mixing rates in agitated vessels," *AIChEJ,* 6, 432, 1960.
27. Gupta, S. K.. "Momentum transfer operations," Tata McGraw-Hill Publishing Co. Ltd., New Delhi, 1982.
28. Samant, K. D. and Ng, Ka, M., "Development of liquid-phase agitated reactors: synthesis, simulation and scale-up," *AIChEJ,* Vol. 45, No. 11, pp. 2371–2391, November 1999.
29. Pahl, M. H. and Muschelknautz, "Static mixers and their applications," *Int, Chem. Eng.,* 22, pp. 195–205, 1982.
30. Coker, A. K., "Sizing process piping for single-phase fluids," *The Chemical Engineer,* October 10, 1991.
31. Round, *Can. J. Chem. Eng.,* 58, 122, 1980.
32. Lockhart, R.W. and Martinelli, R. C., *Chem. Eng., Prog.,* 45, pp. 39–46, 1949.
33. Coker, A. K., "Understand two-phase flow in process piping," *Chem. Eng. Prog.,* pp. 60–65, November 1990.
34. Green, A., Johnson, B., and John, A., "Process intensification magnifies profits," *Chem. Eng.,* pp. 66–73, December 1999.
35. Chen, S. J., Kenics Technical data, KTEK-2, 1978.
36. Li, H. Z., Fasol, C., and Choplin, L., "Pressure drop of Newtonian and non-Newtonian fluids across a sulzer SMX static mixer," *Trans. IChemE,* Vol. 75, Part A, pp. 792–798, Nov. 1997.
37. Bakker, A. and La Roche, R., "Flow and mixing with kenics static mixers," *Cray Channels,* 15, 3, Cray Research Inc., Eagan, MN, pp. 25–28, 1993.
38. Bakker, A., et al., "Computerizing the steps in mixer selection," *Chem. Eng.,* 101, 3, pp. 120–129, March 1994.
39. Chopey, N. P., *Handbook of Chemical Engineering Calculations,* pp. 7–28, McGraw-Hill Book Company, 1984.
40. Zwietering, T. N., "Suspending of solid particles in liquid by agitators," *Chem. Eng. Sci.,* Vol. 8, pp. 244–253, 1958.
41. Weisman, J. and Efferding, L. E., *AIChEJ,* 6 (3), 419, 1960.
42. Nienow, A. W., *Chem. Eng. Sci.,* 23, pp. 1453-1459, 1968.
43. Corpstein, R. R., Myers, K. J., and Fasano, J., "The high efficiency road to liquid-solid agitation," *Chem. Eng.,* Oct. 1994.

44. Bakker, A., Fasano, J., and Leung, D. E., "Pinpoint mixing problems with lasers and simulation software," *Chem. Eng.,* pp. 94–100, January 1994.

45. Bondy, F. and Lippa, S., "Heat transfer in agitated vessels," *Chem. Eng.,* pp. 62–71, 1983.

46. Dream, R. F., "Heat transfer in agitated vessels," *Chem. Eng.,* pp. 90–96, Jan. 1999.

47. McEwan, J., "How to predict batch reactor heating and cooling," *Chem. Eng.,* p. 179, May 1989.

48. Walas, M. S., *Chemical Process Equipment Selection and Design,* Butterworths Series in Chemical Engineering, 1988.

49. Falconer, J. L. and Huvard, G. S., "Important concepts in undergraduate kinetics and reactor design courses," *Chemical Engineering Education,* pp. 140–141, Vol. 33, No. 2, Spring 1999.

CHAPTER EIGHT

Residence Time Distributions in Flow Reactors

INTRODUCTION

In the preceding chapters, the performance equations for ideal reactors were reviewed and the optimal temperature progressions of various industrial reactions were considered. Fluid mixing characteristics were also reviewed using various agitator types, and the power consumption, time for heating and cooling in batch systems, and the scale-up of mixers was determined. However, real reactors deviate from ideal ones and the differences are due to a number of factors. These factors include channeling of fluid as it moves through the reactor, the presence of stagnant regions within the reactor, bypassing or short-circuiting of portions of fluids in a packed bed, the longitudinal mixing caused by vortices and turbulence, and the failure of impellers or other mixing devices to provide perfect mixing.

In an ideal continuous stirred tank reactor (CSTR), the reactant concentration is uniform throughout the vessel, while in a real stirred tank, the reactant concentration is relatively high at the point where the feed enters and low in the stagnant regions that develop in corners and behind baffles. In an ideal plug flow reactor, all reactant and product molecules at any given axial position move at the same rate in the direction of the bulk fluid flow. However, in a real plug flow reactor, fluid velocity profiles, turbulent mixing, and molecular diffusion cause molecules to move with changing speeds and in different directions. The deviations from ideal reactor conditions pose several problems in the design and analysis of reactors. In this chapter, we shall establish a basis for examining both qualitatively and quantitatively the effect of departure from idealized flow behavior on the performance of a reactor.

THE RESIDENCE TIME DISTRIBUTION FUNCTIONS AND THEIR RELATIONSHIPS

Consider a homogeneous system as shown in Figure 8-1 through which matter is passing at a steady rate. The inlet and outlet streams are completely mixed, and there is a single exit from which the fluid elements cannot return to the system. If each fluid element is being timed as it enters the system and is stopped when it leaves, then each element at the exit has a time t associated with it, which determines the time spent in the system—this quantity is defined as the residence time for that element. In real systems there will be a spread of residence times leading to a residence time distribution (RTD). To predict the behavior of a system, it must be determined by how long different fluid elements remain in the system.

Mac Mullin and Weber [1] introduced the concept of the RTD in the analysis of chemical reactors, and Danckwerts [2] developed this concept further in his classical paper, which has since formed the basis of various investigations involving flow systems in chemical and biochemical reactors. Levenspiel [3], Levenspiel and Bischoff [4], Himmelblau and Bischoff [5], Wen and Fan [6], and Shinnar [7] have given extensive treatments of this subject.

AGE DISTRIBUTION FUNCTIONS AND RESIDENCE TIME DISTRIBUTION FUNCTION E(t)

The "age" of an element of fluid is defined as the time elapsed since it entered the system. The fraction of fluid having ages between t and t + dt is (uc/m) • dt, where u is the volumetric rate of flow of fluid through the system, m is the quantity of tracer injected, and c is the local concentration of tracer at time t after injection. Danckwerts [8] introduced the concept of a fluid element or "point," meaning a small volume with respect to the reactor vessel size, but still large enough

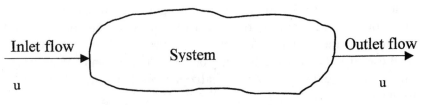

Figure 8-1. Steady-state homogeneous system.

to contain sufficient molecules to define continuous properties such as density and concentration.

Consider a system of constant volume V through which a stream of fluid is passing at a steady volumetric rate u. If t is the age of elements in the stream at a point or over a region in the system, then the element whose age is between t and t + dt is known as the t-element. The relative frequency of the appearance of t-elements at a point or over a region in the system is an age distribution at the point or over that region.

RESIDENCE TIME DISTRIBUTION FUNCTION E(t)

This is defined as the fraction of material in the outlet stream that has been in the system for the period between t and t + dt, and is equal to E(t)dt, where E(t) is called the exit age distribution function of the fluid elements leaving the system. This is expressed as

$$E(t)dt = \begin{pmatrix} \text{fraction of fluid leaving the vessel that has} \\ \text{residence time (exit age) of } (t, t+dt) \end{pmatrix}$$

All fluid elements have some residence time, therefore, over a sufficiently long period, all tracer will eventually come out. This is represented by

$$\int_0^\infty E(t)dt = 1 \qquad (8\text{-}1)$$

The fraction of fluid element in the exit stream with age less than t_1 is

$$\int_0^{t_1} E(t)dt \qquad (8\text{-}2)$$

Figure 8-2 shows the distribution of this property and the fraction of the elements for which the residence time is less than a given value t_1.

E(t) has the unit of inverse time (e.g., min^{-1}), and in probability terms, E(t) is a density rather than a distribution. It is a fundamental indicator of the flow and mixing pattern in a chemical reactor and its

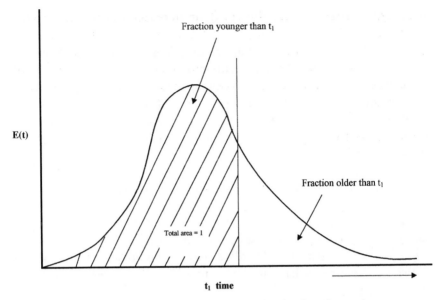

Figure 8-1. Residence time density function.

measurement is an important step in the analysis of non-ideal reactors. If a tracer is introduced into a reactor the concentration of the tracer is monitored at the exit stream, it is possible to ascertain the age distribution of the fluid elements within the reactor.

The residence time distribution function must satisfy the following conditions:

$E(t) = 0$ for $t < 0$ since no fluid can exit before it enters.
$E(t) \geq 0$ for $t > 0$ since mass fractions are always positive.

By definition,

$$\int_0^\infty E(t)dt = 1$$

CUMULATIVE RESIDENCE TIME DISTRIBUTION FUNCTION F(t)

This function is the volume fraction of material in the outlet stream, which has been in the system for times less than t is equal to F(t). If "ages" are assigned to the different fluid elements leaving the system, $F(t)$ would be the volume fraction of the outlet stream having an age less than t. For constant density systems, volume fractions are the same as the weight fractions of the effluent with an age less than t. Therefore, the cumulative residence time distribution function $F(t)$ is expressed by

$$F(t) = \int_0^t E(t')dt' \tag{8-3}$$

$$\text{or } E(t) = \frac{dF(t)}{dt} \tag{8-4}$$

It can also be expressed by

$$F(t) = \frac{\int_0^t C(t)u\,dt}{\int_0^\infty C(t)u\,dt} \tag{8-5}$$

where $C(t)$ = the effluent tracer concentration, (g tracer/l)
$\quad\quad\; u$ = the volumetric flowrate, (l/min)

Replacing the integrals by finite differences yields

$$F(t) = \frac{\sum_0^t (C(t)u\,\delta t)}{\sum_0^\infty (C(t)u\,\delta t)} \tag{8-6}$$

If the data are at evenly spaced time increments and the mass flowrate is constant, this gives

$$F(t) = \frac{\sum_0^t C(t)}{\sum_0^\infty C(t)} \tag{8-7}$$

It is possible to determine the cumulative residence time distribution function F(t) from either a tracer step-change or a tracer impulse response. From its definition, the properties of F(t) are:

- $F(t) = 0$ when t < 0
- $F(t) \geq 0$ when $t \geq 0$
- $F(\infty) = 1$
- $dF(t)/dt = E(t) > 0$

The cumulative exit age distribution is a non-negative, monotone non-decreasing function as shown in Figure 8-3.

INTERNAL AGE DISTRIBUTION FUNCTION I(t)

The fraction of the material within the system for times between t and t + dt is equal to I(t)dt, where t is the age or length of time a fluid element has been in the vessel. I(t) has properties similar to E(t) for a perfect mixer.

$$\int_0^\infty I(t)dt = 1 \tag{8-8}$$

$$\text{and} \int_0^{t_1} I(t)dt = \left(\begin{array}{l}\text{fraction of fluid element in vessel} \\ \text{younger than age } t_1\end{array}\right)$$

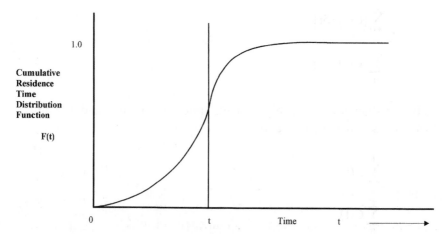

Figure 8-3. Cumulative residence time distribution function.

The exit age distribution function E(t) is obtained from outside the vessel while the internal age distribution function I(t) is obtained from inside the vessel. I(t) can be represented in terms of the RTD or the F-curve as

$$I(t) = \frac{u}{V}\left[1 - \int_0^t E(t)dt\right] \tag{8-9}$$

$$\text{or } \bar{t}\,I(t) = 1 - \int_0^t E(t)dt \tag{8-10}$$

where $\bar{t}$ = V/u = mean residence time

In terms of F(t)

$$I(t) = \frac{u}{V}\left[1 - F(t)\right] \tag{8-11}$$

It also follows from Equation 8-9 that

$$E(t) = -\bar{t}\frac{d}{dt}\,I(t) \tag{8-12}$$

For the perfectly mixed CSTR, the internal age distribution function is

$$\bar{t}\,I(t) = 1 - \int_0^t \frac{1}{\bar{t}}e^{-t/\bar{t}}\,dt$$

$$= 1 + \left[e^{-t/\bar{t}}\right]_0^t$$

$$I(t) = \frac{1}{\bar{t}}\,e^{-t/\bar{t}} = E(t) \tag{8-13}$$

Equation 8-13 shows that in a perfectly mixed vessel, the internal and exit conditions are identical. Figure 8-4 illustrates the internal age distribution characteristics.

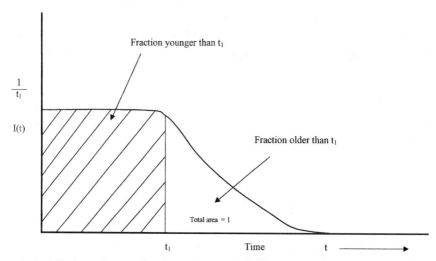

Figure 8-4. Characteristic distribution of residence time, which is equivalent to age distribution of the inlet stream.

THE INTENSITY FUNCTION $\Lambda(t)$

Naor and Shinnar [9] proposed the intensity function, which is defined as the fraction of fluid in the vessel of age t that will leave at a time between t and t+dt. The intensity function $\Lambda(t)$ is related to the E(t) and I(t) functions from

$$\begin{pmatrix} \text{Amount of fluid} \\ \text{leaving between} \\ \text{times}\,(t,\,t+dt) \end{pmatrix} = \begin{pmatrix} \text{Amount not} \\ \text{leaving before} \\ \text{time t} \end{pmatrix}\begin{pmatrix} \text{Fraction of age t} \\ \text{that will leave between} \\ \text{times}\,(t,\,t+dt) \end{pmatrix}$$

Introducing the various definitions gives

$$[uE(t)dt] = [V\ I(t)][\Lambda(t)dt]$$

$$\text{or}\ \ \Lambda(t) = \frac{u}{V}\frac{E(t)}{I(t)} = \frac{1}{\bar{t}}\frac{E(t)}{I(t)} = \frac{-dI(t)}{dt}\bullet\frac{1}{I(t)} \tag{8-14}$$

The intensity function $\Lambda(\theta)$ as shown in Figure 8-5 is useful in detecting the existence of dead space and bypassing, and it allows

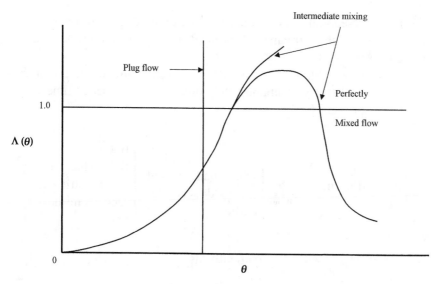

Figure 8-5. Intensity function.

some insight into the basic mechanism of the flow processes. The pertinent features of $\Lambda(\theta)$ are:

- The intensity function of a completely mixed vessel is a constant, $\Lambda(\theta) = 1$, a straight line parallel to the abscissa.
- The intensity function of two or more identical, ideally mixed vessels in series starts from the origin and increases monotonically to an asymptotic value $(\Lambda \rightarrow n)$, which is equal to the number of vessels.
- There are no stagnant zones in the system.

MEAN RESIDENCE TIME

The average time spent by material flowing at volumetric rate u through a volume V is known as the mean residence time $\bar{t}$, expressed as

$$\bar{t} = \frac{V}{u}$$

The mean residence time $\bar{t}$ requires two conditions. First, there must not be a density change in the flowing stream as it passes through the system; secondly, there must be no backmixing of the material in the system either upstream past the feed point, or of the product stream back into the system.

It is also possible to determine $\bar{t}$ from the residence time distribution function $E(t)$ by examining the contents of a vessel at time $t = 0$. This is expressed by

$$
\begin{pmatrix} \text{total volume} \\ \text{of fluid in the} \\ \text{vessel at } t = 0 \end{pmatrix} = \sum_{\text{all time}} \begin{pmatrix} \text{volume of fluid} \\ \text{which had entered} \\ \text{t to } t + dt \text{ seconds} \\ \text{earlier} \end{pmatrix} \begin{pmatrix} \text{fraction of this} \\ \text{fluid which stays} \\ \text{more than t} \\ \text{seconds in the} \\ \text{vessel} \end{pmatrix}
$$

This is expressed mathematically as

$$
V = \int_0^\infty (u \, dt) \left(\int_t^\infty E(s) \, ds \right) \tag{8-15}
$$

where s is a dummy variable, the volumetric flowrate u is a constant, and $E(s)$ is continuous. Therefore,

$$
\frac{V}{u} = \int_0^\infty \left[\int_t^\infty E(s) \, ds \right] dt \tag{8-16}
$$

Applying integration by parts (i.e., $\int w \, dv = wv - \int v \, dw$)

where $w = \int_t^\infty E(s) ds$

$\qquad v = t$

$\qquad dv = dt$

$$
\frac{dw}{dt} = -E(t) \quad \text{(applying Leibnitz's rule)}
$$

$$
dw = -E(t) \, dt
$$

yields

$$\frac{V}{u} = \left[\left(\int\limits_{t}^{\infty} E(s)\,ds \right) t \right]_{0}^{\infty} - \int\limits_{0}^{\infty} t[-E(t)]\,dt$$

$$= 0 - 0 + \int\limits_{0}^{\infty} t\,E(t)\,dt$$

Therefore, $\quad \bar{t} = \dfrac{V}{u} = \displaystyle\int\limits_{0}^{\infty} t\,E(t)\,dt$ \hfill (8-17)

NORMALIZED RTD FUNCTION E(θ)

It is often convenient to use a dimensionless time θ and a corresponding version of the RTD $E(\theta)$. The dimensionless time θ is

$$\theta = \frac{t}{\bar{t}} \quad \text{and} \quad d\theta = \frac{1}{\bar{t}}\,dt$$

The relationship between $E(t)$ and $E(\theta)$ is found from the basis that both represent the same physical entity, the fraction of exit fluid with age θ. Thus, $E(\theta)d\theta = E(t)dt$ or $(1/\bar{t})E(\theta)dt = E(t)dt$. Therefore,

$$E(\theta) = \bar{t}\,E(t) \tag{8-18}$$

For a perfectly mixed CSTR, it is assumed that the vessel contents are perfectly homogeneous and have the same composition as the exit stream. Considering a step input into the CSTR, a macroscopic material balance gives

$$uc_0 = uc + V\frac{dc}{dt} \tag{8-19}$$

$$\text{or} \quad \bar{t}\frac{dc}{dt} + c = c_0 \tag{8-20}$$

Applying the Laplace transform to Equation 8-20, which is of the form $L(dy/dx) = s\bar{y}(s) - y_0$, with the initial condition $y_0 = 0$, gives

$$\bar{t}\{s\bar{c}(s)-c_o\} + \bar{c}(s)=\frac{c_o}{s}$$

or $\bar{c}(s)\{\bar{t}s+1\}=\dfrac{c_o}{s}$ (8-21)

Rearranging Equation 8-21 gives

$$\frac{\bar{c}(s)}{c_o} = \frac{1}{s(s\bar{t}+1)}$$ (8-22)

Converting Equation 8-22 into partial fraction and solving yields

$$\frac{\bar{c}(s)}{c_o}=\frac{1}{s} - \frac{\bar{t}}{s\bar{t}+1}$$ (8-23)

The inverse Laplace transform of Equation 8-23 gives

$$\frac{c(t)}{c_o} = 1 - e^{-t/\bar{t}}$$ (8-24)

$$= F(\theta) \quad \text{(by definition)}$$

Consequently, since

$$E(t)=\frac{dF(\theta)}{dt}$$

$$E(t)=\frac{1}{\bar{t}}e^{-t/\bar{t}}$$ (8-25)

Substituting Equation 8-25 into Equation 8-18 gives

$$E(\theta) = e^{-\theta}$$ (8-26)

The internal age distribution function in terms of $F(\theta)$ gives

$$I(t) = \frac{1}{\bar{t}}\left[1 - F(\theta)\right]$$

$$= \frac{1}{\bar{t}}e^{-t/\bar{t}} \tag{8-27}$$

$$I(\theta) = e^{-\theta} = E(\theta) \tag{8-28}$$

The intensity function $\Lambda(t)$ in terms of $E(t)$ and $I(t)$ gives

$$\Lambda(t) = \frac{1}{\bar{t}}\frac{E(t)}{I(t)} = \frac{1}{\bar{t}} \tag{8-29}$$

or $\Lambda(\theta) = 1$ \hfill (8-30)

During plug flow, all material passes through the vessel without any mixing, and each fluid element stays in the vessel for exactly the same length of time. For a step input, the front or interface between the tracer and non-tracer fluids traverses down the vessel and exits at the other end in a time equal to the mean residence time $\bar{t}$. Therefore, the $F(\theta)$ curve is a step function and is expressed as

$$F(\theta) = U(t - \bar{t}) \tag{8-31}$$

where $U(t - \bar{t}) = \begin{cases} 0 & t < \bar{t} \\ 1 & t > \bar{t} \end{cases}$

$$I(t) = \frac{1}{\bar{t}}\left[1 - F(\theta)\right] = \frac{1}{\bar{t}}\left[1 - U(t - \bar{t})\right] \tag{8-32}$$

or $E(t) = \dfrac{dF}{dt} = \dfrac{d}{dt}U(t - \bar{t}) = \delta(t - \bar{t})$ \hfill (8-33)

$$I(\theta) = 1 - U(\theta - 1) \tag{8-34}$$

$$E(\theta) = \delta(\theta - 1) \tag{8-35}$$

The intensity function for plug flow is

$$\Lambda(t) = \frac{\delta(t - \bar{t})}{1 - U(t - \bar{t})} \tag{8-36}$$

which by a limiting process is shown to be equivalent to

$$\Lambda(t) = \begin{cases} 0 & 0 \le t < \bar{t} \\ +\infty & t = \bar{t} \end{cases} \tag{8-37}$$

or $\Lambda(\theta) = \begin{cases} 0 & 0 \le \theta < 1 \\ +\infty & \theta = 1 \end{cases}$ $\tag{8-38}$

Table 8-1 gives the relationships between the age distribution functions and Figure 8-6 shows the age distribution functions of ideal reactors.

In using the normalized distribution function, it is possible to directly compare the flow performance inside different reactors. If the normalized function $E(\theta)$ is used, all perfectly mixed CSTRs have numerically the same RTD. If $E(t)$ is used, its numerical values can change for different CSTRs.

MOMENTS OF RESIDENCE TIME DISTRIBUTIONS

A recommended method of characterizing the RTD in flow systems is by using their moments. These are known as the mean, variance, and skewness. The mean value or the centroid of distribution for a concentration versus time curve is

$$\mu = \bar{t} = \frac{\int\limits_0^\infty tC(t)dt}{\int\limits_0^\infty C(t)dt} \tag{8-39}$$

If the distribution curve is only known at a number of discrete time values, t_i, then the mean is expressed by

$$\mu = \bar{t} = \frac{\sum\limits_{t=0}^{t=\infty} t_i C(t_i)\Delta t_i}{\sum\limits_{t=0}^{t=\infty} C(t_i)\Delta t_i} = \frac{\sum\limits_{t=0}^{t=\infty} tC(t)}{\sum\limits_{t=0}^{t=\infty} C(t)} \quad \text{(if } \Delta t_i \text{ is constant)} \tag{8-40}$$

Table 8-1
Relationships between the age distribution functions

Functions	Definitions
$C(\theta) = \dfrac{dF(\theta)}{d\theta}$	$E(\theta) = \bar{t}E(t) = \dfrac{-dI(\theta)}{d\theta} = -\bar{t}^2 \dfrac{dI(t)}{dt}$
$F(\theta) = \displaystyle\int_0^\theta C(\theta')d\theta'$	$1 - I(\theta) = 1 - \bar{t}I(t) = \displaystyle\int_0^\theta E(\theta')d\theta' = \int_0^t E(t')dt'$
	$F(\theta) + I(\theta) = 1$
	$F(\theta) = 1 - I(\theta) = \displaystyle\int_0^\theta E d\theta = 1 - \sum E\Delta\theta$
	$\Lambda(\theta) = \bar{t}\Lambda(t) = \dfrac{E(\theta)}{I(\theta)} = \dfrac{-d\ln I(\theta)}{d\theta} = \dfrac{E(t)}{\bar{t}I(t)} = -\dfrac{1}{I(t)} \cdot \dfrac{dI(t)}{dt}$
$I(\theta)$	$1 - \displaystyle\int_0^\theta E(\theta)d\theta = 1 - F(\theta) = I(\theta) = W(\theta)$
$E(\theta)$	$E(\theta) = dF(\theta)/d\theta = -dW(\theta)/d\theta = -dI(\theta)/d\theta$
$W(\theta)$	$1 - \displaystyle\int_0^\theta E(\theta)d\theta = 1 - F(\theta) = I(\theta)$
H	$\displaystyle\int_0^1 F(\theta)d\theta$

In terms of dimensionless time $\theta = t/\bar{t}$. To convert to E(t), F(t) etc., use $dt = \bar{t}\,d\theta$; $E(\theta) = \bar{t}\,E(t)$; $F(\theta) = F(t)$; $I(\theta) = \bar{t}\,I(t)$; $W(\theta) = W(t)$. E, F, and W are RTD functions. I and H are not RTD functions.

$$\text{or } \mu = \bar{t} = \frac{\displaystyle\int_0^\infty tE(t)\,dt}{\displaystyle\int_0^\infty E(t)\,dt} = \frac{\sum t_i E(t_i)\Delta t_i}{\sum E(t_i)\Delta t_i} \qquad (8\text{-}41)$$

The second moment is taken about the mean and is referred to as the variance or square of the standard deviation σ^2, defined by

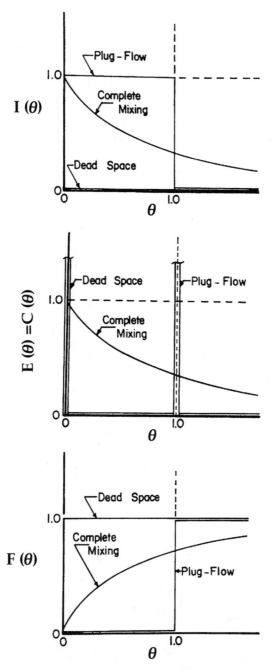

Figure 8-6. Age distribution functions of ideal reactors. *(Source: Wen and Fan [6]).*

$$\sigma^2 = \frac{\int\limits_0^\infty (t-\bar{t})^2 C(t)\,dt}{\int\limits_0^\infty C(t)\,dt} = \frac{\int\limits_0^\infty t^2 C(t)\,dt}{\int\limits_0^\infty C(t)\,dt} - \bar{t}^2 \qquad (8\text{-}42)$$

In discrete form σ^2 is

$$\sigma^2 = \frac{\sum (t_i - \bar{t})^2 C(t_i)\Delta t_i}{\sum C(t_i)\Delta t_i}$$

$$= \frac{\sum t_i^2 C(t_i)\Delta t_i}{\sum C(t_i)\Delta t_i} - \bar{t}^2 \qquad (8\text{-}43)$$

$$\text{or } \sigma^2 = \int\limits_0^\infty (t-\bar{t})^2 E(t)\,dt = \int\limits_0^\infty t^2 E(t)\,dt - \bar{t}^2$$

$$= \frac{\sum t_i^2 E(t_i)\Delta t_i}{\sum E(t_i)\Delta t_i} - \bar{t}^2 = \sum t_i^2 E(t_i)\Delta t_i - \bar{t}^2 \qquad (8\text{-}44)$$

The dimensionless variance σ_θ^2 is expressed as

$$\sigma_\theta^2 = \frac{\sigma^2}{\bar{t}^2} = \int\limits_0^\infty (\theta - 1)^2 E(\theta)\,d\theta \qquad (8\text{-}45)$$

The magnitude of the variance σ represents the square of the distribution spread and has the units of $(\text{time})^2$. The greater the value of this moment, the greater the spread of the RTD. The variance is particularly useful for matching experimental curves to one of a family of theoretical curves.

The third moment is taken about the mean and is known as the skewness, γ. It is defined by

$$\gamma = \int\limits_0^\infty (t-\bar{t})^3 E(t)\,dt \qquad (8\text{-}46)$$

The magnitude of this moment measures the extent that the distribution is skewed in one direction or another in reference to the mean.

DETERMINING RTD FROM EXPERIMENTAL TRACER CURVES

Previously, the RTD functions for flow systems were described. In practice, the RTD is determined experimentally by injecting a tracer at the inlet of the system and monitoring the response at the outlet; information about the system is derived from these results. In a tubular reactor, the RTD measurements are used to determine how closely the plug flow assumption is obeyed.

Wen and Fan [6] have provided a comprehensive listing of various tracers and experimental techniques for determining the RTD in flow systems. Recent studies [10,11,12] have been performed employing an impulse tracer to determine the RTD in bubble columns and an oscillatory flow electrochemical reactor. The author [13,14] has employed both step-change and an impulse to determine the RTD of nozzle type reactors: analysis of the RTD involves an atomic absorption spectro-photometer (AAS), a cine-projector, and a chart recorder. Figures 8-7 and 8-8 show the nozzle-type reactors and the AAS, respectively. Figure 8-9 gives a typical response curve from the AAS.

The methods used in determining the RTD are an impulse signal and a step-change or a periodic input of the tracer. The following reviews these methods of injecting a tracer to analyze the RTD in flow systems.

IMPULSE SIGNAL

Suppose that a quantity m of tracer is injected at the inlet of a system during a period of time, which is very short compared to the mean residence time, $\bar{t}$. The concentration of this tracer material is measured in the outlet stream as a function of time. Assume the following conditions:

1. Constant flowrate u(l/min) and fluid density ρ(g/l).
2. Only one flowing phase.
3. Closed system input and output by bulk flow only (i.e., no diffusion takes place across the system boundaries).
4. Flat velocity profiles at the inlet and outlet.

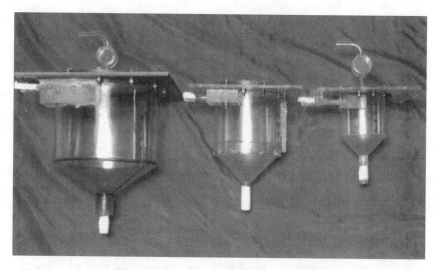

Figure 8-7. Cylindrical nozzle reactors.

Figure 8-8. Atomic absorption spectrophotometer with 16 mm cine projector in place.

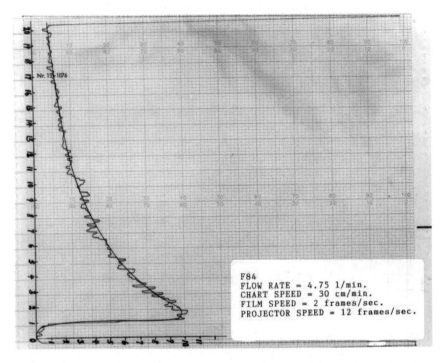

Figure 8-9. A typical RTD response curve from an AA spectrophotometer.

5. Linearity with respect to the tracer analysis, that is, the magnitude of the response at the outlet is directly proportional to the amount of tracer injected.
6. The tracer is completely conserved within the system and is identical to the process fluid in its flow and mixing behavior.

Any tracer can be chosen for the given system, providing it satisfies the following conditions:

- It should not affect the flow.
- It should be injected in a short time and be conveniently analyzed by a suitable method.
- It should not react with or be absorbed by the surface of the system.
- Its molecular diffusivity should be low and it should be conserved (i.e., a mass balance on it must be possible).

• The tracer must be uniformly distributed in the inlet fluid, and the analysis of the exit should give a proper average concentration in the outlet fluid.

In a time period from $t = 0$ to $t = \delta t$ seconds, a quantity m_i(g) of a tracer is introduced at the system inlet, and the tracer concentration $C(t)$ (g/l) is measured in the exit from the system. Subject to the above conditions, the residence time density function from the measured tracer response is:

$$E(t) = \frac{C(t)}{\int_0^\infty C(t)dt} = \frac{\left[\begin{array}{l}\text{Tracer concentration in the outlet}\\ \text{(or a quantity proportional to it)}\\ \text{at time t}\end{array}\right]}{\left[\begin{array}{l}\text{Total area under tracer concentration}\\ \text{(or a quantity proportional to it) curve}\\ \text{versus time as measured at the outlet}\end{array}\right]} \quad (8\text{-}47)$$

Consider the tracer leaving a system between t and $t + \delta t$. From the definition of $E(t)$, the quantity of tracer is

$$m_i E(t_i)\delta t \tag{8-48}$$

From the definition of $C(t)$, the concentration of tracer in the outlet is

$$C(t) \bullet u \bullet \delta t \tag{8-49}$$

Therefore,

$$m_i E(t)\delta t = C(t)u\delta t \tag{8-50}$$

$$\text{or } E(t) = \frac{u}{m_i}C(t) \tag{8-51}$$

Using Equation 8-51 it is possible to determine $E(t)$ from the measured response $C(t)$, if both the volumetric flowrate u and the mass of tracer injected m_i are known. Since the mass of tracer emerging in the interval from t and $t + \delta t$ is $C(t)u\delta t$, then the total quantity of the tracer is:

$$m_i = \int_0^\infty u\, C(t)\, dt \qquad\qquad (8\text{-}52)$$

$$\text{or } \frac{m_i}{u} = \int_0^\infty C(t)\, dt \qquad\qquad (8\text{-}53)$$

Substituting Equation 8-53 into Equation 8-51 gives

$$E(t) = \frac{C(t)}{\int_0^\infty C(t)\, dt} \qquad\qquad (8\text{-}54)$$

If the concentration of tracer in the outlet stream is not measured directly, a quantity $R_I(t)$, which is proportional to $C(t)$, must be measured. For example, $R_I(t)$ may be light absorbance if the tracer is a dye, a conductance if the tracer is an electrolyte, or a counting rate if a radioactive tracer is used. If $C(t) = kR_I(t)$, where k is the proportionality constant, it can be substituted in Equation 8-54 to give the residence time density function $E(t)$ as

$$E(t) = \frac{R_I(t)}{\int_0^\infty R_I(t)\, dt} \qquad\qquad (8\text{-}54)$$

Figure 8-10 shows typical pulse input and output signals.

STEP-CHANGE

If the feed to a system is switched instantaneously from one supply to another, a step-change residence time experiment can be performed. The second feed must be distinguishable from the first, but both should have the same density and viscosity and behave in the same manner within the system. A step signal is often performed by switching from undyed to dyed feed, unless the tracer is expensive or unpleasant (e.g., radioactive tracers); in such a case, an impulse tracer is used.

Pulse Analysis of a Tracer

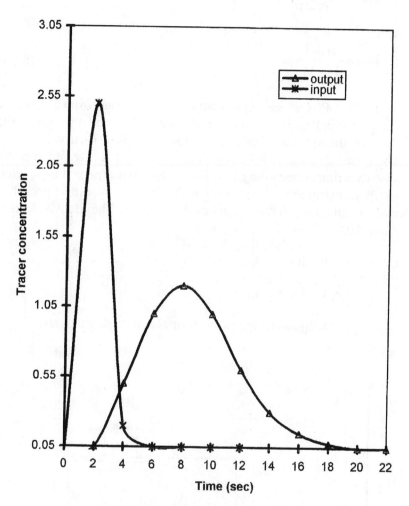

Figure 8-10. Arbitrary pulse input and output signals.

Consider the effluent from a system at time t after the step-change to a marked feed has taken place. The concentration of the tracer in the effluent is related to the F-function of the system because all unmarked material in the effluent must be in the system for a time greater than t. If P is some property that varies linearly with the volume fraction of feed one in a mixture of feed two, then

$$1 - F(t) = \frac{P_2 - P(t)}{P_2 - P_1}$$

$$\text{or} \quad F(t) = \frac{P(t) - P_1}{P_2 - P_1} \tag{8-56}$$

A graph of $P(t)$ versus time would show a change from P_1 (before the step-change) to P_2 (a long time afterwards), and this gives the F-curve of the system. Figure 8-11 shows a typical F-curve.

The F-curve can also be determined from the E-curve obtained by a pulse experiment according to Equation 8-3. For a plug flow reactor, the step is extremely sharp, and in the limit it would approach a Heaviside function at the mean residence time. The Heaviside unit function $H(t - t_o)$ is

$$H(t - t_o) = 0 \quad \text{for } t < t_o$$

$$= 1 \quad \text{for } t > t_o$$

It enables the step-change to be formulated mathematically

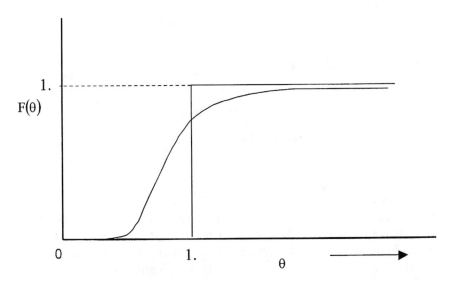

Figure 8-11. Step change analysis.

$$\delta(t-t_o) = \frac{dH(t-t_o)}{dt}$$

(8-57)

PERIODIC INPUT

A periodic variation (e.g., a sinusoidal variation) of the concentration of the tracer in some property of the feed into a system can also be used. Measurement of the attenuation and phase shift of such an input signal is then related to the RTD. The advantage of this method is that it averages out the slight variations in both the feed rate and mixing as the effect of noise is minimized. This method of input is rather complex and, therefore, is only used in laboratory investigation. Figure 8-12 shows a graphical representation of the sinusoidal response of a linear system. From the input signal (i.e., as a function of various sinusoidal frequencies of the input signal), the periodic response analysis makes it possible to determine the amplitude change and the phase shift of the response of the system. This is then compared with the response of the mathematical model under consideration.

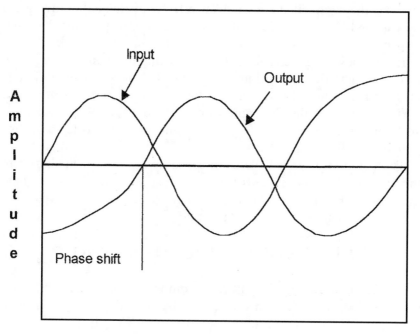

Figure 8-12. Sinusoidal response.

INTERPRETATION OF TRACER EXPERIMENTS

Previously, the forms of the age distribution curves for the various ideal flow reactors were reviewed. However, the actual experimental curves produced by the tracer techniques will often deviate from these curves depending on the extent and type of mixing encountered in real systems. The degree of mixing is related to the spread and peak of the E-curves. The narrower the degree of mixing, the narrower and sharper the E-curve. The tracer information may need to be assessed as to whether the reactor shows considerable bypassing of the fluid, a large dead space, or gross channeling of the fluid.

For example, a tracer impulse injected at the inlet of a reactor may produce two peaks, the first corresponding to the channeling of the fluid and the second to the bulk fluid. Bypassing, short-circuiting, or channeling occurs when a certain amount of fluid immediately leaves the system and fails to mix with the bulk of the fluid. An imperfect packing in a fixed bed reactor is a possible cause of this phenomenon, and an overly close arrangement of the inlet and outlet of a CSTR is another. Stagnant zones, dead corners, or dead space imply that a certain part of the liquid does not mix or exchange with inlet or outlet streams, thereby reducing the reactor effective volume. Improving the flow is possible by inserting redistributors and through proper baffling, respectively.

Piston flow signifies that some of the liquid passes through the reactor in plug flow. This liquid, unlike that involved in short-circuiting, has a certain residence time in the reactor. In certain operations, it is essential that the flow approach as close as possible some ideal situation, usually plug flow (e.g., in continuous, large-scale chromatographic separations).

These non-idealities enable us to construct useful flow models from the tracer information. Figures 8-13, 8-14, and 8-15, respectively, show the E-curve, F-curve, I-curve, and Λ-curve for reactors with bypassing, dead space (stagnancy), and channeling.

ANALYSIS OF RTD FROM PULSE INPUT

The discrete data can be analyzed from the residence time distribution by using either the histogram method or the trapezoidal rule

(text continued on page 692)

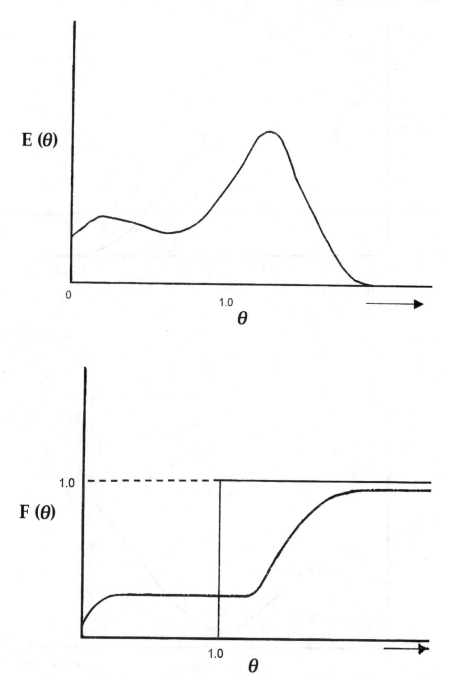

Figure 8-13. E-curve, F-curve, I-curve, and Λ-curve for a typical system with bypassing.

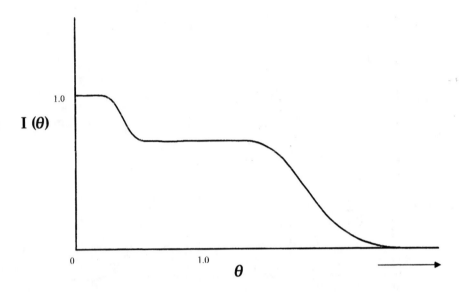

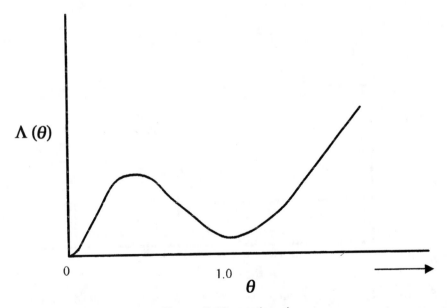

Figure 8-13. continued

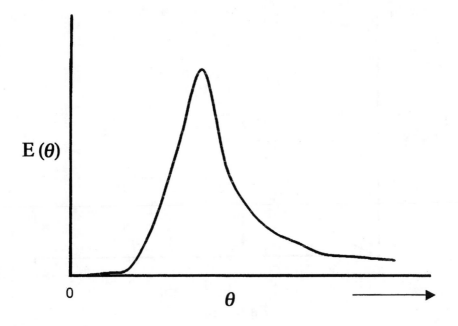

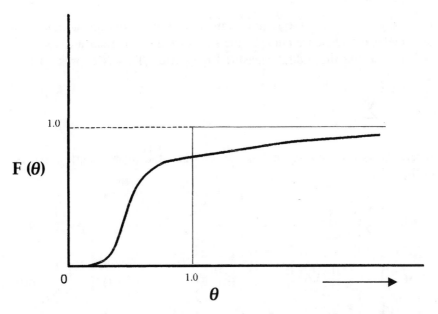

Figure 8-14. E-curve, F-curve, and Λ-curve for a typical system with dead space.

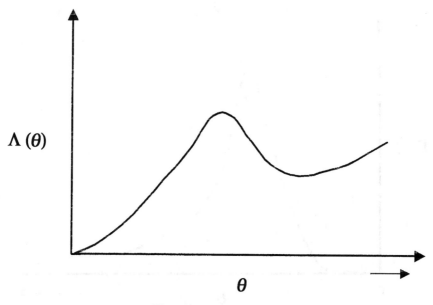

Figure 8-14. continued

(text continued from page 688)

method to determine the required areas. In the histogram method, the area under the response curve is the sum of the rectangular areas whose heights are the individual measured responses. This is expressed by:

$$\text{area} = \sum_{i=1}^{n} C_i(t)\Delta t_i \tag{8-58}$$

where Δt_i is a time interval (or the width of the ith rectangle), and is defined by:

$$\Delta t_i = \frac{t_1 + t_2}{2} \quad (i = 1) \tag{8-59}$$

$$\Delta t_i = \frac{(t_{i+1} - t_i) + (t_i - t_{i-1})}{2} = \frac{t_{i+1} - t_{i-1}}{2} \quad (i = 2 \text{ to } n-1) \tag{8-60}$$

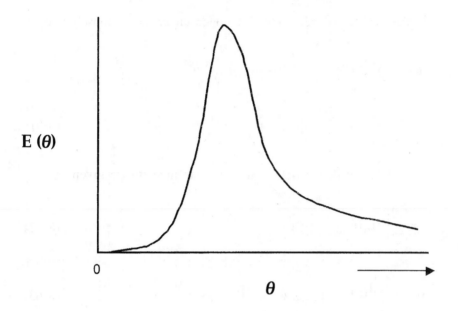

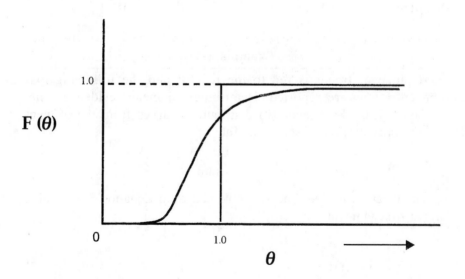

Figure 8-15. E-curve and F-curve for a typical system with channeling.

In the trapezoidal rule method, the area under the response curve is:

$$\text{area} = \sum_{i=1}^{n-1} C_i(t)_{\text{average}} \Delta t_i$$

$$= \sum_{i=1}^{n-1} \left(\frac{C_i + C_{i+1}}{2} \right) (t_{i+1} - t_i) \tag{8-61}$$

The mean residence time $\bar{t}$ and the variance σ_t^2 are given by

$$\bar{t} = \sum_{i=1}^{n-1} \left[t_i E_i(t)_{\text{average}} \right] \Delta t_i \tag{8-62}$$

$$\sigma_t^2 = \left\{ \sum_{i=1}^{n-1} \left[t_i^2 E_i(t) \right]_{\text{average}} \Delta t_i \right\} - \bar{t}^2 \tag{8-63}$$

Missen et al. [15] have developed numerical methods for the step input F(t). Figure 8-16 illustrates a typical method of determining the area under C(t) versus t curve.

Example 8-1

A pulse of tracer is fed to the reactor and the following exit concentrations are reported. Determine the mean residence time, variance, $E(\theta)$, $F(\theta)$, and $I(\theta)$ distribution curves from the effluent tracer concentration as shown in Table 8-2.

Solution

The mean residence time $\bar{t}$ is obtained from Equation 8-39, which is the first moment.

$$\mu = \bar{t} = \frac{\displaystyle\int_0^\infty t\, C(t)\, dt}{\displaystyle\int_0^\infty C(t)\, dt} \tag{8-39}$$

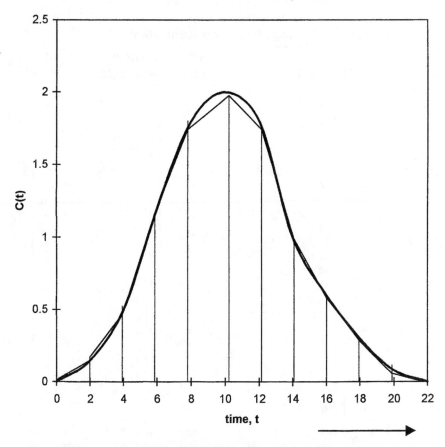

Figure 8-16. Trapezoidal method of area determination.

In discrete form,

$$\mu = \bar{t} = \frac{\displaystyle\sum_{0}^{\infty} t\,C(t)\delta t}{\displaystyle\sum_{0}^{\infty} C(t)\delta t} \tag{8-40}$$

The variance is the second moment, and is defined by

Table 8-2
Effluent tracer concentration

Time (min)	Effluent tracer concentration, g/cm^3
0.0	0.0
2.5	1.5
5.0	3.0
7.5	4.5
10.0	5.0
12.5	5.2
15.0	5.0
17.5	4.5
20.0	4.0
22.5	4.0
25.0	2.0
27.5	1.0
30.0	0.5
32.5	0.2
35.0	0.0

$$\sigma^2 = \frac{\int_0^\infty t^2 C(t)dt}{\int_0^\infty C(t)dt} - \bar{t}^2 \qquad (8\text{-}42)$$

or in discrete form is expressed as

$$\sigma^2 = \frac{\sum_0^\infty t^2 C(t)\delta t}{\sum_0^\infty C(t)\delta t} - \bar{t}^2 \qquad (8\text{-}43)$$

Table 8-3 gives the parameters for calculating the mean residence time, variance, $E(\theta)$, $F(\theta)$, and $I(\theta)$ values. The mean residence time $\bar{t}$ with $\delta t = 2.5$ is

Table 8-3

t	C(t)	tC(t)	t²C(t)	E(t) min⁻¹	θ	E(θ)
0.0	0.0	0.00	0.000	0.000	0.000	0.000
2.5	1.5	3.75	9.375	0.015	0.171	0.219
5.0	3.0	15.00	75.000	0.030	0.342	0.438
7.5	4.5	33.75	253.125	0.045	0.513	0.657
10.0	5.0	50.0	500.000	0.050	0.684	0.731
12.5	5.2	65.0	812.500	0.051	0.856	0.745
15.0	5.0	75.0	1,125.000	0.050	1.027	0.731
17.5	4.5	78.75	1,378.125	0.045	1.198	0.657
20.0	4.0	80.00	1,600.000	0.040	1.369	0.584
22.5	4.0	90.00	2,025.000	0.040	1.540	0.584
25.0	2.0	50.00	1,250.000	0.020	1.711	0.292
27.5	1.0	27.50	756.25	0.010	1.882	0.146
30.0	0.5	15.00	450.00	0.005	2.053	0.073
32.5	0.2	6.50	211.25	0.002	2.225	0.029
35.0	0.0	0.00	0.00	0.000	2.396	0.000
Total	**40.4**	**590.25**	**10,445.625**			

$$\mu = \bar{t} = \frac{\sum\limits_{0}^{\infty} tC(t)\delta t}{\sum\limits_{0}^{\infty} C(t)\delta t} = \frac{(590.25)(2.5)}{(40.4)(2.5)}$$

$$= 14.61 \text{ min}$$

The variance $\sigma^2 = \dfrac{\sum\limits_{0}^{\infty} t^2 C(t)\delta t}{\sum\limits_{0}^{\infty} C(t)\delta t} - \bar{t}^2$

$$= \frac{(10,445.625)(2.5)}{(40.4)(2.5)} - 14.61^2$$

$$\sigma^2 = 45.10 \text{ min}^2$$

The exit age residence time distribution function E(t) is

$$E(t) = \frac{C(t)}{\sum\limits_{0}^{\infty} C(t)\delta t} \tag{8-54}$$

where $\sum\limits_{0}^{\infty} C(t)\delta t = (40.4)(2.5) = 101$.

The dimensionless residence time $\theta = t/\bar{t}$ and $E(\theta) = \bar{t}E(t)$.

Table 8-4 shows the details of the summation, which is performed using the trapezoidal rule. The internal age distribution $I(\theta)$ is obtained from

$$I(\theta) = 1 - F(\theta) = 1 - \sum E\delta\theta$$

The Microsoft Excel spreadsheet (Example8-1.xls) was developed to calculate the moments (mean residence time and variance) and the

Table 8-4

θ	$E(\theta)$	Area present in the interval between i – 1 and i, $E_i(\theta) = \dfrac{(E_i + E_{i-1})(\theta_i - \theta_{i-1})}{2}$	Summing to $\theta_i = -\sum E\delta\theta$ $F(\theta)$	$I = 1 - \sum E\delta\theta$ $I(\theta)$
0.000	0.000	0.000	0.000	1.000
0.171	0.219	0.019	0.019	0.981
0.342	0.438	0.056	0.075	0.925
0.513	0.657	0.094	0.169	0.831
0.684	0.731	0.119	0.288	0.712
0.856	0.745	0.126	0.414	0.586
1.027	0.731	0.126	0.540	0.460
1.198	0.657	0.119	0.659	0.341
1.369	0.584	0.106	0.765	0.235
1.540	0.584	0.100	0.865	0.135
1.711	0.292	0.075	0.940	0.060
1.882	0.146	0.037	0.977	0.023
2.053	0.073	0.019	0.996	0.004
2.225	0.029	0.009	1.005	—
2.396	0.000	0.002	1.007	—

RTD functions, $E(\theta)$, $F(\theta)$, and $I(\theta)$. Figure 8-17 shows plots of $E(\theta)$, $F(\theta)$, and $I(\theta)$ versus θ.

Example 8-2

Levenspiel and Smith (1957)[*] have reported the following data obtained from a residence time experiment involving a length of 2.85-cm diameter pyrex tubing. A volume of potassium permanganate ($KMnO_4$) solution that would fill a 2.54 cm length of the tube was rapidly injected into a water stream with a linear velocity of 35.70 cm/sec. A photoelectric cell 2.74 m downstream from the injection point was used to monitor the local $KMnO_4$ concentration. Determine the mean residence time of the fluid, the variance, $E(\theta)$, $F(\theta)$, and $I(\theta)$ from the following effluent $KMnO_4$ concentrations.

[*]*Chem. Eng. Sci. 6, 227 (1957).*

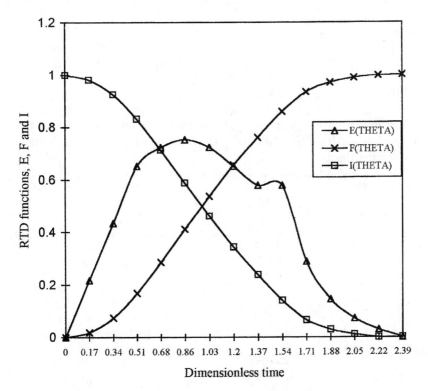

Figure 8-17. Residence time distribution functions $E(\theta)$, $F(\theta)$, and $I(\theta)$ versus θ.

Time (sec)	KMnO$_4$ concentration (arbitrary units)
0.0	0.0
2.0	11.0
4.0	53.0
6.0	64.0
8.0	58.0
10.0	48.0
12.0	39.0
14.0	29.0
16.0	22.0
18.0	16.0
20.0	11.0
22.0	9.0
24.0	7.0
26.0	5.0
28.0	4.0
30.0	2.0
32.0	2.0
34.0	2.0
36.0	1.0
38.0	1.0
40.0	1.0
42.0	1.0

Solution

A table was constructed to determine the mean residence time $\bar{t}$, variance σ^2, $E(\theta)$, $F(\theta)$, and $I(\theta)$ parameters from the effluent tracer versus time data.

The mean residence time $\bar{t}$ with $\delta t = 2.0$ is

$$\mu = \bar{t} = \frac{\displaystyle\sum_0^\infty t\,C(t)\delta t}{\displaystyle\sum_0^\infty C(t)\delta t} = \frac{(4,252)(2.0)}{(386)(2.0)} = 11.016 \text{ sec} \tag{8-40}$$

The variance $\sigma^2 = \dfrac{\displaystyle\sum_0^\infty t^2 C(t)\delta t}{\displaystyle\sum_0^\infty C(t)\delta t} - \bar{t}^2$ \hfill (8-43)

$$= \frac{(65,392.0)(2.0)}{(386)(2.0)} - 11.016^2 = 48.06 \text{ sec}^2$$

The exit age residence time distribution function $E(t)$ is

$$E(t) = \frac{C(t)}{\sum\limits_{0}^{\infty} C(t)\delta t}$$

(8-54)

where $\sum\limits_{0}^{\infty} C(t)\delta t = (386)(2.0) = 772$.

The dimensionless residence time $\theta = t/\bar{t}$ and $E(\theta) = \bar{t}E(t)$.

Table 8-5 shows the details of the summation, which is performed using the trapezoidal rule. The internal age distribution $I(\theta)$ (Table 8-6) is obtained from $I(\theta) = 1 - F(\theta) = 1 - \sum E\delta\theta$. An Excel spreadsheet (Example8-2.xls) was developed and Figure 8-18 shows plots of $E(\theta)$, $F(\theta)$, and $I(\theta)$ versus θ.

Table 8-5

Time, sec	C(t)	tC(t)	$t^2C(t)$	E(t) sec^{-1}	θ	E(θ)
0.0	0.0	0.0	0.0	0.000	0.000	0.000
2.0	11.0	22.0	44.0	0.014	0.182	0.154
4.0	53.0	212.0	848.0	0.069	0.363	0.760
6.0	64.0	384.0	2,304.0	0.083	0.545	0.914
8.0	58.0	464.0	3,712.0	0.075	0.726	0.826
10.0	48.0	480.0	4,800.0	0.062	0.908	0.683
12.0	39.0	468.0	5,616.0	0.051	1.089	0.562
14.0	29.0	406.0	5,684.0	0.038	1.271	0.419
16.0	22.0	352.0	5,632.0	0.028	1.452	0.308
18.0	16.0	288.0	5,184.0	0.021	1.634	0.231
20.0	11.0	220.0	4,400.0	0.014	1.816	0.154
22.0	9.0	198.0	4,356.0	0.012	1.997	0.132
24.0	7.0	168.0	4,032.0	0.009	2.179	0.099
26.0	5.0	130.0	3,380.0	0.006	2.360	0.066
28.0	4.0	112.0	3,136.0	0.005	2.542	0.055
30.0	2.0	60.0	1,800.0	0.003	2.723	0.033
32.0	2.0	64.0	2,048.0	0.003	2.905	0.033
34.0	2.0	68.0	2,312.0	0.003	3.086	0.033
36.0	1.0	36.0	1,296.0	0.0013	3.268	0.014
38.0	1.0	38.0	1,444.0	0.0013	3.450	0.014
40.0	1.0	40.0	1,600.0	0.0013	3.631	0.014
42.0	1.0	42.0	1,764.0	0.0013	3.813	0.014
Total	**386.0**	**4,252.0**	**65,392.0**			

Table 8-6

θ	$E(\theta)$	Area present in the interval between i − 1 and i, $E_i(\theta) = \dfrac{(E_i + E_{i-1})(\theta_i - \theta_{i-1})}{2}$	Summing to $\theta_i = -\sum E\delta\theta$ $F(\theta)$	$I = 1 - \sum E\delta\theta$ $I(\theta)$
0.000	0.000	0.000	0.000	1.000
0.182	0.154	0.014	0.014	0.986
0.363	0.760	0.083	0.097	0.903
0.545	0.914	0.152	0.249	0.751
0.726	0.826	0.158	0.407	0.593
0.908	0.683	0.137	0.544	0.456
1.089	0.562	0.113	0.657	0.343
1.271	0.419	0.089	0.746	0.254
1.452	0.308	0.066	0.812	0.188
1.634	0.231	0.049	0.861	0.139
1.816	0.154	0.035	0.896	0.104
1.997	0.132	0.026	0.922	0.0.78
2.179	0.099	0.021	0.943	0.057
2.360	0.066	0.015	0.958	0.042
2.542	0.055	0.011	0.969	0.031
2.723	0.033	0.008	0.977	0.023
2.905	0.033	0.006	0.983	0.017
3.086	0.033	0.006	0.989	0.011
3.268	0.014	0.004	0.993	0.007
3.450	0.014	0.003	0.996	0.004
3.631	0.014	0.003	0.999	0.001
3.813	0.014	0.003	1.002	—

Example 8-3

Hull and von Rosenberg (1960)[*] injected a pulse of radioactive tracer into the catalyst inlet at the bottom of a reactor. The radioactive tracer concentration was then measured at various points in the reactor. The catalyst rate was 340 lb/hr and the holdup was 18.4 lb. The results of the tracer response were given as follows:

Time, t min	Counts/min × 10^{-3} (smoothed to equidistant points)
0	0
0.5	5
1.0	22
1.5	27

[*]*Ind. Eng. Chem. 52, 989 (1960).*

2.0	26
2.5	22
3.0	19
3.5	15
4.0	10
4.5	7
5.0	4
5.5	3
6.0	3
6.5	(1)
7.0	(0)
	164

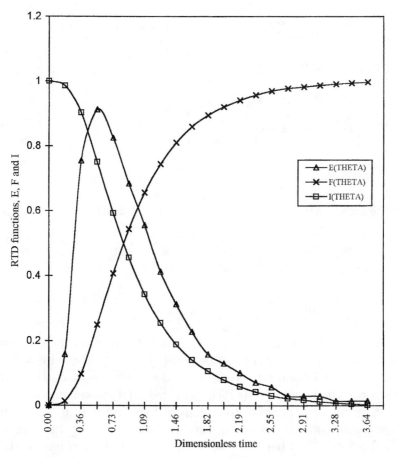

Figure 8-18. Residence time distribution functions $E(\theta)$, $F(\theta)$, and $I(\theta)$ versus θ.

Determine the various distribution functions, namely $E(\theta)$, $F(\theta)$, and $I(\theta)$.

Solution

A Microsoft Excel spreadsheet (Example8-3.xls) was developed to determine the $E(\theta)$, $F(\theta)$, and $I(\theta)$ distribution functions. The results of the spreadsheet calculations give the mean residence time from the distribution curve

$$\bar{t} = \sum_{i=1}^{n-1} \left[t_i E_i(t)_{average} \right] \delta t \qquad (8\text{-}62)$$

$$= 2.53 \text{ min}$$

The variance σ_t^2 is:

$$\sigma_t^2 = \left\{ \sum_{i=1}^{n-1} \left[t_i^2 E_i(t) \right]_{average} \delta t_i \right\} - \bar{t}^2 \qquad (8\text{-}63)$$

$$= 1.7 \text{ min}^2$$

The experimental mean residence time $\bar{t}_{expt}$ is:

$$\bar{t}_{ept} = \frac{(18.4)(60)}{340}$$

$$= 3.25 \text{ min}$$

There is approximately a 22% deviation between the experimental and the distribution mean residence time. However, the main purpose was to use the information from the RTD curve to improve the reactor operation. The results of the RTD provided vital information concerning the effects of operating conditions and structural designs on solid-mixing patterns in fluidized systems. The perfect mixing function was generated by $e^{-\theta}$, where $\theta = t/\bar{t}$. Figure 8-19 shows plots of these functions against dimensionless residence time θ.

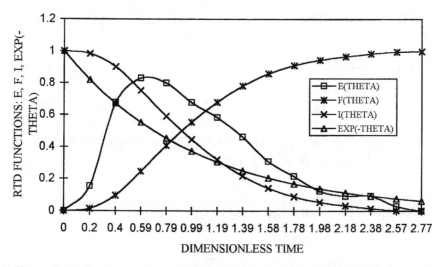

Figure 8-19. Residence time distribution functions E(θ), F(θ), and I(θ) versus θ.

Example 8-4

The response to an impulse input has the shape of a trapezoid with equations

$$C = \begin{vmatrix} t-2 & \text{when } 2 \le t \le 5 \\ 3 & \text{when } 5 \le t \le 8 \\ 11-t & \text{when } 8 \le t \le 11 \\ 0 & \text{elsewhere} \end{vmatrix}$$

Determine the mean residence time $\bar{t}$ and the residence time distribution function E(t).

Solution

The mean residence time $\bar{t}$ is

$$\mu = \bar{t} = \frac{\int_0^\infty t\,C(t)\,dt}{\int_0^\infty C(t)\,dt} \tag{8-40}$$

and the E(t) function is

$$E(t) = \frac{C(t)}{\int\limits_{0}^{\infty} C(t)dt} \tag{8-54}$$

A plot of C(t) for the given conditions gives a trapezoidal shape as shown in Figure 8-20. Alternatively,

$$\int\limits_{0}^{\infty} C(t)dt = \int\limits_{2}^{5} (t-2)dt + \int\limits_{5}^{8} 3dt + \int\limits_{8}^{11} (11-t)dt$$

integrating $\int\limits_{0}^{\infty} C(t)dt$ gives

$$I = \left[\frac{t^2}{2} - 2t \right]_2^5 + 3[t]_5^8 + \left[11t - \frac{t^2}{2} \right]_8^{11}$$

= 18.0, which is the same value as the area of the trapezoidal shape.

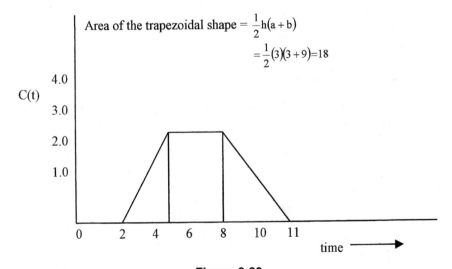

Area of the trapezoidal shape $= \frac{1}{2}h(a+b)$

$$= \frac{1}{2}(3)(3+9) = 18$$

Figure 8-20

To determine the mean residence time $\bar{t}$, it is necessary to find the integral of $\int_0^\infty C(t)dt$ over the given conditions as follows:

$$\int_0^\infty tC(t)dt = \int_2^5 t(t-2)dt + \int_5^8 3tdt + \int_8^{11} t(11-t)dt$$

$$= \int_2^5 \left(t^2 - 2t\right)dt + \int_5^8 3tdt + \int_8^{11} \left(11t - t^2\right)dt$$

Integrating gives

$$I = \left\{ \left[\frac{t^3}{3} - t^2\right]_2^5 + \frac{3}{2}[t^2]_5^8 + \left[\frac{11t^2}{2} - \frac{t^3}{3}\right]_8^{11} \right\}$$

$$= 117$$

The mean residence time $\bar{t}$ is

$$\mu = \bar{t} = \frac{\int_0^\infty tC(t)dt}{\int_0^\infty C(t)dt} = \frac{117}{18} = 6.5 \qquad (8\text{-}40)$$

The E(t) function is $E(t) = C(t)/\int_0^\infty C(t)dt$ for the following conditions.

$$E(t) = \frac{C(t)}{18} \begin{cases} = \dfrac{t-2}{18} & \text{when } 2 \le t \le 5 \\[2mm] = \dfrac{3}{18} & \text{when } 5 \le t \le 8 \\[2mm] = \dfrac{11-t}{18} & \text{when } 8 \le t \le 11 \\[2mm] = 0 & \text{elsewhere} \end{cases}$$

RESIDENCE TIME DISTRIBUTION FOR
A LAMINAR FLOW TUBULAR REACTOR

Consider a steady unidimensional flow in a tubular reactor as shown in Figure 8-21 in the absence of either radial or longitudinal diffusion. The velocity u(r) is the parabolic distribution for a Newtonian fluid at constant viscosity, with the fluid in the center of the tube spending the shortest time in the reactor.

The laminar velocity profile in Figure 8-21a is approximated by a series of annuli, within each of which the velocity is constant as illustrated in Figure 8-21b. Each annulus is considered to be a plug flow tubular reactor having its own space velocity. The velocities of the fluid elements at different radii are given by the parabolic velocity profile for fully developed laminar flow. The velocity is expressed as

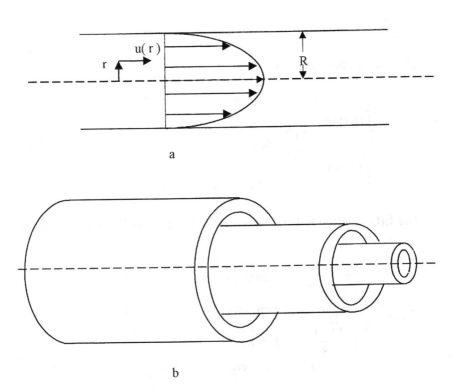

a

b

Figure 8-21. Laminar flow reactor.

$$u(r) = u_O\left[1 - \left(\frac{r}{R}\right)^2\right] \qquad (8\text{-}64)$$

where $u(r)$ = velocity at any radius r
$\qquad u_O$ = centerline velocity
$\qquad r$ = distance from the centerline of the pipe
$\qquad R$ = inside radius of the pipe

The average velocity through the tube is

$$\bar{u} = \frac{u_O}{2} \qquad (8\text{-}65)$$

Because u varies with r, the residence times of the various fluid elements will also vary with r. The time that it will take a fluid element to traverse the reactor of length L is

$$t = \frac{L}{u(r)} = \frac{L}{u_O\left[1 - \left(\frac{r}{R}\right)^2\right]} \qquad (8\text{-}66)$$

The average residence time $\bar{t}$ is

$$\bar{t} = \frac{2L}{u_O} \qquad (8\text{-}67)$$

or the length L in terms of $\bar{t}$ is

$$L = \frac{\bar{t}}{2}u_O \qquad (8\text{-}68)$$

Combining Equations 8-66 and 8-68 gives

$$t = \frac{\bar{t}\,u_O}{2u_O\left[1 - \left(\frac{r}{R}\right)^2\right]}$$

$$\frac{\bar{t}}{2t} = \left[1 - \left(\frac{r}{R}\right)^2\right] \tag{8-69}$$

The fluid at the centerline is moving the fastest, so this material will be the first to leave. It leaves at a time t_{min} given by

$$t_{min} = \frac{L}{u_O} = \frac{L}{2\bar{u}} = \frac{\bar{t}}{2} \tag{8-70}$$

Therefore,

$$F(t) = 0 \text{ for } t < \frac{\bar{t}}{2} \tag{8-71}$$

The fraction of the volumetric flowrate that occurs in the region bounded by $r = 0$ and $r = r$ is equal to $F(t)$ where t is equal to the time necessary for a fluid element to traverse the reactor length at a given r.

$$F(t) = \frac{\text{Volumetric flowrate between } r = 0 \text{ and } r = r}{\text{Total volumetric flowrate}}$$

$$\text{or } F(t) = \frac{\int_0^r u(r) 2\pi r dr}{\int_0^R u(r) 2\pi r dr} \tag{8-72}$$

Substituting Equation 8-64 into Equation 8-72 gives

$$F(t) = \frac{\int_0^r \left[1 - \left(\frac{r}{R}\right)^2\right] r \, dr}{\int_0^R \left[1 - \left(\frac{r}{R}\right)^2\right] r dr} \tag{8-73}$$

Integrating Equation 8-73 gives

$$F(t) = \frac{\left(\dfrac{r^2}{2} - \dfrac{r^4}{4R^2}\right)}{\left(\dfrac{R^2}{2} - \dfrac{R^4}{4R^2}\right)} = \left(\frac{r}{R}\right)^2 \left[2 - \left(\frac{r}{R}\right)^2\right] \qquad (8\text{-}74)$$

From Equation 8-69

$$\left(\frac{r}{R}\right)^2 = 1 - \frac{\bar{t}}{2t} \qquad (8\text{-}75)$$

Substituting Equation 8-75 into Equation 8-74 yields

$$F(t) = \left[1 - \frac{\bar{t}}{2t}\right]\left[2 - \left(1 - \frac{\bar{t}}{2t}\right)\right]$$

$$F(t) = 1 - \left(\frac{\bar{t}}{2t}\right)^2 \qquad (8\text{-}76)$$

Figure 8-22 shows the $F(\theta)$ curves for laminar flow in a tubular reactor and for other idealized flow patterns.

The exit age distribution function $E(t)$ for a laminar flow reactor is

$$
\begin{aligned}
E(t) &= 0 & &\left| t < \frac{\bar{t}}{2} \right. \\
E(t) &= \frac{\bar{t}^2}{2t^3} & &\left| t \geq \frac{\bar{t}}{2} \right.
\end{aligned}
\qquad (8\text{-}77)
$$

The dimensionless form of the RTD function is

$$
\begin{aligned}
E(\theta) &= 0 & &\left| \text{for } \theta < 0.5 \right. \\
E(\theta) &= \frac{1}{2\theta^3} & &\left| \text{for } \theta \geq 0.5 \right.
\end{aligned}
\qquad (8\text{-}78)
$$

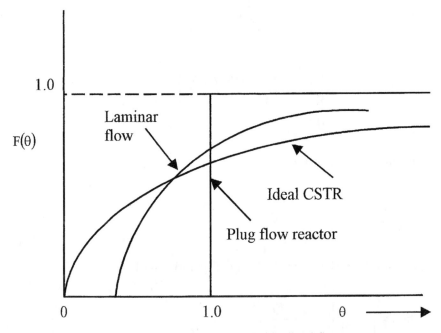

Figure 8-22. Curves for reactors with idealized flow patterns.

Figure 8-23 shows a plot for the exit age RTD.

Note that not all velocity profiles in a tubular reactor are parabolic. Power law fluids have the profile

$$u(r) = u_O \left[1 - \left(\frac{r}{R} \right)^{\left(\frac{n+1}{n} \right)} \right] \tag{8-79}$$

where $n \geq 0$ is known as the power law constant. Newtonian fluids are a special case of power law fluids and where $n = 1$, Equation 8-79 reduces to the parabolic profile. When $n < 1$, the fluid is known as pseudoplastic, and the velocity profile is thus flattened. When $n > 1$, the fluid is known as dilatant, and the velocity profile is sharper than the parabolic profile. The residence time distribution function for the power law fluids in a tubular reactor is

$$F(t) = \left[1 + \frac{2n\bar{t}}{(3n+1)t} \right] \left[1 - \frac{(n+1)\bar{t}}{(3n+1)t} \right]^{\frac{2n}{(n+1)}} \quad \text{for} \quad t > \frac{(n+1)\bar{t}}{3n+1} \tag{8-80}$$

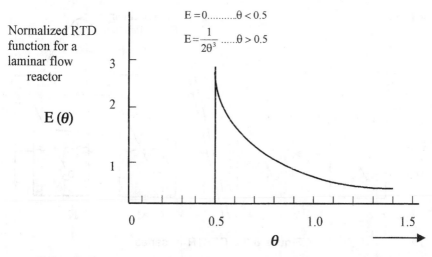

Normalized RTD function for a laminar flow reactor

$E(\theta)$

$E = 0 \ldots \ldots \theta < 0.5$

$E = \dfrac{1}{2\theta^3} \ldots \theta > 0.5$

Figure 8-23. Residence time distribution for tubular laminar flow.

E- AND F-CURVES FOR A SERIES OF STIRRED TANK REACTORS

Consider a series of well-stirred reactors each having the same volume, where the total volume is V_R. Let $C_1, C_2 \ldots C_m \ldots C_n$ be the concentrations leaving the successive stages of the CSTR. The unsteady state tracer balance on the mth stage is

$$uC_{m-1} = uC_m + V_m \frac{dC_m}{dt} \tag{8-81}$$

A model frequently employed to simulate the behavior of an actual reactor is a series of ideal stirred tank reactors as shown in Figure 8-24.

The actual volume V_R can be replaced by N identical stirred tank reactor whose total volume is the same as that of the actual reactor. That is:

$$V_R = NV \tag{8-82}$$

where V is the volume of each stirred tank.

It is possible to determine the value of N that gives the best fit to the response curve of the actual reactor as described below. Consider a steady flow u m^3/sec of fluid in and out of the first reactor volume

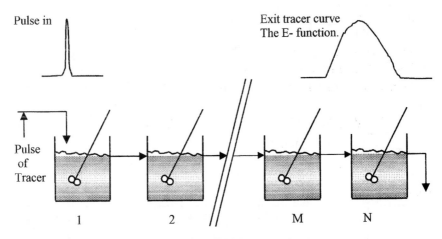

Figure 8-24. CFSTR in series.

V_1. At time $t = 0$, a pulse of tracer is injected into the vessel that, when evenly distributed in the vessel, has a concentration C_0. At any time t after the tracer is introduced, a material balance on the first reactor gives

Input Rate – Output Rate = Rate of accumulation of tracer
in the reactor

$$0 - uC_1 = V_1 \frac{dC_1}{dt} \tag{8-83}$$

where C_1 is the concentration of tracer in tank 1. Separating variables and integrating Equation 8-83 gives

$$\int_{C_0}^{C_1} \frac{dC_1}{C_1} = -\frac{u}{V_1} \int_0^t dt \tag{8-84}$$

leading to

$$\ln\left(\frac{C_1}{C_0}\right) = -\frac{u}{V_1} t$$

or $C_1 = C_0\, e^{-t/\bar{t}_1}$ \hfill (8-85)

where $\bar{t} = V_1/u$.

Since V_i is the volume of a single reactor in the series, $\bar{t}_i$ is the residence time in any one of the reactors, then

$$\bar{t}_i = \frac{V_i}{u} = \frac{V_1}{u} = \frac{V_2}{u} = \frac{V_3}{u} \cdots\cdots\cdots = \frac{V_N}{u} = \frac{V_R}{Nu} \hspace{1cm} (8\text{-}86)$$

The area under the C/C_O versus t curve is $\bar{t}_i$, and from this the E-curve is

$$\bar{t}_1 E_1 = e^{-t/\bar{t}_1} \hspace{1cm} (8\text{-}87)$$

A material balance on the tracer in the second reactor where C_1 enters and C_2 leaves gives

$$uC_1 = uC_2 + V_2 \frac{dC_2}{dt} \hspace{1cm} (8\text{-}88)$$

Substituting Equations 8-85 and 8-86, respectively, into Equation 8-88 yields

$$C_O e^{-t/\bar{t}_i} = C_2 + \bar{t}_i \frac{dC_2}{dt} \hspace{1cm} (8\text{-}89)$$

Rearranging Equation 8-89 gives

$$\frac{dC_2}{dt} + \frac{C_2}{\bar{t}_i} = \frac{C_O}{\bar{t}_i} e^{-t/\bar{t}_i} \hspace{1cm} (8\text{-}90)$$

Equation 8-90 is a first order differential equation, which is of the form $dY/dx + P(x)y = Q(x)$. The integrating factor is $e^{\int P dx}$. In Equation 8-90, the integrating factor is

$$I.F. = e^{\int \frac{1}{\bar{t}_i} dt} = e^{\frac{t}{\bar{t}_i}}$$

The solution after mathematical manipulation gives

$$C_2 = \frac{t}{\bar{t}_i} C_0 e^{-t/\bar{t}_i} \tag{8-91}$$

A material balance on the tracer in the third reactor where C_2 enters and C_3 leaves gives

$$uC_2 = uC_3 + V_3 \frac{dC_3}{dt} \tag{8-92}$$

Substituting Equations 8-86 and 8-91 into Equation 8-92 and rearranging gives

$$\frac{dC_3}{dt} + \frac{1}{\bar{t}_i} C_3 = \frac{1}{\bar{t}_i} \cdot \frac{t}{\bar{t}_i} C_0 e^{-t/\bar{t}_i} \tag{8-93}$$

The integrating factor of Equation 8-93 is

$$\text{I.F.} = e^{\int P dx} = e^{\int \frac{1}{\bar{t}_i} dt} = e^{\frac{t}{\bar{t}_i}}$$

The solution of Equation 8-93 is

$$C_3 = \frac{C_0}{2\bar{t}_i^2} t^2 e^{-t/\bar{t}_i} \tag{8-94}$$

For the exit-age distribution, the fraction of material exiting the system from the three CSTRs that are in the system between t and $t + \delta t$ is $ME(t)\delta t = uC_3(t)\delta t$ or

$$E(t) = \frac{u}{M} C_3 \tag{8-95}$$

where M is the total amount of tracer injected. Substituting Equation 8-94 into Equation 8-95 yields

$$ME(t)\delta t = u \frac{C_0}{2\bar{t}_i^2} t^2 e^{-t/\bar{t}_i} \delta t \tag{8-96}$$

where by definition $M/V = (u/u)C_0$, therefore

$$M = \bar{t}_i u C_O \quad \left(sec \bullet \frac{m^3}{sec} \bullet \frac{kg}{m^3} \right) \tag{8-97}$$

Equation 8-96 then becomes

$$\bar{t}_i u C_O E(t) = \frac{u C_O t^2}{2 \bar{t}_i^2} e^{-t/\bar{t}_i} \tag{8-98}$$

and the exit-age distribution function E(t) becomes

$$E(t) = \frac{t^2}{2 \bar{t}_i^3} e^{-t/\bar{t}_i} \tag{8-99}$$

For a series of CSTRs, the RTD for CSTR in series E(t) is

$$E(t) = \frac{t^{(N-1)}}{\bar{t}_i^N} \frac{1}{(N-1)!} e^{-t/\bar{t}_i} \tag{8-100}$$

Because $V_R = NV_i$, then $\bar{t}_i = \bar{t}/N$, where $\bar{t}$ is the total reactor volume divided by the flowrate u. In terms of dimensionless time θ, $\theta = t/\bar{t}$, $\bar{t} = N\bar{t}_i$. The exit age distribution function $E(\theta)$ is

$$E(\theta) = \frac{N(N\theta)^{N-1}}{(N-1)!} e^{-N\theta} \tag{8-101}$$

Figure 8-25 shows the RTD of different numbers of tanks in series. As the number increases, the behavior of the system approaches that of a plug flow reactor.

NB: C_O for N tanks is usually defined as $\dfrac{M}{NV_i} = \dfrac{M}{V_R}$

Similarly, for the step response, the output F-curve from a series of N ideal stirred tanks is

$$F(t) = 1 - e^{-Nt/\bar{t}_i} \left[1 + \frac{Nt}{\bar{t}_i} + \frac{1}{2!} \left(\frac{Nt}{\bar{t}_i} \right)^2 + \ldots + \frac{1}{(N-1)!} \left(\frac{Nt}{\bar{t}_i} \right)^{N-1} \right] \tag{8-102}$$

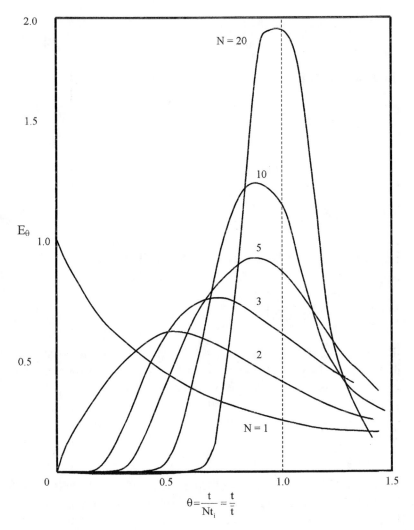

Figure 8-25. RTD curves for the tanks-in-series model.

In terms of dimensionless time θ, $\theta = t/\bar{t}$, $\bar{t} = N\bar{t}_i$, therefore

$$F(\theta) = 1 - e^{-N\theta}\left[1 + N\theta + \frac{(N\theta)^2}{2!} + \ldots + \frac{(N\theta)^{N-1}}{(N-1)!} + \ldots\right] \qquad (8\text{-}103)$$

These curves as N increases are illustrated in Figure 8-26.

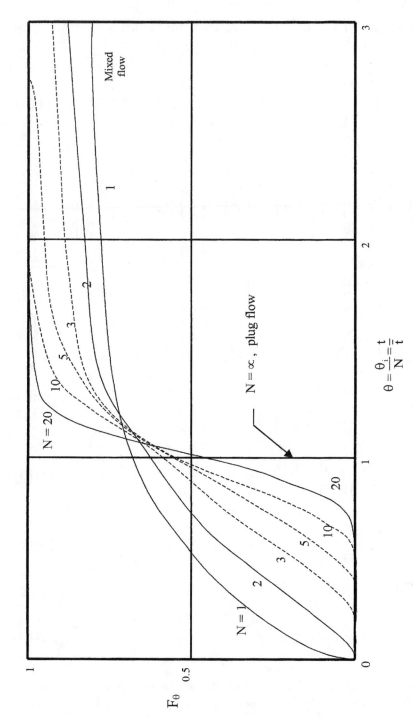

Figure 8-26. The F-curve for the tanks-in-series model.

The number of tanks in series from the dimensionless variance (Equation 8-45) $\sigma_\theta^2 \bullet \sigma_\theta^2$ from the tracer experiment is

$$\sigma_\theta^2 = \frac{\sigma^2}{\bar{t}^2} = \int_0^\infty (\theta - 1)^2 E(\theta) d\theta \qquad (8\text{-}45)$$

Expanding Equation 8-45 gives

$$\sigma_\theta^2 = \frac{\sigma^2}{\bar{t}^2} = \int_0^\infty \theta^2 E(\theta) d\theta - 2\int_0^\infty \theta E(\theta) d\theta + \int_0^\infty E(\theta) d\theta$$

$$\sigma_\theta^2 = \int_0^\infty \theta^2 E(\theta) d\theta - 1$$

$$= \int_0^\infty \theta^2 \frac{N(N\theta)^{N-1}}{(N-1)!} e^{-N\theta} d\theta - 1 \qquad (8\text{-}104)$$

$$\sigma_\theta^2 = \frac{N^N}{(N-1)!} \int_0^\infty \theta^{N+1} e^{-N\theta} d\theta - 1$$

$$= \frac{N^N}{(N-1)!} \left[\frac{(N+1)!}{N^{N+2}} \right] - 1 \qquad (8\text{-}105)$$

$$\sigma_\theta^2 = \frac{1}{N} \qquad (8\text{-}106)$$

The number of tanks in series is

$$N = \frac{1}{\sigma_\theta^2} = \frac{\bar{t}^2}{\sigma^2} \qquad (8\text{-}107)$$

The number of tanks increases as the variance decreases, and the curve width or variance changes from $\sigma_\theta^2 = 1$ for $N = 1$ to $\sigma_\theta^2 = 0$ as

$N \to \infty$. This indicates that the behavior of stirred tanks in series ranges from perfect mixing to plug flow.

RTD FUNCTIONS FOR CSTRs
WHERE N IS NOT AN INTEGER

It is often difficult to use Equations 8-101 or 8-103 to represent the response curve of an actual reactor where N is not an integer. Stokes and Nauman [16] have considered this situation, which is reviewed below.

If the equivalent number of perfectly mixed stages N is not an integer, let i be an integer slightly smaller than N. A fraction $0 \le h < 1$ is defined as

$$N = i + h \tag{8-108}$$

The model consists of $i + 1$ stirred tanks in series, i of which have a common volume, V_{Total}/N, and one of which has a smaller volume hV_{Total}/N. For $N > 1$, the cumulative distribution F-functions are:

$$F(\theta) = 1 - \left(\frac{h}{h-i}\right)^i \exp\left(\frac{-N\theta}{h}\right) - \exp(-N\theta)$$

$$\times \left[\sum_{j=0}^{i-1} \frac{(N\theta)^j}{j!} \left\{1 - \left(\frac{h}{h-1}\right)^{i-j}\right\}\right] \tag{8-109}$$

for $N > 1$ and

$$F(\theta) = 1 - N \exp(-N\theta) \tag{8-110}$$

for $1 > N > 0$. The dimensionless variances are:

$$\sigma_\theta^2 = -1 + 2\left(\frac{-h}{1-h}\right)^i \frac{h^2}{N^2} + \frac{2}{N^2}\sum_{j=0}^{i-1}(j+1)\left[1 - \left(\frac{-h}{1-h}\right)^{i-j}\right] \tag{8-111}$$

for $N > 1$ and

$$\sigma_\theta^2 = \frac{2}{N} - 1 \tag{8-112}$$

for $1 \geq N > 0$,

where i = number of tanks truncated to an integer
   j = index of summation
   N = number of tanks in series
   θ = dimensionless residence time
   σ^2 = dimensionless variance

Equations 8-109, 8-110, 8-111, and 8-112 are reduced to an ordinary tanks-in-series model when N = i and h = 0. For the equivalent number of ideal CSTRs, N is obtained by minimizing the residual sum of squares of the deviation between the experimental F-curve and that predicted by Equation 8-109. The objective function is minimized from the expression

$$\Phi = \sum_{i=1}^{M} \left[F^E(\theta_i) - F^M(\theta_i) \right]^2 \tag{8-113}$$

where $F^E(\theta_i)$ = experimental F-curve data
   $F^M(\theta_i)$ = predicted F-curve from Equation 8-109

A one-dimensional search optimization technique, such as the Fibonacci search, is employed to minimize Equation 8-113. A computer program (PROG81) was developed to estimate the equivalent number of ideal tanks N for the given effluent tracer response versus time data. Additionally, the program calculates the mean residence time, variance, dimensionless variance, dispersion number, and the Peclet number.

The fractional tank extension model is simple to use and consists of only a single parameter to represent the non-ideality. It is useful for calculating reaction yields. If each of the tanks is a perfect mixer, the yield can be calculated from a set of algebraic equations. It is also useful in the direct fitting of experimental data. The model is well suited for visualizing various levels of micromixing. Zwietering [17] gives several examples for N = 2 and 3, which are extended to the fractional tank model. For N > 1, the true limit of maximum mixedness must be evaluated numerically. However, the greatest degree of micromixing corresponding to the physical model arises when each tank is a perfect mixer and the smallest tank is at the discharge end of the system. The drawback is that the individual extent of bypassing and dead regions cannot be estimated quantitatively using this method. Macromixing, micromixing, and maximum mixedness models are reviewed in Chapter 9.

THE DISPERSION MODEL

The design of chemical flow reactors depends on the overall rate at which the reaction proceeds, and the extent of backmixing occurring in the reactors. The first factor relates to the determination of the rate equations as reviewed in Chapter 2. The second factor, backmixing or dispersion, is used to represent the combined action of all phenomena, namely molecular diffusion, turbulent mixing, and non-uniform velocities, which give rise to a distribution of residence times in the reactor. The performance of the reactor and the degree of conversion attainable for a given feed and reactor volume differs from that which would be predicted assuming ideal plug flow (i.e., negligible diffusion), with the magnitude of the difference depending on the relative rates of diffusion and convection.

Suppose a tracer pulse is introduced into the fluid entering a system. If the reactor is an ideal plug flow, the tracer pulse traverses through the reactor without distortion (Figure 8-27), and emerges to give the characteristic ideal plug flow residence impulse response as shown in Figure 8-28. If diffusion occurs, the tracer spreads away from the center of the original pulse in both the upstream and downstream directions. At various times from the injection, the tracer occupies positions in the reactor as illustrated in Figure 8-29.

If the concentration of the tracer molecules in the reactor effluent are known, then the measured response will depend on the length of the reactor, the rate of diffusion, and mean fluid velocity. The response

Tracer cloud   Flat velocity profiles

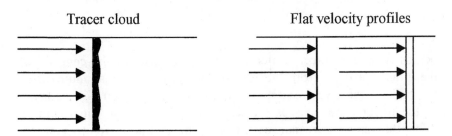

Figure 8-27. Plug flow

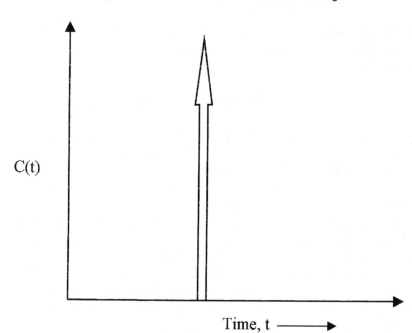

Figure 8-28. RTD response for a plug flow reactor.

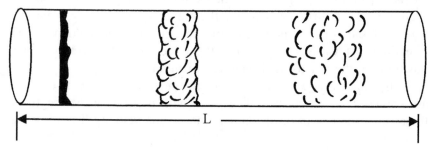

Figure 8-29. Dispersion of the tracer in a plug flow reactor.

begins when the tracer, which has diffused ahead of the centroid of the cloud, emerges from the reactor, builds up to a maximum when the bulk of the tracer emerges, and decreases as the trailing portion of the tracer signal passes the detector at the reactor outlet. Figure 8-30 illustrates the responses that may be measured for reactors of increasing length.

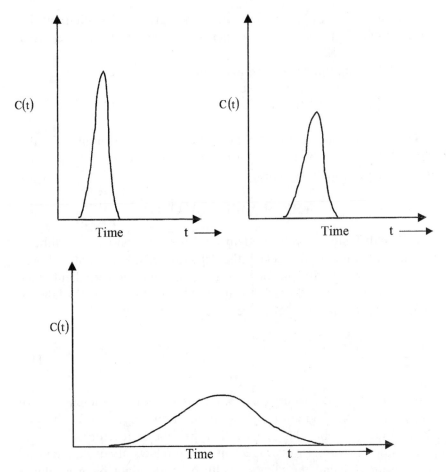

Figure 8-30. RTD curves at increasing tubular reactor length.

The distribution of tracer molecule residence times in the reactor is the result of molecular diffusion and turbulent mixing if the Reynolds number exceeds a critical value. Additionally, a non-uniform velocity profile causes different portions of the tracer to move at different rates, and this results in a spreading of the measured response at the reactor outlet. The dispersion coefficient D (m^2/sec) represents this result in the tracer cloud. Therefore, a large D indicates a rapid spreading of the tracer curve, a small D indicates slow spreading, and D = 0 means no spreading (hence, plug flow).

Flow patterns in the reactor can vary greatly. To characterize backmixing, of the longitudinal dispersion number, D/uL, is often used,

where D = longitudinal dispersion coefficient, m^2/sec
    u = fluid velocity, m/sec
    L = length of reactor, m

In a packed bed or flow in pipes, the dispersion number is also defined as D/ud, where d is the particle size in packed beds or the tube diameter in empty pipes.

THE MODEL EQUATION

Fick's diffusion law is used to describe dispersion. In a tubular reactor, either empty or packed, the depletion of the reactant and non-uniform flow velocity profiles result in concentration gradients, and thus dispersion in both axial and radial directions. Fick's law for molecular diffusion in the x-direction is defined by

$$\frac{\partial C}{\partial t} = D\frac{\partial^2 C}{\partial x^2}$$

(8-114)

where D is the longitudinal or axial dispersion coefficient, and it characterizes the degree of mixing during flow. Figure 8-31 shows a sketch of an annular element of this type tubular reactor.

Consider a steady flow of reactant A to products at constant density through an element of radius r, width δr, and height δl in a tubular reactor at isothermal condition. Suppose that radial and axial mass transfer is expressed by Fick's law, with $(D_e)_l$ and $(D_e)_r$ as effective diffusivities. The rate at which A reacts is $(-r_A)$, mol/m^3 sec. A material balance on a tubular element of radii r and r + δr and height δl is carried out from

$$\frac{\text{Input}}{\text{Rate}} = \frac{\text{Output}}{\text{Rate}} + \frac{\text{Disappearance}}{\text{by reaction}} + \text{Accumulation}$$

A enters element l by convection and dispersion. The output term is the rate of passage of A through the cross-section at 1 + δl. The disappearance by reaction is the volumetric rate of disappearance $(-r_A)$ times the volume of the differential element $2\pi r\delta r\delta l$.

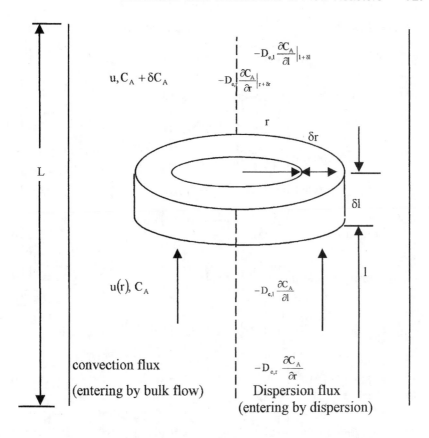

Input rate: $2\pi r \cdot \delta r\left(uC_{A,l} + N_{A,l}\right) + 2\pi r \cdot \delta l \cdot N_{A,r}$

Output rate: $2\pi r \cdot \delta r\left(uC_{A,l+\delta l} + N_{A,l+\delta l}\right) + \left(2\pi r + \delta r\right) \cdot \delta l \cdot N_{A,r+\delta r}$

Reactions term: $2\pi r \cdot \delta r \cdot \delta l\left(-r_A\right)_{\text{over element}}$

Accumulation rate: $2\pi r \cdot \delta r \cdot \delta l\left(\dfrac{\delta C_{A,\text{ over element}}}{\delta t}\right)$

Figure 8-31. Annular element of a tubular reactor.

$$\text{Input rate} = 2\pi r \cdot \delta r(uC_{A,l} + N_{A,l}) + 2\pi r \cdot \delta l \cdot N_{A,r} \qquad (8\text{-}115)$$

The flux N_A is the sum of the diffusive flux J_A with respect to the center of mass plus the contribution to the flux of A caused by the bulk mass movement.

$$N_A = J_A + Nx_A$$

where $J_{Ax} = -D_{AB}(\partial \rho / \partial x)$.

Output rate $= 2\pi r \cdot \delta r(uC_{A,l+\delta l} + N_{A,l+\delta l})$

$$+ (2\pi r + \delta r) \cdot \delta l \cdot N_{A,r+\delta r} \tag{8-116}$$

Rate of disappearance $= (-r_A)_{\text{over element}} \delta V_R$

$$\text{by reaction} = 2\pi r \delta r \delta l \, (-r_A)_{\text{over element}} \tag{8-117}$$

$$\text{Accumulation rate} = 2\pi r \delta l \delta r \frac{\delta C_A}{\delta t} \tag{8-118}$$

Combining Equations 8-115, 8-116, 8-117, and 8-118 gives

$$d\left[uC_A - (D_e)_l \frac{\partial C_A}{\partial l}\right](2\pi r \delta r) - 2\pi(D_e)_r \delta l \, d\left(r \frac{\partial C_A}{\partial r}\right)$$

$$+ 2\pi r \delta l \, \delta r\left[(-r_A) + \frac{\delta C_A}{\delta t}\right] = 0 \tag{8-119}$$

Rearranging Equation 8-119 as $\delta l \to 0$ gives

$$\frac{\partial C_A}{\partial t} = (D_e)_l \frac{\partial^2 C_A}{\partial l^2} + (D_e)_r\left(r \frac{\partial C_A}{\partial r}\right) - u\frac{\partial C_A}{\partial l} - (-r_A) \tag{8-120}$$

The effective axial dispersion coefficient $(D_e)_l$ and radial dispersion coefficient $(D_e)_r$ are assumed independent of concentration and position. Equation 8-120 is used for tubular flow of fluids in pipes and flow through packed beds where voidage is considered when defining $(D_e)_r$ and $(-r_A)$. Equation 8-120, which is analogous to Equations 6-105 and 6-106 was employed with other homogeneous and heterogeneous systems in which axial symmetry can be assumed and where the flow does not deviate significantly from plug flow. However, the model cannot be used when the flow characteristics deviate considerably from plug flow such as in a CSTR, bubbling fluidized bed, and in two-phase flow of gas-liquid systems.

AXIAL DISPERSED PLUG FLOW MODEL

When the effective radial dispersion coefficient $(D_e)_r$ can be neglected in comparison to the effective axial dispersion coefficient $(D_e)_l$, Equation 8-120 reduces

$$\frac{\partial C_A}{\partial t} = (D_e)_l \frac{\partial^2 C_A}{\partial l^2} - u \frac{\partial C_A}{\partial l} - (-r_A) \tag{8-121}$$

This model is referred to as the axial dispersed plug flow model or the longitudinal dispersed plug flow model. $(D_e)_r$ can be neglected relative to $(D_e)_l$ when the ratio of column diameter to length is very small and the flow is in the turbulent regime. This model is widely used for chemical reactors and other contacting devices.

In terms of the normalized coordinates

$$\theta = \frac{t}{\bar{t}} = \frac{tu}{L}, \ \ z = \frac{l}{L}$$

where L is the reactor length. Equation 8-121 then becomes

$$\frac{\partial C_A}{\partial \theta} = \frac{D_{e,l}}{uL} \frac{\partial^2 C_A}{\partial z^2} - \frac{\partial C_A}{\partial z} - (-r_A)\bar{t} \tag{8-122}$$

DIMENSIONAL ANALYSIS OF THE DISPERSION MODEL

It is often useful to write a model equation such as Equation 8-121 in terms of dimensionless variables. This introduces the Peclet number $N_{Pe} = uL/D_{e,l}$, which represents the ratio of characteristic dispersion time to characteristic convection time (residence time), and the Damköhler number,

$$N_{Da} = kC_{AO}^{n-1}\bar{t} = kC_{AO}^{n-1} \frac{L}{u} \tag{8-123}$$

which represents the ratio of characteristic convection time (residence time) to the characteristic process (reaction) time.

Assuming an nth order reaction at steady state condition, $(\partial C/\partial \theta) = 0$, Equation 8-122 becomes

$$\frac{D_{e,l}}{uL} \frac{\partial^2 C_A}{\partial z^2} = \frac{\partial C_A}{\partial z} + \left(-r_A \bar{t}\right) \qquad (8\text{-}124)$$

Where $(-r_A) = kC_A^n$, Equation 8-124 gives

$$\frac{D_{e,l}}{uL} \frac{d^2 C_A}{dz^2} = \frac{dC_A}{dz} + kC_A^n \bar{t}$$

or $\quad \dfrac{1}{N_{Pe}} \dfrac{d^2 C_A}{dz^2} = \dfrac{dC_A}{dz} + kC_A^n \bar{t} \qquad (8\text{-}125)$

Rearranging Equation 8-125 gives

$$\frac{d^2 C_A}{dz^2} = N_{Pe} \left[\frac{dC_A}{dz} + kC_A^n \, \bar{t} \right] \qquad (8\text{-}126)$$

When $f = C_A/C_{AO}$ (the dimensionless concentration), Equation 8-126 gives

$$C_{AO} \frac{d^2 f}{dz^2} = N_{Pe} \left[C_{AO} \frac{df}{dz} + kf^n C_{AO}^n \, \bar{t} \right]$$

or $\quad \dfrac{d^2 f}{dz^2} = N_{Pe} \left[\dfrac{df}{dz} + k\bar{t}C_{Ao}^{n-1} f^n \right] \qquad (8\text{-}127)$

where the Damkhöler number is

$$N_{Da} = k\bar{t}C_{AO}^{n-1} \qquad (8\text{-}128)$$

The dimensionless group $D_{e,l}/uL$ is known as the dispersion number and is the parameter that measures the extent of axial dispersion. The degree to which axial dispersion influences the performance of a chemical reactor is determined by the value of the Peclet number (N_{Pe}). A high value of N_{Pe} corresponds to a slightly dispersed reactor. That is,

$$\lim_{D_{e,l} \to 0} N_{Pe} = \lim_{D_{e,l} \to 0} \frac{uL}{D_{e,l}} = \infty$$

signifying an ideal plug flow. Similarly, a low value of N_{Pe},

$$\lim_{D_{e,l} \to \infty} N_{Pe} = \lim_{D_{e,l} \to \infty} \frac{uL}{D_{e,l}} = 0$$

represents a high degree of backmixing.

The concentration of A is uniform throughout the reaction volume, and the dispersion model becomes a well-stirred tank model. Although the validity of the dispersion model depends to a great extent on the process, as a rule-of-thumb the dispersion model may be applied with confidence as long as $N_{Pe} > 20$. It should be used with great caution when N_{Pe} falls below this value. It has been shown that the tank-in-series (TIS) model gives a good representation of a wide range of mixing phenomena. Small values of N are appropriate for agitated tanks and large values for columns. In contrast, the dispersion plug flow (DPF) model depicts mixing in columns better than the TIS. However, this involves additional mathematical complexity. Generally, both the dispersion model and the tanks-in-series models are roughly equivalent, and either can be employed for representing flow in real vessels.

Boundary Conditions of the Dispersion Model

The two sets of boundary conditions for Equations 8-122 and 8-126 are:

$$\text{Closed ends} \quad uC_{AO} = \left(uC_A - D_{e,l} \frac{\partial C_A}{\partial z} \right)_{z=0} \text{ and } \left(\frac{\partial C_A}{\partial z} \right)_{z=1} = 0 \text{ and}$$

$$\text{Open ends} \quad C_A(\pm\infty, t) = 0, \quad C_A(0,t) = C_A(t) \qquad (8\text{-}129)$$

Closed ends apply to reactor operation and open ends are used with tracer studies when tracer is injected and sampled some distance from the ends of the vessel.

Langmuir [18] first proposed the axial dispersion model and obtained steady state solutions from the following boundary conditions:

$$uC_{z \to 0^-} = uC_{z \to 0^+} - D_{e,1}\left(\frac{\partial C}{\partial z}\right)_{z \to 0^+}$$

$$\left(\frac{\partial C}{\partial z}\right)_{z=L} = 0 \quad \text{for all } t \tag{8-130}$$

Danckwerts [2] also obtained steady state solutions based on the same boundary conditions and various studies have since been performed by Taylor [19], Aris [20], and Levenspiel and Smith [21].

CORRELATION FOR AXIAL DISPERSION COEFFICIENTS

A dimensionless group that frequently occurs in this context is the Bodenstein number expressed as

$$N_{Bo} = N_{Re} \cdot N_{Sc} = \frac{d_t u \rho}{\mu} \cdot \frac{\mu}{\rho D_{e,1}} = \frac{u d_t}{D_{e,1}} \tag{8-131}$$

The dispersion model has been successfully employed in modeling the behavior of packed bed reactors. In this case,

$$N_{Bo} = \frac{u \cdot d_p}{6 \varepsilon D'_{e,1}} \tag{8-132}$$

where u/ε = the average interstitial velocity in the bed
$\quad d_t$ = pipe diameter
$\quad d_p$ = mean particle diameter
$\quad u$ = fluid velocity
$\quad \varepsilon$ = void fraction (porosity) of the bed
$\quad D'_{e,1}$ = effective dispersion coefficient based on actual cross-sectional area available for fluid transport

The inverse of the Bodenstein number is $\varepsilon D'_{e,1}/u \cdot d_p$, sometimes referred to as the intensity of dispersion. Himmelblau and Bischoff [5], Levenspiel [3], and Wen and Fan [6] have derived correlations of the Peclet number versus Reynolds number. Wen and Fan [6] have summarized the correlations for straight pipes, fixed and fluidized beds, and bubble towers. The correlations involve the following dimensionless groups:

$$N_{Pe} = \frac{D_{e,l}}{ud} \quad \text{or} \quad \frac{\varepsilon D'_{e,l}}{ud_p} \quad \text{Peclet number} \tag{8-133}$$

$$N_{Re} = \frac{d_t u \rho}{\mu} \quad \text{Reynolds number (empty tubes)} \tag{8-134}$$

$$= \frac{d_p u \rho}{\mu} \quad \text{Particle Reynolds number (packed beds)}$$

$$N_{Sc} = \frac{\mu}{\rho D_{mol}} \quad \text{Schmidt number} \tag{8-135}$$

where d = the flow channel diameter for an empty tube
        d_p = mean particle diameter for packed beds
        μ = fluid viscosity
        ρ = fluid density
   D_{mol} = molecular diffusivity of the reactant (or tracer) in the fluid
           ($D_{mol} \approx 10^{-5}$ cm²/sec, for liquids, 1 cm²/sec for gases)

Knowing the viscosity and density of the reaction mixture, the flow channel diameter, void fraction of the bed, and the superficial fluid velocity, it is possible to determine the Reynolds number, estimate the intensity of dispersion from the appropriate correlation, and use the resulting value to determine the effective dispersion coefficient $D_{e,l}$ or $D'_{e,l}$. Figures 8-32 and 8-33 illustrate the correlations for flow of fluids in empty tubes and through pipes in the laminar flow region, respectively. The dimensionless group $D_{e,l}/\bar{u}d_t = D_{e,l}/2\bar{u}R$ depends on the Reynolds number (N_{Re}) and on the molecular diffusivity as measured by the Schmidt number (N_{Sc}). For laminar flow region, $D_{e,l}$ is expressed by:

$$D_{e,l} = D_{mol} + \frac{\bar{u}^2 d_t^2}{192 D_{mol}}$$

$$\text{or} \quad \frac{1}{N_{Pe}} = \frac{1}{N_{Re} \cdot N_{Sc}} + \frac{N_{Re} \cdot N_{Sc}}{192} \tag{8-136}$$

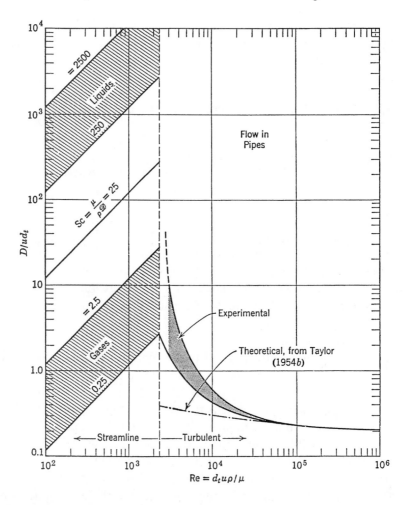

Figure 8-32. Correlation for the dispersion of fluids flowing in pipes. *(Source: Levenspiel, O.,* Ind. Eng. Chem., *50, 343, 1958.)*

The axial dispersion model also gives a good representation of fluid mixing in packed-bed reactors. Figure 8-34 depicts the correlation for flow of fluids in packed beds.

Small Amount of Dispersion ($D_{e,l}/uL < 0.01$)

For a relatively small amount of dispersion or large value of N_{Pe}, the tracer curve is very narrow. Consequently, the spreading of the

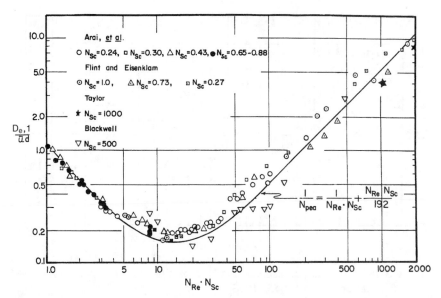

Figure 8-33. Correlation of axial dispersion coefficient for flow of fluids through pipes in laminar flow region ($N_{Re} < 2,000$). *(Source: Wen, C. Y. and Fan, L. T.,* Models for Flow Systems and Chemical Reactors, *Marcel Dekker Inc., 1975.)*

tracer curve does not significantly change as it passes the measuring point. As a good approximation, it is symmetrical and Gaussian as shown in Figure 8-35. The axial dispersion model is expressed as

$$C = \frac{1}{2\sqrt{\pi\left(\dfrac{D_{e,1}}{uL}\right)}} \exp\left[\frac{-(1-\theta)^2}{4\left(\dfrac{D_{e,1}}{uL}\right)}\right] \qquad (8\text{-}137)$$

Levenspiel [3] has given a family of Gaussian curves, and the equations representing these curves are given in Table 8-7.

Large Amount of Dispersion ($D_{e,1}/uL > 0.01$)

For a large amount of dispersion or small value of N_{Pe}, the pulse response is broad, and it passes the measurement point slowly enough for changes to occur in the shape of the tracer curve. This gives a non-symmetrical E-curve.

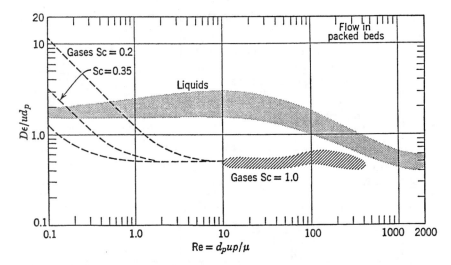

Figure 8-34. Correlation of dispersion to Reynolds number. *(Source: Levenspiel, O., Ind. Eng. Chem., 50, 343, 1958.)*

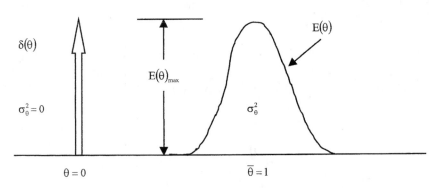

Figure 8-35. Guassian response to pulse input for small dispersion.

Two types of boundary conditions are considered, the closed vessel and the open vessel. The closed vessel (Figure 8-36) is one in which the inlet and outlet streams are completely mixed and dispersion occurs between the terminals. Piston flow prevails in both inlet and outlet piping. For this type of system, the analytic expression for the E-curve is not available. However, van der Laan [22] determined its mean and variance as

Table 8-7

$$E(\theta) = \bar{t}E(t) = \frac{1}{\sqrt{4\pi\left(\dfrac{D_{e,l}}{uL}\right)}} \exp\left[\frac{-(1-\theta)^2}{4\left(\dfrac{D_{e,l}}{uL}\right)}\right]$$

$$E(t) = \sqrt{\frac{u^3}{4\pi D_{e,l}L}} \exp\left[\frac{-(L-ut)^2}{4\dfrac{D_{e,l}}{uL}}\right]$$

$$\bar{t}_E = \frac{V}{\upsilon} = \frac{L}{\upsilon} \text{ or } \bar{\theta}_E = 1$$

$$\sigma_\theta^2 = \frac{\sigma_t^2}{\bar{t}^2} = 2\left(\frac{D_{e,l}}{uL}\right) \text{ or } \sigma^2 = 2\left(\frac{D_{e,l}L}{u^3}\right)$$

$\delta(\theta)$

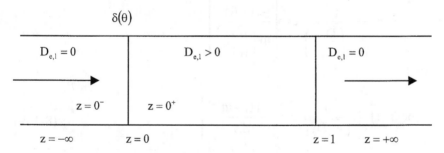

Figure 8-36. Flow conditions for closed-closed system.

$$\bar{t}_E = \bar{t} = \frac{V}{\upsilon} \text{ or } \theta_E = \frac{\bar{t}_E}{\bar{t}} = \frac{\bar{t}_E \upsilon}{V} = 1$$

$$\sigma_\theta^2 = \frac{\sigma_t^2}{\bar{t}} = 2\left(\frac{D_{e,l}}{uL}\right) - 2\left(\frac{D_{e,l}}{uL}\right)^2\left[1 - e^{\frac{-uL}{D_{e,l}}}\right] \qquad (8\text{-}138)$$

For an open vessel (Figure 8-37), diffusion can occur across the system boundaries. It is assumed that the dispersion coefficient $D_{e,l}$

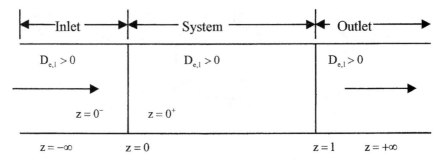

Figure 8-37. Flow conditions for open-open system.

or the Peclet number N_{Pe} is the same from just before the inlet to just beyond the outlet. The open vessel represents a convenient experimental device as a section of a long pipe. Levenspiel and Smith [21] derived the mean and variance in the $E(\theta)$ as:

$$E(\theta) = \frac{1}{\sqrt{4\pi\left(\dfrac{D_{e,l}}{uL}\right)}}\; \exp\left[\frac{-(1-\theta)^2}{4\theta\left(\dfrac{D_{e,l}}{uL}\right)}\right]$$

$$\text{or } E(t) = \frac{u}{\sqrt{4\pi D_{e,l}}}\; \exp\left[\frac{-(L-ut)^2}{4D_{e,l}}\right] \tag{8-139}$$

The moments of the solution are:

$$\theta_E = \frac{\bar{t}_E}{\bar{t}} = 1 + 2\left(\frac{D_{e,l}}{uL}\right)$$

$$\text{or } t_E = \frac{V}{\upsilon}\left[1 + 2\frac{D_{e,l}}{uL}\right]$$

$$\sigma_\theta^2 = \frac{\sigma_t^2}{\bar{t}^2} = 2\frac{D_{e,l}}{uL} + 8\left(\frac{D_{e,l}}{uL}\right)^2 \tag{8-140}$$

If $D_{e,l}/uL$ is small enough so that $(D_{e,l}/uL)^2$ can be neglected, then

$$\sigma_\theta^2 = 2\frac{D_{e,l}}{uL}$$

or $\quad \dfrac{D_{e,l}}{uL} = \dfrac{1}{2}\sigma_\theta^2 \qquad\qquad\qquad\qquad (8\text{-}141)$

Open-Closed and Closed-Open Vessels

van der Laan [22] developed equations for the mean and variance in terms of the dispersion coefficient $D_{e,l}/uL$ as

$$\overline{\theta}_{E_{OC}} = \overline{\theta}_{E_{CO}} = 1 + \frac{D_{e,l}}{uL} \quad \text{or} \quad \overline{t}_{E_{OC}} = \overline{t}_{E_{CO}} = \frac{V}{\upsilon}\left(1 + \frac{D_{e,l}}{uL}\right)$$

$$\sigma_{\theta_{OC}}^2 = \sigma_{\theta_{CO}}^2 = \frac{\sigma_{t_{CO}}^2}{\overline{t}^2} = \frac{\sigma_{t_{OC}}^2}{\overline{t}^2} = 2\left(\frac{D_{e,l}}{uL}\right) + 3\left(\frac{D_{e,l}}{uL}\right)^2 \qquad (8\text{-}142)$$

Table 8-8 summarizes the results of the cases discussed above including the boundary conditions, the expression for $C(\theta)$ at $z = 1$, and the mean and variance for $C(\theta)$.

The criterion for the validity of Equation 8-141 is $N_{Pe} \gg 1.0$. A rough rule-of-thumb is $N_{Pe} > 10$. If this condition is not satisfied, the correct equation depends on the boundary conditions at the inlet and outlet. A procedure for determining dispersion coefficient $D_{e,l}$ is as follows:

1. Measure the outlet response to a tracer impulse injected at the inlet.
2. Calculate the mean μ and the variance σ^2 using Equations 8-40 and 8-43, respectively.
3. Estimate the Peclet number from Equation 8-141. If the value of N_{Pe} is 10 or greater, accept it. Otherwise, use a refinement of the theory, which accounts for the boundary conditions at the outlet, or formulate another model.
4. Determine the mean (superficial) fluid velocity, $\overline{u}$, as the volumetric flowrate divided by the flow channel cross-section.
5. Estimate the dispersion coefficient $D_{e,l}$ from the definition of the Peclet number.

Table 8-8
Comparison of solutions of the axially dispersed plug flow model for different boundary conditions

Case	Boundary conditions	Reduced concentration at z = 1, C(θ) at z = 1	Moments mean $\bar{\theta}$	Variance σ_θ^2
Gaussian	—	$\dfrac{1}{2}\left(\sqrt{\dfrac{N_{Pe}}{\pi}}\right)\exp\left[\dfrac{-N_{Pe}(1-\theta)^2}{4}\right]$ Levenspiel	1	$\dfrac{2}{N_{Pe}}$
+Closed-Open	$z=0,\ \delta(\theta)=C(\theta)-\dfrac{1}{N_{Pe}}\dfrac{dC}{d\theta}$ $z\to\infty,\ C\to 0$ $\theta\le 0,\ C=0$	$\left(\dfrac{N_{Pe}}{\pi\theta}\right)^{0.5}\exp\left[\dfrac{-N_{Pe}(1-\theta)^2}{4\theta}\right]-N_{Pe}\exp(N_{Pe})\,\mathrm{erfc}(A)$ $A=\left(\dfrac{N_{Pe}}{\theta}\right)^{0.5}\left(\dfrac{1+\theta}{2}\right)$ Villermaux and van Swaaij	$1+\dfrac{1}{N_{Pe}}$	$\dfrac{2}{N_{Pe}}+\dfrac{3}{N_{Pe}^2}$
Closed-Closed	$z=0\ \ \delta(\theta)=C(\theta)-\dfrac{1}{N_{Pe}}\dfrac{\partial C(\theta)}{\partial z}$ $z=1\ \dfrac{\partial C(\theta)}{\partial z}=0\quad \theta\le 0,\ C(\theta)=0$	Complex expression Otake and Kunigita	1	$\dfrac{2N_{Pe}-2+2e^{-N_{Pe}}}{N_{Pe}^2}$
Open-Open	$z=-\infty,\ \dfrac{dC}{dz}=0$ $z=+\infty,\ \dfrac{dC}{dz}=0$	$\dfrac{1}{2}\left(\dfrac{N_{Pe}}{\pi\theta}\right)^{0.5}\exp\left[\dfrac{-N_{Pe}(1-\theta)^2}{4\theta}\right]$ Levenspiel and Smith	$1+\dfrac{2}{N_{Pe}}$	$\dfrac{2}{N_{Pe}}+\dfrac{8}{N_{Pe}^2}$

Figure 8-38 shows the residence time distributions of some commercial and fixed bed reactors. These shapes can be compared with some statistical distributions, namely the Gamma (or Erlang) and the Gaussian distribution functions. However, these distributions are represented by limited parameters that define the asymmetry, the peak,

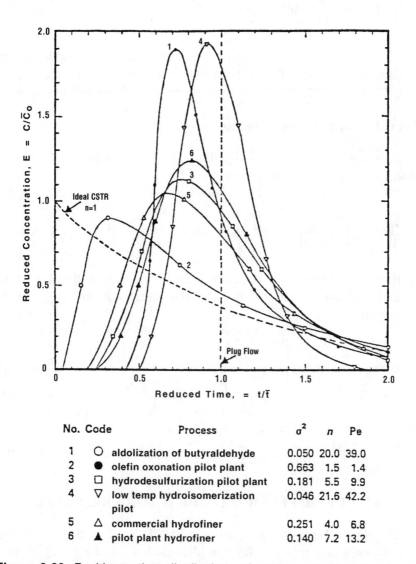

No.	Code	Process	σ^2	n	Pe
1	○	aldolization of butyraldehyde	0.050	20.0	39.0
2	●	olefin oxonation pilot plant	0.663	1.5	1.4
3	□	hydrodesulfurization pilot plant	0.181	5.5	9.9
4	▽	low temp hydroisomerization pilot	0.046	21.6	42.2
5	△	commercial hydrofiner	0.251	4.0	6.8
6	▲	pilot plant hydrofiner	0.140	7.2	13.2

Figure 8-38. Residence time distributions of some commercial and fixed bed reactors. The variance, equivalent number of CSTR stages, and Peclet number are given for each reactor. *(Source: Walas, S. M.,* Chemical Process Equipment—Selection and Design, *Butterworths, 1990.)*

and the shape in the vicinity of the peak. The parameters are the moments, variance, skewness, and kurtosis. Wen and Fan [6] have developed moments of RTDs of various flow models. The Gamma distribution function for non-integral values is given by

$$E(t)_{gamma} = \frac{N^N}{\Gamma(N)} \theta^{N-1} \exp(-N\theta)$$
(8-143)

The value of N is the only parameter in Equation 8-143. It can be computed from the RTD experimental data from its variance (Equation 8-107).

Gaussian Distribution Function

A well-known statistical distribution is the normal or Gaussian distribution and is expressed by

$$f(\theta) = \frac{1}{\sigma\sqrt{2\pi}} \exp\left[\frac{-(\theta-1)^2}{2\sigma^2}\right] \quad -\infty \le \theta \le +\infty$$
(8-144)

Because only positive values of θ are of concern in the RTD, this function is normalized by dividing the integral from 0 to ∞ with the result

$$E(\theta)_{gauss} = f(\sigma)\exp\left[\frac{-(\theta-1)^2}{2\sigma^2}\right]$$
(8-145)

$$\text{where } f(\sigma) = \frac{\sqrt{2/\pi\sigma^2}}{1 + \text{erf}(1/\sigma\sqrt{2})}$$
(8-146)

EFFECT OF DISPERSION ON REACTOR PERFORMANCE

The axial dispersion plug flow model is used to determine the performance of a reactor with non-ideal flow. Consider a steady state reacting species A, under isothermal operation for a system at constant density; Equation 8-121 reduces to a second order differential equation:

$$\frac{D_{e,l}}{uL} \frac{d^2C_A}{dz^2} - \frac{dC_A}{dz} - (-r_A)\bar{t} = 0 \tag{8-147}$$

For a first order reaction $(-r_A) = kC_A$, and Equation 8-147 is then linear, has constant coefficients, and is homogeneous. The solution of Equation 8-147 subject to the boundary conditions of Danckwerts and Wehner and Wilhelm [23] for species A gives

$$\frac{C_A}{C_{AO}} = 1 - X_A = \frac{4a\exp\left[\dfrac{1}{2}\dfrac{uL}{D_{e,l}}\right]}{(1+a)^2\exp\left[\dfrac{a}{2}\dfrac{uL}{D_{e,l}}\right] - (1-a)^2\exp\left[\dfrac{-a}{2}\dfrac{uL}{D_{e,l}}\right]} \tag{8-148}$$

where $a = \left[1 + 4k\bar{t}\left(\dfrac{D_{e,l}}{uL}\right)\right]^{0.5}$

In terms of the Peclet number N_{Pe}, Equation 8-148 becomes

$$\frac{C_A}{C_{AO}} = 1 - X_A = \frac{4a\exp\left[\dfrac{1}{2}N_{Pe}\right]}{(1+a)^2\exp\left[\dfrac{a}{2}N_{Pe}\right] - (1-a)^2\exp\left[\dfrac{-a}{2}N_{Pe}\right]} \tag{8-149}$$

where $a = \left[1 + \dfrac{4k\bar{t}}{N_{Pe}}\right]^{0.5}$

Equations 8-148 and 8-149 give the fraction unreacted C_A/C_{AO} for a first order reaction in a closed axial dispersion system. The solution contains the two dimensionless parameters, N_{Pe} and $k\bar{t}$. The Peclet number controls the level of mixing in the system. If $N_{Pe} \to 0$ (either small u or large $D_{e,l}$), diffusion becomes so important that the system acts as a perfect mixer. Therefore,

$$\frac{C_A}{C_{AO}} = \frac{1}{1 + k\bar{t}} \quad \text{as } N_{Pe} \to 0 \tag{8-150}$$

As $N_{Pe} \to \infty$ (large u or small $D_{e,l}$), the system behaves as a piston flow reactor so that

$$\frac{C_A}{C_{AO}} = e^{-k\bar{t}} \text{ as } N_{Pe} \to \infty \qquad (8\text{-}151)$$

Levenspiel and Bischoff [24] compared the fraction unreacted by the dispersion model to the solution with that for plug flow:

$$\frac{C_A}{C_{AO}} = 1 - e^{-k\bar{t}} = 1 - e^{-kL/u} \qquad (8\text{-}152)$$

They further determined the ratio of the dispersion reactor volume to the plug flow reactor volume required to accomplish the same degree of conversion for several values of the dimensionless dispersion parameter $D_{e,l}/uL$. Figure 8-39 shows the results of Equation 8-147

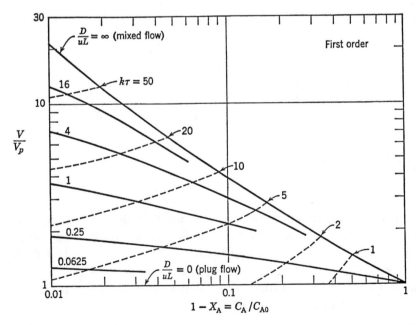

Figure 8-39. Comparison of real and plug flow reactors for the first order reaction A → Products [24].

for a first order reaction. The abscissa is $(C_A)_{outlet}/(C_A)_{inlet}$ or $1 - X_A$, and the ordinate is the ratio of the dispersed plug flow reactor volume to the volume of an ideal plug flow reactor (PFR), which will be required to achieve the same conversion. The extent to which axial dispersion influences the reactor performance is determined by the value of a dimensionless group (Figure 8-32) as $D_{e,l}/ud$. For high conversions and large $D_{e,l}/uL$, significantly larger reactors are required than are predicted by a plug flow analysis. At large $D_{e,l}/uL$, the use of the dispersion model is inappropriate; therefore, a lower segment of Figure 8-39 is feasible for design calculations. Figure 8-39 shows that as the rate of dispersion increases, the reactor volume required to achieve a fixed conversion increases from a minimum value for ideal plug flow ($D_{e,l} = 0$) to a maximum value for a perfect mixer ($D_{e,l} = \infty$). Figure 8-40 shows the results for second order kinetics, which is also most appropriate when $D_{e,l}/uL$ is small. For a small dispersion number ($D_{e,l}/uL \ll 1$) and ($D_{e,l}/uL \cdot k\bar{t} \ll 1$),

$$\frac{C_A}{C_{AO}} = \exp\left[-k\bar{t} + (k\bar{t})^2 \frac{D_{e,l}}{uL}\right] = \exp\left[-k\bar{t} + \frac{k^2\sigma^2}{2}\right] \tag{8-153}$$

$$\text{or } \ln\left(\frac{C_A}{C_{AO}}\right) = -k\bar{t} + (k\bar{t})^2 \frac{D_{e,l}}{uL} \tag{8-154}$$

Using Equation 8-154 for the same reactor space time and volume gives

$$\ln\left(\frac{C_{A,\text{dispersion}}}{C_{A,\text{plug flow}}}\right) = (k\bar{t})^2 \frac{D_{e,l}}{uL} \tag{8-155}$$

Equation 8-155 shows that the conversion in the dispersion reactor will always be less than that of the plug flow reactor ($C_{A,\text{dispersion}} > C_{A,\text{plug flow}}$). In the case where the effluent composition is fixed instead of the reactor size, Equations 8-152 and 8-154 can be manipulated to show that for small $D_{e,l}/uL$,

$$\frac{L_{\text{dispersion}}}{L_{\text{plug flow}}} = \frac{V_{\text{dispersion}}}{V_{\text{plug flow}}} = 1 + (k\bar{t})\frac{D_{e,l}}{uL} \tag{8-156}$$

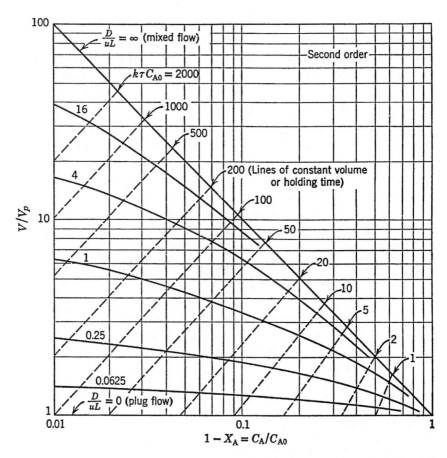

Figure 8-40. Comparison of real and plug flow reactors for second order reactions [24].

COMPARISION OF TANK IN SERIES (TIS) AND DISPERSION PLUG FLOW (DPF) MODELS

Both the tank in series (TIS) and the dispersion plug flow (DPF) models require tracer tests for their accurate determination. However, the TIS model is relatively simple mathematically and thus can be used with any kinetics. Also, it can be extended to any configuration of compartments with or without recycle. The DPF axial dispersion model is complex and therefore gives significantly different results for different choices of boundary conditions.

For a large dispersion (small N_{Pe}), the TIS model is preferred. The TIS model provides a mechanism that is readily visualized and often used. The Gamma and dispersion models can be related by equating the variances of their respective $E(\theta)$ functions. This gives

$$\sigma_\theta^2 = \frac{2\left[N_{Pe} - 1 + \exp(-N_{Pe})\right]}{N_{Pe}^2}$$

$$= \frac{1}{N_{erlang}} \tag{8-157}$$

For both large and small values of N_{Pe},

$$N_{erlang} = \frac{N_{Pe}}{2} \tag{8-158}$$

RESIDENCE TIME DISTRIBUTION IN A STATIC MIXER

Static mixers, as reviewed in Chapter 7, contain mixing elements enclosed in a tubular housing through which radial mixing is achieved. They redistribute fluid across the flow channel and consequently rearrange temperature and composition distributions. They are often used to promote mixing in laminar flow systems thus having a pronounced effect on the RTD.

A major unknown variable in reactor design is the scale of imperfection in the flow pattern. This is caused by various factors such as the existence of stagnant regions, channeling of fluid through the system, or recycling of the fluid within the vessel. It is important that these anomalies are considered during scale-up to avoid errors in the final design. Therefore, complete knowledge of what happens to every molecule as it passes through the vessel is necessary to design an effective mixer. This can be acquired by having a complete velocity distribution of the molecules in the mixer.

Computational fluid dynamics (CFD) is an effective tool in analyzing the velocity distribution and other pertinent parameters in a static mixer. CFD is also being recognized as an effective tool in enhancing the performance of mixers and reactors, which is reviewed in Chapter 10.

A practical method of predicting the molecular behavior within the flow system involves the RTD. A common experiment to test non-uniformities is the stimulus response experiment. A typical stimulus is a step-change in the concentration of some tracer material. The step-response is an instantaneous jump of a concentration to some new value, which is then maintained for an indefinite period. The tracer should be detectable and must not change or decompose as it passes through the mixer. Studies have shown that the flow characteristics of static mixers approach those of an ideal plug flow system. Figures 8-41 and 8-42, respectively, indicate the exit residence time distributions of the Kenics static mixer in comparison with other flow systems.

The dispersed plug flow model has been successfully applied to describe the flow characteristics in the Kenics mixer. The complex flow behavior in the mixer is characterized by the one-parameter. The Peclet number, N_{Pe}, is defined by:

$$N_{Pe} = \frac{uL}{D_{e,l}} \tag{8-159}$$

where u = average linear flow velocity in the mixer
       L = length of the mixer
       $D_{e,l}$ = axial dispersion coefficient

For an ideal plug flow system, there is no axial dispersion, hence $D_{e,l} = 0$ and $N_{Pe} \rightarrow \infty$. In contrast, for an ideal backmix flow system, $D_{e,l} \rightarrow \infty$ and $N_{Pe} = 0$. The Peclet number for all flow systems lies somewhere between these ideal conditions. A flow system with $N_{Pe} > 100$ is considered to be plug flow, and as indicated in Figure 8-42, the step response curve of the Kenics mixer corresponds to an average Peclet number of 145. Therefore, it is inferred that the unit is nearly a plug flow device. Additionally, the plug flow characteristics are attributed to the thorough radial mixing, which reduces the spread of the residence times.

The Kenics mixer has been shown to provide thorough radial mixing. This results in a reduction in radial gradients in velocity, composition, and temperature. Because the unit possesses nearly plug flow characteristics, both temperature and product quality controls are achieved.

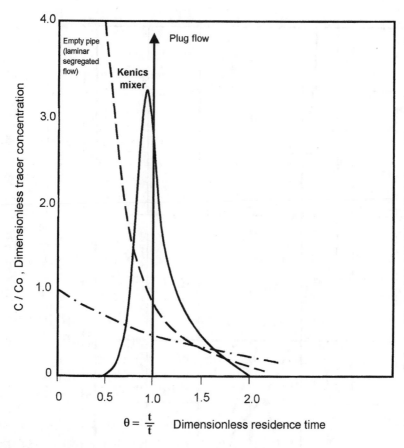

$$\theta = \frac{t}{\bar{t}}$$ Dimensionless residence time

Figure 8-41. E-Function: Exit residence time distrubtion for the Kenics Mixer and other flow systems.

Example 8-5

Using the experimental residence time distribution data of Levenspiel and Smith in Example 8-2, determine the number of ideal tanks N, the variance, dispersion number, and Peclet number.

Solution

The computer program PROG81 estimates the equivalent number of ideal tanks N for the given experimental residence time distribution

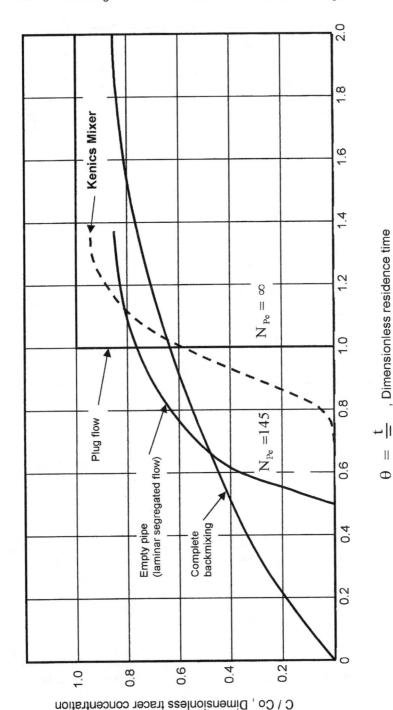

Figure 8-42. F-Function: Exit residence time distrubtion for the Kenics Mixer and other flow systems.

data by minimizing Equation 8-103. The input data are the t versus
E(t) data, upper and lower limits of the parameter, and maximum
number of Fibonacci searches. The lower limit of the parameter is
zero. The upper limit is an arbitrary number as high as 10 or 25,
depending on the shape of the E-curve and how closely it approaches
plug flow behavior. Table 8-9 gives the detailed computer results and
Figure 8-43 shows $F_{Exp}(\theta)$ and $F_{Model}(\theta)$ versus θ.

Table 8-9
Tracer concentration versus time data

TIME	CONCENTRATION
.000	.000
2.000	11.000
4.000	53.000
6.000	64.000
8.000	58.000
10.000	48.000
12.000	39.000
14.000	29.000
16.000	22.000
18.000	16.000
20.000	11.000
22.000	9.000
24.000	7.000
26.000	5.000
28.000	4.000
30.000	2.000
32.000	2.000
34.000	2.000
36.000	1.000
38.000	1.000
40.000	1.000
42.000	1.000

TIME	CONCENTRATION	E(T)
.000	.000	.000
2.000	11.000	.014
4.000	53.000	.069
6.000	64.000	.083
8.000	58.000	.075
10.000	48.000	.062
12.000	39.000	.051
14.000	29.000	.038
16.000	22.000	.028
18.000	16.000	.021
20.000	11.000	.014
22.000	9.000	.012
24.000	7.000	.009
26.000	5.000	.006
28.000	4.000	.005
30.000	2.000	.003
32.000	2.000	.003
34.000	2.000	.003
36.000	1.000	.001
38.000	1.000	.001
40.000	1.000	.001
42.000	1.000	.001

Table 8-9
(continued)

THETA	DIM-ETHETA	FTHETA
.000	.000	.000
.182	.157	.014
.363	.756	.097
.545	.913	.249
.726	.828	.407
.908	.685	.544
1.089	.556	.657
1.271	.414	.745
1.452	.314	.811
1.634	.228	.860
1.816	.157	.895
1.997	.128	.921
2.179	.100	.942
2.360	.071	.957
2.542	.057	.969
2.723	.029	.977
2.905	.029	.982
3.087	.029	.987
3.268	.014	.991
3.450	.014	.994
3.631	.014	.996
3.813	.014	.999

THE MEAN RESIDENCE TIME:	11.016
THE VARIANCE:	48.067
THE DIMENSIONLESS VARIANCE:	.396
THE DISPERSION NUMBER:	.198
PECLET NUMBER:	5.05

MIXING STAGES FOR A NON-IDEAL SYSTEM RTD

FRACTIONAL TANK EXTENSION MODEL

MAXIMUM NUMBER OF SEARCHES:	25
UPPER BOUND:	20.00
LOWER BOUND:	.00

THETA	EXPT. F-CURVE	MODEL F-CURVE
.000	.000	0.000
.182	.014	.024
.363	.097	.116
.545	.249	.252
.726	.407	.396
.908	.544	.529
1.089	.657	.641
1.271	.745	.732
1.452	.811	.802
1.634	.860	.856
1.816	.895	.896
1.997	.921	.926
2.179	.942	.947
2.360	.957	.963
2.542	.969	.974
2.723	.977	.982
2.905	.982	.987
3.087	.987	.991
3.268	.991	.994
3.450	.994	.996
3.631	.996	.997
3.813	.999	.998

EQUIVALENT NUMBER OF IDEALLY MIXED STAGE:	2.31
ROOT MEAN SQUARE DEVIATION:	0.8289E-02

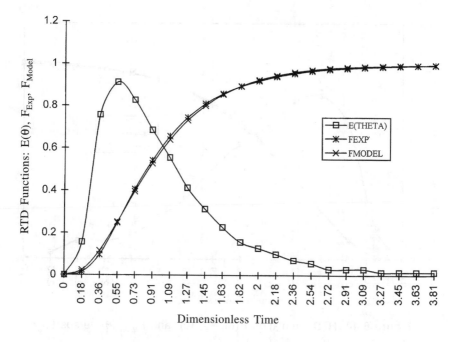

Figure 8-43. RTD simulation $E(\theta)$, $F_{Exp}(\theta)$, and $F_{Model}(\theta)$ versus θ.

Example 8-6

Using the experimental residence time distribution data of Hull and von Rosenberg in Example 8-3, determine the number of ideal tanks N, the variance, dispersion number, and Peclet number.

Solution

Computer program PROG81 determines the number of tanks, the variance, dispersion number, and the Peclet number from Hull and von Rosenberg data. The results of the simulation suggest that about three stirred tanks in series are equivalent to the RTD response curve. Figure 8-44 shows the shows $E(\theta)$, $F_{Exp}(\theta)$, and $F_{Model}(\theta)$ versus θ.

Example 8-7

Tracer impulse data of a commercial hydrodesulfurizer (Sherwood, *A Course in Process Design*, MIT Press, 1963) with 10 mm catalyst pellets are given in the following table.

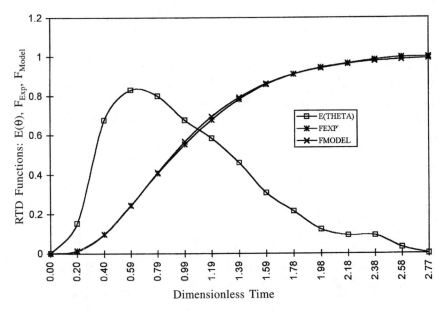

Figure 8-44. RTD simulation $E(\theta)$, $F_{Exp}(\theta)$, and $F_{Model}(\theta)$ versus θ.

Time	Concentration
7.5	0.0
10.0	1.2
12.5	4.5
17.5	14.2
20.0	40.7
22.5	46.4
25.0	32.9
27.5	23.5
30.0	16.6
32.5	11.8
35.0	8.5
37.5	6.0
40.0	4.1
42.5	2.6
45.0	1.5
47.5	0.4
50.0	0.0

Determine the various response functions.

Solution

An Excel spreadsheet (Example8-7.xls) was used to determine the various RTD functions and the computer program PROG81 was used to simulate the model response curve with the experimental data. The results show the equivalent number of ideally mixed stages (nCSTRs) for the RTD is 13.2. The Gamma distribution function from Equation 8-143 is:

$$E(t)_{Gamma} = \frac{N^N}{\Gamma(N)} \theta^{N-1} \exp(-N\theta) \tag{8-143}$$

$$E(t)_{Gamma} = \frac{13.2^{13.2}}{\Gamma(13.2)} \theta^{12.2} \exp(-13.2\theta)$$

The dispersion number is 0.039 and the Peclet number (N_{Pe}) is 26. Figure 8-45 shows $F_{Exp}(\theta)$ and $F_{Model}(\theta)$ versus θ.

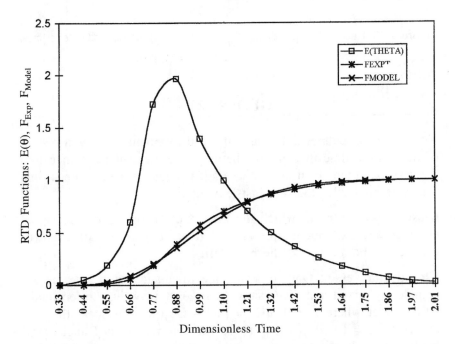

Figure 8-45. RTD simulation $E(\theta)$, $F_{Exp}(\theta)$, and $F_{Model}(\theta)$ versus θ.

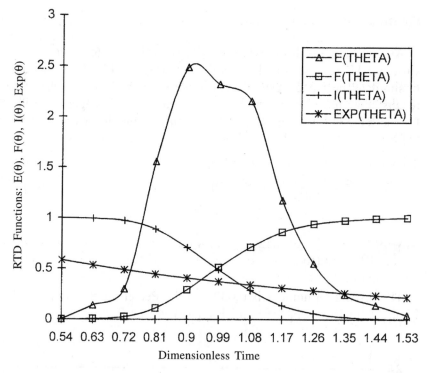

Figure 8-46. Residence time distrubtion functions versus dimensionless time.

GLOSSARY

Age of a fluid element: The age of a fluid element at a given instant of time is the time that elapses between the element's entrance into the vessel and the given instant. The age is equal to the residence time for those molecules that are just leaving the vessel.

Closed vessel: One in which the rate of diffusion across the vessel entrance and exit boundaries is negligible compared to the rate of convective motion across the boundaries.

Closed end vessel: One in which the inlet and outlet streams are completely mixed and dispersion occurs only between the terminals. At the inlet where $z = 0$, $uC_o = [uC - D_{e,1}(\partial C/\partial z)]_{z=0}$; at the outlet where $x = L$, $(\partial C/\partial z)_{z=0}$. These are referred to as Danckwerts' boundary conditions.

Clump: A collection of molecules small enough to occupy no more than a microscopic volume in a reactor, but large enough for the concept of concentration to be meaningful.

Concentration: The main types are: C_δ, the effluent from a vessel with impulse input of tracer; $C^o = m/V_r$, the initial mean concentration resulting from impulse input of magnitude m; C_u, the effluent from a vessel with a step input of magnitude C_f.

Dispersion: The movement of aggregates of molecules under the influence of a gradient of concentration, temperature, etc. The effect is represented by Fick's law with a dispersion coefficient substituted for molecular diffusion coefficient. The rate of transfer = $-D_e(\partial C/\partial z)$.

Impulse: An amount of tracer injected instantaneously into a vessel at time zero. The symbol $m\delta(t - a)$ represents an impulse of magnitude m injected at time t = a. The effluent concentration resulting from an impulse input is C_δ.

Intensity of segregation: A measure of the difference in concentration between neighboring clumps of fluid, $I_S = C_A'^2 / \overline{C}_A(1 - \overline{C}_A)$, where $\overline{C}_A$ is the time-averaged concentration and C_A' is the fluctuating component of concentration A.

Life expectation, λ: Of a molecule or aggregate in a vessel at a particular time is the period that it will remain in the vessel before ultimately leaving. The distribution of life expectancies is identical with that of residence times, $E(\lambda) = E(t)$.

Macromixing: The phenomenon whereby residence times of clumps are distributed about a mean value. Mixing on a scale greater than the minimum eddy size or minimum striation thickness, by laminar or turbulent motion.

Maximum mixedness: Exists when any molecule that enters a vessel immediately becomes associated with those molecules with which it will eventually leave the vessel. This occurs with those molecules that have the same life expectation. A state of maximum mixedness is associated with every residence time distribution (RTD).

Maximum mixedness model: The fluid in a flow reactor that behaves as a micro fluid. Mixing of molecules of different ages occurs as early as possible.

Mean residence time: The average time spent by the molecules in a vessel. It is the first moment of the effluent concentration from a vessel with impulse input and is defined by

$$\bar{t} = \frac{\displaystyle\int_0^\infty t\, C_\delta dt}{\displaystyle\int_0^\infty C_\delta dt}$$

Micromixing: Mixing among molecules of different ages (i.e., mixing between macrofluid clumps). Mixing on a scale smaller than the minimum eddy size or minimum striation thickness by molecular diffusion.

Mixing, ideal or complete: A state of complete uniformity of composition and temperature in a vessel. In a flow system, the residence time distribution is exponential, ranging from zero to infinity.

Open end vessel: One in which there are no discontinuities (abrupt changes) in concentration at the inlet and outlet where both bulk and dispersion flow occur. The boundary conditions are $C = C_o$ when $z = 0$ and $\delta C/\delta z = 0$ when $z = \infty$.

Peclet number: $N_{Pe} = uL/D_e$, where u is a linear velocity, L is a linear dimension, and D_e is the dispersion coefficient. In packed beds, $N_{Pe} = ud_P/D_e$, where u is the interstitial velocity and d_P is the particle diameter.

Perfect tracer: Tracer molecules behave identically with the process fluid molecules within the process unit.

Plug flow: A condition in which all effluent molecules have had the same residence time.

Pulse: A type of input in which the concentration of tracer in the input stream is changed suddenly, maintained at a non-zero value for a definite period, then changed to the original value and maintained that way for the period of interest. When the pulse is of constant magnitude it is called a square pulse.

Residence time distribution (RTD): The time that elapses from when the element enters the vessel to when it leaves.

Residence time distribution functions: These give information about the fraction of the fluid that spends a certain time in a process vessel.

In the case of tracer from a vessel that contained an initial average concentration C^o, the area under a plot of $E(t) = C_{effluent}/C^o$ between the ordinates at t_1 and t_2 is the fraction of the molecules that have residence times in this range. In the case of step constant input of concentration C_f to a vessel with zero initial concentration, the ratio $F(t) = C_{effluent}/C_f$ at t_1 is the fraction of molecules with residence time less than t_1.

Tracer: A substance that is used for measuring the residence time distribution in a vessel. Usually, it is inert and used in small concentrations so as not to change the physical properties of the process fluid appreciably, and analyzable for accuracy.

Scale of segregation: A measure of the average distance between "clumps" of the same component in a mixture.

Segregration: The tendency of the contents of a reactor to behave as a macrofluid.

Segregrated flow: Occurs when all molecules that enter together also leave together. A state of segregation is associated with every RTD. Each aggregate of molecules reacts independently of every other aggregate, as in an individual batch reactor.

Segregrated flow model: The fluid in a flow reactor is assumed to behave as a macrofluid. Each clump functions as a miniature batch reactor. Mixing of molecules of different ages occurs as late as possible.

Skewness: The third moment of a residence time distribution, which is expressed by:

$$\gamma^3(t) = \int_0^\infty (t - \bar{t})^3 E(t)dt$$

It is a measure of asymmetry.

Step Input: An input in which the concentration of tracer is changed to some constant value C_f at time zero and maintained at this level indefinitely. For example, $C_f u(t - a)$ represents a step of magnitude C_f beginning at $t = a$. The resulting effluent concentration is C_u.

Striation thickness: Average distance between adjacent interfaces of materials to be mixed by a laminar mechanism.

Variance: The second moment of the RTD. There are two terms: one of absolute time, $\sigma^2(t)$, and the other of dimensionless time, $\theta = t/\bar{t}$, $\sigma_\theta^2 = \sigma^2(t)/\bar{t}^2$

$$\sigma^2(t) = \frac{\int_0^\infty (t-\bar{t})^2 C_\delta dt}{\int_0^\infty C_\delta dt} = \int_0^\infty (t-\bar{t})^2 E(t)dt$$

$$\sigma_\theta^2 = \frac{\sigma^2(t)}{\bar{t}^2} = \int_0^\infty (\theta-1)^2 E(\theta)d\theta$$

REFERENCES

1. Mac Mullin, R. B. and Weber, M., Jr., *Trans. Am. Inst. Chem. Eng.,* 31, 409, 1935.
2. Danckwerts, P. V., *Chem. Eng. Sci.,* 2, pp. 1–18, 1953.
3. Levenspiel, O., *Chemical Reaction Engineering,* 3rd ed., John Wiley & Sons, New York, 1999.
4. Levenspiel, O. and Bischoff, K. B., *Adv. Chem. Eng.,* 4, 95, 1963.
5. Himmelblau, D. M. and Bischoff, K. B., *Process Analysis and Simulation, Deterministic Approach,* John Wiley & Sons, New York, 1968.
6. Wen, C. Y. and Fan, L. T., *Models for Flow Systems and Chemical Reactors,* Marcel Dekker, Inc., New York, 1975.
7. Shinnar, R., "Use of Residence and Contact Time Distributions in Reactor Design," Chapter 2, pp. 63–149 of *Chemical Reaction and Reactor Engineering,* Carberry, J. J. and Varma, A., Eds., Marcel Dekker, New York, 1987.
8. Danckwerts, P. V., *Chem. Eng. Sci.,* Vol. 9, pp. 78–79, 1958.
9. Naor, P. and Shinnar, R., *I & EC Fundamentals,* pp. 278–286, 1963.
10. Yim, S. S. S. and Ayazi, S. P., "Residence Time Distribution in a Rotary Flow Through Device, Fluid Mixing V," *Inst. ChemE. Symp. Series,* No. 140, pp. 191–201, 1996.
11. Al-Taweel, A.M., et al., *Trans. IChemE,* Vol. 74, Part A, pp. 456–461, May 1996.
12. Carpenter, N. G. and Roberts, E. P. L., *Trans. IChemE,* Vol. 77, Part A, pp. 212–217, May 1999.

13. Coker, A. K., "Mathematical Modelling and a Study of Flow Patterns in Cylindrical Nozzle," M.Sc. Thesis, Aston University, 1979.

14. Coker, A. K., "A Study of Fast Reactions in Nozzle Type Reactors," Ph.D. Thesis, Aston University, 1985.

15. Missen, R. W., Mims, C. A., and Saville, B. A., *Introduction to Chemical Reaction Engineering and Kinetics,* John Wiley & Sons, 1999.

16. Stokes, R. L. and Nauman, E. B., *The Canadian J. of Chem. Eng.,* Vol. 48, pp. 723–725, 1970.

17. Zwietering, Th., N., *Chem. Eng. Sci.,* 11, 1, 1959.

18. Langmuir, I., *J. Am. Chem. Soc.,* 30, 1742, 1908.

19. Taylor, G. I., *Proc. Roy Soc.* (London) 219A, 186, 1953; 225A, 473, 1954a; 223A, 446, 1954b.

20. Aris, R., *Chem. Eng. Sci.,* 9, 266, 1959.

21. Levenspiel, O. and Smith, W. K., *Chem Eng. Sci.,* 6, 227, 1957.

22. van der Laan, E. T., *Chem. Eng. Sci.,* 7, 187 1958.

23. Wehner, J. and Wilhelm, R. H., *Chem. Eng. Sci.,* 6, 89, 1956.

24. Levenspiel, O. and Bischoff, K. B., *Ind. Eng. Chem.,* 51, 1431 1959; 53, 313, 1961.

CHAPTER 9

Models for Non-Ideal Systems

INTRODUCTION

In a non-ideal system, the distribution functions exhibit various forms. Therefore, knowledge of these distributions for a given system is important. By comparing the curves corresponding to ideal systems, it is possible to determine the type of ideal behavior that a system most closely resembles. If the differences are small, the system can be treated as one of the ideal models discussed in Chapter 5. However, a large difference may imply a complex system, in which case modification will be required to reflect the real behavior of the system. The techniques for measuring the RTD by means of a tracer offer a diagnostic approach in analyzing the structure of flow in continuous systems.

The tank-in-series (TIS) and the dispersion plug flow (DPF) models can be adopted as reactor models once their parameters (e.g., N, $D_{e,l}$ and N_{Pe}) are known. However, these are macromixing models, which are unable to account for non-ideal mixing behavior at the microscopic level. This chapter reviews two micromixing models for evaluating the performance of a reactor— the segregated flow model and the maximum mixedness model—and considers the effect of micromixing on conversion.

BASICS OF NON-IDEAL FLOW

Three main factors affect either the interaction of fluids in a system or the flow pattern. These are:

- How early or late mixing of material occurs in the system.
- The state of aggregation of the flowing material, its tendency to clump, and for a group of molecules to move about together.
- The residence time distribution of material that is flowing through the system.

The fluid elements of a single flowing stream can mix with each other either early or late in their flow through the system. This is especially important with two reactants entering a vessel. Two extreme states of aggregation of a process are referred to as macrofluid and microfluid (Figure 9-1). In a macrofluid, molecules move together in clumps. The clumps are distributed in residence times, but all the molecules within a clump have the same age. In contrast, the clumps are dispersed in a microfluid because all the molecules move independently of one another regardless of age.

The RTD in a system is a measure of the degree to which fluid elements mix. In an ideal plug flow reactor, there is no mixing, while in a perfect mixer, the elements of different ages are uniformly mixed. A real process fluid is neither a macrofluid nor a microfluid, but tends toward one or the other of these extremes. Fluid mixing in a vessel, as reviewed in Chapter 7, is a complex process and can be analyzed on both macroscopic and microscopic scales. In a non-ideal system, there are irregularities that account for the fluid mixing of different

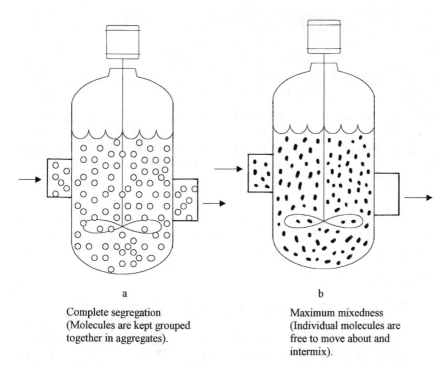

a

Complete segregation
(Molecules are kept grouped
together in aggregates).

b

Maximum mixedness
(Individual molecules are
free to move about and
intermix).

Figure 9-1. Two extreme states of fluid micromixing.

ages at the microscopic level. The RTD does not account for how fluid elements at the microscopic scale interact with each other. There is an obvious distinction between macromixing, or a set of mechanisms associated with the RTD, and micromixing, which results from the interactions between the fluid elements.

Macromixing defines the variation in the residence time experienced by molecules flowing through a flow system, while micromixing relates to the concentration history experienced by the molecules as they traverse through the system. The conversion in a system can be determined from the RTD with knowledge of the macromixing process. Generally, the performance of non-ideal flow systems can be determined from known parameters, such as:

- The kinetics of the system
- The RTD of fluid in the system
- The earliness or lateness of fluid mixing in the system
- The nature of the fluid (i.e., whether a micro or macro fluid) in the system

SEGREGRATED FLOW MODEL

In a mixing process where complete segregation occurs, the fraction of fluid elements with residence times $(t, t + dt)$ is $E(t)dt$, and the average exit composition in the vessel is defined by:

$$
\begin{pmatrix} \text{mean concentration} \\ \text{of reactant in the} \\ \text{exit stream} \end{pmatrix} = \overline{C}_A = \lim_{\Delta t \to 0} \sum_{\substack{\text{all elements} \\ \text{of exit stream}}} \begin{pmatrix} \text{concentration of} \\ \text{reactant remaining} \\ \text{in an element of age} \\ \text{between t and t + dt} \end{pmatrix} \begin{pmatrix} \text{fraction of exit} \\ \text{stream which is} \\ \text{of age between} \\ \text{t and t + dt} \end{pmatrix} \qquad (9\text{-}1)
$$

Consider a segregated flow system as illustrated in Figure 9-2, with a known RTD, in which a reaction A $\rightarrow$ products. The disappearance rate of A is

$$(-r_A) = (r_A)(C_A) \qquad (9\text{-}2)$$

Since the fluid flows in clumps and is completely segregated by age (i.e., each clump behaves like a batch reactor), it is possible to determine the reactant concentration in every clump for a constant density fluid as:

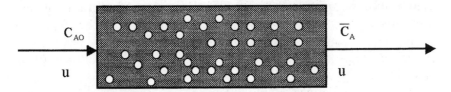

Figure 9-2. Reactor with segregated flow.

$$-\left(\frac{dC_A}{dt}\right)_{clump} = (r_A)$$
(9-3)

Rearranging Equation 9-3 and integrating yields

$$\int_{C_{AO}}^{C_{A,clump}} \frac{-dC_A}{r_A} = t$$
(9-4)

Therefore, $C_{A,clump}$ is a function of time for all nth order reactions. Using Equation 9-3 for the reactant concentration remaining in an element that has been in the reactor for a time t, E(t)dt is the fraction of the exit stream with ages between t and t + dt. Equation 9-1 then becomes

$$\overline{C}_A = \int_0^\infty C_{A,clump}(t)E(t)dt$$
(9-5)

The reactant conversion X_A is $X_A = 1 - \overline{C}_A/C_{AO}$. In terms of the fractional conversion, Equation 9-5 becomes

$$1 - X_A = \frac{\overline{C}_A}{C_{AO}} = \int_0^\infty \left(\frac{\overline{C}_A}{C_{AO}}\right)_{clump} E(t)dt$$
(9-6)

A zero order reaction in a batch reactor gives

$$\frac{C_A}{C_{AO}} = 1 - \frac{kt}{C_{AO}}$$
(9-7)

Substituting Equation 9-7 into Equation 9-6 and integrating gives

$$\frac{\overline{C}_A}{C_{AO}} = 1 - \frac{kt}{C_{AO}}\left(1 - e^{-C_{AO}/kt}\right) \qquad (9\text{-}8)$$

For a first order reaction, we have

$$\left(\frac{\overline{C}_A}{C_{AO}}\right)_{clump} = e^{-kt} \qquad (9\text{-}9)$$

and

$$\frac{\overline{C}_A}{C_{AO}} = \int_0^\infty e^{-kt}\, E(t)dt$$

or

$$\frac{\overline{C}_A}{C_{AO}} \cong \sum e^{-kt}\, E(t)\Delta t \qquad (9\text{-}10)$$

The residence time density function $E(t)$ is

$$E(t) = \tfrac{1}{\bar{t}}e^{-t/\bar{t}} \qquad (8\text{-}25)$$

Substituting Equation 8-25 into Equation 9-10 gives

$$\frac{\overline{C}_A}{C_{AO}} = \int_0^\infty e^{-kt} \bullet \frac{1}{\bar{t}}e^{\frac{-t}{\bar{t}}}\, dt$$

$$= \frac{1}{\bar{t}}\int_0^\infty e^{-\left(k+\frac{1}{\bar{t}}\right)t}\, dt \qquad (9\text{-}11)$$

Integrating Equation 9-11 by parts (i.e., $\int wdv = wv - \int vdw$) yields

$$\frac{\overline{C}_A}{C_{AO}} = \frac{1}{1+k\bar{t}} \qquad (9\text{-}12)$$

Equation 9-12 is identical to that derived in Chapter 5. There appears to be two methods for determining the conversion in a perfectly mixed reactor. In Chapter 5, complete mixing in the reactor was assumed, and no RTD was explicitly used. Here, the first order reaction gives identical results when the RTD is used as specified by Equation 8-25.

For a second order reaction of a single reactant,

$$\left(\frac{C_A}{C_{AO}}\right)_{clump} = \frac{1}{1 + ktC_{AO}} \tag{9-13}$$

The segregated flow model equation gives

$$1 - X_A = \left(\frac{\overline{C}_A}{C_{AO}}\right) = \int_0^\infty \frac{1}{1 + ktC_{AO}} \cdot \frac{1}{\overline{t}} e^{\frac{-t}{\overline{t}}} \, dt \tag{9-14}$$

where $\alpha = 1/ktC_{AO}$ and $\theta = t/\overline{t}$. Equation 9-14 becomes

$$1 - X_A = \left(\frac{\overline{C}_A}{C_{AO}}\right) = \alpha e^\alpha \int_0^\infty \frac{e^{-(\alpha+\theta)}}{\alpha+\theta} \, d(\alpha+\theta) = \alpha e^\alpha \, ei(\alpha) \tag{9-15}$$

where $E_i(x) \equiv$ exponential integral $\int_x^\infty (e^{-y}/y)dy$.

Equation 9-15 gives the conversion expression for the second order reaction of a macrofluid in a mixed flow. An exponential integral, $ei(\alpha)$, which is a function of α, and its value can be found from tables of integrals. However, the conversion from Equation 9-15 is different from that of a perfectly mixed reactor without reference to RTD. An earlier analysis in Chapter 5 gives

$$\frac{C_A}{C_{AO}} = \frac{-1 + \sqrt{1 + 4kC_{AO}\overline{t}}}{2kC_{AO}\overline{t}} \tag{5-245}$$

For an nth order reaction, the conversion in a batch reactor is:

$$1 - X_A = \left(\frac{\overline{C}_A}{C_{AO}}\right) = \left[1 + (n-1)C_{AO}^{n-1} kt\right]^{1/(1-n)} \tag{9-16}$$

For an nth order reaction of a macrofluid, Equation 9-16 is substituted into Equation 9-6 to give

$$1 - X_A = \frac{\overline{C}_A}{C_{AO}} = \int_0^\infty \left[1 + (n-1)C_{AO}^{n-1}\right]^{\frac{1}{(1-n)}} \bullet \frac{e^{\frac{-t}{\overline{t}}}}{\overline{t}} dt \qquad (9\text{-}17)$$

For N-stirred tanks in series, the conversion is expressed by

$$\frac{\overline{C}_A}{C_{AO}} = \frac{1}{\overline{t}^N} \frac{N^N}{(N-1)!} \int_0^\infty t^{N-1} e^{-\left[k + \frac{N}{\overline{t}}\right]t} dt$$

$$\frac{\overline{C}_A}{C_{AO}} = \frac{1}{\left[1 + \frac{k\overline{t}}{N}\right]^N} = \frac{1}{\left[1 + k\overline{t}_i\right]^N} \qquad (8\text{-}160)$$

Complete segregation represents an extreme form of mixing on a molecular scale. Molecules enter mixed and leave mixed, but no mixing takes place within the system. Any RTD is possible in this state. Complete segregated systems can be modeled as piston flow reactors in parallel, as shown in Figure 9-3 or as single piston flow reactor with side exits as in Figure 9-4. Other residence time distributions have some maximum possible level of micromixing, referred to

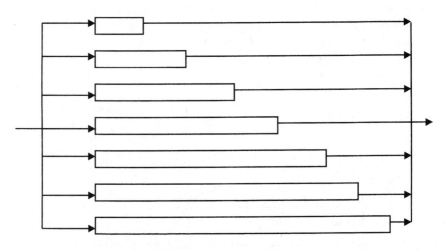

Figure 9-3. Models for segregated reactors: Piston flow elements in parallel.

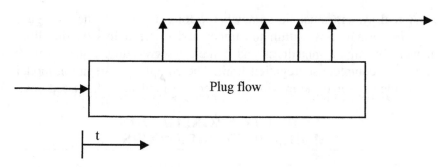

Figure 9-4. Single-piston flow element with side exits.

as maximum mixedness. Figure 9-5 shows an illustration of micro-mixing. The abscissa (macromixing) measures the breadth of the residence time distribution. It is zero for piston flow, fairly broad for the exponential distribution of stirred tanks, and broader yet for situations with stagnancy and bypassing. The ordinate (micromixing) varies from none to complete and it measures the effects of micro-mixing. This can be negligible for piston flow and have maximum importance for stirred tank reactors. Soundly designed reactors will fall in the normal region bounded by three apexes: piston flow, perfect mixing, and complete segregation with an exponential distribution.

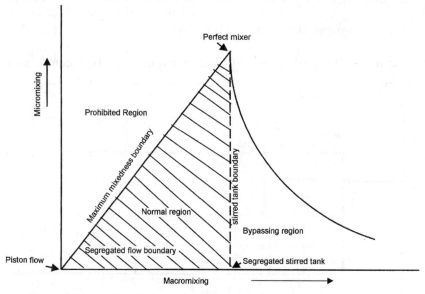

Figure 9-5. Schematic representation of mixing space. *(Source: Nauman, E. G.,* Chemical Reactor Design, *John Wiley & Sons, 1987.)*

A real system must lie somewhere along a vertical line in Figure 9-5. Performance is within the upper and lower points on this line; namely maximum mixedness and complete segregation. Equation 9-10 gives the complete segregation limit. The complete segregation model with side exits and the maximum mixedness model are discussed next.

COMPLETE SEGREGATION
MODEL WITH SIDE EXITS

Consider a plug flow reactor with side exits (Figure 9-6) through which portions of the flow leave. Molecules are classified as they traverse within the system according to their ages λ (i.e., the life expectancy of the fluid in the reactor at that point), implying that mixing occurs as late as possible.

A material balance on an elemental volume ΔV of the reactor between ages λ and $\lambda + d\lambda$ follows.

Input of A by flow
in vessel at $\lambda = u[1 - f(\lambda)]C_A(\lambda)$ $\hspace{2cm}$ (9-18)

Output of A by flow
in vessel at $\lambda + d\lambda = u[1 - F(\lambda + d\lambda)]C_A (\lambda + d\lambda)$ $\hspace{1cm}$ (9-19)

Output from the side exit $= uC_A E(\lambda)d\lambda$ $\hspace{2cm}$ (9-20)

where the volume of fluid with a life expectancy between λ and $\lambda + d\lambda$ is

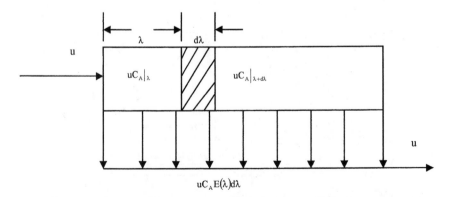

Figure 9-6. A plug flow reactor with side exits.

$$\Delta V = \int_0^\infty u\big[1 - F(\lambda)\big]d\lambda \; - \int_{\lambda+d\lambda}^\infty u\big[1 - F(\lambda)\big]d\lambda = u\big[1 - F(\lambda)\big]\Delta\lambda \qquad (9\text{-}21)$$

The rate of loss of substance A
in this volume is = $(-r_A)\Delta V$

$$= (-r_A)u[1 - f(\lambda)]\Delta\lambda \qquad (9\text{-}22)$$

A balance on component A between λ and $\lambda + d\lambda$ gives

$$\begin{pmatrix} \text{Input of A} \\ \text{by flow in} \\ \text{vessel at } \lambda \end{pmatrix} = \begin{pmatrix} \text{output of A by} \\ \text{flow in vessel} \\ \text{at } \lambda + d\lambda \end{pmatrix} + \begin{pmatrix} \text{output of A by} \\ \text{flow through} \\ \text{sides of vessel} \end{pmatrix} + \begin{pmatrix} \text{rate of loss of} \\ \text{substance A} \\ \text{by reaction} \\ \text{in } \Delta V \end{pmatrix}$$

which is:

$$u[1 - F(\lambda)]C_A(\lambda) = u[1 - F(\lambda + d\lambda)]C_A(\lambda + d\lambda)$$

$$+ \; uC_A E(\lambda)d\lambda + (-r_A)u[1 - F(\lambda)]\Delta\lambda \qquad (9\text{-}23)$$

Rearranging Equation 9-23 and dividing by u as $\Delta\lambda \to 0$ gives:

$$0 = \frac{d\big\{[1 - F(\lambda)]C_A(\lambda)\big\}}{d\lambda} + C_A(\lambda)E(\lambda) + (-r_A)[1 - F(\lambda)] \qquad (9\text{-}24)$$

Equation 9-24 gives

$$\frac{dC_A}{d\lambda} = -(-r_A) \qquad (9\text{-}25)$$

Each fluid element containing molecules of the same age during its passage through the reactor can be considered as a small batch reactor. The mean concentration overall fluid elements at the reactor outlet is the expected concentration of all fluid elements over the RTD.

MAXIMUM MIXEDNESS MODEL (MMM)

Upon entering the reactor, the molecules of the feed immediately mix and become associated with the molecules in whose company they will eventually leave. Figure 9-7 shows a plug flow with multiple inlets whose flow pattern is given by the same RTD. Zwietering [1] developed the MMM as represented by a plug flow in which the feed enters continuously and incrementally along the reactor length. Consider the material balance of reactant A around the differential volume between λ and $\lambda + d\lambda$.

$$\begin{pmatrix} \text{Input of A by} \\ \text{flow in vessel} \\ \text{at } \lambda + d\lambda \end{pmatrix} + \begin{pmatrix} \text{Input of A by} \\ \text{flow through} \\ \text{the side at } \lambda \end{pmatrix} = \begin{pmatrix} \text{output of A} \\ \text{by flow in} \\ \text{vessel at } \lambda \end{pmatrix} + \begin{pmatrix} \text{rate of loss of} \\ \text{substance A} \\ \text{by reaction} \\ \text{in } \Delta V \end{pmatrix}$$

Input of A by flow
in vessel at $\lambda + d\lambda = u[1 - F(\lambda + d\lambda)]C_A(\lambda + d\lambda)$ (9-26)

Input of A by flow
through the side at $\lambda = [uE(\lambda)\Delta\lambda]C_{AO}$ (9-27)

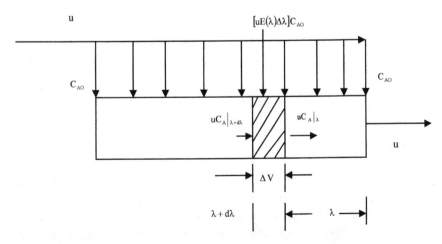

Figure 9-7. A plug flow reactor with side entry.

Output of A by flow
in vessel at λ = $u[1 - F(\lambda)]C_A(\lambda)$ (9-28)

The rate of loss of substance A
in this volume is = $(-r_A)\Delta V$

$$= (-r_A)u[1 - F(\lambda)]\Delta\lambda \qquad (9\text{-}29)$$

The steady-state material balance on A gives

$u[1 - F(\lambda + d\lambda)]C_A(\lambda + d\lambda)$

$+ [uE(\lambda)\Delta\lambda]C_{AO} = u[1 - F(\lambda)]C_A(\lambda) + u(-r_A)[1 - F(\lambda)]\Delta\lambda$ (9-30)

Rearranging Equation 9-30 and dividing by u gives:

$[1 - F(\lambda + d\lambda)]C_A(\lambda + d\lambda) + E(\lambda)\Delta\lambda C_{AO}$

$- [1 - F(\lambda)]C_A(\lambda) - (-r_A)[(1 - F(\lambda)]\Delta\lambda = 0$ (9-31)

Taking the limits as $\Delta\lambda \to 0$ gives:

$$E(\lambda)C_{AO} + \frac{d\left\{[1-F(\lambda)]C_A(\lambda)\right\}}{d\lambda} - (-r_A)[1-F(\lambda)] = 0 \qquad (9\text{-}32)$$

or $C_{AO}E(\lambda)+[1-F(\lambda)]\dfrac{dC_A(\lambda)}{d\lambda}-C_AE(\lambda)-(-r_A)[1-F(\lambda)]=0$ (9-33)

Rearranging Equation 9-33 yields:

$$\frac{dC_A(\lambda)}{d\lambda} + \frac{(C_{AO}-C_A)E(\lambda)}{[1-F(\lambda)]} - (-r_A) = 0 \qquad (9\text{-}34)$$

Equation 9-34 does not use an initial value of C_A as a boundary condition. Instead, the usual boundary condition associated with Equation 9-34 is

$$\lim_{\lambda \to 0} \frac{dC_A}{d\lambda} = 0 \qquad (9\text{-}35)$$

The outlet concentration from a maximum mixedness reactor is found by evaluating the solution to Equation 9-34 at $\lambda = 0$; $C_{Aout} = C_A(0)$. The analytical solution to Equation 9-34 is rather complex; for reaction order $n > 1$, the $(-r_A)$ term is usually non-linear. Using numerical methods, Equation 9-34 can be treated as an initial value problem. Choose a value for $C_{Aout} = C_A(0)$ and integrate Equation 9-34. If $C_A(\lambda)$ achieves a steady state value, the correct value for $C_A(0)$ was guessed. Once Equation 9-34 has been solved subject to the appropriate boundary conditions, the conversion may be calculated from $C_{Aout} = C_A(0)$.

EFFECT OF MICROMIXING ON CONVERSION

As stated earlier, the actual state of micromixing lies between two extremes. However, further details are extremely difficult to use either theoretically or experimentally. Kramers and Westerterp [2] gave results for the extreme case of a perfectly macromixed reactor with various reactions occurring. Figures 9-8 and 9-9 illustrate a comparison between the segregated flow and complete mixing for $n = 0$ and $n > 1$, respectively. It can be seen in Figure 9-9 that the differences between

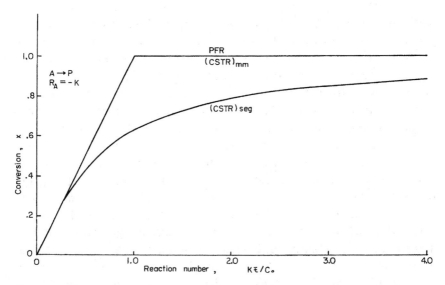

Figure 9-8. Conversion for a zero order reaction in a CSTR and a PFT. *(Source: Wen, C. Y. and L. T. Fan,* Models for Flow Systems and Chemical Reactors, *Marcel Dekker Inc., 1975.)*

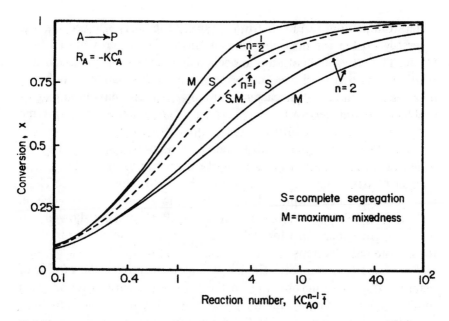

Figure 9-9. Conversion for isothermal elementary reactions in a CSTR [2].

the levels of micromixing are small. For a first order reaction (or a series of first order reactions), micromixing has no effect on the reactor performance. Therefore, conversion is predicted exactly by the segregated flow model (Equation 9-10). Micromixing decreases conversion for reaction orders greater than 1 and increases for orders less than 1. The segregated flow model, consequently, predicts an upper bound on conversion for n > 1 and a lower bound for n < 1.

Several types of models have been developed for intermediate levels of micromixing with arbitrary macromixing RTD. Weinstein and Adler [3], Villermaux and Zoulalain [4], and Ng and Rippin [5] have proposed a model that divides the reactor into two environments: one in a segregated state and the other in a maximum mixedness state. The fraction in each state can be fitted to the actual reactor data, which are then correlated. For bimolecular reaction, the maximum difference between the conversion corresponding to complete segregation and that corresponding to maximum mixedness can be as much as 50%. Tsai et al. [6] have developed the two-environment model, which is an empirical modification of the original two-environment model. It considers material transfers from the entering environment to the leaving environment at a rate that is proportional to the amount of material remaining in the entering environment. They showed that the reversed two-environment is more appropriate for the growth in a flow reactor than other micromixing models. Chen and Fan [7] inferred that a reversed two-environment model is more representative of polymer reactors. In this case, the reactants initially enter the maximum mixedness region, but become segregated as polymerization increases the viscosity and decreases diffusivities.

Industrial reactors generally operate adiabatically. Cholette and Blanchet [8] compared adiabatic plug flow reactor to the $CSTR_{mm}$. For exothermic reactions, they inferred that the performance of a $CSTR_{mm}$ is better than that of a plug flow reactor at low values of conversion, and vice-versa at high values of conversion. They further showed that the design considerations for endothermic reactions are similar to those for isothermal reactions.

For isothermal autocatalytic and adiabatic reactions, the effects of macromixing and micromixing on the conversions are different from those for isothermal elementary reactions. Figures 9-10 and 9-11 compare the performances of these reactions under the two extreme states of micromixing. Worrel and Eagleton [9] performed mixing and segregation studies in a CSTR. They deduced that the maximum mixedness conversion is attained rapidly as impeller speed is increased. However, the effect of micromixing on the conversion was negligible.

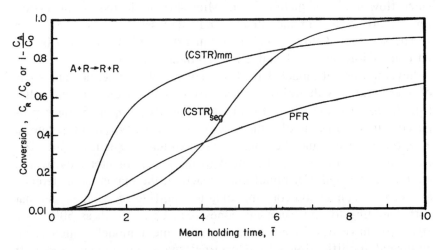

Figure 9-10. Conversion for autocatalytic in a CSTR and a PFT. *(Source: Wen, C. Y. and L. T. Fan, Models for Flow Systems and Chemical Reactors, Marcel Dekker Inc., 1975.)*

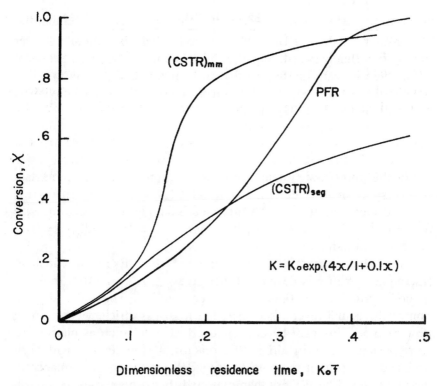

Figure 9-11. Conversion for a first order, adiabatic reaction in a CSTR and a PFT. *(Source: Wen, C. Y. and L. T. Fan,* Models for Flow Systems and Chemical Reactors, *Marcel Dekker Inc., 1975.)*

Their studies also showed the importance of the reacting system kinetics for better design operation of CSTR reactors.

Generally, empirical models involving micromixing should be viewed with caution because it is difficult to obtain valid correlations for the parameters in the absence of sound experimental data. Also, a model may not provide a unique measure of micromixing except for the specific system being studied. Micromixing is a complex phenomenon, which cannot be easily characterized by one or more adjustable parameters. A model with few parameters might provide an excellent fit for one reaction and a poor fit for another reaction that occurs in the same vessel. A comprehensive discussion and analyses on micromixing are provided by Nauman and Buffham [10], and Wen and Fan [11].

Example 9-1

Consider the data of Hull and von Ronsenberg in Example 8-3 for mixing in a fluidized bed. Suppose the solids in the fluidized bed were not acting as a catalyst, but were actually reacting according to a first order rate law $(-r) = kC$, $k = 1.2$ min^{-1}. Compare the actual conversion with that of an ideal plug flow.

Solution

For the plug flow system, the concentration of unreacted reactant with respect to A is determined by $C_A/C_{AO} = e^{-kt}$ where $\bar{t}$ is the mean residence time; $\bar{t} = 2.5$ min (see Example 8-3) and $C_A/C_{AO} = e^{-kt} = e^{-(1.2)(2.5)} = 0.05$. The conversion, $X_A = 1 - C_A/C_{AO} = 0.95$ or 95% conversion.

Using the macromixing Equation 9-10, an Excel spreadsheet (Example9-1.xls) was developed that gives $\overline{C}_A/C_{AO} = 0.11$. In terms of conversion, $X_A = 0.89$ or 89% conversion. Thus, the simulation of the actual reactor suggests that it is less efficient than a plug flow reactor. This may be due to several factors such as channeling or bypassing of the fluid in the reactor. If the reactor had been designed using the usual plug flow model, the actual performance in the plant would be far less than expected. For a higher conversion in the reactor, some modifications would be required to remedy the non-ideal flow.

Example 9-2

The flow through a reactor is 10 dm^3/min. A pulse test gave the following concentration measurements at the outlet.

t(min)	c $\times$ 10^5	t(min)	c $\times$ 10^5
0	0	15	238
0.4	329	20	136
1.0	622	25	77
2.0	812	30	44
3	831	35	25
4	785	40	14
5	720	45	8
6	650	50	5
8	523	60	1
10	418		

1. What is the volume of the reactor? Calculate and plot $E(\theta)$, $F(\theta)$, and $I(\theta)$ versus θ.
2. If the reactor is modeled as a tank-in-series (TIS) system, how many tanks are needed to represent this reactor?
3. If the reactor is modeled by a dispersion model, what is the Peclet number (N_{Pe})?
4. What is the reactor conversion for a first order reaction with $k = 0.1$ min^{-1} for
   a. The tank in series model
   b. The segregation model
   c. The dispersion model
   d. Plug flow
   e. A single CFSTR

Solution

An Excel spreadsheet (Example9-2.xls) was developed for the RTD functions and the parameters. The calculated mean residence time is 9.878 min, and thus the volume of the reactor is

$$V = \bar{t} \cdot u$$

$$= 9.878 \times 10$$

$$= 98.8 \ dm^3$$

Figure 9-12 shows plots of $E(\theta)$, $F(\theta)$, and $I(\theta)$ versus θ.

Assuming that the reactor is modeled as TIS, the number N is determined as

$$N = \frac{1}{\sigma_\theta^2}$$

where σ_θ^2 is the dimensionless variance. The value from the spreadsheet is $\sigma_\theta^2 = 0.765$. Therefore, the number of tanks required to represent this reactor is $N = 1.31$.

Assuming that the reactor is modeled by a dispersion model using the closed boundary conditions, the Peclet number is determined by

$$\sigma_\theta^2 = \frac{2}{N_{Pe}} - \frac{2}{N_{Pe}^2}\left(1 - e^{-N_{Pe}}\right) \tag{8-161}$$

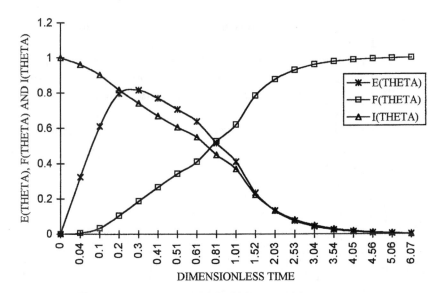

Figure 9-12. Plots of E(θ), F(θ), and I(θ) versus θ.

Expressing Equation 8-161 as a polynomial gives

$$f(N_{Pe}) = \sigma_\theta^2 - \frac{2}{N_{Pe}} + \frac{2}{N_{Pe}^2}\left(1 - e^{-N_{Pe}}\right) \tag{8-162}$$

As an alternative to using the computer program (PROG16), Equation 8-162 can be expressed in a spreadsheet. Using the GOAL-SEEK command from the Tools menu (Appendix B) from the Excel spreadsheet menu to determine the Peclet number gives $f(N_{Pe}) = 0$. This function is incorporated in the spreadsheet. The computed Peclet number is $N_{Pe} = 0.8638$.

For a first order reaction with $k = 0.1$ min^{-1}, the following conversions were determined:

For the TIS model,

$$X_A = 1 - \frac{1}{\left[1 + \dfrac{k\bar{t}}{N}\right]^N} \tag{8-160}$$

From the spreadsheet, $X_A = 0.521$ or 52.1%.

The segregation model gives:

$$\frac{\overline{C}_A}{C_{AO}} \cong \sum e^{-kt}\, E(t)\, \Delta t \tag{9-10}$$

$X_A = 0.47$ or 47%

For the dispersion model, the conversion is determined by Equation 8-149 as

$$\frac{C_A}{C_{AO}} = 1 - X_A = \frac{4a\exp\left[\dfrac{1}{2}N_{Pe}\right]}{(1+a)^2 \exp\left[\dfrac{a}{2}N_{Pe}\right] - (1-a)^2 \exp\left[\dfrac{-a}{2}N_{Pe}\right]} \tag{8-149}$$

where $a = \left[1 + \dfrac{4k\bar{t}}{N_{Pe}}\right]^{0.5}$

Using the value of $N_{Pe} = 0.86$, the calculated conversion from the spreadsheet is $X_A = 0.53$ or 53%.

The conversion for the plug flow is expressed as:

$$X_A = 1 - e^{-k\bar{t}}$$

$X_A = 0.627$ or 62.7%

For a single CSTR

$$X_A = \frac{k\bar{t}}{1 + k\bar{t}}$$

$$= \frac{(0.1)(9.878)}{1 + (0.1)(9.878)}$$

$$= 0.497 \text{ or } 49.7\%.$$

A summary of the conversion is given as:

Flow systems	Conversion, X_A (%)
Plug flow reactor	62.7
Dispersion	52.6
Tank-in-series	52.1
Segregated model	47.0
A single CSTR	49.7

REFERENCES

1. Zwietering, T. N., *Chem. Eng. Sci.,* 11, 1–15, 1959.
2. Kramers, H. and Westerterp, K. R., *Elements of Chemical Reactor Design and Operation,* Academic Press, New York, 1963.
3. Weinstein, H. and Adler, R. J., *Chem. Eng. Sci.,* 22, 65, 1967.
4. Villermaux, J. and Zoulalain, A., *Chem. Eng. Sci.,* 24, 1513, 1969.
5. Ng, D. Y. C. and Rippin, D. W. T., "The effect of incomplete mixing on conversion in homogeneous reactions," Proc. 3rd Europ. Symp. of Chem. Eng., Amsterdam, Netherlands, 1964.
6. Tsai., B. I., Fan, L. T., Erickson, L. E., and Chen, M. S. K., *J. Appl Chem. Biotechnol.,* 21, 307, 1971.
7. Chen, M. S. K. and Fan, L. T., *Can. J. Chem. Eng.,* 49, 704–708, 1971.
8. Cholette, A. and Blanchet, J., *Can. J. Chem. Eng.,* 39, 192, 1961.
9. Worrel, G. R. and Eagleton, L. C., *Can. J. Chem. Eng.,* 39, 254, 1964.
10. Nauman, E. B. and Buffham, B. A., *Mixing in Continuous Flow System,* John Wiley & Sons, 1983.
11. Wen, C. Y. and Fan, L. T., *Models for Flow Systems and Chemical Reactors,* Marcel Dekker, Inc., 1975.

CHAPTER TEN

Application of Computational Fluid Dynamics and Computational Fluid Mixing in Reactors

INTRODUCTION

Computational fluid dynamics (CFD) is the analysis of systems involving fluid flow, energy transfer, and associated phenomena such as combustion and chemical reactions by means of computer-based simulation. CFD codes numerically solve the mass-continuity equation over a specific domain set by the user. The technique is very powerful and covers a wide range of industrial applications. Examples in the field of chemical engineering are:

- Polymerization
- Multiphase flow in reactors (e.g., bubble column)
- Reaction modeling
- Sedimentation
- Separation
- Complex pipeline network
- Mixing

The numerical solution of the energy balance and momentum balance equations can be combined with flow equations to describe heat transfer and chemical reactions in flow situations. The simulation results can be in various forms: numerical, graphical, or pictorial. CFD codes are structured around the numerical algorithms and, to provide easy assess to their solving power, CFD commercial packages incorporate user interfaces to input parameters and observe the results. CFD codes contain three main elements, namely, a pre-processor, a solver, and a post-processor.

PRE-PROCESSOR

The pre-processor consists of inputting a problem into a CFD program using a friendly interface, which is transformed into a suitable format for the solver. The user activities at this stage involve:

- Defining the geometry of the region of interest: the computational domain. Some CFD packages now allow geometries prepared in CAD packages to be imported directly into the CFD code, thus reducing pre-processing time as many plant designs are already available in CAD packages.
- Grid generation by subdividing the domain into a number of smaller non-overlapping subdomains. This creates a grid (or mesh) of cells (or control volumes or elements).
- Selecting the physical and chemical phenomena that need to be modeled.
- Defining fluid properties.
- Specifying appropriate boundary conditions of cells that coincide with or touch the domain boundary.

The accuracy of a CFD program is greatly influenced by the number of cells in the grid. The larger the number of cells, the more accurate the solution. Additionally, the cost of computer hardware and calculation time combined with the accuracy of a solution depends on the preciseness of the grid.

SOLVER

The three modes of numerical solution techniques are finite difference, finite element, and spectral methods. These methods perform the following steps:

- Approximation of the unknown flow variables by means of simple functions.
- Discretization by substitution of the approximations into the governing flow equations and subsequent mathematical manipulations.
- Solution of the algebraic equations.

The commercial CFD codes use the finite volume method, which was originally developed as a special finite difference formulation. The numerical algorithm consists of the following steps:

- Integrating the governing equations of fluid flow over all the finite control volumes of the solution domain.
- Discretizating by substituting the various finite-difference type approximations for the terms in the integrated equation representing flow processes, which converts the integral equations into a system of algebraic equations.
- Solving the algebraic equations by an iterative method.

POST-PROCESSOR

The results of CFD packages are obtained by several means, due to the versatile data visualization tools that are now incorporated in many workstations. The results can be of several forms:

- Vector plots
- Two- or three-dimensional surface plots
- Particle tracking
- Line and shaded contour plots
- Domain geometry and grid display
- View manipulation (translation, rotation, scaling, and color post-script output)

Successful utilization of CFD packages depends on understanding the physical and chemical phenomena that must be considered for a typical problem. Good modeling skills are essential when making necessary assumptions so that the complexity of the problem can be reduced at the same time as retaining the principal features of the problem. The correct assumptions often govern the quality of the information generated by CFD.

An adequate knowledge of the numerical solution algorithm is also needed. The mathematical concepts used to determine the success of such algorithms are convergence, consistency, and stability. Convergence is the property of a numerical method that produces a solution, which approaches the exact solution as the grid spacing, control volume size, or element size is reduced to zero. Consistent numerical schemes are systems of algebraic equations that can be demonstrated to be the same as the original governing equation as the grid spacing tends to zero. Finally, stability deals with damping the errors as the numerical method proceeds. An unstable technique causes round-off errors in the initial data resulting in extraneous oscillations and divergence. The principal steps in the simulation must be outlined. This includes allocating memory, defining the domain, setting the cells, choosing the physical model and boundary conditions, defining the physical constants, saving the case file, running the program, and saving the data.

At the end of the simulation, the user must decide whether the results are adequate, as it is impossible to assess the validity of the models of physics and chemistry that are incorporated in CFD codes. Thus, the final results must always be validated by an experiment. CFD is a powerful additional problem-solving tool, and validation of the code requires highly detailed information concerning the boundary conditions of a problem. Experimental techniques such as laser doppler velocimetry (LDV), digital particle image velocimetry (DPIV), and gamma-ray computed tomography (GRCT) are modern tools employed for flow analysis and visualization. These techniques can be used to validate CFD results.

Mixing of single and multiphase fluids in stirred tanks is a common operation in chemical process industries. Proper knowledge of fluid flow is essential for scale-up, equipment design, process control, and economic factors. Computational fluid mixing (CFM) is a powerful tool that can be used to model fluid flows of different impeller designs in stirred tank reactors. CFM models enable designers to visualize mixing as it occurs in the vessel, thus allowing the designer to select the best agitator to achieve the desired process performance. In this chapter, the theory of fluid flow in three dimensions, fluid mixing in stirred tank reactors employing CFM, and CFD tools are reviewed. The combination of these tools and experimental techniques helps to provide accurate point values of time-averaged and fluctuating velocities, as well as the time-average and semi-instantaneous flow fields in stirred tanks. At the end of this chapter is a review on the use of CFD in enhancing the performance of reactors.

THEORY AND FLUID FLOW EQUATIONS

The governing equations of fluid flow represent mathematical statements of the conservation of mass, known as the continuity equation:

- The mass of fluid is conserved.
- The rate of change of momentum equals the sum of the forces on a fluid particle (Newton's second law).
- The rate of change of energy is equal to the sum of the rate of heat addition to and the rate of work done on a fluid particle (first law of thermodynamics).

The fluid is regarded as a continuum, and its behavior is described in terms of macroscopic properties such as velocity, pressure, density and temperature, and their space and time derivatives. A fluid particle or point in a fluid is the smallest possible element of fluid whose macroscopic properties are not influenced by individual molecules. Figure 10-1 shows the center of a small element located at position (x, y, z) with the six faces labelled N, S, E, W, T, and B. Consider a small element of fluid with sides δx, δy, and δz. A systematic account

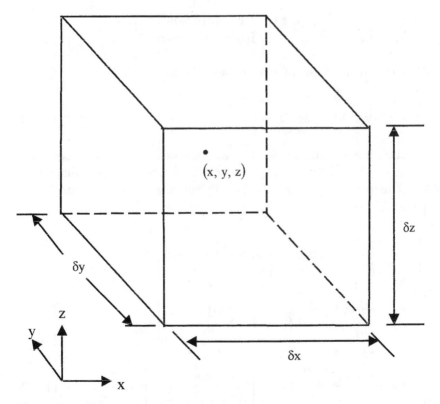

Figure 10-1. Fluid element for the conservation of mass, momentum, and energy.

of changes in the mass, momentum, and energy of the fluid element due to fluid flow across its boundaries and to the action of sources inside the elements where necessary, results in the fluid flow equations.

All fluid properties are functions of space and time, namely $\rho(x, y, z, t)$, $p(x, y, z, t)$, $T(x, y, z, t)$, and $u(x, y, z, t)$ for the density, pressure, temperature, and velocity vector, respectively. The element under consideration is so small that fluid properties at the faces can be expressed accurately by the first two terms of a Taylor series expansion. For example, the pressure at the E and W faces, which are both at a distance $1/2\delta x$ from the element center, is expressed as

$$p - \frac{\partial p}{\partial x} \cdot \frac{1}{2}\partial x \quad \text{and} \quad p + \frac{\partial p}{\partial x} \cdot \frac{1}{2}\delta x$$

Conservation of mass.

The mass balance for the fluid element can be expressed as:

Rate of increase of mass in fluid element $=$ Net rate of flow of mass into fluid element

The rate of increase of mass in the fluid element is:

$$\frac{\partial}{\partial t}\left(\rho\, \delta x\, \delta y\, \delta z\right) = \frac{\partial \rho}{\partial t}\, \delta x\, \delta y\, \delta z \tag{10-1}$$

Accounting for the mass flowrate across a face of the element, which is given by the density product, area, and velocity component normal to the face (Figure 10-2), the net rate of flow of mass into the element across its boundaries is given by:

$$\left[\rho u - \frac{\partial(\rho u)}{\partial x}\frac{1}{2}\delta x\right]\delta y\, \delta z - \left[\rho u + \frac{\partial(\rho u)}{\partial x}\frac{1}{2}\delta x\right]\delta y\, \delta z$$

$$+ \left[\rho v - \frac{\partial(\rho v)}{\partial y}\frac{1}{2}\delta y\right]\delta x\, \delta z - \left[\rho v + \frac{\partial(\rho v)}{\partial y}\frac{1}{2}\delta y\right]\delta x\, \delta z$$

$$+ \left[\rho w - \frac{\partial(\rho w)}{\partial z}\frac{1}{2}\delta z\right]\delta x\, \delta y - \left[\rho w + \frac{\partial(\rho w)}{\partial z}\frac{1}{2}\delta z\right]\delta x\, \delta y \tag{10-2}$$

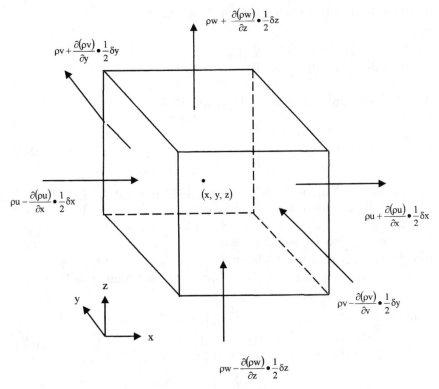

Figure 10-2. Mass flows in and out of the fluid element.

Flows that are directed into the element give an increase of mass in the element and are assigned a positive sign. The flows leaving the element are assigned a negative sign. Equating the rate of increase of mass into the element across its faces (Equation 10-2), yields:

$$\frac{\partial \rho}{\partial t} \delta x\, \delta y\, \delta z = \left[\rho u - \frac{\partial(\rho u)}{\partial x}\frac{1}{2}\delta x\right]\delta y\, \delta z - \left[\rho u + \frac{\partial(\rho u)}{\partial x}\frac{1}{2}\delta x\right]\delta y\, \delta z$$

$$+ \left[\rho v - \frac{\partial(\rho v)}{\partial y}\frac{1}{2}\delta y\right]\delta x\, \delta z - \left[\rho v + \frac{\partial(\rho v)}{\partial y}\frac{1}{2}\delta y\right]\delta x\, \delta z$$

$$+ \left[\rho w - \frac{\partial(\rho w)}{\partial z}\frac{1}{2}\delta z\right]\delta x\, \delta y - \left[\rho w + \frac{\partial(\rho w)}{\partial z}\frac{1}{2}\delta z\right]\delta x\, \delta y \qquad (10\text{-}3)$$

Dividing by the element volume $\delta x \delta y \delta z$ gives

$$\frac{\partial \rho}{\partial t} + \frac{\partial}{\partial x}(\rho u) + \frac{\partial}{\partial y}(\rho v) + \frac{\partial}{\partial z}(\rho w) = 0 \qquad (10\text{-}4)$$

or, in a more compact vector, Equation 10-4 is:

$$\frac{\partial \rho}{\partial t} + \text{div}(\rho u) = 0 \qquad (10\text{-}5)$$

Equation 10-5 is the unsteady, three-dimensional mass conservation or continuity equation at a point in a compressible fluid. The first term on the left side is the rate of change in time of the density (mass per unit volume). The second term describes the net flow of mass leaving the element across its boundaries and is called the convective term.

For an incompressible fluid (i.e., constant density) Equation 10-4 becomes

$$\frac{\partial u}{\partial x} + \frac{\partial v}{\partial y} + \frac{\partial w}{\partial z} = 0 \qquad (10\text{-}6)$$

or $\text{div } u = 0$ \qquad (10\text{-}7)$

An equation for conservation of momentum can be derived by considering Newton's second law, which states that the rate of change of momentum of a fluid particle equals the sum of the forces on the particle. That is,

$$\begin{array}{l}\text{Rate of increase} \\ \text{of momentum of} \\ \text{fluid particle}\end{array} = \begin{array}{l}\text{Sum of forces} \\ \text{on fluid particle}\end{array}$$

The rates of increase of x, y, and z momentum per unit volume of a fluid particle are given by

$$\rho \frac{Du}{Dt} \qquad \rho \frac{Dv}{Dt} \qquad \rho \frac{Dw}{Dt}$$

The two types of forces on fluid particles are classified as:

- *Body forces:* gravity force, centrifugal force, Coriolis force electromagnetic force.
- *Surface forces:* pressure forces, viscous forces.

It is possible to determine the x-component of the momentum equation by setting the rate of change of x-momentum of the fluid particle equal to the total force in the x-direction on the element due to surface stresses plus the rate of increase of x-momentum due to sources, which gives:

$$\rho \frac{Du}{Dt} = \frac{\partial(-p + \tau_{xx})}{\partial x} + \frac{\partial \tau_{yx}}{\partial y} + \frac{\partial \tau_{zx}}{\partial z} + S_{Mx} \tag{10-8}$$

The y-component and z-component of the momentum equation are given by

$$\rho \frac{Dv}{Dt} = \frac{\partial \tau_{xy}}{\partial x} + \frac{\partial(-p + \tau_{yy})}{\partial y} + \frac{\partial \tau_{zy}}{\partial z} + S_{My} \tag{10-9}$$

$$\rho \frac{Dw}{Dt} = \frac{\partial \tau_{xz}}{\partial x} + \frac{\partial \tau_{yz}}{\partial y} + \frac{\partial(-p + \tau_{zz})}{\partial z} + S_{Mz} \tag{10-10}$$

The sign associated with the pressure is opposite to that associated with the normal viscous stress. The usual sign convention assumes that a tensile stress is the positive normal stress so that the pressure, which by definition has compressive normal stress, has a negative sign.

The effects of the surface stresses are accounted for explicitly, and source terms S_{Mx}, S_{My}, and S_{Mz} include contributions due to the body process only. For example, the body force due to gravity would be modeled by $S_{Mx} = 0$, $S_{My} = 0$, and $S_{Mz} = -\rho g$.

The Navier-Stokes equation with viscous stresses for the x, y, and z directions are given by:

$$\rho \frac{Du}{Dt} = \frac{-\partial p}{\partial x} + \frac{\partial}{\partial t}\left[2\mu \frac{\partial u}{\partial x} + \lambda \operatorname{div} u\right] + \frac{\partial}{\partial y}\left[\mu\left(\frac{\partial u}{\partial y} + \frac{\partial v}{\partial x}\right)\right]$$

$$+ \frac{\partial}{\partial z}\left[\mu\left(\frac{\partial u}{\partial z} + \frac{\partial w}{\partial z}\right)\right] + S_{Mx} \tag{10-11}$$

$$\rho \frac{Dv}{Dt} = \frac{-\partial p}{\partial y} + \frac{\partial}{\partial x}\left[\mu\left(\frac{\partial u}{\partial y} + \frac{\partial v}{\partial x}\right) + \frac{\partial}{\partial y}\left[2\mu\frac{\partial v}{\partial y} + \lambda\,\mathrm{div}\,u\right]\right]$$

$$+ \frac{\partial}{\partial z}\left[\mu\left(\frac{\partial v}{\partial z} + \frac{\partial w}{\partial y}\right)\right] + S_{My} \tag{10-12}$$

$$\rho \frac{Dw}{Dt} = \frac{-\partial p}{\partial z} + \frac{\partial}{\partial x}\left[\mu\left(\frac{\partial u}{\partial z} + \frac{\partial w}{\partial x}\right)\right] + \frac{\partial}{\partial y}\left[\mu\left(\frac{\partial v}{\partial z} + \frac{\partial w}{\partial y}\right)\right]$$

$$+ \frac{\partial}{\partial z}\left[2\mu\frac{\partial w}{\partial z} + \lambda\,\mathrm{div}\,u\right] + S_{Mz} \tag{10-13}$$

$$\text{where} \quad \mathrm{div}\,u = \frac{\partial u}{\partial x} + \frac{\partial v}{\partial y} + \frac{\partial w}{\partial z} \tag{10-14}$$

TURBULENCE ON TIME-AVERAGED NAVIER-STOKES EQUATIONS

Equations 10-1 through 10-14 only hold for laminar flow. At values of Reynolds number above $N_{Re_{crit}}$, a complicated situation emerges resulting in a radical change of flow character. In this case, fluid motion becomes intrinsically unsteady even with constant imposed boundary conditions. The velocity and all other flow properties vary in a random and chaotic manner. This regime is known as the turbulent flow. The instantaneous fluid velocity at a certain point is represented by its average value and a superimposed fluctuation. Taking Cartesian coordinates so that the velocity vector, u, has x-component u, y-component v, and z-component w:

$$\mathrm{div}\,u = 0 \tag{10-15}$$

$$\frac{\partial u}{\partial t} + \mathrm{div}(uu) = \frac{-1}{\rho}\frac{\partial p}{\partial x} + v\,\mathrm{div}\,\mathrm{grad}\,u \tag{10-16}$$

$$\frac{\partial v}{\partial t} + \mathrm{div}(vu) = \frac{-1}{\rho}\frac{\partial p}{\partial y} + v\,\mathrm{div}\,\mathrm{grad}\,v \tag{10-17}$$

$$\frac{\partial w}{\partial t} + \text{div}(wu) = \frac{-1}{\rho}\frac{\partial p}{\partial z} + v\,\text{div grad}\,w \qquad (10\text{-}18)$$

where

$$u = U + u'(t) \qquad (10\text{-}19)$$

$$v = V + v'(t) \qquad (10\text{-}20)$$

$$z = W + w'(t) \qquad (10\text{-}21)$$

The resulting momentum balance for the mean velocity components U, V, and W becomes:

$$\frac{\partial U}{\partial t} + \text{div}(UU) = \frac{-1}{\rho}\frac{\partial P}{\partial x} + v\,\text{div grad}\,U$$

$$+ \left[-\frac{\overline{\partial u'^2}}{\partial x} - \frac{\overline{\partial u'v'}}{\partial y} - \frac{\overline{\partial u'w'}}{\partial z} \right] \qquad (10\text{-}22)$$

$$\frac{\partial V}{\partial t} + \text{div}(VU) = \frac{-1}{\rho}\frac{\partial P}{\partial y} + v\,\text{div grad}\,V$$

$$+ \left[-\frac{\overline{\partial u'v'}}{\partial x} - \frac{\overline{\partial v'^2}}{\partial y} - \frac{\overline{\partial v'w'}}{\partial z} \right] \qquad (10\text{-}23)$$

$$\frac{\partial W}{\partial t} + \text{div}(WU) = \frac{-1}{\rho}\frac{\partial P}{\partial z} + v\,\text{div grad}\,W$$

$$+ \left[-\frac{\overline{\partial u'w'}}{\partial x} - \frac{\overline{\partial v'w'}}{\partial y} - \frac{\overline{\partial w'^2}}{\partial z} \right] \qquad (10\text{-}24)$$

The extra stress terms result from six additional stresses, three normal stresses, and three shear stresses:

$$\tau_{xx} = -\rho\overline{u'^2} \quad \tau_{yy} = -\rho\overline{v'^2} \quad \tau_{zz} = -\rho\overline{w'^2}$$

$$\tau_{xy} = \tau_{yx} = -\rho\overline{u'v'} \quad \tau_{xz} = \tau_{zx} = -\rho\overline{u'w'} \quad \tau_{yz} = \tau_{zy} = -\rho\overline{v'w'} \quad (10\text{-}25)$$

These extra turbulent stresses are termed the Reynolds stresses. In turbulent flows, the normal stresses $-\rho\overline{u'^2}$, $-\rho\overline{v'^2}$, and $-\rho\overline{w'^2}$ are always non-zero because they contain squared velocity fluctuations. The shear stresses $-\rho\overline{u'v'}$, $-\rho\overline{u'w'}$, $-\rho\overline{v'w'}$ and are associated with correlations between different velocity components. If, for instance, u' and v' were statistically independent fluctuations, the time average of their product $\overline{u'v'}$ would be zero. However, the turbulent stresses are also non-zero and are usually large compared to the viscous stresses in a turbulent flow. Equations 10-22 to 10-24 are known as the Reynolds equations.

For a turbulence model to be useful in a general-purpose CFD code, it must be simple, accurate, economical to run, and have a wide range of applicability. Table 10-1 gives the most common turbulence models. The classical models use the Reynolds equations and form the basis of turbulence calculations in currently available commercial CFD codes. Large eddy simulations are turbulence models where the time-dependent flow equations are solved for the mean flow and the largest eddies and where the effects of the smallest eddies are modeled.

TIME-DEPENDENT TURBULENT MIXING AND CHEMICAL REACTION IN STIRRED TANKS[1]

Blend time and chemical product distribution in turbulent agitated vessels can be predicted with the aid of Computational Fluid Mixing

---

[1]Written by A. Bakker and J. B. Fasano. Presented at the AIChE Annual Meeting, November 1993, St. Louis, Paper No. 70C.

Table 10-1
Turbulence models

Classical models	Based on (time-averaged) Reynolds equations.
	1. Zero equation model—mixing length model.
	2. Two-equation model k-e model.
	3. Reynolds stress equation model.
	4. Algebraic stress model.
Large eddy simulation	Based on space-filtered equations.

(CFM) models. The blend time predictions show good agreement with the experimental correlation discussed below. Calculations for turbulent, time-dependent mixing of two chemicals exhibiting a competitive pair of reactions, are compared with the experimental results. The effects of the position of the inlet feed stream in the turbulent flow field are also studied below, which will show that process problems with turbulent chemical reactors can be avoided by incorporating the results of CFM simulations in the design stage.

Blending of chemical reactants is a common operation in the chemical process industries. Blend time predictions are usually based on empirical correlations. When a competitive side reaction is present, the final product distribution is often unknown until the reactor is built. The effects of the position of the feed stream on the reaction byproducts are usually unknown. Also, the scale-up of chemical reactors is not straightforward. Thus, there is a need for comprehensive, physical models that can be used to predict important information like blend time and reaction product distribution, especially as they relate to scale and feed position.

The objective of the following model is to investigate the extent to which Computational Fluid Mixing (CFM) models can be used as a tool in the design of industrial reactors. The commercially available program, Fluent^{TM}, is used to calculate the flow pattern and the transport and reaction of chemical species in stirred tanks. The blend time predictions are compared with a literature correlation for blend time. The product distribution for a pair of competing chemical reactions is compared with experimental data from the literature.

MODEL

The flow pattern is calculated from conservation equations for mass and mometum, in combination with the Algebraic Stress Model (ASM) for the turbulent Reynolds stresses, using the Fluent V3.03 solver. These equations can be found in numerous textbooks and will not be reiterated here. Once the flow pattern is known, the mixing and transport of chemical species can be calculated from the following model equation:

$$\frac{\partial}{\partial t}(\rho X_i) + \frac{\partial}{\partial x_i}(\rho u_i X_i) = \frac{\partial}{\partial x_i}\left(\frac{\mu_t}{Sc_t}\frac{\partial X_i}{\partial x_i}\right) + R_i \qquad (10\text{-}26)$$

Here, X_i is the mass fraction of chemical species i, and R_i is the rate of creation or depletion by chemical reaction. For a single-step, first order reaction, such as $A + B \rightarrow R$, the reaction rate is given by:

$$R_i \propto \left(C_A C_B + \overline{C}_A \overline{C}_B \right) \tag{10-27}$$

Here, C_A and C_B (upper case) denote the mean molar concentrations of reactants A and B while C_A and C_B (lower case) denote the local concentration fluctuations that result from turbulence. When the species are perfectly mixed, the second term on the right side containing the correlation of the concentration fluctuations, will approach zero. Otherwise, if the species are not perfectly mixed, this term will be negative and will reduce the reaction rate. Estimating this correlation term is not straightforward and numerous models are available. An excellent discussion on this subject was given by Hannon [1].

The model used here is a slightly modified version of the standard Fluent model [2]. Two possible reaction rates are calculated, the kinetic reaction rate R_{ki} and a second reaction rate R_{mi} that is controlled by the turbulent mixing. The kinetic reaction rate for species i is calculated as:

$$R_{ki} = KM_i \prod_{j \text{ reactants}} \frac{\rho X_j}{M_j} \tag{10-28}$$

The turbulent mixing limited reaction rate for species i is calculated as:

$$R_{mi} = \left(M_i A_{mn} \frac{\varepsilon}{k} \right) \cdot \text{minimum} \left[\left(\frac{\rho X_j}{\upsilon_j M_j} \right)_{\text{reactants } j} \right] \tag{10-29}$$

The "minimum" function gives the minimum value of $(\rho X_j / \upsilon_j M_j)$ for all the reactants j taking part in this reaction. Finally, the reaction rate R_i is calculated as the product of the molar stoichiometry υ_i of species i and the minimum of R_{ki} and R_{mi}:

$$R_i = -\upsilon_i \text{ minimum } (R_{mi}, R_{ki}) \tag{10-30}$$

Here, M_i is the molecular weight of species i and A_{mn} is an empirically determined model constant for reaction n. In this reaction system, υ_i is +1 for reactants and −1 for products. K is the kinetic rate constant of the reaction.

The idea behind this model is that in regions with high turbulence levels the eddy lifetime k/ε will be short, the mixing fast, and as a result the reaction rate is not limited by small-scale mixing. On the other hand, in regions with low turbulence levels, small-scale mixing may be slow, which will limit the reaction rate.

REACTION MODELING RESULTS

The following competitive-consecutive reaction system was studied:

$$A + B \xrightarrow{\ K_1\ } R \tag{10-31}$$

$$B + R \xrightarrow{\ K_2\ } S \tag{10-32}$$

This is the reaction system used by Bourne et al. [3] and Middleton et al. [4]. The first reaction is much faster than the second reaction: $K_1 = 7,300 \ m^3 \cdot mole^{-1} \cdot sec^{-1}$ versus $K_2 = 3.5 \ m^3 \cdot mole^{-1} \cdot sec^{-1}$. The experimental data published by Middleton et al. [4] were used to determine the model constant A_{mn}. Two reactors were studied, a 30-l reactor equipped with a D/T = 1/2 D-6 impeller and a 600-l reactor with a D/T = 1/3 D-6 impeller. A small volume of reactant B was instantaneously added just below the liquid surface in a tank otherwise containing reactant A. A and B were added on an equimolar basis. The transport, mixing, and reaction of the chemical species were then calculated based on the flow pattern in Figure 10-3. Experimental data were used as impeller boundary conditions. The product distribution X_S is then calculated as:

$$X_S = \frac{2C_S}{C_R + 2C_S} \tag{10-33}$$

In the reaction model used here it was assumed that small-scale mixing only affected the first reaction and that once this reaction had occurred, the species were locally well mixed. As a result, small-scale

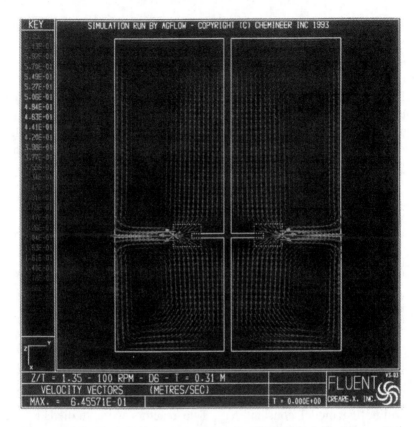

Figure 10-3. Flow field in 30-l reactor (see full-color version on CD).

turbulent mixing did not affect the second reaction. This was achieved using different values of A_{mn} for both reactions. For the second reaction, A_{m2} was set to infinity. The value for A_{m1} was then varied to study the effect on the predicted final product distribution.

Figure 10-4 shows the predicted X_S as a function of A_{m1} for the 30-1 reactor at 100 rpm. Decreasing A_{m1} slows the first reaction and increases the formation of the secondary product S. As a result, the predicted X_S decreases with increasing A_{m1}. It was found that $A_{m2} = 0.08$ gave the best predictions when compared to the experimental data from Middleton et al. [4] Figure 10-5 shows a comparison between the experimental data from Middleton et al. and the current model predictions for both the 30-1 and 600-1 reactors. X_S is plotted as a function

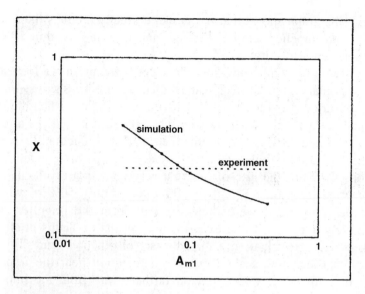

Figure 10-4. Predicted X_S versus A_{m1} for A_{m2} equals infinity. 30-l reactor at 100 rpm.

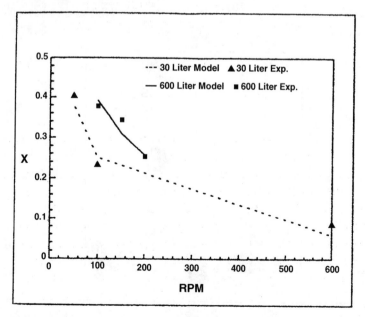

Figure 10-5. X_S as a function of rpm. Model predictions compared with data from Middleton et al. [4].

of rpm. This graph shows that the model correctly predicts the effects of scale and impeller rotational speed, and is usually within 10% of the experimental results.

The effect of the inlet position of the feed stream on the formation of the secondary byproducts S was studied. Figure 10-6 shows values of X_S for various feed locations. X_S varies only slightly when the inlet is located in the fluid bulk. However, when the feed is injected directly above the impeller, such that the feed stream immediately passes through the highly turbulent impeller zone, local mixing is much faster and does not limit the rate of the first reaction. As a result, there is less reaction byproduct S and the final X_S in only 50% of what it would be if the feed were located away from the impeller. This qualitatively agrees with the experimental results of Tipnis et al. [5]. Although Tipnis et al. used a different set of reactions and different tank geometries, it was also found that injection near the impeller results in a lower X_S than injection farther away from the impeller and that the relative differences are similar to those found in this study.

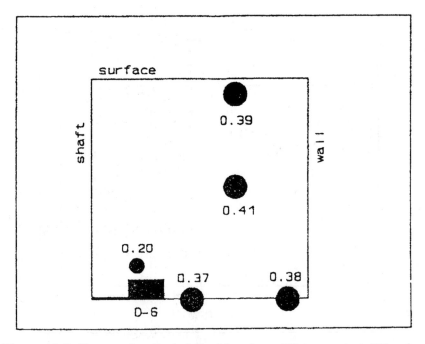

Figure 10-6. X_S as a function of feed location. 600-l vessel at 100 rpm. $A_{m1} = 0.08$ and A_{m2} equals infinity.

Figure 10-7 shows the concentrations of R and S and the product distribution X_S as a function of time for the feed location just above the impeller. The values are normalized with respect to the final values. R and S increase steadily with time. X_S increases at first, reaching a local maximum just before the species are mixed by the impeller. The improved quality of the mixture favors the first reaction and X_S decreases until it reaches a local minimum. At this point there is enough R present to allow the second reaction to occur even in relatively well mixed regions, and X_S increases again until it asymptotically reaches a final value. Figures 10-8a through 10-8h show the local concentrations of species A, R, and S as a function of time for the 600-l tank at 100 rpm.

BLEND TIME

The mixing of two nonreacting species in a tank equipped with a high-efficiency impeller (Chemineer HE-3) was calculated using Fluent

(text continued on page 806)

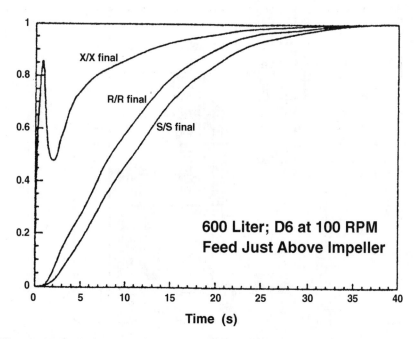

Figure 10-7. Concentrations of R and S and product distribution X_S as a function of time, normalized with final values.

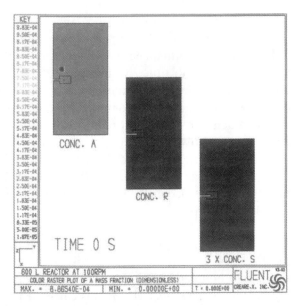

Figure 10-8a. The local mass fractions in the 2-D reaction simulation at Time = 0 s (see full-color version on CD).

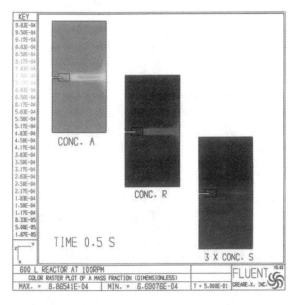

Figure 10-8b. The local mass fractions in the 2-D reaction simulation at Time = 0.5 s (see full-color version on CD).

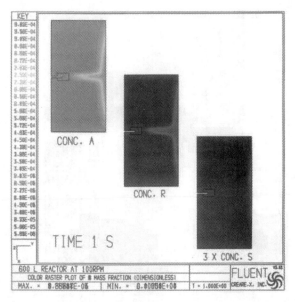

Figure 10-8c. The local mass fractions in the 2-D reaction simulation at Time = 1 s (see full-color version on CD).

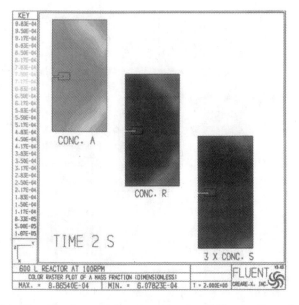

Figure 10-8d. The local mass fractions in the 2-D reaction simulation at Time = 2 s (see full-color version on CD).

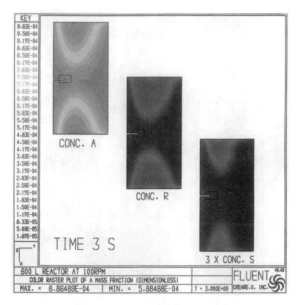

Figure 10-8e. The local mass fractions in the 2-D reaction simulation at Time = 3 s (see full-color version on CD).

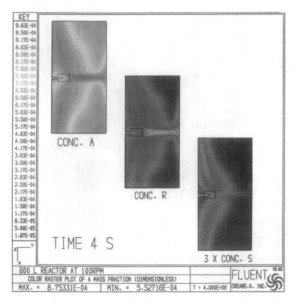

Figure 10-8f. The local mass fractions in the 2-D reaction simulation at Time = 4 s (see full-color version on CD).

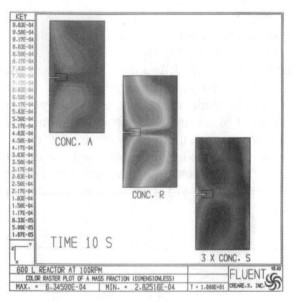

Figure 10-8g. The local mass fractions in the 2-D reaction simulation at Time = 10 s (see full-color version on CD).

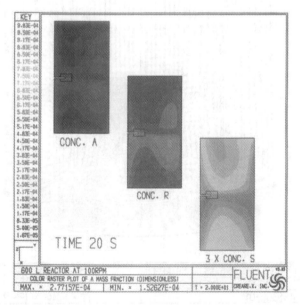

Figure 10-8h. The local mass fractions in the 2-D reaction simulation at Time = 20 s (see full-color version on CD).

(text continued from page 801)

V3.03. The tank diameter was T = 1 m. Furthermore, Z/T = 1, D/T = 0.33, C/T = 0.32, and rpm = 58. The flow pattern in this tank is shown in Figure 10-9. Experimental data were used as impeller boundary conditions. Figure 10-10 shows the uniformity of the mixture as a function of time. The model predictions are compared with the results of the experimental blend time correlation of Fasano and Penny [6]. This graph shows that for uniformity above 90% there is excellent agreement between the model predictions and the experimental correlation. Figure 10-11a shows the concentration field at t = 0 sec. Figures 10-11b through 10-11d show the concentration field at t = 0,

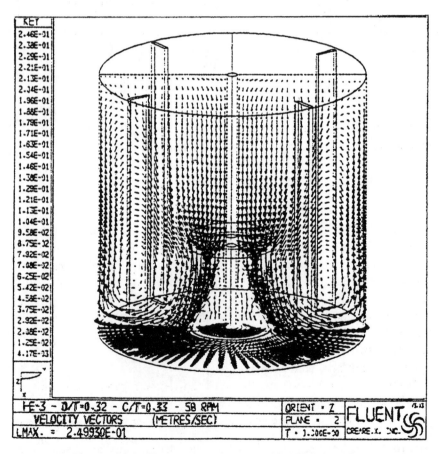

Figure 10-9. Flow pattern in tank with HE-3 impeller.

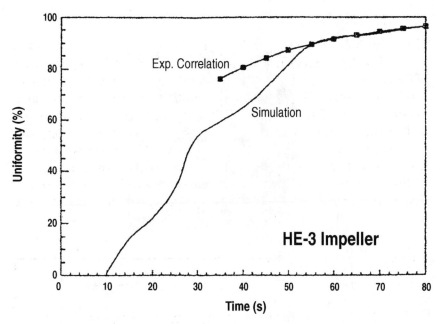

Figure 10-10. Uniformity as a function of time.

4, 10, and 20 sec, respectively. After 80 sec the species are homogeneously mixed.

DISCUSSION

The models presented correctly predict blend time and reaction product distribution. The reaction model correctly predicts the effects of scale, impeller speed, and feed location. This shows that such models can provide valuable tools for designing chemical reactors. Process problems may be avoided by using CFM early in the design stage. When designing an industrial chemical reactor it is recommended that the values of the model constants are determined on a laboratory scale. The reaction model constants can then be used to optimize the product conversion on the production scale varying agitator speed and feed position.

However, the range of validation of the reaction model was limited. Only one impeller type and one reaction system were studied. In the future, the model should be tested for a wider range of geometries and reaction systems and, if necessary, modified to increase its validity.

(text continued on page 810)

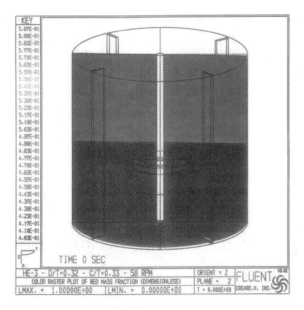

Figure 10-11a. The concentration field in the 3-D blending simulation at Time = 0 s (see full-color version on CD).

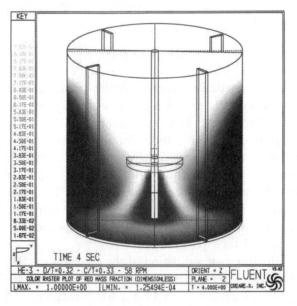

Figure 10-11b. The concentration field in the 3-D blending simulation at Time = 4 s (see full-color version on CD).

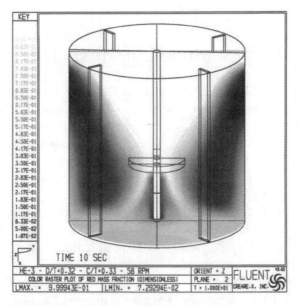

Figure 10-11c. The concentration field in the 3-D blending simulation at Time = 10 s (see full-color version on CD).

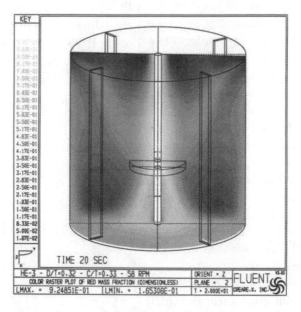

Figure 10-11d. The concentration field in the 3-D blending simulation at Time = 20 s (see full-color version on CD).

(text continued from page 807)

REFERENCES AND RECOMMENDED READING

1. Hannon, J., "Mixing and Chemical Reaction in Tubular Reactors and Stirred Tanks," PhD. Thesis, Cranfield Institute of Technology, U.K., 1992.
2. *Fluent V3.03 Users Manual,* Fluent Inc., Lebanon, NH, 1990.
3. Bourne, J. R., Kozicki, F., and Rys, P., "Mixing and Fast Reaction Reaction –1; Test Reactions to Determine Segregation," *Chem. Eng. Sci.,* 36, (10), 1643, 1981.
4. Middleton, J. C., Pierce, F., and Lynch, P. M., "Computations of Flow Fields and Complex Reaction Yield in Turbulent Stirred Reactors and Comparison with Experimental Data," *Chem. Eng. Res. Des.,* Vol. 64, pp. 18–21, 1986.
5. Tipnis, S. K., Penny, W. R., and Fasano, J. B., An experimental investigation to determine a scale-up method for fast competitive parallel reactions in agitated vessels, AIChE Annual Meeting, St. Louis, November 1993.
6. Fasano, J. B. and Penny, W. R., "Avoid Blending Mix-ups," *Chem. Eng. Prog.,* pp. 56–63, October 1991.
7. Versteeg, H. K. and Malalasekera, W., *An Introduction to Computational Fluid Dynamics—The Finite Volume Method,* Addison Wesley Longman Ltd., 1995.

NOMENCLATURE

A_{mn} = Model constant for reaction n
C_i = Concentration of species i (mole m^{-3})
k = Turbulent kinetic energy density (m^{2}s^{-2})
K = Reaction rate constant (m^3 mole^{-1} s^{-1})
M_i = Molecular weight of species i
R_i = Production/depletion of species i (kg m^{-3} s^{-1})
R_{ki} = Kinetic reaction rate of species i (kg m^{-3} s^{-1})
S_{ct} = Turbulent Schmidt number
t = Time (sec)
u_i = Velocity in directions i (m s^{-1})
X_i = Spatial coordinate in direction i (m)
X_S = Product distribution
ε = Turbulent kinetic energy dissipation rate density (m^2 s^{-3})
μ_t = Turbulent viscosity (kg m^{-1} sec^{-1})

ρ = Liquid density, (kg m^{-3})
υ_i = Stoichiometry species i

IMPROVE REACTORS VIA COMPUTATIONAL FLUID DYNAMICS[2]

INTRODUCTION

Flow modeling is an excellent tool for enhancing the performance of any process vessel. Applying such a technique to reactors can be especially fruitful, because of reactors' central role in chemical processes. In most reactor-design situations, the reactions and the catalysis system are set beforehand. For any given combination of them, reactor performance becomes a complex function of the underlying transport processes. These in turn are governed by the fluid dynamics of the reactor.

Detailed flow analysis and modeling can, accordingly lead to numerous performance improvements. For instance, an engineer can raise the throughput of a fixed-bed reactor by identifying and getting rid of any fluid maldistribution in the vessel. Modifying the feed-pipe locations in a semi-batch reactor can improve its selectivity. As another example, the capacity of an oxidation reactor might be raised by allowing oxygen-enriched instead of ordinary air to be used without risking safety.

The basic strategy for realizing such performance enhancements is summarized in Table 10-2. The principles can be applied to improving

Table 10-2
Strategy for performance enhancement—A summary

Analyze the existing reactor or design	Develop alternative versions	Evaluate and test the new alternatives
• Flow regimes, operability • Residence time distribution • Mixing • Transport processes		• Screening • Optimization • Validation

---

[2]*Written by Vivek V. Ranade and reprinted by special permission from* Chemical Engineering, *May; Copyright © 1997 by Chemical Week Association, New York, NY, 10106.*

the performance of an existing reactor, or to optimizing a reactor that has not yet been specified or built.

FLOW MODELING CHOICES

The engineer is offered a large variety of flow-modeling methods, whose complexity ranges from simple order-of-magnitude analysis to direct numerical simulation. Up to now, the methods of choice have ordinarily been experimental and semi-theoretical, such as cold flow simulations and tracer studies.

Despite their popularity, these methods normally have an inherent limitation—the fluid dynamics information they generate is usually described in global parametric form. Such information conceals local turbulence and mixing behavior that can significantly affect vessel performance. And because the parameters of these models are necessarily obtained and fine-tuned from a given set of experimental data, the validity of the models tends to extend over only the range studied in that experimental program.

Computational fluid dynamics (CFD), which solves fluid-dynamics problems with the aid of a digital computer, avoids these drawbacks while offering other advantages as well.[3] With CFD, the engineer can build a flow model having global and local information alike, simulating the flow and turbulence characteristics of reactors or other process equipment in detail. And for systems at high temperatures or pressures, or having high corrosivity or a high degree of hazard, CFD models may well be the only good tools available for studying the fluid dynamics.

CFD requires relatively few restrictive assumptions, and gives a complete description of the flow field. Complex configurations can be treated. When applied to a reactor that is operating, CFD's diagnostic probing via computer simulation does not disturb the operation. Various alternative configurations can be screened quickly using the validated CFD model. Applying CFD to reactors or other equipment [1,2] entails a number of tasks or activities:

- Formulating the relevant transport equations.
- Establishing the necessary constitutive and closure equations (the former relate fluid stresses with velocity gradients; the latter relate unknown Navier-Stokes-equation correlations with known quantities).

---

[3]Chemical Engineering, *December 1996, pp. 66–72.*

- Formulating appropriate boundary conditions.
- Selecting the most suitable numerical techniques to solve the equations.
- Choosing or developing a suitable computer program (also referred to as a code) to implement the numerical techniques.
- Validating these techniques and programs.
- Developing effective flow-simulation strategies for the equipment.

We illustrate these tasks and activities by applying them to one particular reactor. First, however, we set the stage by summarizing how the steps are carried out for a reactor of any type.[4]

APPLYING CFD

The first step is to define the objectives of the flow model, and to identify those flow aspects that are relevant for the performance of the reactor. Then, the engineer must identify and quantify the various times and space scales involved, as well as the geometry of the system. These actions allow the problem to be represented by a mathematical model. *Creating this model accurately is the most crucial task in the flow modeling project.*

The next step is to map the model onto a CFD computer program. This mainly requires specifying the geometry of the equipment, generating appropriate spatial grids (meshes) and boundary conditions, and setting the required terms in the transport equations included in the program. Sometimes, it is also necessary to generate sub-models, providing such information as physical properties, reaction rates, interphase drag coefficients, or heat or mass-transfer parameters.

Prior knowledge or educated guesses about the various time and space scales and the likely regions of steep gradients help in generating a suitable grid. Complex geometries can be handled by using coordinate systems that are neither rectangular nor polar, but instead are fitted to the shape of the vessel at hand. While generating the grids, extremes of aspect ratio and skewness are avoided. It is not too important that the grids intersect at right angles, so long as the actual angles are greater than 45°. Once the appropriate grid is generated,

---

[4]*Although this article focuses on design of reactors, keep in mind that the principles presented here can also be applied to other equipment through which process streams pass.*

the user specifies the relevant properties of the process fluids. Typically, these include viscosity, density and thermal conductivity.

Then, the model's boundary conditions on the edges and external surfaces of the vessel are specified, taking care to understand and then allow for the influence of the location of outflow and inflow boundaries on the predicted flow results. Wall functions (empirical functions that estimate the effects of walls on flow) can provide boundary conditions near the wall for turbulent flows, a practice that avoids the need for fine grids near the walls. However, this technique is used only with caution when heat or mass transport from the wall is important. Once the problem has been mapped onto the CFD program, the engineer assesses the sensitivity of the predicted results to the chosen grid spacing, time steps, boundary conditions and other parameters of the model. Making some preliminary simulations can help provide such understanding.

Because so many parameters affect the performance of the CFD program and the predicted results, we recommend that the simulations for this step be planned systematically. Such a plan must vary according to the problem at hand and the available information. However, some general comments can still be made.

For instance, grid sequencing is often useful. In this strategy, iterative model results are first obtained, rapidly, using a small number of grids (i.e., a coarse mesh). These results can then be used as initial guesses for iterations yielding more-precise answers from finer meshes. Similarly, it is often desirable to increase the complexity of the problem in steps, after starting with a highly simplified version. Overall convergence rates of CFD models can be controlled to some extent by suitably adjusting under relaxation parameters (parameters that damp the iterative solution of algebraic equations; for suggestions in this regard see reference 3). Furthermore, monitoring the rate of convergence for each equation can also give guidelines on how to enhance the overall convergence rate of the CFD program. Before putting the model to work, a final step remains: Validation exercises must verify that the model has indeed captured the important flow characteristics for the task at hand.

A SPECIFIC REACTOR

The foregoing summary can be understood better by seeing how the procedure is applied to a particular case, namely a radial-flow, fixed-bed reactor. Such reactors appear in a variety of processes, including the catalytic synthesis of ammonia, catalytic reforming of petroleum stocks, xylene isomerization, and various desulfurizations. They handle large gas flowrates with minimum pressure drop, and are especially suitable for processes in which fluids must be contacted with solid particles at high space velocity [4].

The fluid dynamics of radial flow reactors (RFRs) in general is complex, and involves severe changes in the flow direction. In RFRs, feed enters parallel to the reactor axis, either through a center pipe or an annulus, and then flows radially inward through the annular catalyst basket. Perfect radial flow always results in the highest conversion [4]. Any axial component of the flow decreases the conversion, because it mixes fluids of different "ages" within the bed (an effect similar to backmixing in tubular reactors). Therefore, flow maldistribution is the most important performance determinant in a radial flow fixed bed reactor. Flow modeling can be used to screen various design solutions, so as to minimize the maldistribution.

DEFINE THE PROBLEM

One half of a typical radial-flow, fixed-bed reactor is shown in Figure 10-12a. The reactor configuration is axis-symmetric. The annular catalyst bed is supported by cylindrical inner and outer screens and by top and bottom cover plates, the latter actually being also in the form of a screen. Reactants enter at the top. As the reader can visualize, the flow changes direction several times after hitting the cover plate. The reactants enter the catalyst bed from the annular space between the bed and reactor shell, passing through the outer support screen. The product stream goes out through the outlet located at the bottom of the central pipe.

Because the cover plate also acts as a shroud, the active catalyst bed is limited to the annular Zone A as shown in the inset in Figure 10-12a. The extent of flow maldistribution within the active bed depends mainly on the throughput, the configuration details around and within the bed, and the flow resistance offered by the support screens and the bed. It may be possible to control the bed resistance to some extent by appropriately selecting the pellet size. However, the design of support screens and the configuration are the most important parameters governing the fluid dynamics and therefore the performance. For this reactor, our capacity-enhancement attempt in this example consists mainly of two strategies:

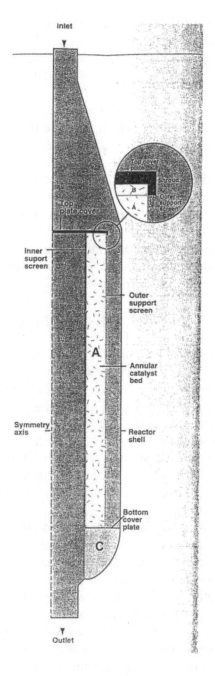

Figure 10-12a. The configuration of this radial fixed-flow reactor provides scope for investigating proposed performance-enhancing modifications.

- Assessing the fluid dynamics of the initial RFR configuration, and identifying the scope of eliminating the flow maldistribution, if any.
- Exploring the possibility of loading more catalyst, by increasing the volume of active catalyst bed. This might be achieved by eliminating the shroud and filling the catalyst up to the top cover plate (thus picking up the space shown as Zone B of Figure 10-12a, inset) and also by filling the catalyst till the bottom of the reactor (Zone C in Figure 10-12a). However, it is important to ensure, by proper screen redesign, that such shroud elimination would not lead to flow maldistribution. Also, the support screens for the catalyst filled into Zone C would need to be redesigned so as to ensure the uniform flow through the catalyst bed. It is therefore essential to develop a detailed flow model to carry out our two strategies.

DEVELOP THE FLOW MODEL

The typical rate of flow (throughput) in RFRs such as the one under consideration is such that the flow is turbulent (feed velocity at the inlet is 40 m/s). Therefore, it is necessary to select an appropriate turbulence model—in particular, we need to simulate the flow in the upper region of the reactor correctly, so that the influences of the aforementioned severe changes in the flow directions can be predicted accurately. Expecting that the recirculating flow in the upper region of the reactor will show spatial variation in velocity and length scales of turbulence, we need at least a two-equation turbulence model. We select a standard model, known as the k-ε[5], that has proved versatile.

The next step is to characterize the resistances offered by the porous catalyst bed and support screens. Several correlations relating the pressure drop through porous beds and velocity and bed characteristics are available. We select an Ergun equation[6] to represent the resistance of catalyst bed:

$$\frac{\Delta P}{L} = \left\{ \left[150 \frac{\mu}{D_P \Phi_P^2} \right] \left[\frac{(1-\varepsilon)^2}{\varepsilon^3} \right] V + \left[1.75 \frac{\rho}{D_P \Phi_P} \right] \left[\frac{(1-\varepsilon)}{\varepsilon^3} \right] V^2 \right\} \quad (10\text{-}33)$$

---

[5]*For more insight into this model, see "CFD Comes of Age in CPI," Chemical Engineering, December 1996, pp. 68–72.*

[6]*See, for instance, Darby, Chemical Engineering Fluid-Mechanics, Marcel Dekker, New York, 1996, pp. 68–72.*

where $\Delta P/L$ = the pressure drop per unit length
$\qquad \mu$ = viscosity
$\qquad D_P$ = equivalent pellet diameter
$\qquad \Phi_P$ = sphericity
$\qquad \rho$ = density
$\qquad V$ = superficial velocity

Our knowledge of the pellet size and shape and the bed voidage are thus sufficient to characterize the bed resistance. For the particular case investigated here, the resistance can be represented as:

$$\frac{\Delta P}{L} = \left(\frac{\mu}{\alpha}\right) V + C_2 \left[\left(\frac{1}{2}\right)\rho\, V^2\right] \qquad (10\text{-}34)$$

with permeability α as 10^{-8} m^2 and inertial resistance coefficient C_2 as 10^4/m. The resistance of the screens can be represented in terms of contraction and expansion losses. The velocity heads lost during flow through the screens can also be expressed in the form similar to that described in Equation 10-34 by setting α equal to very high value (10^{10}) with an appropriated value for C_2. The compressibility of the gaseous feed can be ignored because the pressure drops are low. The physical properties of the feed (viscosity as 10^{-5} Pa-s and density as 1 kg/m^3) can thus be assumed constant. Since we need not simulate any reactions or heat transfer in this exercise, we do not need to develop sub-models.

MAP THE MODEL INTO A PROGRAM

We choose to map the model developed in the preceding section onto a particular CFD software program, Fluent Version 4.31[7]. The reactor geometry is likewise modeled and, an appropriate grid for the problem is generated by using preBFC[7] software. In light of the reactor configuration, we select a geometry that is axis-symmetric and two-dimensional. Preliminary numerical experiments show us that so long as the number of grids in the radial direction is over 40 and the number in the axial direction over 100, the predicted results of the pressure drop and flow maldistribution become insensitive to the actual number of grids.

---

[7]*Of Fluent Inc., Lebanon, NH.*

The mesh used for all the subsequent computations (50 grids in the radial direction and 116 in the axial direction) appears in Figure 10-12b.

The standard k-ε model simulates the turbulence in the reactor. For flow within the porous catalyst bed, however, we suppress the turbulence. We enter the appropriate physical properties of the system, and employ standard boundary conditions at the impermeable walls and the reactor outlet. To represent the turbulence of the feed stream at the inlet, we treat it as pipe-flow turbulence. These model equations can then be solved; for instance, via the well-known Simple algorithm [3]. To facilitate fast convergence, it is useful to make a reasonable initial guess of the pressure drop across the catalyst bed.

We choose to carry out only few numerical experiments to select the solution parameters. Detailed optimization of the solution parameters is difficult and often expensive computationally, so we do not recommend it. Finally, we must validate the model. Though detailed experimental data for the velocity and pressure profiles are not available for this particular RFR, we can employ the data on the overall pressure drop across the bed to validate the model to some extent. We find that the predicted overall pressure drop across the bed (10 kPa) shows good agreement with the available data.

PUT THE MODEL TO WORK

Now the model is ready for optimizing the reactor design. Of particular interest in this example are details on the flow at the locations where the flow direction changes abruptly, and the extent of maldistribution within the active catalyst bed.

First we run the model so as to study the influence of screen resistance on the overall flow patterns and on the maldistribution. The resulting profiles of inward radial velocity at the inner screen across the catalyst bed appear in Figure 10-13 for different screen resistances. It can be seen that higher screen resistance leads to more-uniform flow, as one would expect. The existing screens (with resistance coefficients C_2 of 2×10^5/m) appear to be satisfactory, since the deviations experienced are less than 10%.

From the contours of the stream function and a closeup of the flow field near the shroud and top cover plate, we note that the shroud causes significant recirculation at the top end of the catalyst bed. Also, the downward velocity field in the annular region between catalyst bed and reactor shell is found not to be uniform. The comparative

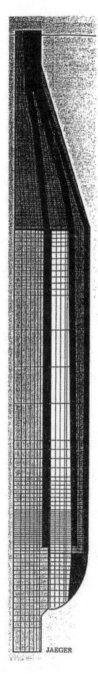

Figure 10-12b. When CFD is the investigator's tool, a grid or mesh is mathematically superimposed over the vessel.

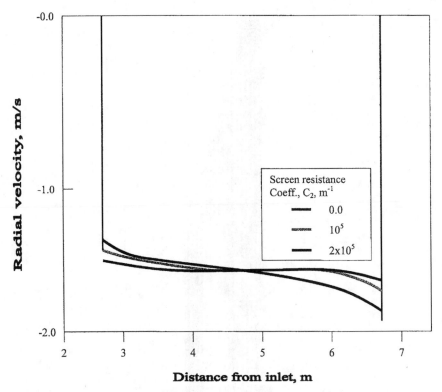

Figure 10-13. The radial velocity at the inner screen of the reactor does not vary much with screen resistance.

uniformity (Figure 10-13) of the flow coming out of the catalyst bed into the central pipe signals to us that we cannot enhance the capacity of the reactor much by modifying the existing screen designs. So we turn to the second possibility, namely removing the shroud and increasing the active catalyst loading by filling the catalyst in Zones B and C as shown in Figure 10-12a.

We simulate such a scenario, specifying support screens for Zones B and C that are similar to the existing ones in Zone A. Then we re-examine the contours of the stream function and the details of flow near the top cover plate. It turns out that the simulated removal of shroud gives more circulation in the upper region of the RFR (Figure 10-14). The predicted profile of the radial velocity at the inner edge of the catalyst bed appears in profile B in Figure 10-15 (the corresponding profile for the earlier case is also shown for comparison

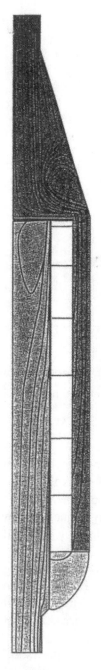

Figure 10-14. Images generated during CFD help assess the desirability of proposed vessel-design modifications.

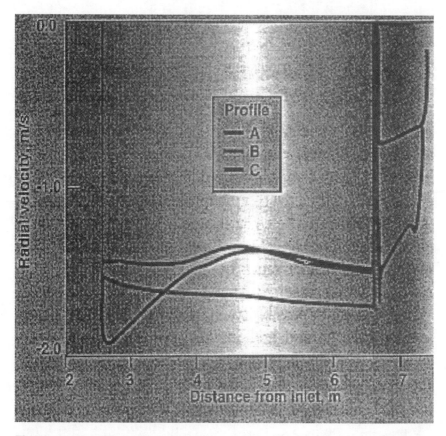

Figure 10-15. The lines show radial velocity at the inner reactor screen for different modification proposals.

purposes as profile A). Profile B confirms that the removal of shroud and filling the catalyst in Zones B and C would make the flow through the catalyst bed much less uniform.

The high resistance offered by the support screens of Zone C leads to very low flow through that zone, and also leads to a recirculating zone in the annular space between the catalyst bed and the reactor shell. The non-uniform flow through the catalyst bed also brings about significant recirculation in the central pipe. In short, the model tells us that the mere removal of shroud and filling the catalyst in Zones B and C may not lead to capacity enhancements, due to the associated problems of maldistribution.

On the other hand, the model may be able to bring out a way to redesign the support screens to improve the flow uniformity. Specifically, we postulate that the shroud might be replaceable by a support screen with (due to its shroudlike role) significantly higher resistance than the existing support screens. To examine this possibility, we adapt the model so as to contain an outer support screen resistance for Zone B five times that for Zone A. Meanwhile, to get more flow through Zone C, the support screen resistance at that area might be reduced. We simulate this possibility by specifying zero resistance for the cover plate (which, as previously noted, is actually a screen) for Zone C. The resistance of the inner support screen for Zone C is set to 5×10^4/m.

The resulting contours of the stream lines, a close-up of a vector plot near the top cover plate and the profile of radial velocity at the inner edge of the catalyst bed (Profile C, Figure 10-15) all show us that these last-named proposed changes would have many benefits. The main ones are:

- Recirculation in the annular space between the catalyst bed and reactor shell is eliminated.
- The size of the recirculating zone in the central pipe is much reduced.
- Flow non-uniformity due to shroud removal is largely eliminated.
- Flow through Zone C also has been considerably increased, and is now of the same order as that through Zones A and B.

In summary, our CFD analysis has let us simulate a variety of changes in the design of a particular RFR. Some of the changes, such as just removing the shroud, prove to offer little or no benefit. Others, notably the support-screen redesign, seem well worth pursuing.

KEEP IN MIND

It must be reemphasized that the value of a flow model's recommendations depends on how well the model represents the real process situation. The reactor and the process streams must be described accurately, as must the relationship between the fluid dynamics and the process performance. Often, process engineers are tempted to rely on commercial CFD programs for the fluid dynamics equations. However, any commercial program may have particular limitations for simulating complex process equipment. On the other hand, almost all of such programs let the user make adaptations that accommodate the situation at hand.

Turbulence, which prevails in the great majority of fluid-flow situations, poses special problems. Due to the wide range of space and time scales in turbulence flow, its exact numerical simulation is possible only at relatively low Reynolds number (around 100 or below) and if the geometry is simple.

For most applications, the engineer must instead resort to turbulence models along with time-averaged Navier-Stokes equations. Unfortunately, most available turbulence models obscure physical phenomena that are present, such as eddies and high-vorticity regions. In some cases, this deficiency may partially offset the inherent attractiveness of CFD noted earlier.

Concluding, it is essential to represent complex, real-life flow situations by computationally tractable models that retain adequate details. As an example, a computational "snapshot" approach that simulates the flow in stirred reactors or other vessels for any arbitrary impeller has been developed [5]. This approach lets the engineer simulate the detailed fluid dynamics around the impeller blades with much less computations that would otherwise be required. Improvements in CFD technique are likely to encourage further work along these lines.

Application of CFD to reactors and other process vessels is sure to grow. It not only permits innovation and performance enhancement for new processes and equipment, but also allows substantial improvement in the operation of mature technologies.

USES OF CFD CODES

Almost all modern CFD codes have a $k - \varepsilon$ model. Advanced models like algebraic stress models or Reynolds stress model are provided FLUENT, PHOENICS and FLOW3D. Table 10-3 summarizes the capabilities of some widely used commercial CFD codes. Other commercially CFD codes can be readily assessed on the web from hptt//www.cfd-online.com This is largest CFD site on the net that provides various facilities such as a comprehensive link section and discussion forum.

(text continued on page 828)

Table 10-3
Particular features of some commercial CFD codes—A comparison

	PHOENICS	FLUENT	FLOW3D	ASTEC	FIDAP
Supplier	Cham, London, U.K. or Hunstville, AL or Creare.x. Inc. Hanover, NH	Fluent Europe, Sheffield, U.K. or Lebanon, NH	CFDS AEA Tech., Oxon, U.K. or Pittsburgh, PA	CFDS, AEA Tech., Oxon, U.K. or Pittsburgh, PA	Fluid Dynamics, International, Inc. Evanston, IL, USA
Numerical Method	Finite volume	Finite volume	Finite volume	Finite volume	Finite Element
Turbulence modeling capability (range of models).	Eddy viscosity k-l, k-ε, and Reynolds stress.	k-ε and Algebraic stress, Reynolds stress and renormalization group theory (RNG) V. 4.2	k-ε, low Reynolds No., Algebraic stress, Reynolds stress and Reynolds flux.	k-ε	Mixing length (user subroutine) and k-ε.
Compressible flow capability	Available (including supersonic)	Available	Available (including supersonic)	Weakly compressible, Mach. No. up to 0.2	Not available
Non-Newtonian modeling capability	Choice of models available, user subroutines.	Power law, and difficult to implement user subroutines.	Choice of models available.	Not available.	Power law, Bingham, Generalized power law, and user subroutines.

Multiphase flow capability.	Available.	Available.	Available.	Available.	Available.
Free surface capability.	Available.	Not available.	Limited.	Available.	Available.
Mesh generation features.	Structured mesh, interactive mesh generation.	Structured mesh.	Structured mesh multiblock.	Unstructured mesh, multiblock interface to PATRAN and IDEAS.	Unstructured.
Combustion modeling capability.	Combustion/chemical reaction.	Combustion/chemical reaction	Combustion/chemical reaction	Combustion/chemical reaction	Simple combustion/Chemical reaction
Target Application	General purpose	General purpose.	General purpose.	Main area of application is in heat and fluid in highly complex geometries.	General purpose.
Code expandability.	User supplied FORTRAN subroutines can be added.	Cannot add FORTRAN subroutines directly. Need full source code for code expansion.	User supplied FORTRAN subroutines can be added.	User supplied FORTRAN subroutines can be added.	User supplied FORTRAN subroutines can be added.

Source: Dombrowski, N., et al., "Know the CFD Codes," Chem. Eng. Prog., September 1993.

(text continued from page 825)

REFERENCES

1. Ranade, V. V., "Computational fluid dynamics for reactor engineering," *Reviews in Chem. Eng.,* 11, 229, 1995.
2. Shyy, W., Application of body-fitted coordinates in transport processes, in "Advances in Transport Processes," Mujumdar, A. S. and Mashelkar, R. A., Eds., Vol. 9, Elsevier, Amsterdam, 1993.
3. Patankar, S. V., Numerical Heat Transfer and Fluid Flow, Hemisphere, Washington, D.C., 1980.
4. Chang, H.-C. and Calo, J. M., An analysis of a radial-flow packed-bed reactors: how are they different?, in "Chemical Reactors," H. S., Fogler, Ed., ACS Symposium series 168, ACS, Washington, D.C., 1981.
5. Ranade, V. V. and Dometti, S. M. S., "Computational snapshot of flow generated by axial impellers in baffled stirred vessels," *Chem. Eng. Res. Des.,* 74A, 476–484, 1996.

Suggested Literature

Computational Fluid Dynamics:
1. Spalding, D. B., Imperial College Report HTS/80/1, 1980.
2. Anderson, D.A., et al., *Computational fluid mechanics and heat transfer,* Hemisphere, New York, 1984.
3. Nullaswamy, M., "Turbulence models and their applications to the predictions of internal flows," *Computers and Fluids,* 15, 151, 1987.
4. Oran, E. S. and Boris, J. P., *Numerical Simulation of Reactive Flows,* Elsevier, New York, 1987.
5. Hutchings, B. and Iannuzzelli, R., "Taking the measure of fluid dynamics software," *Mech. Eng.,* 72, May 1987.
6. Dombrowski, N., et al., "Know the CFD Codes," *Chem. Eng. Prog,* September 1993.

Stirred-Vessels Reactors:
7. Ranade, V. V., Interaction of macro-micromixing in agitated reactors, in "Advances in Transport Processes," Mujumdar, A. S. and Mashelkar, R. A., Eds., Vol. 9, Elsevier, Amsterdam, 1993.
8. Bakker, A., et al., "Pinpoint mixing problems with lasers and simulation software," *Chem. Eng.,* January 1994.

9. Ranade, V. V. and van den Akker, H.E.A., "Modeling of flow in gas-liquid stirred vessels," *Chem. Eng. Sci.,* 49, 5175, 1994.

Bubble-Column Reactors:
10. Svendsen, H. F., et al., "Local flow structures in internal-loop and bubble column reactors," *Chem. Eng. Sci.,* 47, 3297, 1992.
11. Ranade, V. V., "Computational fluid dynamics for reactor engineering," *Reviews in Chem. Eng.,* 11, 229, 1995.

Fixed-Bed Reactors:
12. Foumeny, E. A. and Benyahia, F., "Can CFD improve the handling of air, gas and gas-liquid mixtures?' *Chem. Eng. Prog.,* February 1993.
13. Ranade, V. V., "Modeling of flow maldistribution in a fixed bed reactor using phoenics," *J. of Phonics,* 7, 3, 1994.

Fluidized-Bed Reactors:
14. Ding, J. and Gidaspaw, D., "Bubbling fluidization model using kinetic theory of granular flows," *AIChE J,* 36, 523, 1990.
15. Gidaspaw, D. and Therdtianwong, A., Proceedings of International Conference on Circulating Fluidized Beds, 436, 1993.
16. Patel, M. K., et al., "Numerical modelling of circulating fluidized beds," *Comp. Fluid Dyn.,* 1, 161, 1993.

Flow Metering and Process Piping:
17. Patel, B. R. and Sheikoholeslami, Z., Numerical modelling of turbulent flow through the orifice meter, International Symposium on Fluid Flow Measurement, Washington, D.C., November 1986.
18. Ranade, V. V. and Kumaran, G., Computational fluid dynamics for piping engineering in "Proceeding of First National Seminar on Piping Engineering," MIT, June 1995.
19. Dometti, S. M. S. and Ranade, V. V., Simulation of vortex flow meters, National Workshop on Modelling in Hydraulic Engineering, CWPRS, June 1995.

CHAPTER ELEVEN

Biochemical Reaction

INTRODUCTION

The processing of biological materials and employing biological agents such as cells, enzymes, or antibodies are the principal domain of biochemical engineering. Biochemical reactions involve both cellular and enzymatic processes and the principal differences between the biochemical and chemical reactions lie in the nature of the living systems. Small living creatures known as microorganisms interact in many ways with human activities. Microorganisms play a primary role in the capture of energy from the sun. Additionally, their biological activities also complete critical segments of the cycles of carbon, oxygen, nitrogen, and other elements necessary for life. The cell is the unit of life, and cells in multi-cellular organism function together with other specialized cells. Generally, all cells possess basic common features. Every cell contains cytoplasm, a colloidal system of large biochemicals in a complex solution of smaller organic molecules and inorganic salts. The use of cells or enzymes taken from cells is restricted to conditions at which they operate, although most plant and animal cells live at moderate temperatures but cannot tolerate extremes of pH. In contrast, many microorganisms operate under mild conditions, some perform at high temperatures, others at low temperatures and also pH values, which exceed neutrality. Some can tolerate concentrations of chemicals that can be highly toxic in other cells. Thus, successful operation depends on acquiring the correct organisms or enzymes while preventing the entry of foreign organisms, which could impair the process.

Some variables such as temperature, pH, nutrient medium, and redox potential are favorable to certain organisms while discouraging the growth of others. The major characteristics of microbial processes that contrast with those of ordinary chemical processing include the following [1]:

- The reaction medium is invariably aqueous.
- The products are made in low concentrations (rarely more than 5%–10% for chemicals and much less for particular enzymes).
- Reaction temperatures are low, usually in the range of 10–60°C. Also the optimum range in individual cases may be 5°C or less.
- The processes require a mild and narrow range of pH.
- Generally, the scale of commercial processes (i.e., with the exception of potable ethanol or glucose isomerate) is modest, and for enzymes it is very low, only a few kg/day.
- The mass of microbes often increases simultaneously increases with the production of chemicals.
- Batch reactors are often used, although there are few large-scale continuous processes.
- The air is sterilized to prevent contamination during the process operation.

The reactant is referred to as a substrate. Alternatively it may be a nutrient for the growth of cells or its main function may require being transformed into some desirable chemical. The cells select reactants that will be combined and molecules that may be decomposed by using enzymes. These are produced only by living organisms, and commercial enzymes are produced by bacteria. Enzymes operate under mild conditions of temperature and pH. A database of the various types of enzymes and functions can be assessed from the following web site: *http://www.expasy.ch/enzyme/.* This site also provides information about enzymatic reactions.

This chapter solely reviews the kinetics of enzyme reactions, modeling, and simulation of biochemical reactions and scale-up of bioreactors. More comprehensive treatments of biochemical reactions, modeling, and simulation are provided by Bailey and Ollis [2], Bungay [3], Sinclair and Kristiansen [4], Volesky and Votruba [5], and Ingham et al. [6].

KINETICS OF ENZYME-CATALYZED REACTIONS

ENZYME CATALYSIS

All enzymes are proteins that catalyze many biochemical reactions. They are unbranched polymers of α-amino acids of the general formula

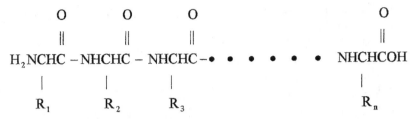

The carbon-nitrogen bond linking the carboxyl group of one residue and the α-amino group of the next is called a peptide bond. The basic structure of the enzyme is defined by the sequence of amino acids forming the polymer. There are cases where enzymes utilize prosthetic groups (coenzymes) to aid in their catalytic action. These groups may be complex organic molecules or ions and may be directly involved in the catalysis or alternately act by modifying the enzyme structure.

The catalytic action is specific and may be affected by the presence of other substances both as inhibitors and as coenzymes. Most enzymes are named in terms of the reactions they catalyze (see Chapter 1). There are three major types of enzyme reactions, namely:

1. Soluble enzyme—insoluble substrate
2. Insoluble enzyme—soluble substrate
3. Soluble enzyme—soluble substrate

The predominant activity in the study of enzymes has been in relation to biological reactions. This is because specific enzymes have both controlled and catalyzed synthetic and degradation reactions in all living cells. Many of these reactions are homogeneous in the liquid phase (i.e., type 3 reactions).

Enzymes are grouped with respect to the type of reaction catalyzed. The largest groups of enzymes are hydrolytic and oxidative enzymes. The former, which catalyze the hydrolysis of esters, glycosidic linkages in polysaccharides, and the hydrolysis of peptide linkages are complex acid-base catalysts, which enhance the transfer of hydrogen ions. Oxidative enzymes are engaged in oxidation-reduction processes, and accelerate electron-transfer processes. There are groups transferring enzymes, which promote the interchange of groups, such as amino and keto groups between two molecules. Other groups of enzymes are lyases, which catalyze the addition of chemical groups to double bonds, and isomerases, which catalyze isomerization reactions. Examples of some common enzyme-catalyzed reactions are:

- Glucose oxidase, which catalyzes the oxidation of glucose to gluconic acid.

$$\text{Glucose} + O_2 + H_2O \xrightarrow{\ E\ } \text{gluconic acid} + H_2O_2$$

where E denotes enzyme.

- L-asparaginase, which catalyzes the hydrolysis of L-asparagine to L-aspartate.

$$L-\text{Asparagine} + H_2 \xrightarrow{\ E\ } L-\text{aspartate} + NH_3$$

This is used in the treatment of some cancer cells, where L-asparaginase is used to remove an essential nutrient, thus inhibiting their growth.

- Decomposition of hydrogen peroxide (H_2O_2) in aqueous solution with catalase.

$$2H_2O_2 \xrightarrow{\ E\ } 2H_2O + O_2$$

The hydrogen peroxide produced in the glucose oxidase catalyzed reaction has an antibacterial action. If the presence of hydrogen peroxide is undesirable in the product, catalase is added to remove the peroxide.

- Hydrolysis of urea with the enzyme (E) urase.

$$(NH_2)_2 CO + H_2O \xrightarrow{\ E\ } CO_2 + 2NH_3$$

- Hydrolysis of starch with amylase to form glucose

$$(C_6H_{10}O_5)_n + nH_2O \xrightarrow{\ E\ } nC_6H_{12}O_6$$

This is an essential step in the conversion of corn to full-grade ethanol.

Kinetic studies involving enzymes can principally be classified into steady and transient state kinetics. In the former, the enzyme concentration is much lower than that of the substrate; in the latter much higher enzyme concentration is used to allow detection of reaction intermediates. In steady state kinetics, the high efficiency of enzymes as a catalyst implies that very low concentrations are adequate to enable reactions to proceed at measurable rates (i.e., reaction times of a few seconds or more). Typical enzyme concentrations are in the range of 10^{-8}M to 10^{-10}M, while substrate concentrations usually exceed 10^{-6}M. Consequently, the concentrations of enzyme-substrate intermediates are low with respect to the total substrate (reactant) concentrations, even when the enzyme is fully saturated. The reaction is considered to be in a steady state after a very short induction period, which greatly simplifies the rate laws.

FACTORS AFFECTING ENZYME CATALYZED REACTIONS

Enzymes possess high specificity and efficiency as catalysts. However, their rates of reactions can be influenced by certain factors, which include the following:

- The concentrations of enzyme and substrate
- The nature of the substrate
- The nature of the enzyme (E)
- The presence of inhibitors and coenzymes
- Temperature
- pH
- External factors (e.g., irradiation) and shear

Some of the main experimental results involving concentration effects are:

1. The rate of reaction $(-r_s)$ or (r_p) is proportional to the total (initial) enzyme concentration C_{ET}.
2. At low (initial) concentration of substrate C_s, the initial rate $(-r_{so})$ is first order with respect to the substrate.
3. At high C_s, $(-r_{so})$ is independent of (initial) C_s.

The effect of temperature satisfies the Arrhenius relationship where the applicable range is relatively small because of low and high temperature effects. The effect of extreme pH values is related to the nature of enzymatic proteins as polyvalent acids and bases, with acid and basic groups (hydrophilic) concentrated on the outside of the protein. Finally, mechanical forces such as surface tension and shear can affect enzyme activity by disturbing the shape of the enzyme molecules. Since the shape of the active site of the enzyme is constructed to correspond to the shape of the substrate, small alteration in the structure can severely affect enzyme activity. Reactor's stirrer speed, flowrate, and foaming must be controlled to maintain the productivity of the enzyme. Consequently, during experimental investigations of the kinetics enzyme catalyzed reactions, temperature, shear, and pH are carefully controlled; the last by use of buffered solutions.

MODELS OF ENZYME KINETICS

Consider the reaction $S \rightarrow P$ occurs with an enzyme as a catalyst. It is assumed that the enzyme E and substrate S combine to form a complex ES, which then dissociates into product P and free (uncombined) enzyme E.

$$E + S \underset{k_2}{\overset{k_1}{\rightleftharpoons}} ES^* \qquad (1\text{-}93)$$

$$ES^* \xrightarrow{\;k_3\;} E + P \qquad (1\text{-}94)$$

The net rate of disappearance of S is:

$$\left(-r_s\right)_{net} = k_1 C_E C_S - k_2 C_{ES^*} \qquad (11\text{-}1)$$

At pseudo equilibrium $(-r_s) = 0$, implying that the steps are very rapid:

$$k_1 C_E C_S = k_2 C_{ES^*} \qquad (11\text{-}2)$$

$$K_m = \frac{k_2}{k_1} = \frac{C_E \, C_S}{C_{ES^*}} \qquad (11\text{-}3)$$

where K_m = the dissociation equilibrium constant for ES^*
  C_E = concentration of the enzyme, E
  C_S = concentration of the substrate, S
  C_{ES^*} = concentration of the complex, ES^*

The concentration of the enzyme-substrate complex from Equation 11-3 is

$$C_{ES^*} = \frac{k_1}{k_2} C_E C_S \qquad (11\text{-}4)$$

Decomposition of the complex to the product and free enzyme is assumed irreversible, and rate controlling:

$$ES^* \xrightarrow{\;k_3\;} P + E \qquad (11\text{-}5)$$

The formation rate of the product P is

$$v \equiv \left(r_P\right) = k_3 C_{ES^*} \qquad (11\text{-}6)$$

C_{ES^*} and C_S are related by a material balance on the total amount of enzyme, C_{ET}.

$$C_E + C_{SE^*} = C_{ET} \tag{1-101}$$

Combining Equations 11-4 and 1-101 gives

$$\frac{k_2}{k_1} \frac{C_{ES^*}}{C_S} + C_{ES^*} = C_{ET} \quad \text{or}$$

$$C_{ES^*} = \frac{k_1 C_S C_{ET}}{k_2 + k_1 C_S}$$

$$= \frac{C_S C_{ET}}{\left(\dfrac{k_2}{k_1} + C_S \right)} \tag{11-7}$$

Combining Equations 11-6 and 11-7 yields

$$v = (r_P) = \frac{k_3 C_S C_{ET}}{\left(\dfrac{k_2}{k_1} + C_S \right)}$$

$$= \frac{k_3 C_S C_{ET}}{(K_m + C_s)} \tag{11-8}$$

where K_m in this instance is referred to as the Michaelis-Menten [7] constant. r_p is expressed as a reaction velocity. Figure 11-1a shows the effect of the substrate concentration C_S on the reaction rate $(-r_S)$ according to simple Michaelis-Menten kinetics.

$$(-r_S) = \frac{V_{max} C_S}{K_m + C_S} \tag{11-9}$$

where $V_{max} = k_3 C_{ET}$ $\tag{11-10}$

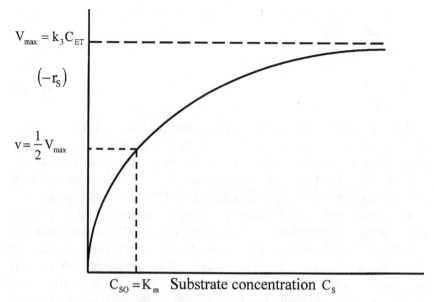

Figure 11-1a. Simple Michaelis-Menten kinetics.

At low substrate concentration

$$K_m \gg C_S \text{ and}$$

$$(-r_S) \approx \frac{V_{max} C_S}{K_m} \qquad (11-11)$$

At high substrate concentration

$$C_S \gg K_m$$

$$(-r_S) \cong V_{max} \qquad (11-12)$$

When the substrate concentration is such that the reaction = $(1/2)V_{max}$

$$(-r_S) = \frac{V_{max} C_S}{K_m + C_S} = \frac{V_{max}}{2} \qquad (11-13)$$

Solving Equation 11-13 for the Michaelis constant gives

$$\frac{2C_{S_{1/2}}}{K_m + C_{S_{1/2}}} = 1$$

$$K_m = C_{S_{1/2}} \qquad\qquad (11\text{-}14)$$

The Michaelis constant is equal to substrate concentration at which the rate of reaction is equal to one-half the maximum rate. The parameters K_m and V_{max} characterize the enzymatic reactions that are described by Michaelis-Menten kinetics. V_{max} is dependent on total enzyme concentration C_{ET} (Equation 11-10), whereas K_m is not.

Enzymatic reactions frequently undergo a phenomenon referred to as substrate inhibition. Here, the reaction rate reaches a maximum and subsequently falls as shown in Figure 11-1b. Enzymatic reactions can also exhibit substrate activation as depicted by the sigmoidal type rate dependence in Figure 11-1c. Biochemical reactions are limited by mass transfer where a substrate has to cross cell walls. Enzymatic reactions that depend on temperature are modeled with the Arrhenius equation. Most enzymes deactivate rapidly at temperatures of 50°C–100°C, and deactivation is an irreversible process.

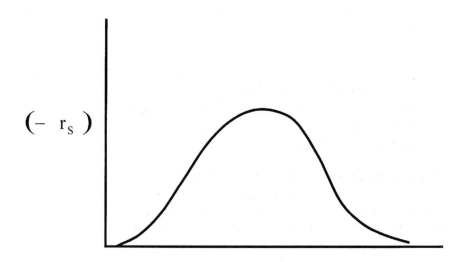

Substrate concentration C_S

Figure 11-1b. Substrate inhibition.

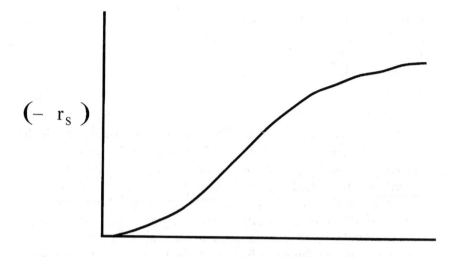

Substrate concentration C_S

Figure 11-1c. Activation.

If V_{max}, the maximum velocity replaces $k_3 C_{ET}$ in Equation 11-8, yielding

$$\left(r_P\right) = \frac{V_{max} C_S}{\left(K_m + C_S\right)}$$

$$(11\text{-}15)$$

Equation 11-15 is known as the Michaelis-Menten equation. It represents the kinetics of many simple enzyme-catalyzed reactions, which involve a single substrate. The interpretation of K_m as an equilibrium constant is not universally valid, since the assumption that the reversible reaction as a fast equilibrium process often does not apply.

PARAMETERS IN THE MICHAELIS-MENTEN EQUATION

The Michaelis-Menten Equation 11-15 is not well suited for estimation of the kinetic parameters V_{max} and K_m. Rearranging Equation 11-15 gives various options for plotting and estimating the parameters.

$$v = \frac{V_{max} C_S}{K_m + C_S}$$

$$(11\text{-}16)$$

Rearranging Equation 11-16 gives

$$\frac{1}{v} = \frac{1}{V_{max}} + \frac{K_m}{V_{max}} \frac{1}{C_S} \tag{11-17}$$

Equation 11-17 is referred to as a *Lineweaver-Burk* equation involving separate dependent and independent variables $1/v$ and $1/C_S$, respectively. Equation 11-17 can be further rearranged to give

$$\frac{C_S}{v} = \frac{K_m}{V_{max}} + \frac{1}{V_{max}} C_S$$

or

$$v = V_{max} - K_m \frac{v}{C_S} \tag{11-18}$$

Equation 11-18 is referred to as the *Eadie-Hofstee* equation, where v is plotted against v/C_S. However, both of these equations are subjected to large errors. Equation 11-18 in particular contains the measured variable v in both coordinates, which is subjected to the largest errors. Figures 11-2 and 11-3 show plots of Lineweaver-Burk and Eadie-Hofstee equations, respectively.

BRIGGS-HALDANE MODEL

Briggs and Haldane [8] proposed a general mathematical description of enzymatic kinetic reaction. Their model is based on the assumption that after a short initial startup period, the concentration of the enzyme-substrate complex is in a pseudo-steady state (PSS). For a constant volume batch reactor operated at constant temperature T, and pH, the rate expressions and material balances on S, E, ES*, and P are

$$(+r_S) = \frac{dC_S}{dt} = k_2 C_{ES^*} - k_1 C_E C_S \tag{11-19}$$

$$(+r_{ES}) = \frac{dC_{ES^*}}{dt} = k_1 C_E C_S - k_2 C_{ES^*} - k_3 C_{ES^*} \tag{11-20}$$

The total concentration of the enzyme and the substrate is

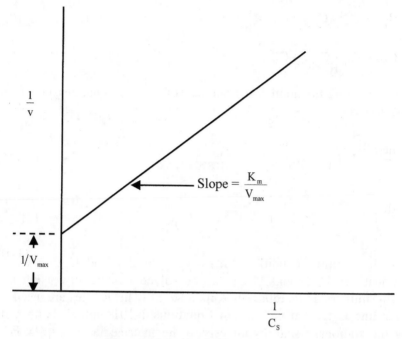

Figure 11-2. Lineweaver-burk plot.

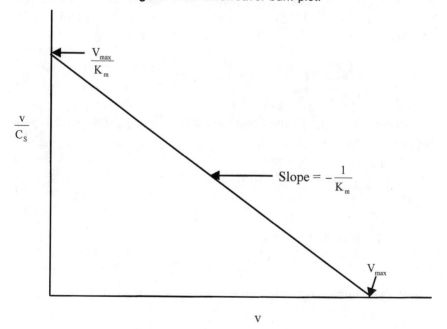

Figure 11-3. Eadie-Hofstee plot.

$$C_{ET} = C_E + C_{ES^*} \tag{1-101}$$

$$C_{ST} = C_S + C_{ES^*} + C_P \tag{11-21}$$

Rearranging Equation 1-101 and substituting in Equations 11-19 and 11-20, gives

$$\frac{dC_S}{dt} = k_2 C_{ES^*} - k_1 C_S \left(C_{ET} - C_{ES^*} \right) \tag{11-22}$$

$$\frac{dC_{ES^*}}{dt} = k_1 C_S \left(C_{ET} - C_{ES^*} \right) - \left(k_2 + k_3 \right) C_{ES^*} \tag{11-23}$$

With the initial conditions, at t = 0, $C_S = C_{ST}$, and $C_{ES^*} = 0$

Equations 11-22 and 11-23 can be solved numerically, using any of the numerical methods in Appendix D. The results are used to determine $C_E(t)$ and $C_P(t)$ from Equations 1-101 and 11-21. Applying the stationary state hypothesis to the intermediate complex ES^*, ($C_{ET} \ll C_{ST}$), (dC_{ES^*}/dt) = 0, Equation 11-23 becomes $0 = k_1 C_S C_{ET} - k_1 C_S C_{ES^*} - (k_2 + k_3) C_{ES^*}$

or

$$C_{ES^*} = \frac{k_1 C_S C_{ET}}{k_1 C_S + k_2 + k_3} \tag{11-24}$$

Substituting C_{ES^*} in the formation rate of the product P in Equation 11-6 gives

$$\left(+r_P \right) = k_3 C_{ES^*} = \frac{k_1 k_3 C_S C_{ET}}{k_1 C_S + k_2 + k_3} \tag{11-25}$$

$$\left(+r_P \right) = \frac{k_3 C_S C_{ET}}{K_m + C_S} \tag{11-26}$$

where the Michaelis-Menten constant K_m is now

$$K_m = \frac{k_2 + k_3}{k_1}$$

If $k_2 \gg k_3$, Equation 11-26 simplifies to the form of Equation 11-8. Substituting $V_{max} = k_3 C_{ET}$ in Equation 11-26 yields

$$r_P = \frac{V_{max} C_S}{K_m + C_S} \tag{11-27}$$

LINEARIZED FORM OF THE INTEGRATED MICHAELIS-MENTEN (MM) EQUATION

For a constant volume batch reactor, the MM equation gives a form of an equation that can be linearized. Using Equation 11-15 gives

$$\left(+r_P\right) = \left(-r_S\right) = \frac{-dC_S}{dt} = \frac{V_{max} C_S}{K_m + C_S} \tag{11-28}$$

Rearranging Equation 11-28 gives

$$\left(K_m + C_S\right)\frac{dC_S}{C_S} = -V_{max}\, dt \tag{11-29}$$

$$-K_m \frac{dC_S}{C_S} - dC_S = V_{max} dt \tag{11-30}$$

Integrating Equation 11-30 with the boundary conditions $C_S = C_{SO}$ at $t = 0$ gives

$$-K_m \ln\left(\frac{C_S}{C_{SO}}\right) - \left(C_S - C_{SO}\right) = V_{max}\, t \tag{11-31}$$

Equation 11-31 can further be rearranged to give

$$\frac{1}{t}\ln\left(\frac{C_{SO}}{C_S}\right) = \frac{V_{max}}{K_m} - \frac{1}{K_m}\left(\frac{C_{SO} - C_S}{t}\right) \tag{11-32}$$

Equation 11-32 shows $1/t\, \ln(C_{SO}/C_S)$ as a linear function of $(C_{SO} - C_S)/t$. The parameters K_m and V_{max} can be estimated from Equation 11-32 using measured values of C_S as a function of t for a given

C_{SO}. Figure 11-4 shows the corresponding plot in terms of substrate concentration.

Example 11-1

The enzyme triose phosphate isomerase catalyzes the interconversion of D-glyceraldehyde 3–phosphate and dihydroxyacetone phosphate.

$$CHO \bullet CH(OH) \bullet CH_2OPO_3^{2-} \Leftrightarrow CH_2OH \bullet CO \bullet CH_2OPO_3^{2-}$$

The following results refer to the initial reaction velocity, v, with glyceraldehyde 3–phosphate (S) as substrate at a total enzyme concentration $C_{ET} = 2.22 \times 10^{-10}$M, pH = 7.42, and 30°C (Putman et al. [9]).

10^3 [S]/M	0.071	0.147	0.223	0.310	0.602	1.47	2.60
10^7 v/Ms^{-1}	1.31	2.45	3.37	3.90	5.63	7.47	8.17

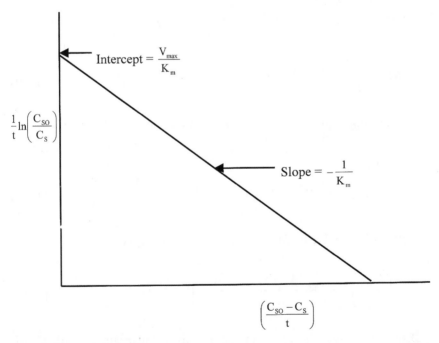

Figure 11-4. Evaluating K_m and V_{max} from 1/t ln(C_{SO}/C_S) versus ($C_{SO} - C_S$)/t plot.

Determine the Michaelis constant K_m and the catalytic constant k_3 for the enzyme under these conditions.

Solution

Using the developed computer program PROG1, the model equation $1/Y = A + B * 1/X$ represents the Lineweaver-Burk plot represented by Equation 11-17 as:

$$\frac{1}{v} = \frac{1}{V_{max}} + \frac{K_m}{V_{max}} \frac{1}{C_S}$$ (11-17)

The results of the regression analysis with the intercept and slope are:

$A = 0.9945 \times 10^{-1}$

$B = 0.4469 \times 10^{-1}$

$$\frac{1}{V_{max}}\left(10^{-7}\right) = A = 0.9945 \times 10^{-1}$$

$V_{max} = 1.0055 \times 10^{-6} \text{ Ms}^{-1}$

$$\frac{K_m}{V_{max}} = \text{Slope} \times 10^4 = B \times 10^4$$

$$= 0.4669 \times 10^{-1} \times 10^4$$

$$K_m = \left(0.4469 \times 10^{-1}\right)\left(1.0055 \times 10^{-6}\right)\left(10^4\right)$$

$$K_m = 4.7 \times 10^{-4} M$$

The catalytic constant k_3 is determined from Equation 11-10.

$$V_{max} = k_3 C_{ET}$$ (11-10)

$$\text{or } k_3 = \frac{V_{max}}{C_{Et}} = \frac{1.0055 \times 10^{-6}}{2.22 \times 10^{-10}}$$

$$k_3 = 4.53 \times 10^3 \, s^{-1}$$

Figure 11-5 shows the Lineweaver-Burk plot of $1/v$ versus $1/C_s$.

Example 11-2

At a particular room temperature, results for the hydrolysis of sucrose, S, catalyzed by the enzyme invertase ($C_{ET} = 1 \times 10^{-5}$ mol L^{-1}) in a batch reactor are given by:

C_S/mmol L^{-1}	1	0.84	0.68	0.53	0.38	0.27
t/h	0	1	2	3	4	5
C_S/mmol L^{-1}	0.16	0.09	0.04	0.018	0.006	0.0025
t/h	6	7	8	9	10	11

Source: Missen, R. W. et al. [10].

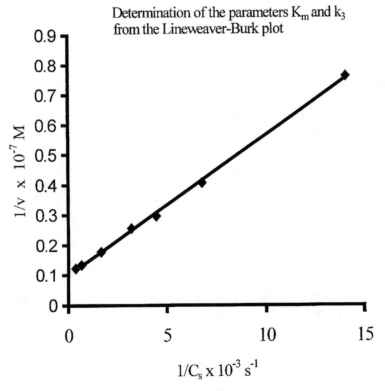

Determination of the parameters K_m and k_3 from the Lineweaver-Burk plot

Figure 11-5. Double reciprocal (Lineweaver-Burk) plot to determine K_m and k_3.

Determine the values of the kinetic parameters V_{max}, K_m, and k_3.

Solution

Using Equation 11-26 for the batch reactor, the dependent variable $1/t \ln(C_{SO}/C_S)$ is plotted as a function of the independent variable $(C_{SO} - C_S)/t$. A linear plot gives a slope $= -1/K_m$ and an intercept $= V_{max}/K_m$. The values of K_m and V_{max} are determined from the slope and the intercept. k_3 is evaluated from $V_{max} = k_3 C_{ET}$, or

$$k_3 = \frac{V_{max}}{C_{ET}} \qquad\qquad (11\text{-}10)$$

Using the Microsoft Excel spreadsheet (EXAMPLE11-2.xls), a regression analysis was carried out giving the slope $= -4.9967$ L/mmol.

$$-\text{Slope} = -\frac{1}{K_m}$$

$$K_m = 0.2 \text{ mmol/L}$$

$$\text{Intercept}\left(\frac{1}{\text{hr}}\right) = \frac{V_{max}}{K_m} = 0.9995$$

$$V_{max} = 0.2 \times 0.9995 \left(\frac{\text{mmol}}{\text{hr} \cdot \text{L}}\right)$$

$$k_3 = \frac{V_{max}}{C_{Et}}$$

$$= \frac{\left(0.2 \times 10^{-3}\right)}{\left(1 \times 10^{-5}\right)} \left\{\frac{\text{mol}}{\text{hr} \cdot \text{L}} \cdot \frac{\text{L}}{\text{mol}}\right\}$$

$$= 20 \text{ hr}^{-1}$$

Figure 11-6 shows a plot of $1/t \ln(C_{SO}/C_S)$ versus $(C_{SO} - C_S)/t$.

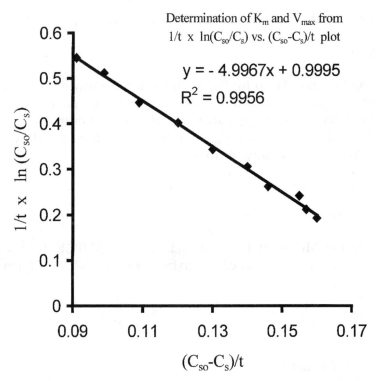

Figure 11-6. $1/t \ln(C_{SO}/C_S)$ versus $(C_{SO} - C_S)/t$ plot.

TREATMENT OF MICHAELIS-MENTEN (MM) EQUATION

A limitation of the linearized forms of the MM equation is that no accurate estimates of K_m and V_{max} can be established. Using the computer program PROG1, it is still impossible to obtain meaningful error estimates of the parameters because linear regression is an inappropriate method of analysis.

There is an increasing trend toward using the integrated rate equation and non-linear regression analysis to estimate K_m and V_{max}. This technique is more complex than the linear regression method and provides additional benefits. First, accurate non-biased estimates of K_m and V_{max} can be determined. Second, non-linear regression may allow the errors (or confidence intervals) of the parameters to be evaluated. Software packages with greater sophistication and graphical displays are now commercially available. This software enables designers and modelers to explore various options or incorporate developed models for analysis. The web sites of these software packages are: *http://www.ebicom.net/~dhyams/cvxpt.htm, http://www.graphpad.com,* and *http://www.mathsoft.com.*

NON-LINEAR REGRESSION USING THE SOLVER

In this chapter, the Microsoft Solver Excel's powerful optimization package is used to perform non-linear least squares curve fitting. Procedures for using the Solver are illustrated in Appendix B. The Solver provides results that are comparable with those from the commercial software packages. The Solver determines the sets of least-squares regression coefficients very quickly and efficiently. The slight differences between the Solver and commercial software packages arise from the fact that the coefficients are found by a search method; the final values will differ depending on the convergence criteria used in each program.

Example 11-3

A non-linear regression analysis is employed using the Solver in Microsoft Excel spreadsheet to determine the values of K_m and V_{max} in the following examples. Example 1-5 (Chapter 1) involves the enzymatic reaction in the conversion of urea to ammonia and carbon dioxide and Example 11-1 deals with the interconversion of D-glyceraldehyde 3–Phosphate and dihydroxyacetone phosphate. The Solver (EXAMPLE11-1.xls and EXAMPLE11-3.xls) uses the Michaelis-Menten (MM) formula to compute v_{cal}. The residual sums of squares between v_{obs} and v_{cal} is then calculated. Using guessed values of K_m and V_{max}, the Solver uses a search optimization technique to determine MM parameters. The values of K_m and V_{max} in Example 11-1 are:

	Lineweaver-Burk method	Non-linear regression method	Percentage deviation (%)
K_m, M	4.7×10^{-4}	4.286×10^{-4}	8.8
$V_{max} Ms^{-1}$	10.055×10^{-7}	9.574×10^{-7}	4.78

The values of K_m and V_{max} in Example 1-5 are:

	Lineweaver-Burk method	Non-linear regression method	Percentage deviation (%)
$K_m, \frac{kmol}{m^3}$	0.0142	0.0539	73.7
$V_{max}, \frac{kmol}{m^3 \bullet s}$	0.852	1.7032	50.0

Figures 11-7 and 11-8 show plots of velocity versus substrate concentration of the interconversion of D-glyceraldehyde 3-Phosphate, and the conversion of urea, respectively.

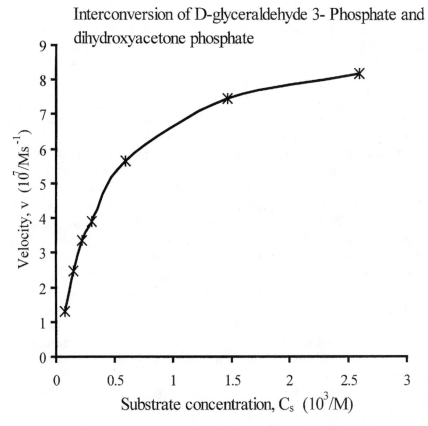

Figure 11-7. Velocity versus substrate concentration plot.

Michaelis-Menten parameters Vmax and Km for the reaction involving urea + urease ------> [urea * urease] ------> 2NH3 + CO2 + urease

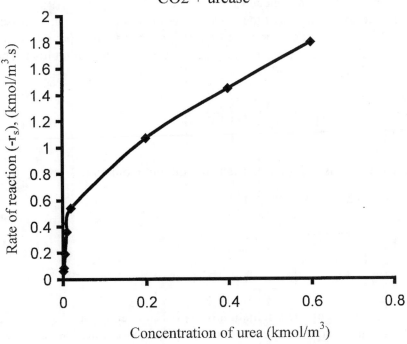

Figure 11-8. Velocity versus substrate concentration plot.

ENZYME KINETICS IN THE PRESENCE
OF AN INHIBITOR

The three most common types of inhibitors in enzymatic reactions are competitive, non-competitive, and uncompetitive. Competitive inhibition occurs when the substrate and inhibitor have similar molecules that compete for the identical site on the enzyme. Non-competitive inhibition results in enzymes containing at least two different types of sites. The inhibitor attaches to only one type of site and the substrate only to the other. Uncompetitive inhibition occurs when the inhibitor deactivates the enzyme substrate complex. The effect of an inhibitor is determined by measuring the enzyme velocity at various substrate concentrations both in the presence and absence of an inhibitor. The inhibitor substantially reduces the enzyme velocity at low substrate concentrations, but does not alter the velocity considerably at very high concentrations of substrate (Figure 11-9).

Figure 11-9 shows that the inhibitor does not alter V_{max}, but increases the observed K_m (i.e., concentration of the substrate that produces half the maximal velocity in the presence of a competitive inhibitor). The observed K_m is defined by

$$K_{mobs} = K_m \left\{ 1 + \frac{[\text{Inhibitor}]}{K_i} \right\} \tag{11-33}$$

where K_i = dissociation constant for inhibitor binding.

Rearranging Equation 11-33 yields

$$K_i = \frac{\text{Inhibitor}}{\left[\dfrac{K_{m,obs}}{K_m} - 1 \right]} \tag{11-34}$$

A more reliable determination of K_i is achieved if the observed K_m is determined at various concentrations of inhibitor. Each curve is then fitted to determine the observed K_m. The inhibitor is entered as the

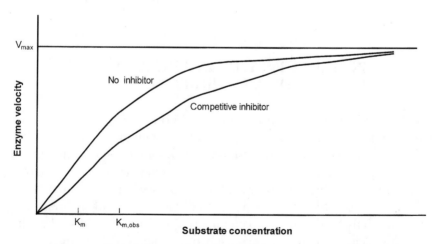

Figure 11-9. Enzyme velocity versus substate concentration.

independent variable and $K_{m,obs}$ as the dependent variable. A linear graph infers that the inhibitor is competitive. The slope and intercept can be determined using linear regression as shown in Figure 11-10.

Competitive and non-competitive inhibitions are easily distinguishable from the Lineweaver-Burk plot. In the case of competitive inhibitors, the intercept on the $1/C_S$ axis increases while the intercept of the $1/v$ axis remains unchanged by the addition of the inhibitor. Conversely, with a non-competitive inhibitor, only the $1/v$ axis intercept increases. The effect of competitive inhibitors can be reversed by increasing the substrate concentration. Where the enzyme or the enzyme substrate complex is made inactive, a non-competitive inhibitor decreases V_{max} of the enzyme, but K_m remains constant.

FERMENTATION

Fermentation* processes use microbiology in producing chemical compounds that are made naturally. Cheap synthetic processes with abundant raw materials are now superseding fermentation processes that have produced commodity chemicals (e.g., ethanol, butanol, and

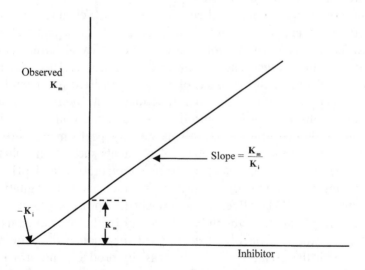

Figure 11-10. $K_{m,obs}$ versus inhibitor plot.

---

*The term "fermentation" is often used loosely by biochemical engineers. Biochemists have a more focused perspective (anaerobic) catabolism of an organic compound in which the compound serves as both an electron donor and acceptor and in which ATP is produced.

acetone). However, fermentation processes are still prominent for producing complex organic chemicals such as citric and lactic acids derived from low-cost carbohydrate sources.

The pharmaceutical industry has employed materials of plant and animal origin as sources of drugs. The industry has utilized the life processes of either plants or animals and microorganisms to produce medicinal and antibiotic products.

Microorganisms are bacteria, yeast, and molds that feed on organic materials. The supply of adequate energy foods coupled with other required nutrients will not only cause these microvegetative organisms to grow and multiply, but will change the food into other chemical substances. Yeast and bacteria are unicellular and of very small dimensions. Yeasts are irregularly oval and around 0.004 to 0.010 mm in diameter. Bacteria are smaller, less than 0.007 mm, and more diverse in shape. Many bacteria are rod-shaped and the bacteria multiply by binary fission. On the other hand, yeast multiplies by budding. Molds are multicellular filaments, which increase by vegetative growth of the filament. The vegetative reproduction cycle of these bacteria and of yeast is short (that is, measured in minutes), and therefore they multiply quite fast.

Fermentation under controlled conditions involves chemical conversions. It is the action of specific microorganism on a substrate to produce the desired chemical compound. The majority of fermentation processes require oxygen and are classified as aerobic. The few processes carried out in the absence of air are classified as anaerobic. Certain anaerobes neither grow nor function in the presence of air. However, conditions can be controlled to encourage both the multiplication of the vegetative organisms and its performance either directly or through the secreted enzymes. Nutrients such as phosphates and nitrogenous compounds at controlled temperature and pH are added during the growth period. pH, temperature, aeration-agitation, pure culture, and yields influence the fermentation process. Optimizing the yield often requires reducing the quantity of the microorganism. A particular strain of an organism is used to produce the required chemical with the greatest yields, the least by-product, and at a low cost. The yeast, bacteria, and molds used in fermentation require both nutrients and specific environments to perform their activities (e.g., sugar concentration or other nutrients), which subsequently affect the product. The most favorable temperature varies from 5°C to 40°C, and pH also exerts a great influence on the process.

Generally, the fermentation process involves the addition of a specific culture of microorganisms to a sterilized liquid substrate or broth in a tank (submerged fermentation), addition of air if aerobic, in a well-designed gas-liquid contactor. The fermentation process is then carried out to grow microorganisms and to produce the required chemicals. Table 11-1 lists examples of the processes used by fermentation.

The fermentation process can be performed batch-wise or continuously at a given temperature and time. The broth is further processed to remove the desired chemical. Figure 11-11 shows a schematic and an abstracted physical model of a fermenter with the liquid phase as the control region.

DESIGN OF BIOLOGICAL REACTORS

TRANSPORT PHENOMENA INFLUENCE IN BIOREACTOR DESIGN

The design basis is the most important consideration when determining the size of a biological reactor. Other pertinent factors are the final product, the microorganism used, its growth rate and oxygen requirement, the product concentration (e.g., expressed in mg/L broth for proteins or 100 mg/L for organic acids), and the type of product (intercellular or extracellular).

Table 11-1
Processes used by fermentation

Oxidation	Reduction	Hydrolysis	Esterification
Alcohol to acetic acid	Aldehydes to alcohols (e.g., acetaldehyde to ethyl alcohol)	Starch to glucose	Hexose phosphate from hexose and phosphoric acid
Sucrose to critic acid	Sulfur to hydrogen sulfide	Sucrose to glucose and fructose and on to alcohol	
Dextrose to gluconic acid			

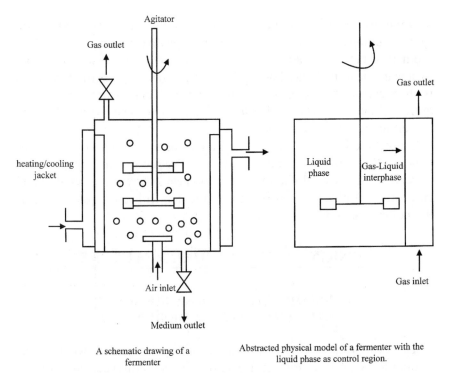

A schematic drawing of a fermenter

Abstracted physical model of a fermenter with the liquid phase as control region.

Figure 11-11. A schematic and an abstracted physical model of a fermenter.

Additionally, three different phenomena are essential factors in the design of a bioreactor. These phenomena are:

- Thermodynamics
- Microkinetics
- Transport

These phenomena principally govern the performance of a bioreactor. The first and second of these phenomena are independent of scale. Neither a typical thermodynamic property (e.g., the solubility of oxygen in a broth) nor microkinetic properties (e.g., the growth and product formation by the microorganism) are dependent on the scale of the bioreactors. However, the actual oxygen concentrations and the kinetic behavior of microorganisms in a bioreactor are dependent on scale. The reason is that oxygen and other nutrients involved in the conversion processes are consumed rapidly, and have to be supplied by transport processes. Scale-up influences these transport processes resulting in the existence of gradients at production scale. When the microorganism travels through the bioreactor, it encounters constantly changing concentrations of nutrients, oxygen, and pressure. Additionally, the microorganisms are always subjected to (turbulent) shear phenomena, which can damage the microorganism, the cell itself, or influence the formation of agglomerates (floc, pellets) of microorganisms.

Scale-up problems are prevalent in aerobic processes for continuous and fed-batch productions than for batch type involving the transport of oxygen and nutrients to the microorganisms. Transport phenomena are governed mainly by flow and diffusion, and these depend on the shear, mixing, mass transfer ($k_L a$), heat transfer, and macrokinetics (i.e., a form of apparent kinetics resulting from a combination of microkinetics and diffusion occurring in immobilized systems, flocs, and pellets). These phenomena will change during scale-up together with the kinetic behavior of microorganism, which depends on local environmental conditions.

Biological reactors play a valuable role in the conversion of substrates by microorganisms and mammalian cells into a wide range of products such as antibiotics, insulin, and polymers. Figures 11-12, 11-13, and 11-14 illustrate various types of biological reactor, and Figure 11-15 shows the physical characteristics of a typical commercial fermentation vessel.

VESSEL DESIGN AND ASPECT RATIO

A vertical cylindrical, and mechanical agitated pressure vessel, equipped with baffles to prevent vortex formation is the most widely used fermenter configuration. The baffles are typically one-tenth of the fermenter diameter in width, and are welded to supports that extend from the sidewall. A small space between the sidewall and the baffle enables cleaning. Internal heat transfer tube bundles can also be used as baffles. The vessels must withstand a 45 psig internal pressure and full vacuum of –14.7 psig, and comply with the ASME code.

Standard components for a bioreactor comprise the vessel, the agitator and impeller, heat exchangers, seals, and valves. Other components include utilities such as clean-in-place (CIP), and steam-in-place (SIP) systems, auxiliary tanks, pH and foam control, inlet tubes,

(text continued on page 862)

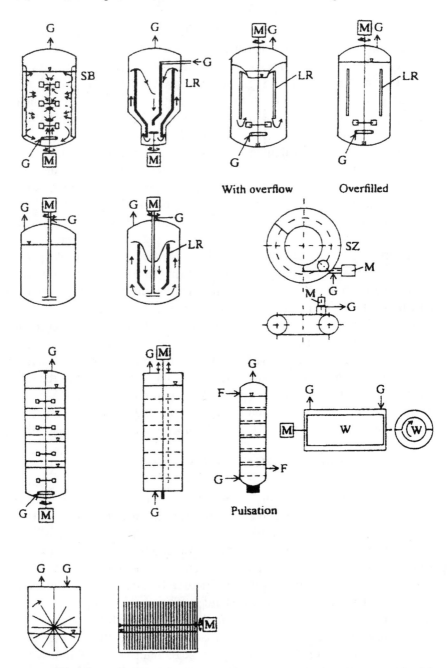

Figure 11-12. Reactors with mechanical stirring: M, moter; G, gas (air); SB, baffles; LR, conduit tube; SZ, foam breaker; W, roller; F, liquid. *(Source: Schuger [11].)*

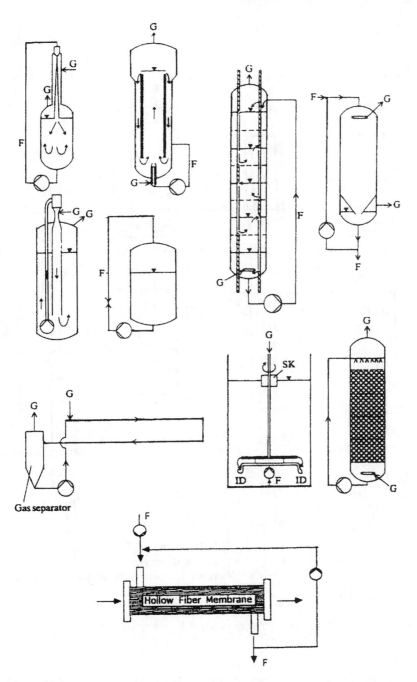

Figure 11-13. Reactors stirred by pumped fluid: ID, Injector nozzle; SK, float; G, gas; F, liquid. *(Source: Schuger [11].)*

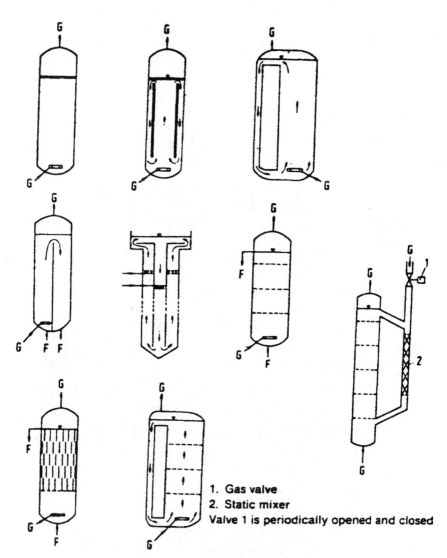

Figure 11-14. Reactors with energy input by compression. *(Source: Schuger [11].)*

1. Gas valve
2. Static mixer
Valve 1 is periodically opened and closed

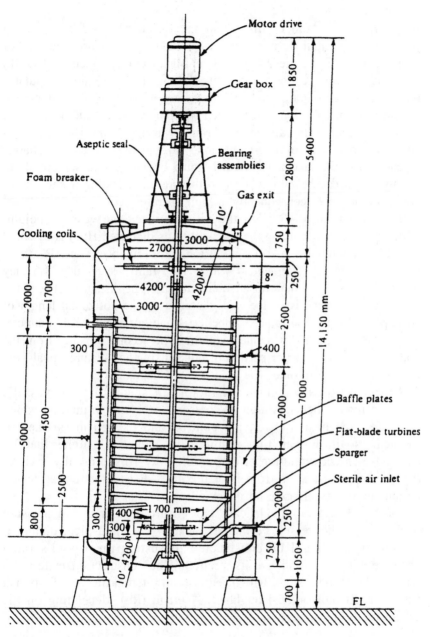

Figure 11-15. Cutaway diagram of a 1,000,000 liter fermenter used for penicillin production. *(Source: Aiba et al. [12].)*

(text continued from page 857)

instrumentation, and control systems. Standard good engineering practice is essential during both design and scale-up procedures. For example, the U.S. requires Current Good Manufacturing Practices (CGMP) as mandated by the Food and Drug Administration, which stipulates minimum design requirements and guidelines in areas such as equipment, documentation, operation, and maintenance. The vessel should be designed to operate monoseptically over a range varying from 18 h–30 days depending on the type of fermentation [13]. Operation must be reliable, repeatable, sterile, and automated. Manual operations are susceptible to variables that can be introduced into the process resulting in the risk of contamination and, consequently, a high operating cost.

The aspect ratio AR (i.e., the ratio between the vessel's height, which is the distance along the sidewall to its diameter, H:D) is essential when specifying the vessel's agitation requirements. The AR varies between 1.2 to 3.0. Fermenters with a H:D ratio greater than 1.5 may require multiple impellers for adequate mixing. The distance between the impellers is two-thirds of the tank diameter or the horizontal length of an impeller blade. The highest impeller is one horizontal blade diameter below the surface of the aerated volume. The second impeller from the bottom provides the major portion of bulk mixing and features axial-flow or high efficiency turbine-type blades (see Chapter 7).

Aspect ratios are significant factors in specifying the vessel's agitation requirements. A high AR presents problems in mixing, which can affect the heat transfer, pH control, and oxygen mass transfer. Fermenters with low air flow requirements and an AR higher that 3:1 can cause an accumulation of carbon dioxide. This is usually toxic to the microbe and results from poor mixing. AR as an essential design parameter in vessel scale-up is reviewed in Chapter 13.

An ASME 2:1 elliptical heads can ensure an increase in pressure resistance of the vessel. Fermenter jackets (e.g., half-pipe, diameter, or true type) should be constructed to sustain the vessel's rated pressure and, thus, enhance its strength. The construction material is type 316L stainless steel, which features an internal mechanical-polish finish of 2B-mill or 25-Roughness Average (Ra)* depending on the nature of the fermentation.

2B mill is a measurement of roughness, which is at the higher end of the scale. Roughness average (Ra) is a biotechnology industry standard. 25Ra is smoother than 2B mill. As a rule, the higher the Ra number, the rougher the finish.

TYPES OF OPERATION

The mode of operation, which can be batch, fed-batch, or continuous, should be considered during the preliminary stages of large-scale design. Batch operation often involves simplicity in maintenance of aseptic conditions and minimizes operating losses due to contamination or equipment failure.

In contrast, fed-batch type involves feeding the reactor with fresh nutrients or carbohydrates or both during fermentation when low glucose concentrations (10–50 mg/L) are required to maintain the productivity of microorganism.

Most fermentations employ a single microorganism, thus it is essential to grow the microbe in a pure culture without contamination from other undesirable microbes. Continuous biological reactors are seldom used. This is due to the difficulty of maintaining monosepsis (i.e., preventing contamination by foreign organisms) coupled with the challenge in controlling the different phases of the microorganism's growth rate. Mathematical models are developed for the batch, fed-batch, and continuous biological reactors later in this chapter.

The desired product in the fermentation process depends on the pertinent condition used, the type of microorganisms (e.g., bacteria, fungi, and mammalian cell lines), and on economic factors. The various types of product are:

- The microorganism manufactures the product, which may be an organic acid, an alcohol, an enzyme or other protein, an amino acid, a vitamin, an antibiotic, or other therapeutic protein.
- The microorganism transforms a substance into one or higher values during a biconversion process.
- The product itself is the microorganism, for example Baker's Yeast and veterinary or human biological or vaccines.

CELL GROWTH

Typical batch fermentation begins with an initial charge of cells referred to as inoculum. Here, the number of living cells varies with time as shown in Figure 11-16. During the initial lag phase, cells are adjusting themselves to the new environment. During this period, the cells perform functions as synthesizing transports proteins for moving

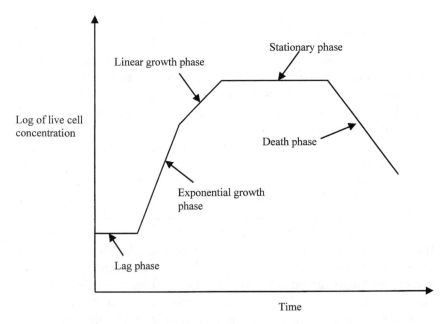

Figure 11-16. Typical growth phase in batch fermentation.

the substrate into the cell. Most microbes and particularly bacteria are adaptable and can utilize various carbon sources. Generally, one or more unique enzymes are required for each source and these must be manufactured by the cell in response to the new environment. These are referred to as induced enzymes. Additionally, the duration of the lag phase depends on the growth medium, such as the reactor from which the inoculum is obtained relative to the reaction medium where it is placed. The manufacture of desirable products is negligible during the lag phase.

After the lag phase, a period of rapid growth develops during which the cell numbers increase exponentially with time. In this phase, the cells are dividing at the maximum rate because all the enzyme's pathways for metabolizing the media are in place. During the exponential growth phase, the total cell mass increases by a fixed percentage during each time interval, and the cells are able to use the nutrients most efficiently. In a closed vessel, the cells cannot multiply indefinitely, and a stationary phase follows the period of exponential growth. At this point, the population achieves its maximum size. In semi-batch fermentations, a linear growth phase occurs when some key reactants are supplied at a fixed rate. If the limiting reactant is oxygen, the supply rate is affected by mass transfer considerations. In fed batch fermentations, the limiting reactant is the substrate (e.g., glucose), which is fed to the system at a fixed rate. A stationary population can be maintained using continuous culture techniques.

During the stationary phase, the growth rate is zero as a result of the depletion of nutrients and essential metabolites. Several important fermentation products (including most antibiotics) are produced in the stationary phase. The stationary phase is followed by a phase where cells die or sporulate. During the death phase, there is a decrease in live cell concentration, which results from the toxic byproducts coupled with the depletion of the nutrient. The number of viable cells usually follows an exponential decay curve during this period.

Each phase in Figure 11-16 is recommended in microbiological processes. For example, a design objective may require minimizing the length of the lag phase or maximizing the rate and length of the exponential phase. The latter objective is achieved by slowing the onset of the transition to stationary growth. Knowledge of parameters that influence the phase of batch growth is essential to attain the largest possible cell density at the end of the process. Microbial cultures are very sensitive systems and, therefore, high demands are required for both control and monitoring of a biochemical reactor. Additionally, off-line analysis of liquid chromatograph or enzymatic assay techniques can also be employed for measurement of substances associated with the process of growth as substrates and products.

The growth of an aerobic organism can be represented by

$$\text{Substrate} \xrightarrow{\text{cells}} \text{more cells} + \text{product}$$

GLUCONIC ACID PRODUCTION

Consider the fermentation of glucose to gluconic acid, which involves the oxidation of the aldehyde group (RCHO) of the sugar to a carboxyl group (RCOOH). The industrial production of gluconic acid is by strains of *Aspergillus* and *Pseudomonas ovalis*. The enzyme that catalyzes the oxidation of glucose is a dehydrogenase (and not an oxidase), which is capable of transforming glucose to gluconolactone. Gluconic acid is produced by the hydrolysis of the glucono-lactone, which can be either enzymatic or a non-enzymatic process. The enzyme that is required for the hydrolysis step is gluconolactonase, although the presence of this enzyme in *Aspergillus* and *Pseudomonas* has not been revealed. Rai and Constantinide [14] considered the hydrolysis stage as a non-enzymatic process. The byproduct of the reaction is decomposed to water and oxygen by the enzyme catalase, which is present in the living cells.

Gluconic acid is extensively used in the food, pharmaceutical, and a wide range of products. In the textile industries, gluconic acid, glucono-δ-lactone, and ammonium gluconates are used in acid catalysts. Gluconates are incorporated in antibiotic complexes (e.g., tetracyline) to improve stability, lower the toxicity, and increase antibiotic levels in the blood. Calcium gluconates are employed for treating calcium deficiencies in both humans and animals.

The hydrogen peroxide produced in the glucose oxidase catalyzed reaction contains an antibacterial action. The addition of a catalase converts the hydrogen peroxide to water and oxygen.

Reaction Mechanisms

The reaction mechanisms in the fermentation of glucose to gluconic acid are:

1. *Cell growth:*

   Glucose + cell $\longrightarrow$ more cells

   For example,

   $$C_6H_{12}O_6 + \text{cells} \rightarrow \text{more cells} \qquad (1\text{-}34)$$

2. *Glucose oxidation:*

   Glucose + O$_2$ $\xrightarrow{\text{Glucose oxidase}}$ gluconolactone + H$_2$O$_2$

   For example,

   $$C_6H_{12}O_6 + O_2 \xrightarrow{\text{Glucose oxidase}} C_6H_{10}O_6 + H_2O_2 \qquad (1\text{-}35)$$

3. *Gluconolactone hydrolysis:*

   Gluconolactone + H$_2$O $\longrightarrow$ gluconic acid

For example,

$$C_6H_{10} + H_2O \rightarrow C_5H_{11}O_5COOH \tag{1-36}$$

4. *Hydrogen peroxide decomposition:*
For example,

$$H_2O_2 \xrightarrow{\text{Catalase}} H_2O + \frac{1}{2}O_2 \tag{1-37}$$

Mathematical Modeling

Rai and Constantinides [14] developed a mathematical model for the fermentation of the bacterium *Pseudomonas ovalis*, which converts the glucose to gluconic acid. The following equations describe the dynamics of the logarithmic growth phase:

The rate of cell growth:

$$\frac{dC_A}{dt} = b_1 C_A \left(1.0 - \frac{C_A}{b_2} \right) \tag{11-35}$$

The net rate of gluconolactone formation:

$$\frac{dC_B}{dt} = \frac{b_3 C_A C_D}{b_4 + C_D} - 0.9082 \, b_5 C_B \tag{11-36}$$

The rate of gluconic acid formation:

$$\frac{dC_C}{dt} = b_5 C_B \tag{11-37}$$

The rate of glucose consumption:

$$\frac{dC_D}{dt} = -1.011 \left(\frac{b_3 C_A C_D}{b_4 + C_D} \right) \tag{11-38}$$

where C_A = concentration of the cell
C_B = concentration of gluconolactone

C_C = concentration of gluconic acid

C_D = concentration of glucose

$b_1 - b_5$ = parameters of the system, which are functions of temperature and pH

Rai and Constantinides [14] performed the fermentation experiments at 30°C and a pH of 6.6. They fitted the dynamic equations (Equations 11-35 to 11-38) to the experimental data using the non-linear regression method developed by Marquardt [15] to evaluate the values of the parameters b_1 to b_5. The values of the parameters are: $b_1 = 0.949$, $b_2 = 3.439$, $b_3 = 18.72$, $b_4 = 37.51$, $b_5 = 1.169$.

At these conditions and known parameters, a computer program using the Runge-Kutta-Gill method was developed (PROG11-1) to simulate the changes in concentrations of the cells, gluconolactone, gluconic acid, and glucose with time for a period of 10 hrs (i.e., $0 \le t \le 10$ hrs).

The initial conditions at the start of this period were:

$C_A(0) = 0.5$ U.O.D./mL (cell optical density)

$C_B(0) = 0.0$ mg/mL

$C_C(0) = 0.0$ mg/mL

$C_D(0) = 50.0$ mg/mL

Table 11-2 gives the results of the computer simulation and Figure 11-17 shows the concentration profiles of the cell, gluconolactone, gluconic acid, and glucose with time. These profiles are in good agreement with the experimental data of Rai and Constantinides [14].

MODELING BIOLOGICAL REACTORS

A model can be defined as a set of relationships between the variables of interest in the system being investigated. A set of relationships may be in the form of equations; the variables depend on the use to which the model is applied. Therefore, mathematical equations based on mass and energy balances, transport phenomena, essential metabolic pathway, and physiology of the culture are employed to describe the reaction processes taking place in a bioreactor. These equations form a model that enables reactor outputs to be related to geometrical aspects and operating conditions of the system.

For example, in a fermentation process, the modeler may be interested in various controlled parameters such as the feed rate, rate and mode

Table 11-2
Concentration of the cell, gluconolactone, gluconic acid, and glucose with time using the Runge-Kutta-Gill method

TIME	CONC. CA	CONC. CB	CONC. CC	CONC. CD
0.00	0.5000	0.0000	0.0000	50.0000
0.50	0.7333	2.5918	0.6112	46.8185
1.00	1.0387	5.0631	2.7153	42.3880
1.50	1.4075	7.5746	6.2800	36.5758
2.00	1.8118	9.9874	11.2925	29.5341
2.50	2.2098	11.8892	17.5868	21.8320
3.00	2.5613	12.7399	24.7193	14.4229
3.50	2.8423	12.1714	31.9763	8.3345
4.00	3.0488	10.3031	38.5688	4.1700
4.50	3.1911	7.7632	43.9202	1.8243
5.00	3.2846	5.3007	47.8353	0.7191
5.50	3.3441	3.3661	50.4601	0.2649
6.00	3.3811	2.0375	52.1093	0.0939
6.50	3.4039	1.1971	53.1014	0.0326
7.00	3.4178	0.6908	53.6822	0.0112
7.50	3.4262	0.3943	54.0166	0.0038
8.00	3.4313	0.2236	54.2073	0.0013
8.50	3.4344	0.1263	54.3154	0.0004
9.00	3.4362	0.0712	54.3764	0.0001
9.50	3.4373	0.0401	54.4108	0.0000
10.00	3.4380	0.0225	54.4301	0.0000

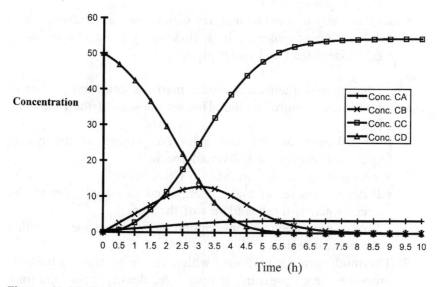

Figure 11-17. Concentration profiles of cell, glucose, gluconolactone, and gluconic acid with time.

of agitation, temperature, and the viability of the microorganism. Thus, constructing the mathematical model of a bioreactor from theoretical and empirical data enables the designer to predict the behavior and optimize the performance of the system.

A fermentation model is an abstracted and generalized description of pertinent aspects of the process. The abstracted physical model (Figure 11-11) can be defined as a region in space throughout which all the variables of interest (e.g., temperature, concentration, pH) are uniform. This is referred to as the control region or control volume. The control region may be a constant volume (e.g., a chemostat or a simple batch fermentation) or it may vary in size (e.g., a fed batch fermenter). The control region may be finite as represented by a well-mixed fermenter or infinitesimal as depicted by a tower fermenter. In this case, the concentrations of substrate and product vary continuously throughout the liquid space, so that only an infinitesimal thin slice can be considered uniform. Such a reactor can be modeled as a series of infinitesimal control regions. The considered boundaries of the control regions are:

- Phase boundaries across which an exchange of mass or energy occurs (e.g., the bubble-liquid interface).
- Phase boundaries across which no exchange occurs (e.g., the walls of the containing vessel).
- Geometrically defined boundaries within one phase across which exchanges occur either by bulk flow or by molecular diffusion (e.g., nutrient inlet and outlet pipes).

Constructing a mathematical model involves developing a set of equations for each control region. This set consists of the following:

1. Balanced equations for each extensive property of the system (e.g., mass, energy, or individual species).
2. Rate equations. These are of two types, namely:
   - Rates of transfer of mass, energy, individual components, or species across the boundaries of the region.
   - Rate of generation or consumption of individual species within the control region.
3. Thermodynamic equations, which relate to thermodynamic properties (e.g., pressure, temperature, density, concentration), either within the control region (e.g., gas laws) or on either side of a phase boundary (e.g., Henry's law).

A balanced equation for every extensive property in each control volume may be written as:

$$\begin{array}{l} \text{rate of accumulation} \\ \text{in control region} \end{array} = \begin{array}{l} \text{rate of input} \\ \text{to control region} \end{array} - \begin{array}{l} \text{rate of output} \\ \text{from control region} \end{array} \quad (11\text{-}39)$$

The input terms can be classified as:

• Bulk flow across geometrical boundaries.
• Diffusion across geometrical boundaries.
• Transfer across phase boundaries.
• Generation within the control region.

The output terms are:

• Bulk flow across geometrical boundaries.
• Diffusion across geometrical boundaries.
• Transfer across phase boundaries.
• Consumption within the control region.

Bulk flow is expressed as the material or energy carried by the bulk flow of fluid into or out of the control region.

MASS TRANSFER ACROSS PHASE BOUNDARIES

Agitation and Oxygen Transfer

The degree of both agitation and aeration must be considered simultaneously. Agitation will affect the transfer of oxygen from the gas phase into the liquid phase. The oxygen uptake rate (OUR), or the amount of oxygen that is required for cell growth per unit of cell mass and maintenance, can be determined from experimental data. This involves measuring the dissolved oxygen in a saturated fermentation broth and, subsequently, determining the rate at which the microbes use up the oxygen. Alternatively, OUR can be determined either by combining the stoichiometric ratio with the growth rate of the micro-organism or from the method described by Roberts, et al. [16].

The oxygen transfer rate (OTR), or the rate at which oxygen is transferred from the gas to the liquid phase, must be a least equal to the peak oxygen load (OUR) at steady state, and can be determined from Equation 11-40:

$$OTR = k_L a \left(C^* - C_L \right)_{mean} \qquad (11\text{-}40)$$

where OTR = oxygen transfer rate, mmoles/L-h

$k_L a$ = mass transfer coefficient, h^{-1}

C^* = oxygen concentration in equilibrium with the partial pressure of oxygen in the bulk gas phase, mmoles/L

C_L = concentration of dissolved oxygen in the fermentation broth, mmoles/L

The generation or consumption terms are defined as follows:

r_x = growth rate of cells

r_p = production rate of some product

r_s = consumption rate of substrate

A convenient method for measuring OTR in microbial systems depends on dissolved oxygen electrodes with relatively fast response times. Inexpensive oxygen electrodes are available for use with open systems, and steam-sterilizable electrodes are available for aseptic systems. There are two basic types: One develops a voltage from an electrochemical cell based oxygen, and the other is a polarographic cell whose current depends on the rate at which oxygen arrives. Measurement of oxygen transfer properties require a brief interruption of oxygen supply [17].

GENERAL MODEL FOR A SINGLE VESSEL

Consider an idealized fermentation process in which growing cells are consuming substrate and producing more cells as represented by the following scheme.

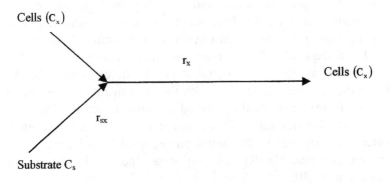

where C_x = cell concentration, kg/m^3
C_s = substrate concentration, kg/m^3
r_{sx} = rate of substrate consumption, kg/m^3•h
r_x = rate of cell growth, kg/m^3•h

Figure 11-18 shows the notations used in modeling and analysis of the fermentation process. The material balance on the microorganism in a CSTR at constant volume, assuming perfect mixing (i.e., concentrations of the cell and substrate inside and at the exit are the same) is expressed by:

$$\begin{matrix} \text{rate of accumulation} \\ \text{in control region} \end{matrix} = \begin{matrix} \text{rate of input} \\ \text{to control region} \end{matrix} - \begin{matrix} \text{rate of output} \\ \text{from control region} \end{matrix} \quad (11\text{-}39)$$

Cell balance: accumulation = $(d/dt)(VC_C)$
Input terms: bulk flow = u_iC_{xi}
       generation = r_xV
Output terms: bulk flow = u_OC_x

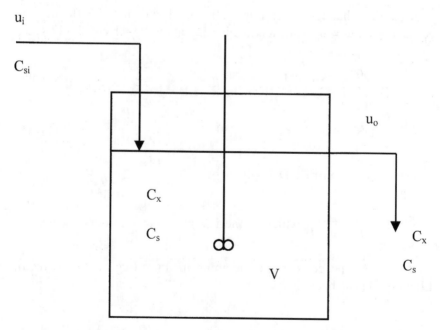

u_i

C_{si}

u_o

C_x

C_s

V

C_x

C_s

Figure 11-18. A chemostat with design parameters.

Substituting these terms in the material balance equation gives

$$V\frac{dC_C}{dt} = u_i C_{xi} + r_x V_R - u_o C_x \qquad (11\text{-}41)$$

Substrate balance: accumulation $= (d/dt)(VC_S)$
Input terms: bulk flow $= u_i C_{Si}$
Output terms: bulk flow $= u_o C_S$
consumption $= r_{sx} V$

The substrate material balance becomes:

$$V\frac{dC_S}{dt} = u_i C_{Si} - u_o C_S - r_{sx} V \qquad (11\text{-}42)$$

Additionally, the change in reactor (liquid) volume can be accounted for by

$$\frac{dV}{dt} = u_i - u_o \qquad (11\text{-}43)$$

Assuming that the reactor is well mixed and that the flowrate is constant (i.e., $u_i = u_o = u$), rearranging Equations 11-41 and 11-42 gives

$$\frac{dC_x}{dt} = D(C_{xi} - C_x) + r_x \qquad (11\text{-}44)$$

$$\frac{dC_s}{dt} = D(C_{si} - C_s) - r_{sx} \qquad (11\text{-}45)$$

where the dilution rate D is given by

$$D = \frac{u}{V}\left(\frac{m^3}{h} \bullet \frac{1}{m^3}\right), \text{ which is the inverse of the residence time.}$$

If there are no cells in the incoming medium (e.g., a sterile feed), Equation 11-44 becomes

$$\frac{dC_x}{dt} = r_x - DC_x \qquad (11\text{-}46)$$

Assuming r_x and r_{sx} are as follows:

$$r_x = \mu C_x \tag{11-47}$$

and $r_{sx} = \mu \dfrac{C_x}{Y_{x/s}}$
$$\tag{11-48}$$

where μ = specific growth rate, h^{-1}
$\quad Y_{x/s}$ = yield coefficient

$$Y_{x/s} = \frac{\text{mass of new cells formed}}{\text{mass of substrate consumed}} \left(\frac{\text{kg of cells}}{\text{kg of substrate}} \right) \tag{11-49}$$

with $Y_{x/s} = 1/Y_{s/x}$. Equations 11-44 and 11-45 now become

$$\frac{dC_x}{dt} = (\mu - D)C_x \tag{11-50}$$

and $\dfrac{dC_s}{dt} = \dfrac{-\mu C_x}{Y_{x/s}} + D(C_{si} - C_s)$
$$\tag{11-51}$$

The specific growth rate will be assumed to take the form

$$\mu(C_s) = \frac{\mu_{max}C_s}{K_s + C_s} \tag{11-52}$$

where μ_{max} = maximum specific growth rate
$\quad K_s$ = the Monod constant

The specific growth rate function $\mu(C_s)$ is written as μ. The various properties of Equation 11-52 are: when the substrate concentration is not limiting (i.e., $C_s \gg K_s$), the specific growth rate $\mu(C_s)$ approaches μ_{max}; the growth rate of the viable cell (r_v) becomes independent of C_s and is proportional to the viable cell concentration (C_{xv}).*

---

*Viable cells mean cells which can grow; non-viable cells mean cells which cannot. Micro-biologists use the same terms but can divide "non-viable" cells into cells, which remain metabolically active in some respects and cells which are effectively dead. They can also distinguish within their category of "viable" cells between cells, which are actively dividing, and cells such as spore cells whose growth activities are potential rather than actual.

$$r_x = \mu(S)C_{xv} = \left(\frac{\mu_{max}\,C_s}{K_s + C_s}\right)C_{xv} \tag{11-53}$$

where C_{xv} is the concentration of viable cell. When C_s is small compared to K_s, the specific growth rate is nearly proportional to the limiting substrate concentration.

At steady state, Equations 11-50 and 11-51 become

$$(\mu - D)C_x = 0 \tag{11-54}$$

and $$D(C_{si} - C_s) - \frac{\mu C_x}{Y_{x/s}} = 0 \tag{11-55}$$

The generalized mass balance equation on a unit volume is:

$$\frac{1}{V}\frac{d(Vy)}{dt} = \sum r_{gen} - \sum r_{cons} + D(y_i - \gamma y) \tag{11-56}$$

where y is the general extensive property, C_x, C_s, C_p, C_v, and C_d. These properties are defined as:

C_x = cell concentration
C_s = substrate concentration
C_p = product concentration
C_v = viable cell concentration
C_d = non-viable cell concentration
V = reactor (liquid) volume
$\sum r_{gen}$ = sum of all the rates of generation
$\sum r_{con}$ = sum of all the rates of consumption
$\gamma = u_o/u_i$

Figure 11-18a shows a flow diagram for constructing the mathematical model of a fermentation process.

THE CHEMOSTAT

The chemostat is an arrangement for continuous fermentation in which the properties of the system are regulated by a controlled supply of some limiting nutrient. It is a powerful research tool in microbial physiology and for evaluating process parameters. The chemostat

BIOCHEMICAL UNIT   MAIN PROCESS DATA LINE   PHYSIOLOGICAL
INPUT

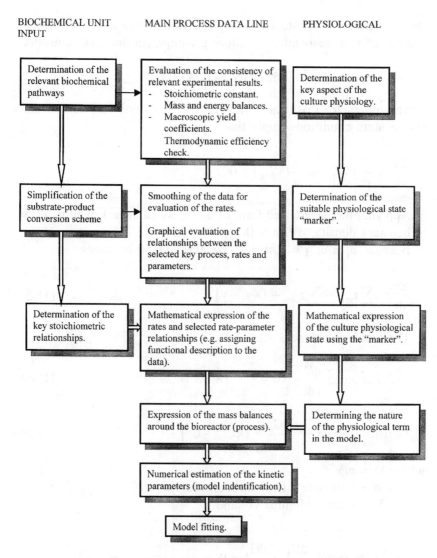

Figure 11-18a. The scheme of constructing the mathematical model of a fermentation process. *(Source: B. Volesky and J. Votruba [4].)*

operates with a constant inflow and outflow, and it is assumed to be well mixed and that concentrations are uniform throughout the whole of the medium volume. Outflow concentrations are the same as those in the fermenter, and constant inflow and outflow medium ensures a constant reactor volume. The fermenter operation mode normally involves sterile feed ($C_x = 0$), and the cell growth rate can be controlled by adjusting the volumetric feedrate (i.e., dilution rate). Figure 11-19 shows a chemostat with associated monitoring equipment and pH controller.

DESIGN EQUATION

The bioreactor involves sterile feed ($C_x = 0$) and constant volume; steady state conditions imply that

$$\frac{dV}{dt} = \frac{d}{dt}(VC_s) = \frac{d}{dt}(VC_x) = 0$$

Applying the generalized form of the mass balance Equation 11-56, at constant density, dV/dt = 0.

Because $u_i = u$, $\gamma = u_i/u = 1$, Equation 11-56 becomes

$$\frac{dy}{dt} = \sum r_{gen} - \sum r_{con} + D(y_i - y) \tag{11-57}$$

The material balances are:

Cell balance: $\quad \dfrac{dC_x}{dt} = 0 + r_x - DC_x$ \hfill (11-58)

Substrate balance: $\quad \dfrac{dC_s}{dt} = D(C_{si} - C_s) - r_s$ \hfill (11-59)

where D = u/V
   u = volumetric flow through the system
   V = reactor (liquid) volume

At steady state, $\quad \dfrac{dC_s}{dt} = \dfrac{dC_x}{dt} = 0$

Therefore, for the substrate balance

$$u(C_{si} - C_s) - r_s V = 0$$

or \hfill (11-60)

$$D(C_{si} - C_s) - r_s = 0 \tag{11-61}$$

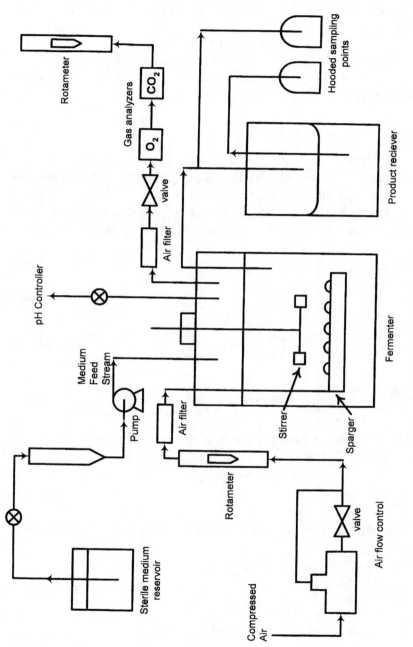

Figure 11-19. Chemostat system.

For the cell balance

$$DC_x = r_x \tag{11-62}$$

Substituting Equation 11-47 into Equation 11-62 gives

$$DC_x = \mu C_x$$

or $D = \mu$ \hfill (11-63)

The dilution rate D is equal to 1/t, where $\bar{t} = u/V$, and is equal to the mean residence time of the tank. Equation 11-63 shows that controlling the dilution rate D (i.e., the flowrate into the reactor) can control the specific cellular growth rate of a microbial species μ.

For the substrate balance, it is known from Equation 11-61 that

$$D(C_{si} - C_s) - r_s = 0$$

or \hfill (11-61)

$$D(C_{si} - C_s) = +r_s = \frac{r_x}{Y_{x/s}} \tag{11-64}$$

$$D(C_{si} - C_s) = \frac{DC_x}{Y_{x/s}} \tag{11-65}$$

The cell concentration in terms of the substrate concentration and the yield coefficient is:

$$C_x = Y_{x/s}(C_{si} - C_s) \tag{11-66}$$

The substrate concentration C_s is obtained from the Monod expression given by

$$\mu = \frac{\mu_{max} C_s}{K_s + C_s} = D \tag{11-67}$$

Rearranging Equation 11-67 yields

$$K_s + C_s = \frac{\mu_{max} C_s}{D}$$

$$\text{or } C_s = \frac{K_s D}{(\mu_{max} - D)} \tag{11-68}$$

Assuming that a single nutrient is limiting, and cell growth is the only process contributing to substrate utilization, substituting Equation 11-68 into Equation 11-66 and rearranging yields

$$C_x = \frac{Y_{x/s}(K_s + C_{si})}{(\mu_{max} - D)} \left[\frac{C_{si}\mu_{max}}{K_s + C_{si}} - D \right] \tag{11-69}$$

WASHOUT IN CONTINUOUS CULTURE

A consequence of applying Equations 11-68 and 11-69 respectively in describing the chemostat, is that the maximum dilution rate is limited to a value that is slightly less than the maximum specific growth rate μ_{max}. This is referred to as the critical dilution rate D_c.

Washout occurs when cells are removed from the reactor at a rate $(D_c \bullet C_{xv})$ that is just equal to the maximum rate at which they can grow. As the flowrate u increases, D increases and causes the steady state value of C_s to increase and the corresponding value of C_x to decrease, respectively. When D approaches μ_{max}, C_x becomes zero, and C_s will rise to the inlet feed value C_{si}. This corresponds to a complete removal of the cells by flow out of the tank, and the condition of loss of all cells at steady state is known as "washout."

The dilution rate at which washout occurs can be determined by setting $C_x = 0$ in Equation 11-69. This becomes

$$\frac{\mu_{max}C_{si}}{K_s + C_{si}} - D_{max} = 0$$

$$\text{or } D_{max} = \frac{\mu_{max}C_{si}}{K_s + C_{si}} \tag{11-70}$$

The rate of cell production per unit reactor volume is DC_x. The maximum cell output rate is determined by solving

$$\frac{d}{dD}(DC_x) = 0 \tag{11-71}$$

Substituting for C_x gives

$$DC_x = DY_{x/s}\left(C_{si} - C_s\right)$$

$$= DY_{x/s}\left(C_{si} - \frac{K_s D}{\mu_{max} - D}\right)$$

$$\frac{d}{dD}(DC_x) = \frac{Y_{x/s}}{dD}\left(C_{si}D - \frac{K_s D^2}{(\mu_{max} - D)}\right) = 0 \qquad (11\text{-}72)$$

Solving for Equation 11-71 yields for the maximum production

$$D_{max\ output} = \mu_{max}\left\{1 \pm \left[\frac{K_s}{(K_s + C_{si})}\right]^{0.5}\right\} \qquad (11\text{-}73)$$

The microbial growth rate (and hence the dilution rate) for maximum production (or consumption) must be below the maximum specific growth rate (and hence the critical dilution rate). Therefore, Equation 11-73 is given by

$$D_{max\ output} = \mu_{max}\left\{1 - \left[\frac{K_s}{(K_s + C_{si})}\right]^{0.5}\right\} \qquad (11\text{-}74)$$

Figure 11-20 shows production rate, cell concentration, and substrate concentration as functions of dilution rate. From Equation 11-74, the maximum production rate can be determined.

Example 11-4

(a) Given that the maximum specific growth rate of an organism is $\mu_{max} = 1\ h^{-1}$ and its $K_s = 1.5\ g/cm^3$. If the input substrate concentration is $145\ g/cm^3$, determine the maximum output rate for this system in a CFSTR.

(b) If the concentration of product at this maximum output rate is $10\ g/cm^3$, determine the maximum output rate of the product.

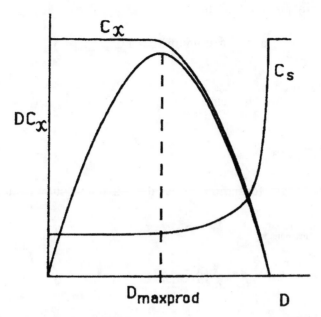

Figure 11-20. Cell concentration and production rate as a function of dilution rate. *(Source: Fogler [19].)*

(c) If the concentration of biomass at this maximum output is 50 g/cm³, determine the cell mass formation rate at the maximum output rate.

Solution

(a) Using Equation 11-74, the maximum output rate in a CFSTR is:

$$D_{\text{max output}} = \mu_{\text{max}}\left\{1 - \left[\frac{K_s}{(K_s + C_{si})}\right]^{0.5}\right\} \qquad (11\text{-}74)$$

$$= 1\left\{1 - \left[\frac{1.5}{1.5 + 145}\right]^{0.5}\right\}$$

$$= 0.8988 \text{ h}^{-1}$$

$$= 0.899 \text{ h}^{-1}$$

(b) The output rate of product is

$$r_P = D \bullet C_{PI}, \text{ at the maximum } r_p = D_{\text{max output}} \bullet C_{PI}$$

$$r_P = 0.899 \bullet 10 \left\{ \frac{1}{h} \bullet \frac{g}{cm^3} \right\}$$

$$r_P = 8.99 \frac{g}{cm^3 h}$$

(c) The cell mass formation rate at the maximum output rate is:

$$r_M = D_{\text{max output}} \bullet C_{MI}$$

$$= 0.899 \bullet 50 \left\{ \frac{1}{h} \bullet \frac{g}{cm^3} \right\}$$

$$= 44.95 \frac{g}{cm^3 h}$$

BATCH FERMENTER

Starting from an inoculum, C_{xi} at t = 0, and an initial quantity of limiting substrate C_{si} at t = 0, the biomass will grow after a short lag phase and will consume substrate. The growth rate slows as the substrate concentration decreases, and becomes zero when all the substrate has been consumed. Simultaneously, the biomass concentration initially increases slowly, then faster until it levels off when the substrate becomes depleted. Figure 11-21 shows a sketch of a batch fermenter.

The mass balances for a constant volume zero feed batch fermentation are as follows:

Using the generalized mass balance equation:

$$\frac{1}{V} \frac{d(Vy)}{dt} = \sum r_{\text{gen}} - \sum r_{\text{con}} + Dy_i - \gamma Dy \tag{11-56}$$

where y is the general extensive property.

The left side of Equation 11-56 becomes

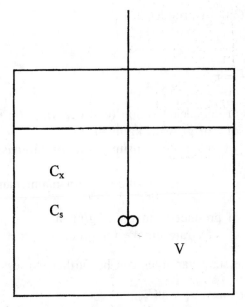

Figure 11-21. The batch fermenter.

$$\frac{1}{V}\frac{d(Vy)}{dt} = \frac{1}{V}\left\{V\frac{dy}{dt} + y\frac{dV}{dt}\right\}$$ (11-75)

For the batch reactor, $u_i = u_O = 0, \dfrac{dV}{dt} = 0$

Equation 11-56 becomes

$$\frac{dy}{dt} = \sum r_{gen} - \sum r_{cons}$$ (11-76)

The equations for the state variables (viable cells, non-viable cells, substrate, and product) are:

Viable cells: $\dfrac{dC_{xv}}{dt} = r_x - r_d$ (11-77)

Non-viable cells: $\dfrac{dC_{xd}}{dt} = r_d$ (11-78)

Substrate: $\dfrac{dC_s}{dt} = -\left(r_{sx} + r_{sm} + r_{sp}\right)$ (11-79)

Product: $\dfrac{dC_p}{dt} = r_p$ (11-80)

where r_d = rate of cell deactivation (cell death), kg/m³•h
$\quad\ r_x$ = rate of cell growth, kg/m³•h
$\quad\ r_{sx}$ = rate of substrate consumption for biomass production, kg/m³•h
$\quad\ r_{sm}$ = rate of substrate consumption for maintenance formation, kg/m³•h
$\quad\ r_p$ = rate of product formation, kg/m³•h
$\quad\ r_{sp}$ = rate of substrate uptake for product formation, kg/m³•h

The rate of these state variables can be further expressed as:

Rate of cell growth: $r_x = \mu C_{xv}$ (11-81)

Rate of cell death: $r_d = k_d C_{xv}$ (11-82)

From Leudeking-Piret kinetics [18],

Rate of product formation: $r_p = \alpha r_x + \beta C_{xv}$ (11-83)

where α = growth-related product formation coefficient (kg product/ kg cells)
$\quad\ \beta$ = non-growth related product formation coefficient (kg product/kg cells)

Rate of substrate consumption for biomass production:

$$r_{sx} = \frac{r_x}{Y'_{x/s}} = \frac{\mu C_{xv}}{Y'_{x/s}}$$ (11-84)

Rate of substrate consumption for product formation:

$$r_{sp} = \frac{r_p}{Y'_{p/s}} = \frac{\alpha \mu C_{xv} + \beta C_{xv}}{Y'_{p/s}}$$ (11-85)

The rate of consumption of substrate to provide the energy for maintenance is:

Rate of substrate consumption for maintenance energy:

$$r_{sm} = m_s C_{xv} \qquad (11\text{-}86)$$

where m_s is the rate constant (kg substrate/kg cells•h). The value of the maintenance energy coefficient, m, depends on the environmental conditions surrounding the cell and on its growth rate.

Substituting the relationships into the material balances gives:

Viable cells: $\quad \dfrac{dC_{xv}}{dt} = \mu C_{xv} - k_d C_{xv} \qquad (11\text{-}87)$

Non-viable cells: $\quad \dfrac{dC_{xd}}{dt} = k_d C_{xv} \qquad (11\text{-}88)$

Substrate: $\quad \dfrac{dC_s}{dt} = -\left\{ \dfrac{\mu C_{xv}}{Y'_{x/s}} + m_s C_{xv} + \dfrac{\alpha \mu C_{xv} + \beta C_{xv}}{Y'_{p/s}} \right\} \qquad (11\text{-}89)$

Product: $\quad \dfrac{dC_p}{dt} = \alpha \mu C_{xv} + \beta C_{xv} \qquad (11\text{-}90)$

Including a set of initial conditions at the time of inoculation, the profiles of these variables with time can be determined by numerical integration with the aid of a computer program (Appendix D).

FED-BATCH REACTOR

The fed batch reactor (FBR) is a reactor where fresh nutrients are added to replace those already used. The rate of the feed flow u_i may be variable, and there is no outlet flowrate from the fermenter. As a consequence of feeding, the reactor volume changes with respect to time. Figure 11-22 illustrates a simple fed-batch reactor. The balance equations are:

For constant density and feed flowrate: $\quad \dfrac{dV}{dt} = u_i \qquad (11\text{-}91)$

Substrate balance: $\quad \dfrac{d(VC_s)}{dt} = u_i C_{si} + r_s V \qquad (11\text{-}92)$

Biomass balance: $\quad \dfrac{d(VC_x)}{dt} = r_x V \qquad (11\text{-}93)$

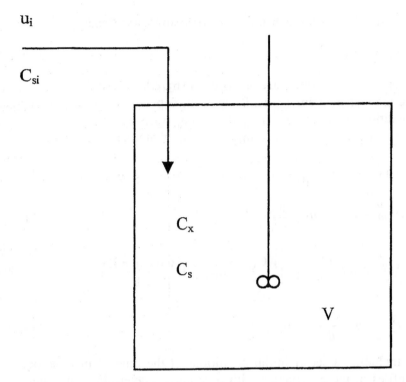

Figure 11-22. The fed-batch reactor.

The quantities VC_x and VC_s represent the mass of biomass and substrate respectively in the reactor. Dividing these masses by the volume V gives the concentrations C_x and C_s as a function of time, which are required in the kinetic relationships to determine r_x and r_s.

A quasi-steady state process is a fed-batch fermenter where $dC_x/dt = 0$ and $\mu = u/V$. Because V increases, μ therefore decreases, and thus the reactor moves through a series of changing steady states for which $\mu = D$, during which C_s and μ decrease and C_x remains constant.

Keller and Dunn [19] have provided a detailed analysis of fed-batch operation. An example of an industrial fed batch process is the production of penicillin by *Penicillium chrysogenum*. In this process, the cellular growth rate of the mycelium during the trophophase (i.e., cell production phase) is high, allowing this phase to be short. Correspondingly, the cellular growth rate during the idiophase (i.e., phase of product formation) is very low, and the emphasis in this phase is on the production of penicillin.

SCALE-UP OF BIOREACTORS

Fermenters are generally designed to provide more oxygen (i.e., for a higher oxygen transfer rate OTR) than is required at the peak oxygen load of the organism. This usually occurs during the log or exponential growth phase. The oxygen uptake rate (OUR) value of a specific fermentation system is required to design the system, and these requirements are different for each microorganism.

Employing the OTR criterion, a high respiration rate of the organism enables successful scale-up of aerobic processes. In such a case, the rate-limiting reaction step is the oxygen transfer resistance at the gas-liquid interface. This can be increased when required by increasing the power as this transfers more oxygen from the gas phase to the liquid phase. Additionally, the back pressure or head pressure can increase the OTR while preventing foam production.

Another essential scale-up parameter is $k_L a$, the mass transfer coefficient at the liquid-air interface. $k_L a$ is determined both by the terms "gassing-in" and "gassing-out," respectively. The former is achieved by saturating the liquid in the vessel with oxygen. This involves measuring the time it takes to increase the oxygen concentration, which is measured with a dissolved oxygen (DO) probe. The latter technique uses nitrogen to purge the oxygen from the liquid. $k_L a$ can also be determined from a mass balance of the gas flow. If scale-up is relied on oxygen transfer alone, be careful not to oversize the fermenter. Scale-up of a batch reactor is reviewed in Chapter 13.

The position of the agitator shaft and motor can affect scale-up of a fermenter. Generally, agitators and shafts are mounted in the top vertical position. However, the development of more reliable shaft seals has enabled the agitator to be mounted in the bottom position, thus increasing accessibility and ease of maintenance. This orientation now permits a shorter shaft, requires simpler physical support structures, creates head space for piping and a cyclone (which captures liquid-entrained gases), and reduces the vertical height requirement [13]. Figure 11-23 shows a piping and instrumentation diagram (P & ID) of a fermenter.

An important factor in scale-up is to maintain constant tip speed (i.e., the speed at the end of the impeller is equal to the velocity in both the model and full-scale plant). If the speed is too high, it can lyse (i.e., kill) off the bacteria, and if the speed is too slow, the reactor contents will not be well mixed.

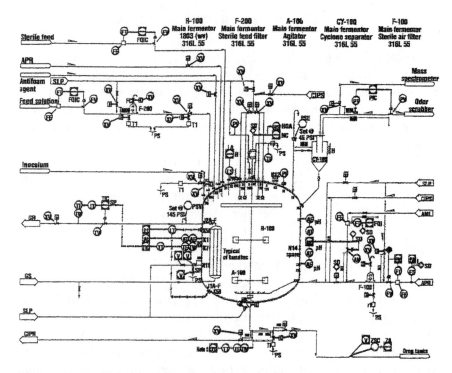

Figure 11-23. Piping and instrumentation diagram of a fermenter. *(Source: Bartow [13], "Supersizing the aerobic fermenter,"* Chemical Engineering, *pp. 70–75, July 1999. Courtesy of Chemical Engineering.)*

An algorithm for the scale-up of a fermenter is as follows [20]:

1. Choose a fermenter volume required based on the desired capacity.
2. Choose an impeller diameter, D_i
3. Determine the reactor's dimension (e.g., D_T = tank diameter) based on geometric similarity, with the impeller diameter being the characteristic length. For example,

$$\frac{D_{T,full-scale}}{D_{T,model}} = \frac{D_{i,full-scale}}{D_{i,model}} \qquad (11\text{-}94)$$

4. Calculate the impeller speed, N.

$$\left(\pi D_i N\right)_{full-scale} = \left(\pi D_i N\right)_{model}$$

$$N_{full-scale} = N_{model} \bullet \frac{D_{i,model}}{D_{i,full-scale}} \qquad (11-95)$$

5. Choose the mass transfer correlation for $k_L a$ (i.e., mass transfer coefficient of the air bubble).
6. Calculate the gas flowrate using the correlation and setting.

$$\left(k_L a\right)_{full-scale} = \left(k_L a\right)_{model}$$

Using the Perez and Sandall [21] correlation for non-Newtonian fluid

$$\frac{k_L a D_i^2}{D_{AB}} = 21.2 \left(\frac{D_i^2 N \rho}{\mu_e}\right)^{1.1} \left(\frac{\mu_e}{D_{AB}}\right)^{0.5} \left(\frac{\mu_e V_s}{\sigma}\right)^{0.45} \left(\frac{\mu_d}{\mu_e}\right)^{0.7} \qquad (11-96)$$

where D_T = tank diameter, m
$\quad D_i$ = impeller diameter, m
$\quad D_{AB}$ = diffusivity, m²/s
$\quad k_L a$ = volumetric mass transfer coefficient, s⁻¹
$\quad N$ = impeller rotational speed, s⁻¹
$\quad Q$ = volumetric flowrate, m³/s
$\quad V_s$ = superficial gas velocity, $(4Q/\pi D_T^2)$, m/s
$\quad \mu_d$ = viscosity of dispersed phase, kg/m•s
$\quad \mu_e$ = effective viscosity, kg/m•s
$\quad \sigma$ = surface tension, N/m
$\quad \rho$ = density, kg/m³

Other relationships for determining $k_L a$ are given in Table 11-3.
    The mass transfer coefficient $k_L a$ depends on the following equipment parameters:

$$k_L a \approx \frac{1}{D_i^2} \left(D_i^2 N\right)^{1.1} \left(\frac{4Q}{\pi D_i^2}\right)^{0.45}$$

$$\approx D_i^{0.2} N^{1.1} \left(\frac{Q}{D_i^2}\right)^{0.45} \qquad (11-97)$$

Table 11-3
Mass transfer coefficients in fermenter

Coalescing media:	$k_L a^{20} = 0.026 \bullet \left(\dfrac{P_g}{V}\right)^{0.4} \bullet (v_s)^{0.5}$
Non-coalescing media:	$k_L a^{20} = 0.0016 \bullet \left(\dfrac{P_g}{V}\right)^{0.7} \bullet (v_s)^{0.2}$
In general:	$k_L a^{20} = c \bullet \left(\dfrac{P_g}{V}\right)^{a} \bullet (v_s)^{b}$
In viscous fermentation broths:	$k_L a = c \bullet \left(\dfrac{P_g}{V}\right)^{a} \bullet (v_s)^{b} \bullet (\eta)^{-0.86}$
For coarse bubble systems:	$k_L a^{20} = 0.32 \bullet (v_s)^{0.7}$

$k_L a^{20}$ = volumetric mass transfer at 20°C
v_s = superficial gas velocity
η = dynamic viscosity

Vessel size (liters)	a	b
5	0.95	0.67
500	0.6 – 0.7	0.67
50,000	0.4 – 0.5	0.50

Source: Bartholomew, Adv. Appl. Microbiol., *2, 289, 1960.*

$$\frac{(k_L a)_{full-scale}}{(k_L a)_{model}} = \frac{\left[D_i^{0.2} N^{1.1} \left(\dfrac{Q}{D_i^2}\right)^{0.45}\right]_{full-scale}}{\left[D_i^{0.2} N^{1.1} \left(\dfrac{Q}{D_i^2}\right)^{0.45}\right]_{model}} = 1 \qquad (11\text{-}98)$$

Therefore,

$$Q_{full-scale} = Q_{model} \left(\frac{D_{i,full-scale}^2}{D_{i,model}^2}\right) \left[\left(\frac{D_{i,model}}{D_{i,full-scale}}\right)^{0.2} \left(\frac{N_{model}}{N_{full-scale}}\right)^{1.1}\right]^{1/0.45} \qquad (11\text{-}99)$$

7. Calculate the power requirement.

An alternative procedure for determining N and Q is to set either Q/ND = constant or V_s = constant, and then determine N from the power or mass transfer correlations.

Table 11-4 gives the different scale-up criterion for the ratio of the full-scale and model for geometrically similar systems with volume of the model (V_m = 10 l) and the volume of the full-scale (V_{fs} = 10 m³). This gives a linear scale-up factor of 10. The table shows that different scale up criteria result in entirely different process conditions at production scale. For example, to give an equal power per unit volume (P/V) value involves an increase in the power input by 10^3. Correspondingly, for an equal value of stirrer speed (N), the power input (P) must be increased by a factor of 10^5. It may be necessary to find out how much the P/V ratio must be changed to achieve the same Reynolds number or how much larger the diameter of the stirrer must be if the stirrer speed is to be maintained. The first answer from Table 11-4 shows that the P/V ratio must be changed by 10^{-4}, and the second answer shows that to use the same N, the value of ND must be increased by 10^2. Subsequently, the diameter would need to be increased by 10^2.

Generally, a constant value of a particular operating/equipment variable is used. Some examples of scale-up based on constant operating variables as frequently used in fermentation are shown in Table 11-4. The properties, which can be kept constant at the different scales, are:

- P/V (for CSTR)
- $k_L a$ for (CSTR and bubble column)
- V_{tip} (for CSTR)
- Mixing times t_m (for CSTR)
- A combination of P/V, V_{tip}, and the gas flowrate (for CSTR)

Table 11-4
Different scale-up criteria and consequences

Scale-up criterion	Ratio of the value at 10 m³ relative to the value at 10 l				
	P	P/V	N (or t_m^{-1})	ND	N_{Re}
Equal P/V	10^3	1	0.22	2.15	21.5
Equal N (or t_m^{-1})	10^5	10^2	1	10^2	10^2
Equal tip speed	10^2	0.1	0.1	1	10
Equal Reynolds number	0.1	10^{-4}	0.01	0.1	1

P = power, V = volume, N = impeller speed, D = stirrer diameter, N_{Re} = Reynolds number.
Source: Biotechnology by Open Learning [22].

When P/V is used as the constant criterion, there is a decrease in power consumption under aerated conditions. In the literature many relationships exist, but in practice the gassed power consumption is about 50% of the ungassed power consumption. The correlations for gassed power consumption (P_g) and the ungassed power consumption (P_s) can be expressed by [22]

$$\frac{P_g}{P_s} = 0.01312 \bullet N_{Fr}^{-0.16} \bullet N_{Re}^{0.064} \left(\frac{\phi_g}{ND^3} \right)^{-0.38} \left(\frac{D_T}{D_i} \right) \tag{11-100}$$

where N_{Fr} = stirrer Froude number
$\quad N_{Re}$ = Reynolds number
$\quad N$ = stirrer speed
$\quad D_T$ = tank diameter
$\quad D_i$ = impeller diameter
$\quad \phi_g$ = volumetric gas flowrate

Other scale-up factors are mixing time, Reynolds number, and shear. Shear is maximum at the tip of the impeller and may be determined by Equation 11-101 [17]:

$$S_{model} = S_{full-scale} \left(\frac{D_{i,model}}{D_{i,full-scale}} \right)^{1/3} \tag{11-101}$$

An excessive shear can cause a breakup of some mycelial fermentation resulting in a low yield. A permissible tip speed is 2.5 to 5.0 m/s. Mixing time has been proposed as a scale-up consideration. Large-scale stirred tank reactors (with a volume larger than 5m³) are poorly mixed compared to small-scale reactors. This is a common cause of a change of regime (e.g., kinetic regime at model scale and transport regime at full-scale), which can present problems for mass and heat transfer involving viscous broths. The mixing time for aerated systems is expressed by [22]

$$Nt_m = \frac{0.6(D_T/D_i)^3(H/D_T)}{3\left\{N_p(H_i/D_i)^2\right\}^{0.5}} \quad \text{for} \quad N_{Re} > 5 \times 10^3 \tag{11-102}$$

where D_i = impeller diameter
  D_T = tank diameter
  N = impeller speed
  N_p = impeller power number
  N_{Re} = Reynolds number
  H = liquid height
  H_i = impeller height
  t_m = mixing time

Table 11-5 shows the mixing times for different sizes of reactors used in penicillin fermentation with $H/D_T = 2.5$. The measured mixing time is compared with the calculated mixing time from Equation 11-102. The observed differences in the measured and calculated t_m may be explained by the following:

- When $H/D_T = 2.5$, then t_m increases more than the Equation 11-102 on the scaling from $1.4 \ m^3$—$190 \ m^3$.
- The viscosity of the broth can influence the mixing time.
- Aerated conditions increase the mixing time.
- There are different criteria for mixing (e.g., for being well mixed).

Therefore, in scale-up, the use of the relationship to produce a larger system with the same t_m using various configurations of D_i, D_T, H, H_i, and N_p can result in an error. Hence, the use of equal mixing times

Table 11-5
Mixing times in different sizes of vessels

$V(m^3)$	$D_T(m)$	$P(kW)$	P/V (kW/m^3)	$t_m(s)$ measured	$t_m(s)$ calculated from Equation 11-102
1.4	1.1	3.8	2.7	29	12
45.0	3.5	120.0	2.7	67	26
190.0	4.4	24.0	1.3	119	31

Source: Biotechnology *by Open Learning [22].*

as determined by theoretical considerations as the criterion for scale-up should be used with caution.

Example 11-5

The antibiotic Tylosin was produced in a CSTR using Streptomyces fradiae in a 5 liter laboratory fermenter. For different substrate flow-rates the concentrations of product and biomass were measured [23].

Flowrate (mL/h)	50	100	150	200	350
Biomass conc. (g DW/L)	39	29	28	22	17
Tylosin conc. (g/L)	0.7	0.4	0.2	0.1	0.01

DW = dry weight.

Determine the specific growth rate and the specific production rate, using the above experimental data.

Solution

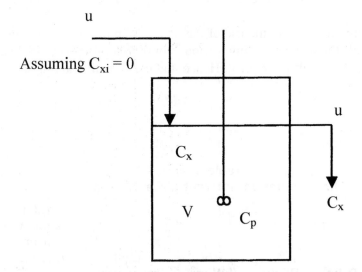

Assuming that the system is operating as steady state condition, the mass balance equation from Equation 11-39 becomes:

Rate of input to control region = Rate of output from control region

The rate of input is the rate at which the product species is formed by the reaction. The rate of output is the rate at which a species leaves the system.

Mass balance of biomass: $V\mu C_x = uC_x$ (11-103)

Mass balance of tylosin: $VqC_x = uC_p$ (11-104)

where μ and q are the specific growth and production rates, respectively. From Equations 11-103 and 11-104, μ and q are:

$$\mu = \frac{u}{V}$$ (11-105)

and $q = \dfrac{uC_p}{VC_x}$ (11-106)

Table 11-6 gives the results of μ and q for different substrate flowrates.

Example 11-6

The growth of Mehylomonas L3 on methanol was observed in a CSTR. Working volume of the tank was V = 0.355 L. The concentration of methanol in fresh medium was 1 g/L. The results of the experiment at a steady state are given in the following table [24].

Flowrate (mL/h)	67	70	120	155	165
Biomass conc. (g DW/L)	0.441	0.47	0.527	0.563	0.563
Methanol conc. (mg/L) in output stream	30	20	23	16	7

DW = dry weight.

Table 11-6
Results of μ and q

Flowrate (mL/h)	50	100	150	200	350
μ(1/h)	0.01	0.02	0.03	0.04	0.07
$q \times 10^3$ (1/h)	0.212	0.276	0.214	0.182	0.041

Evaluate the $Y_{s/X}$ and the specific growth rate μ.

Solution

The mass balances for both biomass and substrate respectively are:

$$V\mu C_x = u C_x \qquad (11\text{-}107)$$

$$u C_{si} = u C_s + Vq C_x \qquad (11\text{-}108)$$

The specific growth rate and the yield coefficient are

$$\mu = \frac{u}{V} \qquad (11\text{-}109)$$

and $Y_{s/X} = \dfrac{(C_{si} - C_s)}{C_x} = \dfrac{q}{\mu} \qquad (11\text{-}110)$

Table 11-7 gives the results for the specific growth rate and the yield coefficient for varying flowrate.

NOMENCLATURE

D_T = tank diameter, m
D_i = impeller diameter, m
D_{AB} = diffusivity, m^2/s
N = impeller rotational speed, s^{-1}
Q = volumetric flowrate, m^3/s
V_s = superficial gas velocity, $(4Q/\pi D_T^2)$, m/s

Table 11-7
Results of μ and $Y_{s/X}$

Flowrate (mL/h)	67	70	120	155	165
μ (1/h)	0.189	0.197	0.338	0.437	0.465
$Y_{s/X}$	2.2	2.13	1.85	1.75	1.76

μ_d = viscosity of dispersed phase, kg/m•s
μ_e = effective viscosity, kg/m•s
σ = surface tension, N/m
ρ = density, kg/m^3

GLOSSARY

Activated sludge: Material containing a very large active microbial population used in the purification of waste water.

Aerobe: Microorganism growing in the presence of air or requiring air for growth.

Aerobic reactor: A reaction vessel operating under aerobic conditions with sterile air being used as the source of oxygen in dissolved state.

Alcoholic (ethanolic) fermentation: A fermentation in which the major products are carbon dioxide and ethanol (alcohol).

Algae: A group of photosynthetic organisms that range from microscopic size to a length of 20 m. They supply oxygen and consume nutrients in several different processes for biological waste treatment.

Amino acids: The building blocks of proteins. There are twenty common ones; methionine, lysine, and trytophan are the ones produced in the greatest volume.

Anaerobe: An organism that grows without air or requires oxygen-free conditions to live.

Anaerobic reactor: A bioreactor in which no dissolved oxygen or nitrate is present and microbial activity is due to anaerobic bacteria.

Antibiotic: A specific type of chemical substance that is administered to fight infections usually caused by bacteria, in humans or animals. Many antibiotics are produced by microorganisms; some are chemically synthesized.

Antibody: A protein (immunoglobulin) produced by humans and higher animals in response to exposure to a specific antigen, and characterized by specific reactivity with its complementary antigen.

Antigen: A substance, usually a protein or carbohydrate, which when introduced in the body of a human or higher animal, stimulates the production of an antibody that will react specifically with it.

Apparent viscosity: Ratio of shear stress to shear rate. It depends on the rate of shear.

Aseptic: Free of infection.

Assay: A technique that measures a specific biological response.

Bacillus subtilis (B. Subtilis): An aerobic bacterium used as a host in rDNA experiments.

Bacteria: Any of a large group of microscopic organisms having round, rod-like, spiral or filamentous, unicellular or non-cellular bodies that are often aggregated into colonies. Bacteria may exist as free living organisms in soil, water, or organic matter, or as parasites on the bodies of plants and animals.

Batch processing: A processing technique in which a bioreactor is supplied with substrate and essential nutrients, sterilized and inoculated with microorganisms, and the process is run to completion followed by removal of products.

Biocatalyst: An enzyme that plays a fundamental role in living organisms or in industry by activating or accelerating a bioprocess.

Biochemical oxygen demand (BOD): The molecular oxygen used in meeting the metabolic needs of aerobic organisms in aqueous medium containing an oxidizable organic molecule.

Biochemical pathway: A sequence of enzymatically catalyzed reactions occurring in a cell.

Bioconversion: Chemical conversion of a naturally occurring biodegradable substance using a biocatalyst.

Biodegradation: The breakdown of substances by microorganisms, mainly by aerobic bacteria.

Biomass: All organic matters including those belonging to the aquatic environment that grow by the photosynthetic conversion of low energy carbon compounds employing solar energy.

Biooxidation: Oxidation (loss of electrons) process accelerated by a biocatalyst.

Biopolymers: Naturally occurring macromolecules that include proteins, nucleic acids, and polysaccharides.

Bioprocess: The process that uses complete living cells or their components (e.g., enzymes, organelles, and chloroplasts) to effect desired chemical and/or physical changes.

Bioreactor: The vessel of defined configuration in which a bioprocess is carried out.

Biosensor: An electronic device that transforms signals produced by biological molecules to detect concentration or transport of specific compounds.

Biosynthesis: Production, by synthesis or degradation, of a chemical compound by living organism, plant or animal cells, or enzymes elaborated by them.

Biotechnology: Commercial processes that use living organisms, or substances from those organisms, to make or modify a product. This includes techniques used for improving the characteristics of economically important plants and animals and for the development of microorganisms to act on the environment.

Broth: Complex fluid mixture in bioreactor, including cells, nutrients, substrate, antifoam, cell products, etc.

Catalase: An enzyme that catalyses the breakdown of hydrogen peroxide to water and oxygen.

Catalysis: A process by which the rate of a chemical reaction is increased by a substance (namely enzymes in biochemical reactions) that remains chemically unchanged at the end of the reaction.

Cell: The smallest structural unit of living matter capable of functioning independently; it is a microscopic mass of protoplasm surrounded by a semipermeable membrane, including one or more nuclei and various non-living substances that are capable, either alone or with other cells, of performing all the fundamental functions of life.

Cell culture: The in vitro growth of cells isolated from multicellular organisms. These cells are usually of one type.

Cellulase: The enzyme that cuts the linear chain of cellulose, a glucose polymer at 1-4-β-linkages into cellodextrins and glucose.

Cellulose: A polymer of six-carbon glucose sugars found in all plant matter; the most abundant biological substance on earth.

Chemical oxygen demand (COD): A measure of the amount of oxygen, expressed in milligrams per liter, required to oxidize organic matter present in a substance using a chemical oxidation method.

Chemostat: A bioreactor in which steady-state growth of microorganisms is maintained over prolonged periods of time under sterile conditions by providing the cells with constant input of nutrients and continuously removing effluent with cells as output.

Clone: A group of genetically identical cells or organisms produced asexually from a common ancestor.

Colony: A visible accumulation of cells that usually arises from the growth of a single microorganism.

Continuous culture: A method of cultivation in which nutrients are supplied and products are removed continuously at volumetrically equal rates maintaining the cells in a condition of stable multiplication and growth.

Culture: A population of microorganisms.

Denitrification: The reduction of nitrates to nitrites and finally to nitrous oxide or even to molecular nitrogen catalyzed by facultative aerobic soil bacteria working under anaerobic conditions.

Deoxyribonucleic acid (DNA): A linear polymer, made up of deoxyribonucleotide units, that is the carrier of genetic information, present in chromosomes and chromosomal material of cell orgenelles such as mitrochondria and chloroplasts, and also in some viruses. Every characteristic inherited trait has its origin in the code of each individual's DNA.

Dextran: Any of several polysaccharides that yield only glucose on hydrolysis.

Dilution rate: A measure of the rate at which the existing medium in continuous cultivation of organisms is replaced with fresh medium.

Diffusion: The net movement of molecules or ions from an area of higher concentration to an area of lower concentration.

Enzyme: A protein that functions as a biocatalyst in a chemical reaction. Any group of catalytic proteins that are produced by living cells mediating and promoting the chemical processes of life without themselves being altered or destroyed. This includes proteases, glucoamylase, alpha amylase, glucose isomerase, and rennin.

Escherichia coli (E. coli): A species of bacteria that inhabits the intestinal tract of most vertebrates. Many non-pathogenic strains are used experimentally as hosts for rDNA.

Exponential growth: The phase of cell growth in which the number of cells or the cell mass increases exponentially.

Facultative: The ability to live under more than one set of conditions, usually in relation to oxygen tension prevailing in the medium, that is aerobic or anaerobic (e.g., facultative anerobes).

Fatty acids: Organic acids with long carbon chains. Fatty acids are abundant in cell membranes and are widely used as emulsifiers, as metallic soaps, and for other industrial uses.

Fed-batch culture: A cell cultivation technique in which one or more nutrients are supplied to the bioreactor in a given sequence during the growth or bioconversion process while the products remain in the vessel until the end of the run.

Fermentation: An anaerobic bioprocess. An enzymatic transformation of organic substrates, especially carbohydrates, generally accompanied by the evolution of gas as a byproduct. Fermentation is used in various industrial processes for the manufacture of products (e.g., alcohols, organic acids, solvents, and cheese) by the addition of yeasts, moulds, and bacteria.

Fermenter: An industrial microbiological reactor in which the addition of nutrients, removal of products, and insertion of measuring sensors and control devices are maintained while accessories like heating, aeration, agitation, and sterilization systems are provided.

Flocculation: Agglomeration of colloidal material by adding a chemical that causes the colloidal particles to produce larger particles.

Fungi: Simple vegetative bodies from which reproductive structures are elaborated. The fungi contain no chlorophyll and therefore require sources of complex organic molecules. Many species grow on dead organic materials and others live as parasites.

Gene: The basic unit of inheritance; a sequence of DNA coding that can be translated by the cell machinery into a sequence of amino acids linked to form a protein.

Glucose: A 6-carbon sugar molecule, which is the building block of natural substances like cellulose, starch, dextrans, xanthan, and some other biopolymers and used as a basic energy source by the cells of most organisms.

Glycolysis: Metabolic pathway involving the conversion of glucose to lactic acid or ethanol.

Growth curve: Graphical representation of the growth of an organism in nutrient medium in a batch reactor under predetermined environmental conditions.

Immobilized cell: Technique used for the fixation of enzymes or cells onto a solid support.

Incubation: The introduction of microorganisms into a culture medium with a view to growing a large number of identical cells all by propagation.

Inhibition: The decrease of the rate of an enzyme-catalyzed reaction by a chemical compound including substrate analogues. Such inhibition may be competitive with the substrate (binding at the active site of the enzyme) or non-competitive (binding at an allosteric site).

Inhibitor: An agent that prevents the normal action of an enzyme without destroying it (e.g., a substance (inhibitor) B causes the slow-down of the enzyme-substrate reaction of A $\rightarrow$ R).

Inoculum: A batch fermentation that starts with an initial charge of cells.

Lineweaver-Burk plot: Method of analyzing kinetic data (growth rates of enzyme catalyzed reactions) in linear form using a double reciprocal plot of rate versus substrate concentration.

Lipid: A large, varied class of water insoluble organic molecules, including steriods, fatty acids, prostaglandins, terpenes, and waxes.

Log phase: Period of bacterial growth with logarithmic increase in cell number.

Mass transfer: Irreversible and spontaneous transport of mass of a chemical component in a space with a non-homogeneous field of the chemical potential of the component. The driving force causing the transport can be the difference in concentration (in liquids) or partial pressures (in gases) of the component. In biological systems, mass transfer may result from diffusion facilitated transport or active transport.

Maximum oxygen transfer rate: Volumetric mass transfer coefficient times the oxygen solubility at constant temperature and pressure. The value, which is often obtained by the sulfite method, wrongly though, has been used in the past for the evaluation of oxygen transfer rate of a chemical reaction as a rough approximation. For bioreactor systems, gassing out dynamic methods and oxygen sensors are employed.

Medium: Mixture of nutrient substances required by cells for growth and metabolism.

Metabolism: The total of all chemical reaction activities in a living organism producing energy and growth.

Michaelis-Menten kinetics: Kinetics of conversion of substrates in enzyme-catalyzed reactions.

Microorganisms: Microscopic living entities, viruses, bacteria, and yeasts.

Mixed culture: Two or more species of microorganisms living in the same medium.

Moulds: Multi-cellular filaments that increase by vegetative growth of the filament.

Monod kinetics: Kinetics of microbial cell growth as a function of substrate concentration proposed by Jacques Monod and widely used to understand growth-substrate relationships.

Mycelium: A mass of filaments composing the vegetative body of many fungi and some bacteria.

Nutrient: A substance used as food.

Oxygen transfer rate (OTR): The product of volumetric oxygen transfer rate $k_L a$ and the oxygen concentration driving force $(C^* - C_L)$, $(ML^{-3}T^{-1})$, where k_L is the mass transfer coefficient based on liquid phase resistance to mass transfer (LT^{-1}), a is the air bubble surface area per unit volume (L^{-1}), and C^* and C_L are oxygen solubility and dissolved oxygen concentration, respectively. All the terms of OTR refer to the time average values of a dynamic situation.

Oxygen uptake rate (OUR): The actual rate at which oxygen utilization takes place and is given by $(\mu/Y_{O2}) C_x$, where μ is specific growth rate of the concerned cells, h^{-1}, Y_{O2} is the oxygen uptake per unit cell gg^{-1}, and C_x is the concentration of cells, gml^{-1}.

Organism: A living biological specimen.

Pasteurization: The process of mild heating to kill particular spoilage organisms or pathogens.

Pathogen: A disease producing agent. The term is usually restricted to a living organism such as a bacterium or virus.

Peptide: A linear polymer of amino acids. A polymer of numerous amino acids is called a polypeptide.

Peroxidase: An enzyme that breaks down the hydrogen peroxide, H_2O_2 + NADH + H^+ → $2H_2O$ + NAD^+.

pH: A measure of the acidity or basicity of a solution on a scale of 0 (acidic) to 14 (basic) [e.g., lemon juice has a pH of 2.2 (acidic), water has a pH of 7.0 (neutral), and a solution of baking soda has a pH of 8.5 (basic)].

Protease: Protein digesting enzyme.

Protein: One or more polypeptides assembled in biologically active units; proteins function as catalysts in metabolism and as structural elements of cells in tissues.

Pure culture: A culture containing only one species of microorganism.

Recombinant: DNA (r-DNA): The hybrid DNA produced by joining pieces of DNA from different organisms together in vitro (i.e., in an artificial apparatus).

Ribonucleic acid (RNA): A nucleic acid that contains the sugar ribose.

Single cell protein: Cells, or protein extracts, of microorganisms grown in large quantities for use as human or animal protein supplements.

Sludge: Precipitated mechanically or biologically separated solid matter produced during water and or sewage treatment or industrial processes. Such solids may be amenable to biological control.

Species: One kind of microorganism.

Specific enzyme activity: Amount of substrate rendered into product per unit dry weight of enzyme protein per unit time.

Specific growth rate coefficient, μ: The rate of change of microbial cell mass concentration per unit cell concentration: $\mu = (1/C_x)(dC_x/dt)$ (where C_x = cell mass concentration, g/l).

Spore: A uni- or multicellular, asexual, reproductive, or resting body that is resistant to unfavorable environmental conditions, and which produces a new vegetative individual when the environment is favorable.

Sporulation: The germination of a spore by a bacterium or by a yeast.

Starch: A polymeric substance of glucose molecules and a component of many terrestrial and aquatic plants used by some organisms as a means of energy storage: starch is broken down by enzymes (amylases) to yield glucose, which can be used as a feedstock for chemical or energy production.

Steriods: A group of organic compounds, some of which act as hormones to stimulate cell growth in higher animals and humans.

Substrate: A chemical substance acted upon by an enzyme or a cell to give rise to a product.

Taxonomy: Classification of organisms.

Thixotropic fluid: A fluid when subjected to a constant shear stress exhibits an apparent viscosity that increases with time.

Total organic carbon (TOC): The total amount of organic carbon in a mixture.

Total oxygen demand (TOD): The total amount of molecular oxygen consumed in the combustion of oxygen-demanding substances at about 900°C.

Toxicity: The ability of a substance to produce a harmful effect on an organism by physical contact, ingestion, or inhalation.

Toxin: A substance produced in some cases by disease-causing micro-organisms, which is toxic to other living cells.

Vaccine: A suspension of attenuated or killed bacteria or virus, or portions thereof, injected to produce active immunity.

Vegetative cell: Cells involved with obtaining nutrients, as opposed to reproduction or resting.

Volumetric mass transfer coefficient, $k_L a$: The proportionality coefficient reflecting both molecular diffusion, turbulent mass transfer, and specific area for mass transfer.

Virus: Any of a large group of submicroscopic agents infecting plants, animals, and bacteria, and unable to reproduce outside the tissue of the host. A fully formed virus consists of nucleic acid (DNA or RNA) surrounded by a protein, or protein and lipid coat.

Vitamin: An organic compound present in variable, minute quantities in natural foodstuffs, and essential for the normal processes of growth and maintenance of the body. Vitamins do not furnish energy, but are necessary for energy transformation and for regulation of metabolism.

Yeast: A fungus of the family Saccharomycetaceae that is used especially in the making of alcoholic liquors and as leavening in baking, and also in various bioprocesses.

REFERENCES

1. Walas, S. M., *Chemical Reaction Engineering Handbook of Solved Problems*, Gordon & Breach Publishers, 1995.
2. Bailey, J. E. and Ollis, D. F., *Biochemical Engineering Fundamentals*, 2nd ed., Mc.Graw-Hill Book Co., 1986.
3. Bungay, H. R., *Computer Games and Simulation for Biochemical Engineering*, John Wiley & Sons, 1985.
4. Sinclair, C. G. and Kristiansen, B., *Fermentation Kinetics and Modelling*, J. D. Bu'Lock (Ed.) Open University Press 1987.
5. Volesky, B. and Votruba, J., *Modeling and Optimization of Fermentation Processes*, Elsevier, 1992.
6. Ingham, J., et al., *Chemical Engineering Dynamics—An Introduction to Modelling and Computer Simulation*, 2nd ed., Wiley-VCH Verlag GmbH, 2000.
7. Michaelis, L. and Menten, M. L., *Biochem. Z.*, 49, 333, 1913.
8. Briggs, G. E. and Haldane, J. B. S., *Brochem, J.*, 19, 338, 1925.
9. Putman, S. J., et al., *Biochem J.*, 129, 301, 1972.
10. Missen, R. W., et al., *Introduction to Chemical Reaction Engineering and Kinetics*, John Wiley & Sons, 1999.
11. Schügerl, K. J., "New bioreactors for aerobic processes," *Int. Chem. Eng.*, 22, 591, 1982.
12. Aiba, S., Humphrey, A. E., and Mills, N. F., *Biochemical Engineering*, 2nd ed., University of Tokyo Press, Tokyo, 1973.
13. Bartow, M. V., "Supersizing the aerobic fermenter," *Chemical Engineering*, pp. 70–75, 1999.

14. Rai, V. R. and Constantinide, A., "Mathematical modeling and optimization of the gluconic acid fermentation," *AIChE Symp. Ser.,* Vol. 69, No. 132, p. 114, 1973.
15. Marquadt, D. W., "An algorithm for least-squares estimation of non-linear parameters," *J. Soc. Indust. Appl. Math.,* 11, 431, 1963.
16. Roberts, R. S., et al., "The effect of agitation on oxygen mass transfer in a fermenter," *Chem. Eng., Ed.,* 26, 142, 1992.
17. Perry, R. H. and Green, D. W., *Perry's Chemical Engineers' Handbook,* 7th ed., McGraw-Hill Book Co., 1999.
18. Leudeking, R. and Piret, E. L., "A kinetic study of the lactic acid fermentation," *J. Biochem. Microbiol, Technol. Eng.,* 1: 393, 1959.
19. Keller, R. and Dunn, I. J., "Computer simulation of the biomass production rate of cyclic fed batch continuous culture," *J. Appl. Chem., Biotechnol.,* 28, 508–514, 1978.
20. Fogler, H. S., *Elements of Chemical Reaction Engineering,* 3rd ed., Prentice Hall Int. Series, 1999.
21. Perez, J. F. and Sandall, O. C., *AIChE J.,* 20, 770, 1974.
22. Biotechnology by Open Learning, *Bioreactor Design and Product Yield,* Butterworth-Heinemann, ISBN 07506-15095, 1992.
23. Gray, P. P. and Bhuwapathanapun, S., *Biotechnol. Bioeng.,* 22: 1785, 1980.
24. Hirt, W., *Biotechnol. Bioeng.,* 23, 235, 1981.

CHAPTER TWELVE

Safety in Chemical Reaction Engineering

INTRODUCTION

In the chemical process industries (CPI), chemical manufacture, especially in the fine, pharmaceutical, and speciality chemical industries, involves the processing of reactive chemicals, toxic or flammable liquids, vapors, gases, and powders. Although the safety records of these industries have improved in recent years, fires, explosions, and incidents involving hazardous chemical reactions still occur. A basis for good engineering practice in assessing chemical reaction hazards is essential, with the goal to help designers, engineers, and scientists responsible for testing and operating chemical plants meet the statutory duties of safety imposed by governmental organizations [e.g., Environmental Protection Agency (EPA), Occupational Safety and Health Administration (OSHA), USA, Health & Safety Executive, UK].

The control of chemical reactions (e.g., esterification, sulfonation, nitration, alkylation, polymerization, oxidation, reduction, halogenation) and associated hazards are an essential aspect of chemical manufacture in the CPI. The industries manufacture nearly all their products, such as inorganic, organic, agricultural, polymers, and pharmaceuticals, through the control of reactive chemicals. The reactions that occur are generally without incident. Barton and Nolan [1] examined exothermic runaway incidents and found that the principal causes were:

- Inadequate temperature control
- Inadequate agitation
- Inadequate maintenance
- Raw material quality
- Little or no study of the reaction chemistry and thermochemistry
- Human factors

Other factors that are responsible for exothermic incidents are:

- Poor understanding of the reaction chemistry resulting in a poorly designed plant.
- Underrated control and safety backup systems.
- Inadequate procedures and training.

HAZARD EVALUATION IN THE CHEMICAL PROCESS INDUSTRIES

The safe design and operation of chemical processing equipment requires detailed attention to the hazards inherent in certain chemicals and processes. Chemical plant hazards can occur from many sources. Principal hazards arise from:

- Fire and explosion hazards.
- Thermal instability of reactants, reactant mixtures, isolated intermediates, and products.
- Rapid gas evolution that can pressurize and possibly rupture the plant.
- Rapid exothermic reaction, which can raise the temperature or cause violent boiling of the reactants and also lead to pressurization.

Earlier reviews have been concerned with energy relationships for a particular chemical process, which are based on two general classifications of chemical processes, conventional and hazardous [2]. The former is used to describe a non-explosive, non-flammable reaction and the latter to describe an explosive and flammable reaction. The division of reactions does not account for the varying degree of safety or hazards of a particular reaction, which may lie between these extremes, and consequently becomes too limiting in the design. Safety is most likely to be neglected with conventional reactions while over design may be expected with hazardous reactions. These limitations can be avoided by introducing a third class of reactions, called "special," that covers the intermediate area between conventional and hazardous where reactions are relatively safe. Shabica [2] has proposed some guidelines for this third classification and Figure 12-1 shows the increasing degree of hazard of a particular process. The Appendix at the end of this chapter lists hazard ratings of chemical reactions and

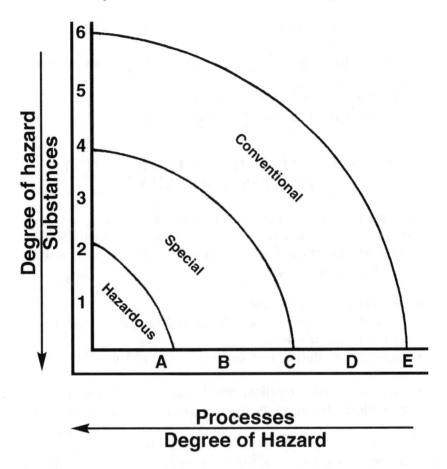

Processes
Degree of Hazard

Figure 12-1. The increasing degree of hazard of matter plotted against the increasing hazard of a process. *(Source: Shabica [2].)*

substances for the three classifications. Figures 12-2, 12-3, and 12-4 show the flowsheets and floor plans of the hazardous, special, and conventional equipment.

Chemical reaction hazards must be considered in assessing whether a process can be operated safely on the manufacturing scale. Furthermore, the effect of scale-up is particularly important. A reaction, which is innocuous on the laboratory or pilot plant scale, can be disastrous in a full-scale manufacturing plant. For example, the heat release from a highly exothermic process, such as the reduction of an aromatic nitro compound, can be easily controlled in laboratory glassware. However,

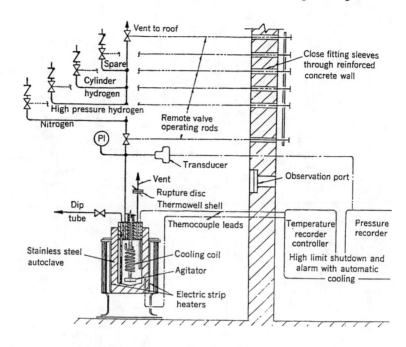

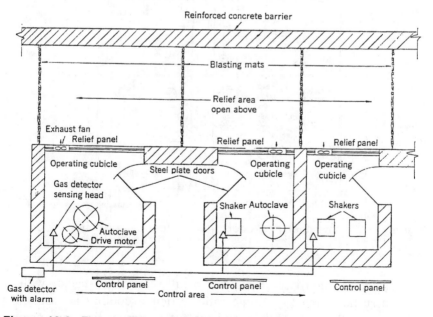

Figure 12-2. Flowsheet and floor plan of hazardous equipment. *(Source: Shabica [2].)*

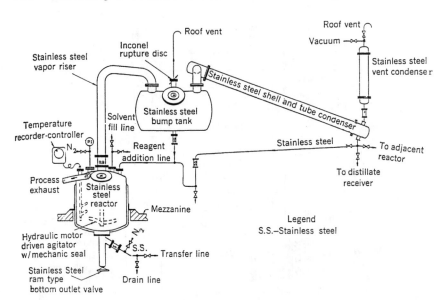

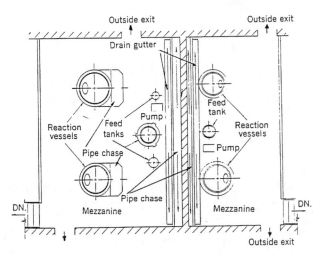

Figure 12-3. Flowsheet and floor plan of special equipment. *(Source: Shabica [2].)*

if the same reaction is performed in a full-scale plant reactor with a smaller surface area to volume ratio, efficient cooling is essential. Otherwise, a thermal runaway and violent decomposition may occur. Similarly, a large quantity of gas produced by the sudden decomposition of a diazonium compound can be easily vented on the laboratory

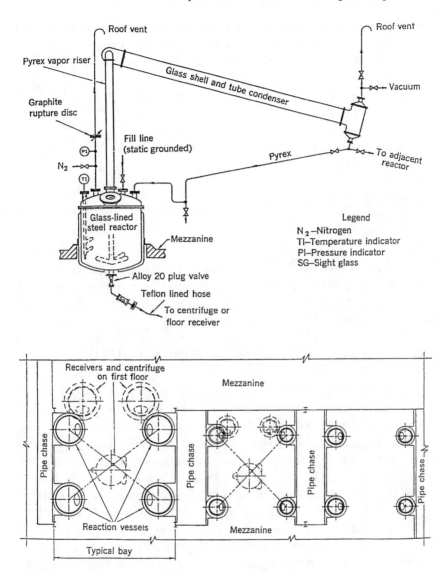

Figure 12-4. Flowsheet and floor plan of conventional equipment. *(Source: Shabica [2].)*

scale, but the same decomposition on a large scale could pressurize and rupture a full-scale plant.

Additionally, consequences of possible process maloperations, such as incorrect charging sequence, contamination of reactants, agitation failure and poor temperature control, adding reactants too quickly, omitting one of the reactants, and incorrect reactant concentration (recycling) must be considered. A number of parameters govern the reaction hazards associated with a process. These include the following:

- Chemical components
- Process temperature
- Process pressure
- Thermochemical characteristics of the reaction
- Reaction rate
- Reaction ratios
- Solvent effects
- Batch size
- Operational procedure

The assessment of the hazards of a particular process requires investigating the effects of these parameters through experimental work, interpretating the results in relation to the plant situation, and defining a suitable basis for safe operation.

HAZARD ASSESSMENT PROCEDURES

Hazard assessments are essential and should be performed on all chemical processes. The reactors such as batch, semi-batch, and continuous as reviewed in Chapter 4, can be used for performing various operations. Many industrial reactions are exothermic (i.e., accompanied by the evolution of heat) and, therefore, overheating can occur. In batch operations, all the reactants are added to the reactor at the start of the reaction. In semi-batch operations, one or more of the reactants are charged to the reactor in the beginning, and others are then metered in, giving some control over the reaction rate and, thus, the rate of heat production. Overheating often results in thermal runaway, which is characterized by progressive increases in the rate of heat generation, and, hence, temperature and pressure.

Thermal runaway is a particular problem in unsteady state batch reactions, where the rate of reaction and, therefore, the rate of heat production varies with time. The consequences of thermal runaway are sometimes severe as in the incidents at Seveso [3]. In this case, a bursting disk ruptured on a reactor. The reactor was used to manufacture trichlorophenol at a temperature of 170–185°C and was heated by steam at 190°C. The batch was dehydrated and left at a temperature of 158°C over the weekend. Ethylene glycol and caustic soda give exothermic secondary reactions producing sodium polyglycolates, sodium oxalate, sodium carbonate, and hydrogen, thus exhibiting an autocatalytic behavior. These reactions caused a temperature rise allowing the production of tetrachlorodioxin and the subsequent vessel pressurization. The Bhopal fatal incident [4] (not a thermal runaway) involved a toxic release of an intermediate, methyl isocyanate (MIC), which resulted in over 2,000 deaths. A runaway polymerization of highly toxic methyl isocyanate occurred in a storage tank. The runaway polymerization caused an uncontrolled pressure rise, which caused a relief valve to lift and discharge a plume of toxic vapor over the city.

The task of specifying the design, operation, and control of a reactor with stirrer, heating, or cooling coils, reflux facilities, and emergency relief venting can pose a problem if all the time-dependent parameters are not considered. The use of batch processing techniques in the fine chemical industry is often characterized by:

- Multi-product plant (must have adaptable safety system)
- Complex developing chemistry
- High frequency of change
- Process control is simpler than continuous processes

These factors are attributed to batch and semi-batch processes rather than continuous processes. However, the use of continuous processes on fine chemical manufacturing sites is limited. It is often preferable to use the semi-batch mode as opposed to batch processes. The Appendix lists hazards of pertinent chemical reactions for toxic and reactive hazards chemicals. Information concerning the safety of various chemicals (e.g., ammonia and others) can be readily obtained from the World Wide Web. Table 12-1 shows how to access a material safety data sheet at the Vermont Safety Information (VIRI) site on the Internet.

EXOTHERMS

Temperature-induced runaways have many causes including loss of cooling, loss of agitation, and excessive heating. During a temperature-

Table 12-1
Accessing a material safety data sheet (MSDS)

1. Type in *http://www.siri.org/.*
2. When the first screen appears, click on SIRI MSDS collection-Vermont site.
3. When the first page appears, type in the chemical you want in the find box.

   Example: Find ambox{ammonia}

   Then click on box{search}

4. The next page shows a list of companies that provide data on ammonia.

   Example: Mallinckrodt Baker—Ammonia Solution Strong
   Mallinckrodt Baker—Ammonia Aqueous

5. Scroll through "Ammonia Solution, Strong" for the required information:

   Product Identification
   Composition/Information on Ingredients
   Hazards Identification
   First Aid Measures
   Fire Fighting Measures
   Accidental Release Measures
   Handling and Storage
   Exposure Controls/Personal Protection
   Physical and Chemical Properties
   Stability and Reactivity
   Toxicological Information
   Ecological Information
   Disposal Considerations
   Transport Information
   Regulatory Information
   Other Information

induced upset, the reactor temperature rises above the normal operating target. When any of the temperature elements senses a high reactor temperature, the programmable logic controller (PLC) software shut-down system automatically puts the reactor on idle (isolates). This action doubles the cooling water flow to the reactor by opening a bypass valve. If these reactions are unsuccessful in terminating the temperature rise, the shut-down system opens the quench valves, dumping the reaction mass into the water-filled quench tank. If the dump valves fail to operate, the reactor contents soon reach a tempera-ture where a violent self-accelerating decomposition occurs. The uncontrolled exotherm generates a large volume of gas and ejects the process material out of the reactor into the containment pot.

ACCUMULATION

It is important to know how much heat of reaction can accumulate when assessing the hazards related to an exothermic reaction. Accumulation in a batch or semi-batch process can be the result of:

- Adding a reactant too quickly
- Loss of agitation
- Performing the reaction at too low a temperature
- Inhibition of the reaction
- Delayed initiation of the desired reaction

Reactants can accumulate when the chosen reaction temperature is too low, and as such the reaction continues even after the end of the addition. In such a case, a hazardous situation could occur if cooling were lost as exemplified by Barton and Rogers [5].

Impurities or the delayed addition of a catalyst causes inhibition or delayed initiation resulting in accumulation in the reactors. The major hazard from accumulation of the reactants is due to a potentially rapid reaction and consequent high heat output that occurs when the reaction finally starts. If the heat output is greater than the cooling capacity of the plant, the reaction will run away. The reaction might commence if an agitator is restarted after it has stopped, a catalyst is added suddenly, or because the desired reaction is slow to start.

THERMAL RUNAWAY CHEMICAL
REACTION HAZARDS

Thermal runaway reactions are the results of chemical reactions in batch or semi-batch reactors. A thermal runaway commences when the heat generated by a chemical reaction exceeds the heat that can be removed to the surroundings as shown in Figure 12-5. The surplus heat increases the temperature of the reaction mass, which causes the reaction rate to increase, and subsequently accelerates the rate of heat production. Thermal runaway occurs as follows: as the temperature rises, the rate of heat loss to the surroundings increases approximately linearly with temperature. However, the rate of reaction, and thus the

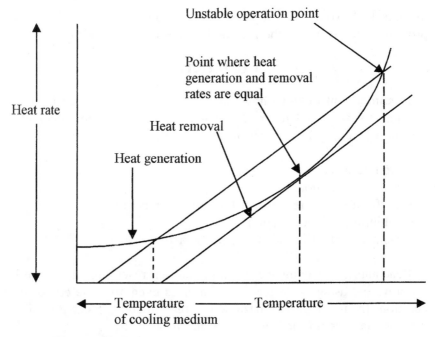

Figure 12-5. A typical curve of heat rate versus temperature.

rate of heat generation, increases exponentially. If the energy release is large enough, high vapor pressures may result or gaseous products may be formed, which can lead to over-pressurization and possible mechanical destruction of the reactor or vessel (thermal explosion).

The released energy might result from the wanted reaction or from the reaction mass if the materials involved are thermodynamically unstable. The accumulation of the starting materials or intermediate products is an initial stage of a runaway reaction. Figure 12-6 illustrates the common causes of reactant accumulation. The energy release with the reactant accumulation can cause the batch temperature to rise to a critical level thereby triggering the secondary (unwanted) reactions. Thermal runaway starts slowly and then accelerates until finally it may lead to an explosion.

THE ϕ-FACTOR

The fraction of heat required to heat up the vessel, rather than its contents, depends on the heat capacity of the vessel (i.e., how much

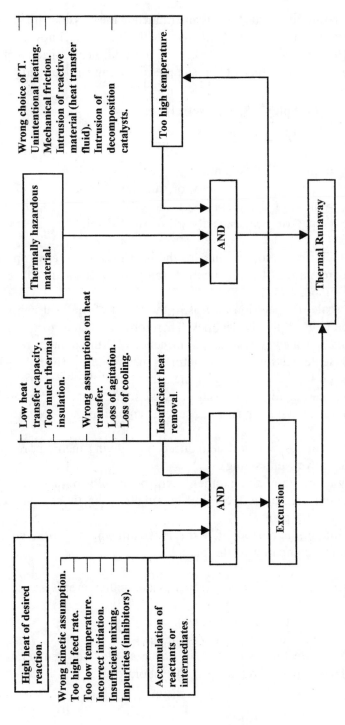

Figure 12-6. Causes of runaways in industrial reactors. *(Source: W. Regenass, "Safe Operation of Exothermic Reactions," Internal Ciba-Geigy Publication, 1984.)*

energy is required to raise the temperature of the vessel with respect to the reaction mass). The heat loss to the vessel is known as the ϕ-factor [6]. The ϕ-factor is the ratio of the total heat capacity of the sample and vessel to that of the sample alone, and is defined by

$$\phi = \frac{M_s C_{ps}(\text{sample}) + M_v C_{pv}(\text{vessel})}{M_s C_{ps}(\text{sample})} \qquad (12\text{-}1)$$

where M_s = mass of the sample (kg)
C_{ps} = specific heat capacity of sample (J/kg•K)
M_v = mass of the vessel (kg)
C_{pv} = specific heat capacity of vessel (J/kg•K)

The ϕ-factor does not account for the heat loss to the environment. It is used to adjust the self-heating rates as well as the observed adiabatic temperature rise.

As the scale of operation increases, the effect of the heat consumption by the plant typically declines. Therefore, the extent to which the kinetics of the runaway reaction is influenced by the plant is reduced. For plant scale vessels, the ϕ-factor is usually low (i.e., 1.0–1.2) depending on the heat capacity of the sample and the vessel fill ratio. Laboratory testing for vent sizing must simulate these low ϕ-factors. If the laboratory ϕ-factor is high, several anomalies will occur:

• The rate of reaction will be reduced (i.e., giving incorrect reaction rate data for vent sizing).
• The magnitude of the exotherm will be smaller by a factor of ϕ.
• Measured pressure effects will be smaller than those that would occur on the plant.
• The induction period of a thermal runaway reaction will be increased compared to the plant.

The consequences of these erroneous anomalies are:

• Inadequate safety design system.
• Undersized emergency relief system.
• Unknown decompositions may occur at elevated temperatures, which may not be realized in the laboratory.

ONSET TEMPERATURE

The reaction onset temperature is the temperature where it is assumed significant fuel consumption begins. The onset temperature is expressed by [7]

$$T_{onset} = \frac{B}{\ln\left[\dfrac{x_o \Delta H_{Rx} \cdot A}{C \cdot \phi \cdot T_{ex}^{\bullet}}\right]} \tag{12-2}$$

where $B = \dfrac{E_A}{R}$ $\qquad\qquad\qquad\qquad$ (12-3)

$\quad$ A = rate constant (pre-exponential factor from Arrhenius equation
$\qquad$ k = A exp $(-E_A/RT)$, sec^{-1} (i.e., for a first order reaction)
$\quad$ B = reduced activation energy, K
$\quad$ C = liquid heat capacity of the product (J/kg•K)
$\quad E_A$ = activation energy, J/mol
$\quad$ R = gas constant (8.314 J/mol•K)
T_{onset} = onset temperature, K
$\quad x_o$ = initial mass fraction
ΔH_{Rx} = heat of decomposition, J/kg
$\quad \phi$ = dimensionless thermal inertial factor for the sample holder
$\qquad$ or product container (ϕ-factor)
$\quad T_{ex}^{\bullet}$ = bulk heat-up rate driven by an external heat source, °C/sec

TIME-TO-MAXIMUM RATE

An onset temperature can be selected based on an arbitrary time-to-maximum-rate from the relation

$$t_{mr} = \frac{C \cdot \phi \cdot T^2 \exp(B/T_{onset})}{x_o \cdot \Delta H_{Rx} \cdot A \cdot B} \tag{12-4}$$

Rearranging Equation 12-4 yields

$$T_{onset} = \frac{B}{\ln\left[\dfrac{t_{mr} \cdot x_o \cdot \Delta H_{Rx} \cdot A \cdot B}{C \cdot \phi \cdot T^2}\right]} \tag{12-5}$$

Equation 12-5 gives an onset temperature T_{onset} that corresponds to a time-to-maximum rate t_{mr} (min) using a successive substitution solution procedure. An initial guess of $T = 350$ K for the right side of Equation 12-5 will give a solution value of T_{onset} on the left side of Equation 12-5 within 1% or on an absolute basis $\pm 3°C$. Convergence is reached within several successive substitution iterations.

MAXIMUM REACTION TEMPERATURE

Once the onset temperature is determined, it is then possible to obtain the maximum reaction temperature by the adiabatic temperature rise and any contribution due to external heat input. The theoretical adiabatic temperature increase is

$$\Delta T_{adia} = \frac{x_o \cdot \Delta H_{Rx}}{C \cdot \phi} \tag{12-6}$$

The contribution to the overall temperature increase from external heat input is defined by

$$\Delta T_{ex} = \frac{T_{ex}^\bullet \cdot t_{mr}}{2} \tag{12-7}$$

where ΔT_{ex} = temperature increase attributed to external heating effects, °C

$T_{ex}^\bullet$ = bulk heat-up rate due to external heating alone, °C/min

t_{mr} = time-to-maximum rate as determined by Equation 12-5 with tempertaure T set equal to T_{onset}, min

From these expressions, the maximum reaction temperature T_{max} is defined by

$$T_{max} = T_{onset} + \Delta T_{adia} + \Delta T_{ex} \tag{12-8}$$

METHODS FOR SCREENING THERMAL
RUNAWAY CONDITIONS

Various techniques have been employed for testing and screening hazardous compounds produced in the CPI. The results obtained from these instruments assist the designers and technologists in scale-up of the plants. The suitability of the various instruments and test methods are illustrated in Figure 12-7. From this figure, it can be inferred that not all the techniques can provide all the answers in different temperature versus time regimes.

The evaluation of chemical reaction hazards involves establishing exothermic activity and/or gas evolution that could give rise to incidents. However, such evaluation cannot be carried out in isolation or by some simple sequence of testing. The techniques employed and the results obtained need to simulate large-scale plant behavior. Adiabatic calorimeters can be used to measure the temperature time curve of self-heating and the induction time of thermal explosions. The pertinent experimental parameters, which allow the data to be determined under specified conditions, can be used to simulate plant situations.

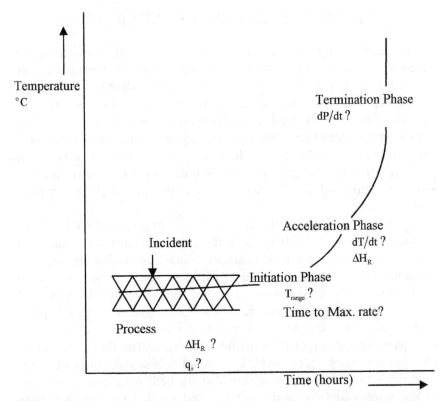

Figure 12-7. Simplified scenario of a thermal runaway. *(Source: T. Hoppe and B. Grob, "Heat flow calorimetry as a testing method for preventing runaway reactions," Int. SYmp. on Runaway Reactions, CCPS, AIChE, March 7–9, 1989.)*

Adiabatic reaction calorimetry has proven to be a successful tool for assessing thermal hazards of reactive chemicals, generating thermodynamic data, and optimizing process design for maximizing safety. It is not only used for screening [8], but also for obtaining kinetic and thermodynamic data [9,10]. The results are used for constructing computer models of chemical processes that may undergo a thermal runaway reaction. Figures 12-8a, b, c, and d show plots of various parameters for an adiabatic system. The various proprietory adiabatic calorimeters and vent sizing instruments used in laboratory investigations or for scale-up of vessels in the CPI are reviewed below.

TEST INSTRUMENTS

ACCELERATING RATE CALORIMETER (ARC™)

The accelerating rate calorimeter (ARC™) is an instrument that measures the temperature, temperature rate, and pressure changes of a reaction sample as a function of time under adiabatic "runaway" conditions. The ARC provides data pertaining to thermochemistry and kinetics. The ARC is used to assess thermal stability (e.g., for safe storage and to detect undesirable secondary reactions), gather total heat release and heat release rate data (to prevent thermal runaways), measure pressure, and pressure rate data due to gas release (to prevent vessel rupture and for the design of emergency relief and mitigation systems).

In the ARC (Figure 12-9), the sample of approximately 5 g or 4 ml is placed in a one-inch diameter metal sphere (bomb) and situated in a heated oven under adiabatic conditions. These conditions are achieved by heating the chamber surrounding the bomb to the same temperature as the bomb. The thermocouple attached to the sample bomb is used to measure the sample temperature. A heat-wait-search mode of operation is used to detect an exotherm. If the temperature of the bomb increases due to an exotherm, the temperature of the surrounding chamber increases accordingly. The rate of temperature increase (self-heat rate) and bomb pressure are also tracked. Adiabatic conditions of the sample and the bomb are both maintained for self-heat rates up to 10°C/min. If the self-heat rate exceeds a predetermined value (~0.02°C/min), an exotherm is registered. Figure 12-10 shows the temperature versus time curve of a reaction sample in the ARC test,

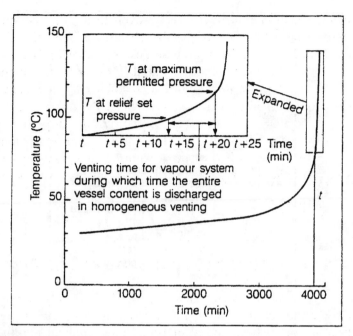

Figure 12-8a. Temperature against time for an adiabatic system.

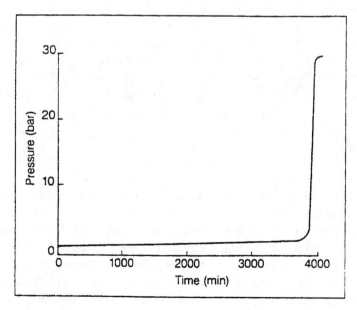

Figure 12-8b. Pressure against time for an adiabatic system.

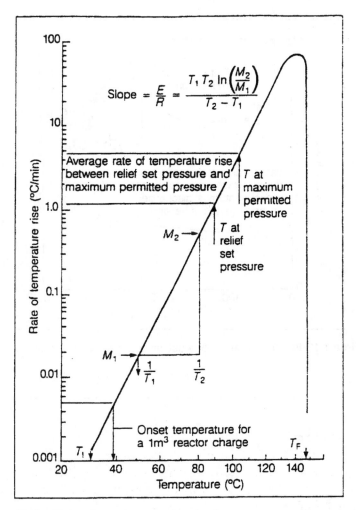

Figure 12-8c. Rate of temperature rise against temperature for an adiabatic system.

and Figure 12-11 illustrates a sketch of the self-heat rate as a function of temperature.

The exotherm starts at ~150°C and ends at a final temperature of ~272°C, resulting in an adiabatic temperature increase of ~122°C. This value is the uncorrected measured temperature increase for both sample and vessel (bomb). The correction involves multiplication by the φ-factor. For example, if the value of the φ-factor from an experiment is 1.68, then the corrected adiabatic temperature is 205°C.

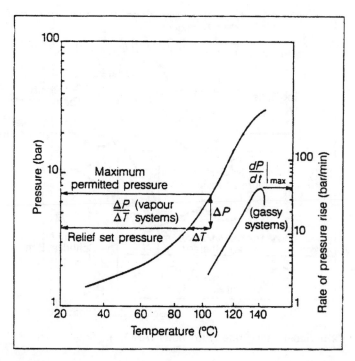

Figure 12-8d. Pressure and rate of pressure rise against temperautre for an adiabatic system.

The ARC analysis has been extended to determine the conditions that may lead to thermal runaway in reactors or storage vessels [10]. The equipment timeline is defined by the ratio of the heat capacity of the reactor and its contents to the reactor heat transfer area and heat transfer coefficient. This is expressed by

$$\theta = \frac{MC_p}{UA} \tag{12-9}$$

where MC_p = heat capacity of the reactor and its contents, J/K
U = overall heat transfer coefficient, W/m²K
A = heat transfer area, m²

The temperature corresponding to the equipment timeline on the time to maximum rate (TMR) plot is the temperature of no return. Above the temperature of no return, the rate of heat generation from

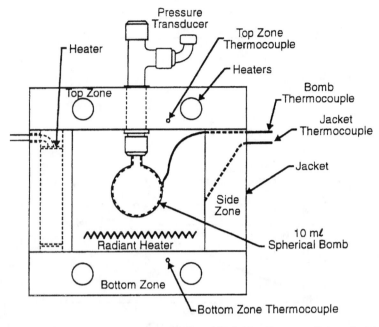

Figure 12-9. Accelerating rate calorimeter (ARC™). *(Source: Arthur D. Little.)*

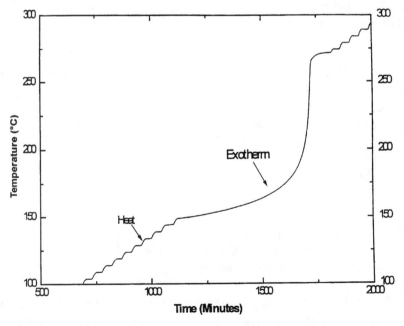

Figure 12-10. ARC temperature versus time plot of a reaction sample.

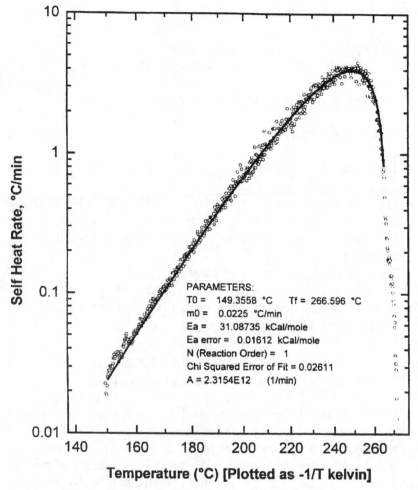

Figure 12-11. Self-heat rate analysis. ARC data are shown along with a fitted model obtained by assuming the following kinetic parameters: reaction order = 1, activation energy = 31.08 kcal/mol, and frequency factor = 2.31E12 min^{-1}.

the reaction will exceed the available cooling rate and the reaction will run away. Limitations of using the ARC for pressure-relief design include having to correct for the ϕ-factor [9,11,12]. The correction can fail in the case of complex reactions. Huff [12] has shown that ARC evaluations can be applied to multiple reactions. He further indicated that the corrected calorimeter data yield conservative approaches of developing detailed computer simulation models.

The advantages of the ARC are [13,14]:

- An excellent onset detection sensitivity.
- A good theoretical background for the experiment kinetic interpretation.
- A means of obtaining pressure information.

The disadvantages are:

- The experiment duration is much longer than for differential thermal analysis (DTA) tests.
- The interpretation of complex kinetics obtained under adiabatic conditions is far from being straightforward.
- The correction of the pressure record to adiabatic conditions is a difficult problem.
- There is no stirring in the original set-up.
- The need for skilled operators and maintenance problems.
- Troubleshooting new software on the personal computer against the old control computer.

AUTOMATED PRESSURE TRACKING ADIABATIC CALORIMETER (APTAC)

The APTAC, developed by Arthur D. Little, is a miniature chemical reactor that measures and analyzes the thermal and properties of exothermic chemical reactions. The results from the APTAC instrument help to identify potential hazards and combat key elements of process safety design including emergency relief systems, process optimization, effluent handling, and thermal stability. The APTAC uses a 130 ml spherical test cell with a 0.51mm-wall thickness situated inside a 4-l pressure vessel as shown in Figure 12-12. The APTAC uses a high-power heater assembly that serves to heat the sample as well as provide temperature control of the chamber surrounding the test cell. Additionally, the APTAC uses a continuous nitrogen supply for pressure balancing. The APTAC is more versatile than the ARC and the principal differences between them are:

- *Data analysis:* Analyzing the basic temperature and pressure data from the APTAC is less complex. The APTAC design gives data that are easier to use than the ARC.

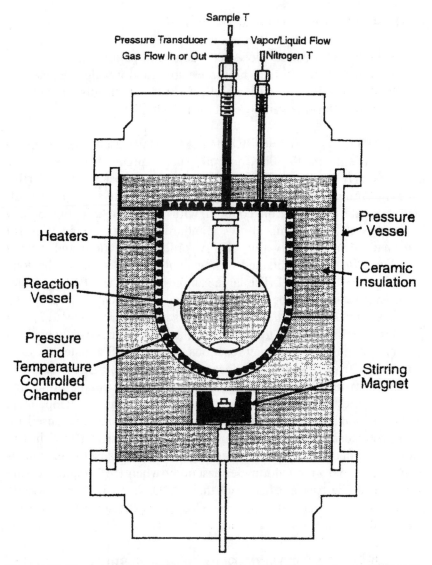

Figure 12-12. Automated pressure tracking adiabatic calorimeter (APTAC). *(Source: Arthur D. Little.)*

- *Reagent delivery:* It is possible to add reagents to the sample while the experiment is running using the APTAC instrument.
- *Sample venting:* The pressure in APTAC's sample container can be automatically released and the material collected into an external vessel.

The APTAC can function as:

- A high-performance, closed-cell adiabatic calorimeter.
- A high-temperature and high-pressure reaction calorimeter.
- An open test cell from which effluent can be discharged and analyzed for emergency relief system studies.

The APTAC is capable of studying exothermic reactions at temperatures from ambient to 500°C and pressures ranging from vacuum to 2,000 psia. It can detect and track exotherms at heat generation rates ranging from 0.04°C/min to 400°C/min. It employs the heat-wait-search mode of operation coupled with adiabatic tracking of any detected exotherm. This allows standard analysis of the data for information on the activation energy, order of reaction, and heat of reaction. The results obtained during cool-down determine the total amount of gas generated. The large sample size of the APTAC enables additional reagents to be added while the experiment is in progress. This enables the instrument to simulate actual process operations for process development or to determine the effectiveness of quenching agents in controlling runaway reactions. The APTAC collates pressure relief data from the normal closed-cell test, and then releases the sample pressure from the vessel into an external sample collection vessel. The test gives information about maximum rates of temperature rise, gas production, and condensate vapors, which are then used for vent sizing using the DIERS simplified method. The APTAC instrument has reduced maintenance and repair costs because no chemicals vent directly into the calorimeter assembly when the sample pressure is released. The typical ϕ-factor of the APTAC is 1.15. This is slightly higher than the vent sizing package (VSP) calorimeter because of the thicker wall.

VENT SIZING PACKAGE (VSP)

The vent sizing package (VSP) was developed by Fauskes & Associates, Inc. The VSP and its latest version VSP2 employ the low thermal mass test cell stainless steel 304 and Hastelloy test cell with a volume of 120 ml contained in a 4-l, high-pressure vessel as shown in Figure 12-13. The typical ϕ-factor is 1.05–1.08 for a test cell wall thickness of 0.127–0.178 mm. Measurements consist of sample temperature T_1 and pressure P_1, and external guard temperature T_2 and

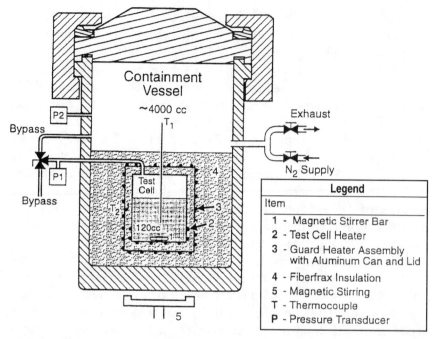

Legend		
Item		
1 - Magnetic Stirrer Bar		
2 - Test Cell Heater		
3 - Guard Heater Assembly with Aluminum Can and Lid		
4 - Fiberfrax Insulation		
5 - Magnetic Stirring		
T - Thermocouple		
P - Pressure Transducer		

Figure 12-13. Vent sizing package (VSP) apparatus. *(Source: Fauskes & Associates Inc.)*

containment pressure P_2. During the runaway or the self-heating period, the guard heater assembly serves to provide an adiabatic environment for the test sample by regulating T_2 close to T_1. For closed (non-vented) test cells, the containment vessel serves to prevent bursting of the test cell by regulating its own pressure P_2 to follow the test cell sample pressure P_1. This pressure-tracking feature makes it possible to use the thin wall (low ϕ-factor) test cell design. Vented or open tests, where the vapor or gas generated is vented either into the containment vessel or to an external container, are unique capabilities of the VSP instruments. The typical onset sensitivity is 0.1°C/min for the VSP and 0.05°C/min for the VSP2.

VSP experiments allow the comparison of various process versions, the direct determination of the wanted reaction adiabatic temperature rise, and the monitoring of the possible initiations of secondary reactions. If no secondary reaction is initiated at the wanted reaction adiabatic final temperature, a further temperature scan allows the determination of the temperature difference between the wanted reaction adiabatic final temperature and the subsequent decomposition reaction onset temperature.

VSP experiments are not suitable to measure the controlling reactant accumulation under normal process conditions. This is obtained using reaction calorimeters. VSP experiments are suitable to assess the consequences of runaway decomposition reactions after their identification using other screening methods [Differential Thermal Analysis (DTA), Differential Scanning Calorimetry (DSC) and Accelerating Rate Calorimeter (ARC)]. The decomposition is then initiated by a temperature scan or an isothermal exposure, depending on the known kinetic behavior of the reaction. In cases where a reliable baseline cannot be determined on DTA thermograms, a VSP experiment enables a better and safer determination of the decomposition exotherm.

VSP experiments provide both thermal information on runaway reactions and information on pressure effects. The type of pressurization, following the DIERS methodology (i.e., vapor pressure, production of non-condensable gases, or both), can be determined from VSP experiments. The following experimental conditions are readily achievable using the VSP [14,15]:

- Temperature up to 350–400°C under temperature scan conditions.
- Pressure up to 200 bar.
- Closed or open test cells. Open test cells are connected to a second containment vessel. Test cells of various materials can be used including stainless steel, Hastelloy C276, and glass. Glass test cells of various sizes and shape are suitable for testing fine chemical and pharmaceutical products/reaction conditions when the process is only in glass vessels or when only small samples are available, or if the samples are very expensive.
- Mechanical stirring is recommended for testing polymerization reactions or viscous reaction mixtures. This requires the use of taller containment vessels than the original to install the electric motor for the agitator.

REACTIVE SYSTEM SCREENING TOOL (RSST)

Fauskes & Associates Inc. developed the RSST as an inexpensive screening tool [15,16]. The RSST (Figures 12-14 and 12-15) consists of a spherical glass reaction vessel and immersion heater (optional),

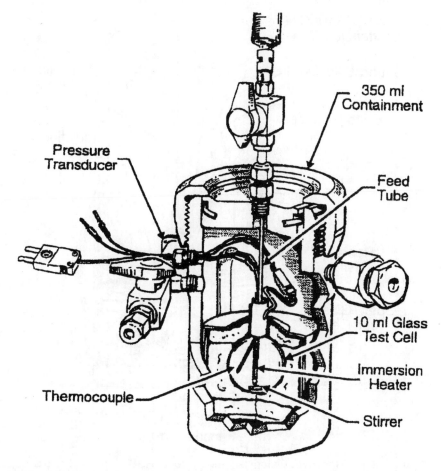

Figure 12-14. Reactive system screening tool (RSST) apparatus. *(Source: Fauskes & Associates Inc.)*

its surrounding jacket heater and insulation, thermocouples and a pressure transducer, and a stainless steel containment vessel that serves as both a pressure simulator and safety vessel. The RSST uses a magnetic stirrer base, a control box containing the heater power supply, temperature/pressure amplifiers, and a data acquisition and control panel. The sample cell volume is 10 ml and the containment volume is 350 ml. The apparatus has a low effective heat capacity relative to that of the sample whose value, expressed as the capacity ratio, is approximately 1.04. This key feature allows the measured data to be directly applied to the process scale.

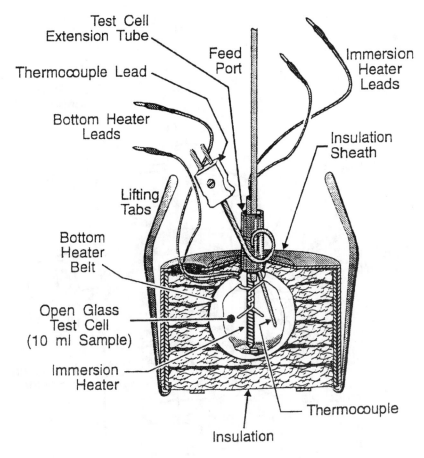

Figure 12-15. Reactive system screening tool (RSST) test cell. *(Source: Fauskes & Associates Inc.)*

A sample in the test cell is heated at a constant rate and the sample self-heat rate is obtained as a function of sample temperature. This heat-up rate may be varied and is controlled by feedback from the sample temperature measurement. The imposed linear ramp rate can be varied from 0.1°C/min to ramp rates approaching those required to simulate fire exposure by using the immersion heater option. The heater compensates for losses, and the self-heating rate of an exothermic system is adjusted for heater input. Reagents may be added to a sample during a test. Data handling programs give results of pressure versus temperature, temperature versus time, pressure versus time, and Arrhenius plots.

The RSST can rapidly and safely determine the potential for runaway reactions. It also measures the temperature rates and, in the case of gassy reactions, pressure increases to allow reliable determinations of the energy and gas release rates. This information is combined with analytical tools to evaluate reactor vent size requirements. This is extremely useful when screening a large number of different chemicals and processes.

The advantages of the RSST are:

- The small glass test cell (10 ml) is useful when only small samples are available or if the sample is expensive.
- The possible installation of the RSST on site when the samples cannot be transported.
- The low cost of the RSST.

The disadvantages of the RSST compared to the VSP are:

- The small test cell makes the study of process reaction steps more difficult.
- Only open cell tests are possible using the RSST.
- Multiphase or viscous reaction mixtures cannot be stirred well.

Figure 12-16 gives a recommended test sequence for identification of reactive system type and vent sizing.

PHI-TEC

The PHI-TEC II adiabatic calorimeter as shown in Figure 12-17 was developed by Hazard Evaluation Laboratory Ltd. (UK). The PHI-TEC can be used both as a high sensitivity adiabatic calorimeter and as multi-purpose vent sizing device [17,18]. The PHI-TEC employs the principles established by DIERS and includes advanced features compared to the VSP. It also provides important information for storage and handling and provides useful insight into the options suitable for downstream disposal of vented material.

The device consists of a low thermal mass test cell of 110 ml volume. This is made of stainless steel with a 0.015-mm wall thickness, but test cells made from other materials can be used. The contents of the cell are mixed using either a magnetically driven stirring bar or a direct drive agitation test cell. The test cell is

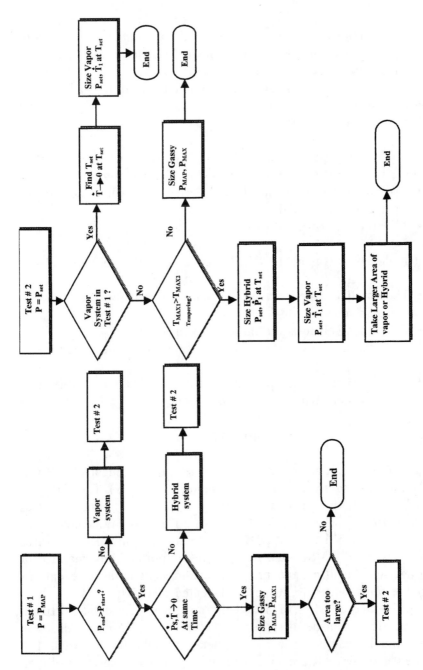

Figure 12-16. Recommended test sequence for indentification of reactive system and vest sizing.

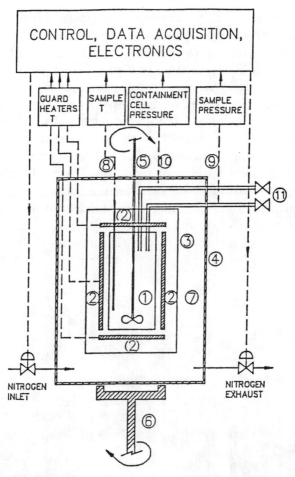

1. ADIABATIC TEST CELL, UP TO 120cm^3.
2. 3 RADIANT GUARD HEATERS.
3. CALORIMETRY ASSEMBLY
4. PRESSURE VESSEL, 4 LITERS
5. DIRECT STIRRING (OPTIONAL).
6. MAGNETICALLY DRIVEN AGITATOR

7. INSULATION.
8. SAMPLE THERMOCOUPLE.
9. SAMPLE PRESSURE TRANSDUCER.
10. VESSEL PRESSURE TRANSDUCER
11. CONNECTIONS FOR INJECTION
    VENTING, DISPOSAL SYSTEM
    (OPTIONAL).

Figure 12-17. PHI-TEC adiabatic calorimeter. *(Source: Hazard Evaluation Laboratory Ltd.)*

surrounded by three independent guard heaters, which are controlled to match the sample temperature that is measured within the test cell. These reduce heat losses from the sample down to a very low level, and thereby maintain the sample adiabatic condition. The guard heaters can track the sample temperature up to 200°C/min. The test cell and heaters are mounted in a calorimetry assembly, which sits within a high-pressure containment vessel. A computer handles the temperature control duty and also activates a pressure compensation system. High-pressure nitrogen is admitted to, or bled off from, the pressure vessel so that the pressure in the vessel closely matches the measured sample pressure within the test cell as the pressure in the test cell increases. This capability allows very thin-walled metal cells to be used, which minimizes the thermal capacity of the cell in relation to that of the test sample. This also enables the test cell to be used at pressures up to 138 bara despite its lack of mechanical strength. The computer linked with the instrument is used for data acquisition and control. Singh [19] has provided mathematical models to describe a closed thermal runaway reaction and the effect of venting from data obtained from PHI-TEC experiments. Figure 12-18 shows an overall view of the PHI-TECII adiabatic calorimeter and Figure 12-19 depicts the test

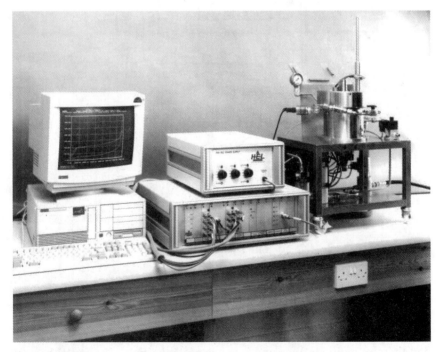

Figure 12-18. An overall view of the PHI-TEC adiabatic calorimeter. *(Source: Hazard Evaluation Laboratory Ltd.)*

Figure 12-19. The PHI-TEC II adiabatic calorimeter: Test cell and calorimeter in main pressure vessel rated for use to 138 bara. *(Source: Hazard Evaluation Laboratory Ltd.)*

cell and calorimeter in the main pressure vessel. Figures 12-20, 12-21, 12-22, and 12-23 show plots of various variables involving adiabatic decomposition of 20 w/w % di-tertiary butyl peroxide (DTBP) in toluene using the PHI-TEC II calorimeter.

(text continued on page 866)

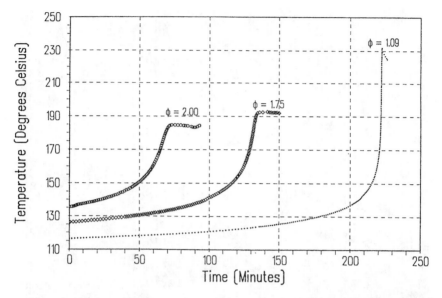

Figure 12-20. Temperature versus time plot of 20 w/w% di-tertiary butyl peroxide (ϕ = 1.09, ϕ = 1.75, and ϕ = 2.00). *(Source: Hazard Evaluation Laboratory Ltd.)*

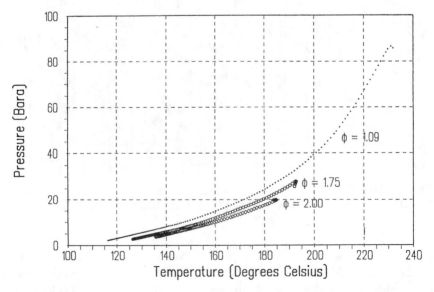

Figure 12-21. Pressure versus temperature plot of 20 w/w% di-tertiary butyl peroxide (ϕ = 1.09, ϕ = 1.75, and ϕ = 2.00). *(Source: Hazard Evaluation Laboratory Ltd.)*

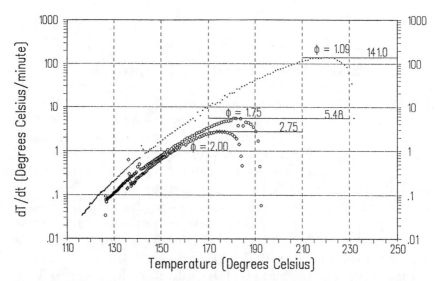

Figure 12-22. dT/dt versus temperature plot of 20 w/w% di-tertiary butyl peroxide (ϕ = 1.09, ϕ = 1.75, and ϕ = 2.00). *(Source: Hazard Evaluation Laboratory Ltd.)*

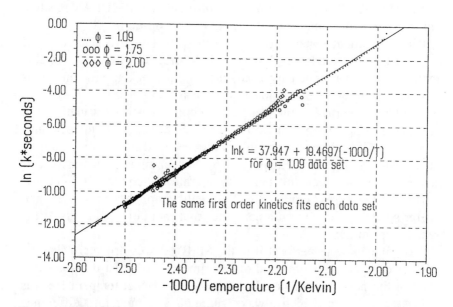

Figure 12-23. ln(k* seconds) versus −1,000/Absolute temperature plot of 20 w/w% di-tertiary butyl peroxide (ϕ = 1.09, ϕ = 1.75, and ϕ = 2.00). *(Source: Hazard Evaluation Laboratory Ltd.)*

(text continued from page 863)

SIMULAR REACTION CALORIMETER

Many chemical reactions in the CPI are performed in a semi-batch process under isothermal conditions. The DIERS methodology in sizing the relief system for runaway reactions is based on adiabatic conditions using the batch process. There are instances where a semi-batch process under isothermal condition is advantageous in significantly reducing the vent size or even possibly eliminating runaway reaction and the need for pressure relief. Singh [20] performed studies involving the esterification reaction between methanol (CH_3OH) and acetic anhydride ($CH_3CO)_2O$ at a constant temperature of 40°C in a calorimeter known as the SIMULAR.

The SIMULAR, developed by Hazard Evaluation Laboratory Ltd., is a chemical reactor control and data acquisition system. It can also perform calorimetry measurements and be employed to investigate chemical reaction and unit operations such as mixing, blending, crystallization, and distillation. Figure 12-24 shows a schematic detail of the SIMULAR, and Figure 12-25 illustrates the SIMULAR reaction calorimeter with computer controlled solids addition.

The instrument consists of a reaction vessel with a size range of 0.2–5.0 l and is made of glass, stainless steel, or any other suitable material and surrounded by a heating/cooling jacket. Silicone oil is circulated through the reaction vessel jacket and a heater/chiller unit before being returned to the jacket. The cooling/heating circulation unit is capable of operating from –50 – 300°C, and controlling temperatures to within 0.1°C. A glass condenser can be added for reflux operation at atmospheric pressure (Figure 12-26). Sensor signals such as reactor temperature, oil inlet and outlet temperatures, condenser inlet and outlet temperatures, condenser flow, calibration heater power, and stirrer speed are logged on a computer that monitors and controls the reactor performance.

Experiments were performed in the SIMULAR calorimeter using the "power compensation" method of calorimetry (note that it can also be used in the heat flow mode). In this case, the jacket temperature was held at conditions, which always maintain a temperature difference (~20°C) below the reactor solution. A calibration heater was used to

(text continued on page 870)

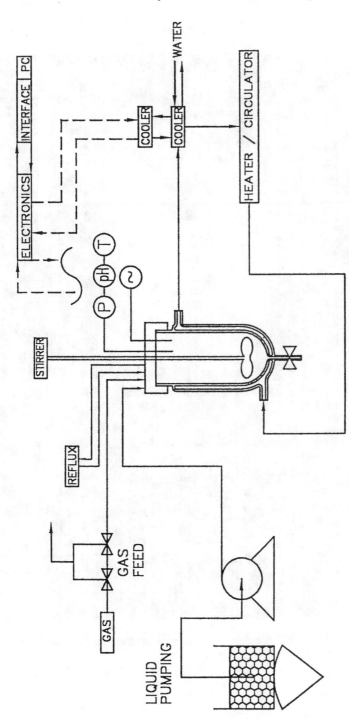

Figure 12-24. Schematic overview of a typical calorimeter. (*Source: Hazard Evaluation Laboratory Ltd.*)

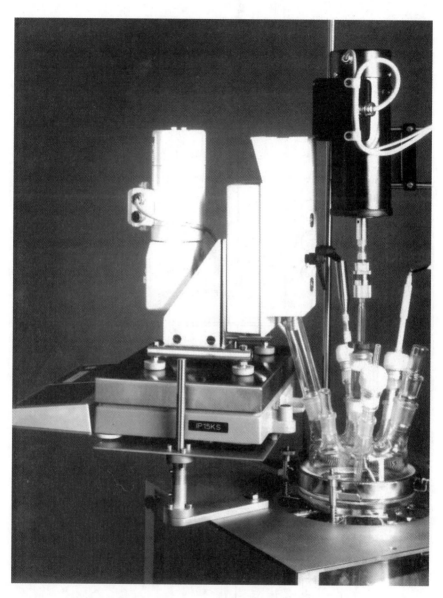

Figure 12-25. SIMULAR reaction calorimeter: Computer controlled solids feed. *(Source: Hazard Evaluation Laboratory Ltd.)*

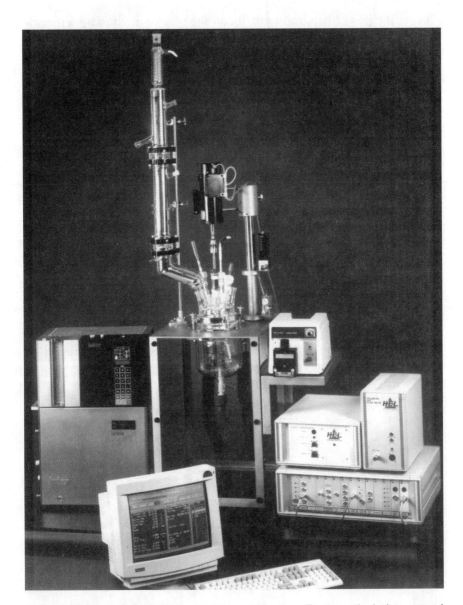

Figure 12-26. The SIMULAR reaction calorimeter. Features include pumped liquid feed, gas mass flow control, gas evolution measurement, and distillation equipment. *(Source: Hazard Evaluation Laboratory Ltd.)*

(text continued from page 866)

provide the power required to maintain the reactor at constant temperature. As the chemical reaction proceeded and heat was generated, the calibration power was varied to maintain constant conditions. Figures 12-27, 12-28, 12-29, and 12-30 give plots of various parameters using the SIMULAR reaction calorimeter.

TWO-PHASE FLOW RELIEF SIZING FOR RUNAWAY REACTION

Many methods have been used to size relief systems: area/volume scaling, mathematical modeling using reaction parameters and flow theory, and empirical methods by the Factory Insurance Association (FIA). The Design Institute for Emergency Relief Systems (DIERS) of the AIChE has performed studies of sizing reactors undergoing runaway reactions. Intricate laboratory instruments as described earlier have resulted in better vent sizes.

A selection of relief venting as the basis of safe operation is based on the following considerations [21]:

- Compatibility of relief venting with the design and operation of the plant/process
- Identifying the worst-case scenarios

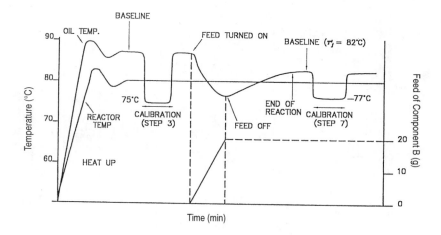

Figure 12-27. Temperature versus time plot from a semi-batch heat flow experiment. *(Source: Hazard Evaluation Laboratory Ltd.)*

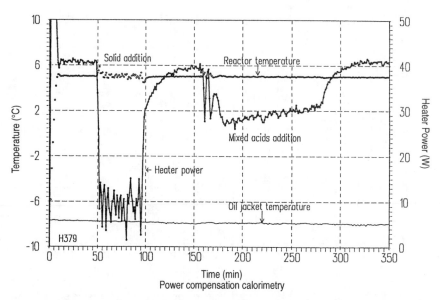

Figure 12-28. Acid plus solid gives product 1; product 1 plus mixed acids gives product 2. *(Source: Hazard Evaluation Laboratory Ltd.)*

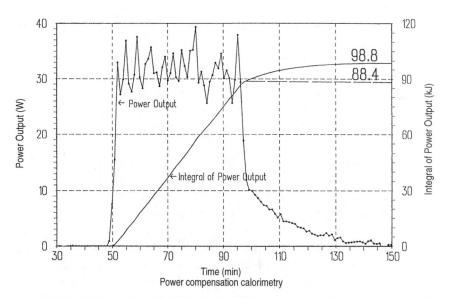

Figure 12-29. Acid plus solid gives product 1 (power output and its integral versus time). *(Source: Hazard Evaluation Laboratory Ltd.)*

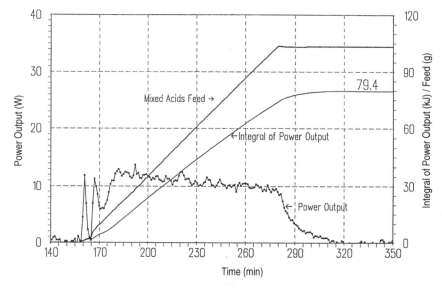

Figure 12-30. Product 1 plus mixed acid gives product 2 (power output and its integral versus time). *(Source: Hazard Evaluation Laboratory Ltd.)*

- Type of reaction
- Means of measuring the reaction parameters during the runaway reaction
- Relief sizing procedure
- Design of the relief system including discharge ducting and safe discharge area

RUNAWAY REACTIONS

A runaway reaction occurs when an exothermic system becomes uncontrollable. The reaction leads to a rapid increase in the temperature and pressure, which if not relieved can rupture the containing vessel. A runaway reaction occurs because the rate of reaction, and therefore the rate of heat generation, increases exponentially with temperature. In contrast, the rate of cooling increases only linearly with temperature. Once the rate of heat generation exceeds available cooling, the rate of temperature increase becomes progressively faster. Runaway reactions nearly always result in two-phase flow reliefs. In reactor venting, reactions essentially fall into three classifications:

- Vapor pressure systems
- Hybrid (gas + vapor) systems
- Gassy systems

The significance of these categories in terms of relief is that, once the vent has opened, both vapor pressure and hybrid reactions temper, by losing enough heat through vaporization, to maintain temperature and pressure at an acceptable level. In a gas generating system, there is negligible or sometimes no control of temperature during venting, such that relief sizing is based on the peak gas generation rate. Experimental studies conducted by the testing methods must not only be able to differentiate between the reaction types, but must also simulate large-scale process conditions. The results of the experimental studies should greatly enhance the design of the relief system. The reactor venting categories are reviewed below.

Vapor Pressure Systems

In this type of reaction, no permanent gas is generated. The pressure generated by the reaction is due to the increasing vapor pressure of the reactants, products, and/or inert solvent as the temperature rises.

It is the rate of temperature increase (i.e., power output) between the set pressure and the maximum allowable pressure, which determines the vent size and not the peak rate. Boiling is attained before potential gaseous decomposition (i.e., the heat of reaction is removed by the latent heat of vaporization). The reaction is tempered, and the total pressure in the reactor is equal to the vapor pressure. The principal parameter determining the vent size is the rate of the temperature rise at the relief set pressure.

Systems that behave in this manner obey the Antoine relationship between pressure P and temperature T as represented by

$$\ln P = A + B/T \tag{12-10}$$

where A and B are constants and T is the absolute temperature. An example is the methanol and acetic anhydride reaction.

Gassy Systems

Here, the system pressure is due entirely to the pressure of non-condensible gas rather than the vapor pressure of the liquid. The gas is the result of decomposition. The exothermic heat release is largely retained in the reaction mass because the cooling potential of volatile materials is not available. As such, both the maximum temperature and maximum gas generation rate can be attained during venting. Gaseous decomposition reactions occur without tempering. The total pressure in the reactor is equal to the gas pressure. The principal parameter determining the vent size is the maximum rate of pressure rise. Unlike vapor-pressure systems, the pressure is controlled (and reduced) without cooling the reaction.

A survey within the Fine Chemical Manufacturing Organization of ICI has shown that gassy reaction systems predominate due to established processes such as nitrations, diazotizations, sulphonations, and many other types of reactions [22]. Very few vapor pressure systems have been identified that also generate permanent gas (i.e., hybrid type).

Hybrid Systems

These are systems that have a significant vapor pressure and at the same time produce non-condensible gases. Gaseous decomposition reaction occurs before boiling—the reaction is still tempered by vapor stripping. The total pressure in the reactor is the summation of the gas partial pressure and the vapor pressure. The principal parameters determining the vent size are the rates of temperature and pressure rise corresponding to the tempering condition. A tempered reactor contains a volatile fluid that vaporizes or flashes during the relieving process. This vaporization removes energy through heat vaporization and tempers the rise in temperature rate due to the exothermic reaction.

In some hybrid systems, the generated vapor in a vented reaction is high enough to remove sufficient latent heat to moderate or "temper" the runaway (i.e., to maintain constant temperature). This subsequently gives a smaller vent size.

Richter and Turner [23] have provided systematic schemes for sizing batch reactor relief systems. They employed logic diagrams that outlined the various decisions to produce a model of the system. Figure 12-31 reviews the reaction kinetics and thermodynamics required to create a reactor model. Figure 12-32 shows a sequence of steps used to model flow from the reactor, assuming relief is occurring as a homogeneous vapor-liquid mixture. The DIERS program has supported the use of a homogeneous vapor-liquid mixture (froth) model, which

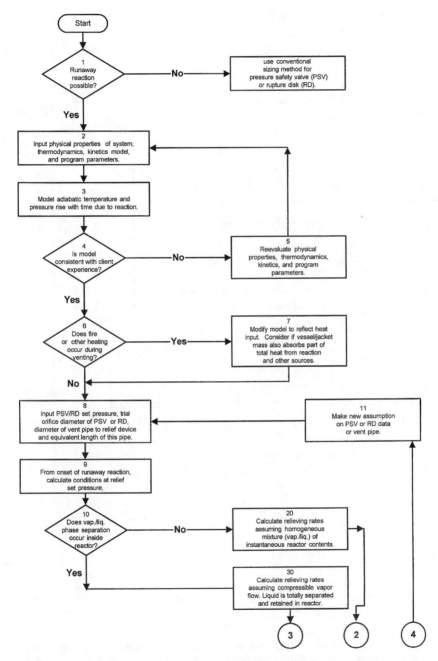

Figure 12-31. Block flow diagram showing the reaction kinetics and thermo-dynamics needed to create a reactor model. *(Source: Reproduced with permission of the AIChE. Copyright © 1996 AIChE. All rights reserved.)*

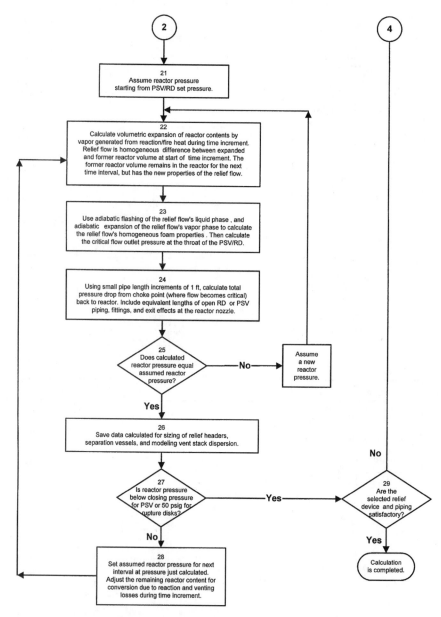

Figure 12-32. Steps used to model flow from the reactor, assuming relief is occuring as a homogeneous vapor/liquid mixture. *(Source: Reproduced with permission of the AIChE. Copyright © 1996 AIChE. All rights reserved.)*

relies on the assumption that the vapor phase is in equilibrium with the batch liquid phase [24]. Figure 12-33 shows the steps used to model compressible vapor venting from the reactor. Table 12-2 lists formulae for computing the area of the three systems.

SIMPLIFIED NOMOGRAPH METHOD

Boyle [25] and Huff [26] first accounted for two-phase flow with relief system design for runaway chemical reactions. Computer simulation approaches to vent sizing involve extensive thermokinetic and thermophysical characterization of the reaction system. Fisher [27] provided an excellent review of emergency relief system design involving runaway reactions in reactors and vessels. The mass flux through the relief device representing choked two-phase flow through a hole is expressed as:

$$G = \frac{Q_m}{A} = \frac{\Delta H_V}{v_{fg}} \left(\frac{g_c}{C_P T_S} \right)^{0.5} \tag{12-11}$$

For two-phase flow through pipes, an overall dimensionless discharge coefficient, ψ, is applied. Equation 12-11 is referred to as the equilibrium rate model (ERM) for low-quality choked flow. Leung [28] indicated that Equation 12-11 be multiplied by a factor of 0.9 to bring the value in line with the classic homogeneous equilibrium model (HEM). Equation 12-11 then becomes

$$G = \frac{Q_m}{A} = 0.9\psi \frac{\Delta H_V}{v_{fg}} \left(\frac{g_c}{C_P T_S} \right)^{0.5} \tag{12-12}$$

where A = area of the hole, m^2
$\quad g_c$ = correction factor 1.0 (kg•m/sec^2)/N
$\quad Q_m$ = mass flow through the relief, kg/sec
$\quad \Delta H_V$ = heat of vaporization of the fluid, J/kg
$\quad v_{fg}$ = change in specific volume of the flashing liquid, m^3/kg
$\quad C_P$ = heat capacity of the fluid (J/kg•K)
$\quad T_S$ = absolute saturation temperature of the fluid at the set pressure, K

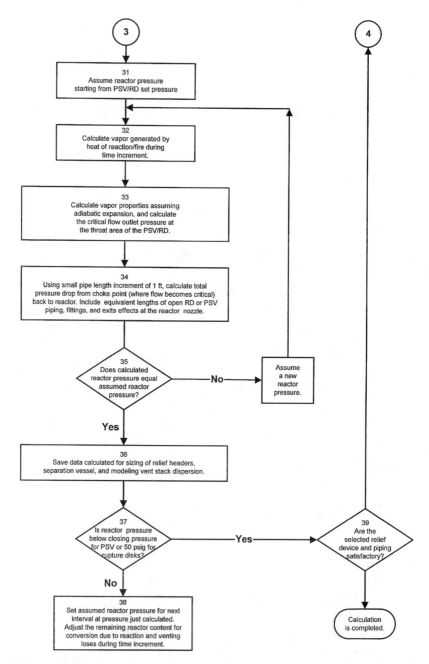

Figure 12-33. Steps used to model compressible vapor venting from the reactor. *(Source: Reproduced with permission of the AIChE. Copyright © 1996 AIChE. All rights reserved.)*

Table 12-2
Vent areas and diameters of the vapor,
gassy, and hybrid systems

Vapor system	Gassy system	Hybrid system
$A = 1.5 \times 10^{-5} \left(\dfrac{m_o \dfrac{dT}{dt}}{F \cdot P_s} \right)$	$A = 3 \times 10^{-6} \left(\dfrac{1}{F} \right) \left(\dfrac{m_o}{m_t} \right) \left(\dfrac{\dfrac{dP}{dt}}{P_{MAP}^{1.5}} \right)$	$A = 5.6 \times 10^{-6} \left(\dfrac{1}{F} \right) \left(\dfrac{m_o}{m_t} \right) \left(\dfrac{\dfrac{dP}{dt}}{P_s^{1.5}} \right)$
$d = \left(\dfrac{4A}{\pi} \right)^{0.5}$	$d = \left(\dfrac{4A}{\pi} \right)^{0.5}$	$d = \left(\dfrac{4A}{\pi} \right)^{0.5}$
L/D=0, F=1.00	L/D=0, F=1.00	L/D=0, F=1.00
L/D=50, F=0.85	L/D=50, F=0.70	L/D=50, F=0.70
L/D=100, F=0.75	L/D=100, F=0.60	L/D=100, F=0.60
L/D=200, F=0.65	L/D=200, F=0.45	L/D=200, F=0.45
L/D=400, F=0.50	L/D=400, F=0.33	L/D=400, F=0.33

The result is applicable for homogeneous venting of a reactor (low quality, not restricted just to liquid inlet condition). Figure 12-34 gives the value of ψ for L/D ratio. For example, for a pipe length of zero, $\psi = 1$, and as the pipe length increases, the value of ψ decreases. Equation 12-12 can be further rearranged in terms of a more convenient expression as follows:

$$\frac{\Delta H_V}{\upsilon_{fg}} = T_S \frac{dP}{dT} \tag{12-13}$$

Substituting Equation 12-13 into Equation 12-12 gives

$$G = 0.9 \psi \frac{dP}{dT} \left(\frac{g_c T_S}{C_P} \right)^{0.5} \tag{12-14}$$

The exact derivative is approximated by a finite difference derivative to yield

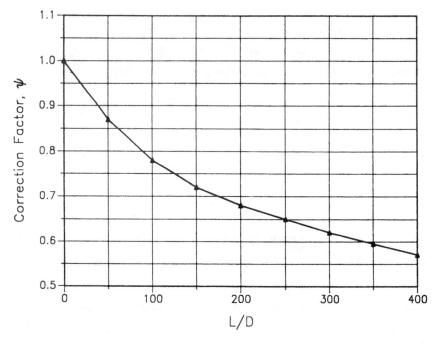

Figure 12-34. Correction factor versus L/D for two-phase flashing flow through pipes. *(Source: J. C. Leung and M. A. Grolmes, "The Discharge of Two-Phase Flashing Flow in a Horizontal Duct," AIChE Journal, 33(3), p. 524, 1987.)*

$$G \cong 0.9 \, \psi \, \frac{dP}{dT} \left(\frac{g_c T_s}{C_P} \right)^{0.5} \qquad\qquad (12\text{-}15)$$

where ΔP = overpressure

ΔT = temperature rise corresponding to the overpressure

Fauske [29] developed a simplified chart for the two-phase calculation. The relief area was expressed as:

$$A = \frac{V \rho}{G \, \Delta t_V} \qquad\qquad (12\text{-}16)$$

where A = relief vent area, m^2

V = reactor volume, m^3

ρ = density of the reactants, kg/m^3

G = mass flux through the relief (kg/sec•m²)
Δt_V = venting time, sec

Boyle [25] developed Equation 12-16 by defining the required area as the size that would empty the reactor before the pressure could rise above some allowable pressure for a given vessel. The mass flux G is given by Equations 12-12 or 12-15, and the venting time is given by

$$\Delta t_V = \frac{(\Delta T)(C_P)}{q_S} \tag{12-17}$$

where ΔT = temperature rise corresponding to the overpressure ΔP
    T = temperature
    C_P = heat capacity
    q_S = energy release rate per unit mass at the set pressure of the relief system

Combining Equations 12-16, 12-17, and 12-11 yields

$$A = V\rho(g_C C_P T_S)^{-0.5} \frac{q_S}{\Delta P} \tag{12-18}$$

Equation 12-18 gives a conservative estimate of the vent area, and the simple design method represents overpressure (ΔP) between 10%–30%. For a 20% absolute overpressure, a liquid heat capacity of 2,510 J/kg • K for most organics, and considering that a saturated water relationship exists, the vent size area per 1,000 kg of reactants is:

$$A = \left(\frac{m^2}{1,000\,kg}\right) = \frac{0.00208\left(\dfrac{dT}{dt}\right)}{P_S} \frac{°C/min}{bar} \tag{12-19}$$

Figure 12-35 shows a nomograph for determining the vent size. The vent area is calculated from the heating rate, the set pressure, and the mass of reactants. The nomograph is used for obtaining quick vent sizes and checking the results of the more rigorous computation. Crowl and Louvar [30] expressed that the nomograph data of Figure 12-35 applies to a discharge coefficient of $\psi = 0.5$, representing a discharge L/D of 400.0. However, use of the nomograph at other discharge pipe lengths and different ψ requires a suitable conversion.

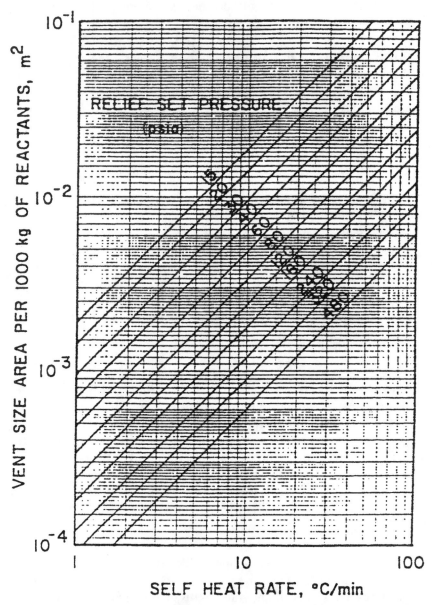

Figure 12-35. A vent sizing monogram for tempered (high vapor pressure) runaway chemical reactions. *(Source: H. K. Fauske, "Generalized Vent Sizing Monogram for Runaway Chemical Reactions,"* Plant/Operations Progress, *Vol. 3, No. 4, 1984. Reproduced with permission of the American Institute of Chemical Engineers, Copyright © 1984 AIChE. All rights reserved.)*

VENT SIZING METHODS

Vents are usually sized on the assumption that the vent flow is:

- All vapor or gas
- All liquid
- A two-phase mixture of liquid and vapor or gas

The first two cases represent the smallest and largest vent sizes required for a given rate at increased pressure. Between these cases, there is a two-phase mixture of vapor and liquid. It is assumed that the mixture is homogeneous, that is, that no slip occurs between the vapor and liquid. Furthermore, the ratio of vapor to liquid determines whether the venting is closer to the all vapor or all liquid case. As most relief situations involve a liquid fraction of over 80%, the idea of homogeneous venting is closer to all liquid than all vapor. Table 12-3 shows the vent area for different flow regimes.

VAPOR PRESSURE SYSTEMS

These systems are called "tempering" (i.e., to prevent temperature rise after venting) systems because there is sufficient latent heat available to remove the heat of reaction and to temper the reaction at the set pressure. The vent requirements for such systems are estimated from the Leung's Method [31,32]:

$$A = \frac{M_O q}{G\left[\left(\frac{V}{M_O} T_S \frac{dP}{dT}\right)^{0.5} + \left(C_V \Delta T\right)^{0.5}\right]^2} \tag{12-20}$$

Table 12-3
Vent areas for different flow regimes

Type of flow	Required vent area as a multiple of all vapor vent area
All vapor	1
Two-phase: churn turbulent	2–5
Bubbly	7
Homogeneous	8
All liquid	10

Alternatively, the vent area can be expressed as:

$$A = \frac{M_O q}{G\left[\left(\frac{V}{M_O} \frac{\Delta H_V}{\upsilon_{fg}}\right)^{0.5} + \left(C_V \Delta T\right)^{0.5}\right]^2} \tag{12-21}$$

where M_o = total mass contained within the reactor vessel prior to relief, kg

q = exothermic heat release rate per unit mass, J/(kg•sec), W/kg

V = volume of the vessel, m^3

C_V = liquid heat capacity at constant volume (J/kg•K)

ΔH_V = heat of vaporization of the fluid, J/kg

υ_{fg} = change in specific volume of the flashing liquid, $(\upsilon_g - \upsilon_f)$, m^3/kg

The heating rate q is defined by

$$q = \frac{1}{2} C_V \left[\left(\frac{dT}{dt}\right)_s + \left(\frac{dT}{dt}\right)_m\right] \tag{12-22}$$

The first derivative of Equation 12-22, denoted by the subscript s, corresponds to the heating rate at the set pressure, and the second derivative, denoted by subscript m, corresponds to the temperature rise at the maximum turnaround pressure. Both derivatives are determined from an experiment (e.g., in the PHI-TEC or VSP).

The above equations assume the following:

- Uniform froth or homogeneous vessel venting occurs.
- The mass flux, G, varies little during the relief.
- The reaction energy per unit mass, q, is treated as constant.
- Constant physical properties C_V, ΔH_V, and υ_{fg}.
- The system is a tempered reactor system. This applies to the majority of reaction systems.

It is important to use consistent units in applying the above two-phase equations. The best procedure is to convert all energy units to their mechanical equivalents before solving for the relief area, especially when Imperial (English) units are used. To be consistent, use the SI unit.

The vapor pressure systems obey the Antoine relationships:

$$\ln P = A + \frac{B}{T} \tag{12-10}$$

Differentiating Equation 12-10 yields

$$\frac{1}{P}\frac{dP}{dT} = -\frac{B}{T^2} \tag{12-23}$$

$$\frac{dP}{dT} = -\frac{B}{T^2}P \tag{12-24}$$

An equation representing the relief behavior for a vent length L/D < 400 is given by [33]

$$M_O = \frac{\left(D_p\right)^2 (\Delta P_S)}{2.769 \left(\dfrac{dT}{dt}\right)_s} \cdot \left(\frac{T_S}{C_p}\right)^{0.5} \tag{12-25}$$

where M_o = allowable mass of the reactor mixture charge (kg) to limit the venting overpressure to P_P (psig)

D_P = rupture disk diameter, in

ΔP_S = allowable venting overpressure (psi), that is, the maximum venting pressure minus the relief device set pressure

P_P = maximum venting pressure (psig)

P_S = relief device set pressure (psig). Note that the relief device set pressure can range from the vessel's MAWP to significantly below the MAWP.

T_S = equilibrium temperature corresponding to the vapor pressure where the vapor pressure is the relief device set pressure (K)

$(dT/dt)_s$ = reactor mixture self-heat rate (°C/min) at temperature $T_S(K)$ as determined by a DIERS or equivalent test

C_P = specific heat of the reactor mixture (cal/g-K or Btu/lb-°F)

Equation 12-25 is a dimensional equation, therefore, the dimensions given in the parameters must be used.

FAUSKE'S METHOD

Fauske [32] represented a nomograph for tempered reactions as shown in Figure 12-35. This accounts for turbulent flashing flow and requires information about the rate of temperature rise at the relief set pressure. This approach also accounts for vapor disengagement and frictional effects including laminar and turbulent flow conditions. For turbulent flow, the vent area is

$$A = \frac{1}{2}\frac{M_O\left(\dfrac{dT}{dt}\right)_s (\alpha_D - \alpha_O)}{F\left(\dfrac{T_S}{C_S}\right)^{0.5} \Delta P (1 - \alpha_O)} \quad \text{for } 0.1P_S \leq \Delta P \leq 0.3P_S \quad (12\text{-}26)$$

where M_O = initial mass of reactants, kg
   $(dT/dt)_s$ = self-heat rate corresponding to the relief set pressure, K/sec
   $(dT/dt)_m$ = self-heat rate at turnaround temperature, K/sec
   P_S = relief set pressure, Pa
   α_D = vessel void fraction corresponding to complete vapor disengagement
   α_O = initial void fraction in vessel
   T_S = temperature corresponding to relief actuation, K
   C_S = liquid specific heat capacity (J/kg•K)
   ΔP = equilibrium overpressure corresponding to the actual temperature rise, Pa
   ΔT = temperature rise following relief actuation, K
   F = flow reduction correction factor for turbulent flow (L/D = 0, F ≈ 1.0; L/D = 50, F ≈ 0.85; L/D = 100, F ≈ 0.75; L/D = 200, F ≈ 0.65; L/D = 400, F ≈ 0.55, where L/D is the length-to-diameter ratio of the vent line)

Figure 12-36 shows a sketch of the temperature profile for high vapor pressure systems.

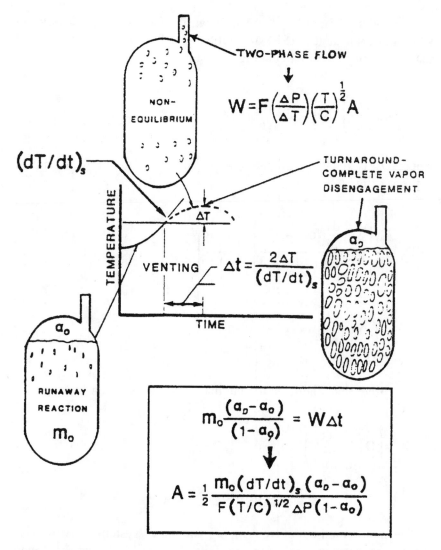

Figure 12-36. Vent sizing model for high vapor pressure systems; due to non-equilibrium effects turnaround in temperature is assumed to coincide with the onset of complete vapor disengagement.

The PHI-TEC or VSP bench scale apparatus can be employed to determine information about the self-heat rate and vapor disengagement when this is not readily available. Additionally, the VSP equipment can be used for flashing flow characteristics using a special bottom vented test cell. Here, the flowrate, G_O (kg/sm²), is measured in a simulated vent line (same L/D ratio) of diameter D_O using the vent sizing package (VSP) apparatus. The following recommended scale-up approach in calculating the vent size is:

$$\text{If} \quad G_O\left(\frac{D_T}{D_O}\right) \geq G_T \cong F\left(\frac{\Delta P}{\Delta T}\right)\left(\frac{T_S}{C_{P,S}}\right)^{0.5}$$

where D_T is the vent diameter required for turbulent flow, and is expressed by

$$D_T \cong \frac{3}{2}\left[\frac{M_O\dfrac{dT}{dt}(\alpha_D - \alpha_O)}{F P_S(1 - \alpha_O)}\right]^{0.5}\left(\frac{C_{P,S}}{T_S}\right)^{0.25}$$

$$\text{If} \quad G_O\left(\frac{D_T}{D_O}\right) \leq G_T \cong F\left(\frac{\Delta P}{\Delta T}\right)\left(\frac{T_S}{C_{P,S}}\right)^{0.5}$$

the required vent diameter for laminar flow D_L is given by

$$D_L = \left(D_T^2 D_O \frac{G_T}{G_O}\right)^{0.33}$$

GASSY SYSTEMS

The major method of vent sizing for gassy system is two-phase venting to keep the pressure constant. This method was employed before DIERS with an appropriate safety factor [34]. The vent area is expressed by:

$$A = \frac{Q_g(1 - \alpha)\rho_f}{G_1} = \frac{Q_gM}{GV} \tag{12-27}$$

where Q_g = volumetric gas generation rate at temperature and in reactor during relief, m³/sec

M = mass of liquid in vessel, kg

ρ_f = liquid density, kg/m^3

G, G$_1$ = mass vent capacity per unit area (kg/sec•m^2)

α = void fraction in vessel

V = total vessel volume, m^3

Unlike systems with vapor present, gassy systems do not have any latent heat to temper the reaction. The system pressure increases as the rate of gas generation with temperature increases, until it reaches the maximum value. It is possible to underestimate the vent size if sizing depends on the rate of gas generated at the set pressure. Therefore, it is more plausible to size the vent area on the maximum rate of gas generation. Homogeneous two-phase venting is assumed even if the discharge of liquid during venting could reduce the rate of gas generation even further. The vent area is defined by

$$A = 3.6 \times 10^{-3} Q_g \left(\frac{M_O}{VP_m} \right)^{0.5} \tag{12-28}$$

The maximum rate of gas generation during a runaway reaction is proportional to the maximum value of dP/dt and can be calculated from

$$Q_{g1} = \frac{M_O}{M_t} \left(\frac{V_t}{P_m} \frac{dP}{dt} \right) \tag{12-29}$$

An equation representing the relief behavior for a length L/D < 400 as in the tempered system is given by [33]

$$M_O = \left[V_P \, \Delta P_P \left\{ \frac{(D_P)^2 (M_S)(P_P)}{(2.07)(T_T)(dP/dt)} \right\}^2 \right]^{1/3} \tag{12-30}$$

where V$_P$ = vessel total volume, gal

P$_P$ = maximum allowable venting pressure, psia

M$_S$ = sample mass used in a DIERS test or equivalent test, g

dP/dt = pressure rise in test, psi/sec

T$_T$ = maximum temperature in test, K

P_{amb} = ambient pressure at the end of the vent line, psia

$\Delta P_P = P_P - P_{amb}$, psi

Equation 12-30 is a dimensional equation and the dimensions given in the parameters must be used.

HOMOGENEOUS TWO-PHASE VENTING UNTIL DISENGAGEMENT

ICI [34] developed a method for sizing a relief system that accounts for vapor/liquid disengagement. They proposed that homogeneous two-phase venting occurs and increases to the point of disengagement. Their derivations were based on the following assumptions:

• Vapor phase sensible heat terms may be neglected.
• Vapor phase mass is negligible.
• Heat evolution rate per unit mass of reactants is constant (or average value can be used).
• Mass vent rate per unit area is approximately constant (or safe value can be used).
• Physical properties can be approximated by average values.

The vent area is given by

$$A = \frac{qV(\alpha - \alpha_O)}{Gv_f \left\{ \dfrac{h_{fg}v_f(\alpha - \alpha_O)}{v_{fg}(1 - \alpha_O)(1 - \alpha)} + C\Delta T \right\}} \tag{12-31}$$

where A = vent area, m^2

C = liquid specific heat (J/kg•K)

G = mass vent capacity per unit area, kg/m^2s

h_{fg} = latent heat, J/kg

q = self-heat rate, W/kg

V = total vessel volume, m^3

ΔT = temperature rise corresponding to the overpressure, K

α_O = initial void fraction

α = void fraction in vessel

v_f = liquid specific volume, m^3/kg

v_{fg} = difference between vapor and liquid specific volumes, m^3/kg

Equation 12-31 is valid if disengagement occurs before the pressure would have turned over during homogeneous venting, otherwise it gives an unsafe (too small) vent size. Therefore, it is necessary to verify that

$$q > \frac{GAh_{fg}\upsilon_f^2}{V\upsilon_{fg}(1-\alpha)^2}$$ (12-32)

is satisfied at the point of disengagement.

TWO-PHASE FLOW THROUGH AN ORIFICE

Sizing formulae for flashing two-phase flow through relief devices were obtained through DIERS. It is based on Fauske's equilibrium rate model (ERM) and assumes frozen flow (non-flashing) form a stagnant vessel to the relief device throat. This is followed by flashing to equilibrium in the throat. The orifice area is expressed by

$$A = \frac{W}{C_D}\left\{\left(\frac{xV_G}{kP_1}\right) + \left[\frac{(V_G - V_L)^2 C_L T_1}{\lambda^2}\right]\right\}^{0.5}$$ (12-33)

where A = vent area, m^2
    C_D = actual discharge coefficient
    C_L = average liquid specific heat (J/kg•K)
    k = isentropic coefficient
    P_1 = pressure in the upstream vessel, N/m^2
    T_1 = temperature corresponding to relief actuation, K
    x = mass fraction of vapor at inlet
    V_G = specific volume of gas, m^3/kg
    V_L = specific volume of liquid, m^3/kg
    W = required relief rate, kg/s
    λ = latent heat, J/kg

The theoretical rate is given by

$$W = A_f(2\Delta P \cdot \rho)^{0.5}$$ (12-34)

For a simple sharp-edged orifice, the value of C_D is well established (about 0.6). For safety relief valves, its value depends on the shape of the nozzle and other design features. Additionally, the value of C_D varies with the conditions at the orifice. For saturated liquid at the inlet, Equation 12-33 simplifies to yield

$$A = \frac{W(V_G - V_L)(C_L T_1)^{0.5}}{C_D \lambda} = \frac{W}{C_D \left(\dfrac{dP}{dT}\right)\left(\dfrac{T_1}{C_L}\right)^{0.5}} \qquad (12\text{-}35)$$

These equations are based on the following assumptions:

- Vapor phase behaves as an ideal gas
- Liquid phase is an incompressible fluid
- Turbulent Newtonian flow
- Critical flow—this is usually the case since critical pressure ratios of flashing liquid approach the value of 1

It is recommended that a safety factor of at least 2.0 be used [34]. In certain cases, lower safety factors may be employed. The designer should consult the appropriate process safety section in the engineering department for advice. Table 12-4 lists the applicability of both Leung and Fauske's methods.

Conditions of Use

- If Fauske's method yields a significantly different vent size, then the calculation should be reviewed.
- The answer obtained from the Leung's method should not be significantly smaller than that of Fauske's method.
- ICI recommends a safety factor of 1–2 on flow or area. The safety factor associated with the inaccuracies of the flow calculation will depend on the method used, the phase nature of the flow, and the pipe friction. For two-phase flow, use a safety factor of 2 to account for friction or static head.
- Choose the smaller of the vent size from the two methods.

A systematic evaluation of venting requires information on:

- The reaction—vapor, gassy, or hybrid
- Flow regime—foamy or nonfoamy

- Vent flow—laminar or turbulent
- Vent sizing parameters—dT/dt, $\Delta P/\Delta T$, and ΔT

It is important to ensure that all factors (e.g., long vent lines) are accounted for, independent of the methods used. Designers should ascertain that a valid method is chosen rather than the most convenient or the best method. For ease of use, the Leung's method for vapor pressure systems is rapid and easy. Different methods should give vent sizes within a factor of 2. Nomographs give adequate vent sizes for long lines up to a L/D of 400, but sizes are divisible by 2 for nozzles. The computer software VENT was developed to size two-phase relief for vapor, gassy, and hybrid systems.

DISCHARGE SYSTEM

DESIGN OF THE VENT PIPE

It is necessary to know the nature of the discharging fluid in determining the relief areas. DIERS and ICI techniques can analyze systems that exhibit "natural" surface active foaming and those that do not. DIERS further found that small quantities of impurities can affect the flow regime in the reactor. Additionally, a variation in impurity level may arise by changing the supplier of a particular raw material. Therefore, care should be taken when sizing emergency relief on homogeneous vessel behavior, that is, two-phase flow. In certain instances, pressure relief during a runaway reaction can result in three-phase discharge, if solids are suspended in the reaction liquors. Solids can also be entrained by turbulence caused by boiling/gassing in the bulk of the liquid. Caution is required in sizing this type of relief system, especially where there is a significantly static head of fluid in the discharge pipe. Another aspect in the design of the relief system includes possible blockage of the vent line. This could occur from the process material being solidified in cooler sections of the reactor. It is important to consider all discharge regimes when designing the discharge pipework.

Safe Discharge

Reactors or storage vessels are fitted with an overpressure protection vent directly to roof level. Such devices (e.g. relief valves) protect only

(text continued on page 896)

Table 12-4
Applicability of recommended vent sizing methods

Leung's method	Fauske's method
1. The overpressures are limited in the range of 0%–50% of absolute set pressure, (i.e., $0 \leq \Delta P \leq 0.5P_s$).	The overpressures are limited in the range of 10%–30% of absolute set pressure, (i.e., $0.1P_s \leq \Delta P \leq 0.3P_s$). It assumes that flow is turbulent and that the vapor behaves as an ideal gas.
2. The method includes the mass unit vent flow capacity per unit area, G. This allows using any applicable vent capacity calculation method.	The method incorporates the equilibrium rate model (ERM) for vent flow capacity when friction is negligible. Additionally, a correction factor is used for longer vent lines of constant diameter and with negligible static head change.
3. The heat evolution rate per unit mass, the vent capacity per unit area, physical properties (e.g., latent heat of liquid, specific heat, and vapor/liquid specific volumes) are constant.	It allows for total vapor-liquid disengagement of fluids that are not "natural" surface active foamers.
4. A hand calculation method that can be used to take into account two-phase relief when the materials in the vessel are "natural" surface active foamers.	To account for disengagement, the vessel void fraction at disengagement should be evaluated (i.e., the point at which the vent flow ceases to be two-phase and starts to be vapor only).

5. The method uses the drift-flux level swell calculation models to take into account there being more vapor in the inlet stream to relief device than average for the vessel.

The method is potentially unsafe if early disengagement (i.e., before the pressure would otherwise have turned over during two-phase venting) occurs.

6. Methods are given for the bubbly and churn-turbulent flow regimes within the vessel.

The criterion for safe use of the method takes into account that disengagement is

$$q < \frac{G A h_{fg} v_f^2}{V v_{fg} (1 - \alpha_D)^2}$$

However, when disengagement is not accounted for, then $\alpha_D = 1$ and the above equation is not required.

A = vent area, m^2
G = mass vent capacity per unit area ($kg/m^2 s$)
h_{fg} = latent heat, J/kg
v_f = liquid specific volume (m^3/kg)
v_{fg} = differences between vapor and liquid specific volumes (m^3/kg)
V = total vessel volume, m^3
α_D = void fraction in vessel at disengagement
q = heat release rate per unit mass of reactor contents (W/kg)

(text continued from page 893)

against common process maloperations and not runaway reactions. The quantity of material ejected and the rate of discharge are low, resulting in good dispersion. The increased use of rupture (bursting) discs can result in large quantities (95% of the reactor contents) being discharged for foaming systems.

The discharge of copious quantities of chemicals directly to the atmosphere can give rise to secondary hazards, especially if the materials are toxic and can form a flammable atmosphere (e.g., vapor or mist) in air. In such cases, the provision of a knockout device (scrubber, dump tank) of adequate size to contain the aerated/foaming fluid is required.

The regulatory authorities can impose restrictions on discharging effluent from the standpoint of pollution. Therefore, reliefs are seldom vented to the atmosphere. In most cases, a relief is initially discharged to a knockout system to separate the liquid from the vapor. The liquid is collected and the vapor is then discharged to another treatment unit. The vapor treatment unit depends on the hazards of the vapor and may include a vent condenser, scrubber, incinerator, flare, or a combination of these units. This type of system is referred to as the "total containment system" as shown in Figure 12-37. The knockout drum is sometimes called a catchtank or blowdown drum. The horizontal knockout serves as both a vapor-liquid separator as well as a holdup vessel for the disengaged liquid. These are commonly used where there is greater space such as in petroleum refineries and petrochemical plants. The two-phase mixture enters at one end and the vapor leaves at the opposite end. Inlets may be provided at each end with a vapor outlet at the center of the drum to minimize vapor velocities involving two-phase streams with very high vapor flowrates. When there is limited space in the plant, a tangential knockout drum is employed as illustrated in Figure 12-38. Coker [35] has given detailed design procedures of separating gas-liquid separators.

Direct Discharge to the Atmosphere

Careful consideration and analysis by management are essential before flammable or hazardous vapors are discharged to the atmosphere. Consideration and analysis must ensure that the discharge can be performed without creating a potential hazard or causing environmental problems. The possible factors are:

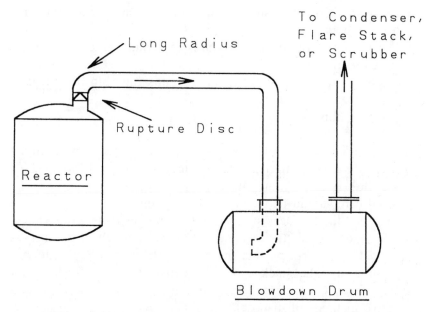

Figure 12-37. Relief containment system with blowdown drum. The blowdown drum separates the vapor from the liquid. *(Source: Grossel [39].)*

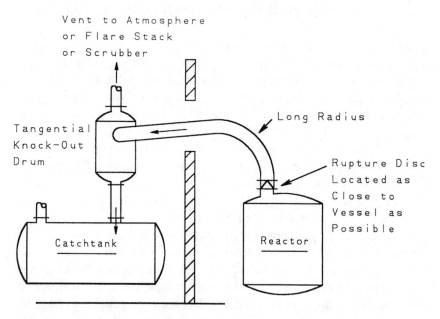

Figure 12-38. Tangential inlet knockout drum with separate liquid catchtank. *(Source: Grossel [39].)*

- Exposure of plant personnel and/or the surrounding population to toxic vapors or corrosive chemicals
- Formation of flammable mixtures at ground level or elevated structures
- Ignition of vapors at the point of emission (blowdown drum vent nozzle)
- Air pollution

These factors and methods for evaluating their effects are discussed in greater detail in American Petroleum Institute publications (API RP521) [36], Lees [37], and Wells [38].

Special care should be taken when designing the vapor vent stack to ensure that the tip faces straight up (i.e., no goose-neck) to achieve good dispersion. The stack should not be located near a building so as to avoid vapor drifting into the building. However, if the drum is near a building, the stack should extend at least 12 ft above the building floor. Grossel [39] has provided various descriptions of alternative methods of disposal.

RUPTURE (BURSTING) DISC

Some method of pressure relief is required on all pressure vessels and for other process equipment where increasing pressure might rupture the vessel. Much of the piping used in modern chemical operations also requires overpressure protection. Safety relief valves or rupture discs are employed for pressure relief. In many cases, either a rupture disc or a safety relief valve can be used. Safety relief valves are usually used for process protection and rupture discs are used for vessel protection. The safety relief valve or rupture disc must be designed to operate at a known pressure and prevent the pressure within the system from increasing. Therefore, it is important to consider the flowrate the valve can handle.

A rupture disc is a simple device that essentially consists of a thin material held in place between two flanges. The disc is usually made of metal, although it may be made of other materials. The choice of material is important because the rupture disc must be designed to close tolerances to operate properly. During emergency venting, the disc ruptures when the pressure level rises to a chosen level. The vessel is then vented and the pressure in the vessel eventually drops to atmospheric pressure. The rupture disc should be large enough to vent the vessel at the required maximum rate. Safety relief devices (i.e., safety relief valves) may also be designed to open at a given temperature. Sometimes, a combination of relief devices are used that fail at either a given temperature or a given pressure.

An important application of a rupture disc device is at the inlet of a pressure relief valve. The sizing of the pressure relief valve or rupture disc device combination requires that the pressure relief valve first be sized to meet the required relieving capacity. The normal size of the rupture disc device installed at the inlet of the pressure relief valve must be equal to or greater than the nominal size of the inlet connection of the valve to permit sufficient flow capacity and valve performance. The failure modes of rupture discs are [40]:

(a) Failure to rupture at maximum bursting pressure.
(b) Failure to open fully at maximum bursting pressure.
(c) Premature rupture below minimum bursting pressure.
(d) Leakage through the disc to the vent (through cracks or pinholes).
(e) Leakage through the holder to the atmosphere.

Cases (a) and (b) are the critical failure modes for protecting the equipment, and the reliability of the disc in these modes is referred to as the primary reliability. Case (c) is inconvenient and sometimes costly, but it is not usually dangerous unless the containment or disposal system for the discharge is inadequate. Cases (d) and (e) are comparatively rare.

The design of a rupture disc depends on the vented material. If a gas or liquid alone is to be vented, the design is relatively straight-forward. The relief system can be designed on the basis of single-phase fluid flow [41]. However, in many systems containing liquids under their own vapor pressure, emergency venting can be a combination of liquid and vapor. In such cases, the two-phase fluid must be taken into account. The methods (DIERS) used to design such relief systems are substantially different than those for single-phase fluids.

Rupture discs should be removed from service at predetermined intervals for visual inspection. Depending on the condition of the disc and recommendations by the manufacturers, they are either replaced or returned to service. The most common mode of failure is case (c), premature rupture below the minimum bursting pressure. An analysis of this mode of failure indicates that this can be the result of:

- Incorrect specification of the disc. This includes overlooking intermittent vacuum conditions and other pressure/temperature transients, failure to predict corrosion, or operating at a pressure too close to the rupture disc pressure, resulting in fatigue failure.
- Damage to the disc by faulty handling before installation or during maintainance.
- Incorrect installation of the disc, especially by faulty fitting.
- Damage of flawed disc is undetected during installation.
- Disc left in service beyond its recommended service life.
- Incorrect disc fitting.

ASME Code Rupture Disc Terminology

The following provide rupture disc terminology definitions:

- A *rupture disc device* is a non-reclosing pressure relief device actuated by inlet static pressure and designed to function by the bursting of a pressure containing disc.
- A *rupture disc* is the pressure containing and pressure sensitive element of a rupture disc device.
- A *rupture disc holder* is the structure that encloses and clamps the rupture disc in position.
- The *manufacturing design range* is a range of pressure within which the marked burst pressure must fail to be acceptable for a particular requirement as agreed upon between the rupture disc manufacturer and the user or his agent.
- The *specified disc temperature* supplied to the rupture disc manufacturer is the temperature of the disc when the disc is expected to burst.
- A *batch of rupture discs* are those discs manufactured of a material at the same time and the same size, thickness, type, heat, and manufacturing process including heat treatment.
- The *minimum net flow area* is the calculated net area after a complete burst of the disc with appropriate allowance for any structural or support members that may reduce the net flow area through the rupture disc device. The net flow area for sizing purposes should not exceed the nominal pipe size area of the rupture disc device.
- The *certified flow resistance factor,* K_R, is a dimensionless factor used to calculate the velocity head loss that results from the

presence of a rupture disc device in a pressure relief system. The relationships between the relief terms in the glossary are illustrated in Figure 12-39. The design of safety relief valves for single-phase fluids such as vapor, liquid, steam, air, and fire conditions is described elsewhere [41].

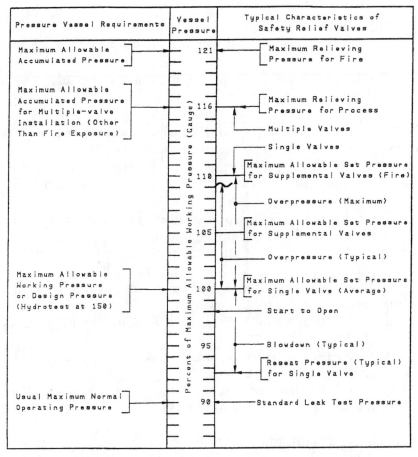

Pressure Vessel Requirements	Vessel Pressure	Typical Characteristics of Safety Relief Valves
Maximum Allowable Accumulated Pressure	121	Maximum Relieving Pressure for Fire
Maximum Allowable Accumulated Pressure for Multiple-valve Installation (Other Than Fire Exposure)	116	Maximum Relieving Pressure for Process / Multiple Valves
		Single Valves
	110	Maximum Allowable Set Pressure for Supplemental Valves (Fire)
		Overpressure (Maximum)
	105	Maximum Allowable Set Pressure for Supplemental Valves
		Overpressure (Typical)
Maximum Allowable Working Pressure or Design Pressure (Hydrotest at 150)	100	Maximum Allowable Set Pressure for Single Valve (Average) / Start to Open
	95	Blowdown (Typical) / Reseat Pressure (Typical) for Single Valve
Usual Maximum Normal Operating Pressure	90	Standard Leak Test Pressure

Notes:
1. The operating pressure may be any lower pressure required.
2. The set pressure and all other values related to it may be moved downward if the operating pressure permits.
3. This figure conforms with the requirements of the ASME Boiler and Pressure Vessel Code, Section VIII, "Pressure Vessels," Division 1.

Figure 12-39. Guidelines for relief pressures. *(Source: API RP 521 Guide for Pressure-Relieving and Depressurizing systems, 2nd ed. Washington, DC, American Petroleum Institute, 1982, p. 23.)*

Rupture Disc Sizing Methodologies

The coefficient of discharge method ($K_d = 0.62$) was specified to calculate the capacity of the rupture disc device. However, the validity of this method is limited to a disc mounted close to the pressure vessel and the discharging to atmosphere. The ASME Code provides guidance for the limited use of this method:

- The disc must discharge to the atmosphere.
- The disc must be installed within 8 pipe diameters of the vessel nozzle.
- The length of discharge piping must not be greater than 5 pipe diameters.

The resistance to flow method (K_R), which has been employed by industry for years, has now been adopted by the ASME Code for rupture disc. Sizing is performed on a relief system basis and not by capacity of individual components. The key elements of this method are:

- The rupture disc is treated as another component in the relief system that provides some resistance to flow.
- System relief capacity must be multiplied by a factor of 0.90.
- ASME Code default resistance to flow factor $K_R = 2.4$.
- Certified resistance factors K_R can be provided by the rupture disc manufacturer for each disc model.
- Based on Darcy formula (Crane 410), $h_L = f(L/D \cdot v^2/2g)$ ∴ $K_R = f(L/D)$.
- API recommendation for rupture disc has been $L/D = 75$ ∴ $K_R = 0.02(5) = 1.5$.
- Certified values based on tests of specific disc models in accordance with ASME Section VIII and PTC25.

Manufacture Certification

Significant changes by the ASME Code have been employed that now have direct impact on the use of rupture disc devices. The change to the Code affecting rupture disc manufacturers is the application of the UD Code symbol to rupture discs that meet the requirements of the ASME Code. Authorization to use the UD stamp is based on an audit by an ASME representative of various manufacturing, testing, and quality assurance systems. Additionally, validation tests are performed to ensure close agreement between production discs and certified discs.

Rupture Disc Performance Requirements

The ASME Code provides requirements for rupture disc performance. The rupture tolerance at the specified disc temperature should not exceed ± 2 psi for marked pressure up to and including 40 psi and $\pm 5\%$ for marked burst pressures above 40 psi.

The rupture disc must be marked at a pressure within the manufacturing range. The manufacturing ranges available are defined in the product literature for each rupture disc model.

Rupture discs are typically manufactured to order, where each order represents a lot. The ASME Code defines three methods of acceptance testing for rupture discs. The most common method requires that at least two discs from the lot be burst tested at the specified disc temperature. The results of these tests must fall within the rupture tolerance. Table 12-5 shows the changes between the old requirements and the new sizing methodology and certification process.

Table 12-5
Rupture disc requirements

Old requirements	New requirements
Manufacturer's name	Manufacturer's name
Model or part number	Model or part number
Lot number	Lot number
Disc material	Disc material
Nominal size	Nominal size
Stamped burst pressure	Marked burst pressure
Specified disc temperature	Specified disc temperature
Capacity	Minimum net flow area
	Certified flow resistance, K_R
	ASME UD code symbol

INHERENTLY SAFE PLANTS IN REACTOR SYSTEMS

Hazards should be considered and if possible eliminated at the design stage or in the process development where necessary. This involves considering alternative processes, reducing or eliminating hazardous chemicals, site selection or spacing of process units. It is essential to consider inherently safer principles in the design stage because designers may have various constraints imposed on them by the time the process is developed. Englund [42] has given details of inherently safer plants of various equipment in the CPI. He emphasized the importance of user friendly plants as originally expatiated by Kletz [43]. The following reviews the practical application to reactor systems.

REACTOR SYSTEM

The following should be considered in designing an inherently safer plant involving reactor systems:

1. A good understanding of reaction kinetics is required to establish safe conditions for operation of exothermic reactions.
2. Use continuous reactors if possible. It is usually easier to control continuous reactors than batch reactors. If a batch reaction system is required, minimize the amount of unreacted hazardous materials in the reactor. Figures 12-40 and 12-41 show typical examples.

   Methods have been developed for improving batch process productivity in the manufacture of styrene-butadiene latex by the continuous addition of reactants so the reaction occurs as the reactor is being filled. These are not continuous processes even though the reactants are added continuously during most of a batch cycle. The net result is that reactants can be added almost as fast as heat can be removed. There is relatively little hazardous material in the reactor at any time because the reactants, which are flammable or combustible, are converted to non-hazardous and non-volatile polymer very quickly.
3. If possible, produce and consume hazardous raw materials in situ. Some process raw materials are so hazardous to ship and store that it is very desirable to minimize the amount of these materials on hand. In some cases, it is possible to achieve this

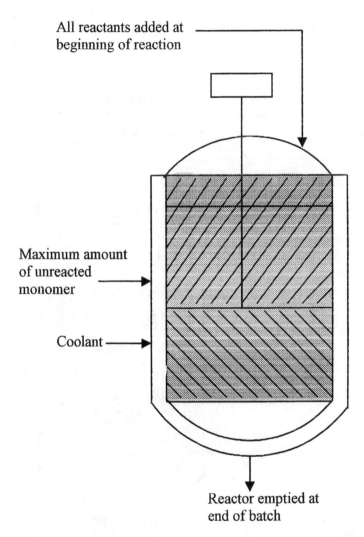

All reactants added at
beginning of reaction

Maximum amount
of unreacted
monomer

Coolant

Reactor emptied at
end of batch

Figure 12-40. Process A: Batch reaction with all reactants added at the beginning of the reaction. There is a considerable amount of flammable and hazardous material in the reactor at the beginning. *(Source: S. M. Englund, "Inherently Safer Plants: Practical Applications," Process Safety Progress, Vol. 14, No. 1, pp. 63–70, AIChE, 1995.)*

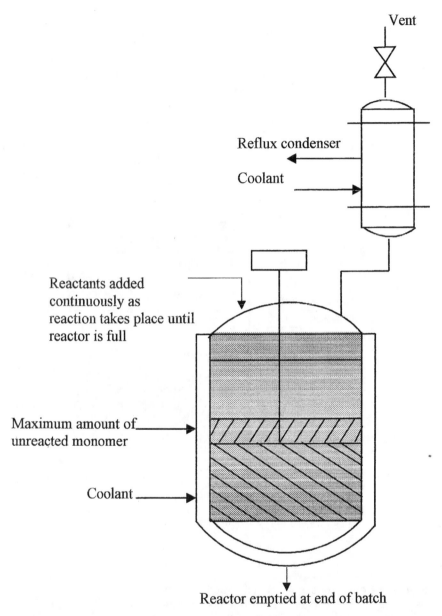

Figure 12-41. Process B: Batch reaction with reactants added during the reaction. Little flammable and hazardous material are present at any time. A reflux (or knockback) condenser is used to provide additional heat transfer. *(Source: S. M. Englund, "Inherently Safer Plants: Practical Applications,"* Process Safety Progress, *Vol. 14, No. 1, pp. 63–70, AIChE, 1995.)*

by using less hazardous chemicals, so there is only a small amount of the hazardous material in the reactor at any time.

4. Liquid-phase of solid-phase reactors contains more material than vapor-phase reactors and thus contains more stored energy than vapor-phase reactors.

5. Using high purity raw materials and products can reduce the amount the waste material that must be handled.

6. Consider designing the reactor for the highest possible pressure in case of a runaway reaction to reduce the possibility of releasing material to the environment. For example, for a certain process in the manufacture of polystyrene, the composition in the reactor included monomer, polymer, and solvent. The maximum pressure that could be reached by adiabatic polymerization of the mixture in the reactor, beginning at the reaction temperature of 120°C (248°F), was about 300 psig or 2,068 kPa (gauge). With this knowledge, it was possible to design polymerization equipment that will withstand this pressure, plus a reasonable safety factor, with considerable confidence that a runaway will not cause a release of material through a pressure relief system or because of an equipment rupture.

7. Limit the total possible charge to a batch reactor by using a precharge or feed tank of limited capacity. Alternatively, limit the addition rate by selecting a pump with a maximum capacity lower than the safe maximum addition rate for the process, or by using restriction orifices.

8. The maximum or minimum temperature attainable in a vessel can be limited by properly designed jacket heating systems. If steam heating is used, maximum temperatures can be limited by controlling steam pressure. A steam desuperheater may be needed to avoid excessive temperature of superheated steam from a pressure letdown station.

9. Tubular reactors often offer the greatest potential for inventory reduction. They are usually simple, have no moving parts, and a minimum number of joints and connections that can leak.

10. Mass transfer is often the rate-limiting step in gas-liquid reactions. Novel reactor designs that increase mass transfer can reduce reactor size and may also improve process yields.

SCALE-UP

In scale-up, runaway exothermic chemical reactions can be prevented by taking appropriate safety measures. The onset or critical temperature for a runaway reaction depends on the rate of heat generation and the rate of cooling, which are closely linked to the dimensions of the vessel.

This can be represented by the rate of heat generation being proportional to the volume of the reaction mixture. In other words,

Rate of heat generation $\propto$ volume

The rate of natural cooling is proportional to the surface area of the vessel and is represented by

Rate of cooling $\propto$ surface area

This shows that as the vessel size increases, the volume increases at a faster rate than the surface area. Thus, care must be taken before scale-up is implemented as even a small increase in volume can result in insufficient cooling and subsequent loss of control.

The chemical reaction rate is generally a function of a reactant concentration and temperature. In the case of an exothermic reaction, unless the heat of reaction is removed, an increase in temperature may result in a runaway reaction. For most homogeneous reaction, the rate is increased by a factor of 2 or 3 for every 10°C rise in temperature. This is represented by

$$Q_G \; \alpha e^{-E/RT} \text{ or} \tag{12-36}$$

$$Q_G = Ae^{-E/RT} \tag{12-37}$$

where A = pre-exponential (Arrhenius) factor
   E = activation energy
   Q_G = rate of heat generation
   R = universal gas constant
   T = temperature, K

The rate of heat generation depends on four main factors:

1. Reaction temperature
2. Whether the reactor operates in batch or semi-batch mode

3. Whether reactant accumulation occurs
4. The occurrence of thermal events (e.g., precipitation, decomposition, gas evolution, phase change)

If an external jacket or internal coil is used through which coolant flows at temperature T_C to remove the heat, then the rate of heat removal is represented by

Rate of heat removal $\propto (T_r - T_C)$

$$Q_c = UA(T_r - T_c) \qquad\qquad (12\text{-}38)$$

where A = surface area of heat transfer
$\quad\quad\;\; Q_c$ = rate of heat removal
$\quad\quad\;\; T_c$ = coolant temperture
$\quad\quad\;\; T_r$ = reactant temperture
$\quad\quad\;\; U$ = overall heat transfer coefficient

The rate of heat removal also depends on four main factors:

1. The effective surface area for heat transfer
2. The temperature difference between the reactant mass and coolant
3. The thermal properties of the reaction mixture, vessel walls, and coolant
4. The nature of the coolant

Hence, the rate of heat generation is exponential with reaction temperature T_r, but the heat removal rate is approximately linear because U is a weak function of T (Chapter 6). Therefore, a critical value of T_r will exist at which control is lost.

The rate of heat generation from an exothermic reaction is directly related to the mass of reactants involved. This and the ability to remove the heat, is an essential consideration in the scale-up of reactors. In a conventional vertical cylindrical reaction vessel of diameter D,

The volume (and hence mass) of reactants $\propto D^2 L$

where L is the height. In a convective heat transfer to a coolant in an external jacket, heat transfer area $\propto D \cdot L$

The ratio is

$$\frac{\text{heat transfer area}}{\text{potential heat release}} \; \alpha \; \frac{D \cdot L}{D^2 \cdot L}$$

$$\alpha \; \frac{1}{D} \tag{12-39}$$

As the size of the vessel increases, the ratio becomes increasingly unfavorable. Thus, there is a critical value for D for which an exothermic reaction can easily be controlled on a laboratory scale and may be hazardous on a larger scale.

Heat Transfer Film Coefficient

Empirical dimensionless group correlations have been used in the scale-up process. In particular, the correlation for the inside film heat transfer coefficient for agitated, jacketed vessels has been employed for the scale-up to a larger vessel. Reaction calorimeters are often used to give some indication of heat transfer coefficients compared to water in the same unit. Correlation for plant heat transfer is of the general form

$$\left(\begin{array}{c}\text{Nusselt} \\ \text{Number}\end{array}\right) = f(\text{Prandtl Number})^a (\text{Reynolds Number})^b \left(\begin{array}{c}\text{Viscosity} \\ \text{factor}\end{array}\right)^c$$

This is expressed mathematically as

$$\frac{h_i D}{k} = f \left(\frac{C_p \mu}{k}\right)^{0.33} \left(\frac{\rho N d^2}{\mu}\right)^{0.67} \left(\frac{\mu_b}{\mu_w}\right)^{0.14} \tag{12-40}$$

where h_i = inside film coefficient heat transfer coefficient (W/m²•K)
$\quad\;\; f$ = geometric factor
$\quad\;\; k$ = thermal conductivity of wall material (W/m•K)
$\quad\;\; \rho$ = density, kg/m³
$\quad\;\; N$ = speed of agitator (1•sec⁻¹)
$\quad\;\; d$ = diameter of agitator, m
$\quad\;\; D$ = reactor diameter, m
$\quad\;\; \mu$ = viscosity, Nsm⁻²
$\quad\;\; C_p$ = specific heat at constant pressure (J/kg•°C)

The overall heat transfer coefficient U is defined by

$$\frac{1}{U} = \frac{1}{h_i} + \frac{x}{k} + \frac{1}{h_o}$$ (12-41)

where h_o = jacket side heat transfer coefficient
    k = thermal conductivity
    x = wall thickness

The inside convective heat transfer coefficient h_i is the only element of the overall heat transfer coefficient U that varies with the agitation speed N. If heat is removed from an agitated reactor using an internal coil or external jacket, the overall heat transfer rate depends on the rotation speed of the agitator N and if the process side offers the major resistance. This is expressed by

$$U \propto (N)^{2/3}$$ (12-42)

Therefore, a failure of agitation may result in hot spots and a runaway reaction.

HAZARD AND OPERABILITY STUDIES (HAZOP)

The design and operation of a process plant form an integral part of safety and systematic procedures and should be employed to identify hazards and operability and, where necessary, should be quantified. During the design of a new plant, the hazard identification procedure is repeated at intervals. This is first performed on the pilot plant before the full-scale version as the design progresses. Potential hazards whose significance can be assessed with the help of experiments are often revealed by this study.

The hazard and operability study (HAZOP) identifies a potential hazard. It provides little information on risk and consequences or its seriousness. However, by assessing the potential hazard, sometimes the designer may decide that the consequences of the hazard are either trivial or unlikely to be ignored. In certain instances, the solution is obvious and the design is modified. A fault tree analysis is useful where the consequences of the hazard are severe, or where its causes are many. The fault tree indicates how various events or combinations of events can give rise to a hazard. It is employed to identify the most likely causes of the hazard and, thus, show where additional safety precautions will be most effective. Keltz [44] and Barton and Rogers [5] have provided excellent texts with case studies on HAZOP.

A HAZOP study is a structured review of the plant design and operating procedures. The main goals are to:

- Identify potential maloperations
- Assess their consequences
- Recommend corrective actions

The recommended actions may eliminate a potential cause or interrupt the consequences. They involve:

- Pipework or other hardware changes
- Changes in operating conditions
- More precise operating instructions
- Addition of alarms with prescribed operator responses
- Addition of automatic trip systems

The procedure for a HAZOP study is to apply a number of guide words to various sections of the process design intention. The design intention informs what the process is expected to carry out. Table 12-6 shows these guide words, and Figure 12-42 summarizes the entire procedure. Common property words are:

- Concentration
- React
- Flowrate or amount
- Separate
- Level
- Temperature
- Heat transfer
- Particle size
- Pressure

STUDY CO-ORDINATION

HAZOP is a structured review exercise performed by a team of between three and six people, one of whom acts as a chairman. Another member of the team acts as a secretary and records the results of the proceeding. A typical team comprises:

- *Chairman:* study leader, experienced in hazop.
- *Project or design engineer:* usually a mechanical engineer, responsible for keeping the costs within budget.

Table 12-6
List of guide words used for the HAZOP procedure

Guide word	Meaning	Explanation
No	Complete negation of the design intention.	Application to flow, concentration, react, heat transfer, separate and similar functions. No level means an empty vessel or a two-phase interface is lost.
More or less	Quantitative increase or decrease.	Applicable to all property words.
As well as	Design intention achieved together with something else.	Flow as well as describes contamination in a pipeline. React as well as covers side reactions.
Part of	Design intentions only partly achieved.	It is more precise to use the guide word "less" wherever possible. However, fluctuations in a property word are covered by part of.
Reverse	Logical opposite of the intention.	Applicable principally to flow.
Other than	Complete substitution.	Applicable where the wrong material flows in a line or the wrong reagent is charged to a reactor or the required reaction does not occur but others do.

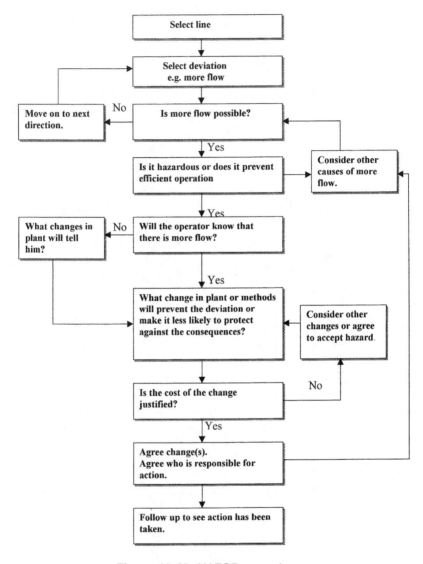

Figure 12-42. HAZOP procedure.

- *Process engineer:* usually the chemical engineer who developed the flowsheet.
- *Commissioning manager:* usually a chemical engineer, who will have to start up and operate the plant.
- *Control system design engineer:* modern plants contain sophisticated control and trip systems and HAZOPs often result in the addition of yet more instrumentation.

- *Research chemist:* if new chemistry is involved.
- *Independent team leader:* an expert in the HAZOP technique whose job is to ensure that the team follows the procedure.

The team should have a wide range of knowledge and experience. If a contractor designs the plant, then the HAZOP team should include people from both the contractor and client organizations. On a computer-controlled batch plant, the software engineer should be a member of the HAZOP team, which should include at least one other person who understands the computer logic.

In an existing plant, the team should include several people with experience of the plant. A typical team comprises:

- *Plant manager:* responsible for plant operation.
- *Process foreman:* knows what actually happens rather than what is supposed to happen.
- *Plant engineer:* responsible for mechanical maintenance such as testing of alarms and trips, as well as installation of new instruments.
- *Process investigation manager:* responsible for investigating technical problems and for transferring laboratory results to plant-scale operations.
- *Independent team leader.*

HAZOP OF A BATCH PROCESS

A HAZOP study of a batch processing plant involves a review of the process as a series of discrete stages, as the conditions in a single reactor change with time in a given cycle. Variations in the rate of change, as well as changes in the duration of settling, mixing, and reacting are important deviations. A HAZOP on a batch process includes the flowrates of service fluids—it is more common to look at the reactants in terms of amounts charged or discharged rather than flowrates.

Sequential operating instructions are applied to identify the design intent through various stages of the process. HAZOP guide words can then be applied to each design intent instead of the process line. For example, for the instruction "raise concentration or temperature to a specified value for a given time," the HAZOP would consider the consequences of not achieving or exceeding the desired concentration or temperature in the specified time. To illustrate further, if an instruction states that "1 tonne of A has to be charged to a reactor," then the team should consider deviations such as

```
DON'T CHARGE          A
CHARGE MORE           A
CHARGE LESS           A
CHARGE AS WELL AS     A
CHARGE PART OF        A (if A is a mixture)
CHARGE OTHER THAN     A
REVERSE CHARGE        A (that is, can flow occur from the
                         reactor to the A container?)
```

This can be the most serious deviation if

```
A IS ADDED EARLY
A IS ADDED LATE
A IS ADDED TOO QUICKLY
A IS ADDED TOO SLOWLY
```

Many accidents occur because process materials flow in the wrong direction. For example, ethylene oxide and ammonia were reacted to make ethanolamine. Some ammonia flowed from the reactor in the opposite direction, along the ethylene oxide transfer line into the ethylene oxide tank, past several non-return valves and a positive displacement pump. It got past the pump through the relief valve, which discharged into the pump suction line. The ammonia reacted with $30m^3$ of ethylene oxide in the tank, which ruptured violently. The released ethylene oxide vapor exploded causing damage and destruction over a wide area [5]. A hazard and operability study might have disclosed the fact that reverse flow could occur.

HAZAN

Hazard analysis (HAZAN) is a quantitative way of assessing the likelihood of failure. Other names associated with this technique are risk analysis, quantitative risk assessment (QRA), and probability risk assessment (PRA). Keltz [44] expressed the view that HAZAN is a selective technique while HAZOP can be readily applied to new design and major modification. Some limitations of HAZOP are its inability to detect every weakness in design such as in plant layout, or miss hazards due to leaks on lines that pass through or close to a unit but carry material that is not used on that unit. In any case, hazards should generally be avoided by changing the design. Assessing hazard by HAZAN or any other technique should always be the alternative choice.

A small team, similar to that used in HAZOP, carries out hazard analysis. The three steps in HAZAN are:

1. Estimate how frequently the incident will occur.
2. Estimate the consequences to employees, members of the public, plant, and profits.
3. Compare the results of the first two steps with a target or criterion and decide whether it is necessary to act to reduce the incident's frequency or severity.

FAULT TREE ANALYSIS

Fault tree analysis is used to assess the frequency of an incident. A fault tree is a diagram that shows how primary causes produce events, which can contribute to a particular hazard. There are several pathways in which a single primary cause can combine with other primary causes or events. Therefore, a single cause may be found in more than one hazard and may occur at different locations in the fault tree.

The graphical structure of the fault tree enables the primary causes and secondary events that produce the hazards to be combined. It is then possible to compare the relative contributions of the different events to the probability of the hazardous outcome by employing the probability of occurrence of causes and events on the fault tree.

Example 12-1

An 800-gal reactor containing a styrene mixture with a specific heat of 0.6 cal/gm °C has a 10-in. rupture disk and a vent line with equivalent length = 400. The vessel MAWP is 100 psig and the rupture disk set pressure is 20 psig. The styrene mixture had a self-heat rate of 60°C/min at 170°C as it is tempered in a DIERS venting test. Determine the allowable reactor mixture charge to limit the over-pressure to 10% over the set pressure.

Solution

Using Equation 12-25 gives

$$M_O = \frac{\left(D_p\right)^2 \left(\Delta P_S\right)}{2.769 \left(\dfrac{dT}{dt}\right)_s} \cdot \left(\frac{T_S}{C_p}\right)^{0.5}$$ (12-25)

where M_O = allowable mass of the reactor mixture charge (kg) to limit the venting overpressure to P_P (psig)

D_p = 10 inches

P_S = relief device set pressure = 20 psig

P_P = maximum venting pressure = 1.10(20) = 22.0 psig

ΔP_S = allowable venting overpressure (psi) (i.e., the maximum venting pressure minus the relief device set pressure); (22.0 − 20) = 2 psi

T_S = equilibrium temperature corresponding to the vapor pressure where the vapor pressure is the relief device set pressure (K); = 170°C (170 + 273.15 = 443.15 K)

$(dT/dt)_s$ = reactor mixture self-heat rate (°C/min) at temperature T_S (K); = 60°C/min

C_P = specific heat of the reactor mixture (cal/g-K) = 0.6 cal/gm•K

M_O = [(10²)(2.0)]/[(2.769)(60)] • (443.15/0.6)$^{0.5}$

M_O = 32.72 kg

The density of styrene is 0.9 g/cm³. Therefore, the quantity charged in the reactor is:

$$\frac{(32.72)\,kg\left(10^3\right)g}{1\,kg} \times \frac{1\,cm^3}{0.9\,g} \times \frac{1l}{10^3\,cm^3} \times \frac{1\,gal}{3.785l} = 9.6 \text{ gal } (\approx 10 \text{ gal})$$

The reactor charge is quite small for an 800-gal reactor. If the pressure is allowed to rise to 10% above MAWP, then

ΔP_S = 1.1(100) − 20

= 90 psi

The amount charged is:

$$M_O = \frac{\left(10^2\right)(90)}{(2.769)(60)} \cdot \left(\frac{443.15}{0.6}\right)^{0.5}$$

$M_O = 1,472.2$ kg

$$= \frac{1,472.2}{(0.9)(3.785)} \text{ gal}$$

$$= 432 \text{ gal}$$

This shows that a much larger initial charge will be required. An Excel spreadsheet program (EXAMPLE12-1.xls) was developed for this example.

Example 12-2

A 700-gal reactor with a net volume of 850 gal containing an organic mixture has a 10-in. rupture disk and a vent line with an equivalent length L/D = 400. The vessel MAWP is 100 psig and the rupture disk set pressure is 20 psig. A venting test that was performed showed that the reaction was "gassy." The test mass was 30 g, the peak rate of pressure rise was 550 psi/min, and the maximum test temperature was 300°C. Determine the allowable reactor mixture charge to limit the overpressure to 10% of the MAWP.

Solution

Using Equation 12-30 for a "gassy" reaction, the charge amount is:

$$M_O = \left[V_P \, \Delta P_P \left\{ \frac{(D_P)^2 (M_S)(P_P)}{(2.07)(T_T)(dP/dt)} \right\}^2 \right]^{1/3} \qquad (12\text{-}30)$$

where V_P = vessel total volume (gal) = 850 gal

P_P = maximum allowable venting pressure (psia) = 1.10 (100) + 14.7 = 124.7 psia

M_S = sample mass used in a DIERS test or equivalent test (g) = 30 g

dP/dt = pressure rise in test (psi/sec) = 550 psi/min = 9.17 psi/sec

T_T = maximum temperature in test (K) = 300°C = 573.15 K

$\Delta P_P = P_P - P_{amb}$ = 124.7 – 14.7 = 110 psi

P_{amb} = ambient pressure at the end of the vent line = 14.7 psia

$$M_O = \left[(850)(110) \left\{ \frac{(10)^2(30)(124.7)}{(2.07)(573.15)(9.17)} \right\}^2 \right]^{1/3}$$

M_O = 479.9 kg

The charge amount to the reactor is 480 kg. A Microsoft Excel spreadsheet (EXAMPLE12-2.xls) was developed for this example.

Example 12-3

Determine the vent size for a vapor pressure system using the following data and physical properties.

Reactor Parameters:

Volume, V=10 m^3
Mass, M = 8,000 kg
Vent opening pressure = 15 bara (217.5 psia)
Temperature at set pressure T_s = 170°C (443.15 K)
Overpressure allowed above operating pressure = 10%
= 1 bar (10^5 Pa)

Material Properties:

Specific heat, C_p=3,000 J/kg•K
Slope of vapor pressure and temperature curve dP/dT=20,000 Pa/K

Rate of Reaction

ΔT = dP/20,000 = 10^5/20,000 = 5 K
Rate of set pressure, $(dT/dt)_s$ = 6.0 K/min
Rate of maximum pressure = 6.6 K/min
Average rate = 6.3 K/min = 0.105 K/s

Solution

Using Fauske's nomograph from Figure 12-35 at a self-heat rate of 6.3 K/min and a set pressure of 217.5 psia, the corresponding vent size area per 1,000 kg of reactants = 0.0008 m^2.
The vent area of 8,000 kg reactants = 0.0064 m^2. The vent size

$$d = \left(\frac{4\ \text{Area}}{\pi}\right)^{0.5} = 90.27\,\text{mm}\,(3.6\,\text{in.})$$

If Figure 12-35 is applicable for F = 0.5, then for F = 1.0, the area is

$$A = \left(0.0064\,\text{m}^2\right)\left(\frac{0.5}{1.0}\right) = 0.0032\,\text{m}^2$$

The area assumes a 20% absolute overpressure. The result can be adjusted for other overpressures by multiplying the area by a ratio of 20/(new absolute percent overpressure).

Using the Leung's method and assuming L/D = 0 and F = 1.0, two-phase mass flux from Equation 12-14 gives

$$G = (0.9)(1.0)(20{,}000)\left(\frac{1.0 \times 443.15}{3{,}000}\right)^{0.5}$$
$$= 6{,}918.1\,\frac{\text{kg}}{\text{ms}^2}$$

Rate of heat generation: q

$$q = C_p \frac{dT}{dt}\left(\frac{\text{J.K}}{\text{kg}\bullet\text{K}\bullet\text{s}}\right) = (3{,}000)(0.105)\frac{\text{W}}{\text{kg}} = 315\frac{\text{W}}{\text{kg}}$$

From Equation 12-20, the vent area A is:

$$A = \frac{8{,}000 \times 315}{6{,}918.1\left\{\left(\dfrac{10}{8{,}000} \times 443.15 \times 20{,}000\right)^{0.5} + (3{,}000 \times 5.0)^{0.5}\right\}^2}$$

$A = 7.024 \times 10^{-3}\ \text{m}^2$

D = 94.6 mm (3.72 in.)

The vent size is about 4.0 in.

Using the Fauske's method and assuming $\alpha = 1.0$, $\alpha_O = 0.0$, and F = 1.0, $\Delta P = 1$ bar $= 10^5$ N/m^2.

From Equation 12-25, the vent area is

$$A = \frac{1}{2} \cdot \frac{(8,000)(0.105)(1-0)}{(1.0)\left(\dfrac{443.15}{3,000}\right)^{0.5}(10^5)(1-0)}$$

A= 10.927×10^{-3} m^2

D = 117.95 mm (4.64 in.)

Example 12-4

A 3,500-gal reactor with styrene monomer undergoes adiabatic polymerization after being heated inadvertently to 70°C. The maximum allowable working pressure (MAWP) of the reactor is 5 bar absolute. Determine the relief vent diameter required. Assume a set pressure of 4.5 bara and a maximum pressure of 5.4 bara. Other data and physical properties are given as follows [12]:

Data:

Volume (V)	13.16 m^3 (3,500 gal)
Reaction mass (m_o), kg	9,500
Set temperature (T_s)	209.4°C = 482.5 K

Data from VSP:

Maximum temperature (T_m)	219.5°C = 492.7 K
$\left(\dfrac{dT}{dt}\right)_s$	29.6°C/min = 0.493 K/s (sealed system)
$\left(\dfrac{dT}{dt}\right)_m$	39.7°C/min = 0.662 K/s

Physical property data:

	4.5 bar set	5.4 bar set
V_f, m^3/kg	0.001388	0.001414
V_g, m^3/kg; ideal gas assumed	0.08553	0.07278
C_P, kJ/kg•K	2.470	2.514
ΔH_V, kJ/kg	310.6	302.3

Solution

The heating rate q is determined by Equation 12-22

$$q = \frac{1}{2}C_V\left[\left(\frac{dT}{dt}\right)_s + \left(\frac{dT}{dt}\right)_m\right]$$ (12-22)

Assuming $C_V = C_P$:

$$q = \frac{1}{2}(2.470\,\text{kJ/kg} \cdot \text{K})[0.493 + 0.662](\text{K/s})$$

$$= 1.426\,\frac{\text{kJ}}{\text{kg} \cdot \text{s}}$$

The mass flux through the relief G is given by Equation 12-12, assuming L/D = 0 and $\psi = 1.0$

$$G = \frac{Q_m}{A} = 0.9\psi\frac{\Delta H_V}{\upsilon_{fg}}\left(\frac{g_C}{C_P T_S}\right)^{0.5}$$ (12-12)

$$G = \frac{(0.9)(1.0)(310,660\,\text{J/kg})[1(\text{Nm})/\text{J}]}{(0.08553 - 0.001388)\,\text{m}^3/\text{kg}}\left\{\frac{\left[1(\text{kg}\,\text{m/s}^2)/\text{N}\right]}{(2,470\,\text{J/kg}\,\text{K})(482.5\,\text{K})[1(\text{Nm})/\text{J}]}\right\}^{0.5}$$

$$= 3,043.81\,\frac{\text{kg}}{\text{m}^2\text{s}}$$

The relief area is determined by Equation 12-21

$$A = \frac{M_O q}{G\left[\left(\frac{V}{M_O}\frac{\Delta H_V}{\upsilon_{fg}}\right)^{0.5} + (C_V\,\Delta T)^{0.5}\right]^2}$$ (12-21)

The change in temperature ΔT is $T_m - T_s$

$\Delta T = 492.7 - 482.5 = 10.2 \text{ K}$

$$A = \frac{(9,500\,\text{kg})(1,426\,\text{J/kg}\bullet\text{s})\left[1(\text{Nm})/\text{J}\right]}{(3,043.81\,\text{kg}/\text{m}^2\text{s})\left[\begin{array}{c}\left\{\left(\dfrac{13.16\,\text{m}^3}{9,500\,\text{kg}}\right)\left(\dfrac{310,660\,\text{J/kg}\left[1(\text{Nm})/\text{J}\right]}{0.08414\,\text{m}^3/\text{kg}}\right)\right\}^{0.5}\\[4mm]+\left[\left(2,470\dfrac{\text{J}}{\text{kg}\bullet\text{K}}\right)(10.2\,\text{K})\left[1(\text{Nm})/\text{J}\right]\right]^{0.5}\end{array}\right]^2}$$

$A = 0.0839 \text{ m}^2$

The required relief diameter is

$$d = \left(\frac{4A}{\pi}\right)^{0.5}$$

$$= \left(\frac{(4)(0.0839\,\text{m}^2)}{3.14}\right)^{0.5} = 0.327\,\text{m}$$

$d = 327 \text{ mm } (12.87 \text{ in.})$

The following considers a situation that involves all vapor relief. The size of a vapor phase rupture disk required is determined by assuming that all of the heat energy is absorbed by the vaporization of the liquid. At the set temperature, the heat release rate q is

$$q = C_V\left(\frac{dT}{dt}\right)_s = \left(2.470\frac{\text{kJ}}{\text{kg}\bullet\text{K}}\right)\left(0.493\frac{\text{K}}{\text{s}}\right)$$

$$q = 1.218\frac{\text{kJ}}{\text{kg}\bullet\text{s}}$$

The vapor mass flow through the relief is then

$$Q_m = \frac{q\,m_o}{\Delta H_v}$$

$$= \frac{(1,218\,J/kg)(9,500\,kg)}{(310,660\,J/kg)}$$

$$= 37.25\,\frac{kg}{s}$$

The required relief area for vapor is determined by

$$A = \frac{Q_m}{C_o P}\left\{\frac{R_g T}{\gamma g_c M}\left(\frac{2}{\gamma+1}\right)^{\left(\frac{\gamma+1}{\gamma-1}\right)}\right\}^{0.5}$$

where Q_m = discharge mass flow, kg/s
$\quad\quad C_o$ = discharge coefficient
$\quad\quad A$ = area of discharge
$\quad\quad P$ = absolute upstream pressure
$\quad\quad \gamma$ = heat capacity ratio for the gas
$\quad\quad g_c$ = gravitational constant
$\quad\quad M$ = molecular weight of the gas (M = 104 for styrene)
$\quad\quad R_g$ = ideal gas constant (8,314 Pa m³/kg mol K)
$\quad\quad T$ = absolute temperature of the discharge

Assuming $C_o = 1$ and $\gamma = 1.32$

$$A = \frac{(37.25\,kg/s)}{(1.0)(4.5\,bar)(10^5\,Pa/bar)\left[1(N/m^2)/Pa\right]}$$

$$\times\left\{\frac{(8,314\,Pa\,m^3/kg-mol)(482.5\,K)\left[1(N/m^2)/Pa\right]}{(1.32)\left[1(kg\,m/s^2)/N\right](104\,kg/kg-mol)}\right\}^{0.5}\times\left\{\left(\frac{2}{2.32}\right)^{\left(\frac{2.32}{-0.32}\right)}\right\}^{0.5}$$

$$A = 0.0242\ m^2$$

The relief diameter = 0.176 m

d = 176 mm (6.9 in.)

Thus, the size of the relief device is significantly smaller than for two-phase flow. Sizing for all vapor relief will undoubtedly give an incorrect result, and the reactor would be severely tested during this runaway occurrence. Table 12-7 gives the results of the VENT software program of Example 12-4.

Example 12-5

---

(1) Show that in a continuous flow stirred tank reactor the critical temperature above which heat is generated faster than it can be removed is of the form

$$T_M - T_C = \frac{RT_M^2}{E}$$

where T_M = the temperature at which heat is produced faster than it can be removed, K

T_C = coolant temperature, K

Table12-7
Vent sizing for two-phase (runaway reactions)
flow vent sizing for a tempered system

MASS OF REACTANT IN THE VESSEL, kg:	9500.000
REACTOR VOLUME, m^3:	13.160
SLOPE OF VAPOR PRESSURE TEMPERATURE CURVE: N/m^2.K:	7649.734
LATENT HEAT OF VAPORIZATION, kJ/kg.:	.31060000E+03
SPECIFIC HEAT CAPACITY OF LIQUID, kJ/kg.K:	2.470
SET TEMPERATURE, oC:	209.400
MAXIMUM TEMPERATURE, oC:	219.500
SPECIFIC VOLUME OF GAS, m^3/kg:	.085530
SPECIFIC VOLUME OF LIQUID, m^3/kg:	.001388
DIFFERENCE BETWEEN GAS AND LIQUID SPECIFIC VOLUMES m^3/kg:	.084142
SELF HEAT RATE AT SET TEMPERATURE, K/s:	.4930
SELF HEAT RATE AT MAXIMUM TEMPERATURE, K/s:	.6620
MASS FLUX PER UNIT AREA, kg/m^2.s:	3043.067
HEAT RELEASE RATE PER UNIT MASS, kW/kg:	1.426
VENT AREA, m^2:	.84579520E-01
VENT SIZE, m:	.32816150E+00
VENT SIZE, mm:	.32816150E+03
VENT SIZE, inch:	.12919750E+02

E = activation energy, cal/gmol
R = gas law constant, cal/gmol•K

(2) If the reaction has an activation energy of 30,000 cal/gmol and the cooling water is 15°C, determine the maximum temperature the reaction can operate under without reaching the runaway condition.

Solution

(1) Assuming a first order exothermic reaction (A $\xrightarrow{\ k\ }$ Products) in a CFSTR with the inlet and outlet flows controlled at steady and equal rates, the rate at which energy is generated is given by

$$Q_g = \left(\frac{-\Delta H_R}{a}\right)(-r_A) V_R \qquad (6\text{-}115)$$

The rate of reaction is given by

$$(-r_A) = kC_A = k_O e^{\frac{-E}{RT}} C_A \qquad (12\text{-}43)$$

where Q_g = rate of heat generation, cal/sec
$(-\Delta H_R/a)$ = heat of reaction, cal/g-mol
$\quad k_O$ = frequency factor, sec^{-1}
$\quad C_A$ = concentration of reaction, g-mol/dm^3
$\quad V_R$ = reaction volume, dm^3
$\quad E$ = activation energy, cal/gmol
$\quad R$ = gas constant, 1.987 cal/gmol•K
$\quad T$ = absolute temperature, K

Substituting Equation 12-43 into Equation 6-115 gives

$$Q_g = \left(\frac{-\Delta H_R}{a}\right) k_O C_A e^{\frac{-E}{RT}} V_R \qquad (12\text{-}44)$$

Because the reaction is exothermic, heat must be removed from the reactor to keep the temperature from increasing out of control. The heat transfer rate of removal is:

$$Q_r = UA(T - T_C) \qquad (12\text{-}45)$$

where Q_r = rate of heat removal, cal/sec
U = overall heat transfer coefficient, cal/cm^2•sec•K
A = heat transfer area, cm^2
T_C = coolant temperature, K

If heat can be removed as fast as it is generated by the reaction, the reaction can be kept under control. Under steady state operating conditions, the heat transfer rate will equal the generation rate (see Figure 6-26). If the heat removal rate Q_r is less than the heat generation rate Q_g (e.g., a condition that may occur because of a cooling water pump failure), a temperature rise in the reactor is experienced. The net rate of heating of the reactor content is the difference between Equations 12-44 and 12-45.

$$Q_n = Q_g - Q_r \tag{12-46}$$

$$Q_n = \left(\frac{-\Delta H_R}{a} \right) k_O C_A e^{\frac{-E}{RT}} V_R - UA(T - T_C) \tag{12-47}$$

If Q_n is positive, the reaction will have an increasing temperature. If the rate of increase of Q_n is positive (i.e., $dQ_n/dT > 0$), the reaction has the potential to accelerate and become uncontrollable. From Equation 12-47,

$$C_1 = \left(\frac{-\Delta H_R}{a} \right) k_O C_A V_R$$

and $C_2 = UA$. Equation 12-47 then becomes

$$Q_n = C_1 e^{\frac{-E}{RT}} - C_2(T - T_C) \tag{12-48}$$

If $Q_n > 0$ and $dQ_n/dT > 0$, the reaction has the potential to runaway. Therefore,

$$\frac{dQ_n}{dT} = C_1 \left(\frac{E}{RT^2} \right) e^{\frac{-E}{RT}} - C_2 > 0 \tag{12-49}$$

At the point where $Q_g = Q_r$ (i.e., when the reaction begins to go out of control), $T = T_m$, the maximum temperature for reaction control, which yields,

$$C_1 e^{\frac{-E}{RT_M}} = C_2 (T_M - T_C) \tag{12-50}$$

Therefore,

$$C_2 (T_M - T_C) \left(\frac{E}{RT_M^2} \right) = C_2 \tag{12-51}$$

Rearranging Equation 12-51 gives

$$T_M - T_C = \frac{RT_M^2}{E} \tag{12-52}$$

(2) The maximum temperature T_M in which the reaction can operate without reaching the runaway condition is given by Equation 12-52. Thus, for E = 30,000 cal/gmol, T_C = 15°C = 288.15 K and R = 1.987 cal/gmol•K,

$$T_M - 288.15 = \frac{1.987 \, T_M^2}{30,000} \left\{ \frac{cal}{gmol \cdot K} \cdot K^2 \cdot \frac{gmol}{cal} \right\} \text{ or}$$

$$T_M^2 - 15,098 \, T_M + 4,350,528 = 0 \tag{12-53}$$

Equation 12-53 is a quadratic equation of the form $aX^2 + bX + c = 0$, and the roots are:

$$X = \frac{-b \pm \sqrt{b^2 - 4ac}}{2a}$$

The plausible root is

$$X = \frac{-b - \sqrt{b^2 - 4ac}}{2a}$$

The maximum temperature T_M = 293.87 K (20.72°C). Therefore, for any temperature greater than 293.87 K or 20.72°C, the reactor will begin to generate heat faster than the heat can be removed and the reaction can runaway.

An approximate 6°C-temperature rise is assumed safe. To ensure safe operation, the followings actions should be taken:

- Reduce the feed rate of the reactant.
- Provide adequate emergency venting for the reactor.
- Inject an inhibitor into the reactor or quench the reaction when it begins to go out of control.
- Add diluents to reduce the reaction rate. These can be separated from the products afterwards.
- Increase the size of the heat exchanger so that all the energy can be removed.

GLOSSARY

Accumulation: The buildup of unreacted reagent or intermediates, usually associated with reactant added during semi-batch operations.

Activation energy E_a: The constant E_a in the exponential part of the Arrhenius equation, associated with the minimum energy difference between the reactants and an activated complex (transition state that has a structure intermediate to those of the reactants and the products), or with the minimum collision energy between molecules that is required to enable a reaction to occur.

Adiabatic: A system condition in which no heat is exchanged between the chemical system and its environment.

Adiabatic induction time: Induction period or time to an event (spontaneous ignition, explosion, etc.) under adiabatic conditions, starting at operating conditions.

Adiabatic temperature rise: Maximum increase in temperature that can be achieved. This increase occurs when the substance or reaction mixture decomposes or reacts completely under adiabatic conditions. The adiabatic temperature rise follows from:

$$\Delta T_{adia} = x_0 \, (\Delta H_{RX})/C\phi$$

where x_o = initial mass fraction; ΔH_{RX} = heat of reaction, J/kg; C = liquid heat capacity J/kg•K; ϕ = dimensionless thermal inertial factor (Phi-factor).

Autocatalytic reaction: A reaction, the rate of which is increased by the catalyzing effect of its reaction products.

Autoignition temperature: The autoignition temperature of a substance, whether solid, liquid, or gaseous, is the minimum temperature required to initiate or cause self-sustained combustion (e.g., in air, chlorine, or other oxidant) with no other source of ignition.

Back pressure: The static pressure existing at the outlet of a pressure relief device as a result of the pressure in the discharge system. It is the sum of the superimposed and built-up back pressure.

Blowdown: The difference between actual popping pressure of a pressure relief valve and actual re-seating pressure expressed as a percentage of set pressure.

Blowdown pressure: The value of decreasing inlet static pressure at which no further discharge is detected at the outlet of a safety relief valve of the resilient disk type after the valve has been subjected to a pressure equal to or above the popping pressure.

Boiling-Liquid-Expanding-Vapor-Explosive (BLEVE): The violent rupture of a pressure vessel containing saturated liquid/vapor at a temperature well above its atmospheric boiling point. The sudden decrease in pressure results in explosive vaporization of a fraction of the liquid and a cloud of vapor and mist, with accompanying blast effects. The resulting flash vaporization of a large fraction of the liquid produces a large cloud. If the vapor is flammable and an ignition source is present at the time of vessel rupture, the vapor cloud burns in the form of a large rising fireball.

Built-up back pressure: Pressure existing at the outlet of a pressure relief device caused by flow through that particular device into a discharge system.

Burst pressure: The value of inlet static pressure at which a rupture disc device functions.

Chatter: The abnormal rapid, reciprocating motion of the movable parts of a pressure relief valve in which the disk contacts the seat.

Combustible: A term used to classify certain liquids that will burn on the basis of flash points. Both the National Fire Protection Association (NFPA) and the Department of Transportation (DOT) define "combustible liquids" as having a flash point of 100°F (37.8°C) or lower.

Condensed phase explosion: An explosion that occurs when the fuel is present in the form of a liquid or solid.

Confined explosion: An explosion of a fuel-oxidant mixture inside a closed system (e.g., a vessel or building).

Confined Vapor Cloud Explosion (CVCE): A condensed phase explosion occurring in confinement (equipment, building and/or congested surroundings). Examples of CVCE include explosions in vessels and pipes and processing or storing reactive chemicals at elevated conditions. The excessive buildup of pressure in the confinement results in this type of explosion leading to high overpressure, shock waves, and heat load (if the chemical is flammable and ignites). The fragments of exploded vessels and other objects hit by blast waves, become airborne, and act as missiles.

Containment: A physical system in which no reactants or products are exchanged between the system and its environment under any conditions.

Decomposition energy: The maximum amount of energy that can be released upon decomposition. The product of decomposition energy and total mass is an important parameter for determining the effects of a sudden energy release (e.g., in an explosion).

Deflagration: The chemical reaction of a substance in which the reaction front advances into the unreacted substance at less than sonic velocity. Where a blast wave is produced that has the potential to cause damage, the term "explosive deflagration" is used.

Detonation: A release of energy caused by an extremely rapid chemical reaction of a substance in which the reaction front advances into the unreacted substance at equal to or greater than sonic velocity.

Design Institute for Emergency Relief Systems (DIERS): Institute under the auspices of the American Institute of Chemical Engineers funded to investigate design requirements for vent lines in the case of two-phase venting.

Disk: Is the pressure containing movable element of a pressure relief valve that effects closure.

Dow Fire and Explosion Index (F&EI): A method (developed by Dow Chemical Company) for ranking the relative fire and explosion risk associated with a process. Analysts calculate various hazard and explosion indexes using material characteristics and process data.

Exotherm: A reaction is called exothermic if energy is released during the reaction.

Explosion: A release of energy that causes a pressure discontinuity or blast wave. Explosions in the storage or process units can be categorized in three main groups, according to mode of occurrence and damage potential. These explosions are initiated either by the thermal stratification of the liquid and vapor or by such high explosion shock waves, which have sufficient strength to rupture the reaction/storage vessels or conduits. An explosion may or may not be accompanied with fire; it depends on the type of explosion and the chemical involved in the explosion.

Explosion rupture disk device: A rupture disk device designed for use at high rates of pressure rise.

Fail-safe: Design features that provide for the maintenance of safe operating conditions in the event of a malfunctioning control device or an interruption of an energy source (e.g., direction of failure of a motor-operated valve on loss of motor power).

Failure: An unacceptable difference between expected and observed performance.

Fire point: The temperature at which a material continues to burn when the ignition source is removed.

Flammability limits: The range of gas or vapor compositions in air that will burn or explode if a flame or other ignition source is present. *Important:* The range represents an unsafe gas or vapor mixture with air that may ignite or explode. Generally, the wider the range the greater the fire potential.

Flammable: A "flammable liquid" is defined as a liquid with a flash point below 100°F (37.8°C). Flammable liquids provide ignitable vapor at room temperatures and must be handled with caution. Flammable liquids are Class 1 liquids and are subdivided as follows:

Class 1A: Those having flash points below 73°F and having a boiling point below 100°F.

Class 1B: Those having flash points below 73°F and having a boiling point at or above 100°F.

Flares: Flares are used to burn the combustible or toxic gas to produce combustion products, which are neither toxic nor combustible. The diameter of a flare must be suitable to maintain a stable flame and prevent a blowdown (when vapor velocities are greater than 20% of the sonic velocity).

Flash fire: The combustion of a flammable vapor and air mixture in which flame passes through that mixture at less than sonic velocity, such that negligible damaging overpressure is generated.

Flash point: The lowest temperature at which vapors above a liquid will ignite. The temperature at which vapor will burn while in contact with an ignition source, but which will not continue to burn after the ignition source is removed.

Gassy system: In gassy systems, the pressure is due to a permanent gas that is generated by the reaction.

Hazard: An inherent chemical or physical characteristic that has the potential for causing damage to people, property, or the environment.

Hazard analysis: The identification of undesired events that lead to the materialization of a hazard, the analysis of the mechanisms by which these undesired events could occur, and usually the estimation of the consequences.

Hazard and Operability Study (HAZOP): A systematic qualitative technique to identify process hazards and potential operating problems using a series of guide words to study process deviations. A HAZOP is used to question every part of the process to discover what deviations from the beginning of the design can occur and what their causes and consequences may be. This is done systematically by applying suitable guide words. This is a systematic detailed review technique for both batch and continuous plants, which can be applied to new or existing processes to identify hazards.

Hazardous chemical reactivity: Any chemical reaction with the potential to exhibit an increase in temperature and/or pressure too high to be absorbed by the environment surrounding the system. These include both reactive and unstable materials.

Hybrid mixture: A suspension of dust in air/vapor. Such mixtures may be flammable below the lower explosive limit of the vapor and can be ignited by low energy sparks.

Hybrid system: Hybrid systems are those in which the total pressure is due to both vapor pressure and permanent gas.

Inherently safe: A system is inherently safe if it remains in a non-hazardous situation after the occurrence of non-acceptable deviations from normal operating conditions.

Inhibition: A protective method where the reaction can be stopped by an addition of another material.

Interlock system: A system that detects out-of-limits or abnormal conditions or improper sequences and either halts further action or starts corrective action.

Isothermal: A system condition in which the temperature remains constant. This implies that temperature increases and decreases that would otherwise occur are compensated for by sufficient heat exchange with the environment of the system.

Likelihood: A measuring of the expected frequency which an event occurs. This may be expressed as a frequency (e.g., events per year), a probability of occurrence during a time interval (e.g., annual probability), or a conditional probability (e.g., probability of occurrence, given that a precursor event has occurred).

Lower Explosive Limit (LEL) or Lower Flammable Limit (LFL): The lowest concentration of a vapor or gas (the lowest percentage of the substance in air) that will produce a flash of fire when an ignition source (heat, arc, or flame) is present.

Maximum Allowable Working Pressure (MAWP): The maximum allowed pressure at the top of the vessel in its normal operating position at the operating temperature specified for that pressure.

Mitigation: The lessening of the risk of an accidental event. A sequence of action on the source in a preventive manner by reducing the likelihood of occurrence of the event, or in a protective manner by reducing the magnitude of the event and for the exposure of local persons or property.

Onset temperature: The temperature at which the heat released by a reaction can no longer be completely removed from the reaction vessel and, consequently, results in a detectable temperature increase. The onset temperature depends on detection sensitivity, reaction kinetics, vessel size, and on cooling, flow, and agitation characteristics.

Oxidant: Any gaseous material that can react with a fuel (either gas, dust, or mist) to produce combustion. Oxygen in air is the most common oxidant.

Overpressure: A pressure increase over the set pressure of the relief device, usually expressed as a percentage of gauge set pressure.

Phi-factor, ϕ: A correction factor that is based on the ratio of the total heat capacity (mass × specific heat) of a vessel and the total heat capacity of the vessel contents.

$$\phi = \frac{\text{Heat capacity of sample} + \text{Heat capacity of vessel}}{\text{Heat capacity of sample}}$$

The ϕ factor enables temperature rises to be corrected for heat lost to the container or vessel. The ϕ factor approaches the value of one for large vessels and for extremely low-mass vessels.

Pressure relief device: A device designed to open to prevent a rise of internal fluid pressure in excess of a specified value due to exposure to emergency or abnormal conditions. It may also be designed to prevent excessive internal vacuum. It may be a pressure relief valve, a non-reclosing pressure relief device, or a vacuum relief valve.

Process safety: A discipline that focuses on the prevention of fires, explosions, and accidental chemical releases at chemical process facilities. Excludes classic worker health and safety issues involving working surfaces, ladders, protective equipment, etc.

Purge gas: A gas that is continuously or intermittently added to a system to render the atmosphere nonignitable. The purge gas may be inert or combustible.

Quenching: Rapid cooling from an elevated temperature (e.g., severe) cooling of the reaction system in a short time (almost instantaneously), which "freezes" the status of a reaction and prevents further decomposition.

Relieving pressure: A set pressure plus overpressure.

Risk: The likelihood of a specified undesired event occurring within a specified period or in specified circumstances.

Risk analysis: A methodical examination of a process plant and procedure that identifies hazards, assesses risks, and proposes measures that will reduce risks to an acceptable level.

Runaway: A thermally unstable reaction system, which shows an accelerating increase of temperature and reaction rate. The runaway can finally result in an explosion.

Rupture disk device: A non-reclosing pressure relief device actuated by inlet static pressure and designed to function by the bursting of a pressure containing disk.

Safety relief valve: A pressure relief valve characterized by rapid opening pop action or by opening generally proportional to the increase in pressure over the opening pressure. It may be used for either compressible or incompressible fluids, depending on design, adjustment, or application.

Set pressure: The inlet pressure at which the relief device is set to open (burst).

Stagnation pressure: The pressure that would be observed if a flowing fluid were brought to rest along an isentropic path.

Superimposed back pressure: The static pressure existing at the outlet of a pressure relief device at the time the device is required to operate. It is the result of pressure in the discharge system from other sources.

Temperature of no-return: Temperature of a system at which the rate of heat generation of a reactant or decomposition slightly exceeds the rate of heat loss and possibly results in a runaway reaction or thermal explosion.

Thermally unstable: Chemicals and materials are thermally unstable if they decompose, degrade, or react as a function of temperature and time at or about the temperature of use.

Thermodynamic data: Data associated with the aspects of a reaction that are based on the thermodynamic laws of energy, such as Gibbs' free energy, and the enthalpy (heat) of reaction.

Time to maximum rate (TMR): The time taken for a material to self-heat to the maximum rate of decomposition from a specific temperature.

Unconfined Vapor Cloud Explosion (UCVE): Occurs when a sufficient amount of flammable material (gas or liquid having high vapor pressure) is released and mixes with air to form a flammable cloud such that the average concentration of the material in the cloud is higher than the lower **limit of explosion**. The resulting explosion has a high potential of damage as it occurs in an open space covering large areas. The flame speed may accelerate to high velocities and produce significant blast overpressure. Vapor cloud explosions in densely packed plant areas (pipe lanes, units, etc.) may show accelerations in flame speeds and intensification of blast.

Upper Explosive Limit (UEL) or Upper Flammable Limit (UFL): The highest concentration of a vapor or gas (the highest percentage of the substance in the oxidant) that will produce a flash or fire when an ignition source (heat, arc, or flame) is present.

Vapor specific gravity: The weight of a vapor or gas compared to the weight of an equal volume of air, which expresses of the density of the vapor or gas. Materials lighter than air have vapor specific gravity less than 1.0 (e.g., acetylene, methane, hydrogen). Materials heavier than air (e.g., ethane, propane, butane, hydrogen, sulphide, chlorine, sulfur dioxide) have vapor specific gravity greater than 1.0.

Vapor pressure: The pressure exerted by a vapor above its own liquid. The higher the vapor pressure, the easier it is for a liquid to evaporate and fill the work area with vapors that can cause health or fire hazards.

Vapor pressure system: A vapor pressure system is one in which the pressure generated by the runaway reaction is solely due to the increasing vapor pressure of the reactants, products, and/or solvents as the temperature rises.

Venting (emergency relief): Emergency flow of vessel contents out of the vessel. The pressure is reduced by venting, thus avoiding a failure of the vessel by overpressurization. The emergency flow can be one-phase or multiphase, each of which results in different flow and pressure characteristics. Multiphase flow (e.g., vapor and or gas/liquid flow) requires substantially larger vent openings than single-phase vapor (and/or gas) flow for the same depressurization rate.

REFERENCES

1. Barton, J. A. and Nolan, P. F. "Incidents in the Chemical Industry due to Thermal-runaway Chemical Reactions, Hazards X; Process

Safety in Fine and Speciality Chemical Plants," IChemE, Symp., Ser. No. 115 pp. 3–18, 1989.

2. Shabica, A. C., "Evaluating the Hazards in Chemical Processing," *Chem. Eng. Prog.,* Vol. 59, No. 9, pp. 57–66, 1963.

3. Mumford, C. J., "PSI Chemical Process Safety," Occupational Health and Safety Training Unit, University of Portsmouth Enterprise Ltd., version 3, 1993.

4. Wells, G., *Major Hazards and Their Managements,* Institution of Chemical Engineer (IChemE), 1997.

5. Barton, J. and Rogers, R., *Chemical Reaction Hazards,* 2nd ed., Institution of Chemical Engineers, 1997.

6. Townsend, D. I. and Tou, J. "Thermal Hazard Evaluation by an Accelerating Rate Calorimeter," *Thermochimica Acta,* 27, pp. 1–30, 1980.

7. Grolmes, M. A., "Pressure Relief Requirement for Organic Peroxides and Other Related Compounds," Int. Symp. on Runaway Reactions, Pressure Relief Design and Effluent Handling, AIChE, pp. 219–245, March 11–13, 1998.

8. Fauske, H. K., Clare, G. H., and Creed, M. J., "Laboratory Tool for Characterizing Chemical Systems," Proceeding of the Int. Symp. on Runaway Reactions, AIChE/CCPS, Boston, MA, March 7–9, 1989.

9. Leung, J. C., Fauske, H. K., and Fisher, H. G., "Thermal Runaway Reactions in a Low Thermal Inertia Apparatus," *Thermochimica Acta,* 104, pp. 13–29, 1986.

10. Ahmed, M., Fisher, H. G., and Janeshek, A. M., "Reaction Kinetics from Self-heat Data Correction for the Depletion of Sample," Proceeding of the Int., Symp. on Runaway Reactions, AIChE/CCPS, Boston, MA, March 7–9, 1989.

11. Wilberforce, J., "The Use of the Accelerating Rate Calorimeter to Determine the SADT of Organic Peroxides," ARC technical document, 1981.

12. Huff, J., "Emergency Venting Requirements," *Plant/Operations Progress,* 1 (4), pp. 211–219, 1982.

13. Townsend, D., Ferguson, H., and Kohlbrand, H., "Application of ARC™ Thermokinetic Data to the Design of Safety Schemes for Industrial Reactors," *Process Safety Progress,* 14 (1), pp. 71–76, 1995.

14. Gustin, J. L., "Choice of Runaway Reaction Scenarios for Vent Sizing Based on Pseudo-adiabatic Calorimetric Techniques," Int. Symp. on Runaway Reaction, Pressure Relief Design, and Effluent Handling, AIChE, pp. 11–13, 1998.

15. Creed, M. J. and Fauske, H. K., "An Easy Inexpensive Approach to the DIERS Procedure," *Chem. Eng. Prog.,* 86 (3), 45, 1990.
16. Fauske, H. K., "The Reactive System Screen Tool (RSST): An Easy, Inexpensive Approach to the DIERS Procedure," Int. Symp. on Runaway Reaction, Pressure Relief Design, and Effluent Handling, AIChE, pp. 51–63, March 11–13, 1998.
17. Singh, J., "PHI-TEC: Enhanced Vent Sizing Calorimeter Application and Comparison with Existing Devices," Proceeding of the Int. Symp. on Runaway Reactions, AIChE/CCPS, pp. 313–330, Boston, MA, March 7–9, 1989.
18. Waldram, S. P., "Toll Manufacturing: Rapid Assessment of Reactor Relief Systems for Exothermic Batch Reactions," Hazards XII European Advances in Process Safety, IChemE, Symp., Series No. 134, pp. 525–540, 1994.
19. Singh, J., "Vent Sizing for Gas-Generating Runaway Reactions," *J., Loss Prev. Process Ind.,* Vol. 7, No. 6, pp. 481–491, 1994.
20. Singh, J., "Batch Runaway Reaction Relief: Re-evaluation of 'credible' Scenario," *Process Safety Prog.,* Vol. 16, No. 4, pp. 255–261, 1997.
21. Coker, A. K., "Size Relief Valves Sensibly Part 2," *Chem. Eng. Prog.,* pp. 94–102, November 1992.
22. Gibson, N., Maddison, N., and Rogers, R. L., "Case Studies in the Application of DIERS Venting Methods to Fine Chemical Batch and Semi-batch reactors, Hazards from Pressure: Exothermic Reactions, Unstable Substances, Pressure Relief and Accidental Discharge," IChemE, Symp, Ser. No. 102, EFCE Event No. 359, pp. 157–169, 1987.
23. Ritcher, S. H. and Turner, F., "Properly Program the Sizing of Batch Reactor Relief Systems, *Chem. Eng. Prog.,* pp. 46–55, 1996.
24. Fisher, H., et al., "Emergency Relief System Design Using DIERS Technology," AIChE's Design Institute for Emergency Relief Systems, DIERS, AIChE, New York, 1992.
25. Boyle, W. J., "Sizing Relief Area for Polymerization Reactors," *Chem. Eng. Prog.,* 63 (8), p. 61, 1967.
26. Huff, J. E., *CEP Loss Prevention Technical Manual,* 7, 1973.
27. Fisher, H. G., "An Overview of Emergency Relief System Design Practice," *Plant/Operations, Prog.,* 10 (1), pp. 1–12, January 1991.
28. Leung, J. C., "Simplified Vent Sizing Equations for Emergency Relief Requirements in Reactors and Storage Vessels, *AIChE Journal,* Vol. 32, No. 10, pp. 1622–1634, 1986.

29. Fauske, H. K., "Generalized Vent Sizing Monogram for Runaway Chemical Reactions, *Plant/Operation, Prog.*, Vol. 3, No. 3, October 1984.
30. Crowl, D. A. and Louvar, J. F., *Chemical Process Safety Fundamental with Applications,* Prentice-Hall, Inc., New Jersey, 1990.
31. Leung, J. C. and Fauske, H. K., *Plant/Operation Progress,* 6 (2), pp. 77–83, 1987.
32. Fauske, H. K., "Emergency Relief System Design for Runaway Chemical Reaction: Extension of the DIERs Methodology, *Chem. Eng. Res. Des.,* Vol. 67, pp. 199–202, 1989.
33. Noronha, J. A., Seyler, R. J., and Torres, A. J., "Simplified Chemical Equipment Screening for Emergency Venting Safety Reviews Based on the DIERS Technology," Proceedings of the Int. Symp. on Runaway Reactions, pp. 660–680, Cambridge, MA, March 7–9, 1989.
34. Duxbury, H. A. and Wilday, A. J., "The Design of Reactor Relief System," *Trans. IChemE.,* Vol. 68, Part B, pp. 24–30, February 1990.
35. Coker, A. K., "Computer Program Enhances Guidelines for Gas-Liquid Separator Designs," *Oil & Gas Journal,* May 1993.
36. American Petroleum Institute Recommended Practice 521, "Guide for Pressure-Relieving and Depressurizing Systems," 2nd ed., 1982.
37. Less, F. P., *Loss Prevention in the Process Industries,* 2nd ed., Butterworth-Heinemann, 1996.
38. Wells, G. L., *Safety in Process Plant Design,* John Wiley & Sons, New York, 1980.
39. Grossel, S. S., "Design and Sizing of Knock-out Drums/Catchtanks for Reactor Emergency Relief Systems," *Plant/Operations Prog.,* Vol., 5, No. 3, pp. 129–135, 1986.
40. Parry, C. F., *Relief Systems Handbook,* Institution of Chemical Engineers, 1992.
41. Coker, A. K., "Size Relief Valves Sensibly," *Chem. Eng. Prog.,* pp. 20–27, 1992.
42. Englund, S. M., "Inherently Safer Plants, Practical Applications," *Process Safety Prog.,* Vol. 14, No. 1, pp. 63–70, 1995.
43. Kletz, T., "Friendly Plants," *Chem. Eng. Prog.,* pp. 18–26, 1989.
44. Kletz, T., "Hazop and Hazan—Identifying and Assessing Process Industry Hazards," 4th ed., IChemE, 1999.

CHAPTER 12: APPENDIX

Table 12-8
Hazard rating chemical reactions

Reaction	Degree of hazard	Reaction	Degree of hazard
A. Reductions		*G. Hydrolysis, aqueous nitriles, esters*	E
1. Clemmensen	D		
2. Sodium-amalgam	D		
3. Zinc-acetic acid	E		
4. Zinc-hydrochloric acid	E		
5. Zinc-sodium hydroxide	E		
6. Ferrous ammonium sulfate	E		
7. Lead tetraacetate	E		
8. Meerwein-Pondorff	D		
9. Lithium aluminum hydride	B		
10. Dialkyl aluminum hydride	B		
11. Rosenmund	A		
12. Catalytic high pressure	A		
13. Catalytic low pressure	B		
B. Oxidations		*H. Simple metathetical replacements*	D
1. Hydrogen peroxidedil ute aqueous	E		
2. Air or I_2 (mercaptan to disulfide)	D		
3. Oppenauer	D		
4. Selenium dioxide	D		
5. Aqueous solution nitric acid, permanganate, manganic dioxide, chromic acid, dichromate	E		

6. Electrolytic — B
7. Chromyl chloride — C
8. Ozonolysis — A
9. Nitrous acids — A
10. Peracids-low molecular weight or two or more positive groups — A
11. Peracids-high molecular weight — B
12. t-butyl hypochlorite — C
13. Chlorine — C

C. *Alkylations: Carbon-carbon*
1. Jarousse — E
2. Alkali metal — C
3. Alkali metal alcoholate — D
4. Alkali metal amides and hydrides — C
5. Reformatsky — E
6. Michael — E
7. Grignard — B
8. Organo metallics, such as dialkyl zinc or cadmiun-alkyl or aryl lithium — B
9. Alkali acetylides — A
10. Diels-Alder — D
11. Arndt-Eistert — A
12. Diazoalkane and aldehyde. — A
13. Aldehydes or ketones and hydrogen cyanide — C

I. *Preparation and reaction of peroxides and peracids*
1. Concentrated — A
2. Dilute — D

J. *Pyrolysis*
1. Atmospheric pressure — D
2. Pressure — B

Carbon-oxygen
1. Williamson — D
2. Formaldehyde-hydrochloric acid — E
3. Ethylene oxide. — C

Table 12-8
(continued)

Reaction	Degree of hazard	Reaction	Degree of hazard
4. Dialkyl sulfate	D		
5. Diazoalkane	A		
		K. Schmidt reaction	B
Carbon-nitrogen			
1. Cyanomethylation	C		
2. Chloromethylation	D		
3. Ethylenimine	C		
4. Ethylene oxide	C		
5. Quarternization	D		
		L. Mannich reaction	D
D. Condensations			
1. Erlenmeyer	D		
2. Perkin	D		
3. Acetoacetic ester	D		
4. Aldol	D		
5. Claisen	D		
6. Knoevenagel	D		
7. Condensations using catalysts such as phosphoric acid; $AlCl_3$; $KHSO_4$; $SnCl_4$; H_2SO_4; $ZnCl_2$; $NaHSO_3$; $POCl_3$; HCl; $FeCl_3$	E		
8. Acyloin	C		
9. Diketones with hydrogen sulfide	C		

10. Diketones with Diamines—quinazolines	D
11. Diketones with NH$_2$OH—isoxazolines	D
12. Diketones with NH$_2$NH$_2$—pyrazoles	D
13. Diketones with semi-carbazide—pyrazoles	D
14. Diketones with ammonia—pyrazoles	D
15. Carbon disulfide with aminoacetamide—thiazolone	A
16. Nitriles and ethylene diamines—imidazolines	D

E. Aminations

1. Liquid ammonia	B
2. Aqueous ammonia	E
3. Alkali amides	C

F. Esterifications

1. Inorganic	E
2. Alkoxy magnesium halides	B
3. Organics:	
Alcohol and acids or acid chlorides or acid anhydride	D
Alkyl halide and silver salts of acids	E
Alkyl sulfate and alkali metal salt of acids	D
Alkyl chlorosulfates and alkali salts of carboxylic acid	D
Ester-exchange	D
Carboxylic acid and diazomethane	A
Acetylene and carboxylic acid-vinyl ester	A

M. Halogenations

SO$_2$X$_2$, SOX$_2$, SX	D
POX$_3$, PX$_3$, HX	D
Cl$_2$, Br$_2$	C

N. Nitrations

1. Dilute	D
2. Concentrated	B

Source: A. C. Shabica, Chem. Eng. Prog., *Vol. 59, No. 9, Reproduced with permission of the AIChE. Copyright © 1963. All rights reserved.*

Table 12-9
Hazard rating substances

1. Hazardous—highly dangerous; will explode owing to heat, flame, shock

Acetylides	Hydrogen (high pressure)	Phenyl diazosulfide
Acetyl peroxides	Hydrogen peroxide (< 35% water)	Picric acid and Cu, Pb and Zn salts, (dry)
Aluminum methyl	Iodine azide	Radio isotopes, gamma emitters
Ammonium chlorate	Lead azide	Silver azide
Diazoethane	Manganese heptoxide	Tetracene
Diazomethane	Mannitol hexanitrate	Tetracetylene dicarbonic acid
Dichloroacetylene	Mercury acetylide	Tetranitromethane
N,N-diethyl carbanilide	Mercury azide	Trinitroaniline
Ethylene oxide	Methyl isocyanide	Trinitrobenzene
Formyl peroxide	Nitrocellulose, dry	Trinitrochlorbenzene 2, 4, 6 trinitro-m- cresol
Fulminates	Ozonides (dry)	Trinitrotoluene
Glycerol trinitrate	Parathion	2, 4, 6-trinitroxylene
Hexane hexanitrate	Perchloric acid (less than 10% water)	Zinc peroxide

2. Highly dangerous—gives off highly toxic fumes on heat, flame, shock

Carbon disulfide	Ethyl nitrite	Thionyl chloride–fluoride
Carbon monoxide	Hyponitrous acid	Thiophosgene
Carbonoxysulfide	Mercuric perchlorate	Vinyl chloride
Dibromoacetylene	Methyl nitrite	Vinyl ether
Diethyl ether	Methyl phosphine	Vinylidene chloride
Dimethyl ether	Phosgene	

3. Special—dangerous when exposed to heat or flame may explode or is spontaneously flammable in air

Acetyl benzoyl peroxide	Furan	Inorganic salts of alkyl nitrates
Acetylene	Hydrazine,	Nitroso guanidine
Acetylene chloride	Anhydrous	Nitrosomethyl urea (dry)

Alkyl ethers < C_4
Allylene
Aluminum chlorate
Ammonium chromate
Benzoyl peroxide, dry
Butadiene-1,3,
Chlorates
Dialkyl phosphines
Diazoacetic ester
Diazoamidobenzol
Diazobenzene chloride
Diethyl carbonate
Diisopropyl & higher alkyl ethers
Ferrous perchlorate

Hydrazoic acid
Hydrides, volatile
Hydrogen cyanide (unstabilized)
Hydrogen (low pressure)
Hydrogen peroxide (> 35% water)
Magnesium peroxide
Mercurous azide
Methyl acetylene
Methyl lactate
Nickel hypophosphite
Nitriles > ethyl
Nitrogen bromide

Ozone
Ozonides (solution)
Pentaborane
Perbenzoic acid
Perchlorates
Perchloric acid (10% or more water)
Performic acid
Peroxides (organic)
Phosphorus (white)
Phenylazoimide
Silicon hydride
Silver oxalate
Sodium chlorite
Trinitrobenzaldehyde

4. Dangerous—will react with water or steam to produce hydrogen or toxic fumes or highly flammable gases

Acetone
Alkali cyanides, hydrides, & metals
Aluminum borohydride, carbide, chloride, hydride & nitride
Benzoyl chloride
Boron compounds
Carbonyls
Chlorine
Cyanogen
Cyanogen bromide
Cyanogen chloride

Diakyl carbamyl chlorides
Diakyl aluminum hydrides
Diborane
Dibromoketone
Dichloromethyl chloroformate
Diphosgene
Fuming nitric acid
Grignard reagents
Hydrides nonvolatile
Hydrogen cyanide (stablized)
Hydrogen fluoride

Lithium aluminum hydride
Magnesium metal
Nitric acid
Oleum
Phosphenyl Chloride
Phosphides
Phosphorus nitride,
Oxyhalides,
Penta-halides, & tri-halides
Radio isotopes, beta emitters
Sodium azide

Table 12-9
(continued)

5. Conventional—when heated to decomposition gives off highly toxic fumes and/or products that may be explosion hazards when exposed to flame

Acetic Acid	Amino pyrine	Epichlorohydrin
Acetic anhydride	2-amino thiazole	Esters
Acetoacetanilide	Ammonia	Ethylene imine
Acetone cyanhydrin	Ammonium nitrate.	Hydrazine hydrate
Acetyl chloride	Amyl nitrite	Hydrogen sulfide.
Acrolein	Aniline	Hydroxylamine salts
Acrylonitrile	Benzene sulfonyl chloride	Inorganic Hg. compounds
Alcohols	Benzoyl peroxide (wet)	Iodates
Alkaloids	Bromine & iodine	Iodides
Alkyl & aryl acids,	N-bromo acetamide	Ketone $> CH_3$
Alcohols, amines,	Chlorobutanol	Methyl acrylate
Esters, ethers $>$	Chloroform	Methylene chloride
C_3, halides, hydrocarbons, ketones,	Chloroform	Nitriles
mercaptans & sulfides	2-chloro pyridine	Nitrosomethyl urea (wet)
Alkyl dihalides	Cresols	Petroleum ether
Alkyl nitrates	Cyanates	Piperidine

Allyl amine	N, N-diakyl amino	Propargly bromide
Allyl cyanide	Ethylamine	Pyridine
Allyl ether	Dibutylphosphite	Sodium alkoxide
Allyl halide	Dicyandiamide	Sodium azide
Amines	Diethylamino ethanol	Sodium borohydride
Aminoacetophenone	Diethyl phosphite	Tetrabromoethane
p-aminoazobenzene	Diethyl sulfate	Tetrachloroethane
p-aminophenol	Diketene	

6. Moderately stable when heated gives off acrid fumes

Alcohol	Charcoal	Inorganic acids
Alkyl acids > C_2	Chlorphenyl carbamate.	Methyl diacetoacetate
Alumina	Diethylene glycol	Phenyl cellosolve
Ammonia	Dimethyl formamide	Silica gel
Amyl alcohol	Formic acids	Tetradecane
Benzoic anhydride	Hydrocarbon	Tetrahydro phthalic anhydride
Benzophenone		

Table 12-10
Toxic hazards from incompatible chemicals

Substances in Column I should be stored/handled so that they cannot contact corresponding substances in Column II because toxic materials (Column III) would be produced.

Column I	Column II	Column III
Arsenical materials	Any reducing agent	Arsine
Azides	Acids	Hydrogen azide
Cyanides	Acids	Hydrogen cyanide
Hypochlorites	Acids	Chlorine or hypocholorous acid
Nitrates	Sulfuric acid	Nitrogen dioxide
Nitric acid	Copper, brass and heavy metals	Nitrogen dioxide (nitrous fumes)
Nitrites	Acids	Nitrous fumes
Phosphorus	Caustic Alkalis or reducing agents	Phosphine
Selenides	Reducing agents	Hydrogen selenide
Sulfides	Acids	Hydrogen sulfide
Tellurides	Reducing agents	Hydrogen telluride

Source: C. J. Mumford, PS1 Chemical Process Safety, October 1993, Reproduced with permission of the Occcupational Health and Safety Training Unit, University of Portsmouth, U.K.

Table 12-11
Reactive hazards with incompatible chemicals

Substances in Column I must be stored/handled so that they cannot contact corresponding substances in Column II under uncontrolled conditions, or violent reactions may occur.

Column I	Column II
Acetic acid	Chromic acid, nitric acid, hydroxyl-containing compounds, ethylene glycol, perchloric acid, peroxides, and permanganates.
Acetone	Concentrated nitric and sulfuric acid mixtures.
Alkali and alkaline earth metals, eg. sodium, magnesium, calcium, powdered aluminum.	Carbon dioxide, carbon tetrachloride, and other chlorinated hydrocarbons. (Also prohibit water, foam and dry chemical on fires involving these metals—dry sand should be available).
Anhydrous ammonia	Mercury, chlorine, calcium hypochlorite, iodine, bromine and hydrogen fluoride.
Ammonium nitrate	Acids, metal powders, flammable liquids, chlorates, nitrites, sulfur, finely divided organics or combustibles.
Aniline	Nitric acid, hydrogen peroxide.
Bromine	Ammonia, acetylene, butadiene, butane and other petroleum gases, sodium carbide, turpentine, benzene, and finely divided metals.
Calcium oxide	Water.
Carbon, activated	Calcium hypochlorite.
Chlorates	Ammonium salts, acids, metal powders, sulfur, finely divided organics or combustibles.
Chromic acid and chromium trioxide	Acetic acid, naphthalene, camphor, glycerol, turpentine, alcohol and other flammable liquids.

Table 12-11
(continued)

Column I	Column II
Chlorine	Ammonia, acetylene, butadiene, butane and other petroleum gases, hydrogen sodium carbide, turpentine, benzene and finely divided metals.
Chlorine dioxide	Ammonia, methane, phosphine and hydrogen sulfide.
Copper	Acetylene, hydrogen peroxide
Fluorine	Isolate from "everything"
Hydrazine	Hydrogen peroxide, nitric acid, any other oxidant
Hydrocarbons (benzene, butane, propane, gasoline, turpentine, etc).	Fluorine, chlorine, bromine, chromic acid, peroxide
Hydrocyanic acid	Nitric acid, alkalis.
Hydrofluoric acid, anhydrous (hydrogen fluoride).	Ammonia, aqueous or anhydrous
Hydrogen peroxide	Copper, chromium, iron, most metals or their salts, any flammable liquid, combustible materials, aniline, nitromethane.
Hydrogen sulfide	Fuming nitric acid, oxidizing gases.
Iodine	Acetylene, ammonia (anhydrous or aqeous).
Mercury	Acetylene, fulminic acid, ammonia.

Nitric acid (conc)	Acetic acid, acetone, alcohol, aniline, chromic acid, hydrocynanic acid, hydrogen sulfide, flammable liquids, flammable gases, and nitratable substances, paper, cardboard and rags.
Nitroparaffins	Inorganic bases, amines.
Oxalic acid	Silver, mercury.
Oxygen	Oils, grease, hydrogen, flammable liquids, solids or gases.
Perchloric acid	Acetic anhydride, bismuth and its alloys, alcohol, paper, wood, grease, oils.
Peroxides, organic	Acids (organic or mineral), avoid friction, store cold.
Phosphorus (white)	Air, oxygen.
Potassium perchlorate	Acids (see also chlorates).
Potassium perchlorate	Acids (see also perchloric acid).
Potassium permanganate	Glycerol, ethylene glycol, benzaldehyde, sulfuric acid
Silver	Acetylene, oxalic acid, tartaric acid, fulminic acid, ammonium compounds.
Sodium	See alkali metals (above)
Sodium nitrite	Ammonium nitrate and other ammonium salts.
Sodium peroxide	Any oxidisable substance, such as ethanol, methanol, glacial acetic acid, acetic anhydride, benzaldehyde, carbon disulfide, glycerol, ethylene glycol, ethyl acetate, methyl acetate and furfural.
Sulfuric acid	Chlorates, perchlorates, permanganates.

Source: C. J. Mumford, PS1 Chemical Process Safety, October 1993, Reproduced with permission of the Occupational Health and Safety Training Unit, University of Portsmouth, U.K.

CHAPTER THIRTEEN

Scale-Up in Reactor Design

INTRODUCTION

Scale-up and dimensionless analysis are indispensable mathematical tools for designers engaged in industrial processes, where chemical or biochemical conversion of feeds, together with various transport processes such as momentum, mass, or heat, take place. These processes are scale-dependent implying different behavior on laboratory, model, or full-scale plants. Examples of these processes are homogeneous and heterogeneous reactions. Optimizing chemical processes often involves a deep knowledge of their physical and technical elements. The effects of each parameter that influence the process must be examined for which sometimes mathematical solutions do not exist. In such cases, the designer must depend on both model experiments and the theory of similarity between the model and the full-scale plant. This theory enhances the planning and execution of experiments and the analysis of the data to produce adequate and reliable information on the size and process parameters of the full-scale plant.

Chemical reactions obey the rules of chemical kinetics (see Chapter 2) and chemical thermodynamics, if they occur slowly and do not exhibit a significant heat of reaction in the homogeneous system (microkinetics). Thermodynamics, as reviewed in Chapter 3, has an essential role in the scale-up of reactors. It shows the form that rate equations must take in the limiting case where a reaction has attained equilibrium. Consistency is required thermodynamically before a rate equation achieves success over the entire range of conversion. Generally, chemical reactions do not depend on the theory of similarity rules. However, most industrial reactions occur under heterogeneous systems (e.g., liquid/solid, gas/solid, liquid/gas, and liquid/liquid), thereby generating enormous heat of reaction. Therefore, mass and heat transfer processes (macrokinetics) that are scale-dependent often accompany the chemical reaction. The path of such chemical reactions will be identical on a pilot plant scale and the full-scale plant if both transfer processes are similar and the chemistry is identical.

Pilot plant experiments represent an essential step in the investigation of a process toward formulating specifications for a commercial plant. A pilot plant uses the microkinetic data derived by laboratory tests and provides information about the macrokinetics of a process. Examples include the interaction of large conglomerates of molecules, macroscopic fluid elements, the effects of the macroscopic streams of materials and energy on the process, as well as the true residence time in the full-scale plant.

In a continuous reaction process, the true residence time of the reaction partners in the reactor plays a major role. It is governed by the residence time distribution characteristic of the reactor, which gives information on backmixing (macromixing) of the throughput. The principal objectives of studies into the macrokinetics of a process are to estimate the coefficients of a mathematical model of the process and to validate the model for adequacy. For this purpose, a pilot plant should provide the following:

- Identify the flow pattern of the prototype system by subjecting it to an impulse, step, or sinusoidal disturbance by injection of a tracer material as reviewed in Chapter 8. The result is classified as either complete mixing, plug flow, and an option between a dispersion, cascade, or combined model.
- Investigate and determine thermal parameters involving conditions under which heat transfer takes place and an accurate value of the heat transfer surface area. Additionally, a review of the process is required for thermal stability (see Chapter 6).
- Make an accurate estimate of the effects produced by diffusion parameters and arrange the process to proceed in a kinetic or a diffusion region. The reactor efficiency is determined on the basis of the chosen model.
- Investigate the system under dynamic conditions with disturbances acting over many circuits and monitor how the process variables change with time, resulting in a judicious recommendation regarding the control of the system.
- Develop a mathematical model of the system, which is tested and fitted to the system with the aid of a computer.
- Optimize the mathematical model with the sole aim of enhancing the performance of the system.

The following reviews scale-up of chemical reactors, considers the dimensionless parameters, mathematical modeling in scale-up, and scale-up of a batch system.

DEVELOPMENT AND SCALE-UP
OF REACTORS

The objective of scale-up in reactor design is to determine a criterion or criteria on which to base the transfer of the laboratory scale into a full-scale commercial unit. Before proceeding from a laboratory to an industrial scale, additional investigations are required. However, it is difficult to define these additional steps to gather all the information as promptly as possibe and at minimum cost. The methodology of process development leading to scale-up becomes the principal factor for the success of the operation. In achieving this purpose, experiments are classified into three main types: laboratory, pilot plant, and demonstration units.

In laboratory-type experiments, certain aspects of the process are investigated by handling relatively small amounts of raw materials to reduce the material constraints to a minimum. In laboratory experiments, a series of measurements are taken concerning all the mechanisms that are independent of size (thermodynamics and chemical kinetics). A number of physical properties, such as densities, viscosities, specific heats, and phase equilibria, involved in the model must be ascertained throughout the operating conditions of the process.

Pilot plant experiments vary over a wide range, accounting for industrial constraints (e.g., duration of operation, control parameters, equipment reliability, and impurities in the raw materials). Scale-up problems are investigated during pilot plant experiments. A pilot plant is an experimental rig, which displays the part of the operation that corresponds to an industrial plant. It allows for simultaneous analysis of the physical and chemical mechanisms. A pilot plant is indispensable for measuring the extent of the possible interactions between these two types of mechanisms. It can be small to minimize extraneous costs such as the total operation cost as well as other constraints.

Experiments at the level of a demonstration unit apply to the construction of a first industrial unit, but on a modest scale (e.g., one-tenth of full-scale production). This step can be very costly, but has been proven to be indispensable.

SIMILARITY CRITERIA

Some physical processes occurring in a single phase may be scaled-up using the principles of physical modeling. This is based on the criteria of geometric and chemical similarity derived from differential equations, which describe the process, or from dimensional analysis of the process variable (Chapter 7). In physical modeling, the process of interest is reproduced on different scales, and the effect of physical features and linear dimensions is analyzed. Experimental data are reduced to relationships involving dimensionless groups composed of various combinations of physical quantities and linear dimensions. The relationships can be classified into dimensionless groups or similarity criteria.

Physical modeling involves searching for the same or nearly the same similarity criteria for the model and the real process. The full-scale process is modeled on an increasing scale with the principal linear dimensions scaled-up in proportion, based on the similarity principle. For relatively simple systems, the similarity criteria and physical modeling are acceptable because the number of criteria involved is limited. For complex systems and processes involving a complex system of equations, a large set of similarity criteria is required, which are not simultaneously compatible and, as a consequence, cannot be realized.

The objectives are not realized when physical modeling are applied to complex processes. However, consideration of the appropriate differential equations at steady state for the conservation of mass, momentum, and thermal energy has resulted in various dimensionless groups. These groups must be equal for both the model and the prototype for complete similarity to exist on scale-up.

SCALE-UP IN RELATION TO VARIOUS FACTORS

In many cases, two identical reaction systems (e.g., a pilot plant scale and a full-scale commercial plant) exhibit different performances. This difference in performance may result from different flow patterns in the reactors, kinetics of the process, catalyst performance, and other extraneous factors.

As discussed in Chapters 8 and 9, the fluid elements in various reactors reside in different intervals of time and, consequently, the residence time is different. This is determined from the response curve of the system. Thus, if the model and the full-scale plant are to give the same yield, both should have the same residence time distribution or the same response curve. In the case of a tubular reactor, the effect of longitudinal mixing may be minimized. Additionally, a smoother velocity profile may be obtained through the use of higher flow velocities or packing. Swirl in the mixing may be avoided by providing baffles, which additionally secure hydrodynamic reproducibility (Chapter 7).

In the case of exothermic or endothermic reactions, scale-up may impair conditions for heat input or removal because the ratio of the heat transfer surface area to the reactor volume is reduced. Identical conditions for heat transfer in both the model and full-scale plants may be achieved in exothermal reactions if both have the same thermal stability coefficient. This requirement is obtained by introducing external heat exchangers. Alternatively, a reactor with a strong exothermic reaction can be divided into several small size reactors. In this manner, the ratio of the external heat transfer surface area to the reactor volume is increased, thereby avoiding an excessive temperature rise in the reactor.

HEAT EFFECT

The dominant feature of many reactor scale-ups is the enormous thermal effects that accompany reactions. In many reactor designs, the salient issue is that the reactor is capable of transferring adequate heat compared to attaining a certain degree of reaction. Many commercial reactors operate adiabatically, implying there is negligible heat exchange with the surroundings compared to the enormous flows of heat within the system. The heat generated or consumed by the reaction should be defined so that the temperature history of the reacting fluid is known. Temperature runaways (Chapter 12) and reduced conversions (Chapter 6) relative to that possible at equilibrium are two problems that are prevalent in exothermic reactions. In a reactor with isothermal operation, it is necessary to know how much heat is being exchanged with the surroundings to maintain isothermality.

For example, consider the synthesis of ammonia at two different temperatures. Table 13-1 shows values for the heat of reaction. The negative enthalpy changes upon reaction show that the products of the reaction contain less enthalpy than did the reactants. This implies that heat must have been generated, which is characteristic of an exothermic

Table 13-1
Heat of reaction in ammonia synthesis [1]

$N_2 + 3H_2 \rightleftharpoons 2NH_3$
$\Delta H^O_{25°C} = -22,100 \, \text{cal/mol} \, N_2$ reacted $(-92,500 \, \text{J/mol})$
$\Delta H^O_{528°C} = -25,800 \, \text{cal/mol} \, N_2$ reacted $(-108,000 \, \text{J/mol})$

reaction. Another common feature is the decrease in the total number of moles. Conversely, reactions where the number of moles increases are usually endothermic, hence, energy must be added to the system to break the bonds. Observing changes in the total number of moles involving endothermic or exothermic reaction can assist in predicting reactor performance. However, a quantitative expression of the heat of reaction is essential and can be obtained whether or not the reaction involves a change in the total number of moles. Table 13-1 shows a negligible change in the heat of reaction over a 500°C temperature range. The enormous heat effects indicate that ammonia synthesis reactors must be designed with particular attention given to heat exchange [1].

COEFFICIENTS OF PROCESS STABILITY

The scale-up of exothermic processes is greatly enhanced through the use of the coefficient of thermal stability. Kafarov [2] defined this as the ratio of the slope ($\tan \alpha_2$) of the line representing the heat removal (due to the heat transfer medium and changes in enthalpy) to the slope ($\tan \alpha_1$) of the line representing heat generation (by the reaction) at the intersection of the two lines when plotted on the T versus Q coordinates. This is expressed as

$$\beta = \frac{\tan \alpha_2}{\tan \alpha_1} \tag{13-1}$$

At $\beta > 1$, the process is stable. At $\beta > 1$, the process is unstable. Furthermore, if β^* is designated as the stability of the prototype reactor and β the stability of the model, the transition stability coefficient is defined by

$$\delta = \frac{\beta^*}{\beta} \qquad\qquad\qquad (13\text{-}2)$$

At $\delta = 1$, both the model and the prototype are similarly stable. At $\delta > 1$, the prototype has a greater stability than the model. At $\delta < 1$, the model has a greater stability than the prototype.

In applying the transition stability coefficient for scaling up, the designer requires systematic process steps in a pilot plant scale reactor with a satisfactory performance stability over a temperature interval of interest at stable component concentrations and other variables that may be used to control the process.

In heterogeneous reactions, knowledge of the process control by diffusion, kinetics, or both is essential. When a chemical process proceeds in a kinetic region, the reaction rate controls the entire process, while the velocities and physical properties of the streams are of secondary significance. When a chemical process takes place in a diffusion region, the controlling step is the rate at which the reaction components diffuse toward the reaction zone. The process is subsequently controlled by the velocities and physical properties of the streams. A method by which the effect of diffusional resistance may be eliminated from laboratory kinetic data should be employed in scaling up heterogeneous reactors. In certain instances, a process may be restricted to the desired region by controlling hydrodynamic conditions.

DIMENSIONAL ANALYSIS AND
SCALE-UP EQUATIONS

Dimensionless groups for a process model can be easily obtained by inspection from Table 13-2. Each of the three transport balances is shown (in vector/tensor notation) term-by-term under the description of the physical meanings of the respective terms. The table shows how various well-known dimensionless groups are derived and gives the physical interpretation of the various groups. Table 13-3 gives the symbols of the dimensions of the terms in Table 13-2.

Tables 13-2 and 13-3 elucidate how the common dimensionless groups are derived. The boundary conditions governing the differential equations combined with the relative size of the system should be considered when determining dimensionless parameters. Using Table 13-2 to determine the dimensionless groups for any of the three equations, divide one set of the dimensions into all the others including the boundary conditions.

Table 13-2
Development of the relations among dimensionless groups

MASS BALANCE	Rate of change of mass per unit volume	Rate of change of mass by convection per unit volume	Rate of change of mass by molecular transfer (diffusion) per unit volume	Generation per unit volume (chemical reaction)		Boundary Condition(s) (Interphase Transfer)		
						Empirically determined flux specified (3)a	Concentration specified (1, 2b)a	Mass flux specified (2a, 4)a
Symbols	$\frac{\partial}{\partial t} c_i$	$+ [\nabla \cdot c_i v] =$	$- [\nabla \cdot J_i]$	$+ R_i$		$N_A\|_{x=0} = K\Delta c$	$c = c_0$ or $c_i = c_2$	$N_1 = N_2$ or $N_A = N_0$
Dimensions b	$\frac{C}{\theta} = \frac{CV}{L}$	$\frac{VC}{L}$	$\frac{DC}{L^2}$	$R_i\ddagger$		$\frac{1}{L}(KC)$		

$\xleftarrow{\frac{LV}{D}} =$ Mass Peclet $\longrightarrow$

$\frac{KL}{D} =$ Sherwood $\longrightarrow$

$\frac{R_i\ddagger L}{VC} =$ Damköhler I

$\frac{R_i\ddagger L^2}{DC} =$ Damköhler II

MOMENTUM BALANCE	Rate of change of momentum per unit volume	Rate of change of momentum by convection per unit volume	Rate of change of momentum by molecular transfer (viscous transfer) per volume	Generation per volume (External forces) (Ex: gravity)		Empirically determined flux specified (3)a	Velocity specified (1, 2b)a	Momentum flux specified (2a, 4)a
Symbols	"Inertial Forces" $\frac{\partial}{\partial t} \rho v$	$+ [\nabla \cdot \rho v v] =$	"Viscous Forces" $- [\nabla \cdot \tau]$	$+ \rho g$	Pressure gradient $- \nabla p$	$\tau\|_{x=0} = \tau$ or $\tau = \frac{\sigma}{L}$	$v = 0$ or $v_1 = v_2$	$\tau_1 = \tau_2$ or $\tau = \tau_0$
Dimensions b	$\frac{\rho V}{\theta} = \frac{\rho V^2}{L}$	$\frac{\rho V^2}{L}$	$\frac{\mu V}{L^2}$	ρg	$\frac{P}{L}$	$\frac{1}{L}(\tau\ddagger)$		

$\xleftarrow{\frac{LV\rho}{\mu}} =$ Reynolds $\longrightarrow$

$\frac{V^2}{gL} =$ Froude $\longrightarrow$

$f = \left(\frac{\Delta P}{L}\right)\left(\frac{L}{\rho V^2}\right) =$ Friction Factor $\longrightarrow$

$\frac{\tau\ddagger L}{\mu V} =$ Bingham

$\frac{\rho V^2 L}{\sigma} =$ Weber $\qquad (\sigma \to \tau L)$

ENERGY BALANCE	Rate of change of energy per unit volume	Rate of change of energy by convection per unit volume	Rate of change of energy by diffusion (conduction) per unit volume	Generation per volume (Ex: electrical, chem. rxn, etc.)	Other terms: rev. and irrev. transfer, viscous dissipation, etc.	Empirically determined flux specified (3)a	Temperature (1, 2b)a	Heat flux specified (2a, 4)a
Symbols	$\frac{\partial \rho C_p T}{\partial t}$	$+ (\nabla \cdot \rho C_p T v) =$	$- (\nabla \cdot q)$	S_R		$q\|_{x=0} = h\Delta T$	$T = T_0$ or $T_1 = T_2$	$q_1 = q_2$ or $q = q_0$
Dimensions b	$\frac{T\ddagger \rho C_p}{\theta} = \frac{T\ddagger V \rho C_p}{L}$	$\frac{T\ddagger V \rho C_p}{L}$	$\frac{kT\ddagger}{L^2}$	$S_R\ddagger$		$\frac{1}{L}(hT)$		

$\frac{hL}{k} =$ Nusselt

$\frac{h}{\rho C_p V} =$ Stanton

$\frac{V\rho C_p L}{k} =$ Heat Peclet

$\frac{S_R\ddagger L}{\rho V C_p T\ddagger} =$ Damköhler III

Left side vertical labels: Lewis $= \frac{k}{\rho C_p D}$; Schmidt $= \frac{\mu}{\rho D}$; $\frac{\nu}{D} = \frac{\mu}{\rho D}$; $\frac{\alpha}{D} = \frac{k}{\rho C_p D}$; Prandtl $= \frac{C_p \mu}{k}$; $\frac{Pe}{Re}$; $Le \cdot Pr$

Source: Himmelblau, D. M., and Bischoff, K. B., Process Analysis and Simulation, *John Wiley & Sons,* New York, 1965.

Table 13-3
Notation for dimensionless quantities in Table 13-2

Variable	Equivalent characteristic quantity
c	C
t	θ
Length	L
R_t	$R_t^{\pm}$
v	V
p	P
τ	$\tau^{\pm}$
T	$T^{\pm}$
S_R	$S_R^{\pm}$

Source: Himmelblau, D. M. and Bischoff, K, Process Analysis and Simulation, *John Wiley, New York, 1965.*

In Table 13-2, the numerator of a well-known group is identified by the "arrowhead," while the denominator is assigned by the "tail" of the arrow. For example, the Damköhler I represents the division of the dimensions representing the generation per unit volume by the dimensions representing the transport of mass by convection. This is defined as:

$$\frac{R_i^{\pm}}{\dfrac{VC}{L}} = \frac{R_i^{\pm} L}{VC} = \text{Damköhler I} \tag{13-3}$$

The Damköher II is determined by dividing the dimensions representing the generation per unit volume by the dimensions representing the transport of mass by molecular process as:

$$\frac{R_i^{\pm}}{\dfrac{DC}{L^2}} = \frac{R_i^{\pm} L^2}{DC} = \text{Damköhler II} \tag{13-4}$$

The Damköhler III is obtained by dividing the dimensions representing generation per unit volume of chemical reaction by the dimensions representing the transport of energy by convection. This is defined by

$$\frac{S_R^{\pm}}{\dfrac{T^{\pm} V \rho C_p}{L}} = \frac{S_R^{\pm} L}{T^{\pm} V \rho C_p} = \text{Damköhler III} \tag{13-5}$$

Other dimensionless parameters, namely the Reynolds number, Prandtl number, and Nusselt number can be represented as follows:

$$\frac{\dfrac{\rho V^2}{L}}{\dfrac{\mu V}{L^2}} = \frac{\rho VL}{\mu} = \text{Reynolds number } (N_{Re}) \qquad (13\text{-}6)$$

$$\frac{N_{Pe}}{N_{Re}} = \frac{\dfrac{V\rho C_p L}{k}}{\dfrac{\rho VL}{\mu}} = \frac{C_p\mu}{k} = \text{Prandtl number } (N_{Pr}) \qquad (13\text{-}7)$$

$$\frac{\dfrac{1}{L}(hT)}{\dfrac{kT}{L^2}} = \frac{hL}{k} = \text{Nusselt number } (N_{Nu}) \qquad (13\text{-}8)$$

It is also possible to interpret similarity in terms of dimensionless groups and variables in a process model. Kline [3] expressed that the numerical values of all the dimensionless groups should remain constant during scale-up. However, in practice this has been shown to be impossible [4].

Generally, it is rarely possible to satisfy all of the various similarity criteria in scale-up, especially those related to chemical and thermal dimensionless groups. An alternative solution is to determine the effect of each group and perform intermediate scale experiments to determine the effect of the critical groups.

REACTOR SIZE AND SCALE-UP EQUATIONS

In a preliminary consideration, the following scale-up relationships between the model and the prototype may be used in determining the reactor size. In attaining the same residence time in the model and the prototype, the dimensionless parameter (Damkhöler I) may be applied. This is expressed by

$$\left(Da_I\right)_m = \left(Da_I\right)_p \qquad (13\text{-}9)$$

or
$$\left(\frac{R_i^{\pm}L}{VC}\right)_m = \left(\frac{R_i^{\pm}L}{VC}\right)_p \qquad (13\text{-}10)$$

where the subscripts m and p represent the model and the prototype, respectively. Canceling out the physical constants, and noting that the concentration (initial or final) is the same for the model and the prototype yields

$$\left(\frac{L}{V}\right)_m = \left(\frac{L}{V}\right)_p \qquad (13\text{-}11)$$

This shows that the residence time in the model and the prototype must be the same.

MATHEMATICAL MODELING

One of the limitations of dimensional similitude is that it shows no direct quantitative information on the detailed mechanisms of the various rate processes. Employing the basic laws of physical and chemical rate processes to mathematically describe the operation of the system can avert this shortcoming. The resulting mathematical model consists of a set of differential equations that are too complex to solve by analytical methods. Instead, numerical methods using a computerized simulation model can readily be used to obtain a solution of the mathematical model.

PROCEDURES IN SCALE-UP USING MATHEMATICAL MODELING

There is no strict procedure for scale-up of a chemical reactor by mathematical modeling. Himmelblau [4] has reviewed the mathematical modeling approach, and others specific examples are expatiated by Rase [5] and Carberry [6]. The following describe the steps applied to a scale-up problem as they relate to a catalytic reactor having a fixed bed of solid catalysts.

Development of Mathematical Expressions

Various processes involved in fixed bed catalytic reactions are represented by mathematical descriptions derived from basic laws of physical and chemical rate processes. A process model for a single catalyst pellet is first derived because the pore structure and inner surfaces of the catalyst pellets strongly influence the reaction mechanism. Laboratory experimental data are obtained for many particle sizes in combination with pore-structure analysis. The optimum pore structure, size, and shape of the catalyst pellet are selected based on factors such as the cost and activity of the catalyst. The mathematical description of the process for a layer of catalyst pellet is then formulated. This depends on the process model for one catalyst pellet and incorporates corrections for non-uniformity of temperature and concentrations of the reaction mixture throughout the layer.

Process Optimization

In this step, theoretical optimum conditions for the entire catalyst bed involving a number of pertinent parameters, such as temperature, pressure, and composition, are determined using mathematical methods of optimization [7,8]. The optimum conditions are found by attainment of a maximum or minimum of some desired objective. The best quality to be formed may be conversion, product distribution, temperature, or temperature program.

Preliminary Selection of Reaction Types

In general, the optimum conditions cannot be precisely attained in real reactors. Therefore, the selection of the reactor type is made to approximate the optimum conditions as closely as possible. For this purpose, mathematical models of the process in several different types of reactors are derived. The optimum condition for selected parameters (e.g., temperature profile) is then compared with those obtained from the mathematical expressions for different reactors. Consequently, selection is based on the reactor type that most closely approaches the optimum.

Consider the production of ethylene oxide by oxidation of ethylene with air or oxygen:

$$C_2H_4 + \frac{1}{2}O_2 \xrightarrow[\substack{250-300^\circ C \\ 4-5 \text{ atms}}]{\text{AgO catalyst}} \underset{\underset{O}{\diagdown \ \diagup}}{CH_2 - CH_2} + (CO_2 + H_2O) \quad (\Delta H_R = -29.2 \text{ kcal})$$

side reaction

The optimum temperature programming is attained by increasing the temperature in accordance with the degree of conversion. This condition is imposed to minimize the complete combustion of ethylene that requires high activation energy. A fixed bed catalytic reactor type that can approximate this temperature program is the reactor with stagewise catalyst beds having internal heat exchangers between the stages.

Process Stability and Parametric Sensitivity

Stability of a chemical reactor for an exothermic process refers to the condition in which the rate of heat removal is equal to or greater than that of heat generated by the reaction, and for which both temperature and concentration do not significantly vary (Chapter 6). Stability is one of the principal factors in the selection of a reactor. Bilous and Amundson [9] have reviewed parametric sensitivity for the condition of a chemical reactor involving thermal behavior and its sensitivity to small variations in process parameters. For example, a slight variation in the feed concentration or the reactor inlet temperature results in a significant change in the temperature profile in the reactor. Parametric sensitivity of the reactor may be determined through the mathematical description of the product.

Final Selection of Reactor System

In this case, economic and technical considerations are incorporated with the results from the preceding steps to determine the final reactor system with respect to the size of the experimental reactor and its operating conditions. The data from the experimental reactor are used to make appropriate corrections for the mathematical model derived in the preceding steps. At this stage, it is essential to review the previous steps for revision of earlier results.

DIMENSIONAL SIMILITUDE AND MATHEMATICAL MODELING

A combination of dimensional similitude and the mathematical modeling technique can be useful when the reactor system and the processes make the mathematical description of the system impossible. This combined method enables some of the critical parameters for scale-up to be specified, and it may be possible to characterize the underlying rate of processes quantitatively.

For example, Hamilton et al. [10] employed dimensional similitude in combination with mathematical modeling in the design of a pilot plant and in evaluating the results to provide the basis for scale-up to a commercial scale plant involving a reaction of the type

$$A \underset{k_{-1}}{\overset{k_1}{\rightleftharpoons}} B \underset{k_{-2}}{\overset{k_2}{\rightleftharpoons}} C$$

A differential equation describing the material balance around a section of the system was first derived, and the equation was made dimensionless by appropriate substitutions. Scale-up criteria were then established by evaluating the dimensionless groups. A mathematical model was further developed based on the kinetics of the reaction, describing the effect of the process variables on the conversion, yield, and catalyst activity. Kinetic parameters were determined by means of both analogue and digital computers.

Rase [5] has presented several case studies of different scale-up methods involving industrially important reactions. Murthy [11] has provided a general guideline for the scale-up of slurry hydrogenation reactors, and other scale-up processes are illustrated elsewhere [12].

SCALE-UP OF A BATCH REACTOR

In scaling-up a batch reactor plant to a full-scale commercial plant, every full-size reactor in the plant will have a working volume, which is some multiple of the capacity of a pilot plant reactor. The production of the full-scale plant is usually predicted from cycle times experienced in the small-scale plant (see Chapter 4).

Generally, it is not difficult to maintain the same cycle time when charging the feeds to the reactor. However, problems can arise during temperature control, such as the times required to complete temperature adjustment steps in the larger unit. When comparing both model and full-scale plants with the same jacket configurations having the same operating conditions, the same steps take longer at the larger scale because heat transfer area does not scale-up at the same rate as the volume.

Steve [13] introduced the concept of the "aspect ratio" of a reactor (defined as the tangent -to- tangent length of the reactor divided by the reactor diameter). It was shown that when the aspect ratio (R) is the same for model and prototype reactors, then the increase in the heat transfer area is only a fraction of the increase in the volumetric capacity. The following reviews the method developed by Steve, and determines the aspect ratio required to make the heat transfer area increase proportionally to the volume.

THE MATHEMATICS OF SCALE-UP

Consider the reactor in Figure 13-1 with defined geometric configurations, the working volume W_V is expressed as:

$$W_V = F\left[aD^3 + 7.48\left(\frac{\pi}{4}\right)D^2L\right] \tag{13-12}$$

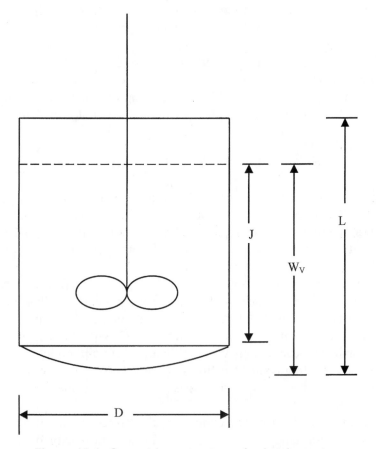

Figure 13-1. Geometric parameters of a batch reactor.

Conversion: 7.48 gal = 1 ft^3

The aspect ratio R is:

$$R = \frac{L}{D} \tag{13-13}$$

The constant d is given by

$$d = 7.48\left(\frac{\pi}{4}\right) = 5.875 \tag{13-14}$$

Substituting Equations 13-13 and 13-14 into Equation 13-12 yields

$$W_V = F[aD^3 + dRD^3] \tag{13-15}$$

$$D = \left[\frac{\left(\frac{W_V}{F}\right)}{(a + dR)}\right]^{1/3} \tag{13-16}$$

where F is the fraction of the total volume occupied by the working volume.

Assume that the total area of the bottom head is effective for heat transfer. Therefore, the total surface area of the reactor in Figure 13-1 available for heating or cooling the working volume is defined by

$$A = tD^2 + \pi DJ \tag{13-17}$$

where J = straight side length covered by the working volume (W_V), ft
t = factor for determining the surface area (ft^2) of the bottom head = 0.93 for ASME standard F&D heads

The working volume can be defined using Equation 13-12 or it can be defined as related to the straight side length that it occupies. Making this expression in J equal to Equation 13-12 with constants as expressed by Equation 13-14 gives

$$W_V = F[aD^3 + dD^2L] = aD^3 + dD^2J \tag{13-18}$$

or $\quad J = (F-1)\dfrac{aD}{d} + FL$ $\hfill$ (13-19)

Substituting Equation 13-19 into Equation 13-17 gives

$$A = tD^2 + \left[(F-1)\frac{aD}{d} + FL\right]\pi D$$

$$= tD^2 + \left[\frac{4a(F-1)D}{7.48\,\pi} + FL\right]\pi D$$

$$A = tD^2 + \frac{4a(F-1)D^2}{7.48} + \pi FLD \qquad (13\text{-}20)$$

where $L = RD$ $\hfill$ (13-13)

Substituting Equation 13-13 into Equation 13-20 and rearranging yields

$$A = \left[t + \frac{4a(F-1)}{7.48}\right]D^2 + \pi FRD^2 \qquad (13\text{-}21)$$

where $\quad f = t + \dfrac{4a(F-1)}{7.48}$

and $g = \pi F$

The heat transfer area A becomes:

$$A = fD^2 + gRD^2 \qquad (13\text{-}22)$$

DEFINING THE SCALE-UP FACTORS

The scale-up factors are defined as follows [13]:

SUF_V = volume scale-up factor

$$= \frac{W_{V2}}{W_{V1}} = \left(\frac{\text{large scale reactor working volume}}{\text{pilot plant reactor working volume}}\right) \qquad (13\text{-}23)$$

SUF_V = heat transfer area scale-up factor

$$= \frac{A_2}{A_1} = \left(\frac{\text{large scale heat transfer area}}{\text{pilot plant heat transfer area}} \right) \tag{13-24}$$

Substituting Equation 13-22 into Equation 13-24 gives

$$SUF_A = \frac{D_2^2 (f + gR_2)}{D_1^2 (f + gR_1)} \tag{13-25}$$

where $D = \left[\dfrac{W_V/F}{(a + dR)} \right]^{1/3}$ $\tag{13-16}$

Assume that the scaled-up reactor has the same bottom head type (e.g., ASME standard F&D) as the pilot plant scale reactor. Equation 13-25 then becomes

$$SUF_A = \left[\frac{\dfrac{W_{V2}/F}{(a + dR_2)}}{\dfrac{W_{V1}/F}{(a + dR_1)}} \right]^{2/3} \left(\frac{f + gR_2}{f + gR_1} \right) \tag{13-26}$$

$$= \left(\frac{W_{V2}}{W_{V1}} \right)^{\frac{2}{3}} \left[\frac{a + dR_1}{a + dR_2} \right]^{\frac{2}{3}} \left(\frac{f + gR_2}{f + gR_1} \right)$$

SUF_A in terms of SUF_V becomes

$$SUF_A = \left[SUF_V \cdot \left(\frac{a + dR_1}{a + dR_2} \right) \right]^{\frac{2}{3}} \left(\frac{f + gR_2}{f + gR_1} \right) \tag{13-27}$$

Equation 13-27 is a general expression that describes how the heat transfer area increases when increasing the volume of a reactor from the pilot plant to the full-scale plant.

MAINTAINING CONSTANT ASPECT RATIO: R

Because there are advantages in maintaining geometric similarity between the pilot plant and the full-scale plant reactors, the larger unit often has the same aspect ratio as the small unit. That is, $R_2 = R_1$. Equation 13-27 then becomes

$$SUF_A = SUF_V^{2/3} \qquad (13\text{-}28)$$

For example, if the larger reactor is twice the size of the pilot unit (i.e., $SUF_V = 2$), and the two units share the same aspect ratio, then the heat transfer area only increases by 1.59 from Equation 13-28:

$$SUF_A = 2^{\frac{2}{3}} = 1.59$$

The SUF_A would, therefore, be 79.5% of the SUF_V. Table 13-4 gives the required aspect ratio for the larger unit for $SUF_A = SUF_V$ at various R_1 values. Table 13-4 shows that SUF_A becomes an increasingly smaller percentage of SUF_V as the latter becomes larger at a constant aspect ratio R.

LARGER ASPECT RATIO: $SUF_A = SUF_V$

In determining the aspect ratio required when $SUF_A = SUF_V$ (although it may not be a practical solution), substitute SUF_V for SUF_A on the left side of Equation 13-27:

$$SUF_V = SUF_V^{\frac{2}{3}} \cdot \left(\frac{a + dR_1}{a + dR_2} \right)^{\frac{2}{3}} \left(\frac{f + gR_2}{f + gR_1} \right) \qquad (13\text{-}29)$$

$$SUF_V^{1/3} = \left(\frac{a + dR_1}{a + dR_2} \right)^{2/3} \left(\frac{f + gR_2}{f + gR_1} \right)$$

$$\text{or } SUF_V = \left(\frac{a + dR_1}{a + dR_2} \right)^{2} \left(\frac{f + gR_2}{f + gR_1} \right)^{3} \qquad (13\text{-}30)$$

Table 13-4
Required R_2 values for $SUF_A = SUF_V$ at various R_1 values

R_1	SUF_V	Required R_2 value	
		Rounded	Raw value
1	2	3	3.11
	3	5	5.13
	5	9	9.14
1.5	2	4	4.01
	3	6.5	6.47
2	2	5	4.96
	3	8	7.89
	4	11	10.81
3	2	7	6.92
	3	11	10.81
4	2	9	8.90

Source: Steve, E. H., *Reactor Design Considerations*, Chemical Engineering, *pp. 96–98, December 1997.*

The following constants are defined as:

$$M = f + gR_1 \tag{13-31}$$

$$N = a + dR_1 \tag{13-32}$$

Equation 13-30 becomes:

$$SUF_V = \frac{N^2}{M^3} \cdot \frac{(f + gR_2)^3}{(a + dR_2)^2} \tag{13-33}$$

Rearranging Equation 13-33 yields

$$\left(SUF_V \cdot \frac{M^3}{N^2}\right)(a + dR_2)^2 = (f + gR_2)^3 \tag{13-34}$$

where $\Psi = SUF_V \cdot \dfrac{M^3}{N^2}$ $\qquad (13\text{-}35)$

Equation 13-34 becomes:

$$\Psi\left(a^2 + 2adR_2 + d^2R_2^2\right) = f^3 + 3f^2gR_2 + 3fg^2R_2^2 + g^3R_2^3 \qquad (13\text{-}36)$$

Rearranging Equation 13-36 into a cubic equation becomes

$$g^3R_2^3 + 3fg^2R_2^2 - \Psi d^2R_2^2 + 3f^2gR_2 - 2adR_2\Psi + f^3 - a^2\Psi = 0$$

$$\text{or } R_2^3 + \left[\frac{3fg^2 - \Psi d^2}{g^3}\right]R_2^2 + \left[\frac{3f^2g - 2ad\Psi}{g^3}\right]R_2 + \left[\frac{f^3 - a^2\Psi}{g^3}\right] = 0 \quad (13\text{-}37)$$

where the coefficients are:

$$C_1 = \frac{3fg^2 - \Psi d^2}{g^3}$$

$$C_2 = \frac{3f^2g - 2ad\Psi}{g^3}$$

$$C_3 = \frac{f^3 - a^2\Psi}{g^3} \qquad (13\text{-}38)$$

Substituting Equation 13-38 into Equation 13-37 gives

$$R_2^3 + C_1R_2^2 + C_2R_2 + C_3 = 0 \qquad (13\text{-}39)$$

Equation 13-39 is a cubic equation in terms of the larger aspect ratio R_2. It can be solved by a numerical method, using the Newton-Raphson method (Appendix D) with a suitable guess value for R_2. Alternatively, a trigonometric solution may be used. The algorithm for computing R_2 with the trigonometric solution is as follows:

1. Design choice inputs: R_1 = 4, SUF_V = 2, F = 0.8
2. Constants: a = 0.606, t = 0.931
3. Calculate: d, f, and g.
4. Calculate: M, N, ψ.
5. Compute the coefficients of the cubic equation, C_1, C_2, and C_3.

6. Calculate: $P = \frac{1}{3}\left(3C_2 - C_1^2\right)$ $\qquad (13\text{-}40)$

7. Calculate: $QV = \dfrac{1}{27}\left(2C_1^3 - 9C_1C_2 + 27C_3\right)$ (13-41)

8. Calculate: $DV = \left(\dfrac{P}{3}\right)^3 + \left(\dfrac{QV}{2}\right)^2$ (13-42)

   If $DV < 0$, the roots are real and unequal and a trigonometric solution is preferred.

9. Compute the angle Φ (ABV = the absolute value of the parameter P).

$$\Phi = \arccos\left[\dfrac{QV^2\!\big/4}{ABV^3\big/27}\right]^{0.5}$$ (13-43)

10. Calculate the three roots: YY_1, YY_2, and YY_3

$$YY_1 = +2\left[\dfrac{ABV}{3}\right]^{0.5}\cos\left(\dfrac{\Phi}{3}\right)$$ (13-44)

$$YY_2 = -2\left[\dfrac{ABV}{3}\right]^{0.5}\cos\left(\dfrac{\Phi+\pi}{3}\right)$$ (13-45)

$$YY_3 = -2\left[\dfrac{ABV}{3}\right]^{0.5}\cos\left(\dfrac{\Phi-\pi}{3}\right)$$ (13-46)

11. Compute the possible roots of R_2.

$$(R_2)_i = YY_i - \dfrac{C_1}{3}$$ (13-47)

12. Select the positive value of R_2 as the larger aspect ratio.

An Excel spreadsheet (EXAMPLE 13-1.xls) was developed to determine R_2 for given values of R_1, SUF_V, and other required parameters.

A Rule-of-Thumb

Steve [13] suggested a rule-of-thumb for calculating R_2 required to make SUF_A equal to SUF_V. Using Table 13-4, it is possible to determine the full-scale pilot plant aspect ratio by

$$R_2 = \left[(R_1+1)SUF_V\right] - 1$$ (13-48)

Using Equation 13-48, the scale-up aspect ratio (R_2) from R_1 is determined as follows:

Pilot plant aspect ratio R_1	Scale-up aspect ratio R_2
1	3
1.5	4
2	5
3	7
4	9

Example 13-1

Determine the aspect ratio of a full-scale batch reactor for $SUF_A = SUF_V$ and at various values of R_1 and SUF_V as shown in Table 13-5.

Solution

An Excel spreadsheet (EXAMPLE 13-1.xls) was developed to determine R_2 for given values of R_1 SUF_V, and other required parameters. The results of the simulation exercise are shown in Table 13-6.

Table 13-5

R_1	SUF_V
1	2
	3
	4
	5
1.5	2
	3
	4
2	2
	3
	4
3	2
	3
	4
4	2
	3
	4

Table 13-6
Effect of scaling-up volume at constant aspect ratio

R_1	SUF_V	R_2	SUF_A	$SUF_A/SUF_V \times 100$
1	2	3.11	1.58	79.4
	3	5.13	2.08	69.3
	4	7.14	2.52	63.0
	5	9.14	2.92	58.5
1.5	2	4.01	1.59	79.4
	3	6.47	2.08	69.3
	4	8.92	2.52	63.0
2	2	4.97	1.59	79.4
	3	7.89	2.08	69.3
	4	10.81	2.52	63.0
3	2	6.92	1.59	79.4
	3	10.81	2.08	69.3
	4	14.70	2.52	63.0
4	2	8.89	1.59	79.4
	3	13.77	2.08	69.3
	4	18.65	2.52	63.0

HEAT TRANSFER MODEL

PREDICTING THE TEMPERATURE
AND TIME FOR SCALE-UP

In predicting the time required to cool or heat a process fluid in a full-scale batch reactor for unsteady state heat transfer, consider a batch reactor (Figure 13-2) with an external half-pipe coil jacket and non-isothermal cooling medium (see Chapter 7). From the derivation, the time θ to heat the batch system is:

$$\theta = \left(\frac{X}{X-1}\right)\left(\frac{Mc}{W_h C_h}\right)\ln\left(\frac{t_1 - T_1}{t_1 - T_2}\right) \tag{13-49}$$

where $X = K_1 = e^{\frac{UA}{W_h C_h}}$

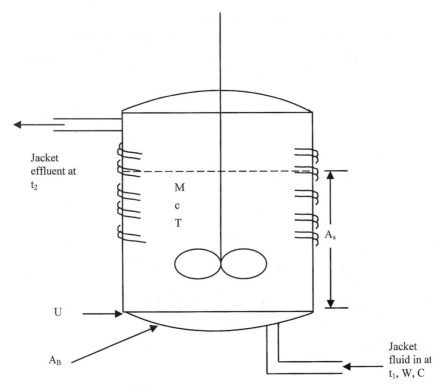

Figure 13-2. A jacketed batch reactor.

Alternatively, for batch cooling with a non-isothermal cooling medium, the time required is:

$$\theta = \left(\frac{X}{X-1}\right)\left(\frac{Mc}{W_cC_c}\right)\ln\left(\frac{T_1-t_1}{T_2-t_1}\right)$$

(13-50)

where $X = K_2 = e^{\frac{UA}{W_cC_c}}$

TIME FOR SCALE-UP

Following Steve's method [13], for a reactor at any scale

$$X_i = e^{\frac{UA_i}{W_iC}}$$

(13-51)

where X_i, A_i, and W_i are the specific values for the scale being considered. For example, when scaling-up from W_{V1} to W_{V2}

$$X_1 = e^{\frac{UA_1}{W_1C}} \tag{13-52}$$

$$\text{and } X_2 = e^{\frac{UA_2}{W_2C}} \tag{13-53}$$

If the larger reactor has a larger size half-pipe jacket, the jacket flow increase is linear, hence

$$W_2 = kW_1 \tag{13-54}$$

Rearranging Equation 13-24 gives

$$A_2 = SUF_A \bullet A_1 \tag{13-55}$$

Substituting Equations 13-55 and 13-54 into Equation 13-53 gives

$$X_2 = e^{\frac{UA_1 SUF_A}{kW_1C}} \tag{13-56}$$

If both scale reactors have the same aspect ratio (L/D) then:

$$SUF_A = SUF_V^{2/3} \tag{13-28}$$

X_2 in terms of SUF_V then becomes:

$$X_2 = e^{\frac{UA_1 SUF_V^{2/3}}{kW_1C}} \tag{13-57}$$

Applying the algebraic law of powers in Equation 13-57 gives

$$X_2 = e^{a_1 b} \tag{13-58}$$

$$\text{where } a_1 = \frac{UA_1}{W_1C} \tag{13-59}$$

$$\text{and } b = \frac{SUF_V^{2/3}}{k} \tag{13-60}$$

Jacket Outlet Temperature

Using Equation 7-120 for batch cooling with an internal coil and a non-isothermal cooling medium,

$$
\begin{array}{cccc}
\mathrm{I} & \mathrm{II} & \mathrm{III} & \mathrm{IV}
\end{array}
$$

$$
\frac{dq'}{d\theta} = -Mc\frac{dT}{d\theta} = W_c C_c (t_2 - t_1) = UA\Delta t_{LMTD} \qquad (7\text{-}120)
$$

where $X = e^{\dfrac{UA}{W_c C_c}}$

The log mean temperature difference Δt_{LMTD} is determined by

$$
\Delta t_1 = T - t_2 \qquad\qquad \Delta t_2 = T - t_1
$$

$$
\begin{aligned}
\Delta t_{LMTD} &= \frac{\Delta t_2 - \Delta t_1}{\ln\left(\dfrac{\Delta t_2}{\Delta t_1}\right)} \\[2em]
&= \frac{(T - t_1) - (T - t_2)}{\ln\left(\dfrac{T - t_1}{T - t_2}\right)} \qquad\qquad (13\text{-}61)
\end{aligned}
$$

Equating III and IV of Equation 7-120 gives

$$
W_c C_c (t_2 - t_1) = UA\left[\frac{(T - t_1) - (T - t_2)}{\ln\left(\dfrac{T - t_1}{T - t_2}\right)}\right] \qquad (13\text{-}62)
$$

$$\text{or } (t_2 - t_1) = \frac{UA}{W_cC_c} \left[\frac{t_2 - t_1}{\ln\left(\dfrac{T - t_1}{T - t_2}\right)} \right] \tag{13-63}$$

$$\ln\left(\frac{T - t_1}{T - t_2}\right) = \frac{UA}{W_cC_c} \tag{13-64}$$

Rearranging Equation 13-64 gives

$$T - t_1 = e^{\frac{UA}{W_cC_c}} (T - t_2) \tag{13-65}$$

$$\text{or } T - t_1 = X(T - t_2) \tag{13-66}$$

where $X = e^{\frac{UA}{W_cC_c}}$

The outlet fluid cooling temperature t_2 can be expressed by

$$t_2 = \overset{\text{I}}{\frac{T(X-1) + t_1}{X}} = \overset{\text{II}}{T + \left(\frac{t_1 - T}{X}\right)} \tag{13-67}$$

The values of t_1 and X in Equation 13-67 are constant. Temperature T varies from the temperature at the start of the cooling cycle T_1 to the temperature at the end of the cooling cycle T_2. When cooling $T_2 < T_1$, the I form of Equation 13-67 clearly shows that the predicted t_2 at the beginning of the cycle will be larger than the predicted t_2 at the end of the cycle.

Alternatively, for batch heating with a non-isothermal heating medium, the log mean temperature difference Δt_{LMTD} is determined by

$$\Delta t_{LMTD} = \frac{[(t_1 - T) - (t_2 - T)]}{\ln\left[\dfrac{t_1 - T}{t_2 - T}\right]} \tag{13-68}$$

Substituting X_2 into Equations 13-67, 13-49, and 13-50, it is possible to determine the effect of scale-up on the jacket outlet temperature and the time required in completing the temperature adjustment step. In directly predicting the expected value for the outlet temperature t_2 from the jacket of a reactor scaled-up by some volume scale-up factor SUF_V, substitute Equation 13-58 into form II of Equation 13-67. This gives

$$t_2 = T + \left(\frac{t_1 - T}{e^{a_1 b}}\right) \tag{13-69}$$

For the full-scale reactor:

$$M_2 = M_1 \cdot SUF_V \tag{13-70}$$

and $W_{C2} = kW_{C1}$ (13-71)

Subscript 2 indicates the full-scale reactor, which yields

$$\frac{M_2 c}{W_{C2} C_C} = \frac{SUF_V M_1 c}{kW_{C1} C_C} \tag{13-72}$$

In directly predicting the expected value for the time required to complete a cooling step in a reactor scaled-up by some volume scale-up factor SUF_v, substituting Equations 13-58 and 13-72 into Equation 13-50 yields

$$\theta = \left(\frac{M_2 c}{W_{C2} C_C}\right)\left(\frac{X_2}{X_2 - 1}\right) \ln\left(\frac{T_1 - t_1}{T_2 - t_1}\right)$$

$$\text{or } \theta = \left(\frac{SUF_V \cdot M_1 \cdot c}{kW_{C1} C_C}\right)\left(\frac{e^{a_1 b}}{e^{a_1 b} - 1}\right) \ln\left(\frac{T_1 - t_1}{T_2 - t_1}\right) \tag{13-73}$$

Alternatively, for a heating condition, the heating time required for scale-up is:

$$\theta = \left(\frac{SUF_V \cdot M_1 \cdot c}{kW_{h1} C_h}\right)\left(\frac{e^{a_1 b}}{e^{a_1 b} - 1}\right) \ln\left(\frac{t_1 - T_1}{t_1 - T_2}\right) \tag{13-74}$$

Example 13-2

Consider an agitated batch reactor having the following parameters:

A, ft^2	265
C, Btu/lb °F	0.52
C_C, Btu/lb °F	1.0
M, lb	29,962
t_1, °F	104
T_1, °F	221
T_2, °F	158
U, Btu/hr-ft^2 °F	74.0
W_C, lb/hr	28,798

Determine the time required to cool twice the volume of the pilot plant batch reactor.

Solution

The time required to cool the pilot plant batch reactor is 0.85 hr. Applying form II of Equation 13-67 and $W_C C_C (t_2 - t_1)$ for the heat removal in the pilot plant reactor yields

$$t_2 = T + \frac{(t_1 - T)}{X} \qquad (13\text{-}67)$$

$$X = e^{\frac{UA}{W_C C_C}}$$

$$= \exp\left(\frac{74 \times 265}{28,798 \times 1.0}\right)$$

$$= 1.976$$

Heat removal, $Q = W_C C_C (t_2 - t_1)$
This results in the following:

	t_2 °F	Heat removal, Q Btu/hr	Tons^f
Start of cycle	161.79	1.66×10^6	138.3
End of cycle	130.67	0.77×10^6	64.2

^{f} 1 ton of refrigeration = 3,517 W = 12,000 Btu/hr

The results quantitatively show the significance of considering both boundaries of the unsteady state operation. The load required at the end of the cycle is less than one-half the load required at the start of the cycle. Steve [13] inferred that reporting the value at only one boundary may be misleading. For example, the total heat load (provided by chillers and heaters) from multiple reactors with overlapping temperature adjustment cycles may be skewed if the changes within those cycles are not considered.

For the full-scale plant, assume:

A constant overall heat transfer coefficient, $U_2 = U_1$

$$SUF_V = \frac{W_{V2}}{W_{V1}} = 2$$

The size of the half-pipe jacket is identical at both scales, that is, $k = 1.0$

Therefore,

$$W_{C2} = kW_{C1} = 1.0 \times W_{C1}$$

$$a_1 = \frac{U_1 A_1}{W_{C1} C_C} = \frac{74 \times 265}{28,798 \times 1.0} = 0.681$$

$$b = \frac{SUF_V^{2/3}}{k} = \frac{2^{2/3}}{1} = 1.587$$

where $X_2 = e^{a_1 b}$

$$= \exp(0.681 \times 1.587)$$

$$= 2.95$$

Applying form II of Equation 13-67 and $W_C C_C (t_2 - t_1)$ for the heat removal in the full-scale plant reactor yields

	t_2 °F	Heat removal, Q Btu/hr	Tons[f]
Start of cycle	181.3	2.23×10^6	185.8
End of cycle	139.68	1.03×10^6	85.8

There are increased heat loads in the full-scale batch reactor because the heat transfer area is 1.59 times larger; the cooling rate W_C remains constant.

The time required to cool the batch reactor in a full-scale batch reactor is determined by Equation 13-73

$$\theta = \left(\frac{SUF_V \bullet M_1 \bullet c}{kW_{Cl}C_C} \right)\left(\frac{e^{a_1 b}}{e^{a_1 b}-1} \right) \ln\left(\frac{T_1 - t_1}{T_2 - t_1} \right) \qquad (13\text{-}73)$$

$$\theta = \left(\frac{2 \times 29,962 \times 0.52}{28,798 \times 1.0} \right)\left(\frac{2.95}{2.95-1} \right) \ln\left(\frac{221-104}{158-104} \right)$$

$$= 1.27 \, hr$$

The cooling time in a full-scale batch reactor is 1.27 hr, which is approximately 49% more than the 0.85 hr predicted at the smaller scale. Multiple temperature adjustment steps occur in a chemical reactor production cycle. The sum of these cycle time increases may be significant, and the plant capacity at the larger scale may be adversely affected if the cycle times are not corrected [13]. An Excel spreadsheet (EXAMPLE 13-2.xls) was developed for Example 13-2.

JACKET ZONING OF A BATCH REACTOR

The heat transfer performance of a batch reactor system equipped with a partial pipe jacket can be improved using multiple zoning. This involves dividing the jacket into two or more zones. In this instance, each independent zone receives the heat transfer medium at the cooling or heating supply temperature at the inlets. The outlets from these zones are combined, and the flows of the medium blend to create a stream of some mixture temperature before the combined flow returns to the cooling or heating utility equipment. The zoning of the reactor enhances the heat transfer without attaining an increase in the jacket pressure drop [14].

Consider a single-zone jacket where there is an increase in the jacket flow, and a corresponding increase in the outside film coefficient because $h_j = f\left(N_{Re}^{-0.2}, G\right)^*$. Therefore, a two-fold increase in the jacket flow results in an increase in h_{j2} by $2^{0.8} h_{ji}^*$. The overall heat transfer coefficient $U = 1/[F_{FOL} + 1/h_j]$, and a larger outside coefficient subsequently increases the overall heat transfer coefficient. The overall heat flux will increase due to the combined effects of the increased flow and lower jacket outlet temperature. The net result is an increase in the pressure drop.

To reduce the pressure drop, a batch reactor with a half-pipe jacket of length L and flowrate W can be partitioned into a two-zone jacket, each with a length L/2 and each supplied with W jacket flowrate. This doubles the jacket flow at a lower pressure drop in each zone. The flow in each zone can then be increased to increase the outside and overall heat transfer coefficients, which is similar to those of the single-zone jacket.

MATHEMATICAL MODELING OF THE BATCH REACTOR

Consider a batch reactor with a partial jacket that is divided into three zones (Figure 13-3). The flow to each zone is the same as the single-zone flow. The fluid velocity is maintained constant to maintain the same outside heat transfer coefficient as a single zone jacket.

The heat balance to the three-zone jacket can be expressed by

$$W_J C_C(t_m - t_1) = W_z C_C(t_{z1} - t_1) + W_z C_C(t_{z2} - t_1)$$

$$+ W_z C_C(t_{z3} - t_1) \tag{13-75}$$

$^*h_j = 0.023\, N_{Re}^{-0.2} N_{Pr}^{-2/3} C_P G \bullet k$ or $h_j = f\left(N_R^{-0.2}, G\right)$. $G = v\rho$, $N_{Re} = \rho v D/\mu$. $\dfrac{h_1}{h_2} = \dfrac{G_1/N_{Re_1}^{0.2}}{G_2/N_{Re_2}^{0.2}}$.

Therefore $\dfrac{h_1}{h_2} = \dfrac{N_{Re_1}^{0.2}}{N_{Re_2}^{0.2}} \cdot \dfrac{G_2}{G_1} \left(\dfrac{N_{Re_1}}{N_{Re_2}}\right)^{0.2} = \left(\dfrac{v_1}{v_2}\right)^{0.2}$. But $\dfrac{G_2}{G_1} = \dfrac{v_2}{v_1}$. Therefore $\left(\dfrac{N_{Re_1}}{N_{Re_2}}\right)^{0.2} \cdot \dfrac{G_2}{G_1} =$

$\dfrac{v_1^{0.2}}{v_2^{0.2}} \cdot \dfrac{v_2}{v_1} = \left(\dfrac{v_2}{v_1}\right)^{0.8}$. If $v_2 = 2v_1$, $\dfrac{h_2}{h_1} = 2^{0.8}$.

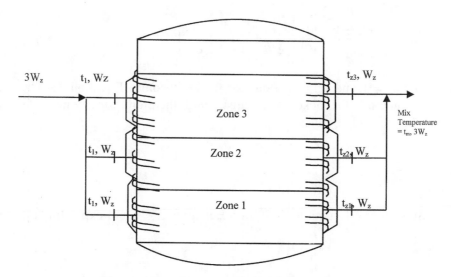

Figure 13-3. A zoned jacketed batch reactor.

Rewriting Equation 13-67 in a generalized form that shows the outlet temperature from any zone gives

$$t_{zi} = T + \frac{t_1 - T}{e^{ai}} \qquad (13\text{-}76)$$

where $a_i = \dfrac{UA_i}{W_z C_C}$

and subscript i is for the individual zone.

Substituting Equation 13-76 into Equation 13-75, where $W_j = 3W_z$, gives

$$t_m = T + \left[\frac{t_1 - T}{3} \right] \left[\frac{1}{e^{a1}} + \frac{1}{e^{a2}} + \frac{1}{e^{a3}} \right] \qquad (13\text{-}77)$$

Equation 13-77 is modified to give a more generalized equation for t_m for any number of zones as:

$$t_m = T + \left[\frac{t_1 - T}{n_z} \right] \left[\frac{1}{e^{a1}} + \frac{1}{e^{a2}} + \ldots + \frac{1}{e^{az}} \right] \tag{13-78}$$

For $n_z = 1$ and $t_m = t_2$, Equation 13-78 becomes Equation 13-67.

Assume that the jackets have equal areas, therefore, for any multiple number of jacket zones,

$$A_i = \frac{A_T}{n_z} \tag{13-79}$$

Substituting Equation 13-79 into Equation 13-78 and expanding a_1, a_2, $\ldots$ a_z, gives

$$t_m = T + \left[\frac{t_1 - T}{n_z} \right] \left[\frac{1}{e^{\frac{UA_T}{n_z W_z C_C}}} + \frac{1}{e^{\frac{UA_T}{n_z W_z C_C}}} + \ldots + \frac{1}{e^{\frac{UA_T}{n_z W_z C_C}}} \right] \tag{13-80}$$

Since $a_1 = a_2 = \ldots = a_z$, Equation 13-80 is expressed by

$$t_m = T + \left[\frac{t_1 - T}{n_z} \right] \left[\frac{n_z}{e^{\frac{UA_T}{n_z W_z C_C}}} \right]$$

$$= T + \frac{t_1 - T}{\left[e^{\frac{UA_T}{W_z C_C}} \right]^{\upsilon}} \tag{13-81}$$

where $\upsilon = \dfrac{1}{n_z}$

Because U and A_T are constants over the cooling cycle, Equation 13-81 can be used to predict the temperature of the mixed outlet streams as:

$$t_m = T + \left[\frac{t_1 - T}{X^{\upsilon}}\right] \qquad (13\text{-}82)$$

TEMPERATURE ADJUSTMENT TIME

Consider the jacket zoning of a batch reactor to determine the times required for a temperature adjustment step [14]. W_C is replaced by $n_Z W_Z$ and the jacket outlet temperature t_2 by t_m in Equation 13-50. The heat balance for cooling is expressed by:

$$\underset{\text{I}}{\frac{dq'}{d\theta}} = \underset{\text{II}}{-Mc\frac{dT}{d\theta}} = \underset{\text{III}}{n_z W_z C_C (t_m - t_1)} = \underset{\text{IV}}{UA\Delta t_{LMTD}} \qquad (13\text{-}83)$$

For heating, the unsteady state heat balance is given by

$$\frac{dq'}{d\theta} = \underset{\text{V}}{Mc\frac{dT}{d\theta}} = \underset{\text{VI}}{n_z W_z C_h (t_1 - t_m)} = \underset{\text{VII}}{UA\Delta t_{LMTD}} \qquad (13\text{-}84)$$

The log mean temperature difference Δt_{LMTD} for cooling is:

$$\Delta t_{LMTD} = \frac{(T - t_1) - (T - t_m)}{\ln\left[\dfrac{T - t_1}{T - t_m}\right]} \qquad (13\text{-}85)$$

For heating, Δt_{LMTD} is expressed by:

$$\Delta t_{LMTD} = \frac{(t_1 - T) - (t_m - T)}{\ln\left[\dfrac{t_1 - T}{t_m - T}\right]} \qquad (13\text{-}86)$$

Combining forms II and III of Equation 13-83 for cooling and substituting t_m in Equation 13-82 gives

$$-Mc\frac{dT}{d\theta} = n_z W_z C_C(t_m - t_1)$$

$$= n_z W_C C_C \left\{ \left[T + \left(\frac{t_1 - T}{X^\upsilon} \right) \right] - t_1 \right\}$$

$$= n_z W_z C_C \left\{ (T - t_1) \left(\frac{X^\upsilon - 1}{X^\upsilon} \right) \right\} \tag{13-87}$$

Rearranging Equation 13-87 and integrating between the boundary conditions, $\theta = 0 \rightarrow \theta$, $T = T_1 \rightarrow T_2$, gives

$$-\int_{T_1}^{T_2} \frac{dT}{T - t_1} = \left(\frac{n_z W_z C_C}{Mc} \right) \left(\frac{X^\upsilon - 1}{X^\upsilon} \right) \int_0^\theta d\theta$$

$$\ln \left[\frac{T_1 - t_1}{T_2 - t_1} \right] = \left(\frac{n_z W_z C_C}{Mc} \right) \left(\frac{X^\upsilon - 1}{X^\upsilon} \right) \theta \tag{13-88}$$

The cooling time θ is given by

$$\theta = \left(\frac{Mc}{n_z W_z C_C} \right) \left(\frac{X^\upsilon}{X^\upsilon - 1} \right) \ln \left[\frac{T_1 - t_1}{T_2 - t_1} \right] \tag{13-89}$$

Alternatively, the time required for heating the batch system is:

$$\theta = \left(\frac{Mc}{n_z W_z C_h} \right) \left(\frac{X^\upsilon}{X^\upsilon - 1} \right) \ln \left[\frac{t_1 - T_1}{t_1 - T_2} \right] \tag{13-90}$$

THE OUTLET TEMPERATURE OF
A SCALED-UP BATCH SYSTEM

Consider the scale-up of a batch reactor from a pilot plant reactor to a full-scale reactor. Rewriting Equation 13-82 to the full-scale reactor yields:

$$t_m = T + \frac{t_1 - T}{X_2^P} \qquad (13\text{-}91)$$

where $P = \dfrac{1}{n_{z2}}$

The subscript 1 refers to the process parameters for the reactor at the smaller scale, and the subscript 2 refers to those parameters for the reactor at the larger scale.

Because the pilot plant and full-scale reactors have the same aspect ratio (L/D), this gives

$$A_{T2} = SUF_A \bullet A_{T1} = SUF_V^{2/3} \bullet A_{T1} \qquad (13\text{-}28)$$

Also, $W_2 = kn_{z2} \, W_{z1}$ $\qquad (13\text{-}92)$

Combining Equations 13-28 and 13-92, and the definitions of X_1 and X_2, gives

$$X_2 = e^{\frac{UA_{T2}}{W_2 C_C}} = e^{\frac{U\left(SUF_V^{2/3}\right)A_{T1}}{kn_{z2}W_{z1}C_C}} \qquad (13\text{-}93)$$

$$\text{or} \quad X_2 = X_1^{\left(\frac{SUF_V^{2/3}}{kn_{z2}}\right)} \qquad (13\text{-}94)$$

Substituting Equation 13-94 into Equation 13-91 gives a general expression for predicting the temperature resulting from mixing the outlet streams from multiple jacket zones. This is expressed by

$$t_{m2} = T + \left(\frac{t_1 - T}{X_1^s}\right) \qquad (13\text{-}95)$$

where $s = \dfrac{SUF_V^{2/3}}{kn_{z2}}$

The time required to complete a cooling temperature adjustment step for the full-scale reactor is:

$$\theta = \left(\frac{M_2 c}{W_{C2} C_C} \right) \left(\frac{X_2}{X_2 - 1} \right) \ln \left[\frac{T_1 - t_1}{T_2 - t_1} \right] \tag{13-96}$$

where $M_2 = SUF_V \bullet M_1$ $\tag{13-67}$

Substituting Equations 13-70 and 13-92 into Equation 13-96 gives for cooling:

$$\theta = \left(\frac{SUF_v}{kn_{z2}} \right) \left(\frac{M_1 c}{W_{z1} C_C} \right) \left(\frac{X_2}{X_2 - 1} \right) \ln \left[\frac{T_1 - t_1}{T_2 - t_1} \right] \tag{13-97}$$

Alternatively, the time required for heating a full-scale batch reactor is:

$$\theta = \left(\frac{SUF_v}{kn_{z2}} \right) \left(\frac{M_1 c}{W_{z1} C_h} \right) \left(\frac{X_2}{X_2 - 1} \right) \ln \left[\frac{t_1 - T_1}{t_1 - T_2} \right] \tag{13-98}$$

Example 13-3

The jacket of a pilot plant batch reactor is divided into three zones. Using the data in Example 13-2, determine the outlet temperature, the heat removed, and the cooling time for both pilot and full-scale batch reactors.

Solution

An Excel spreadsheet program (EXAMPLE13-3.xls) was developed to determine the mixed outlet temperature at the start and end of the cooling cycles, the heat removal, and the cooling time for the pilot and full-scale batch reactors. Table 13-7 gives the results for the number of zones. The results show that segmenting the jacket with accompanying higher jacket flows reduces the batch cooling time.

Table 13-7

Effect of jacket zoning in a scale-up batch reactor

Case	Number of jacket zones	Total jacket flowrate, lb/hr	Flow in each zone, lb/hr	Factors s	X_i^s	T_{m2} (or t_{z2}) at Start °F	End °F	Cooling load, tons Start	End	Cooling step time, hr
1	1	28,798	28,798	1.587	2.947	181.3	139.68	185.5	85.6	1.27
2	2	57,596	28,798	0.794	1.717	152.86	126.55	234.5	108.2	1.0
3	3	86,394	28,798	0.529	1.434	139.41	120.34	254.9	117.6	0.92
4	4	115,192	28,798	0.397	1.31	131.69	116.78	265.8	122.8	0.88

ASPECT RATIO (R) IN JACKET ZONING
AND SCALE-UP OF A BATCH REACTOR

A combination of jacket zoning and aspect ratio adjustment is used to provide a scale-up reactor that performs temperature adjustment steps in the same time as the pilot scale unit [14]. From Equation 13-27

$$SUF_A = \left[SUF_V \bullet \left(\frac{a + dR_1}{a + dR_2} \right) \right]^{\frac{2}{3}} \left(\frac{f + gR_2}{f + gR_1} \right) \qquad (13\text{-}27)$$

Using $\beta = \left(\frac{a + dR_1}{a + dR_2} \right)^{\frac{2}{3}} \left(\frac{f + gR_2}{f + gR_1} \right)$ $\qquad (13\text{-}99)$

Equation 13-27 becomes

$$SUF_A = SUF_V^{\frac{2}{3}} \bullet \beta \qquad (13\text{-}100)$$

If the pilot plant and the full-scale reactors have the same aspect ratio, then $\beta = 1$ and

$$SUF_A = SUF_V^{\frac{2}{3}} \qquad (13\text{-}28)$$

In specifying the number of jacket zones and the aspect ratio for a full-scale reactor, there is a limitation on the temperature adjustment time. This implies that it must be of the same duration as experienced in the pilot plant reactor. Combining Equations 13-89 and 13-97 yields

$$\left[\frac{SUF_V}{kn_{z2}} \right] \left[\frac{X_2}{X_2 - 1} \right] = \frac{X^{\upsilon}}{X^{\upsilon} - 1} \qquad (13\text{-}101)$$

Rearranging the variables to establish a series of constants gives

$$H = \frac{SUF_V}{kn_{z2}} \qquad (13\text{-}102)$$

$$r = \frac{X^{\upsilon}}{X^{\upsilon} - 1} \qquad (13\text{-}103)$$

$$K = \frac{r}{H} \qquad (13\text{-}104)$$

Substituting H, r, and K into Equation 13-101 gives

$$r = H\left(\frac{X_2}{X_2 - 1}\right) \qquad (13\text{-}105)$$

$$K = \frac{X_2}{X_2 - 1}$$

$$\text{or } X_2 = \frac{K}{K - 1} \qquad (13\text{-}106)$$

From Equation 13-53

$$X_2 = e^{\frac{UA_2}{W_2 C}} \qquad (13\text{-}53)$$

$$\text{and } W_2 = n_{z2}kW_1 \qquad (13\text{-}107)$$

$$A_2 = SUF_A \cdot A_1$$

$$= SUF_V^{2/3} \cdot A_1 \cdot \beta \qquad (13\text{-}108)$$

Substituting Equations 13-107 and 13-108 into Equation 13-53 gives

$$X_2 = e^{\left(\frac{U A_1 SUF_V^{2/3} \beta}{W_1 C_C k n_{z2}}\right)} \qquad (13\text{-}109)$$

Equating Equations 13-106 and 13-109 gives

$$e^{\left(\dfrac{UA_1 SUF_V^{2/3}\beta}{W_1 C_C kn_{z2}}\right)} = \dfrac{K}{K-1} \tag{13-110}$$

or $$\dfrac{UA_1\, SUF_V^{2/3}\, \beta}{W_1\, C_C\, kn_{z2}} = \ln\left(\dfrac{K}{K-1}\right)$$

$$\beta = \dfrac{W_1\, C_C kn_{z2}}{UA_1 \left(SUF_V\right)^{2/3}} \ln\left[\dfrac{K}{K-1}\right] \tag{13-111}$$

From Equation 13-99

$$M = f + gR_1 \tag{13-31}$$

$$N = a + dR_1 \tag{13-32}$$

Equation 13-99 becomes

$$\beta = \left(\dfrac{N}{a + dR_2}\right)^{2/3}\left(\dfrac{f + gR_2}{M}\right) \tag{13-112}$$

where $$Q = \dfrac{f + gR_2}{\left(a + d R_2\right)^{2/3}} \tag{13-113}$$

Equation 3-112 becomes

$$\beta = \dfrac{N^{2/3}}{M} \bullet Q \tag{13-114}$$

or $$Q = \beta \dfrac{M}{N^{2/3}} \tag{13-115}$$

Rearranging Equation 13-113 gives

$$Q(a + dR_2)^{2/3} = f + gR_2 \tag{13-116}$$

Expanding and rearranging Equation 13-116 yields

$$Q^3(a + dR_2)^2 = (f + gR_2)^3$$

$$R_2^3 + \left(\frac{3fg^2 - Q^3d^2}{g^3}\right)R_2^2 + \left(\frac{3f^2g - 2adQ^3}{g^3}\right)R_2 + \left(\frac{f^3 - a^2Q^3}{g^3}\right) = 0 \quad (13\text{-}117)$$

Equation 13-117 is a cubic equation in R_2 where

$$C_1 = \frac{3fg^2 - Q^3d^2}{g^3}$$

$$C_2 = \frac{3f^2g - 2adQ^3}{g^3}$$

$$C_3 = \frac{f^3 - a^2Q^3}{g^3} \quad (13\text{-}118)$$

Equation 13-117 becomes

$$R_2^3 + C_1R_2^2 + C_2R_2 + C_3 = 0 \quad (13\text{-}119)$$

Using the Microsoft Excel spreadsheet EXAMPLE13-4.xls, the aspect ratio (i.e., the root R_2) of Equation 13-119 of the zoned full-scale reactor can be determined. R_2 was reviewed earlier for a single-zone batch reactor.

Example 13-4

Using the data in Example 13-2, determine the aspect ratio for the same cooling time (0.85 hr) as the larger unit, with the working volume twice that of the pilot plant scale of a segmented batch reactor.

Solution

Table 13-8 shows the results of the simulation exercise for R_2 (L/D_2) at varying number of zones. When $n_{z2} = 1$, the value of R_2 is quite high, which relates to a taller reactor with smaller size. Correspondingly,

Table 13-8

Aspect ratio R_2 $(L/D)_2$ for zoning scale-up of a batch reactor using Equation 13-119

n_{z2}	r	H	K	K − 1	X_2	SUF_A	C_1	C_2	C_3	$(L/D)_2$
1	2.025	2	1.013	0.013	77.92	6.397	−129.67	−26.6	−1.35	129.87
2	2.025	1	2.025	1.025	1.976	2	−2.96	−0.47	0.00	3.11
3	2.025	0.667	3.036	2.036	1.491	1.76	−1.69	−0.21	0.01	1.8
4	2.025	0.5	4.05	3.05	1.328	1.667	−1.28	−0.12	0.02	1.36

the value of X_2 in this case is also high. The results show that R_2 decreases as the number of zones n_{z2} increases. However, an increase in the aspect ratio from 1 to 3 would require an additional blade on the agitator, which could present problems in achieving the same mixing pattern as the pilot scale reactor.

NOMENCLATURE

a	Factor for determining the capacity (gal) for the bottom head = 0.606 for ASME Standard flanged and dished heads
A	Total heat transfer surface area, ft^2
A_B	Heat transfer area on the bottom head of the reactor, ft^2
A_i	Heat transfer area of a zone = A_T/n_z, ft^2
A_S	Heat transfer area on the straight side of the reactor, ft^2
A_T	Total heat transfer surface area = $A_S + A_B$, ft^2
c	Heat capacity of mass of material in the batch reactor, Btu/lb • °F
C_c	Heat capacity of coolant jacket fluid, Btu/lb • °F
C_h	Heat capacity of heating jacket fluid, Btu/lb • °F
C_1, C_2, C_3	Coefficients of the cubic equation
D	Reactor inside diameter, ft.
F	Fraction of the total volume occupied by the working volume.
G	Mass flowrate, lb/hr • ft^2
h_j, h_{j1}, h_{j2}	Outside film heat transfer coefficients, Btu/hr • ft^2 • °F
J	Straight side length covered by the working volume, W_V, ft
k	Scale-up term for jacket flow, dimensionless
L	Reactor straight side (tangent-to-tangent) length, ft
L′	Flow length of a half-pipe jacket, ft
LMTD	Log mean temperature difference, °F
M	Mass of material in the batch reactor, lb
n_z	Number of zones, dimensionless
P	$1/n_{z2}$, dimensionless
R	Aspect ratio, L/D dimensionless
SUF_A	Heat transfer area scale-up factor = A_2/A_1
SUF_V	Volume scale-up factor = W_{V2}/W_{V1}
t	Factor for determining the surface area, ft^2 of the bottom head = 0.931 for ASME standard F&D heads
t_1	Jacket inlet temperature, °F
t_2	Jacket outlet temperature, °F
t_m	Mixed outlet temperature, °F

t_{zi} Outlet temperature from zone i, °F
T Temperature of mass of material in the batch reactor, °F
T_1 Initial batch temperature, °F
T_2 Final batch temperature, °F
U Overall heat transfer coefficient, Btu/hr • ft^2 • °F
W Jacket flowrate, lb/hr
W_C Coolant jacket flowrate, lb/hr
W_J n_z • W_Z, lb/hr
W_h Heating jacket flowrate, lb/hr
W_Z Flow rate to a zone, lb/hr
W_V Working volume, gal
X $e^{UA/WC} = e^{a_i}$
z A generalized subscript used for counting total zone numbers
YY_i Roots of the cubic equation
θ Time required to complete the cooling or heating cycle, hr

REFERENCES

1. Kabel, R. L., *Reaction Kinetics*, Bisio, A. and Kabel, R. L., "Scaleup of chemical processes conversion from laboratory scale tests to successful commercial size design," John Wiley & Sons, 1985.
2. Kafarov, V., *Cybernetics Methods in Chemistry & Chemical Engineering*, MIR Publishers, Moscow, 1976.
3. Kline, S. J., *Similitude and Approximation Theory*, McGraw-Hill, New York, 1965.
4. Himmelblau, D. M., *Mathematical Modeling*, Bisio, A., and Kabel, R. L., "Scaleup of chemical processes," John Wiley & Sons, 1985.
5. Rase, H. F., "Chemical reactor design for process plants, vol II: Case studies and design data," John Wiley & Sons, New York, 1977.
6. Carberry, J. J., *Chemical and Catalytic Reaction Engineering*, McGraw-Hill, New York, 1976.
7. Edgar, T. F. and Himmelblau, D. M., *Optimization of Chemical Processes*, McGraw-Hill, 1988.
8. Beveridge, G. S. G. and Schechter, R. S., *Optimization: Theory and Practice*, McGraw-Hill, 1970.
9. Bilous, O. and Amundson, N. R., *Am. Inst. Chem. Eng., J.*, 2, 117, 1965.
10. Hamilton, L. and Harrington, *Chem., Eng., Prog.*, 58 (2), pp. 55–59, 1962.

11. Murthy, A. K. S., "Design and scaleup of slurry-hydrogenation systems," *Chemical Engineering,* pp. 94–107, September 1999.
12. Perry, R. H. and Green, D. W. (Eds.) *Perry's Chemical Engineers Handbook,* 6th ed., McGraw-Hill, New York, 1984.
13. Steve, E. H., "Reactor design considerations (Parts 1 and 2)," *Chemical Engineering,* pp. 96–102, 1997.
14. Steve, E. H., "Jacket zoning in reactor scale-up (Parts 3 and 4)," *Chemical Engineering,* pp. 92–98, January 1998.
15. Steve, E. H., private communications, 2000.

Nomenclature

A	cross sectional area; heat transfer area of heating or cooling coil, m^2
A, B	reactants
$a, b, \ldots c, d$	substances $A, B, \ldots C, D \ldots$
A_v	cross-sectional area of tank, m^2
B	number of blades on impeller
C^*	oxygen concentration in equilibrium with the partial pressure of oxygen, kg/m^3
C, D	products of reaction
C'_{AO}	initial concentration, mol/unit mass
C_A	concentration of key reactant species with stoichiometric coefficient, mol/m^3
C_{AO}	initial concentration, mol/m^3
C_d	nonviable cell concentration, kg/m^3
C_E	enzyme concentration, kg/m^3
C_{ES}	concentration of the complex, ES^*, kg/m^3
C_{ET}	total enzyme concentration, kg/m^3
C_L	concentration of dissolved oxygen, kg/m^3
C_M	Monod constant, mol/m^3
C_p	heat capacity, $kJ/kg \bullet K$
C_p	specific molar heat capacity or heat capacity at constant pressure, $J/mol \bullet K$
C_p	product concentration, kg/m^3
C_s	substrate concentration, kg/m^3
C_v	specific molar heat capacity or heat capacity at constant volume, $J/mol \bullet K$
C_v	viable cell concentration, kg/m^3
C_x	cell concentration, kg/m^3
d	diameter, mm

D	dilution rate, h^{-1}
D_A	diffusion coefficient of species A in a bulk fluid, m^2/s
D_A	impeller diameter, m
D_{max}	maximum rate of cell production, h^{-1}
D_l	axial diffusivity, m^2/s
D_p	effective particle diameter, mm
D_r	radial diffusivity, m^2/s
d_t	tube diameter, mm
D_T	tank diameter, m
E	exit age residence time distribution function (min^{-1})
E	height of the agitator from the bottom of tank
E_A	activation energy, J/mol
F	cumulative residence time distribution function
F_{AO}	feed molar rate, mol/s
FF_j	fouling factor, inside vessel
f_i	fugacity of pure i
G	Gibbs free energy, J/mol A
G	mass flow rate, kg/s; superficial mass velocity, lb/hr•ft^2, kg/m^2•s
g_c	dimensional constant, 1 kg•m/s^2•N
GHSV	gas hourly space velocity, h^{-1}
h	enthalpy per unit mass, kJ/kg
H	enthalpy, J/mol
h	heat transfer coefficient, W/m^2•K
H	liquid height, m
h_j	film coefficient on inside surface of jacket, W/m^2•K
ΔH_v	heat of vaporization, J/mol
ΔH_R	heat of reaction, J/mol
J	baffle width
k	reaction rate constant, $(mol/m^3)^{1-n} s^{-1}$
K	equilibrium constant for the stoichiometry
k	thermal conductivity, W/m•K
k_d	death rate coefficient, s^{-1}
k, k^i, k^{ii}, k^{iii}, k^{iv}	reaction rate constants based on r, r^i, r^{ii}, r^{iii}, r^{iv}
K_I	dissociation constant for inhibitor binding, kg/m^3
k_l	axial thermal conductivity, W/m•K
k_{La}	mass transfer coefficient, min^{-1}

K_m	Michaelis-Menten constant, $kmol/m^3$
k_o	pre-exponential factor of single reaction
k_p	specific rate constant with product, s^{-1}
k_r	radial thermal conductivity, $W/m \cdot K$
K_s	The Monod constant, kg/m^3
L	length of tubular reactor; unit depth of packed bed; agitator blade length
LHSV	liquid hourly space velocity, h^{-1}
M	mass, kg
M_A	molecular weight of A, kg/mol of A
M_j	molecular weight, kg/mol
M_L	molar mass of species, kg/kmol; mass of solid or catalyst, kg
$\bar{m}_i$	mass fraction of component i
n	overall order of reaction
N	number of equal-sized CFSTR; impeller rotational speed, s^{-1}
N_A	moles of compound A
n_A	number of molecules A
N_o	initial mole
N_Q	pumping number
N_T	total mole
OTP	optimum temperature progression, $^{\circ}C$
OTR	oxygen transfer rate
OUR	oxygen uptake rate
ΔP	pressure drop, Pa
P	total pressure, Pa
p_A	partial pressure of compound A, Pa
Q	heat duty, J
Q_g	heat generation, W
Q_p	effective pumping capacity
Q_r	heat removal, W
R	ideal gas law constant
	8.314 $J/mol \cdot K$
	1.987 $cal/mol \cdot K$
	0.08206 $lit \cdot atm/mol \cdot K$
R	recycle ratio
R	number of baffles
$-r_A$	rate of reaction of species A, $mol/m^3 \cdot s$
$-r'_A$	molar rate of conversion per unit mass of the solid, $kmol/kg \cdot s$

$r, r^i, r^{ii}, r^{iii}, r^{iv}$	rate of reaction, an intensive measure
r_p	production rate of some product, $kg/m^3 \cdot h$
r_s	consumption rate of substrate, $kg/m^3 \cdot h$
r_{sx}	rate of substrate concentration, $kg/m^3 \cdot h$
r_x	growth rate of cells, $kg/m^3 \cdot h$
Σr_{con}	sum of all the rates of consumption, $kg/m^3 \cdot h$
Σr_{gen}	sum of all the rates of generation, $kg/m^3 \cdot h$
S	cross-sectional area of reactor, m^2
S_A	linear function of the characteristic velocity
ST	space time, s
SV	space velocity, s^{-1}
T	temperature, ^oC
t	time, s
t_b	total batch cycle time, h
T_c	coolant temperature, ^oC
T_h	heating coil temperature, ^oC
$t_{holding\ time}$	holding time, s
T_{opt}	optimum temperature, ^oC
ΔT_{LMTD}	log mean temperature difference, ^oC
t_r	reaction time, s
t_t	turnaround time, s
$\bar{t}$	reactor holding time or mean residence time of a fluid in a flow system = V/u, s
$t_{plug\ flow}$	plug flow mean residence time, s
u	volumetric flowrate, m^3/s
$\bar{u}$	internal energy per unit mass of mixture, kJ/kg
U	overall heat transfer coefficient, $W/m^2 \cdot K$
V	reactor volume, m^3
v	characteristic velocity, m/s
v_L	velocity, m/s
V_{max}	maximum rate of reaction (MM kinetic parameter) $kmol/m^3 \cdot s$
V_o	initial volume, m^3
V_p	volume of catalyst solid, m^3
V_R	reactor volume; final volume, m^3
W	mass of solids in the reactor; mass of catalyst, kg
W	agitator blade width
W_S	shaft work, J
X_A	fraction of A converted, the conversion
x_i	mole fraction of i in the mixture
x_w	wall thickness of vessel, mm

Z	gas compressibility factor
Z_{AA}	Boltzmann constant = 1.30×10^{-16} erg/K
y	general extensive property concentration (kg y)/m^3
$Y_{x|s}$	yield coefficient, (kg x)/(kg s)

DIMENSIONLESS GROUPS

$$\frac{D}{vL}$$ vessel dispersion number

$$\frac{D}{vd}$$ intensity of axial dispersion

$$Re = \frac{dv\rho}{\mu}, \frac{\rho ND^2}{\mu} \left(\frac{\text{inertia forces}}{\text{viscous forces}} \right)$$ Reynolds number

$$Pr = \frac{C_p \mu}{k} \left(\frac{\text{hydrodynamic boundary layer}}{\text{thermal boundary layer}} \right)$$ Prandtl number

$$Nu = \frac{hD}{k} \left(\frac{\text{total heat transfer}}{\text{heat transfer by conduction}} \right)$$ Nusselt number

$$Sc = \frac{\mu}{\rho D} \left(\frac{\text{hydrodynamic boundary layer}}{\text{mass transfer boundary layer}} \right)$$ Schmidt number

$$Bo = \frac{ud}{D} = (Re)(Sc) \left(\frac{\text{total momentum transfer}}{\text{molecular heat transfer}} \right)$$ Bodenstein number

$$Pe = \frac{vL}{D} \left(\frac{\text{mass transfer by convection}}{\text{mass transfer by diffusion}} \right)$$ Peclet number

$$P_o = \frac{Pg_c}{\rho N^3 D_A^5} \left(\frac{\text{total dissipated power}}{\text{power due to inertia}} \right)$$ Power number

$$Fr = \frac{N^2 D_A}{g}, \frac{v^2}{gL} \left(\frac{\text{inertia forces}}{\text{gravitational forces}} \right)$$ Froude number

$$Da_I = \frac{rL}{vC} \left(\frac{\text{chemical reaction rate}}{\text{mass transport by convection}} \right)$$ Damköhler I number

$$Da_{II} = \frac{rL^2}{DC} \left(\frac{\text{chemical reaction rate}}{\text{mass transport by diffusion}} \right)$$ Damköhler II number

GREEK LETTERS

α, β	orders of reaction
ρ	fluid density, kg/m^3
δ	Dirac delta function, an ideal pulse; change
$\delta(t - t_o)$	Dirac delta function occurring at time t_o (s^{-1})
ε_A	expansion factor, fractional volume change on complete conversion of A
θ	dimensionless time units $[t/\bar{t}]$ $(-)$
μ_i	chemical potential of species i
μ	fluid viscosity, $(kg/m{\cdot}s)$, $Pa{\cdot}s$
μ	specific growth rate, s^{-1}
μ_{max}	maximum specific growth rate, s^{-1}
π	total pressure (Pa)
σ^2	variance of a tracer curve or distribution functions (s^2)
τ	space time $C_{AO}V/uC_{AO}$ (s)
υ	stoichiometric coefficient ratio
ϕ	overall fractional yield
λ	Life expectancy, s
Δ	difference
σ_θ^2	dimensionless variance $\left[\sigma^2/\bar{t}^2\right]$ $(-)$
γ_i	fugacity coefficient $(= f_i/p)$
ϕ	Factor
Σ	summation
$ei(\alpha)$	exponential integral (page 767)

GLOSSARY OF ABBREVIATIONS

ASME	American Society of Mechanical Engineers
CFSTR	Continuous flow stirred tank reactor
CFD	Computational fluid dynamics
CFM	Computational fluid mixing
DIERS	Design Institute for Emergency Relief Systems
exp	exponential
IR	Infrared (spectroscopy)
HAZAN	Hazard analysis
HAZOP	Hazard and operability studies
MM	Michaelis-Menten
MMM	Maximum-mixedness model

NMR   Nuclear magnetic resonance
OTP   Optimum temperature progression
OTR   Oxygen transfer rate
OUR   Oxygen uptake rate
PFR   Plug flow reactor
RTD   Residence time distribution
TIS   Tank in series

Index

About the Author

A. Kayode Coker, Ph.D., a senior lecturer and Course Director at Jubail Industrial College, and consultant for AKC Technology in England, is a chartered chemical engineer, corporate member of the Institution of Chemical Engineers in the U.K., and member of the American Institute of Chemical Engineers. He holds a B.Sc. honors degree in Chemical Engineering, an M.Sc. in Process Analysis and Development, and Ph.D. in Chemical Engineering, all from Aston University, Birmingham, U.K.

Dr. Coker has authored *Fortran Programs for Chemical Process Design, Analysis and Simulation,* and written a topic section in the *Encyclopedia of Chemical Processing and Design* (Vol. 61). He has directed and conducted short courses for Procter & Gamble in the U.K., Saudi Basic Industries Corporation (SABIC) and Saudi Aramco Shell Refinery Company (SASREF) in Saudi Arabia. His articles have been published in several international journals.

ABOUT THE CD

INSTALLATION INSTRUCTIONS:
Copy the file "Coker Companion.exe" from this CD-ROM to your hard drive. Launch the program on your hard drive. When installation is complete, all the applications and files mentioned in the text will be inside the resulting folder, "Coker Companion."

To acquire Adobe Acrobat reader, go to http://www.adobe.com/products/acrobat/readstep.html for free download of the program.